FUNDAMENTALS
OF FOREST
BIOGEOCOENOLOGY

V. Sukachev
Vladimir Nikolaevich
N. Dylis

FUNDAMENTALS
OF FOREST
BIOGEOCOENOLOGY

Translated by
Dr J. M. MACLENNAN

OLIVER & BOYD
EDINBURGH AND LONDON

This is an English translation of
Основы лесной биогеоценологии
(Osnovy lesnoi biogeotsenologii)
first published in 1964 by
'Nauka' Publishing Office, Moscow

SBN 05 001637 7

OLIVER AND BOYD LTD
Tweeddale Court
Edinburgh 1
39A Welbeck Street
London W.1

Printed in Great Britain by
Robert Cunningham and Sons Ltd, Alva

Authors' Foreword to English Edition

We believe the following views are almost universally held: (1) all natural phenomena on the earth's surface are interrelated and interdependent and (2) areas of the earth's surface with homogeneity of organic life (plants, animals, and micro-organisms) and of the environment in which it exists (atmosphere, rock strata, hydrological conditions, and soil) form complex, internally-contradictory units. Since the concept of the existence and specific characteristics of these units arose independently, and more or less simultaneously, in several countries, the units were given a variety of names. The best-known of these are 'ecosystem', used mainly in Western Europe and America, and 'biogeocoenose', used mainly in the Soviet Union. Although these terms denote closely-similar concepts they are, as stated in this book, not fully identical. Apart from some differences in detail, the term 'ecosystem' is understood in varying senses by different authors, from an ant-hill with its population to the entire biosphere of the globe. The term 'biogeocoenose' has a more definite meaning and is generally understood to refer to a single system. On that account, and also since etymologically *bio* denotes the organic world and *geo* the abiotic environment in which it exists, and these can be united in a single total coenose—*koinos* (Greek), the authors prefer the term 'biogeocoenose' to 'ecosystem'. We must point out that in the Soviet Union (especially among botanists) the term 'ecology' has a much narrower meaning than in America and in some countries of Western Europe, namely, 'autecology'; this concept does not include biocoenology; ecology and biocoenology are regarded as independent branches of biology. In the U.S.S.R. this branch of knowledge is therefore called 'biogeocoenology' instead of the 'science of ecosystems'.

Data gathered by scientists in the various biological and physico-geographical disciplines and suitable for biogeocoenotic interpretation and use are dispersed through the most diversified publications of various countries, not all of which are easily accessible. For that reason and also because of limitations of space in this book, the authors have been unable to make use of all the literature that to a greater or lesser degree relates to biogeocoenology, but have been obliged to restrict themselves to only part of it, which (as the appended Bibliography shows) is still very extensive. The authors have been able to use data published up to the beginning of 1964.

Although the subject of this book is only forest biogeocoenology, we have also had to discuss general problems of biogeocoenology, since a forest is a combination of the most complex biogeocoenoses. Because of the complexity of forest biogeocoenoses almost no study has as yet been given to the transformation and balance of matter and energy in them, expressed in numerical quantities. Therefore this important aspect of the subject is only partially discussed in the present book.

Since the compilation of a book such as this is a new project it still has many shortcomings, of which the authors themselves are well aware. In publishing this volume, which is still far from perfect, the authors hope that their labours will stimulate further work on the problems of bio-geocoenology, which are not only of general theoretical but also of practical importance as the scientific foundation for more complete and rational utilisation of natural resources.

V. SUKACHEV

N. DYLIS

Contents

Introduction

Proper organisation of the exploitation of natural resources and of their development and conservation for future human generations calls for extensive knowledge of their nature and characteristics. This applies in the fullest measure to forest products, which are extremely varied and at the same time affected by the most diverse aspects of forest life, embracing not only standing timber and its plant life as a whole but also its animal inhabitants and the environment in which the organic life of the forest develops.

In our time a special branch of knowledge has been created for studying the complex natural history of the forest, named 'forest biogeocoenology'. Together with forest economics, it constitutes the main scientific basis of sylviculture and forest economy.

Forests occupy more territory than any other type of terrestrial landscape and play a major role in the life of man, being used by him in many different ways. Recently much attention has been devoted to conservation problems, leading to replanting of cut-over areas and extension of the area of 'man's green friend'. The fullest and most rational utilisation of all forest resources, however, can be achieved only on the foundation of thorough study of all aspects of forest life. The so-called 'cosmic role' of the forest—its role in the entire life of our planet—is not yet clear enough. Correct understanding of that role will assist in the development of the dialectic-materialistic concept of creation as a whole.

This contribution sets forth the fundamentals of forest biogeocoenology and shows its theoretical and practical significance, directing the attention of wide groups, not only of forestry officials but also of biologists and geographers, to the necessity of extensive and intensive field and laboratory studies of the complex life of the forest.

The book has been written by a group of collaborators in the Forestry Laboratory of the U.S.S.R. Academy of Sciences and the V. L. Komarov Botanical Institute of the U.S.S.R. Academy of Sciences. The work of writing separate chapters has been distributed among different specialists, as follows:

Chapter I. Basic concepts in forest biogeocoenology, by V. N. Sukachev, with use of some material by N. E. Kabanov.

Chapter II. The atmosphere as a component of a forest biogeocoenose, by A. A. Molchanov.

Chapter III. A phytocoenose as a component of a forest biogeocoenose, by N. V. Dylis, Yu. L. Tsel'niker, and V. G. Karpov.

1

Chapter IV. Animal life as a component of a forest biogeocoenose, by P. M. Rafes, L. G. Dinesman, and T. S. Perel'.

Chapter V. Micro-organisms as a component of a forest biogeocoenose, by S. A. Egorova, M. G. Enikeeva, and V. S. Bol'shakova.

Chapter VI. Soil as a component of a forest biogeocoenose, by S. V. Zonn.

Chapter VII. Dynamics of forest biogeocoenoses, by V. N. Sukachev.

Chapter VIII. Principles of construction of a classification of forest biogeocoenoses, by N. V. Dylis.

Chapter IX. The possibility of applying the ideas and methods of cybernetics to forest biogeocoenology, by V. D. Aleksandrova.

Chapter X. The theoretical and practical significance of forest biogeocoenology, by V. N. Sukachev.

Most of the chapters were discussed at group meetings of the authors and of other colleagues in the Forestry Laboratory. The general editing of the book was done by V. N. Sukachev and N. V. Dylis.

Much effort has been devoted towards obtaining the greatest possible harmony of form and style of presentation in all chapters. The task was complicated by the need to achieve uniform treatment of a number of new concepts and to interpret previously-known facts from the biogeocoenological point of view. Since the book gives much attention to the inter-relationships of the various components of the forest, and is mainly composed of chapters corresponding to these components, it was inevitable that different chapters should repeat the discussion of these inter-relationships, although generally from a different viewpoint.

Because of the newness of this discipline, some of the principles of biogeocoenology developed in this book have to some extent a controversial character, which is expressed in the views of the various authors.

All of these features lead to a certain lack of harmony in different parts of the book and to shortcomings in it, which are known to the authors themselves. They believe it necessary, however, to publish their work, considering that even in its present form it will be useful to Soviet forestry. The authors also hope that through their book they will help to extend the application of biogeocoenological principles to the study of the earth's biosphere, which, in their opinion, will not only make possible a more profound approach to the discovery of the fundamental laws of its existence but also will assist in proper utilisation of all natural biological resources.

To date, the study by different sciences of various aspects of forest biology has accumulated many data which, by appropriate analysis from a definite point of view, may assist substantially in developing our knowledge of forest biogeocoenology. That task demands much time and effort. The authors have been unable to carry it out within the scope of the present work, but it is very important that it should be done in the future. The authors have used only those published sources considered most important and available to them. They have, therefore, dealt mainly with the temperate zone. Nevertheless, the list of publications used is a long one, and the authors believe it useful to include it in the book.

The authors will be grateful for critical observations of any kind concerning the contents of the book, which will assist in its improvement in the future.

CHAPTER I

Basic concepts in forest biogeocoenology

The biogeocoenose as an expression of the interaction of animate and inanimate natural phenomena on the earth's surface. Inter-relationships of the concepts of biogeocoenose, ecosystem, geographical landscape, and facies.

Forests cover one-third of the earth's surface to-day. Among the countries with a very large forest area is the Soviet Union, whose state forests cover 1131 million hectares, or about half of its territory, with a timber reserve of 75,000 million cubic metres; thus the U.S.S.R. possesses 39 per cent of the world's timber resources. While the role of the forest is a major one in the world's economy, it is particularly important in the national economy of the U.S.S.R. The forest area, however, is seriously diminishing every year both in the world as a whole and in the Soviet Union. This change might not worry humanity if the view sometimes expressed were correct—that the need for wood will constantly decrease because of the constant increase in the use of structural materials other than wood and substitution of these for wood in house-building and the manufacture of furniture and paper, and with the use of various synthetic materials where wood was formerly used. On the contrary, however, we see both in the U.S.S.R. and in other countries that with the expansion and intensification of the national economy the role of wood is increasing.

In this connection the well-known biochemist D. Bernal (1960, p. 131) recently wrote that the least efficient use of wood is as fuel, and that by present-day methods wood, and cellulose in general, may be converted into food and nutritive products.

We must realise that continually in the future forest products are going to be put to wider and more varied uses. It is well known that at present forest technologists are devising methods for the fullest possible utilisation of all by-products obtained from timber during logging and industrial operations. There is every reason to believe that the importance of wood as a raw material for chemical production will constantly increase. Even in the structural field, although it is being replaced to an ever-growing extent by other materials, wood has not ceased to be important. For instance, wood continues to be the best material for railway sleepers and for sashes, doors, and various other parts of buildings.

We may assert that on the whole no type of vegetation produces so much that is valuable to man as does the forest. Cereal plants provide a basic food product; but in variety of uses they fall far behind forests, apart from the fact that not only trees but forest plants in general play a substantial part in food production. Human utilisation of timber and its chemical derivatives, and of all forest vegetation, especially in the future, may be regarded as without limit. For instance, leaves are still very little used. We should also take into account the value of forest animal life to the national economy. Moreover, forests are of great importance to health; in this respect no other vegetational type approaches them. Finally, the proximity of certain types of forest, by the influence that they exert on surrounding territory, is of primary significance in the conservation of soil and water, and the increase of fertility in our fields (the soil-conservation role of forest belts). Therefore the problems of conservation of forests, of speeding up their growth, of increasing their productivity, of improving the quality of their produce, and of fuller use of them and of all other wealth produced by them, are urgent now and doubtless will continue to be so.

The tasks of forestry include organising and making the most rational use of all forest resources and at the same time conserving them for the future, raising their productivity, increasing their usefulness and changing their geographical extent in accordance with the requirements of the national economy. It is obvious that these tasks can be fulfilled only on the basis of all available science, especially forestry sciences. Among these sciences, if we do not count forest economics, sylviculture is of outstanding importance. Sylviculture is the main theoretical foundation of forestry and forest economy. This, as is well known, was most fully expressed in the works of Prof. G. F. Morozov at the beginning of the present century. Morozov especially stressed that for the successful development of forest economy one must know all aspects of nature in the forest. In this connection, according to Morozov, one should understand by 'forest' not only the trees but the whole of their environment, i.e., all other vegetation, fauna, and micro-organisms dwelling with trees on a given area of the earth's surface, with the soil and water conditions pertaining to the latter, and the atmosphere.

Actually, the selection of specific forestry measures and the times and ways in which they are applied depend on the varied biological and physico-geographical features of a forest as well as on economic conditions.

If we consider only basic forestry measures (complete and selective felling, logging, assistance to natural regeneration, artificial planting, protection from pests, fire precautions, etc.), the rational application of these requires knowledge not only of the biological features of the arboreal and other layers of forest vegetation but also of climatic (including micro-climatic), soil, and water conditions, as well as of the faunal and microbial life of the forest. The more complete our knowledge of these complex components of the forest, the more soundly-based will be the forestry measures taken and the greater their practical effect.

It is especially important to stress that for the proper valuation of all these factors in forest economy we must bear in mind (a) that all forest components that play a direct role in forest economy interact on each other, affect each other (e.g., soil conditions depend on climate; micro-climatic and soil conditions on all vegetation; the composition and activities of the fauna on the factors just mentioned; and so on), and (b) that the biological and physico-geographical features of the forest not directly affecting forestry measures affect other factors on which the latter depend directly. Therefore the entire combination of biological and physico-geographical features of a given area of forest are of concern to forest economy in their interaction and inter-relationships. Thus from the sylvicultural point of view any area of forest must be regarded as a specific natural unit, where all vegetation, fauna and micro-organisms, soil and atmosphere exist in a state of close inter-relationship and interaction.

Such interaction, of course, is a feature not only of forests but of any part of the earth's surface. In no other type of vegetation, however, are they displayed so clearly and with so many aspects as in the forest. One must take account of natural units resulting from such interaction in every type of economic use of natural resources (plant, animal, soil, etc.).

The concept of these units has arisen independently at different times in different countries, and they have received different names. We shall use the term 'biogeocoenose' for such a unit, and call the field of know-ledge of biogeocoenoses 'biogeocoenology'. Before discussing these concepts in detail, we must touch briefly on the history and development of the idea of inter-relationships of objects and phenomena in nature, remembering that Aristotle and other Greek philosophers had already suggested that for fuller knowledge of natural phenomena one must know the history of their development.

The idea of the interaction of all natural phenomena, as is well known, is one of the basic premises of materialistic dialectics, well proved by the founders of the latter, K. Marx and F. Engels.

The view that a new branch of science was required, devoted to specific study of the interaction and inter-relationships of all natural phenomena on the earth's surface, was first advanced by V. V. Dokuchaev (1898).

The ideas of Dokuchaev later found expression mainly in geography, especially in the study of geographical landscapes.

In Western Europe the development of that study is linked principally with the name of Passarge (1912, 1920, 1921, 1929). In the U.S.S.R. it began to be developed at the same time by L. S. Berg, who greatly clarified the concept of 'landscape' itself. Therefore we quote the definition of 'landscape' given by him in 1931: 'A geographical landscape is that com-bination or grouping of objects and phenomena in which features of relief, climate, water, soil, plant cover, and animal life, and also human activity, combine into one harmonious whole, typically repeated through-out a given area of land' (Berg 1931, p. 5). In 1946 Berg proposed replace-ment of the term 'geographical landscape' by 'geographical aspect', which, however, has not been accepted.

Among subsequent definitions of 'landscape' we must note that of N. A. Solntsev: 'We should give the name "geographical landscape" to a genetically-homogeneous territory in which a regular and typical re-currence of the same inter-related combinations is observed: of geological structure, of forms of relief, of surface and underground water, of micro-climates, of soil varieties, of phyto- and zoocoenoses' (1948, p. 258).

We may also quote the definitions of 'landscape' given by S. V. Kalesnik, F. N. Mil'kov, and A. G. Isachenko, who did much for the development of landscape-science. S. V. Kalesnik's definition is: 'A geographical landscape is that dialectical whole combination of relief, geological structure, climate, soil, water, organic life, and human activity, typically recurring over a significant area of the geographical envelope of the earth' (1940, p. 13). F. N. Mil'kov defines the concept thus: 'A landscape is an assemblage of interacting and inter-related natural objects and phenomena, presented to us in the form of historically-created and con-tinuously-developing geographical complexes' (1959, p. 11). At the same time Mil'kov uses the term 'landscape' in the same general sense in which we speak of climate, relief, soil, etc. A. G. Isachenko writes: 'A landscape is a genetically-isolated part of the landscape domain, character-ised by both zonal and azonal homogeneity, i.e., by physico-geographical homogeneity as a whole, and possessing individual structure and individual morphological organisation' (1953, p. 252),

From these definitions by our geographers, and also generally from their works on landscape-science, we may see that in spite of their differ-ences in formulation, 'landscape' as understood by them is either a general concept or, more often, a geographical phenomenon, territorially rather extensive, a component of which they consider to be relief whose form remains homogeneous within the landscape. A characteristic feature of landscape, as is usually stated, is a historically and genetically conditioned inter-relationship of all components of the landscape. In that connection Berg included human activity among landscape components and described a landscape as 'one harmonious whole'. Later he modified the definition considerably.

At the same time, when we examine the literature of landscape-science we cannot fail to see that the concept of 'landscape' is often very vague and nebulous, as some geographers themselves have noted. I. P. Gerasi-mov (1954, pp. 69-70) especially points out that even the most ardent and convinced advocates of landscape-science (e.g., A. G. Isachenko 1953, p. 82) remark on the vagueness of the subject. The situation regarding landscape-science has not changed, as is evident from the recent works of N. A. Solntsev.

To make the meaning of the concept 'landscape' more precise there have been proposals to add various qualifications to the word 'landscape', e.g., elementary landscape (Polynov 1953), microlandscape (Larin 1926), micro- meso- macro-landscape (Chetyrkin 1947), landscape of the first and second orders, etc. It has been suggested that the term 'landscape' should be discarded altogether and replaced by another word, such as 'aspect' (Berg), *kraiobraz* (form of a region) (Il'inskii); and abroad it has been proposed to discard even the concept of landscape (Gellert 1961).

Some geographers distinguish two kinds of landscape-classification, 'typological' and 'regional' (Isachenko 1953). The first involves identifying on the earth's surface homogeneous natural complexes of biogenic, inanimate, and soil (bio-inanimate, Vernadskii's term) phenomena; these are further grouped, according to their similarity to a greater or lesser extent, into taxonomic units of different orders, regardless of territorial distribution. The second classification divides up the earth's surface into territorial units of different taxonomic significance according to similarities in the objects under observation: it results in the division of land into districts, which may be based on various criteria such as vegetation, animal life, soil, etc. Classification may also be done on the basis of a complex of natural features. Whereas the first kind of classification proceeds from the smallest homogeneous unit, regardless of what it is called—biogeocoenose, facies, etc.—the second classification also cannot dispense with such small units in dividing the land into districts. Although the *type of biogeocoenose* is the taxonomic unit in the first kind of classification for typology of the earth's surface (more precisely, for typology of the biogeocoenotic cover of the earth), the smallest taxonomic unit for division of territory into districts must naturally be different. The latter method is in general use, but has no name as yet. A. G. Isachenko (1953) suggests giving it the name of 'geographical landscape' or 'elementary physico-geographical district', considering these terms to be synonymous. When biogeocoenoses are used as primary units, i.e., when division into districts is done by combination of all biogeocoenotic indicators, then one must speak of a *biogeocoenotic district*.

The ideas of Dokuchaev about the inter-relation of all phenomena and objects on the earth's surface have been echoed in so-called geobotany.

The science of geobotany, as developed in the U.S.S.R., is characterised by study of the close relationships between plant cover and environmental conditions, especially the soil. This science was very clearly expounded in the works of our greatest sylviculturists G. F. Morozov and G. N. Vysotskii, who were direct disciples of V. V. Dokuchaev. G. N. Vysotskii (1925) expressed these ideas in his 'science of plant cover' ('phytostromatology'), and G. F. Morozov built up the science of forestry on that foundation; later, becoming acquainted with L. S. Berg's study of geographical landscape, he advocated the concept of the forest as a geographical landscape.

role of the biocoenose in the life of that complex. Soon, however, Sukachev discovered that the term 'geocoenose' and its synonym 'syngeocoenose' had been used earlier by S. I. Medvedev (1936) on the suggestion of V. V. Stanchinskii—though in a somewhat wider sense, it is true. Since the term 'biogeocoenose' corresponds more closely to the idea of a natural phenomenon, it was thereafter used in a number of works on the subject (Sukachev 1947, 1948 a, b, 1949, 1951, 1954 a, b, 1955, 1957, 1958, 1960 a, b, 1961 a, Sukatschew 1954, 1960).

Morozov's concept of 'biocoenose', Abolin's of 'epimorph', and Sukachev's of 'biogeocoenose' are much more sharply delimited than the 'geographical landscape' of the above-mentioned geographers. The need to have a more clearly defined unit of land surface had been recognised by several geographers, and even more decisively by pedologists and biologists. On that account, as has been already stated, the term 'elementary landscape' was proposed; B. B. Polynov, our greatest pedologist, defined it thus: 'An elementary landscape in its typical form should represent one definite element of relief, formed from a single stratum or deposit and covered at each moment of its existence by a definite plant community. All of these conditions create a definite type of soil and are evidence of uniform development of interactions between mineral strata and organisms throughout the elementary landscape' (1925, p. 73). Here Polynov, as a pedologist, pays particular attention to the study of soil and the migration of chemical elements in the elementary landscape, about which more will be said later.

The need to have a concept in landscape-science corresponding to the accepted idea of 'elementary landscape' was also stressed in a number of works by A. D. Gozhev, who named it 'microlandscape' (1929), 'type of territory' (1930), 'microtype of territory' (1934), or 'variant' of type of territory (1945, 1948).

The study of so-called 'geochemical landscapes' is of value for forest biogeocoenology. The origin of that study was linked, as stated above, with the name of Polynov, who, using V. V. Dokuchaev's ideas about soil and those of V. I. Vernadskii and A. E. Fersman about geochemistry and biogeochemistry, laid the foundations of the science of landscape geochemistry, now energetically and fruitfully developed by A. I. Perel'man. Out of numerous works by the latter we may mention his 1957 and 1960 articles and especially the book *Landscape geochemistry*, published in 1961. In that book he writes: 'From the geochemical viewpoint landscape is a part of the earth's surface in which migrations of chemical elements of the atmosphere, hydrosphere, and lithosphere take place, by means of solar energy. As a result of these migrations, parts of the earth's crust undergo alteration, becoming increasingly alike so that characteristic natural formations arise, composed of living organisms, soils, a weathered crust, and natural bodies of water. In studying the migration of elements we discover the links between atmosphere and vegetation, and between vegetation, soil, and water, i.e., between all the principal components of landscape. Therefore we may say that landscape geochemistry is the history of atoms in landscape.' Perel'man takes as the basic lowest taxonomic unit in landscape 'elementary landscape' in Polynov's sense, which in essence is close to our concept of biogeocoenose.

Some time ago V. G. Nesterov (1954, 1962) began to advocate his theory of diatopes and bioecoses. He writes: 'The concept of diatopes has been developed for those cases where a single item, forming a unit with another, cannot exist without it although it is contradictory to it.' As an example of a diatope he cites: 'Pine trees with lichens, mosses, cowberries, bilberries, with their appropriate mammals, birds, and insects, are characteristic of sandy soils, with which they constitute a complex unit, a bioecos—*biooikos* (bio–*bios*, organism; eco–*oikos*, environment)—or diatope. The latter, translated literally, means two in one unit (*dia*, a doubling; *topos*, a common place).'[1] 'Alders form a diatope with river flats.' 'A herd of reindeer form a diatope with forest-tundra' (1962, p. 5). 'A line of machine-tools may normally operate on their appropriate bases, under one roof, when electric power of the proper voltage is supplied. . . . In diatopes of communities of organisms and environment, as in diatopes of machines in corresponding circumstances, there is an exchange of energy and material' (1962, p. 6).

To define the relation between the terms 'bioecos' and 'diatope' proposed by Nesterov, we may quote the following from the same book: 'The term "bioecos" reflects an objective reality as a union of two different items, the contradiction between which determines their development. In this sense every natural complex of organisms and environment constitutes a diatope.' Without dwelling on these and many other—often undefined, indecipherable, and occasionally ill-written—statements by that author, we may note that his concept of 'bioecos' corresponds in essence, in many cases, to the concept of 'biogeocoenose'. Remarking, however, on the difference between his views and those of Clements, Morozov, and Michurin on the unity of organisms and environment, he writes that 'we differ substantially from the theory of biogeocoenoses, which regards nature as a complex of many combined items, without defining the two components in it that unite all the constituent parts, and which asserts the moving force of development to be intraspecific competition (at least for single-species phytocoenoses)'. 'The concept of complexes without proper definition of the central combining and guiding contradiction is inadequate. As we see it, nature consists of mutually-linked bioecoses, i.e., units composed of communities of organisms and their living conditions, the contradictions between which determine their development; of these the first determining and combining factor is

[1] Nesterov's translation and transliteration of foreign words is retained. Elsewhere in his book he translates *topos* simply as 'common' and as 'a unit', which of course is incorrect (*topos* = a place).

environment and the second active factor is organisms.' In connection with these doubly-vague statements we must first note that nature certainly does not consist only of mutually-linked bioecoses or biogeocoenoses. It is therefore incorrect to state that the theory of biogeocoenoses 'regards nature as a complex of many combined items, without defining the two components in it that unite all the constituent parts'. The theory of bio-geocoenoses does unite its constituent parts into an assemblage of organisms, a biogeocoenose, and into an assemblage of environmental factors, an ecotope or biotope. The theory of biogeocoenoses considers, as has frequently been written, the main moving force of biogeocoenose development to be the contradictory interaction between biocoenoses and biotopes, i.e., between organisms and the environment in which they live.[1] If one discards all the nebulous and imprecise statements made by Nesterov in his treatment of natural phenomena observed by him, there are no essential differences between his bioecoses and biogeocoenoses.

We have thought it necessary to dwell on our critical analysis of these views of Nesterov—which are the consequence (as is seen from the works cited here and also from other works of his, including his textbook *General forestry*) of his inadequate comprehension of the scientific problems surveyed by him—because these views are expressed by a sylvi-culturist and may introduce great confusion into our forest science.

We must also discuss the recently-coined term 'biocoenotope' (Ioganzen 1962, p. 103). Ioganzen, incorrectly explaining the meaning of 'biogeo-coenose', proposes to substitute the term 'biocoenotope', thinking it possible to use it to 'denote a natural phenomenon, representing a union of biocoenose and biotope'. He writes 'in biocoenotopes there appears a real unity of organisms and their whole environment—hydrosphere, atmosphere, lithosphere, and biosphere, and also of all abiotic, biotic, and anthropic factors operating in nature. In different biocoenotopes there predominate different aspects of the natural complex—abiotic (a lake), biotic (a forest), or anthropic (a village).' From these statements and from all the rest of the article it is clear that the author does not distinguish between the components of a natural unit and the factors that affect it. Man himself is a powerful factor, at present affecting all observed natural phenomena, but in no way to be considered a component of them. This serious methodological error of the author appears clearly at the end of his article, where he writes 'man creates a whole series of cultural biocoenotopes—from gardens to villages'.

It must be noted that from the beginning of the twentieth century

---

[1] V. G. Nesterov does not specify in what sense he understands 'development in nature'. As will be shown below, different authors give different interpretations to phytocoenoses and biogeocoenoses. But nobody regards intraspecific competition of organisms as the moving force in development of phytocoenoses or biogeocoenoses. Intraspecific struggle for existence, leading to natural selection, is regarded as the main factor in species-formation, which is of course quite different from the development of phytocoenoses, biogeocoenoses, or 'bioecoses'.

foreign scientists have been working not only on the concept of geographical landscape but also on that of ecosystem, closely related to biogeocoenose. In 1935 Tansley wrote that, although organisms might appear to claim our main attention, when we look deeply into the matter we cannot separate them from their environment, in combination with which they constitute a single physical system. From the ecologist's point of view such systems are the basic natural units on the earth's surface. These systems Tansley called 'ecosystems', looking on them both as one item in the catalogue of multiform physical systems ranging from the universe down to the atom, and also as an abstract concept.

Although the concept of 'ecosystem' was put forward by Tansley about thirty years ago, for a long time it did not receive any widespread acceptance, and the very idea of inter-relationships and unity of all phenomena and objects on the earth's surface found inadequate expression in foreign literature. Only in recent years has it begun to get much attention, and now ecosystem literature is very extensive.

To show how the term 'ecosystem' is now understood, I shall quote Villee on that subject (Russian translation, 1964, p. 101). 'By the term "ecosystem" ecologists understand a natural unit, representing a combination of animate and inanimate elements; as a result of interaction of these elements a stable system is created, in which matter circulates between the living and non-living parts. Ecosystems may be of various sizes.' Later, examples of ecosystems are presented—a lake; a forest massif; an aquarium with fish, green plants, and molluscs.

In Odum's well-known work on ecology (1962) an essentially similar definition of 'ecosystem' is given (pp. 10, 11).

From the statements of Tansley and Villee it may be seen that the ecosystem concept itself is imprecise and does not always have the same meaning. We must add that other terms have been proposed in foreign countries, corresponding to, or close to, the meaning of 'ecosystem' (see Troll 1950, Odum 1959)—for instance, 'microcosm' (Forbes 1887); 'biosystema' (Thienemann 1941); 'holocen' (Friedrichs 1927, Thienemann 1941, Schmidt 1944, Ylin 1948, Schmithuesen 1961); 'biochora' (Palmann 1948, Etter 1954); 'ecotope' (Troll 1950). In the German Democratic Republic, in dividing the territory into districts on a natural basis, Siegel (quoted by Gellert 1961) took as his basic smallest unit *Fliese*, but it is broader than the landscape concept of Soviet geographers.

The symposium organised by the Ninth International Botanical Congress in Canada in 1959 was very helpful in solving problems of forest ecosystem, when scientists of various countries presented seventeen papers devoted in varying degree to these problems, with the term 'biogeocoenose' and several others being used in different papers, as well as 'ecosystem'. Papers were presented by the following scientists: Arnborg, Sweden; Daubenmire, U.S.A.; Ellenberg, Switzerland; Hills, Linteau, and Rowe, Canada; Kalela, Finland; Medwecka-Kornas, Poland; Ovington, England; Puri, India; Scamoni, German Democratic Republic; Sukachev, u.s.s.r.; and Webb, Australia.

These papers were published in Finland in the English language, in the journal *Silva fennica*, 1960, No. 105.

All the above authors pointed out more or less definitely that representation of the forest as an ecosystem might be considered most fruitful from the biological point of view. Since forestry practice needs a classification of forests that is well known, the most rational is one based on forest ecosystem or biogeocoenose. Krajina (1960), who delivered a summing-up paper at the above symposium, stated that all forest classifications suggested up to that time might be grouped in three categories: (*a*) ecotopic, in which chief attention is given to local habitat (ecotope) factors and which are in turn divided into the classifications macroclimatic (climatopic) and soil (edaphotopic); (*b*) biocoenotic, based on the properties of biocoenoses, which may be divided into phytosociologic (phytocoenotic) and zoosociologic (zoocoenotic); and (*c*) ecosystematic, or biogeocoenotic (holocoenotic), combining the basic features of both the preceding. Briefly surveying forest classifications of these categories presented by different authors, including Soviet scientists, Krajina remarked that in the Soviet Union there had recently been advanced 'the interesting, and holocoenotically more complete, concept of "biogeocoenose", which apparently is ideal for classifying ecosystems of any part of the biosphere' (1960, p. 4). Stating that the concept and the term 'forest biogeocoenose' had already been used in the works of some Western European scientists, but that acceptance of that point of view was meeting with difficulties, the author himself gave, for the time being, only a definition of a plant community as an integral part of a biogeocoenose (of forest type). He also remarked upon the importance of such a complex approach to the solution of practical forestry problems.

We must note that the authors who proposed ecotypic or phytocoenotic classifications (e.g., Arnborg 1960) still did not deny the great value of ecosystematic or biogeocoenotic classifications, but merely stressed the point that vegetation (phytocoenose) is a good indicator of the ecosystematic (biogeocoenotic) characteristics of the forest. Most authors also agreed on the similarity of the concepts 'ecosystem' and 'biogeocoenose'.

Another concept presented at the Ninth International Botanical Congress in Canada was that of 'site', proposed by the Canadian scientist Hills (1960), who had already developed it in detail in a number of earlier works. In one of these he had written that climate, relief, soil, plants, and animals constitute the most important parts of the environmental whole, called 'site'. Literally 'site' means 'location', but in this case Hills understands by the word the total of all that interests the forester. Therefore he writes further that 'site' is the combination of external factors with which the forester should be concerned in developing the forest and exploiting forest productivity. He also holds that evaluation of each component of 'site' is inadequate for forest classification; the principal classification should be by the inter-relationships of all components. Such an approach is holostic (*holos* = whole). In this case one arrives at a whole that is greater than the mere sum of its parts.

Analysing the publications of Hills we may see that he has worked out, in the greatest detail, the study of soil as a component of 'site'. Generally he considers 'site' to be a particular case of 'ecosystem'.

Whittaker (1962) in his new great collated work on classification of plant communities, in which he lists extensive literature on ecosystems, gives this definition: 'An ecosystem is a functional system that includes an assemblage of interacting organisms—plants, animals, and saprobes—and their environment, which acts on them and on which they act.'

In an outstanding work on ecosystems and the biosphere, also published in 1962 and compiled by Prof. P. Duvigneaud of the University of Brussels and his colleagues of the Laboratory of Systematic Botany and Phytogeography of the University and National Centre of Ecology, there is a similar but shorter definition: 'An ecosystem is a functional system that includes an assemblage of living creatures and their environment.' It is added that the term 'ecosystem' can be applied to biocoenoses and their environments of widely differing dimensions. Therefore three categories of ecosystems are distinguished: (*a*) micro-ecosystems (e.g., a tree stump); (*b*) meso-ecosystems (e.g., a forest association); and (*c*) macro-ecosystems (e.g., an ocean). The integration of all the earth's ecosystems creates the gigantic ecosystem of the terrestrial globe—the biosphere. 'The study of an ecosystem, be it a virgin forest or a cultivated field, should always consist in definitive answering of the problems of elaboration, circulation, accumulation, and transformation of material (potential energy) through the interaction of living organisms and their metabolism.'

From the above brief survey of the history of the development of biogeocoenology we may draw the following conclusions.

1. The view that all natural objects and phenomena on the earth's surface are in a state of interaction and interdependence, and that one may speak of the existence of internally-inter-related units formed by them, arose a long time ago and independently in several countries and among scientists studying different aspects of nature, which is evidence of the need for the concept, resulting from the advance of science. It is worthy of note that it arose most readily among pedologists and sylviculturists. That is easily understood. Soil is a particularly obvious example of the total result of interaction of animate and inanimate nature, and a forest particularly clearly illustrates all of the inter-relationships of its components. It is quite natural that geographers also arrived at the same concept, having developed the study of landscapes and proceeding from

the need to divide territory into districts for purposes of the national economy. Those geobotanists, or more precisely phytocoenologists (more often called phytosociologists abroad), who have closely studied the dependence of vegetation on environment also were among the first to produce these ideas and give special attention to the exchange of matter and energy both within these units and between them; and they concentrated their efforts on the delimitation of smaller areas in which the character of these units remained practically uniform.

2. For the above reason a very large number of names for that concept have appeared: 'microcosm', 'epimorph', 'elementary landscape', 'micro-landscape', 'biosystema', 'holocen', 'biochora', 'ecotope', 'geocoenose', 'biogeocoenose', 'ecosystem', 'facies', 'epifacies', 'diatope', 'bioecos', etc. These terms are not fully synonymous, but they are all applied to natural objects with a general similarity. Although the observed natural units are a concrete expression of a well-known general proposition of dialectical materialism about the inter-relationship of all phenomena and objects on the earth's surface, there are considerable divergences among many authors in the understanding of their nature and in the representation of their composition and chief characteristics, indicating to a remarkable degree the varying philosophic approaches to them, which is also partly displayed in the different names. Of the various terms quoted, doubtless the most widespread abroad is 'ecosystem', and in this country 'biogeocoenose', which is used by many geobotanists, pedologists, zoologists, geneticists, geographers, agronomists, sylviculturists, and other specialists. Geographers, however, also use the term 'facies'. It is therefore necessary to dwell only on the relation of these three terms to each other, pointing out why we think it most suitable to use the term 'biogeocoenose'.

Although the term 'ecosystem' is widely used in foreign countries and is sometimes used in ours, it is inexpedient to recommend it because, as we have seen, it is used abroad in widely differing senses. Even where its meaning is close to that of 'biogeocoenose' it is not identical with it.

The following considerations also speak against the term 'ecosystem'. In the first place, among botanists in the Soviet Union 'ecology' is mostly understood as 'autecology', i.e., as the study of the relation of individual plants to their environment, in opposition to Anglo-American botanists, who include in ecology all biocoenology, in particular phytocoenology. Therefore if the term 'ecosystem' corresponds abroad to representation of the inter-relationships of a biocoenose and its environment, in our usual understanding of ecology the term loses that meaning. In the second place, with few of the foreign biologists does the term bring out to the proper degree the fact that here we have a unit of organisms closely inter-related with their environment.

We must point out that in general not only is the concept of 'system' itself not linked to the dialectical unity characteristic of 'biogeocoenose', but it does not include all the inter-relationships, all the interdependences of its abiotic components (atmosphere, mineral strata, water, and soil), which of course are not included in the meaning of ecological components.

With regard to the term 'facies', proposed, as we have seen, by L. S. Berg (1945) and applied by a number of geographers in a sense close to, and sometimes identical with, 'biogeocoenose', I have already written critically (Sukachev 1949, 1960b). But since other ideas have been put forward on this subject I consider it necessary to discuss it here also, giving some other objections besides the previous ones. The term 'facies', as is well known, is widely used in geology. Some very valuable works of a general character on the study of facies have recently appeared (Nalivkin 1956, Markevich 1957, Rukhin 1962). Even in geological literature, however the term 'facies' does not always have the same interpretation. This is well demonstrated by D. V. Nalivkin in the above-mentioned two-volume work. V. P. Markevich also remarks on it. Analysing different definitions, Markevich writes: 'Taking into account the basic meaning of the term "facies" and also the meaning given to the concept by most investigators, we propose the following definition: a facies, or geological facies, is a certain extent of deposit or strata, characterised by a similar complex of palaeontological, petrological, and physico-chemical characteristics leading to tectonic, physico-chemical, biotic, and geological conditions of formation of the deposit.' D. V. Nalivkin also (1956, p. 7) points out that a geological facies is closely linked in its formation with a definite landscape, with its own biocoenose; which is also recognised in one form or another by other geologists. But from the fact that the formation of a geological facies, in which the main role is played by sedimentation processes, is linked with a definite landscape, and in some cases also with a definite biogeocoenose, it does not follow that a geological facies is a biogeocoenose or that these terms are synonyms. This is seen still more clearly from the definition given by L. B. Rukhin. He writes: 'By "facies" one means deposits laid down on a specific area in identical conditions differing from those operating in adjacent districts' (1962, p. 93).

Neither can we agree with the point of view of V. B. Sochava (1959), who writes that the term 'facies' has not been retained in science with regard to the larger categories of physico-geographical divisions. According to Berg (1946), it is expedient to apply it to the finest typological subdivisions of landscape, so facilitating comparison with palaeogeographical facies, which include biocoenoses, biotopes, and sedimentary deposits. A biogeocoenose is a physico-geographical facies, but in the sense of an energetic system. With the biogeocoenotic approach as described by Sukachev (1948 a, b, 1955), a facies is examined from the point of view of exchange of material between organisms and their inanimate environment, transformation of energy, and its migration within the facies.

With regard to Sochava's statement we must first point out that Nalivkin's remark about the inclusion in facies of biocoenoses, biotopes, and sedimentary deposits should be looked upon, as we have said above, in the sense that at the time of formation of a facies it is linked with a specific biocoenose and biotope. This follows from the comprehensive work of Nalivkin (1956) and from other general works mentioned above.

Furthermore, it is scarcely proper to give the same natural phenomenon different names according to whether it is regarded from the geographical (chorological) point of view or as an energetic system. A phenomenon should naturally be given a single name, but it may be necessary to observe it and study it from different aspects.

V. B. Sochava writes: 'Territorially a biogeocoenose and an elementary landscape are the same physico-geographical facies, but treated from different aspects. Representation of a physico-geographical facies in a more universal way requires not only analysis of the inter-relationships between the biotic and abiotic components of a complex, but also study of the links within an abiotic facies' (1961, p. 7). That presentation, however, is incorrect. As has always been stressed in exposition of the theory of biogeocoenoses, in biogeocoenology the interactions of all components are essentially of equal value (in particular, those between the abiotic components of a biogeocoenose) and should be studied equally.

In general it must be remarked that the concept of a facies does not have quite the same meaning and the same content even with those geographers who look upon it as an elementary and even, as it were, indivisible unit of landscape-science; usually it is of wider scope than a biogeocoenose. But in Sochava's opinion, especially judging by his later works, 'facies' is identical with 'biogeocoenose' and the methods of studying both are the same.

Since among the very geographers who cling to the term 'facies' its meaning is ill-defined, we may follow the example of V. I. Prokaev (1961), who, observing that a facies is generally defined as the smallest landscape complex, formed from a single rock stratum, having a uniform microclimate, a single soil type, and a single biocoenose (Solntsev 1949, Kalesnik 1959, and others), points out that that definition has many defects, preventing uniform identification of facies in the field. In the first place, it uses 'elementary partial complexes, the contents of which in the appropriate sciences cannot yet be held to be determined' (e.g., a phytocoenose). In the second place, the definition practically tells the investigator that natural conditions are fully homogeneous within a facies, which in Prokaev's opinion is not consonant with the facts. As an example he cites the lower flood-plain of a stream in the Central Urals, which in his opinion is in no way higher taxonomically than a facies, although it contains areas differing in vegetation and soil and even parts of the old river-bed that contain stagnant water. In the third place, within a facies the boundaries of partial complexes (in the sense of separate components of the facies, *V.S.*) may not coincide. In the fourth place, in Prokaev's opinion the definition can be applied only to original unaltered facies.

Although Prokaev does not give a perfectly clear definition of the term 'facies', from the above and also from the fact that he finds it possible to define within a facies (facies of the first order) facies of the second and even of the third order, it follows that in many cases he understands 'facies' in a wider sense than do several other geographers (e.g., V. B. Sochava). It is certainly clear that in Prokaev's opinion 'facies' is far from always coinciding with 'biogeocoenose'.

The term 'facies' was proposed for use in botany by S. I. Korzhinskii (1888) in a very wide sense (e.g., a facies of coniferous forests). Later, in Western European literature, minor varieties of plant associations were called facies. On the whole the term 'facies' has been, and is being, used with widely differing meanings. Recently there has been an attempt to use it also in pedology. There, however (E. N. Ivanova and Rozov 1959, Gorshenin 1960) 'facies' is understood in a very wide sense, e.g., the central chernozem facies embraces the chernozem belt of Western Siberia and the European part of the U.S.S.R., with different climatic conditions, vegetation, and animals.

Recently, while several of our geographers have been insisting on using the word 'facies' in the narrow landscape-science sense, the botanist B. P. Kolesnikov (1956) has again returned to the wider meaning of the term.

Moreover, even by geographers the expression 'facies' has been used in a very wide sense. V. M. Chetyrkin, in presenting his classification system for dividing Central Asia into districts, writes: 'We shall begin our analysis with one of the highest ranks in our taxonomic system—the geofacies. That is the basic unit for groups of macrolandscapes' (1960, p. 95). Chetyrkin divides almost all of Central Asia into only three geofacies: (*a*) Turanian, (*b*) Dzungaro-Tian-Shan, and (*c*) Central Kazakhstan.

The view has also been expressed that, since the same term often has different meanings in different fields of knowledge, the term 'facies' may be understood differently in landscape-science, in geology, in pedology, and in phytocoenology. That is, in general, undesirable. In an extreme case it is admissible, but only when the fields of knowledge are far apart. In the present case, where the fields concerned are closely linked together and are partly interdependent, use of the same term in different senses is entirely unsuitable.

The convenience of the term 'biogeocoenose' is evidenced by the fact that A. G. Isachenko, while retaining the term 'facies' for the lowest taxonomic unit, considers it more appropriate for himself to declare that 'biogeocoenotic' changes in forest vegetation are extremely multiform (1953, p. 116); and, speaking of the phytocultural landscapes of Yu. P. Byallovich, he says that in that case 'it would be more precise to say "phytocultural biogeocoenose" ' and that the latter is generally 'a category of complex rather of biogeocoenotic than of landscape order' (pp. 145, 146).

F. N. Mil'kov (1959, p. 48) also agrees that the term 'facies' is unsuitable for denoting the phenomenon under discussion; and, citing its use in widely differing senses, in particular its broad geological meaning, considers that the term 'biogeocoenose' is 'without doubt satisfactory to us to a great extent'. But he is disturbed by the 'three-storey structure' of the word. Therefore he recommends returning to the term 'geocoenose' (which I had used earlier), considering it more convenient because the term 'biogeocoenose' can hardly be applied to areas of scree, shifting sands, deserts almost devoid of vegetation, ploughed fallow, etc., where one cannot distinguish a biogeocoenose because it is impossible to discern a phytocoenose there.

In my time I have proposed to replace the term 'geocoenose' by 'biogeocoenose' in order to emphasise the major role played by living organisms in that complex, having in mind phyto-, zoo-, and micro-biocoenoses, i.e., biocoenoses as a whole. Although I have written that the boundaries of a biogeocoenose are as a rule defined by the boundaries of a phytocoenose, at the same time I stated that in some cases other components of such a complex take first place in it. In cases where there are no tall plants at all in a territory, there are still usually micro-organisms and often also lower animals. Therefore the term 'biogeocoenose' is appropriate in such cases. The term 'geocoenose' might in an extreme case be retained only for cases where there are actually no organisms of any kind and consequently no soil, i.e., for mineral outcrops that have just been exposed and have not yet been colonised by any organisms, even micro-organisms. Such a surface condition of mineral strata occurs very rarely, and in any case lasts only for a short time. It may be regarded as the very beginning of a formation, as an 'embryo' biogeocoenose.

The 'three-storeyed' structure of the word presents scarcely any serious inconvenience. We readily use the term 'biogeochemistry', which is also 'three-storeyed'.

Recently N. A. Solntsev described a facies thus: 'A facies, as is well known, represents the simplest natural territorial complex. It is situated on a single element of mesorelief, although it does not always occupy it entirely.' 'Facies, as is well known, comprise complexes over whose entire extent there is uniform lithology of mineral strata and a uniform type of humidification. They have a single microclimate and a single soil type, and are occupied by a single biocoenose (phytocoenose and zoocoenose)' (1961, pp. 53, 54). In a footnote he writes: 'Recently some landscape-scientists, unaware that the chief characteristic of a facies is uniformity of all its components, give the name "facies" to very much larger and more heterogeneous natural territorial complexes that actually consist of several facies. Having done that, they try to prove that uniformity of all components cannot serve as a distinguishing feature of a facies.' He strongly criticises that procedure; later, however, he writes: 'If we have in the forest zone a perfectly regular surface, formed of boulder clay and occupied by a spruce stand of uniform age, then from the landscape viewpoint the whole area is a single facies.' 'If in half of that area the forest is cut down, the forest conditions in the cut-over area are greatly changed.' Therefore he proposes to distinguish two types of facies: (*a*) original and (*b*) derivative.

With regard to these statements by Solntsev we may make the following observations. If we consider the terms 'facies' and 'biogeocoenose' synonymous, as do Solntsev and several other landscape-scientists, then not all of an equal-aged spruce stand on a regular surface formed of boulder clay is necessarily a single facies (biogeocoenose). In such circumstances, for various reasons, there may exist a spruce-and-mountain-sorrel stand without underbrush and a spruce-linden stand with linden underbrush, which will constitute different facies (biogeocoenoses). On the other hand, equality of age is far from being a necessary characteristic of a single facies. It is clear that Solntsev does not define the concept 'facies' precisely enough. His proposal to separate the above two categories of facies, as being something new, produces a rather strange impression. Anticipating a little, we must point out that sylviculturists and geobotanists have long distinguished and described original and derivative types of forest, forest communities, and forest biogeocoenoses. These concepts are more than half a century old; only the term 'facies' has not been applied to them.

We must pay further attention to the following statements by Solntsev. Noting that sometimes we find in nature units intermediate between facies and natural divisions of land, which some landscape-scientists have called 'formations', he writes: 'The term "formation" is, in our opinion, an unfortunate one. Being borrowed from geobotanists, it is at the same time very far from the meaning in which they use it for plant formations, and therefore geographical "formations" and geobotanical "formations" cannot be linked together—they denote quite different things' (p. 55). Here Solntsev adds a footnote: 'From this example it is seen with what care and thought one must approach the selection of landscape terms. . . .' We may agree unreservedly with these statements of Solntsev. But then the question arises: why does he use the term 'facies' when that term, as we have seen, has different meanings in geobotany, in pedology, and in geology?

Recently V. B. Sochava, trying to define the concept of 'landscape', writes: 'Landscape is a complex system of facies that form within the landscape territorial groupings, dynamic and differing in factorial rank. In other words, it is a whole, representing a unit of regularly-distributed, mutually-linked parts' (1962, p. 16). 'Landscape, being a basic taxonomic unit of geographical environment, is in equal degree a category of systematic natural complexes and of territorial division.' 'Species and genera of landscapes within a physico-geographical province represent steps in landscape-classification, for which the concept of landscape is fundamental. That concept is even more important for constructing a system of finer categories of geographical environment, since its elementary subdivisions, facies, are classified within landscapes.' 'When we speak of a certain facies, we think of a generalised representation of an assemblage of homogeneous (but not identical) areas, in which a corresponding natural complex is known.' 'In that understanding of "facies" there exists a type of elementary physico-geographical (landscape) complex, but the expression "type of facies" is superfluous.' 'Facies combine into groups of facies—a higher-rank category of landscape classification' (pp. 20, 21). 'A facies is a unique type of molecule of geographical environment' (p. 17). Sochava suggests calling the study of facies 'molecular geography'.

From these (not very clearly expressed) definitions of 'facies' we may see that Sochava, as previously, identifies it essentially with 'biogeocoenose', not explaining, however, why he prefers such a variously-interpreted term as 'facies' to the clearly-defined term 'biogeocoenose'. We cannot agree with Sochava that the term 'facies' can be used also with the meaning 'type of facies'. G. F. Morozov had already clearly shown that both theoretically and for use in practical forest typology it is necessary to separate the concepts 'stand' and 'type of stand'. In our country's geobotanical literature it has been considered expedient, and has become general practice, to use the term 'plant community' ('phytocoenose') for a concrete group of plants, homogeneous over a specific area, and 'type of plant communities' ('type of phytocoenoses' or 'plant association') as a generalised concept, uniting plant communities or phytocoenoses that are homogeneous in all their specific features. The same holds for the concepts 'biogeocoenose' and 'type of biogeocoenoses', which we shall discuss in more detail below. In all these cases 'type' is the first (lowest-rank) classification unit of these natural objects, just as in the systematics of plant species it, as the basic taxonomic unit, unites homogeneous plant organisms. The same, of course, should apply to the study of facies, although, as shown above, the use of that term is inexpedient.

To complete our discussion of the question as to which of the terms proposed for basic units of the earth's surface may be recommended for use, we must again point out that although the terms may be conventional expressions, preference should be given to that one of them that emphasises the essentials of the concept.

Therefore we cannot recommend use of the term 'facies' for the most elementary subdivision of landscape instead of the term 'biogeocoenose'. Of all the many above-quoted names for that taxonomic unit, 'facies' is the least suitable. Its continued use will cause great confusion and impede the development of both landscape-science and biogeocoenology.

It should remain only in geology, where it is widely used and has no synonyms.

The view has often been expressed that, since 'biocoenose' indicates unity with environment, there is no need to introduce the special term 'biogeocoenose', and 'biocoenose' should be used in the latter sense. Indeed, G. F. Morozov's term 'biocenosa' was understood by him in that sense. The same view was expressed by A. P. Petrov (1947). P. D. Yaroshenko (1953) was at first disinclined to accept the expediency of the concept of 'biogeocoenose', but later fully agreed with the necessity of developing the study of it. In his book *Geobotany* (1961) he devoted several pages to exposition of that concept and its relation to other concepts of similar type, and also said that the term 'biocoenose' should be kept only for an association of organisms. A number of other zoologists have objected to that point of view (Beklemishev 1931, Kashkarov 1945, and others). That objection was developed in greatest detail by K. V. and L. V. Arnol'di (1962, 1963). We must discuss their work, which in many respects is very important and interesting, at somewhat greater length.

The authors, noting that the concept of biogeocoenoses is a sound theoretical basis for subdivision of the 'film of life' of our planet, which V. N. Beklemishev called 'geomerida', give the following definitions of a biogeocoenose: 'A biogeocoenose is that grouping of organisms that in definite conditions of geographical environment and in interaction with it is capable of existing and maintaining its integrity on the basis of the life-activity of a certain group of autotrophic plants, usually successive from generation to generation, and the life-activity of a number of generations of other components linked with it, without the necessity of the introduction from outside of organic material, soil, new stocks of plants and animals, etc.' (1963, p. 66). Developing the definition further, the authors write: 'Although a biotope (or conditions of inanimate environment) may be conditionally analysed apart from the plants and animals of a biocoenose, it is essentially meaningless without the biocoenose, just as the biocoenose is meaningless without the biotope.' 'The concept of the unity of biotope and biocoenose becomes constantly more widespread in our literature; it is methodologically correct, since without that unity there generally does not exist any complete grouping of plants, micro-organisms, and animals, but only an artificial and purely mechanical combination of certain organisms or individual coenological links or pieces of chain, separated from the whole.' 'All of these representations speak in favour of the possibility or even, as we think, the expediency of using instead of the term "biogeocoenose" the simpler and more usual term "biocoenose", or "coenose", if one always understands by it what V. N. Sukachev included in the term, although that author (1961) objected to such assimilation' (p. 168).

The authors' next paragraph, however, shows in effect how difficult it is to dispense with the term 'biogeocoenose'. In that paragraph the authors write of 'the environment of a biocoenose', having in mind the environment that they call 'internal' (i.e., what botanists call 'phyto-environment') and which enters entirely into a biogeocoenose as one of its components. Later they often speak separately of a biocoenose and its internal environment, which of course cannot be done if 'biocoenose' is held to be a synonym of 'biogeocoenose'. If we pay attention to the above-quoted definition of a biogeocoenose given by K. V. and L. V. Arnol'di, it is flatly stated therein that a biogeocoenose is a grouping of organisms that exists in interaction with environmental conditions. But, as is always emphasised, a biogeocoenose is not a grouping of organisms but a complex of organisms and the environment in which they live. Organisms (plants, animals, and micro-organisms) are merely components of that assemblage, into which enter equally the components of atmosphere, mineral strata, soil, and water conditions of that part of the earth's surface occupied by the biogeocoenose. Therefore between K. V. and L. V. Arnol'di's concept of a biocoenose (incorrectly equated by them to a biogeocoenose) and ours there is a fundamental difference.

The correct statement by these authors that a biocoenose is meaningless without a biotope, and a biotope without a biocoenose, still does not mean that we cannot speak separately of a biocoenose and a biotope in describing a biogeocoenose. Having a picture of a biotope presented implies the necessity of having also a specific picture of a biocoenose. That is evident from all of the authors' articles. Even when we have a complete organism in which the dependence of parts and organs on each other and on the whole is even closer than in a biogeocoenose, we examine their inter-relationships separately.

Approximately the same view is held by G. A. Viktorov (1962), who states that the term 'biogeocoenose' does not belong to the field of biology, since in it 'biotic and abiotic factors enter as equal members'.

The above essential difference between the biocoenose of K. V. and L. V. Arnol'di and our biogeocoenose is of great importance—not only theoretical, but also doubly practical. In many cases it is only by equal study of all components of a biogeocoenose that we can discover scientifically its fullest and most rational method of utilisation for the national economy. More than ten years ago the 'Preliminary programme of stationary complex biogeocoenotic investigations' (compendium in *Zemlevedenie*, No. 3, 1950) was compiled. It is, of course, out of date for the present day, but from it we may see that even then the approach to study of biogeocoenoses was different from the approach to study of biocoenoses, regardless of the degree to which the dependence of a biocoenose on its environment was recognised.

All these considerations give us grounds for concluding that, besides the biocoenose concept, there should also exist the concepts of phyto-, zoo-, and micro-biocoenoses,[1] although K. V. and L. V. Arnol'di, like

[1] Recently N. I. Nomokonov (1963) has also strongly expressed the same view.

several other zoologists, consider it incorrect to speak of these separate coenoses, as they are interlinked and form a single organism-complex—a biogeocoenose. The same applies to the other components of a biogeo-coenose, such as atmosphere and soil. Nobody, however, will deny the necessity of separating atmosphere and soil as independent components of a biogeocoenose. When we study and analyse a biogeocoenose we cannot dispense with the names of its separate components, including its vege-table, animal, and microbic populations, which it is necessary to call coenoses (as is usually done by botanists and microbiologists). If we consider the most important characteristic of a coenose to be the inter-actions and inter-relationships of its members, then plants among them-selves, animals also among themselves, and micro-organisms in a bio-coenose are in definite inter-relationships, and these inter-relationships have a specific character for the plant, for the animal, and for the microbic components of biogeocoenoses. While the animals and micro-organisms in a biogeocoenose are not in such close proximity to each other as are the plants, they still interact upon each other to some extent, directly or indirectly.

Taking into account these facts, and also the fact that each component of a biogeocoenose is studied by a specific branch of science and that specific knowledge and specific methods are required of the investigator, it becomes perfectly justified and necessary to differentiate in a biogeo-coenose not only a biotope with its individual components (local atmo-sphere, local mineral strata, local soil, and local hydrological conditions), but also a biocoenose with its components (phyto-, zoo-, and micro-biocoenoses). But we cannot regard these, like soil, as specific biogeo-coenoses. We stress that point because Villee (1964, p. 135) considers them to be specific ecosystems. That again indicates the difference between a biogeocoenose and an ecosystem.

In foreign, and sometimes also in Soviet, literature, the term 'biota' is used for organisms as components of an ecosystem.[1] It is doubtful whether there is need for that term to be introduced. It refers to living organisms, but does not emphasise that together they form a single whole, a coenose. Therefore for the total of all living organisms in a biogeo-coenose (plants, animals, micro-organisms) it is preferable to use the long-established term 'biocoenose'. For the same reason it is also un-suitable to use in biogeocoenology the expression 'living matter', proposed by V. I. Vernadskii, who looked upon it from the biogeochemical point of view.

The question may arise whether it is proper to speak of a microbio-coenose, since micro-organisms are either animals or plants. Certain micro-organisms, however, play a substantial and at the same time a

[1] In American literature the term 'biome' is also used, with a different content. Ap-parently it is more widely used as meaning a very large terrestrial union of ecosystems —tundras, taigas, steppes, deserts, etc. (E. Odum 1959, p. 384, et seq.)

unique role in the life of a biogeocoenose, and require special methods for their study; therefore it is proper to differentiate them in a special coenose (a microbiocoenose) and to look upon them as a specific component of a biogeocoenose.

DEFINITION OF THE CONCEPT OF FOREST BIOGEOCOENOSE; ITS COMPONENTS AND CHIEF CHARACTERISTICS

Reviewing all that has been said above about biogeocoenoses, we may give the following definition of the term: *A biogeocoenose is a combination on a specific area of the earth's surface of homogeneous natural phenomena (atmosphere, mineral strata, vegetable, animal, and microbic life, soil, and water conditions), possessing its own specific type of interaction of these components and a definite type of interchange of their matter and energy among themselves and with other natural phenomena, and representing an internally-contradictory dialectical unity, being in constant movement and development.*

N. P. Naumov writes in the compendium *Outline of the dialectics of living nature* (1963, p. 123) that 'biogeocoenose' as an expression of the close and deep connections between a biocoenose and its environment represents without doubt a unity, at the base of which lies a contradiction and adaptation to concrete local conditions.

Another characteristic feature of a biogeocoenose is the inter-relationships of its components among themselves. Although the terms 'inter-relationships', 'interaction', and 'interdependence' are often considered to have the same meaning (Armand 1949) and are used as synonyms, they are in fact quite different. Not every interaction is an inter-relationship. Trees growing side by side in a forest interact on each other, but, strictly speaking, one cannot say that they are inter-related.

N. V. Timofeev-Resovskii holds that a characteristic of an individual biogeocoenose is that within that 'biochorological unit' 'there exist no biocoenotical, geomorphological, hydrological, climatic, or pedological-geochemical boundaries' (1961, p. 25). We may fully agree with that supplement to the definition, if by 'biochorological unit' we understand not a biocoenotic unit of soil cover but a biogeocoenotic unit in the sense described above.

Sometimes a phytocoenose (and also a biocoenose and a biogeocoenose) is spoken of as if it were a complete biological system, just as one speaks of an organism as a whole. We cannot accept that as correct. By a complete biological system we must understand a system in which the component parts have no independent significance, cannot exist independently, and cannot function outside the system. This specific feature is characteristic of an organism. It is absent in a phytocoenose, a biocoenose, and a biogeocoenose. Many of their component parts, when separated from the whole, can as a rule exist and continue to fulfil their chief life functions, although altering the latter in accordance with the conditions of the different environment. But phytocoenoses, like biocoenoses and bio-geocoenoses, are typical dialectical self-contradictory units.

By *forest biogeocoenose* we may understand *any part of a forest, homogeneous over a specific area in the structure, composition, and characteristics of its components and in the inter-relationships among them, i.e., homogeneous in plant cover, in the animal and microbic life inhabiting it, in the surface of mineral strata, in hydrological, microclimatic (atmospheric), and soil conditions and interaction among them, and in the type of exchange of matter and energy between its components and other natural phenomena.* The branch of biogeocoenology that studies forest biogeocoenoses may be called *forest biogeocoenology*.

In Fig. 1 are shown diagrammatically the directions of the chief inter-actions of the components of a forest, which include *vegetation* (trees, shrubs, herbage, mosses, lichens, water plants, fungi), *animal life* inhabiting the forest (mammals, birds, and other vertebrates, insects, worms, molluscs, and other invertebrates), and *soil and subsurface mineral strata* as far below ground level and *atmosphere* as far above ground level as mutual influence is exerted between them and other components of the

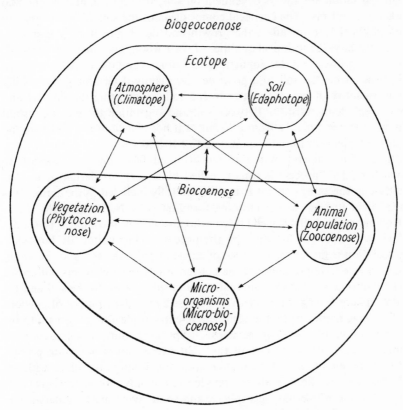

Fig. 1. Diagram of interactions of components of a biogeocoenose.

forest. In view of the fact that microscopic organisms (plant and animal), both by the methods of studying them and by their interaction with other components of the forest, have very specific features, it is expedient to regard *micro-organisms* (bacteria, fungi, actinomycetes, infusoria, amoebae, etc.) as a specific component of the forest, a microbiocoenose.

The interactions of these six forest components are extremely varied and complex. And since all interactions of the components are connected and interdependent, the more fully and deeply they are studied the sounder will be the measures applied by forest management. Therefore forest biogeocoenology includes knowledge of all the characteristics of forest components, determining their interactions, and establishing the laws that govern the latter and the development of forest biogeocoenoses as a whole (Fig. 2).

The various aspects of forest life will be discussed in special chapters below in more detail, and an attempt will be made on the basis of present knowledge to describe the forest components from the biogeocoenotic point of view and to demonstrate the principal inter-relationships among them. We can make the following general statement. Trees and other tall vegetation are always dependent on soil, atmosphere, animal life, and micro-organisms. The chemical composition of the soil, its moisture, and its physical features affect the growth and development of tree species, their fertility, the technical qualities of their wood, and their regeneration, and the growth and development of all other vegetation. All vegetation in turn affects the soil to a great degree, mainly determining the quality and quantity of organic matter in the soil, so affecting its physical and chemical characteristics. Between soil and vegetation there is a constant flow of mineral matter from different soil horizons into the underground parts of plants, with later conversion of it into soil in the form of plant litter. Thus mineral matter is redistributed from one soil horizon to another. This process is usually called biological circulation of matter, although the same content is not always fully included in the term. It is expedient to keep to the meaning formulated by N. P. Remezov, L. E. Rodin, and N. I. Bazilevich (1963), who also give well-compiled methodological instructions for study of the process. They write: 'By biological circulation is meant the entrance of elements from the soil and the atmosphere into living organisms; the conversion of these elements within the latter into new complex combinations and their return to the soil and the atmosphere during the life-activity process by the daily output of portions of organic matter or by the complete re-entry of dead organisms into the composition of the biogeocoenose.' Speaking thus of circulation of matter, we must bear in mind that that expression is not quite precise. Complete circulation of all matter does not usually take place within a biogeocoenose. Some matter always leaves not only a given biogeocoenose but also the whole phytogeosphere, and some living matter enters a biogeocoenose from outside.

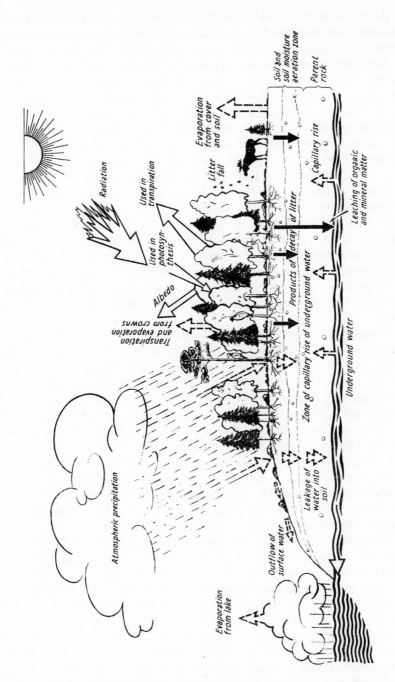

Fig. 2. Diagram of mutual influence of a forest biogeocoenose and environmental conditions (from A. A. Molchanov and N. V. Dylis).

A very significant aspect of biological circulation of matter is that it is inevitably accompanied by the process of transfer and transformation of energy, the balance of which it is most important to take into account.

This problem, which in general has as yet been little studied, is relatively better explained by hydrobiologists who have studied aquatic biogeo-coenoses. Particularly significant in this connection are the works of G. G. Vinberg, in which he uses the term 'ecosystem' in the sense of 'biogeocoenose'. He quite correctly writes: 'Knowledge of the intensity of exchange by components of an ecosystem, their biomass and coefficients of use of food energy in growth enables us to describe quantitatively the rate of biotic circulation, its degree of convertibility, and the actual and potential productivity and type of the ecosystem. Only by quantitatively expressing the inter-relationships among components of an ecosystem are we enabled to determine in what cases a specific link in the food chain or trophic level is limited by energy resources when these are fully used, and in what cases the degree of development of a given trophic level with the presence of an adequate energy base is limited by other factors, which is of primary importance in determining ways of managing ecosystems' (1962, p. 15). (See also the works of the hydrobiologist V. S. Ivliev 1962, etc.)

Considerable attention has been given recently in foreign literature to the energy aspect of biological circulation of matter in aquatic biogeo-coenoses; but little has actually yet been done in this respect for forest biogeocoenoses. This is particularly evident in the well-known ecology textbook of E. Odum (1959), although the importance of studying the energy balance of an ecosystem is discussed in it. It is now the task of students of forest biogeocoenology to devise methods of studying the energy aspect of 'biological circulation' of matter in forest biogeocoenoses and to carry out appropriate investigations in different types of these.

An important part is played in that process by shedding and by the so-called 'forest litter', i.e., accumulation on the soil surface of remains of leaves, twigs, fruits, and other parts of plants. Such plant remains undergo decomposition and mineralisation in forest litter.

Vegetation also plays a large part in the water regime of the soil, ab-sorbing moisture from certain soil horizons and later passing it into the atmosphere by transpiration, by modifying evaporation from the soil surface, and by affecting surface flow and underground movement of water. The effects of vegetation on soil conditions depend on its com-position, age, height, size, and density.

Although water has long been regarded as a most important ecological factor and the effects of vegetation on it have long been studied, S. D. Muraveiskii (1948) drew special attention to the fact that in determining the significance of effects of vegetation the role of water flow (both surface and underground) had usually been underestimated; water flow is parti-cularly important as the source of water and biogenic salts to the vegetation, and consequently regulates the distribution of vegetation. At the same time vegetation, especially forests, has a very great effect on all types of water flow.

No less complex are the interactions observed between vegetation and the atmosphere. While the growth and development of vegetation depend on the temperature, humidity, movement, and composition of the atmosphere, in turn the composition, height, layering, and density of vegetation affect these atmospheric conditions.

We must also remark in this connection that any natural factor, apparently of little significance, may play a large and versatile role in a biogeocoenose. For instance, among atmospheric factors the effect of the wind in breaking twigs off the crowns of trees is often underestimated. M. I. Sakharov (1947) demonstrated that inter-crown and intra-crown breaking of twigs strongly affects the formation of tree crowns, the productivity and technical qualities of the trunks, the development of root systems, and the health of the trees and consequently their resistance to pests and their fruitfulness; and, by increasing the distance between trees and lessening the depth and density of the canopy, causes substantial changes in the phytoclimate, litter, and soil processes, which ultimately affect the development of flora and fauna under the crowns.

Every biogeocoenose has its own climate—its microclimate, sometimes called phytoclimate—by which we understand those atmospheric conditions that are changed by the vegetation itself.

No less close is the inter-relationship between vegetation and the animal life inhabiting a biogeocoenose. Animals, in the process of their life-activity, affect vegetation in many ways: directly by eating it, trampling it, constructing their nests, refuges, etc., in or with it, pollinating flowers, and distributing seeds and fruits; and indirectly by altering the soil, manuring it, loosening it, and generally changing its chemical and physical properties, and to some degree also affecting the atmosphere.

Zoologists studying biogeocoenoses have long since analysed the so-called food chains of organisms, and in that connection speak of the so-called 'pyramid of numbers' rule formulated by C. Elton (1927), to which Allee, Park, et al. (1949) gave universal significance. Viktorov (1960, p. 403), following Elton, drew an important general biogeocoenotic conclusion from that rule, that prey organisms have greater biomass than their predators: without that relation a biocoenose could have no stable existence. Biogeocoenoses, as a rule, consist of four categories of interacting components: (a) abiotic parts; (b) producers—autotrophic organisms that create organic matter, (c) consumers (requiring organic matter) —macroconsumers, heterotrophic organisms, mainly animals, and (d) reducers (microconsumers, saprobes, and saprophytes)—heterotrophic organisms, mainly bacteria and fungi (E. Odum 1959).

The interactions between vegetation and micro-organisms are extremely important. Apart from the fact that micro-organisms (bacteria, fungi, viruses, etc.) often parasitise the higher plants, their role in the soil is a large one. Not counting nodule bacteria, which live on the roots of leguminous and other plants and fix free nitrogen, to be later used by higher plants, various other micro-organisms exert positive and negative influences on the growth and development of higher plants by means of their exudates in the soil, their participation in the distribution of organic matter, their absorption of various gaseous components of the atmosphere, and their transformation of matter in the soil. On the other hand, root exudates of higher plants affect the microbic population of the soil so that within the rhizosphere of these plants, depending on the plant species, the composition of that population varies greatly.

At the same time micro-organisms exist in direct or indirect interaction with animals, both vertebrates and invertebrates.

It has long been known that in a plant community all other plants (as well as micro-organisms) also have interactions, which are often multi-form. In a forest community these interactions are particularly many-faceted and complex. The inter-relationships between plants (phanero-gams and higher sporophytes) are expressed either in the direct action of one on another (parasitism, e.g., by mistletoe on pine, linden, etc.; inter-growth of roots, permitting use by one plant of food material from the root system of another; breaking off or damaging of twigs of one tree by another, e.g., spruce by birch; direct pressure of root systems on each other; etc.); through other organisms (micro-organisms, animals); through exudation of matter (liquid, gaseous, or solid) by above-ground or underground parts, so affecting other plants positively or negatively; or through competition of root systems for water and minerals. Often merely the closeness of above-ground parts leads to certain plants affecting others by altering their environmental conditions, e.g., modification of wind action and protection from windfalls and storm damage are ob-served, and changes occur in the blossoming conditions, water regime, etc., of plants. Finally, forest litter accumulates from parts of plants that die off and are shed (leaves, twigs, fruits, seeds, etc.) and, as stated above, not only affects plants indirectly by altering soil processes but also creates specific conditions for germination of seeds and development of shoots. Thus these mutual influences of plants may be either favourable or un-favourable for their growth and development. In the first case we may speak conventionally of 'mutual aid', in the second of 'struggle for existence' among plants (in the broad Darwinian sense) or of competition. It is to be understood that all these interactions among organisms in a biogeocoenose play at the same time a great part in the biogeocoenose as a whole. They may be observed between individuals both of the same species and of different species, i.e., they may be intraspecific or inter-specific.

S. A. Il'inskaya (1963, p. 9), using a supplementary classification of phytocoenotypes proposed by G. I. Poplavskaya (1924) and V. N. Sukachev (1926), subdivided plants according to their roles in phytocoenoses as follows: (a) edifiers—plants that most actively and profoundly transform factors of the environment of the plants and create a specific internal coeno-environment; (b) associates—plants of secondary, subordinate structural formations; (c) destructors—species that by their ecologo-phytocoenotic features are opposed to associates: they adapt themselves to the circumstances created by the edifiers. According to their coeno-genetic significance, destructors may be progressive or regressive. Progressive destructors often break down existing communities. Regressive destructors are species that survive in a given community from earlier stages of coenogenesis; (d) neutral species, which play no substantial role in the formation of a coenose—their presence in a community is accidental. That classification also has biogeocoenotic significance.

Not only does vegetation interact with other components of a biogeocoenose, but the latter interact with each other. Climatic conditions (atmosphere) affect the soil-formation process, and soil processes, by determining so-called soil respiration (exudation of CO_2 and other gases) alter the atmosphere. The soil affects animal life, not only its own inhabitants but also all other animals. Animal life affects the soil. V. V. Dokuchaev includes animals among soil-formers. Micro-organisms, by their activity in the soil, break down certain compounds, organic and inorganic, and create new substances (including gases) that affect the atmosphere, animal life, and, as stated above, vegetation. We might go still more deeply into the inter-relationships of components of a biogeocoenose, but the above is sufficient to show how multiform they are. These inter-relationships have a dual character, direct and indirect. Indirect interaction always exists, because all components are mutually inter-related and interdependent.

We must also, however, consider the fact that each of the above components of a biogeocoenose is in turn a complex phenomenon, all the component parts and features of which interact with each other.

All of the above-mentioned interactions of the components of a forest biogeocoenose, as of any other biogeocoenose, are displayed, as has been noted above, in the exchange of matter and energy between them. The chief source of energy is solar energy, and green vegetation is outstanding as an accumulator of it. That energy enters other organisms and the soil through vegetation. In addition, heat energy directly and indirectly affects processes in both animate and inanimate components.

Moreover, in one way or another every biogeocoenose affects other biogeocoenoses and natural phenomena in general, whether adjacent to it or at some distance. Exchange of matter and energy takes place not only between components of a given biogeocoenose but also between them and other natural phenomena. While green plants act as the chief accumulator of energy received from the sun, the same plants, as well as animals and plants other than green ones, liberate it. These basic processes are accompanied by various other processes of transformation of matter and energy and exchange of them between the components of a biogeocoenose.

The atmosphere, bedrock, soils, and subsoils are, as it were, the primary materials of a given biogeocoenose; and plants, animals, and micro-organisms are predominantly transformers and exchangers of matter and energy. But soil and forest litter have special significance. They summarise, as stated above, the results of complex biogeocoenotic processes, and are an obvious and very complete expression of the end-products of such processes.

The above, of course, is only an outline of the biogeocoenotic process that reflects the basic functions of components of biogeocoenoses. In fact, each of these components (atmosphere, living organisms, soil, forest litter, and often the underlying mineral strata) is to some extent also material for the biogeocoenotic process, a transformer, a means of exchange, and an end-product of that process (Vernadskii 1927, 1940, Perel'man 1961).

From that viewpoint great interest attaches to data compiled by E. M. Lavrenko, V. N. Andreev, and V. L. Leont'ev (1955) on the gross mass of above-ground vegetation and the annual growth of that mass (in air-dried condition) in different plant associations of 'plakor'[1] habitats in the European part of the U.S.S.R. Whereas the gross mass of herbage and shrub communities amounts to tens of centners per hectare (e.g., meadow-steppe 15 c/ha, birch-green-moss tundra 73 c/ha), for forest communities it is expressed in thousands (e.g., green-moss-bilberry-spruce central taiga 1300 c/ha, oak grove grown from seed 2600 c/ha). The annual growth of the above-ground mass in air-dried condition in centners per hectare for the same communities is given in the following figures: meadow steppe 15 c/ha, birch-green-moss tundra 14 c/ha, green-moss-bilberry-spruce taiga 30 c/ha, and seeded oak grove 56 c/ha. These figures show how much forest exceeds other types of vegetation in accumulation of organic matter and energy, and in the amount of its action on the biological circulation of matter and energy in a biogeocoenose. These figures do not include the mass of underground parts of plants, and of animals living in these phytocoenoses. Although the mass of the latter is very small relatively to the vegetation mass, inclusion of the underground parts of plants in the above figures would increase still more the total mass of living matter in the forest biogeocoenoses as compared with the others.

PLANETARY AND COSMIC ROLE OF BIOGEOCOENOSES IN GENERAL AND OF FOREST BIOGEOCOENOSES IN PARTICULAR

A biogeocoenose as a whole is a laboratory in which a process of accumulation and transformation of energy takes place, as the result of a com-

---

[1] Plakor—flat or slightly rolling areas above valley rims; slopes of watersheds in steppe and forest-steppe zones. (Tr.)

bination of many multiform physiological, physical, and chemical processes, which also interact among themselves.

To evaluate the complete significance of the processes that take place in biogeocoenoses, we must keep in mind the study of the biosphere devised by V. I. Vernadskii. Vernadskii writes that the substance of the biosphere, on account of cosmic radiations (among which solar radiation is the most important), 'is permeated with energy; it becomes active, and collects and distributes through the biosphere energy received in the form of radiation, which in the terrestrial environment is freed and is capable of doing work' (1927, p. 9). Later Vernadskii shows that the biosphere fulfils the following main chemical functions: '(a) the gas function: all gases of the biosphere (N_2, O_2, CO_2, CH_4, H_2, NH_3, H_2S) are created and altered by the biogenic method; (b) the oxygen function: atmospheric oxygen is the result of photosynthesis, performed by chlorophyll-bearing plants; (c) the oxidising function, performed mainly by autotrophic bacteria; (d) the calcifying function: mainly formation of $CaCO_3$; (e) the restoration function: creation of H_2S, FeS_2, etc.; (f) the concentration function: displayed in the accumulation of certain chemical elements distributed through the environment; (g) the function of breaking down organic compounds: performed mainly by bacteria (and fungi); and other functions.' Therefore V. I. Vernadskii considers that 'from the geochemical point of view the whole existence of the earth's crust, at least the mass of its matter, is substantially conditioned by life' (Vernadskii 1934 a, pp. 191 et seq.; Vassoevich 1963, p. 48). 'On the earth's surface there is no chemical force more constantly acting, and therefore more powerful in its ultimate results, than living organisms taken as a whole. And the more we study the chemical phenomena of the biosphere, the more we are convinced that there are no items in it not dependent on life' (Vernadskii 1934, p. 25).

These ideas of Vernadskii, as is well known, have been further widely developed both here in the Soviet Union and abroad. In that respect much work has been done here, as noted above, by A. I. Perel'man.

The works of V. V. Koval'skii are also of great interest: he developed a special branch of ecology, which he called 'geochemical ecology'. It consists in study of the role of chemical factors in the natural environment of living organisms and in their life processes. Although the statement of that problem was not new, and botanists and agronomists in particular had given much study to the physiological and ecological roles of different chemical elements and their compounds, Koval'skii is credited with the direction of work in that field and with a broad geochemical approach to the problem. He asserts that 'as a result of geological and geochemical processes (including the vast activity of living organisms) that have taken place, and are taking place, in different ways in different zones of the earth, the chemical composition of components of the biosphere—soils, natural waters, organisms—displays geographical heterogeneity and has a mosaic character' (1963, p. 830). That mosaic character of the biosphere is also a result of the fact that the chemistry of natural phenomena linked with it reflects the circumstance that it is composed of different biogeocoenoses.

Another important point to which Koval'skii gives special attention is the role played in the life of organisms by trace elements (micro-elements) (see his work of 1962, in which are listed other works of his in this field). Among the general laws on which geochemical ecology is based, of particular value to biogeocoenology is that 'in nature a single chemical element never acts in isolation; the relation between elements is significant, and therefore in organisms we observe not only the strengthening or weakening of certain processes, but even disruptions of the process of exchange of matter' (Koval'skii 1962, p. 17).

Hence we may see that geochemical ecology as a special scientific discipline enters, as it were, on account of its object of study and its specific nature, into the composition of biogeocoenology; little, however, has yet been done in that field. The value of such work, of course, is beyond doubt.

It is widely known that plants and plant cover in natural conditions do much work in the accumulation and vertical distribution (in the soil) of various chemical elements, including trace elements. But we must remember that they do the same with regard to the horizontal distribution of elements, different species of plants acting differently in that respect. That is well demonstrated by the use of radioactive isotopes of micro-elements (Rachkov 1963).

If we consider that the biogeocoenose is the basic unit in which take place the processes that ultimately determine the vast geochemical (on a global scale) and cosmic role of the biosphere, which since the work of V. I. Vernadskii has been constantly drawing increasing attention from scientists of the most varied disciplines, then it becomes clear how valuable is the study of all aspects of life in a biogeocoenose, i.e., biogeocoenology. The latter, by synthesising the results of study of biogeocoenose components by different biological and physico-geographical sciences that apply their own study methods to biogeocoenoses as dialectical unities, produces ultimately a generalisation of the widest, universal scientific significance.

If we bear in mind that the forest is the largest 'film of nature', that nowhere in nature is the action of living matter on inanimate nature so great and so varied as in the forest, we may see how important, theoretically and practically, is the study of forest biogeocoenoses, i.e., forest biogeocoenology.

Biogeocoenoses are found everywhere on the earth's surface where plant or animal organisms exist, regardless of whether or not that surface is covered with water. The aggregate of all biogeocoenoses on the earth's surface constitutes the *biogeocoenotic cover of the earth*. It enters into the composition of the biosphere in Vernadskii's sense, a concept that we have already used. The biosphere is a relatively thick (vertically) envelope of the earth, including: (*a*) the troposphere, (*b*) the hydrosphere, (*c*) the superficial part of the dry land of continents and islands, and (*d*) the part of the lithosphere inhabited by life. E. M. Lavrenko has proposed the name 'phytogeosphere' for that part of the biosphere that corresponds to Vernadskii's 'film of life'. It includes 'not only the plant cover together with the animal and microbic population linked to it, but also the environment penetrated by organisms, i.e., soil and subsoil, and the lower layers of the atmosphere (more precisely, of the troposphere), both these penetrated by plants and those somewhat (up to a few metres) higher than the topmost canopy of plant cover. The phytogeosphere of dry land is the main sphere of agricultural and forestry activities of man. The unit of division of the phytogeosphere is the biogeocoenose' (Lavrenko 1962, pp. 13-14). Thus the concept of the phytogeosphere corresponds to biogeocoenotic cover. Mil'kov (1959) distinguishes the 'landscape geosphere', embracing all landscapes on earth and having a vertical thickness of some hundreds of metres. Although he likens it to Lavrenko's phytogeosphere, his 'landscape geosphere' obviously exceeds the latter in dimensions. I. M. Zabelin (1963, pp. 22-23) distinguishes a special envelope of the earth: the biogenosphere, i.e., the sphere in which life arises, the upper boundary of which is at an altitude of 8 to 18 km (average altitude slightly over 10 km), and the lower boundary under the continents at a depth of about 5 km. From the biogeocoenotic point of view the most interesting of all these envelopes of the earth is the 'phytogeosphere' of E. M. Lavrenko.

Since the upper and lower boundaries of the phytogeosphere (biogeocoenotic cover) are determined, as stated above, mainly by the upper and lower limits of plant cover, and by a higher layer of the atmosphere and a layer of mineral strata that are markedly affected by vegetation, naturally these boundaries depend on the amount of plant cover, and in different types of biogeocoenotic cover they will be located at different altitudes and depths. Therefore we must define precisely the upper and lower limits of biogeocoenoses. We must include in a biogeocoenose that layer of atmosphere that is so altered in its physical and chemical features by the other components of the biogeocoenose that it becomes qualitatively different. The same must be said also of the lower boundary of the biogeocoenose; namely, that a biogeocoenose extends downward to the lower limit of the zone in which the mineral strata have so changed their physical and chemical features under the influence of other components that they have become qualitatively different.

Therefore the existence of a small number of micro-organisms at a considerable altitude in the atmosphere, or the penetration there of some small effect of the plant cover or of gaseous effluents from the soil, does not give grounds for including these horizons of the atmosphere, or, more precisely, of the troposphere, in a biogeocoenose.

Although the above criterion (qualitative alteration of the atmosphere and mineral strata) defines the upper and lower boundaries of biogeocoenoses as well as their horizontal boundaries, it still does not exclude frequent difficulties in determining them, since a certain subjectivity exists in evaluation of the qualitative alterations in these components. Sometimes persons who are insufficiently experienced in the study of natural phenomena reproach both biogeocoenologists and phytocoenologists for not giving criteria for defining separate biogeocoenoses and phytocoenoses in nature precise enough to exclude any vagueness in distinguishing them and drawing their boundaries. They do not take into account the fact that in nature certain phenomena and objects are sharply delimited, whereas between others there are often transitions that cause the drawing of boundaries between them to be arbitrary and to some degree subjective, e.g., in establishing species of organisms, varieties of soil, types of strata and minerals, types of climate, etc.

Different biogeocoenoses have, of course, different vertical dimensions, e.g., forests, steppes, deserts, etc. As a rule, however, we may consider that the upper limit of a biogeocoenose lies a few metres above the plant cover, and the lower one a few metres below the soil surface. Individual roots that occasionally penetrate deeper need not be taken into account.

Thus the vertical dimension of a biogeocoenose, i.e., the distance between its upper and lower boundaries, is usually less than that dimension of a landscape. Even an elementary landscape in A. I. Perel'man's sense will be of greater vertical depth, since its upper limit 'is determined by the zone of dispersion of terrestrial dust (from that or an adjacent landscape), which is occupied by birds and insects', and its lower limit by the water-bearing horizon (Perel'man 1961, pp. 21-22).

Every biogeocoenose that occupies a definite place in nature is adapted to some element of relief. Therefore, as we have seen, most landscape-scientists look upon relief, as well as the level of underground water, as components of landscape, and some also include therein man and all his activities.

We include among the components of a biogeocoenose only the material elements that compose it. Therefore the components of a biogeocoenose comprise the atmospheric layer adjacent to the earth, bedrock, soil, water in the soil and the atmosphere, and biocoenoses, which include plants, animals, and micro-organisms. Relief and the level of underground waters, like terrestrial gravitation, are only factors of the existence of a biogeocoenose.[1] Material bodies, interacting among themselves and deter-

[1] A. G. Isachenko (1953, p. 71) writes, however: 'Artificial exclusion of relief from any territorial complex, however small, contradicts the long-established fact of adaptation of soils and vegetation, in their details, primarily to elements of relief', quoting Larin, Ponomarev, and Ramenskii. We may, of course, supplement that list of authors with the name of V. N. Sukachev, as I have long ago and often (Sukachev 1904, 1915, 1930, 1931, 1934) pointed out the great influence of relief on the distribution of plant communities and consequently of biogeocoenoses, and the adaptation of these types

mining the nature of a biogeocoenose, may also be components of a biogeocoenose. Some biogeocoenose components that exist outside the boundaries of a given biogeocoenose may at the same time create conditions affecting that biogeocoenose to some degree. For instance, take a herbaceous biogeocoenose in a glade in a forest. Trees composing the forest and surrounding the herbaceous biogeocoenose may affect both the movement and the composition of the air that constitutes one of the components of the herbaceous biogeocoenose.

Man also is a very powerful factor affecting natural biogeocoenoses. Human society, which depends to a great extent on natural biogeocoenoses affects them in many different ways.

Since nowadays there are no forest biogeocoenoses that are not subject to the economic, and sometimes the non-economic, activity of man, it is absolutely necessary to take account of that factor. Man himself, however, as has been noted above, should not be included among biogeocoenose components. He is only in the highest degree a powerful factor, able not only to alter forest biogeocoenoses to a greater or lesser extent but also through his culture to create new ones, forest culture-biogeocoenoses.

The question has often been asked: can the terms phytocoenose and biogeocoenose be applied to those parts of the earth where man has artificially created new stands of vegetation? Without dwelling here on that question as far as it concerns phytocoenoses (Sukachev 1950), we must not overlook the views expressed recently by K. V. and L. V. Arnol'di (1962) regarding biocoenoses. These authors write that complete identity of the terms agrobiocoenose and biocoenose cannot be properly admitted, because the principal characteristic of a biocoenose, by right of succession, is repeated monotypy, throughout many generations, of the processes of production and general circulation of organic matter, based on constancy of species-composition of the chief producers and consumers; and the capacity for auto-regulation of that composition is lacking in agrocoenoses. They also remark that while climatic conditions remain unchanged a natural biogeocoenose is the most stable form of utilisation of natural conditions by organisms, and that a natural assemblage strives to return to that form after any short-term destructive action affecting a biocoenose. That feature also is lacking in an agrocoenose. Analysing these and other differences between a natural biocoenose and an agrobiocoenose, they write that in summing up it must be recognised that agrocoenoses are special groupings of organisms, characterised by the

---

to definite forms of relief. From the fact that I have always written that relief does not enter into the concept of 'biogeocoenose', i.e., it is not a component of a biogeocoenose, it does not follow at all that I do not count relief as a factor affecting distribution and also, indirectly, all components of a biogeocoenose and its characteristics. Moreover, I have always stressed that the interaction of the components of a biogeocoenose may often lead to alterations in relief and to the creation of specific forms of microrelief and also (in known cases) of mesorelief.

above specific features; they have much in common, but are not identical, with biocoenoses. The discussion of the problem by these authors is worthy of close attention, but it does not prevent us from looking on formations created by man as biocoenoses and biogeocoenoses. The principal feature of every biogeocoenose—specific interaction of its components— exists there, but it goes without saying that natural and cultural bio- geocoenoses must be distinguished from each other, being regarded as two different categories of biogeocoenoses. (*Translator's note:* in this paragraph use of the terms 'agrocoenose' and 'agrobiocoenose', which may appear slightly inconsistent, is strictly in accordance with the original —as is also, of course, use of the terms 'biocoenose' and 'biogeocoenose'.)

Since a basic feature of every biogeocoenose is the presence of specific inter-relationships of its components, belonging to it alone, the natural and pertinent question arises whether these inter-relationships are of equal value in the life of the biogeocoenose, in its dynamics, in its changes. It is of particular importance to discover whether there is equality of value and significance in the inter-relationships of components belonging to 'live' and to 'dead' nature. On the answer to this question depends largely the answer to the question whether biogeocoenology should be considered as a biological or a physico-geographical discipline.

There are different points of view about this. An article by Solntsev (1960) is devoted to it. The author gives the opinion that 'in the inter- actions of "live" and "dead" nature the principle of unequal significance of interacting factors operates', that the leading factors are the lithogenic ones, followed by the weaker (and dependent thereon) group of hydro- climatic factors; and, finally, the 'weakest', depending on both the litho- genic and the hydroclimatic groups, are the biogenic factors. Therefore to the question: in the interactions of 'live' and 'dead' nature which of the two is stronger, which causes the deeper (more radical) changes in the other? Solntsev answers that (contrary to the views of most Soviet geographers, biologists, and pedologists) one may boldly state that in the interactions of 'live' and 'dead' nature the latter proves to be stronger, as is shown by its decisive influence on the characteristics of 'live' nature of to-day. Summing up his article, Solntsev stresses his principal conclusion, that 'in the interactions of "live" and "dead" nature the leading role is taken by the latter'.

To evaluate the correctness of that conclusion, let us first consider the following circumstance. When there is a part of the earth's surface not yet colonised by plants and animals, there is already interaction between the lithogenic components (mineral strata) and the hydroclimatogenic (elements of the hydrosphere and the atmosphere). We may then speak with a definite meaning of a 'geocoenose'. Both the rate of colonisation of that surface by organisms and their species-composition depend directly on the above factors, but a part is played in that process by so- called inspermation, i.e., the presence in that place of certain plant embryos (their diaspores) and the degree of their adaptation to distribution, to invasion of that area of the earth's surface. The leading role, however, in that inchoate biogeocoenosis and its life is played by factors of the ecotope, i.e., 'dead' or (as Vernadskii says) 'inanimate' nature. If under the influence of geological processes or changes in climate, i.e., exogenic factors, the features of the ecotope alter, the leading role in changes in the biogeocoenose is played by abiogenic factors. If, however, there is insignificant change in the exogenic factors, if conditions for the growth and development of plants and animals are favourable, if organic life rapidly develops and there is a considerable accumulation of organic matter in 'live' or 'dead' form, then its action on 'dead' nature in the ecotope becomes more and more powerful, and in further evolution of the biogeocoenose as an aggregate of plants, animals, and components of the ecotope, 'live' nature (i.e., the biocoenose) plays the leading part. Changes in the original substrate under the influence of vegetation are so considerable that different substrates are much levelled down and smoothed, and on the whole entirely new characteristics of the ecotope are created, expressed in the appearance of a new component of the biogeocoenose, soil. The more vigorous the development of the organic world and the greater the mass formed by it, the more clearly shown is its leading role in the life of biogeocoenoses and their successions. If the characteristics of an ecotope are unfavourable for development of organic life (e.g., in deserts or on poor substrates), the leading role in further development of the biogeocoenose is reserved for 'dead' nature.

Therefore it would be imprecise to say that in the interactions of 'live' and 'dead' nature the leading role is always played by the latter. In different biogeocoenoses the components that are most important in their interactions may be different. Therefore a biogeocoenose is neither a biological nor a physico-geographical concept, but is quite specific and differs essentially from these.

The processes of transformation of matter and energy in biogeocoenoses and their exchange with other natural bodies depend on the following: (a) on the characteristics of their primary components (plants, animals, micro-organisms, atmosphere with its climate, and mineral strata, including hydrological conditions) and their distribution in the biogeocoenoses; (b) on such conditions of the manifestation of action of these components as relief and the time of their existence; (c) on the character of the surroundings of a given biogeocoenose (its environment); and (d) on soil, which, being a biogeocoenose component of secondary, bio-inanimate (Vernadskii) character, is the final result of the interaction of all of them.

That conclusion was well expressed thirty-five years ago by our greatest pedologist, B. B. Polynov. Using the term 'elementary landscape' in a sense close to 'biogeocoenose', Polynov (1925, p. 82) wrote: 'Soil is . . . wholly a product of other elements of landscape; it is substantially different from plants, animals, and mineral strata in that it does not, strictly speaking, contain its own beginning. It does not appear from without, in order to become adapted in some way to landscape: it is from the first moment of its formation a product of landscape and therefore naturally expresses its character to a much greater degree than does any other element of it.'

If the final aim of biogeocoenology is to discover all the laws governing the processes of transformation of matter and energy in biogeocoenoses and between them, in order to control these processes in the interests of man, evidently the study of soil is of great importance in the solution of that problem, especially study of the energy aspect of the process of soil formation (see the very interesting works of V. R. Volobuev).

PRINCIPLES OF DEFINITION OF BIOGEOCOENOSES IN NATURE AND THEIR CLASSIFICATION. CONCEPT OF FOREST TYPOLOGY

The question arises, what are the simplest criteria to use in defining different biogeocoenoses in nature? In this connection primary assistance may be given by analysis of relief. Although relief, as stated above, does not enter into the list of components of biogeocoenoses, being a very important factor in their existence it still may play a very great role in one's preliminary orientation within forest biogeocoenoses and in their delimitation. Within the limits of homogeneous relief the most indicative sign of homogeneity of a biogeocoenose is homogeneity of soil and of plant cover. Of these two indicators for delimitation of biogeocoenoses, homogeneity of plant cover is especially convenient because of its ease of observation. Therefore in delimiting biogeocoenoses in nature it is expedient to use phytocoenoses (plant communities). The boundaries of each separate biogeocoenose are determined, as a rule, by those of a phytocoenose. This is explained by the fact that, among the components of a biogeocoenose, phytocoenoses usually have the leading biogeo-coenose-forming role.

In determining the horizontal limits of a biogeocoenose it is useful also to consider the criteria proposed by A. I. Perel'man (1961) for elementary landscape. He writes: 'In referring any part of the earth's surface to ele-mentary landscape one must consider the possibility (even if only imagin-ary) of a given elementary landscape extending over a large territory. . . . Therefore a patch of solonchak 10 m^2 in extent is an elementary landscape, but, for instance, a single tree, an ant-hill, a heap scraped out of a burrow by a shrew, a tussock in a bog are not elementary landscapes.' That criterion can also be applied in full to biogeocoenoses.

Although every biogeocoenose, as already stated, may be characterised by a definite type of transformation of matter and energy (in particular, within its boundaries) and a definite migration of chemical elements, a biogeocoenose, like an elementary landscape, is not a geochemical land-scape in the sense given to that concept by Polynov (1946b, 1952) and

Perel'man (1961, p. 26). 'A geochemical landscape' may be 'defined as a paragenetic association of combined elementary landscapes, linked together by migration of elements and adapted to a single type of meso-relief.' Therefore watersheds, slopes, valleys, lakes are not separate sections of territory, isolated in nature, but closely-linked, interdependent parts of one whole (Perel'man 1961, p. 26). This quotation is adduced because it stresses the exchange of matter and energy between biogeo-coenoses.

Although a biogeocoenose constitutes a union of all its inter-related, interdependent components, it is not structurally homogeneous in its vertical and horizontal extensions. In the first place, its above-ground and underground parts are sharply distinguished from each other by the character of all their components. The same must be said regarding its horizontal extension. Microrelief, or more often nanorelief, creates heterogeneity in that respect. That may be due to curtain or patchwork distribution of plants. A vegetation of lichens, mosses, and other epi-phytes on the surface of a tree-trunk also interacts specifically with its environment. Such structural subdivisions of a biogeocoenose may be called *biogeocoenotic synusiae*. Originally the term 'synusia' was applied by Gams (1918) to structural parts of a phytocoenose. He distinguished synusiae of different orders. At present there is no universally-accepted meaning of 'synusia'. In Soviet phytocoenotic literature a synusia usually means that structural part of a phytocoenose that is characterised by a specific plant species-composition, by specific ecologo-biological features of the species in it, i.e., by specific life-forms, and by their environment, which is usually conditioned by changes in its plant aggregate; that is, a synusia is characterised also by a specific phyto-environment. If in that definition of 'synusia' we also include the feature that a synusia often has its own specific animal life and usually its own specific microbic popula-tion, we obtain the following definition of a biogeocoenotic synusia: *every structural part of a biogeocoenose that is characterised by specific composition and features of its components and their specific internal interactions, while still maintaining the unity of the biogeocoenose, the community of the interactions of its components, and the exchange of matter and energy between them and its surroundings.* Giving this meaning to synusiae, we must include in them both the above-ground and the under-ground layers of a biogeocoenose, which are determined mainly by the layers of the above-ground parts of plants and their root systems. The layering of a biogeocoenose may also, however, be conditioned by the vertical layering of the atmosphere and by the so-called genetic horizons of the soil. We must place in special synusiae certain parts of tree-trunks that are distinguished by their special features as substrates for develop-ment of the life of organisms differing systematically and ecologically (lichens, mosses, algae, and insects and other animals). Leaves may also constitute individual synusiae with the organisms that inhabit them.

There has been a suggestion to distinguish the phyllosphere, combining the population of all leaves. We must also regard the fallen leaf, with its individual flora and fauna, as a separate synusia.

While the study of synusiae, even in phytocoenoses, is as yet little developed, even less has been done in the study of biogeocoenotic synusiae. Not dwelling in detail, therefore, on that question, and bearing in mind that the structure of biogeocoenoses will be discussed later, we still must not fail to mention the treatment of the subject by K. V. and L. V. Arnol'di, especially since (to the best of my knowledge) they were the first zoologists to give it particular attention.

In contrast to what has been said above about layers as a special kind of synusiae, these authors establish two basic structural units: layering and synusiality. Giving 'layering' approximately the same meaning as do botanists, the authors define a synusia of a biocoenose as 'a grouping of organisms, formulated spatially and consisting of close, or inter-related and mutually adapted, living forms' (1962, p. 172). It seems to us that that definition wholly includes layering. Therefore we must not consider these two terms as proved to be in contrast to each other. That also leads the authors to give finer subdivisions of a biocoenose and a biotope, merely to try to make them harmonise. Therefore they propose roughly the following concordance of subdivisions of biocoenoses and biotopes: (a) a dynamic variant of a biocoenose corresponds to a 'microfacies' (facies in the sense of R. Hesse and N. P. Naumov) of a biotope; (b) a synusia of a biocoenose corresponds to a 'merotope'[1] or 'element of a biotope' in the narrow sense, without including plants or animals in it; (c) subdivisions of a synusia, synusiae of the second and third orders, correspond to the same subdivisions of a 'merotope'. If we use the term 'biogeocoenotic synusia' in the sense given above, there is no need to look into such 'correspondences'.

Without piling up new special terms, it will be sufficient to mention briefly the various kinds of synusiae. Synusiae include, for example, the lower part of a tree-trunk, a definite part of a root system, a layer or horizon (Byallovich 1960) of a biogeocoenose, a decaying leaf in litter with its microbic population, etc. Synusiae may differ widely in extent. Horizontal structural disarticulation of a biogecoenose is so extensive that it affects all of the components and is expressed in a special type of biogeocoenotic synusia. Such synusiae are called 'parcels' by N. V. Dylis (Dylis et al. 1964).

---

[1] The term 'merotope' was proposed by Tischler for subdivisions of a biocoenose (e.g., roots, stems, and leaves of a living plant), and the term 'element of a biotope' by N. P. Naumov (quoted by K. V. and L. V. Arnol'di 1962, p. 174). Taking into account the meaning given above to a biogeocoenotic synusia, we do not see the necessity of keeping the special terms 'merotope' and 'element of a biotope'. We must also remark that the above quotations from the article of these authors once more show the necessity of having, in opposition to their opinion, the specific concepts 'biotope' and 'biocoenose' and combining them in the concept 'biogeocoenose'.

Phytocoenologists have developed the study of change of aspects of phytocoenoses, caused in the first place by the fact that during the vegetative period (more precisely, throughout the year) the appearance of a phytocoenose changes markedly because of changes in the phenological phases of development of the plants composing it, and in the second place by the fact that different years may differ widely in their climatic features and consequently in the appearance of the vegetation. Therefore we may distinguish phenological (or seasonal), meteorological, and in a certain sense also daily, aspects; these correspond to biogeocoenotic aspects. Aspects are sometimes regarded as a particular case of synusiality. It is more correct, however, to distinguish synusiality and rhythmicality as specific features of biogeocoenoses.

Now we pass to the very important problem of typology of forest biogeocoenoses. We must first remark that the classifications of biogeocoenoses and phytocoenoses are not the same, as a phytocoenose is merely a component of a biogeocoenose. Therefore naturally the principles of classification of each must be different. V. B. Sochava writes: 'We must draw a line between the classification of vegetation as such and the classification of natural complexes (facies, biogeocoenoses, ecosystems)' (1963, p. 8). We cannot fully agree, however, with Sochava's view that 'classification of plant communities is a necessary basis for the construction of classification of natural complexes'. While that statement may be accepted for biogeocoenoses, and especially for their lowest ranks, for units of higher ranks the guiding criteria of classification may be the features of other components of the biogeocoenoses.

As is well known, both knowledge of any natural phenomena and rational practical utilisation of them require systematic classification of these phenomena. As soon as forestry even began to be formed as a branch of human activity, a branch of the national economy, it required a classification of forests. Not only to orientate ourselves in the multiformity of our forests, but also to plan forestry measures intelligently, it is imperative to have a classification, a system, of forests. In the most primitive forms of forestry foresters already differentiated forests by their composition and age. As the forest economy became more intensive it could not be satisfied with such a simple classification, and the latter gradually became more complex.

There have been many disputes, which are not fully settled even to-day, about the principles on which forests should be classified, their typology. The All-Union Congress on Forest Typology, called by the Forestry Institute of the u.s.s.r. Academy of Sciences in 1950, harmonised the various points of view to a considerable extent. After studying the problem it came to the following conclusions.

Forest typology should aid the sylviculturist in organising the forest economy and in the most rational application of forestry measures, i.e., in establishing principles on which to build forest typology one must first determine what material features of the forest may affect the operation of various forestry measures. When from that point of view we survey the most important of the latter (e.g., cutting, logging, aiding natural regeneration, clearing felling-areas, control of fires and forest pests, maintenance and improvement of the water-retaining and protective features of the forest, increasing areas of quick-growing species, and other measures for increasing the productivity of our forests), we see that each of these measures is linked not only with the dominant tree species, which is usually the main object of exploitation of the forest, but also with all other forest vegetation (undergrowth, herbage, and moss cover) and micro-organisms in the soil, with the pedological-hydrological conditions of the environment, and also in many cases with the animal life. Therefore types of forest as classification units, defined in the interests of the forest economy, should be homogeneous regarding content of all forest components (tree species, other vegetation, climate, soil, water conditions, animal life). In view of these conclusions, and further developing the ideas of G. F. Morozov (the founder of our scientific forest typology) and the decisions of the above-mentioned All-Union Congress on Forest Typology, we may give the following definition of that basic forest-classification unit:

A forest type is a union of parts of a forest (i.e., separate forest biogeocoenoses) homogeneous in their tree-species composition, in other layers of vegetation, in fauna, in microbic population, in climatic, soil, and water conditions, in the inter-relationships between plants and environment in intra-biogeocoenotic and inter-biogeocoenotic exchange of matter and energy, in regeneration processes, and in the direction of change in them. That homogeneity in the features of biogeocoenose components and of biogeocoenoses as a whole, united in a single type, requires the application of identical forestry measures in identical economic conditions. Forest type, regarded as a type of forest biogeocoenose, has the highest practical significance.

In the above definition of type of forest biogeocoenose there is no need for homogeneity of origin (genesis) of the biogeocoenoses united in a single type. As has been noted earlier, however, in landscape-science some geographers consider genetic homogeneity of a type over its whole extent to be one of its principal characteristics. It is also well known that in forest typology some scientists give great weight to the genesis of types (Aichinger 1951, Kolesnikov 1956, 1958 a, b). Therefore we must dwell on that problem.

The term 'genesis' is not understood uniformly in all sciences. As is well known, in plant systematics genetic plant systems (as well as the so-called natural plant systems) have been created. Whereas a natural system is constructed from the greatest possible number of common characteristics, a genetic system expresses their genetic relationships in the process of evolutionary development of the organic world. Therefore genetic classification is usually shown graphically in the so-called phylogenetic tree. Since any taxonomic unit of plant cover, as has often been stated in the literature (especially Soviet literature), should not be regarded as an organism (as is done by Clements and his followers) or even as a quasi-organism (as is done by Tansley), naturally one should not speak of the genesis of phytocoenoses in the sense used in the systematics of organisms. Since a biogeocoenose also should not be regarded as an organism, there are no grounds for speaking of the genesis of biogeocoenoses in the above sense.

In pedological literature, beginning with Dokuchaev, the term 'genesis' is used with another meaning. Soil being regarded as a product of the action of organisms and atmospheric agents on mineral strata, all soils similarly affected by these factors are held to be closely related genetically. Similarly in forest science G. F. Morozov and several other authors speak of forest types as being genetically close when they are similar in their formative factors. They speak also of the genesis of forest associations, having in mind a series of plant associations preceding a given one in the succession process.

The well-known Austrian sylviculturist Aichinger, after observing plant successions in nature, in describing a forest type (regarded by him as a forest association) points out its links with the preceding forest type in a succession series. In one of Aichinger's latest works (1960) he writes that in determining the type of a forest with significance for practical work one must start from its general physiognomy and base one's conclusions on its floristic, botanico-geographical, ecological, and syngenetic character, taking special account of the influence of man, thus looking on a forest type as a type of vegetational development. 'In such a type of vegetational development,' he writes, 'I include all physiognomically homogeneous plant communities that are similar both in their floristic and phytocoenotic features and in their ecology (*Haushalt*) resulting from the conditions of their environment, and which belong to the same stage in a developmental series' (1960, p. 10). To that Aichinger adds that in his opinion the ultimate scientific aim of survey of a biocoenose is understanding it as a biogeocoenose in Sukachev's sense (p. 12).

One cannot object to Aichinger's statement that in determining the type of a forest one should consider its place in a succession series, but it is far from being always possible or convenient to express it in his terms.

B. P. Kolesnikov also began to devise genetic principles in forest typology in 1946. In two of his works (1958 a, b) he set out his views clearly on that subject. He writes: 'By genetic classification of forest types is meant a classification based on the laws of processes of forest origin and development, which covers all stages of development of a stand and can be used to prognosticate its future condition.' Using the term 'forest-formation process', which can be regarded as a particular case of biogeocoenotic process, Kolesnikov asserts that in the composition

of a single forest type, which he treats as the basic unit for genetic classification of forests, 'naturally one must include stands in all stages of a
single cycle of successions of growth, or of short-term re-establishment,
or of alluvial changes, taking place within an area that is uniform in
location and characteristics' (1958b, p. 117). 'The place of a forest in a
definite stage of the forest-formation process is its basic and most important characteristic' (1958b, p. 118). 'As the basic unit of such a
(genetic—*V.S.*) system,' Kolesnikov writes, 'we have taken a forest type,
large in extent and complex in composition, regarded as the final stage
in the forest-formation process, and equal in duration to the minimum
period of life of one generation of the dominant species. A forest type
is composed of stand types, corresponding to one of the age or short-
term re-establishment stages of development of the forest type during
which external morphological features remain homogeneous in the combined areas of the forest. That lowest and elementary unit of classification
is equal in rank to the "timber type (stand type)" of Ukrainian typologists, to the "forest type" of the 1950 Congress, and to V. N. Sukachev's
"type of forest biogeocoenose" ' (1958b, p. 120).

Kolesnikov's analysis of the problem of development of forest types is
of great value. That problem has great theoretical and practical significance, as has often been stated in the literature (Sukachev 1928, and
others). As is well known, the series of phytocoenotic diagrams given by
me (sometimes incorrectly called classification diagrams) represent genetic
series of forest-type successions.

We have always held, however, that biogeocoenotic classifications of
forest types should be constructed on another principle, namely, on the
degree of similarity of the processes of transformation of matter and
energy in forest biogeocoenoses and exchange of them within biogeocoenoses (between their components) and between biogeocoenoses and
other natural phenomena, expressed in similarity of all components of
the biogeocoenoses and of their composition and structure, as that is the
essence of biogeocoenoses and on it depend primarily the methods of
their practical utilisation. In many cases, but not always, Kolesnikov's
genetic classification of forest types coincides with their biogeocoenotic
classification. Biogeocoenotic classification is the most natural forest
classification.

We can fully agree with Kolesnikov's remark that 'the existence of a
natural classification is an essential prerequisite for devising a genetic
classification' (p. 121). We cannot, however, consider Kolesnikov's
establishment of another elementary classification unit (besides the basic
unit) to be desirable. For forestry purposes it is necessary to have an
elementary unit that is also the basic unit. As the foundation for concrete
forestry measures these elementary (basic) units can and should (as G. F.
Morozov had already pointed out) be composed of various groups, in
harmony with forestry purposes. For the elementary (basic) unit it is

expedient to keep the name 'forest type', which in recent decades has become generally used both in the Soviet Union and abroad. 'Forest type' in that sense, or 'type of forest biogeocoenose', will not always correspond (as will be seen later) to Kolesnikov's 'stand type', although in most cases it will coincide with it. We may also object to the term 'stand type'. Although Morozov used it (following Gutorovich, Serebryannikov, and others of our leading forest typologists), in his lectures he always remarked that the term 'stand' (literally 'plantation': Russian 'nasazhdenie', Tr.) was not a good one. In the first half of the 1930s, as is well known, there arose a strong movement among sylviculturists against use of that term and for its replacement by 'standing timber' (Russian 'drevostoi', Tr.). Our greatest sylviculturist, M. E. Tkachenko, was a strong advocate of the change. It is true that our sylviculturists still often apply the term 'nasazhdenie' to natural stands; but that introduces some confusion. The expression 'nasazhdenie', in the correct sense of the word, is suitable for use only with regard to artificial stands planted by man. Even less justified is use of the term 'nasazhdenie' to mean not only a stand but even a part of a forest. If we have in mind only the forest growth therein we may use the term 'forest community' or 'forest phytocoenose'; if we mean a part of a forest with all its components, then we must speak of a forest biogeocoenose. Therefore it would be more suitable to speak of 'forest type' in the sense accepted by the All-Union Forest Typology Congress of 1950 instead of B. P. Kolesnikov's 'stand type'. But another term is required for Kolesnikov's concept of 'forest type'.

Returning to the question of genesis of forest cover as discussed by Aichinger and Kolesnikov, we must point out that the term 'genesis' in this case means something quite different from its meaning in plant and animal systematics. Neither in phytocoenoses nor in biocoenoses is there 'genesis' as understood in the phylogeny of the organic world; there is no such hereditary transmission of features as is characteristic of organisms. In any succession of biogeocoenoses, when one biogeocoenose is replaced by another some or all of its components are altered and new elements of some components enter from outside. In biogenetic succession there is no inheritance in the sense used in genetics.

If the term 'genesis' is to be used in the dynamics of biogeocoenoses, this should be done only where syngenetic and endogenetic successions are concerned. These successions, which will be discussed in more detail later, must be included in the development—or, more precisely, the self-development—of biogeocoenotic cover. We must bear in mind, however, that biogeocoenoses adjacent in series of syngenetically or endogenetically successive biogeocoenoses may, by the nature of biogeocoenotic processes, as a result of internal and external exchange of matter and energy, differ more from each other than do biogeocoenoses farther apart in the series. For instance, when by syngenetic succession herbaceous are succeeded by forest biogeocoenoses, from the viewpoint of similarity in their biogeocoenotic processes and in the above-mentioned features of organisation they may differ more from each other than do separate forest biogeocoenoses, although the latter may have been widely separated both territorially and in succession series. That is very important both in the theory of forest biogeocoenology and in practical forest economy.

Taking all the above into account, we must remark that although investigation of the origin of a forest type and the history of its formation is of importance for the understanding of that type and its nature, as with any natural phenomenon, use of genetic principles alone in the classification of forest biogeocoenoses does not fulfil the task of forest typology either theoretically or practically. I would stress, however, that investigation of the genesis of any type is necessary for fuller understanding both of its characteristics and of its course of future development.

Forest types, of course, can be determined only in forest-covered areas. But we may evaluate areas not covered with forest from the point of view of their suitability for planting with certain tree species. In that case we shall be interested in the forest-growth conditions of the areas, which may be classified by their suitability for tree-planting into *types of forest-growth conditions*. The latter can also be determined, of course, in forest-covered areas.

As a rule, correlation may be observed between forest types and types of forest-growth conditions. Cases are possible, however, both in nature and in cultivation, where there are different forest types in the same type of forest-growth conditions, on account of different species-composition of the trees and of vegetation in general. Therefore one cannot use the concepts of forest type and type of forest-growth conditions interchangeably.

In this connection, as is well known, the so-called Ukrainian forest-typology school has developed somewhat different methods of forest typology. Although at the All-Union Forest Typology Congress of 1950, in which leading representatives of that school took part, the point of view just mentioned regarding forest types and types of forest-growth conditions was adopted unanimously, P. S. Pogrebnyak and D. V. Vorob'ev, in their works published since that date, have continued to hold the views on forest typology that they had developed earlier. Since these views are even now supported by several Ukrainian sylviculturists, we must examine them critically. The views of P. S. Pogrebnyak and D. V. Vorob'ev have recently been very clearly expounded in D. V. Vorob'ev's book on methods of forest typology investigations, published in 1959. In that book it is stated that the basic classification units, called 'types', are as follows:

'1. *Forest area type* or edatope (type of habitat, type of local growth conditions) combines areas with similar soil and water conditions. Climatic conditions may be different. In different climatic and geographical districts a forest area type is represented by different forest types, and in unforested districts by types of meadow, steppe, etc.

'2. *Forest type* combines forested and unforested areas similar not only in soil and water conditions but also in climatic conditions. Forest types are easily distinguished by the composition of their basic associations. A synonym of "forest type" is "type of forest-growth conditions" in the sense adopted by the Forest Typology Congress (1950). In unforested areas forest types are replaced by types of meadow, steppe, tundra, etc. Each forest type is divided into timber types and (where forest vegetation is absent) herbage types.

'3. *Timber type* combines forest areas similar not only in soil, water, and climatic conditions (i.e., belonging to a single forest type) but also in composition of the tree stand. Within one timber type, however, stands may differ in age and density and may vary in composition (up to 40%) and in productivity.'

'Variation in forest-growth conditions within a forest area type leads to establishment of a variety of types: sub-types, variants, and morphs.' 'Forest types are grouped into a *family of forest types* according to the predominance of various tree species in their principal stands.'

With regard to these taxonomic units we must first say that soil and water conditions must not be regarded as separate from climatic conditions, since they are largely determined by climatic factors. An edatope may be discerned only within a definite climatic region, but local growth conditions include both climatic and edaphic conditions. Incidentally, we must also say that usually the concept of habitat is different from the concept of local growth conditions. The concept 'habitat' corresponds to the concept of range, and that of local growth conditions to the concrete living conditions of certain plants or their communities. Therefore there is generally no need for a forest-typology concept of 'forest area type'. 'Edatope', according to the meaning of that word, should be applied to an assemblage of soil and water conditions corresponding to a definite phytocoenose, or, more precisely, biogeocoenose. It is also illogical and unsuitable to apply the concept 'forest type' to both forested and unforested areas.

Forest type, according to Vorob'ev, is a type of local growth conditions in the sense adopted by the Forestry Typology Congress. We cannot agree to the suitability of the term 'timber type' as understood by Vorob'ev, who looked upon it as a combination of forest areas similar not only in soil and water conditions but also in composition of the tree stand. If 'stand' is understood in Morozov's sense, i.e., as a synonym of 'forest community', or, to be more precise, of 'forest biogeocoenose', then we cannot say that that is a type of standing timber. A number of forest areas grouped together merely by uniformity of their standing timber would correspond to what is customarily called a plant formation or a bio-geocoenotic formation. When we consider that, according to Vorob'ev, forest areas within a timber type may vary in productivity, that breaks down the basic requirement of forestry for the smallest forest-typology unit, that within it forest areas should be uniform in productivity. A family of forest types, according to Vorob'ev, represents in some respects a particular case of biogeocoenotic formation.

Sometimes its ecological character is singled out as a specially positive feature of the approach to forest typology devised by P. S. Pogrebnyak and D. V. Vorob'ev, but that is a misapprehension. If one speaks of an 'ecological character', then, independently of how the term 'ecology' is understood, the biogeocoenotic approach is even more ecological, since biogeocoenology includes ecology, even in the wide Anglo-American sense. Considering all these facts, and also the known internal contra-dictions in the above definitions of taxonomic forest-typology units proposed by Vorob'ev, we must now accept as correct the decisions of the Forest Typology Congress of 1950, which adopted only two main con-cepts: 'type of forest growth conditions' (type of local growth conditions) and 'forest type' in the sense explained by us. That completely fulfils the requirements of forest economy in the field of forest typology.

Recently V. G. Nesterov (1961) proposed a special forest typology, which he called phytocoeno-ecological, holding it to be the most suitable classification of forest for forestry purposes. He gives this definition of 'forest type': 'By "forest type" we must understand an assemblage of forest areas, homogeneous in tree-species composition and environmental conditions', adding 'thus the concept of forest type is taken from G. F. Morozov and not from V. N. Sukachev or P. S. Pogrebnyak.' That statement is quite incomprehensible. In fact, the definition of 'forest type' given by Nesterov does not differ from its meaning as a type of forest biogeocoenose, but his definition is so compressed that it actually be-comes very indefinite in content. That is especially clearly seen from Nesterov's subsequent words: 'Together with the "forest type" concept we think it proper to distinguish within it the type of forest-vegetation community or of stands, by which we must understand an assemblage of standing timber with uniform composition of the chief species in the arboreal layer and in the ground cover (e.g., a pine-cowberry assemblage).' Even if we ignore the incorrectness of the phrase 'type of forest-vegetation community or of stands' and the vagueness of the term 'a pine-cowberry assemblage', we may see that in this case the author has in mind nothing else but what is customarily called a forest association. From that fact, and also from his term 'type of forest formation', it follows that the author has generally failed to grasp sufficiently the now-accepted concepts and terms in forest classification. Evidence of that is also given by the author's statements in the entire chapter 'Study of forest types', where he gives imprecise descriptions both of phytocoenotic and biogeocoenotic teachings and of the teachings of the Ukrainian forest-typology school. The special features of the latter remain, evidently, uncomprehended by him. Therefore we should not be surprised that V. G. Nesterov, having proposed his classification of forests, in which (in his own words) 'primarily and mainly Morozov's teaching about forest types has been followed, and the generalised classification of V. N. Sukachev and the edaphic network of P. S. Pogrebnyak have been taken into account', presents the following general conclusion about his classification: 'The above classification of forest types with regard to pine, spruce, birch, poplar, oak, and alder is not a new departure in forest typology: it is derived from Morozov's teaching. This classification of forest types should be used in conjunction with Sukachev's classification of plant communities and Pogrebnyak's classification of local growth conditions. In combination they should raise our forest economy to a higher level.' But since Nesterov's classification of forests represents an unsuccessful eclectic combination of several of Morozov's suggestions, as he himself points out, in both phytocoenotic and edaphic disciplines, it is obvious that there can be no advantage in combining his classification with those of Sukachev and Pogrebnyak.

Since the main task of forest typology is co-operation in the proper selection, direction, extent, and techniques of forestry measures in accordance with the natural features of a forest, it is clear that the characteristics of the types on which these measures are based or depend should be well known. On the extent of knowledge of the sylvicultural characteristics of types on the one hand, and on understanding of the links between economic measures and the natural features of a forest on the other hand, depends fuller utilisation of forest typology data in practice.

Although in ordinary investigations of forests, where it is not possible to make prolonged and detailed study of them, forest types are distinguished on the basis of easily-observed features, we must not lose sight of the fact that all components of a forest biogeocoenose are of significance in determining types and all interact on one another. These ideas should guide the investigational plan and the study of forest types on a transect. For more thorough study of the characteristics of types, however, one should not limit oneself to their study on transects. It is necessary to organise year-round studies, even studies lasting several years, of these characteristics at field stations.

Recently, especially in American literature, much has been written about the continuity of plant cover, its so-called 'continuum'. Some are even inclined to regard the study of that continuity as a new branch of geobotany. That is obviously an error. In fact, every diligent investigator of plant cover is almost always able to observe transitions either between a number of plant communities situated in different associations or in different areas of a sufficiently extensive territory examined by him.

Sometimes the transitions between the communities of different associations are so gradual and cover so much area that they make it difficult to delimit the associations. Therefore it has long been emphasised that the determination of associations, especially of their boundaries, is often provisional, and that generally associations represent, in a certain sense, an abstract concept. At the same time associations, being characterised by quite definite features, can be defined in nature and shown on a plan or a vegetation map. It has also been demonstrated that, in order to investigate the variety of plant cover and to devise economic measures for its utilisation, it is absolutely necessary to delimit associations or other taxonomic units of vegetation. Therefore when L. G. Ramenskii in 1910 and later years produced the idea of continuity of plant cover, he was in fact 'breaking through an open door'; and when, starting from these premises of his, he spoke against the establishment of a hierarchical series of units of plant cover, he met considerable opposition. But nobody objected, of course, to the careful study of transition zones or of communities lying between different established associations. On the other hand, the necessity of thorough analysis of such transitions together with the environmental conditions, especially in the study of the processes of succession of plant communities, has always been stressed (Sukachev 1930, Ponyatovskaya 1959, Rabotnov 1963).

We must bear in mind, however, that, although changes in vegetation usually correspond to changes in the conditions of its growth, i.e., of the ecotope, the boundaries between communities of different associations are very clear and even abrupt, while the conditions of their ecotopes gradually merge into each other. Long ago Du Rietz brought out that point and correctly explained it by the fact that distribution of plants in communities depends not only on environmental conditions but also on the concrete relations between plant species.

In view of these facts we must conclude that, although we apply the continuum (continuity) principle to biogeocoenotic cover also, in many cases the boundaries between different types of biogeocoenoses (especially forest biogeocoenoses) are much more sharply defined than those between their ecotopes.

Since in the final analysis biogeocoenology should give special attention to the processes of transformation of matter and energy and their exchange within and between biogeocoenoses, and since every effect of one natural phenomenon on another is accompanied by these processes, for further successful development of this branch of knowledge it is necessary, firstly, to use methods of precise quantitative evaluation of all these processes wherever possible, applying mathematical analysis to the data obtained; secondly, to make extensive use of the newest tools and methods provided by physics, chemistry, biophysics, biochemistry, plant physiology, and cybernetics; thirdly, to make the maximum use of the experimental method, organising experiments mainly in natural forest environment.

The undertaking of such biogeocoenotic investigations on a large scale is still, to a great extent, a novel proceeding. Much remains to be done in devising methods and an organisational plan for it.

THE EXPERIMENTAL METHOD IN FOREST BIOGEOCOENOLOGY AND ITS PLACE IN THE SYSTEM OF DISCIPLINES OF NATURAL SCIENCE

In recent years much attention has been given to application of the experimental method in geobotany, and one even speaks of the appearance of a new scientific discipline—experimental geobotany.

The experimental method is also advocated in biogeocoenology, and sometimes one speaks of a special experimental biogeocoenology. These terms can be used only conditionally in view of their conciseness, and we must remember that in all natural sciences experimentation increases every year but we are not therefore entitled to speak of the creation of new branches of knowledge, new sciences. The experimental method in biogeocoenology, as in geobotany (phytocoenology), is only one of the methods used in these sciences, and it should constantly develop further in them. In biogeocoenology experiments may be of many kinds. In particular, they may be set up directly in natural biogeocoenoses, making various alterations in their composition or structure, or in general with experimental study of the features and interactions of their components. The creation of new biogeocoenoses constructed on a specific plan is of great value.

The very important biogeocoenotic problem of the mutual influence of plants in a phytocoenose can be successfully solved mainly by the experimental method.

In recent years many works have been devoted to the mutual influence, through chemical secretions, of plants growing close together. In 1944 G. F. Gauze proposed for that branch of science the name 'chemical biogeocoenology', but even earlier Molisch (1937) had called it 'allelopathy'. The latter name is still occasionally used (e.g., Grümmer 1953 (Russian translation 1957), Winter 1960, and others). G. B. Gortinskii perfectly correctly holds that 'allelopathy is a branch of biogeocoenology, studying the forms and the extent of specific biochemical exchange of matter within biogeocoenoses, with calculation of the effects of such matter on the components of a biogeocoenose' (1963, p. 105).

In biogeocoenological experiments the use of tagged atoms and ionising irradiation is of great value, as was first definitely proved by N. V. Timofeev-Resovskii, who developed work extensively in that direction. He suggests calling that branch of experimental biogeocoenology 'radiation biogeocoenology', the tasks of which he sees as follows: 'By using ionising radiation and radioactive isotopes of different elements, one can (a) study quantitatively the effects on the biomass and structure of a biocoenose caused by such a non-specific and easily-measured factor as ionising radiation, and also determine the relative role of a biocoenose in the distribution of introduced elements through the components of a bio-geocoenose; (*b*) study quantitatively the role of different species of living organisms in concentration and accumulation (and also of distribution) of different chemical elements (mostly scattered and existing in micro-concentration) from the environment, thereby determining the relative role of these species and different groups of organisms in the geochemical processes that take place in biogeocoenoses (with identification of specific accumulators of certain chemical elements), depending on the physico-chemical conditions and the composition of biocoenoses' (Timofeev-Resovskii 1962, pp. 13, 14).

The wide scope of the work done by Timofeev-Resovskii and many of his colleagues in radiation biogeocoenology has shown how fruitful studies of that kind are. They have not yet been applied to forest bio-geocoenoses. An urgent task facing forest biogeocoenology is the ex-tensive application of that method.

Among the large number of questions that Timofeev-Resovskii has posed for radiation biogeocoenology, of special theoretical and practical importance for forest biogeocoenology would be 'combining in one system population-genetic and biogeocoenotic experiments; that would enable us to bring precise criteria to study of the links of a specific popula-tion in the micro-evolutionary plan not only with the conditions of its physico-geographical environment, but also with the entire complex of biotic and geochemical components of biogeocoenoses. The precise determination of elementary phenomena on the biochorological level, and of the links between biogeocoenotic and population-genetic processes, will hasten the introduction of cybernetic principles and concepts into the population-genetic and biogeocoenotic-biochorological levels of life.... That, in turn, creates the possibility of almost unlimited use of mechanical models in the analysis of the most complex biological pheno-mena' (Timofeev-Resovskii 1962, p. 44).

In fact, both sylviculture and agriculture may be regarded from a definite point of view as objects of experimental study not only by phyto-coenology but also by biogeocoenology. In that respect a special branch of phytocoenology is the so-called 'culture-phytocoenology', which covers sylviculture of every kind. The latter is particularly favourable for it because of the long life of trees and their continual changes during growth and development, although they are also a specific subject for experimental study of several general laws of phytocoenology, especially the inter-relations of plants growing close together. At present the prob-lems of culture-phytocoenology and (as a special case of it) of forest culture-phytocoenology are attracting the attention of many scientists. But we must remember that the creation of culture-phytocoenoses is always linked with changes in soil conditions, and to a certain extent also in water and climatic conditions and in fauna (zoocoenoses), and therefore is in fact the creation of new biogeocoenoses or *culture-biogeocoenoses*.

The method of experimental solution of biogeocoenological problems by creation of artificial biogeocoenoses may be called, after V. B. Sochava (1963, p. 7), 'the experimental-modelling of biogeocoenoses method'. But there is no need for that, as it does not facilitate use of the method. Also hardly worth while is re-naming of the test-plot method (long used in forestry, and introduced into geobotany at the beginning of this century) 'the method of natural models of plant associations'. It is necessary to perfect these methods, but one cannot do so by thus changing their names to expressions borrowed from cybernetics; even the term 'model' is scarcely appropriate here.

Solution of the problems of culture-biogeocoenology, especially of forest culture-biogeocoenology, is of exceptional importance for both sylviculture and agriculture, particularly for field-protective and other windbreak plantations. Being the theoretical basis of practical forestry, forest culture-biogeocoenology solves at the same time general problems of biogeocoenology.

We have noted above the great influence of man on biogeocoenotic cover. The more advanced the development of human society, the stronger is that influence, which is constantly being more and more directed to specific ends.

Following the French scientist E. LeRoi, V. I. Vernadskii has proposed the name 'noosphere' (from the Greek nous, mind) for the new evolutionary change in the biosphere that is taking place during the epoch in which we live, representing the reconstruction of the biosphere 'in the interests of free-willed mankind as a unique whole' (1944, p. 118).

When man cultivates new areas (including the planting of trees) or by sylvicultural measures reconstructs natural forests to some degree, he contributes to creation of the noosphere. And since in all his activities, chiefly industrial and agricultural, he keeps making gradual changes in the atmosphere, soil, and natural waters, not to mention the fact that new artificial biocoenoses alter their environment, he actually creates a new biogeocoenotic cover for the earth as the most substantial part of the noosphere.

It is therefore clear that the noosphere, created by man, will provide him with answers to its most extensive and profound problems when the new biogeocoenotic cover will also fulfil these requirements.

The ultimate aim of biogeocoenology—discovery of all the laws that govern the processes of transformation of matter and energy, as stated above—is extremely extensive and complex. Nature, as is well known, is inexhaustible in the variety of its forms, its content, and its processes. Therefore there is no limit to work in that direction. Nevertheless each step forward enriches us in knowledge of the laws that interest us, gives us new equipment for controlling and directing them, and consequently opens new possibilities for fulfilling practical, productive tasks. Biogeocoenology, which is concerned with the whole assemblage of inter-

FFB E

acting natural phenomena with their contradictory dialectical unity, offers the most promising method of fulfilling these tasks. Therein lies its great progressive significance. And since the forest is a natural pheno-menon unique in its complexity, multiformity, and number of inter-relationships, only by this method can solution of the above problems in forest science be achieved. This also has a bearing on the fact that forest biogeocoenology and forest economics serve as the foundation for forest management.

Having surveyed the content and problems of biogeocoenology and especially the methods of its application, we find it natural to reflect on the place it holds in the system of our knowledge. In this respect various points of view have been expressed, which can on the whole be summar-ised by saying that some refer it to biological, some to geographical, disciplines. Including biogeocoenology among biological sciences, some authors, e.g., I. I. Shmal'gauzen (1958) and N. V. Timofeev-Resovskii (1962), have regarded it as the highest level of study of biological pheno-mena, placing it at the culmination of the following series (commencing at the bottom): (a) molecular, (b) cellular, (c) organism, (d) population, (e) species, (f) biocoenotic, and (g) biogeocoenotic phenomena. Other authors, giving 'ecology' a very broad meaning, include biogeocoenology in it. The majority of foreign authors and some Soviet zoologists hold that ecosystem (like its synonym biogeocoenose) is the basic object of study in ecology. Many of our geographers, considering 'biogeocoenose' to be a synonym for 'facies', regard biogeocoenology as a branch of geography, landscape science (N. A. Solntsev, A. G. Isachenko, V. B. Sochava, and others).

As for the place of biogeocoenology in the system of natural sciences, it is appropriate to remember the history of phytocoenology. It had its origins in many early works on botanical geography, from which it separated in our country at the end of the nineteenth and the beginning of the twentieth centuries under the name of phytosociology, later chang-ing its name to phytocoenology (Sukachev 1959a). The history of bio-geocoenology has been essentially similar. The idea of it was conceived almost simultaneously in the wombs of biology and geography. As the discipline developed, however, and the object of its study and its content became more clearly defined, its independent nature became more evident, and it was separated from landscape-science and, generally, from geo-graphy. Originally I also was inclined to consider biogeocoenology a branch of geography (Sukachev 1947, 1948a), but later I came to the con-clusion that it cannot be included among either biological or geographical sciences.

In this connection we must note the remarkable discernment of V. V. Dokuchaev, who, while studying soil-formation, arrived at the concept of 'a genetic, perpetual, and always regular linkage existing among forces, bodies, and phenomena, between animate and inanimate nature, and among the vegetable, animal, and mineral kingdoms'. He went on to express the opinion that a special science should be created to study that interdependent complex of matter and phenomena on the earth's surface, and that it 'will occupy a completely independent and reputable place, with its own strictly-defined tasks and methods, not merging into existing divisions of natural sciences and still less into geography, which is separated from it on all sides' (1898, pp. 6-7).

At present several of our geographers are inclined to regard landscape-science as that special science whose creation was foretold by Dokuchaev. As we have seen, however, the content of landscape-science is not distinguished by great exactitude. Moreover, not all geographers hold that view. Mil'kov, for instance, considering 'geocoenose' to be the same as 'biogeocoenose', writes: 'Geocoenology is an independent science, closely connected with landscape-science. It has its own object of study and its own characteristic methods of investigation. A geocoenose is not an object of study for landscape-geography' (1959, p. 50).

We have very substantial grounds for believing that biogeocoenology and not landscape-science was the special science whose birth was predicted by Dokuchaev and which is classified among neither geographical nor biological sciences. Therefore also the biogeocoenotic level of study of natural phenomena is not one of the range of biological levels: it is a level of a special order.

Nowadays the view (quite correct) is often expressed that 'the mutual interpenetration of sciences and their constantly closer and more varied interactions are characteristic features of the contemporary stage of development of knowledge' (*Interaction of earth-study sciences*, 1913, p. 5). The consequent appearance of entirely new sciences is equally characteristic of the period in which we live. Biogeocoenology is one of these sciences.

CHAPTER II

The atmosphere as a component of a forest biogeocoenose

The biogeosphere includes, as one of its more important components, the lowest part of the earth's atmosphere (more precisely, of the troposphere), which is distinguished by having the greatest density of gases, the most dynamic physical condition, and the closest inter-relationships with plant and animal life, mineral strata, bodies of water, and soil.

The atmosphere is a complex gaseous mass, a conductor of energy, a centre of vast material resources for organisms, and the region of formation of the earth's climate. The effect of the atmosphere on the other components of the biogeosphere takes place by way of many factors, the chief of these being light, heat, water, gaseous composition, and air movement. The effect of these atmospheric factors is particularly powerful on vegetation, which over vast areas of dry land plays the role of the main covering and receptive layer of the planet's surface. These atmospheric factors are modified in turn by the other components of a biogeocoenose (especially vegetation), as a result of which the relationship between the atmosphere and the other biogeocoenose components is one of very close organic interaction (Fig. 3). Although the interaction between the atmosphere and other biogeocoenose components reaches a maximum in the zone of their immediate contact, the results are distributed far beyond that zone, possibly as far as the ozone layer (Vernadskii 1927).

Recently, for instance, it has been established that there is a connection between the weather and the state of the ionosphere, since atmospheric pressure in the troposphere is linked with the concentration of ions at an altitude of 100 to 200 km. As soon as there is a change in pressure, atmospheric processes and weather also change.

The state of the atmosphere in the layer nearest to the earth, i.e., a layer of air 30 to 50 metres in depth immediately adjacent to the earth's surface, is especially important from the biogeocoenotic standpoint. That layer is distinguished by a number of special properties, of which the principal one is that the nearer it is to the surface cover the less is its vertical turbulence.

Within plant communities all properties of the atmosphere undergo profound alteration. This is most evident in forest biogeocoenoses. Within forest communities the effect of forest vegetation is to change the heat regime of the air, its gaseous composition, its humidity, and the amount of organic matter present. Forest vegetation also substantially changes the state of the air in adjacent unforested areas, helping to purify it from dust and to enrich it with oxygen, and moderating extremes of variation in heat and humidity.

We shall now discuss in more detail the significance of the principal components of the atmosphere with relation to other biogeocoenose components, and their effects on the course and the results of biogeocoenotic exchange.

GASEOUS COMPOSITION OF THE ATMOSPHERE

The gaseous composition of the atmosphere is almost uniform over the whole surface of our planet, in spite of the constant absorption of some of its components (e.g., oxygen) by organisms and by various abiotic

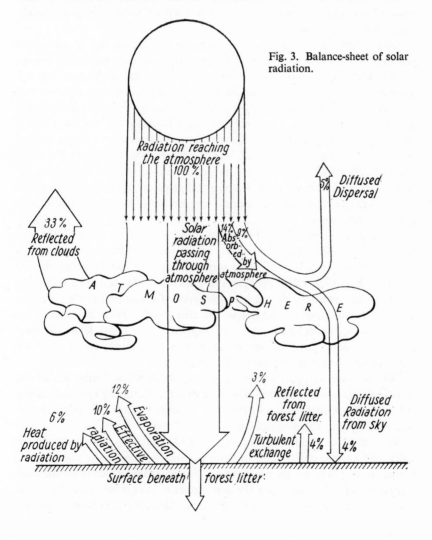

Fig. 3. Balance-sheet of solar radiation.

oxidising reactions. The atmosphere changes little in composition because, while some of its gases are being absorbed in certain reactions,
equivalent quantities are being released by other processes. As a result of
the systematic 'turbulent exchange' between different atmospheric layers
the atmosphere is in constant movement, its composition is equalised, and
at every altitude its components occur in approximately uniform proportions: 78·08% nitrogen, 20·95% oxygen, 0·94% argon, and 0·03%
carbon dioxide and various other gases.

Gaseous nitrogen is of relatively little significance in biogeocoenotic
processes, as it is not directly used by the majority of organisms and is
brought into circulation only in comparatively small quantities by certain
micro-organisms, or in some cases after oxidation (e.g., by lightning).

According to M. A. Bobritskaya (1962) the soil receives annually from
atmospheric precipitation from 3 to 4·5 kg of fixed nitrogen (mainly
ammonia) per hectare.

Among atmospheric gases oxygen and CO_2 are of the greatest biogeocoenotic importance. They take part in biogeocoenotic metabolism
in vast quantities, and are the basis of matter exchange between the
atmosphere and the other components of a biogeocoenose.

Oxygen is accepted directly by all biogeocoenose components, making
possible the respiration of plants and animals in the air, soil, and water,
the transformation of organic matter, and the course of many chemical
processes in mineral strata, soil, and the hydrosphere. Except for green
plants, with which atmospheric oxygen is concerned in a two-way reaction—being absorbed in respiration and released in photosynthesis—all
components of a biogeocoenose use oxygen only in various irreversible
oxidising reactions.

The oxygen content of the atmosphere is always sufficient for any
requirements of plants and animals, but in soils and bodies of water it is
often extremely small, leading to serious shortages both for organic life
and for physico-chemical processes in mineral strata, soil, and water. In
forest biogeocoenoses, except in some types of communities that are well
adapted to shortage of oxygen in the soil (bald cypress and mangrove
communities), inadequate supply of oxygen to the roots leads to lower
production of organic matter by trees and to diminution of the total
metabolism of matter and energy in the forest.

CO_2 is taken up only by green plant communities, being one of the
principal ingredients in the synthesis of organic matter by cells containing
chlorophyll. Other biogeocoenose components, on the other hand,
release CO_2 into the atmosphere.

As a result of the fluctuating rate of CO_2 usage by plants during the
daily and annual cycles, and in spite of equalisation of the composition
of the atmosphere by its movements, considerable variation in the concentration of CO_2 occurs in the atmospheric layer nearest to the earth.

In summer the amount of CO_2 in the surface air is less than in winter, and by day (especially towards evening) it is less than during the night. According to our observations, in oak forests the CO_2 content at the surface of the earth fluctuates as follows: if the content at 7 a.m. is taken as 100%, at 10 a.m. it is 71%, at noon 68%, and at 6 p.m. 57%.

The concentration of CO_2 in the forest also varies with height above the soil surface. Maximum concentration occurs at ground level and minimum at the level of the tree crowns. For instance, in an oak grove according to our observations the CO_2 content at ground level is 0·68 mg per litre, at a height of 5·25 metres 0·46 mg/litre, at a height of 11·5 metres 0·46 mg/litre, and at crown level 0·44 mg/litre.

The CO_2 content also varies with climatic conditions, being higher in wet and warm years than in dry and cold years (Table 1).

Table 1. CO_2 content in the air (in mg/litre) in stands of two different forest types and in years of different humidity. Average during growing period.*

Year	Oak stand in dark-grey soil		Oak stand in slightly-podzolised solonetz (alkali soil)	
	Not thinned	Thinned	Not thinned	Thinned
1954 (dry)	0·518	0·510	0·506	0·524
1955 (wet)	0·599	0·600	0·547	0·562

* Tellerman experimental forestry station, Voronezh province.

There are undoubted differences in the CO_2 content of the air in different types of plant communities. Under a forest canopy the concentration of CO_2 is usually rather greater than in a field, and it is greater in oak than in pine stands; the maximum apparently occurs in tropical rain forests (Richards 1961).

The fluctuations in CO_2 content are due to photosynthesis by green plants. When the CO_2 content increases in the air around plants, assimilation is speeded up—not quite proportionately, it is true, and generally within a certain limited range, determined by light intensity.

The chief sources of CO_2 in the air are animal and plant respiration and decomposition of organic matter in the soil by micro-organisms (soil respiration). Release of CO_2 by the soil becomes more vigorous in proportion to the soil warmth, to the approach of its humidity to the optimum, and to the amount of contained organic matter.

As a result of the breakdown of organic matter by the processes of respiration and decomposition, the CO_2 content of the atmosphere as a whole is maintained at a constant level.

CO_2 gas, which is characterised by a high specific heat capacity, acts as a unique screen permitting heat rays to penetrate to the earth but checking heat rays coming from the earth. Thus CO_2 serves as a temperature regulator for the air and the earth's surface.

SOLAR RADIATION

The atmosphere is a conductor of solar radiation to the biogeosphere of the earth and all processes taking place on the earth's surface are subject to the influence of solar radiation. On the latter depend light, heat, air-mass movement, humidity, and the course of chemical reactions and physical transformations, as well as biological phenomena.

It is customary to call the volume of flow of radiation energy 'intensity of radiation' and to measure it in calories per cm^2 per minute.

In passing through the atmosphere part of the solar radiation is absorbed by it and transformed into other kinds of energy, part is dispersed by clouds, and part is dispersed by gaseous molecules and by other particles suspended in the atmosphere. More than half of the radiation reaching the upper part of the atmosphere is thus used up (Fig. 3).

The intensity of radiation reaching the earth's surface fluctuates widely, depending on the height of the sun above the horizon and therefore on the geographical latitude (Table 3), the density of the atmosphere, its depth, cloudiness, atmospheric content of water vapour, time of day, and local relief.

Table 2. *Intensity of solar radiation in relation to the sun's height (in constant atmospheric conditions).* (*From Tverskoi 1948*)

Height of sun, degrees	Intensity of solar radiation (kcal/cm²)	Height of sun, degrees	Intensity of solar radiation (kcal/cm²)
5	0·39	30	1·11
10	0·60	40	1·21
15	0·82	50	1·27
20	0·95	60	1·31

Table 3. *Daily totals of solar radiation (in kcal/cm²) at different lattitudes.* (*From L. A. Ivanov 1929*)

Degrees of N. latitude	Month					Annual
	May	June	July	August	May-August	
80	4·4	3·8	3·5	1·8	13·5	16·8
60	8·4	8·7	8·2	5·0	30·3	43·6
50	9·4	9·7	9·5	7·9	36·5	54·7
45	9·1	10·8	11·9	9·0	40·8	81·9

The higher the sun, the less is the depth of atmosphere through which its rays pass, and the better they warm the earth's surface and, from the latter, the adjacent layer of air (Table 2).

The depth of the atmosphere affects solar radiation through dispersal of direct rays by gas molecules contained in the atmosphere, and by absorption of rays belonging to certain parts of the spectrum, with consequent alteration of the composition of the solar radiation. In particular, ozone (which exists in the upper atmospheric layers) completely absorbs the shortest ultraviolet rays, which therefore do not reach the earth's surface. Water vapour also alters the spectral composition of solar radiation, absorbing up to 20% of the energy of incident rays.

The total of daily solar radiation received varies greatly according to the exposure of slopes and their angle of inclination, as a result of which every part of the earth's surface (except in tropical regions) contains some ecotopes with radiation normal for a given locality and others with higher or lower radiation (Table 4).

Table 4. *Daily totals of solar radiation (in kcal/cm²) for an average day with average cloudiness.**

Total radiation	Months											
	I	II	III	IV	V	VI	VII	VIII	IX	X	XI	XII
Horizontal surface	20	41	105	196	278	317	267	220	160	79	25	15
Southern slope 30°	57	93	163	251	307	330	285	260	230	149	61	46
Western slope	18	41	94	179	247	287	240	203	151	75	24	14
Eastern slope	20	40	90	173	240	280	237	195	146	74	22	16
Northern slope	—	—	—	11	85	170	220	180	121	48	—	—

* Tellerman experimental forestry station, Voronezh province.

The radiant energy of the sun is the sole source of energy reaching green plants for synthesis of organic matter. Not all the energy reaching the surface of green leaves is photosynthetically active.

Of direct solar radiation only about 35% is active in photosynthesis. That proportion fluctuates slightly during daytime hours when the sun is high above the horizon but falls sharply as the sun's height decreases.

Orange-red rays (wave-length 600-700 μ) are absorbed by chlorophyll and show maximum physiological activity. When there is an adequate amount of rays of that wave-length, maximum production of organic matter occurs in plants.

Blue-violet rays (400-500 μ) are absorbed by chlorophyll, carotenoids, and other cell components, but are only half as effective as orange-red rays; absorption of the blue-violet rays delays flowering of plants, promotes synthesis of proteins, and apparently affects the chemical composition of plants.

Ultraviolet rays (300-400 μ) prevent excessive growth in plants. Rays up to 315 μ in wave-length are particularly effective in this respect.

Short ultraviolet rays (below 300 μ) are lethal to living organisms, but are absorbed by the upper layers of the atmosphere.

Infra-red rays (750-1000 μ) absorbed by plant pigments have negligible effect on physiological processes. Infra-red rays of wave-length over 1000 μ are absorbed mainly by the water in leaf tissues. They take part in the regulation of the heat regime of plants and, no doubt, affect the rate at which physiological processes take place, positively at temperatures up to 20°C and negatively at higher temperatures (Kleshnin 1954).

Green rays (500-600 μ) are the least active.

The composition of radiation, especially its content of physiologically-active rays (physiological radiation), depends on the kind of natural light. There is less physiological radiation in direct sunlight than in diffused light, which contains about 50-60% of physiologically-active rays. The weakening of illumination intensity in diffused light is compensated for by improvement in the physiological quality of the light.

The composition of radiation is of great importance, not only in the photosynthetic activity of plant communities, but also in their transpiration.

Experiments carried out by L. A. Ivanov and E. V. Yurina (1961) show that natural white light has the most favourable effect on transpiration by plants. Replacement of it by monochromatic light tends to lower the rate of transpiration. Different plants react differently to the composition of incident light. Transpiration rate increases in birch, acacia, and pine with red light, and in spruce and mountain ash with green light. The transpiration rates of spruce and mountain ash in green light are lower than for the other plants in red light.

The percentage of physiological radiation actually used by plants is not high. In a forest with optimum density of trees, only 3% of the energy is employed in the formation of organic material, and only about half as much (i.e., about 1·5%) of the full radiation.

RADIATION WITHIN PLANT COMMUNITIES

Some of the incident solar radiation on a plant community with dense foliage is absorbed by the plants, some is reflected, and some penetrates into the plant mass. Naturally, radiation that has passed through a layer of green vegetation differs greatly from that incident on the upper surface, not only having a lower intensity but also being different in quality.

In a forest the light beneath the tree canopy consists of: (a) diffused light reflected from sky and clouds; (b) diffused light reflected from the tree crowns; (c) diffused light that has penetrated through the foliage; and (d) direct sunlight, which on clear days passes between the leaves to form light patches. The proportions of these types of light in a plant community depends on the compactness and density of the canopy, the characteristics of the crowns, and the species-composition of the vegetation.

According to observations made in forest communities, the intensity of direct light is much less beneath a tree canopy, decreasing in proportion to the square of the distance from a gap in the canopy to the sunlit patch beneath. Radiation in sunlit patches under a tree canopy is from 25-50% less intense than direct radiation in an open space. The duration of illumination in any sunlit patch in a forest is much less than that in an open space, since the sunlit patch is constantly moving and illuminates the same part of the soil surface for only 15 to 30 minutes, or even less. Even on clear sunny days the duration of direct illumination beneath the tree canopy is not more than half of that outside the forest. In a spruce-deciduous forest, for example, even with moderate crown density (0·5), the duration of illumination by direct sunlight is less than half of the possible amount. In dense forest the duration of illumination does not exceed 10% of a full day's sunlight.

Diffused light also has its special characteristics in a forest. According to L. A. Ivanov's observations, diffused light in a forest is weaker in physiologically-active rays. Green rays predominate because of reflection and absorption by the tree crowns. Whereas in an open area active rays constitute 48-49% of the diffused light under a cloudy sky, in a pine forest of density 0·5 to 0·6 the proportion is not higher than 30%. In an oak forest, depending on the stand density, the proportion of active rays fluctuates from 2-13%, and in an ash forest from 12-18%.

The filtration of light by the tree canopy is clearly demonstrated by analysis of the spectral composition of radiation in the heart of a stand, in spring during the leaf-formation period and in summer with full foliage (Table 5).

Table 5. *Quality of light (percentage, by wave-lengths, of the total radiation falling on the upper surface of a tree stand) beneath the canopy of an oak forest, in relation to the stage of leaf development.**

Date of observation, phenophase	Wave-length						
	0·730 red	0·615 orange	0·580 yellow	0·530 green	0·463 blue	0·420 violet	0·330 ultra-violet
18 March. Buds not open	63	57	53	49	48	47	45
15 April. Buds beginning to open	61	44	39	38	37	38	32
10 May. Leaves out	21	7	7	6	6	6	5
4-25 June. Full foliage	15	5	5	4	3	3	3

* Tellerman experimental forestry station, Voronezh province. 23-year-old tree stand, with crown density 0·9-1·0.

As the table shows, with development of the leaves radiation beneath the forest canopy constantly decreases, the decrease being much more marked at the blue (short-wave) end than at the red end of the spectrum.

The amount of solar energy reaching the litter surface determines the intensity of transfer of heat and moisture, and also the daily changes in meteorological conditions in the atmospheric layer adjacent to the earth. The principal item in the radiation balance-sheet is the total radiation, consisting of direct and diffused light. Only a small part of the total radiation penetrates the tree canopy, the amount varying in forests of different species, different types, and different ages. We found that, taking the radiation in an open space as 100%, the percentages reaching the surfaces beneath the canopy in different forest communities are as follows: pine-lichen-moss 40%, pine-sphagnum 36%, pine-cowberry 25%, pine-molinia 16%, pine-bilberry 20%, pine-moss 16%, and spruce-mountain-sorrel 5%.

From our observations in the Tellerman oak groves the intensity of radiation beneath the tree canopy markedly increases as leaf-formation conditions deteriorate (Table 6). For instance, radiation beneath the canopy of an oak grove on alkali soil with very low productivity was three times as high as that beneath the canopy of an oak-goutwort grove with the highest productivity.

Radiation beneath the tree canopy varies greatly according to height above the soil surface. If on the surface of the canopy of a 100-year-old pine stand the total radiation is 100%, at the base of the crowns it is 30%, at 1 metre above ground level 25%, on the grass cover 10%, and within the grass cover, on the soil surface, less than 5%.

Table 6. *Total radiation in different types of forest (average data from observations on sunny days about the 15th of each month).**

Type of forest	Radiation beneath tree canopy			Radiation above tree canopy, cal/cm²/min	
	Total†	Diffused‡	Reflected§	Total	Diffused
Oak stands:					
Goutwort	2·9	10·0	30·0	0·69	0·15
Sedge-linden	4·0	19·5	24·5	0·82	0·21
Ash-sedge-goutwort	4·4	18·0	31·2	0·69	0·22
Hedge maple	4·04	10·1	27·1	0·66	0·17
Euonymus	7·16	33·2	33·3	0·75	0·18
Alkali soil	8·7	40·4	25·4	0·64	0·11

 * Tellerman experimental forestry station, Voronezh province.
 † Percentage of radiation above tree canopy.
 ‡ Percentage of diffused radiation above tree canopy.
 § Percentage of total radiation for given forest type.

There are also substantial variations in the radiation beneath the tree canopy according to the age of the stand. In a pine-cowberry stand, total radiation at the age of 15 years is about 36%, at 30 years 16%, at 70 years 23%, at 100 years 31% (Sakharov 1948). In a sedge-goutwort oak grove, according to our observations, total radiation at the age of 20 years is about 2% of that in an open space; at 35, 50, and 70 years 4%; and at 230 years 6%.

As we see, the radiation regime in a forest is extremely variable and depends on species-composition, age, layering, and stand density.

It is not only in tree communities, of course, that radiation varies. Its variations have also been studied in herbaceous phytocoenoses, e.g., in cut-over areas, as is shown by the following data (in cal/cm² per minute):

Bare soil	1·11
Beneath herbaceous canopy, 100 cm above surface of soil	1·11
Within herbaceous growth, at height above soil of:	
50 cm	0·92
10 cm	0·25
0 cm	0·17

Thus, even within an herbaceous community there is a complex radiation system, caused by absorption, transmission, and reflection of parts of the radiation by the above-ground organs of green plants.

The radiation entering a forest biogeocoenose is used in the total evaporation from the surface of the herbaceous layer, in the turbulent heat-exchange between the forest and the atmosphere, in warming the

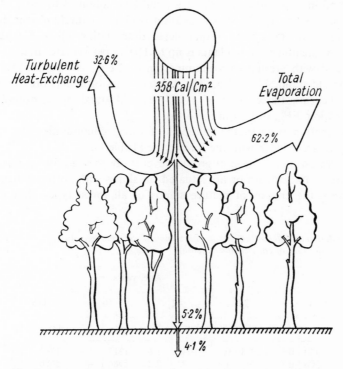

Fig. 4. Average heat balance-sheet for a day from June to August in a 30-year-old oak stand.

soil and the vegetation, and also in evaporation from the surface of the soil and the herbaceous cover. This is graphically shown in Fig. 4, which illustrates the distribution of the items of expenditure of the incident energy in a 30-year-old oak forest at the Tellerman forestry station in Voronezh province in July and August.

ILLUMINATION WITHIN PLANT COMMUNITIES

Illumination beneath the forest canopy, like radiation, varies at different heights above the soil surface and in stands of different species, ages, crown densities, etc. (Table 7).

As shown in Table 7, in any forest the amount of illumination decreases from the tree crowns down to the soil. Illumination at ground level is at a minimum in young dense stands, where there are neither living plants nor living tree branches up to a height of 3 metres. The topmost parts of the crowns of all trees are well lighted, and growth in height there is not restricted at any phase of development of the stand.

The variation in illumination beneath the tree canopy is of very great significance for the development of the lower forest layers—regeneration growth, undershrubs, herbaceous cover. Beneath the canopy of a 13-year-old stand in which illumination at ground level amounts to 0·6% of that in an open space herbaceous vegetation is almost entirely absent, but in a 22-year-old stand with ground-level illumination of 0·8% the dry weight of the understorey plants is up to 120 kg per hectare; in a 42-year-old stand with ground-level illumination of 1·6%, 264 kg/ha; and in a 56-year-old stand, 475 kg/ha. The greatest mass of understorey plants is found beneath the canopy of a 135-year-old stand, with ground-level illumination 2·1% of that in the open.

Illumination beneath the canopies of forest communities also varies widely in accordance with the forest type (Table 8).

The lowest illumination at ground level is observed in forest types that have dense underbrush or are situated on slopes with northern exposure.

Among the crowns of mature stands of all forest types illumination

Table 7. *Illumination at different heights in oak stands of different ages (average from observations on sunny days about the 15th of each summer month).**

Height above ground	Age, years											
	13		22		42		56		135		220	
	lux	%	lux	%	lux	%	lux	%	lux	%	lux	%
Ground level	350	0·6	454	0·8	1038	1·6	1317	1·9	1383	2·1	1143	1·7
1·3 metres	506	0·8	796	1·3	1393	2·2	2986	4·1	3686	5·7	5119	7·5
⅓ height of stand	385	0·6	1293	2·2	2893	4·1	3468	5·1	4208	6·6	7167	10·6
0·5 height of stand	625	1·2	1660	2·8	3693	5·8	7695	10·1	11,563	18·0	14,921	21·0
Among the crowns	1925	3·2	3069	5·1	6394	10·0	12,000	17·7	38,493	59·0	48,659	72·5
Above the crowns	64,210		62,377		64,318		68,508		65,000		68,217	100·0

* Tellerman experimental forestry station, Voronezh province.

Table 8. *Illumination (in lux) in different types of forest (average of observations on sunny days about the 15th of each summer month).**

Type of forest	Illumination in forest at height in metres			
	Ground level	0·5	1·3	10
Oak groves:				
Goutwort	628	884	1676	20,689
Linden-sedge	1248	1676	2347	21,689
Goutwort-sedge	1543	1994	5119	24,689
Hedge maple	790	1624	3454	31,424
Euonymus	676	2654	6277	36,983
Open space	65,000	65,000	65,000	65,000

* Tellerman experimental forestry station, Voronezh province.

varies from 45,000-50,000 lux and is fully favourable for uniform growth in diameter.

In mature and over-mature stands variation in lighting conditions does not produce any noteworthy effect on growth. On the other hand, in young and maturing stands the variations in illumination may exert substantial influence on tree growth (light-produced increment after improvement cutting). Light-demanding species are especially responsive.

Determination of the amount of light absorbed and used in the process of photosynthesis is of great value in forest economy.

In photosynthesis, or the manufacture of organic matter, only 0·5-1·0%, with a maximum of up to 5%, of solar energy is used. These quantities are usually called coefficients of use of solar energy for crop or plant-mass production.

The space occupied by a forest can be regarded as a green screen using a small part of the absorbed light in the synthesis of organic matter. The rest of the light is reflected from leaf surfaces or is absorbed by them and produces more rapid evaporation of water.

According to L. A. Ivanov (1946), the amount of light absorbed by a single tree is determined by the following equation:

$$Q = (1-L) N,$$

where Q is the amount of solar energy falling on one hectare, or the unit of intensity of light falling on the crowns; L is that part of it penetrating the canopy; $(1-L)$ is the intensity of light retained by the crowns (co-efficient of shade); N is the number of trees per hectare.

The amount of light absorbed by a single tree depends on the density of the stand (when the crowns are fully developed). Absorption may be greatest with 1000 to 2000 trees per hectare; at higher densities, as the number of trees increases the amount of absorption per tree decreases. Leaves of different species absorb different amounts of light. At the same time the crowns of different species have similar values for absorption of active rays, as a result of which the total production of organic matter by equal-aged stands of different species (pine, spruce, or birch) is almost the same.

In Leningrad province all these species produce annually about 5·5 to 5·6 tons per hectare of dry organic matter.

This does not mean, however, that all the tree species mentioned produce equal amounts of wood, since larger amounts of photosynthetic products go to form wood in some species and smaller amounts in others. The synthate not going into wood is also unequally distributed among leaves, roots, bark, etc. Comparing the quantities of absorbed light with the quantities of wood produced by different species, L. A. Ivanov (1946) discovered that pine uses light most economically and productively, requiring half as much light to form a given amount of wood as oak does.

With optimum density of pine and oak stands the economic coefficient of light use is equal to 3% of active light and about 1·5% of full radiation. If we consider only wood, and not the total organic mass, the coefficient of light use falls to 2%.

In determining the annual production of organic matter by tree stands, it is important to know not only the total production by photosynthesis but also the production of economically-valuable parts of the trees. To express the relation between the amount of useful wood produced and the amount of work done by the photosynthetic apparatus of the trees, one must take into calculation the coefficient of production of usable wood, which depends on the age of the stand. This coefficient represents the proportion of dry weight as usable wood in the stand. It will probably vary considerably in biogeocoenoses of different types, but definite data are not available.

TEMPERATURE REGIME OF THE ATMOSPHERE

The temperature regime of the atmosphere has a profound effect on the functioning of the biogeosphere, since there are no processes that are not affected by temperature conditions. Solar radiation is the chief source of heat on the earth's surface. Only the highest layers of the atmosphere, however, are heated directly by the sun. The troposphere is heated mainly from the earth's surface as a result of turbulent exchange and heat radiation from the earth. Both direct and diffused radiation take part in heating the air. In different parts of the earth's surface the flow of heat and the heating of the atmosphere vary considerably, since the proportions of direct and diffused radiation are very unequal in different places, and the transparency of the atmosphere and the height of the sun, and consequently the intensity of its radiation, are not constant. Atmospheric circulation also plays a major role, and so (in more limited areas) do relief and exposure of slopes. Variation in the effects of the latter is due to the characteristics of solar heating of slopes with different compass orientation and to the conditions of flow of air-mass from them. In the northern hemisphere the greatest amount of heat falls on southern slopes in spring and autumn, i.e., when the sun is not high above the horizon. At 60°N, for instance, in April southern slopes with an inclination of 30° receive 50% more solar heat, and at 50°N similar slopes receive 28% more solar heat, than do level areas. During the period when the sun is highest in the sky, differences in reception of direct solar radiation are lessened, both for different latitudes and for slopes of different exposures. At that time even steep northern slopes receive up to 80% of the radiation falling on level ground.

Except for the tropic zone, where heat is distributed almost evenly throughout the year, the temperature regime of the atmosphere over the earth's surface is very unequal, and during the annual cycle warm periods and cold periods alternate irregularly. The arrival of cold weather is marked by a decrease in the life-activity of plants, animals, and micro-organisms, in water circulation, and in chemical reactions; the lower the temperature and the more prolonged the cold part of the year, the more pronounced (even to the extent of complete cessation) is the decrease in activity. The variations in local heat conditions observed from year to year and during a single day also substantially affect the activity of the components of a biogeocoenose and their inter-relationships.

The relationships between heat and the different components of a bio-geocoenose are quite specific, and call for separate analysis.

The relationships between heat and phytocoenoses are of primary bio-geocoenotic significance, since heat and water determine the scale, the forms, and the rhythms of the biogeocoenotic work of phytocoenoses. We must remember that the heat regime of the atmosphere determines the horizontal and vertical extent of plant cover, and directly and in-directly affects photosynthesis, transpiration, respiration, mineral and water absorption, plant growth, and accumulation and decomposition of organic matter, now accelerating and now delaying the course of these processes.

Deficiency of atmospheric heat and shortness of the warm period of the year determine the northern and altitudinal boundaries of forest bio-geocoenoses and the limits of many forest formations and of individual tree species, and also retard the rate of tree growth in Arctic regions and high mountains. Excessive falls in temperature and extreme duration of cold seasons have their own specific effects on phytocoenoses, especially forest phytocoenoses. During severe frosts, especially if these alternate with thaws, frost-cracks often appear on the trunks and branches of trees, young shoots and flower-buds are killed off, and sometimes even whole plants of seemingly frost-resistant species perish; this was observed, for instance, for spruce during the Archangel winter of 1939-40, when the temperature fell to −45°C and lower.

Falls in air temperature to below 5°C at the beginning and the end of the vegetative season also strongly affect forest phytocoenoses. Such falls in temperature may be due either to irruptions of cold masses of air (advective frosts) or to cooling of the soil surface by radiation at night (radiational frosts). Spring frosts often nip flower-buds, young leaves, and young tree shoots, as a result of which the productivity and growth in height and diameter of plants are considerably reduced. Early autumn frosts seriously damage young shoots that have not yet matured, causing changes in terminal shoots that result in deformation of the trunks or forking of tops.

Frosts have the greatest effects in topographic basins, in small valleys, and in forest clearings (especially with dense undergrowth). In such places so-called 'lakes of cold' or basins of winter-killing are often formed, seriously reducing the frost-free period, as shown in Table 9.

Table 9. *Average duration of frost-free period at a height of 1·5-2 metres above the soil surface, with relation to location of field (Gol'tsberg 1957).*

Location	Frost-free period, days		
	Spring	Autumn	Total
Tops and upper parts of slopes	+10	+10	+20
Broad valleys (over 1 km)	0	0	0
Valleys in rolling relief	−5	−7	−12
Valleys over 50 metres deep in hilly country	−6, −10	−10, −15	−16, −25
Landlocked valleys more than 50 metres deep and mountain ravines	−12, −18	−18, −22	−30, −40
Mountain plateaux	−5	−10	−15
Valleys of large rivers, shores of large lakes and reservoirs	+5	+10	+15

The effect of heat on a phytocoenose is well-marked in the transpiration of plants. L. A. Ivanov (1946) discovered a direct relation between air temperature and transpiration in tree species, which enabled him to formulate the thermometrical method of calculating the latter. According to A. A. Molchanov (1952), the relation between transpiration and air temperature is more complex. With average monthly temperature below 10°C the curve of the relation between air temperature and transpiration rises slightly above the horizontal; at 10-12°C and higher it rises steeply up to a temperature of 25-30°C, after which it begins to fall sharply.

The links between phytocoenoses and the temperature regime of the air are very dynamic, and the same temperature regime may have widely differing effects on phytocoenoses on account of variation in other coincident factors and differences in phenology of plants in a community, since different heat conditions are required by plants at different phases of growth.

Together with the direct action of heat on plant communities there exists a wide range of multiform, indirect effects of it through the medium of other atmospheric factors (light, air humidity), soil, and the activity of micro-organisms. Finally, in evaluating the links of atmospheric heat with terrestrial phytocoenoses we must remember that the temperature regime in the lower zone of the troposphere is to a great extent conditioned by plant communities.

A forest strongly affects temperature conditions, not only beneath and within the crowns but also above them, because of heat radiation from their upper surfaces. The greater the density and the height of the trees, the more marked are these effects. Tree crowns act as a screen both for light falling upon them and for heat penetrating through them. Therefore the denser and taller the stand, the less heat reaches the air and soil beneath.

The moderating action of the arboreal layer on heat conditions varies according to the type of the forest, to its species-composition, to its situation with regard to relief, and to the forest-vegetational features of the soil. Within a single forest type air temperature beneath the tree layer varies substantially according to the age of the trees; differences in air temperature in stands of different age within a single forest type often exceed those between different forest types.

According to our observations at the Tellerman forestry station, the highest average and maximum temperatures are recorded in 4-year-old stands on cut-over areas (Table 10). Minimum temperatures are also lowest in cut-over areas. As the stand increases in age the air temperature (measured at a height of 2 metres) begins to fall markedly and reaches its lowest point in 23-year-old stands. The difference between the average daily temperatures in 23-year-old and 4-year-old stands may reach 3·2°C. At later ages further increases in temperature are observed. In winter there is almost no difference between air temperatures in stands of different ages.

Table 10. *Average and extreme monthly air temperatures (°C) at a height of 2 metres above ground level in oak stands of different ages, in 1954; forest type: sedge-goutwort oak.**

Month	Age, years														
	4			23			45			58			220		
	Average	Maximum	Minimum	Average	Maximum	Minimum	Average	Maximum	Minimum	Average	Maximum	Minimum	Average	Maximum	Minimum
May	16·3	34·1	−2·8	14·1	32·1	1·8	15·3	32·3	−1·5	16·5	32·4	−2·0	14·7	32·8	−2·2
June	19·4	25·6	4·1	16·2	25·1	6·5	19·7	28·7	6·5	19·9	28·9	6·4	17·4	29·6	6·4
July	21·1	33·5	5·5	19·2	26·2	7·5	20·9	31·6	7·7	20·3	31·7	7·4	19·9	32·2	7·2
August	18·6	32·2	4·5	17·1	23·4	8·5	18·1	32·0	8·8	18·2	32·1	8·9	17·7	32·3	8·8
September	13·4	29·0	0·5	12·6	20·5	1·5	13·7	25·3	1·9	13·8	25·4	0·9	13·1	25·5	0·7

* Tellerman experimental forestry station, Voronezh province.

The extent to which air temperature is modified by tree stands naturally is not the same at different heights above ground level, as may be well seen in forests of any age (Table 11).

Table 11. *Average air temperature during the vegetative period in 1954 at different heights above ground level in oak stands of different ages.**

Height of recording in metres	Temperature (°C)			Relative humidity of air (%)
	Average	Maximum	Minimum	
6-year-old stand, height 2·2 metres				
4·2	17·0	25·0	5·3	74
2·2	17·3	33·4	2·8	77
1·0	16·9	27·9	4·5	86
0·05	15·4	22·9	4·7	94
0·00	15·3	22·0	4·8	98
27-year-old stand, height 9·2 metres				
11·2	17·0	25·8	5·1	75
9·2	17·2	33·6	2·6	77
4·0	16·3	26·7	4·4	85
2·0	15·9	24·9	4·4	86
0·05	15·6	22·7	4·7	92
0·00	15·3	21·9	5·8	98
220-year-old stand, height 31 metres				
33	17·3	25·9	5·1	76
31	17·4	30·8	3·8	78
20	17·2	28·9	4·4	81
10	16·4	26·8	4·4	86
2	16·2	28·9	4·5	85
0·00	15·5	22·9	5·8	98

* Tellerman experimental forestry station, Voronezh province.

The greatest climatic contrasts in a forest biogeocoenose occur at the top of the tree layer. It is there that the greatest daily average and maximum temperatures are recorded, and also the smallest minima.

In winter the temperature gradients disappear in leafless stands in a forest. On cloudy days differences in temperature beneath the canopy of stands of different types and ages are greatly reduced.

The effect of a forest on the temperature of the air is distinctly evident at some horizontal distance away from it.

In particular, a forest considerably moderates frosts and lessens their duration in cut-over areas, and also reduces high temperatures in adjacent fields. At the Tellerman experimental forestry station, during a frost with temperature of −4°C, in a 300-metre clearing the sub-zero temperature lasted 8 hours, in a 100-metre clearing 5·5 hours, in a 50-metre clearing 3·7 hours, and in a 30-metre clearing only 2·3 hours. On the steppes high temperatures are substantially less nearer forest edges; the higher the air temperature, the more marked the effect (Table 12).

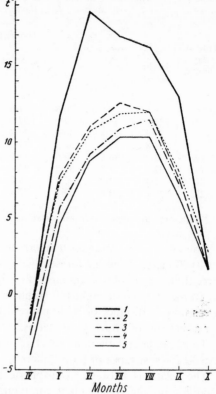

Fig. 5. Soil temperature from April to October at a depth of 5 cm in different types of forest. 1—solonetz (alkali soil) glade; 2—clearing; 3—*Euonymus* oak stand; 4—sedge-goutwort oak stand; 5—hazel-elm stand.

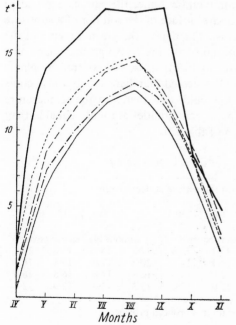

Fig. 6. Soil temperature from April to November at a depth of 50 cm in different types of forest.

Table 12. *Air temperature at a height of 2 metres above ground level in sunny weather, on the steppe at different distances from the forest edge (Molchanov 1961).**

Time of day (hours)	Temperature (°C)							
	In the forest	At forest edge	On steppe, at distance from forest edge (metres)					
			5	10	20	50	240	300
9	29·6	29·6	30·8	31·6	30·6	30·2	30·1	30·1
11	35·0	34·4	37·0	37·0	35·6	35·1	34·7	34·8
13	37·2	37·8	40·0	39·2	37·8	37·7	37·5	37·6
15	35·8	35·8	36·8	37·2	36·9	36·5	36·2	36·4
17	35·9	36·6	37·0	37·6	37·0	36·8	36·6	36·5
19	32·8	33·5	33·8	34·4	34·0	33·7	33·4	34·4
21	31·4	32·4	32·3	34·2	32·8	32·1	32·0	32·0
23	30·1	30·9	29·1	27·9	26·8	26·3	25·8	25·8

* Tellerman experimental forestry station, Voronezh province.

The effect of a forest on temperature fluctuations in clearings extends to 40-50 metres from the forest edge by night and to 20-30 metres by day. Tree thinning enables warm air to flow into full stands to a distance of up to 50 metres, and at night full stands show a warming effect on thinned areas up to a distance of 30-40 metres. Clearings warm the air in a forest for a distance of up to 60-80 metres.

Thus the lateral thermal effect of one biogecoenose on another may extend for a distance of 50-100 metres.

The relations between atmospheric heat and soil are in most cases modified by intervening plant communities, since the surface exposed to solar radiation is usually not the bare surface of the soil nor of a substrate in general, but a green plant mass. The denser the green screen and the thicker the atmospheric layer occupied by the above-ground mass of a phytocoenose, the greater is that modification. A soil covered by plants warms more slowly and becomes less heated than bare soil, but also cools off more slowly and becomes less cold, and therefore the temperature conditions of soils underneath plant communities are more equable than those in areas without vegetation (Table 13).

Table 13. *Average monthly temperature of soil in a clearing (°C).**

Month	Temperature at depth in cm							
	5	10	15	30	5	10	15	30
	with herbaceous cover				without herbaceous cover			
June	14·5	13·8	13·4	12·8	17·8	16·9	15·8	14·8
July	15·7	15·3	14·9	14·6	20·2	17·9	16·8	16·3
August	15·7	15·6	15·0	14·8	18·8	17·6	16·8	16·6
September	11·9	11·9	12·0	12·0	13·3	13·0	12·9	12·8

* Tellerman experimental forestry station, Voronezh province.

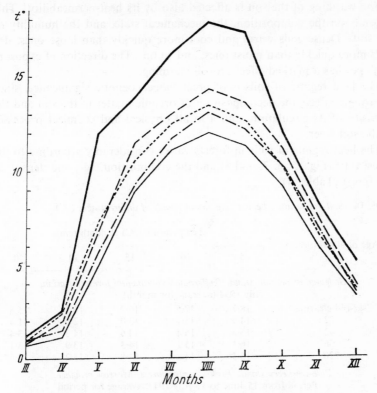

Fig. 7. Soil temperature from March to December at a depth of 100 cm in different types of forest.

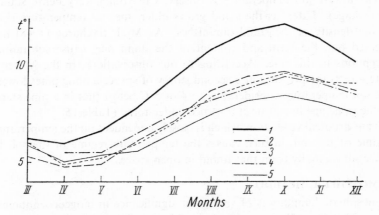

Fig. 8. Soil temperature from March to December at a depth of 320 cm in different types of forest.

The warming of the soil is affected also by its heat-permeability. This depends on the composition, the mechanical state, and the humidity of the soil. Dense soils warm and cool more quickly than loose ones, dry soils more quickly than moist ones, and so on. The direction of exposure of slopes has a marked effect on soil warming.

The heat regime of soils is of great biogeocoenotic significance, since it strongly affects the decomposition of organic matter in the soil and the intensity of its circulation, and also all physical and chemical processes in the soil layer.

The heat regime of soils in forests varies considerably according to the forest type (Figs. 5, 6, 7, and 8) and the composition, age, and density of the forest (Table 14).

Table 14. *Soil temperature beneath canopy of stands of different ages (°C).*

Age of stand (years)	Temperature (°C) at depth (cm)				
	5	10	15	30	50
Sedge-goutwort oak stand. Tellerman experimental forestry station. July 1954 (average for month)					
2-yr-old clearing	18·7	17·5	16·9	16·5	15·8
22	14·9	13·7	13·0	12·6	11·4
42	16·3	15·4	14·6	14·3	13·4
56	16·1	15·1	14·3	13·9	13·1
220	15·5	14·3	13·8	13·2	12·6
Pine-cowberry stand. Prokudin pine forest, Moscow province. Period from 15 June to 15 July 1947, average for period					
Clearing	19·6	18·0	16·8	16·1	
33	12·6	12·3	11·9	10·9	
65	16·6	16·0	14·8	14·0	
150	17·9	16·9	15·6	14·7	

The lowest soil temperature is observed in young, very dense stands (pole stage). Later, as the stand grows older, the soil temperature rises as the density of the stand diminishes. As M. I. Sakharov (1948) has pointed out, the form and content of the stand may cause substantial exceptions to this rule. According to our observations, in the Moscow pine forests the presence of a second storey of spruce among pines lowers the soil temperature considerably, to 4 or 5°C below that in a pine stand of the same age but without a spruce understorey (Table 15).

Thus arboreal vegetation, which is closely dependent on the temperature regime of the soil, in turn causes the temperature regime of the soil to differ substantially from that found in open spaces.

ATMOSPHERIC HUMIDITY

Atmospheric humidity is of very great significance in biogeocoenotic exchange between all components of a biogeocoenose, but its connection with phytocoenoses and soils is especially marked. While it affects all

Table 15. *Average soil temperature during the period 15 June to 15 July 1947, in stamps of different kinds.**

Description of stand	Temperature (°C) at depth in cm			
	5	10	20	30
Clearing	19·6	18·0	16·8	16·1
10 Pine, 150 yrs, density 0·8	17·9	16·9	15·6	14·7
10 Pine, 150 yrs, with dense spruce lower storey	12·8	11·4	10·5	10·0

* Prokudin pine forest, Moscow province.

aspects of plant life and of soil in one way or another, it most strongly affects their exchange of moisture with the atmosphere, sometimes stimulating and sometimes retarding it.

Most important in this respect is the amount of water vapour deficit in the atmosphere. The greater the deficit, the more active are the physical evaporation of water and its transpiration by plant communities. With favourable atmospheric humidity the transpiration mechanism of plants works normally, and moisture-exchange proceeds optimally. Lack of moisture in the atmosphere causes a decrease in the turgor of plant cells, and in combination with high temperature also produces irreversible wilting of plants.

Dryness of the air is most harmful to plants during drought periods, when the sun scorches the earth with its burning rays and the wind helps to speed up transpiration.

A 'sukhovei'[1] begins when the relative humidity of the air falls below 30%, wind velocity exceeds 5 metres per second, and air temperature exceeds 25°C.

Sukhoveis may be divided into several classes according to their harmful effects on plants (Table 16).

Table 16. *Types of sukhovei-drought weather* (Fel'dman 1950).

Temperature (°C)	Relative humidity of air (%)	Moisture deficit (millibars)	Intensity of effect
22·5-27·5	21-40	23	Mild
27·5-32·5	41-60	23	,,
22·5-27·5	0-20	29	Medium
27·5-32·5	21-40	—	,,
32·5-37·5	41-60	29	,,
27·5-32·5	0-20	40	Strong
32·5-37·5	21-40	—	—
32·5-37·6	0-20	50	Severe
37·5-42·5	0-20	64	Very severe

[1] *Translator's note:* As the Russian word 'sukhovei' has a more restricted meaning than our 'drought' and is here defined in the text, it is used henceforth in this translation.

We must remember, however, that the harmfulness of a sukhovei also depends on the amount of moisture in the soil before the sukhovei begins.

The wetter the soil before the sukhovei, the less the damage. M. S. Kulik (1957) suggests that if the amount of available moisture in the soil layer from 0 to 20 cm is over 20 mm, the harmful effects of a sukhovei are mainly restricted to more rapid soil desiccation, but if the amount of available moisture in that soil layer is less than 10 mm, a sukhovei produces direct injury to plants.

In a forest the effects of a sukhovei appear first on the foliage of the trees. Leaves affected by a sukhovei turn red, wilt, and fall off. The death of leaves often causes the withering of individual trees, not only in forests with insufficient soil moisture but even in those where soil moisture is optimal. Trees of age classes IV and V suffer most. According to observations in oak groves in the forest-steppe zone, during droughts the drying of leaves is especially marked in 40- to 60-year-old stands, as the soil is drier there than in younger or older stands. In stands 15 and 135 years old, leaves dry only during sukhoveis of high intensity. Different species suffer to different extents. After sukhoveis, the greatest numbers of leaves are lost by linden trees (about 50%), then ash (25%), hedge-maple (20-25%), elm, and, least of all, oak (10-15%). Desiccation of leaves is most marked in dense stands. Severe and prolonged droughts, by disrupting the water and temperature regimes of vegetation, lead to crop failures in fields and meadows and to reduction of growth of trees in the forests, especially in forest types with inadequate stocks of soil moisture.

A high content of water vapour in the atmosphere also strongly affects biogeocoenotic exchange in the forest. It lowers, to the extent of complete cessation, both the physical evaporation of water and transpiration, and also impedes other forms of exchange between phytocoenoses and their environment. To overcome that danger in regions of very high humidity, guttation of drops of liquid takes place on a large scale. Moisture-exchange by vegetation is thus maintained even in tropical rain-forests, where the atmosphere is saturated with water vapour.

As with light and heat, air humidity is strongly affected by vegetation, and its regime within plant communities differs greatly from that in open areas. In a forest the relative humidity of the air before sunrise is the same at all height levels, since most of the dew is deposited on the tops of the crowns. As the sun rises and the wind becomes stronger, the rate of intermingling of external and internal air in the forest increases and the humidity gradually falls. The effects of the decrease in humidity are felt at constantly lower and lower levels in the tree cover. The continuous flow of water vapour from the soil and the crowns moderates the drying effect of the external air, and although the humidity of the intracoenotic air falls it remains higher than that of the external air.

In phytocoenoses of herbaceous plants air humidity also varies vertically. For instance, in clearings overgrown by herbaceous plants the relative humidity at the soil surface is 98%; at a height of 10 cm it is 94%, at 50 cm 59%, and at 100 cm 56%. During the day air humidity naturally varies, being lowest about 2 to 3 p.m. and rising towards morning and evening. Air humidity is least stable at the upper boundary of plant cover. The denser the herbaceous cover, the greater is the relative humidity within it. According to our observations, with a leaf area of 1·93 cm² the relative humidity in 1 cm³ of air is 75%; with a leaf area of 0·89 cm² it is 67%, with 0·35 cm² 48%, and on an area without herbaceous cover the relative humidity reaches 40%. In clearings the relative humidity of the air varies considerably according to the width of the clearing. In oak groves of the forest-steppe zone, in narrow clearings that are 50 metres across the relative humidity rarely falls below 50% even by day, and then only for a short period (about an hour), whereas in clearings 100 metres wide it falls below 40% and remains at that level for not less than 6½ hours. At the same time the humidity in clearings 600 metres wide falls to 30% and remains at that level for 8 hours on a strip more than 180 metres wide in the centre of the clearing. In forest-steppe conditions these findings are of great significance in the organising of forest-restoration measures in clear-cut areas.

ATMOSPHERIC MOVEMENT

Movements of air-masses in the troposphere are of great importance to the biogeosphere. They originate from uneven heating of the atmosphere at different latitudes, above the continents and above the oceans, and modify the contrasts between foci of heat and cold and between foci of humidity and dryness, equalise the gaseous content of the air, create the climates of the earth, and determine weather and its variations.

Naturally there is no feature and no process in the biogeosphere that is not affected either directly or indirectly, but always profoundly and in many ways, by air-mass movements.

In the life of forest biogeocoenoses the role of atmospheric movements is displayed in a very evident and forceful manner. Forest and wind—this is one of the oldest and best-studied relationships of forestry. N. S. Nesterov (1908) places wind in the same rank as light and moisture with respect to its effects on the life of the forest.

While dispersing the water vapour from leaves, the wind brings drier masses of air into the forest and so speeds up evaporation and transpiration. In other cases the wind may bring air saturated with water vapour, and then transpiration by forest vegetation is lowered. The wind aids the exchange of gases in the air, carrying air with reduced CO_2 away from the leaves. It takes a considerable part in variations in air temperature. It strongly affects the development of tree roots; increases the taper of trunks by swaying; aids crown formation, pollination of flowers, and dispersal of fruits and seeds. Pollen of some tree species is carried by the wind for 100 or 200 km (pine), and small numbers of seeds are carried for kilometres or even tens of kilometres (e.g., seeds of willow, poplar, elm, birch, pine, spruce). In forest areas seeds are, of course, scattered less widely, and even in clearings the seeds of several species travel no more than 100 to 150 metres from their source (Table 17). Even with such short distances the carrying of seeds by wind is one of the most constant forms of lateral exchange of matter between adjacent biogeocoenoses. Such wind-aided exchange takes place on a still larger scale with the blowing of leaves. Birch leaves may be blown for 60 metres, oak and maple leaves 40 metres, larch needles 30 metres, pine needles 25 metres, and spruce needles 20 metres, the volume of blown matter crossing the boundary between two adjacent biogeocoenoses being fairly substantial. Thus if 50 grammes of dry leaves per m^2 fall from the canopy of a birch stand, leaves fall on the soil 10 m away at the rate of 3·22 g/m^2; 20 m away, 1·0 g/m^2; 30 m away, 0·6 g/m^2; and 50 m away, 0·3 g/m^2.

Table 17. *Wind-blowing of seeds from the forest edge* (%).

Species	Distance from forest edge, metres									
	0	12·5	25	50	75	100	125	150	175	200
Ash*	100	25	6	—	—	—	—	—	—	—
Pine	100	50	36	19	9	2·7	1·6	0·8	0·3	0·1
Spruce	100	58	42·2	15·3	9·3	4·7	1·9	1·1	0·8	0·4
Larch	100	69·1	28·1	12·1	6·1	2·6	0·8	0·1	—	—
Birch	100	95	85	79	74	68	61	54	50	43

* Tellerman experimental forestry station; all other species, Archangel province.

In winter the principal biogeocoenotic role of wind is the moving of snow. In forest areas this leads to the piling up of high snow drifts on the outskirts of the forest and in clearings; in open spaces snow piles up in ravines, gullies, and other depressions in the land surface.

The depth of snow cover affects the freezing of the soil and the wintering of snowed-under plants or parts of plants, and (in spring) the depth to which the soil is soaked and the amount of soil moisture.

That role of the wind is especially important in southern arid districts and in the forest-tundra. In spring the wind often moderates frosts by moving away cold masses of air. In dry regions strong winds in spring and summer generate dust storms, damage shoots, lay bare plant roots, and often completely blow away plantings and sowings of forest and farm crops.

High-velocity winds, by increasing transpiration, impair assimilation conditions and consequently lessen the growth of trees in both height and diameter (L. A. Ivanov 1956).

According to our observations, wind velocity lowers the increment of dry matter in plants by increasing their water deficit. Whereas with a wind velocity of 0·5 km/hr the water deficit in pine saplings does not exceed 4·5%, with a wind velocity of 15 km/hr it rises to 12·5%; at 38 km/hr to 18%; and at 60 km/hr to 20%. Correspondingly the increment of dry matter in a pine sapling falls from 23 g with wind velocity of 0·5 km/hr to 19 g at 15 km/hr and to 9 g at 38 km/hr. In certain hydrothermal conditions wind may not only slow down the growth of plants, but actually kills them.

Strong winds often cause heavy mechanical damage to a forest by breaking branches and trunks. In the north of the European part of the U.S.S.R., spruce stands on heavy compact soils are very susceptible to wind damage. The worse the growing conditions for a forest, the more resistant the trees are to wind (Molchanov and Preobrazhenskii 1957). Thus in dense herbage-mountain-sorrel-spruce stands, windblown trees were seen over 78% of the area; in mountain-sorrel-spruce, over 66%; in bilberry-spruce, over 36%; in cowberry-spruce, over 24%; in long-moss-spruce, over 17%; and in riverside spruce stands, over only 4% of the area.

Within forest biogeocoenoses wind conditions are always much different from those in open spaces. A forest is a serious obstacle to low-level winds. The forest breaks the force of the wind, checks its flow by means of the tree crowns and trunks, and changes its direction. As wind passes through a forest much of its force is spent in friction against trunks and branches and is converted into heat or into the mechanical work of swaying trunks and branches, moving leaves, etc.

Forests that differ in composition and in crown density reduce wind velocity unequally, as shown in Table 18.

Table 18. *Wind velocity within forest areas, 200 metres from edge (% of wind velocity in open spaces).* *

Stand	Wind velocity in open space, metres/sec						
	1·2	2·2	2·7	3·5	5·2	6·2	7·0
10 Pine, 150 yrs old, density 0·8, with second storey spruce, dense	0	2	4	8	15	17	19
10 Pine, 65 yrs old, density 0·9	8	10	12	17	28	39	46
10 Pine, 65 yrs old, density 0·4	12	18	20	29	44	61	—

* Prokudin pine forest, Moscow province.

Dense spruce forests break wind force very quickly. At a distance of only 40 metres from the forest edge the wind velocity in a spruce stand was reduced to 1-1·5% of its velocity in a field, which was 2·8 m/sec.

Wind velocity in a forest varies with height. As is seen from Table 19, the smallest velocity is observed at ground level, with irregular increases upwards to the crowns.

Table 19. *Wind velocity in and above forest, according to Geiger (Tkachenko 1939).*

Place of measurement	Height above soil, metres	Average wind velocity, metres/sec
Above crowns	16·9	1·61
Upper surface of crowns	13·7	0·90
Within crown canopy	10·6	0·69
Beneath crowns	7·4	0·67
Halfway between canopy and soil	4·3	0·69
Above soil	1·1	0·60
On soil surface	0	0

In deciduous forests wind velocity decreases markedly at all levels in proportion to the development of foliage on the trees and the under-growth (Table 20).

Table 20. *Wind velocity in an oak stand 25 metres tall at different heights above ground level, with wind velocity on the steppe and above the crowns 2·4 m/sec.**

Height above soil surface (metres)	21 April		15 May		21 May		15 June	
	m/sec	%	m/sec	%	m/sec	%	m/sec	%
1·5	0·59	100	0·56	100	0·35	100	0·21	100
5·0	0·83	141	0·63	113	0·52	148	0·31	148
10·0	0·93	189	0·90	159	0·78	211	0·51	241
18·0	1·16	195	1·09	196	1·15	328	0·65	307

* Tellerman experimental forestry station, Voronezh province.

The change of wind velocity by a forest is not restricted to the zone of air within the forest, but affects considerable areas both to windward and to leeward.

According to the observations of N. S. Nesterov (1908) and others, when the wind is blowing towards a forest, near the forest margin (be-ginning at a distance of about 60 metres) wind velocity falls by 20-30%, and sometimes 50%, from its original figure, depending on the density and height of the forest, the presence of undergrowth, and the velocity of the wind itself. In the Prokudin pine forest, even in wide clearings (600 metres) surrounded by forest, wind velocity is only 58% of that in an open field; in a 200-metre-wide clearing, 37%; in a 100-metre-wide clearing, 15%; and in a 50-metre-wide clearing, only 5-8% of its open-field velocity. The moderating effect of the forest on wind extends as far as 100 or 200 metres over broad open spaces, depending on the force of the wind. This substantially affects the evaporation of moisture from the soil, lowering it by 20% when wind velocity is reduced by 30%.

ATMOSPHERIC PRECIPITATION

Precipitation falling on any area is derived partly from water vapour carried there by movement of air-masses from outside and partly from water vapour arising from local evaporation. Since the air, with the water vapour that it contains, is in constant movement and the horizontal velocity is considerably greater than the rate of vertical movement of water vapour, precipitation on small areas is derived as a rule mostly from moisture brought from outside. Precipitation derived entirely from local sources on land occurs only in limited areas of the humid tropics (equatorial rain forests). In this connection it is essential to determine the coefficient of local turnover, indicating how many times moisture brought from outside and falling as precipitation on a given territory is evaporated and re-precipitated.

The most recent investigations show the average wind velocity at an altitude of 3 km above Europe to be over 30 km/hr. Taking into account the extent of Western and Central Europe, a certain volume of air, moving in one direction at that velocity, can traverse the territory in two days. Therefore the idea (formerly prevalent) that the amount of externally-derived precipitation in Western and Central Europe is only one-third of the total is incorrect.

According to the observations of M. I. Budyko and O. A. Drozdov (1950) and of K. I. Kashin and Kh. B. Pogosyan (1950), precipitation in river basins of fairly large size is a comparatively small proportion of the total amount of water carried above them by air-masses in the course of a year. Precipitation in the European part of the u.s.s.r. is only 37% of the total amount of moisture transported across it by air-masses. Only 13% of that precipitation is derived from local moisture turnover.

During occasional brief periods the proportion of local evaporation in the total amount of precipitation may be greater than stated above, especially in summer during cyclonic activity, when the air circulation conditions of a given area cause little moisture to be carried beyond its boundaries, and a large part of the moisture goes to form clouds.

The annual amount of precipitation falling on any area of the earth's surface varies from 0 to 5000 (10,000) mm, and in some cases is even higher. Even in forested areas variation in the amount of precipitation is very great. In the u.s.s.r., for instance, forests grow in areas receiving little more than 150 mm of precipitation per annum (some districts of Yakutia); in contrast, other forests grow in areas receiving up to 2000 mm (Black Sea coast of Transcaucasia). The forms of atmospheric precipitation (rain, dew, snow, hail, hoarfrost, glare ice) vary, as do their relative proportions, their seasonal distribution, and their amounts. They have very specific effects on terrestrial biogeocoenoses.

The course of precipitation during the year or the growing period is important for all biogeocoenotic processes. Precipitation in the form of rain or snow is most important, as it constitutes the greater part of the moisture received by the earth's surface and fills a major role in interactions between all biogeocoenose components.

Atmospheric precipitation falling on the soil dissolves various mineral substances, disperses them radially and laterally, enables them to interact chemically with vegetation, fulfils the moisture requirements of plants and micro-organisms, and affects temperature conditions and aeration in the soil. In addition, atmospheric precipitation carries into the soil a certain amount of organic and inorganic matter collected by raindrops

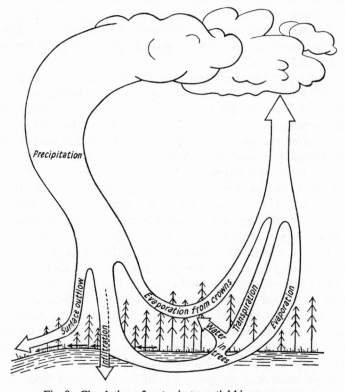

Fig. 9. Circulation of water in terrestial biogeocoenoses.

both from the atmosphere and from the surface of above-ground parts of plants in a biogeocoenose.

Precipitation entering terrestrial biogeocoenoses is partly retained by above-ground parts of plants and litter, and later re-evaporated from them into the atmosphere, and partly penetrates into the soil, from which some moisture is absorbed by plant roots to replace water used in transpiration, some is used by micro-organisms, and some seeps into underground bodies of water and flows away into rivers. A diagram of the transformation of atmospheric water in biogeocoenoses is presented in Fig. 9. The actual magnitude of different stages of moisture-circulation in nature fluctuates widely, and depends on the amount of precipitation, the intensity of rainfall, temperature conditions, wind, physical features of the soil, relief, and the composition and structure of plant communi-

Table 21. *Precipitation retained on the crowns of stands of different ages (annual average in mm).*

Species	Annual precipitation (mm)	Density 1·0 = complete cover	Age of stand (years)												
			20	40	60	80	100	120	140	160	180	200	220	240	260
Pine (near Archangel)	525	1·0	55	86	98	120	118	117	116	112	108	104	100	97	98
		0·8	—	68	—	—	93	—	—	92	—	—	81	—	—
		0·6	—	56	—	—	78	—	—	80	—	—	68	—	—
Pine (near Moscow)	550	1·0	127	150	140	135	120	105	100	97	—	—	—	—	—
Pine (in Khrenovsk forest)	521	1·0	75	—	136	—	130	—	—	120	—	—	—	—	—
		0·8	—	—	—	—	—	—	—	108	—	—	—	—	—
		0·6	—	—	—	—	—	—	—	90	—	—	—	—	—
Pine (in Archadinsk leskhoz)	375	1·0	70	78	73	—	—	—	—	—	—	—	—	—	—
Spruce (Vologda province, Kharovsk)	600	1·0	106	178	175	170	166	160	158	—	—	—	—	—	—
Spruce (near Moscow)	575	1·0	158	175	174	180	170	160	145	130	—	—	—	—	—
Larch	520	1·0	65	86	84	78	74	68	63	59	—	—	—	—	—
Poplar (Tellerman)	500	1·0	88	86	66	51	—	—	—	—	—	—	—	—	—
Ash (Tellerman)	500	1·0	103	99	85	65	—	—	—	—	—	—	—	—	—
Oak (Tellerman)	523	1·0	49	63	64	60	58	59	60	60	64	—	—	—	—
Birch (near Moscow)	550	1·0	62	58	50	45	—	—	—	—	—	—	—	—	—

ties. In forest biogeocoenoses the amount of precipitation reaching the soil surface, with uniform intensity and duration of rainfall, varies according to the tree-species composition, the density and age of the stand, and the development of undergrowth, herbage, and mosses, since all these features of a forest have a substantial effect on the retention of precipitation by the above-ground parts of plant communities (Table 21). The denser the vegetation stand and the more compact the foliage of leaves or needles in the crowns, the greater is the amount of precipitation so retained. This accounts for moisture retention in forests usually being greatest at the period of culmination of growth of tree species, and moisture-retention being greater in stands of the same species at the northern and southern limits of its range. It also leads to great differentiation in water conditions in forests of different tree species. The trunks of trees, as well as the crowns, return some of the precipitation, although to a much smaller extent; smooth-barked trunks retain less than the trunks of species with rough and fissured bark.

Part of the moisture reaching the soil surface is absorbed by litter, part flows along the surface into various depressions and water-courses, and part infiltrates deep into the soil. The amount and depth of infiltration depend on the amount of precipitation reaching the soil surface, on the physical features of the soil (mechanical composition, structure), and on the age of the tree stands (Table 22).

Table 22. *Infiltration of moisture (mm) into dark-grey forest loam under oak stands of different ages, to a depth of more than 5 metres.**

Item	Age of stand (years)						
	4	15	25	45	50	138	220
1953							
Precipitation	346·8	353·8	351·3	338·3	333·9	346·1	339·8
Total moisture absorbed by soil	245·9	179·8	208·0	288·0	260·1	253·5	239·0
Infiltration to stated depth	100·9	174·6	143·3	49·9	73·8	92·6	100·8
1954							
Precipitation	171·4	167·4	162·2	163·7	155·3	161·6	154·3
Total moisture absorbed by soil	168·9	144·8	147·2	148·6	149·0	135·8	119·0
Infiltration to stated depth	25·0	22·0	15·0	15·1	6·3	25·7	35·3
1955							
Precipitation	398·4	381·5	373·8	370·4	374·3	384·0	338·1
Total moisture absorbed by soil	289·2	243·7	205·0	283·0	292·5	279·5	259·5
Infiltration to stated depth	109·5	134·8	168·8	187·4	94·8	107·5	128·7

* Tellerman experimental forestry station, Voronezh province.

In temperate climates the greater part of moisture infiltration into the deeper horizons of the soil results from the melting of snow. An exception is found in sandy soils, where—because of their high permeability—intensive infiltration of water takes place both in spring (during snow-melting) and in summer (during rainstorms).

The amount of water flowing downhill on the surface is extremely variable and depends on relief, on the physical features of the soil, on the intensity of rain or the rate of melting of snow, on the degree of saturation of the soil with water, and on the character of the vegetation. On level ground, surface flow reaches its lowest magnitude above coarse-grained substrates; on sandy soils it does not exceed 1%. Surface flow of water is highest on bare loam and clayey-loam soils. Flow is considerably less in forests, since arboreal vegetation, because of the forest litter and well-developed root systems, favours absorption of water by the soil layers. Different forests, of course, act differently in this respect. In different oak stands in the Tellerman experimental forestry station, for instance, surface flow in a goutwort community reaches 1·6%, in sedge-goutwort (with more turfy cover) 4%, in hedge-maple 18%, and on alkali soil 32% annually. In spruce stands on turf-podzol loamy soil annual surface water flow varies from 15-26%, and in pine stands it is about 7%.

A forest regulates water flow not only within itself but also in adjacent treeless sections of water-courses. This effect is shown in the dependence of flow on the percentage of afforestation of a given water-course basin, especially during the spring thaw, which increases surface flow and produces floods (Table 23).

Table 23. *Coefficient of flow during spring thaw in various forest-vegetational zones in relation to percentage of afforestation.*

Forest-vegetational zone	Species	Afforestation (%)						
		About nil	10	20	40	60	80	100
Coniferous-deciduous forest zone	Spruce	0·92	0·84	0·59	0·38	0·33	0·29	0·26
Coniferous-deciduous forest zone	Pine	0·75	0·40	0·23	0·18	0·17	0·15	0·15
Forest-steppe (Voronezh province)	Oak	0·65	0·25	0·14	0·09	0·07	0·07	0·05
Steppe (Lugansk province)	Oak	0·75	0·15	0·09	0·07	0·07	0·06	0·06

As shown in Table 23, in southern districts of the country an increase in the percentage of afforestation reduces the flow coefficient more rapidly than in northern districts. For correct evaluation of the hydrological role of forests in sparsely-forested and arid districts, we must remember that the examples of smallest consumption of water in outflow and transpiration were observed on small areas of less than one hectare.

As the area increases evaporation and outflow rise gradually at first, and then (with the area of the field 500×600 m) more rapidly (Table 24). Thus the optimum area of a field from the hydrological point of view should be not more than 500×600 m.

Ability to regulate water flow is found not only in extensive continuous forest areas but also in disconnected patches of forest on slopes and forest belts, as Table 25 convincingly shows.

In mountains and hilly areas, forest litter and forest density strongly affect water absorption by the soil. In dense forest the outwash of fine soil by atmospheric precipitation is almost nil, whereas in stands thinned by cutting it may reach 3 to 4 tons per hectare, and in clear-cut areas 10 to 15 tons per hectare, and even more if the soil is broken up by machinery.

Outflow of atmospheric precipitation, and the carrying away thereby of organic matter and fine soil from elevated parts of the biogeosphere into various depressions and water-courses, constitute an important channel of exchange of matter not only between separate terrestrial biogeocoenoses but also between land and water sectors of the biogeosphere, taking a major part in the great geological process of circulation of mineral matter and water.

Water that is absorbed by the soil and does not percolate into subsoil and underground water is sooner or later re-evaporated into the atmosphere, except for colloidal water, which is very firmly held by colloids in the soil. In places covered by vegetation, most soil moisture is lost by transpiration. The rate of transpiration is a function containing many variables. It depends on temperature, dryness of the air, wind, light, amount and state of soil moisture, and, finally, on the nature of the plant, its age, its condition, etc.

The effects of these diverse factors are not proportional and not synchronous, so that the actual amount of transpiration by plants is not always what might be expected. For instance, the rate of transpiration by tree species is less on the steppes than near Moscow, although the reverse would seem more natural, since on the steppes the temperature is

Table 24. *Total evaporation and surface flow in relation to size of treeless areas in the forest-steppe zone.**

Item of water consumption	Size of treeless area (m²)								
	100×100	200×200	300×300	400×400	500×500	600×600	800×800	1000×1000	2000×2000
Annual coefficient of outflow	0·02	0·03	0·04	0·05	0·06	0·11	0·13	0·20	0·40
Total evaporation and outflow	75	79	82	83	84	88	97	97	97

* Tellerman experimental forestry station, Voronezh province.

Table 25. *Coefficient of flow in composite areas with different widths of water-absorbing belts.*

Soil; composition and age of stand	Depth to which soil freezes (cm)	Length of forested slope	Width of forest belt (metres)								
			0	10	20	30	40	60	80	100	120
Coniferous-deciduous subzone (Moscow province)											
Loam, sod-podzolic; 8 pine +2 spruce; 110 yr	30	500	0·96	0·77	0·48	0·32	0·23	0·14	0·10	0·04	0·03
		350	0·87	0·59	0·37	0·25	0·18	0·08	0·05	0·03	0·02
		180	0·57	0·48	0·23	0·16	0·11	0·05	0·03	0·04	0·01
Loam, sod-podzolic; 10 spruce + birch; 100 yr	30	500	0·97	0·88	0·62	0·46	0·27	0·14	0·13	0·11	0·10
		350	0·88	0·68	0·48	0·35	0·21	0·11	0·09	0·08	0·08
Loam, sod-weakly-podzolic; 10 silver linden; 100 yr	20–40	190	0·80	0·56	0·32	0·22	0·12	0·06	0·02	0·02	0·02
Forest-steppe (Voronezh province, Tellerman)											
Dark-grey forest soil; 8 oak +1 linden +1 Norway maple; 75 yr	60–110	500	0·63	0·46	0·26	0·17	0·15	0·11	0·06	0·05	0·05
		380	0·46	0·33	0·19	0·12	0·10	0·08	0·04	0·04	0·04
		200	0·31	0·21	0·12	0·08	0·06	0·05	0·03	0·02	0·02
Alkali meadow; oak stand	20–30	180	0·72	0·56	0·40	0·30	0·28	0·21	0·19	0·16	0·14
Steppe (Lugansk province)											
Chernozem; 7 oak + 2 linden + 1 Tatarian maple +caragana; 20 yr	10–15	500	0·70	0·34	0·22	0·08	0·05	—	—	—	—
		350	0·44	0·21	0·14	0·05	0·03	—	—	—	—
		160	0·27	0·13	0·08	0·03	0·02	—	—	—	—

considerably higher than in Moscow province, the air is drier, and the wind stronger.

Similar unexpected results are obtained when computing the output of moisture in transpiration and evaporation from the crowns of trees per unit of land area. This is explained by differences in leafiness of the tree stands, which varies according to the zones and subzones of forest growth, to the types of forest, and (within the latter) to the composition and age of stands (Fig. 10).

According to our observations, the greater the mass of leaves, the higher is the intensity of transpiration by tree stands and the output of moisture through their transpiration. For the same reason, the lower the

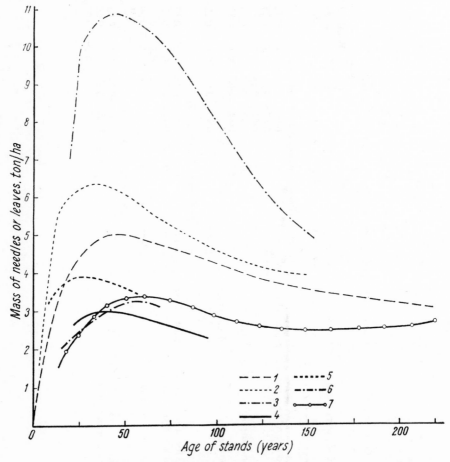

Fig. 10. Mass of leaves (or needles) in tree stands of different ages (tons/ha, oven-dry). 1—in pine stands in Archangel province; 2—in pine stands in Moscow province; 3—in spruce stands in Archangel province; 4—in birch stands in Moscow province; 5—in poplar stands in Voronezh province; 6—in ash stands in Voronezh province; 7—in oak stands in Voronezh province.

productivity of a forest type and of the tree stands, the smaller is the output of moisture through transpiration (Tables 26, 27).

For example, output by transpiration is less in a lichen pine forest than in a cowberry pine forest, and it is less in the latter than in the more productive bilberry pine forest. The same is observed in the subzone of coniferous-deciduous forests around Moscow. There the greatest output of moisture by transpiration is observed in the most productive pine forests. In the long-moss-bilberry, and especially in the sphagnum, pine forests the output of moisture by transpiration is sharply reduced. The same feature is characteristic of forests of other species (Table 27).

The output of moisture by transpiration within a single forest type reaches a maximum during the peak period of growth, which for pines occurs at the age of 60 in the north and of 30-40 in the coniferous-deciduous forest subzone.

Moisture output varies with the structure and density of the stand. In pure pine stands it is higher, and in mixed stands it depends on the species-composition. For instance, in a 150-year-old pine stand with a dense 30- to 40-year-old spruce understorey, transpirational output of moisture is almost twice as high as in pure pine stands of the same age. When the stand density is less its moisture output is also less, e.g., in the coniferous-deciduous forest subzone, with canopy density 0·9 the moisture output of a pine stand by transpiration is 243 mm, and with canopy density 0·4 it is 180 mm.

The same rule is also observed in forests of other species.

The hydrological effects of transpiration by tree stands are not always

Table 26. *Output of moisture by transpiration (mm) in pine forests.*

Age of stand (years)	Density of stand	Output of moisture in different types of forest					
		Lichen-moss	Cowberry	Bilberry	Mixed	Long-moss-bilberry	Sphagnum
Northern taiga (from 15 April to 1 October)							
35	1·0	—	88	—	—	—	—
60	1·0	—	183	197	—	—	—
160	0·8	115	168	176	—	—	—
270	0·7	—	97	114	—	—	—
Coniferous-deciduous forest subzone (from 15 April to 1 October)							
13	1·0	162	201	221	—	—	—
33	1·0	171	250	273	218	—	—
40	1·0	188	242	256	268	—	—
60	0·9	173	223	230	245	165	119
65	0·9	—	236	—	—	—	—
80	0·9	—	—	—	215	—	—
150	0·8	161	135	165	—	—	—
150, with dense 2nd storey of spruce	1·0	—	260	—	—	—	—

Table 27. *Total evaporation in forests of different types and ages, by forest zones.*

Total evaporation, according to P. S. Kuzin	Types of pine forest								Types of spruce forest							
	Age	Density	Lichen-moss	Cowberry	Bilberry	Composite	Long moss-bilberry	Sphagnum	Age	Density	Cowberry	Bilberry	Mountain Sorrel	Dropwort	Grass-sphagnum	Sphagnum
Northern taiga																
20-25	34	1·0	—	211	—	—	—	—	—	1·0	—	248		370		
	60	1·0	—	—	332											
	160	0·8	250	310	340											
	270	0·7	—	—	274											
Central taiga																
									7	1·0		201				
									18	1·0		387				
									37	1·0		449				
									110	1·0		334			334	330
									135	1·0		323	440			
Southern taiga																
									65	1·0		440	500	—		
Mixed forests																
	20	1·0	355	392	430	435	400	385	20	1·0	335	410	521			
	40	1·0	405	455	490	495	455	438	40	1·0	398	456	521			
	40	0·8	356	405	461	463	410	458	—	—	—	—	—			
	60	1·0	393	429	464	470	475	351	55	1·0	445	501	530			
	60	0·7	351	395	399	451	463	343	80	0·9	500	510	518			
	150	0·5	391	432	484	—	—	—	120	1·0	—	—	428			
	150	1·0	328	360	391	395	365	—	150	0·9	331	390	405			
	150*	1·0	—	500												
Forest-steppe																
35-40	60	1·0	312	376	427	411										
	90	1·0	287	327	415	—										
	150	0·9	—	317	405	425										
Steppe																
30-25																
Steppe																
3-25	12	1·0		305												
	40	0·9		298	310											
	60	0·9		279	—											

* With dense spruce understorey.

	Types of oak forest						Birch			Poplar		Ash	
Age	Density	Goutwort	Sedge-goutwort	Hedge maple	*Euonymus* alkali soil	Alkali soil	Age	Bilberry	Bilberry-mtn sorrel	Age	Sedge	Age	Sedge
Northern taiga													
Central taiga													
							1	201					
							18	337					
							37	449					
Southern taiga													
Mixed forests													
							20	330	330				
							25	455	—				
							35	364	—				
							40	—	379				
							60	—	339				
							80	—	295				
							94	—	260				
Forest-steppe													
20	1·0	344	327	320	270	260	10	—	—	10	450	20	450
20	0·8	314	301	—	—	—	30	—	—	30	500	40	450
20	0·5	309	311	—	—	—	40	—	—	60	450	65	400
45	1·0	455	450	379	—	—	60	—	—	60	450	70	400
45	0·8	417	420	360	—	—							
45	0·4	380	418	379	—	—							
Steppe													
60	1·0	476	455	432	380	340							
135	1·0	405	395	—	—	—							
220	1·0	430	413	392	352	325							
Steppe													
30	1·0	—	465	367									
40	1·0												
60													

NOTE: In the forest-steppe zone there are grass-lichen pine groves on dune-type hills, mixed-herbage pine groves on slopes, *Molinia* pine groves in deep depressions, and mixed pine groves on level ground. In the steppe zone all pine groves have 'dead cover'.

the same. In very damp ecotopes transpiration helps to prevent swampiness, but where there is shortage of moisture it may cause drying-up to a point where withering takes place; finally, it may be lower than the output of water by evaporation and surface flow, and then the forest, as compared with open spaces, will help to conserve soil moisture. It all depends on the composition, age, and density of the stands. This dependence suggests a possibility of controlling soil moisture conditions by improvement cutting to regulate the composition and density of stands, especially in districts of fluctuating water conditions. According to our observations, clearing and thinning at low and medium intensity lowers the total evaporation output of moisture in oak stands in the forest-steppe zone during the vegetative period by 32-43 mm as compared with control stands (Table 28). Intensive thinning is less effective: it lowers the total evaporation output of moisture by only 10-20 mm, and then mainly by decreasing evaporation in the second year after cutting. The after-effects of cutting on the total evaporation last for several years, although they gradually diminish, especially after intensive thinning.

Table 28. *Total evaporation (mm) by stands from 15 April to 1 October after clearing and thinning in oak stands in the forest-steppe zone.**

Degree of thinning	Density	Year of observation							Average
		1952	1953	1954	1955	1956	1957	1958	
		Precipitation (mm)							
		421	581	304	607	594	554	603	523
Thinning in spring of 1952 at age of 13 years									
Control	1·0	354	416	286	410	379	429	391	409
Slight	0·8	304	342	331	362	357	405	397	357
Medium	0·7	368	358	333	424	392	389	384	378
Heavy	0·5	432	392	323	415	366	394	391	388
Thinning in spring of 1954 at age of 23 years									
Control	1·0	—	—	—	383	391	429	491	423
Slight	0·85	—	—	—	335	338	410	438	380
Medium	0·75	—	—	—	338	420	395	410	391
Heavy	0·65	—	—	—	337	413	433	458	410
Very heavy	0·4	—	—	—	352	401	415	445	403
Open	0·2	—	—	—	270	429	435	445	368

* Tellerman experimental station, Voronezh province.

Similar changes are observed after cutting in older stands (Table 29).

In clear-cutting the amount of total evaporation depends closely on the width of the clearing: the wider the clearing, the greater the evaporation.

We must remember that total evaporation at different points of a cut-over area is not uniform, varying substantially according to the distance of the point from the forest edge. Maximum output of moisture by evaporation takes place in the centre of a clearing. From the centre towards the forest edge there is a decrease in moisture output, evaporation being less in the eastern half of the clearing than in the western. If we take the evaporation in the centre of a 100-metre clearing as 100, the following figures express the amounts at different distances from the forest edge:

In the western half of the clearing:	(metres)	0	10	20	50
	(%)	94	90	98	100
In the eastern half of the clearing:	(metres)	0	10	20	50
	(%)	56	50	96	100

The above variation is due to the great changes made by a forest on heat and light conditions in the parts of a clearing near the forest edge. The effects of forest proximity on water conditions in adjoining areas are even more marked where forests and fields meet. As is seen from Table 30, the infiltration of precipitation in the soil of a field to a depth of more than 5 metres is sharply increased near forest outskirts, and the effect of the forest in this respect is still evident at a distance of 150 metres from the edge. Also at the forest edge there is much less blowing away of snow and less surface flow of water, but on the other hand more piling up of snowdrifts occurs. In the arid steppes, undoubtedly, a deterioration in hydrological conditions caused by the forest occurs over a fairly wide strip of the adjacent field.

These data indicate the advisability of creating an open system of water-regulating forest belts on the steppe fields; if these belts are spaced

Table 29. *Total moisture output (mm) by evaporation in 45- to 48-year-old oak stands, produced by improvement cutting in 1954.**

Item of output	Control stand, density 1·0	Management by biogroups, intensity of cutting 15%, density 0·85	Improvement cutting by combined method, intensity 24%, density 0·75	Improvement cutting by undercutting method, intensity 50%, density 0·5	Management by Nesterov's method, intensity of cutting 23%, density 0·75
Surface flow	13	13	14	13	11
Evaporation from cover	77	72	81	96	72
Evaporation from crowns	70	48	43	23	36
Transpiration	421	379	392	379	401
Total evaporation	581	512	530	511	523
Infiltration of moisture deeper than 5 metres	8	75	59	78	66
Average precipitation during 4 years (mm)	589	589	589	589	589

* Tellerman experimental forestry station, Voronezh province.

Table 30. *Moisture output from 15 April to 1 October in sedge-goutwort oak stand and in adjacent field (average for 7 years, in mm).**

Item of moisture output	Oak stand 70-77 yr, density 0·7	Distance from forest edge (metres)			
		0	50	150	1000
Surface flow	14	14	25	40	60
Snow drifted in from field	0	72	0	0	0
Snow drifted away	0	0	17	43	77
Total evaporation	412	380	394	369	381
Total output	426	394	436	452	518
Infiltration	95	187	85	69	3
Precipitation and snow drifted in	521	593	521	521	521

* Tellerman experimental forestry station, Voronezh province.

500 metres apart, they will help to conserve 40-77 mm more water in the areas between them than on the open steppe.

For proper biogeocoenotic appraisal of the hydrological conditions of any area, one must discover the location and dynamics of the water table.

The effect of underground water on the soil depends closely on the depth at which it lies. If the underground water is at a considerable depth, neither the water-bearing layer nor the capillary fringe ever enter into the soil profile, and the soil-moistening regime is entirely dependent on atmospheric precipitation and its transformation and distribution within the biogeocoenose. With deep underground water, moisture in the soil does not exceed its minimum water capacity. This type of moisture regime is widespread in soils of the steppe provinces.

In cases where the water-bearing layer and the capillary fringe enter the soil profile periodically but do not remain there long, with consequent cessation of absorption of water by roots, of physical evaporation, and of underground outflow, the moisture in the soil varies from capillary capacity (which, as A. A. Rode points out, is itself variable) to its full water capacity. This type of water regime is found chiefly in the taiga zone.

When the capillary fringe is constantly within the soil profile, although in its lower parts, the water table enters the soil from time to time. In this case soil moisture fluctuates relatively abruptly from full to minimum capacity, but generally there is a surplus of moisture, and the type of regime of such soils is called 'semi-boggy'.

When the capillary fringe is constantly within the soil profile and reaches the surface, and the water table fluctuates mostly within the soil profile, occasionally reaching its surface later, we have a 'boggy' type of regime.

Closely related to the type of soil-moisture regime are the direction of exchange of water and mineral matter (in both the atmosphere-soil and the soil-subsoil systems) and the supply of moisture to phytocoenoses. Supply of moisture to phytocoenoses, in its relation to soil moisture, can be divided into at least six types.

1. With a boggy moisture regime, although there is no shortage of water, lack of oxygen impedes development of the root system and supply of water to the crowns. A physiological drought is thus created, preventing use of water in transpiration by the stand. Output of water is mainly from the surface of mosses. It is either less than, or equal to, the input of water from outside. Drainage operations are required in order to create optimal conditions for tree growth.

2. With a semi-boggy moisture regime, resulting from periodical falling-off in excessive soil moisture, the gross output of moisture from the soil in the form of tree transpiration is increased; evaporation from the crowns is also increased because of the higher stand density; evaporation from the soil surface is notably diminished. In general, moisture circulation is more active than in the preceding type, but the amount of water is still so great that it requires to be reduced either by biological methods (planting trees with high water requirements) or by drainage.

3. With optimum moisture supply, in the taiga type of moisture regime, moisture requirements depend entirely on intensity of plant growth. In this type of moisture circulation the output by transpiration differs little from that by evaporation. It does not increase excessively, because of the limited amount of heat available and the effects of the microclimate.

This type of moisture circulation by vegetation is typical of districts with evenly-distributed precipitation. It is found mainly in the forest zone, to a lesser degree in the forest-steppe zone, and is still less common (mostly in depressions in the relief) in the steppe zone.

4. Another type of moisture supply is distinguished by periodical saturation. Here the gross output of moisture from the land is much less than the possible evaporation. Therefore measures are sometimes necessary to assist in accumulating water in the soil. In these conditions it is important to regulate the density of stands by improvement-cutting, since tree growth is somewhat impaired by shortage of moisture during the summer. In the fields snow-retention, mulching, and supplementary water-supply measures are required. This type of moisture regime occurs in forest-steppe districts.

5. The fifth type of moisture supply is usually associated with an unsaturated moisture regime, the water table being considerably deeper than the lower limit of the soil profile. During the second half of the summer moisture falls to the level at which plants wither, and below it. Actual evaporation is much below the possible level. Regulation of soil moisture in forests should be accompanied by improvement-cutting.

6. The final type of moisture supply results from very small precipita-

tion combined with high air temperature. Such extremely arid conditions of plant environment have produced a number of morphological and physiological adaptations in the vegetation, aimed at protection from evaporation by reducing the total leaf surface (reducing the left lamina). The possible level of evaporation is extremely high, whereas annual precipitation does not exceed 150-200 mm.

Sometimes, as a result of transpiration by forest vegetation, even with the third type of moisture supply (not to mention the fourth) soil moisture may fall to the level at which plants wilt during the second half of the summer. This phenomenon is observed during the period of peak growth; afterwards the annual increment of growth decreases and moisture supply increases. The variation in soil moisture as a forest ages has been described by G. N. Vysotskii (1912). He states that the luxuriant growth of young plantations on steppe soils during the first years after planting is explained by the fact that at first the soil moisture is fully adequate for the young saplings, but later, as they grow older and their water requirements consequently increase, the soil moisture begins to decrease and ultimately reaches a level where it cannot fulfil the requirements of the trees during the period of highest increment of organic matter, and the young forest begins to die off. An entirely different phenomenon is observed when mineral fertilisers are used in the forest. After they are introduced into the soil, consumption of water per unit of growth-increment is considerably less than without fertilisation. We have demonstrated in oak stands on the forest-steppe that the water regime and moisture circulation in trees depend on their mineral nutrition. By changing the mineral-nutrition conditions, we can to a considerable extent alter the water regime and reduce the unproductive use of water in transpiration.

The effect of mineral fertilisation on moisture circulation varies with the chemical composition of the fertiliser and the soil-moisture conditions.

More economical consumption of soil moisture occurs when phosphoric or nitrogenous fertilisers are applied to dark-grey forest soils occupied by oak stands. Supplementary application of nitrogenous fertilisers after the shoots have ceased to grow in length, but before maximum diameter has been attained, has beneficial effects on the content of colloidal water. Nitrogenous fertilisers avert summer leaf-shedding during sukhovei winds and drought. Thus application of mineral fertilisers markedly offsets the unfavourable effects of drought and leads to increased growth in tree stands. This is possibly due to the fact that mineral fertilisers increase the amount of colloidal water in plants, with negligible increase in the osmotic pressure of the cell sap and in the suction pressure of the roots.

Atmospheric precipitation reaching the soil is not pure water. While still in the atmosphere, moisture comes into contact with dust and absorbs various mineral and organic substances. The greater the dust content of the air the greater the mineral content of the precipitation. When a litre of precipitation reaches the earth it contains up to 230 mg of suspended particles and up to 285 mg of dissolved matter. As atmospheric precipitation passes through the crowns of trees and runs down the trunks it acquires much additional mineral matter, the amount and composition of which depend on the composition of the tree stand and its density.

The amount of atmospheric precipitation also affects its chemical content. The heavier and more frequent the precipitation, the weaker is its concentration of dissolved chemicals.

The mineral content of precipitation varies according to the time of year and depends on the amount of precipitation during the year, and especially during the vegetative period.

According to analyses by V. A. Gubareva (1962), the chemical content of rain-water in oak stands depends to a great degree on the age and density of the stands. Rain-water passing through the crowns of dense stands is richest in mineral matter. A high concentration of mineral matter is found in water running down the trunks. Rain falling through the crowns of a 28-year-old stand is most saturated with dissolved chemicals; that falling through the crowns of a 58-year-old stand contains much less, and in last place comes rain falling through the crowns of a 220-year-old stand. Precipitation falling through the crowns of leafless trees contains much smaller amounts of ions.

The chemical content of precipitation varies with the tree-species composition. Precipitation falling through the canopy of an ash stand contains more dissolved matter than that in an oak stand.

The greatest fluctuations are observed in the content of ions of calcium, magnesium, sodium, potassium, and ammonium, and also of sulphate ions and hydrocarbons.

The composition of water flowing down tree trunks depends on the roughness of the bark. The older the stand, the more rugose is the bark and the higher the chemical content of the water flowing down the trunks. The chemical content of water resulting from spring snow-melting is also variable.

Therefore rain-water entering the soilis rich in mineral matter and considerably increases the concentration of mineral matter in soils, especially in their upper horizons.

ATMOSPHERIC DUST AND HARMFUL GASES

The earth has relationships with cosmic bodies, and in consequence there is an exchange, cosmic in extent, of various forms of both matter and energy. The exchange of matter takes place in many different ways (meteorites, cosmic dust, gaseous matter, individual atoms), but has been studied much less than the exchange of energy.

Atmospheric dust is an accumulation of aerosols and other minute solid particles, which exist in an extremely diffuse state in the air. It may be considered a negligible element of the environment if there is little of it, but a very important one with extensive effects (e.g., on vegetation) if it is abundant. When there is heavy production of industrial gases, and also when there are dust storms, vegetation suffers detrimental effects from air-borne dust. The dust content of the air varies, and includes both organic and inorganic matter. Inorganic matter represents from two-thirds to three-quarters of all mechanically-suspended substances, its chief source being light powdery soil. Dust rises wherever vegetation is absent because of adverse conditions. The heaviest dust production is observed in arid (desert and semi-desert) conditions.

Besides dust, the atmosphere contains a considerable amount of meteoric particles and nuclei of condensation.

In spite of the small mass of individual dust particles, the total amount of dust in the air is so great that the annual increase in the earth's mass due to dust amounts to between 1,000,000 and 3,700,000 tons.

The lower layers of the troposphere are particularly dust-laden, as may be seen from Table 31.

Table 31. *Vertical distribution of dust (Fett 1961).*

Height above soil surface (metres)	Average no. of dust particles per cm³	
	Winter 1942-43	Summer 1943
0	98	19·5
100	63	11·8
200	60	11·3
400	165	—
500	7·2	11
600	12·0	9·5
700	12·4	—
800	9·3	11·8
900	4·8	—
1000	6·3	7·1
1200	3·1	5·8
1800	—	3·7
2000	0·5	4·0

The atmosphere contains volcanic dust, dust of marine and vegetational origin, and also dust arising from forest fires, wind-erosion of soil, road traffic, burning of fuel in various industrial operations, etc.

The atmospheric content of dust and nuclei of condensation depends on the relative humidity of the air. The lower the relative humidity, the smaller is the number of dust particles, but the higher the number of nuclei of condensation; and conversely, the higher the humidity, the lower the number of nuclei of condensation and the higher the number of dust particles in 1 cm³.[1] With increase in humidity, nuclei of condensa-

---

[1] Particles with a radius of 10^{-4} to 10^{-3} are classified as dust, and particles with an average radius of 5×10^{-6} as nuclei of condensation. (The unit of measurement is not given in the original. *Tr.*)

tion grow in size by cohesion and so come within the range of conimeter measurement. As the content of nuclei of condensation decreases, the dust content increases. A rise in dust content and a fall in content of nuclei of condensation before rainfall is also due to coagulation.

Ashes, soot, various solid mineral particles, and some gases, when present in the atmosphere, affect the life processes of plants. Dust containing injurious substances causes plant leaves to turn brown or yellow and to wilt. In Fett's opinion, carbon particles do not harm plants. Similarly road dust and cement dust have no harmful effects. Soot, which does not cause wilting of leaves or needles during forest fires, is one of the factors impeding the growth of conifers in large cities.

During dust storms vegetation not only suffers mechanical damage but also becomes clogged with dust.

Soluble matter contained in dust is absorbed by the soil, and later by plant roots; at the same time various water-soluble compounds contained in dust cause plants to wilt. Sometimes they injure chlorophyll-containing cells and lower their chlorophyll content. Insoluble matter contained in dust may have an unfavourable effect on the physical properties of the soil, so harming vegetation.

Dust also has considerable effects on the soil. It lessens the heat output of the soil and the intensity of radiation reaching the earth, thereby impairing the warming of the soil. Dust strongly affects the morphology of the earth's surface and the formation of loess.

Atmospheric dust also contains substances of organic origin: minute seeds, pollen, and spores, micro-organisms, algae, and spore-forming moulds, with particles of plants themselves, etc.

Rainfall during the pollinating period of tree species aids the deposition of pollen on the ground. Because of its pollen content rain-water from the first summer showers usually begins to putrefy quickly, and in summer 'blooms' because of its abundant content of organic matter, pollen, and spores.

Besides dust the atmosphere contains a large number of other admixed substances, in particular various gases. The gases most widely distributed in the atmosphere are the result of volcanic eruptions, decay of organic residues, and burning of various kinds of fuel, and waste-products of industrial processes such as carbon monoxide (CO), carbon dioxide (CO_2), sulphur dioxide (SO_2), and sulphuric anhydride (SO_3). Less common are chlorine (Cl_2) and hydrochloric acid (HCl), fluorine (F_2) and hydrogen fluoride (HF), oxides of nitrogen (NO, NO_2, NO_3), ammonia (NH_3), and hydrogen sulphide (H_2S).

These gases are especially abundant near industrial plants. At a distance of 0.2 km from a factory the average amount of them in 1 m^3 of air is 32 mg; between 0.5 and 1 km, 28 mg; at 1.5 km, 27 mg; from 1.5 to 3 km, 5 mg; and at 3 km or more, up to 1.5 mg per m^3.

Sulphur dioxide has very harmful effects on forest phytocoenoses.

FFB H

When the air contains up to 260 mg/m^3 of SO_2 coniferous trees die within a few hours; SO_2 content of 5·2 to 26·0 mg/m^3 does much harm to coniferous and deciduous trees. SO_2 content of 1·04 to 1·82 mg/m^3 has harmful effects on only the most sensitive tree species with prolonged or frequently-repeated action on them; and SO_2 content of 0·26 to 0·52 mg/m^3 has no harmful effect, or almost none, on trees.

In trees strongly affected by gas the opening of buds is delayed by 9 days and the development of leaves by 36 days, and the duration of the growing period is reduced to 20-40 days.

The extent of damage done by SO_2, like that done by other harmful ingredients of the air, depends on the composition, age, density, and environmental conditions of the forest. Coniferous forests are more sensitive to smoke and gases than are deciduous forests. Among deciduous species, English oak and European white birch display greatest resistance to gases and dust. Dense stands are less resistant than thinned stands, and old trees less resistant than young. In particular, when harmful ingredients of the air affect old stands the life of needles is halved and they are smaller (shorter and thinner), the crowns of trees become much more meagre, and naturally their assimilative ability is lowered. All species suffer less from air pollution and live longer with more favourable soil conditions. Water conditions similarly cause differences in the effects of harmful atmospheric ingredients. The resistance of forest biogeocoenoses is lowered if there is an abundance of stagnant water or a high level of underground water.

A forest aids greatly in the purification of the air from dust and gases and prevents their wider diffusion. Whenever wind velocity is low only insignificant amounts of dust penetrate through the crown canopy into a forest area. Most dust falls at the forest outskirts and quickly settles because of the lack of wind within the forest. In a mature forest the dust content of the air is almost constant and always less than that outside the forest. The number of particles per unit of volume increases only with strong gusty winds. The dust content of forest air falls in the direction of the wind, gusts of which penetrate not more than 500 metres into the heart of the forest. According to observations in the Tellerman oak groves, transportation of dust at a height of 23 metres amounts to 65% at a distance of up to 100 m from the forest edge, to 38% at 400 m, to 25% at 1000 m, to 10% at 2000 m, and to 5% at 3 km from the edge.

The lower dust content of air in the forest results from the cleansing action of the crown canopy.

The amount of dust lying on the leaves of different tree species varies. The rugose leaves of elm retain 6·3 times as much dust as the smooth glossy leaves of poplar and balsam poplar, and 2·3 times as much as oak leaves.

The lower parts of crowns become more dusty than the upper. For instance, on an elm there was 8 times as much dust at a height of 1·5 metres as at the top of the same tree at a height of 13 m, in consequence of the blowing of dust from tree-tops by the wind and its washing down by rain.

Dust is naturally washed away from leaves and twigs by rain and enters the forest litter and the soil, increasing the amount of various substances contained in the soil; the soluble portions of these substances enter into new biological circulation.

CHAPTER III

A phytocoenose as a component of a forest biogeocoenose

BIOMASS OF A PHYTOCOENOSE, ITS STRUCTURE AND WORK

As was mentioned in the first chapter, vegetation (phytocoenose) is an indispensable and substantial part of every biogeocoenose. A phyto-coenose includes all species of higher and lower plants living in a given biogeocoenose (the majority of plants, or at least a large proportion, are autotrophic organisms adapted to the phytocoenose), and serves as the sole source of primary organic matter and the main accumulator of energy. Consequently, the phytocoenose is the basis for the existence of heterotrophic organisms and for most of the chemical and physical reactions in the biogeocoenose. Although the phytocoenose acts on all the features and inter-relationships of other components of the biogeo-coenose, the chief result of its metabolism is the accumulation of biomass (more precisely, phytomass). The biomass of a biogeocoenose is com-posed of animals and micro-organisms as well as of plants. Phytomass, as a rule, considerably exceeds the masses of animals and micro-organisms in both volume and weight, and it is usually the chief object of economic exploitation in a biogeocoenose. This is especially true in a forest bio-geocoenose, in which phytomass is very large and consists mostly of timber, the principal national-economic resource in the forest.

Therefore we begin this chapter with a discussion of the biomass (phytomass) of the plant component of a biogeocoenose.

BIOGEOCOENOTIC WORK OF A PHYTOCOENOSE[1]

In comparison with the work of other biogeocoenose components the biogeocoenotic activity of a phytocoenose is more complex, multiform, and specific. This leads to the most varied and far-reaching effects of world-wide significance, and ensures the central place of the phyto-coenose in the system of biogeocoenose components and in the trans-

[1] By the biogeocoenotic work of a phytocoenose is meant:

1. absorption of various forms of matter and energy by the phytocoenose from other biogeocoenose components, and synthesis of them into organic matter;

2. release into the environment of products of its life-activity (O_2, CO_2, H_2O, etc.) and of part of the absorbed energy, in respiration, transpiration, and other excretory processes (guttation of drops of water, exhalation of aromatic substances, etc.);

3. return of part of the absorbed matter and energy through shedding and de-composition of dead organs and the fall of dead organisms;

4. transformation of the characteristics and composition of other components of the biogeocoenose through the growth of plant bodies and also through 1, 2, and 3.

formation of matter and energy on the earth's surface. As a result of synthesis of organic matter by green plants (which constitute the greater part of the biomass of a phytocoenose), of the growth of plant bodies, of their respiration and transpiration, and of their absorption of mineral matter and water, a phytocoenose accelerates, extends, and complicates the total migrations of matter and energy in the envelopes surrounding the earth; it forms new chains of transformation of matter and energy (biological accumulation of matter, biological circulation); it creates new natural processes (soil formation, peat formation), structures (soil, peat), gases (release of oxygen in the atmosphere, and determination of the atmosphere's total gaseous composition), and the climate of the biogeocoenose; lessens entropy of solar radiations in cosmic space; and permits the life-activity of heterotrophic organisms—animals, man, and the majority of micro-organisms. Besides, and as a result of all that activity of plant communities on the earth's surface, the nature of the relationships between other components of the biogeocoenose is radically altered: from being direct and comparatively simple they become complex, multiform, and intimate. The more extensively the phytomass develops, the more active becomes the role of transforming biogeocoenotic relationships. In this the phytomass of a biogeocoenose differs sharply from its animal and micro-organism components, whose biogeocoenotic activity and role are proportional rather to their numbers and rate of reproduction than to their biomass.

The scale, rhythm, and results of the biogeocoenotic work of phytocoenoses in nature vary within wide limits. On the one hand, they depend on the biochemical and biophysical properties of the phytomass of a biogeocoenose, which vary greatly from place to place because of the different species- and race-compositions of the vegetation, the life forms, the age of the population, the stage of ontogenesis, etc.; on the other hand, they depend on the composition of the other biogeocoenose components that accompany the phytocoenose and govern its work, and primarily on the quantity, quality, and rhythm of the supply of solar radiation and moisture to that part of the earth's surface. This accounts for the zonal and provincial characteristics of the phytomass of a terrestrial biogeocoenose.

The natural multiformity of the biogeocoenotic activity of phytocoenoses is greatly increased in the many kinds of cultural phytocoenoses, the metabolic processes of which have important peculiarities as compared with those of natural communities. These peculiarities are due to the nature of cultivated plants and the application of various economic measures—fertilisation, irrigation, snow-conservation, soil-liming, ploughing and hoeing of soil, thinning of seedlings, dusting with toxic chemicals, etc. Consequently artificial phytocoenoses often greatly exceed natural plant communities in the intensity of their biogeocoenotic activity and produce phenomenal amounts of organic matter (e.g. maize, sugar beet, sugar cane).

AMOUNTS OF PHYTOMASS

Among natural terrestrial phytocoenoses the highest biogeocoenotic activity, both in the extent of absorption of solar radiation and of synthesis of organic matter and in the extent of transformation of other biogeocoenose components, is produced by forest communities. Because

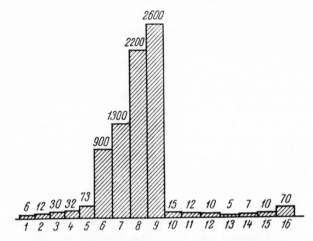

Fig. 11. Gross amount of above-ground plant mass in 'plakor' zonal plant communities in the European part of the u.s.s.r. and Central Asia (centners/ha, oven-dry) (Lavrenko, Andreev and Leont'ev 1955). 1—Arctic 'semi-desert' (mixed-herb-dryad Arctic 'semi-desert'); 2—Arctic tundra (mixed-herb-sedge tundra); 3—northern tundra (osier-herb-moss tundra); 4—southern tundra (birch-green-moss tundra); 5—sparse-forest tundra (birch-spruce sparse-forest); 6—northern taiga (lichen-moss spruce forest); 7—central taiga (green-moss-bilberry spruce forest); 8—southern taiga (*Oxalis* spruce forest); 9—broad-leaved forest (seed oak stand); 10—meadow-steppe (mixed-herb-turf-grass meadow-steppe); 11—typical mixed-herb–turf-grass steppe (typical mixed-herb steppe); 12—typical turf-grass steppe (*Festuca-Stipa* steppe); 13—desert subshrub-turf-grass steppe (*Artemisia-Festuca-Stipa* steppe); 14—steppe-like (northern) and typical subshrub desert (turf-grass-*Artemisia* and *Artemisia* desert); 15—ephemer-subshrub (southern) desert (ephemer-*Artemisia* desert); 16—black saxaul (*Haloxylon*) desert (southern desert subzone).

of the long life of the dominant arboreal vegetation in forests, an enormous quantity of organic matter and solar energy is accumulated in those parts of the earth occupied by forests, in spite of the systematic return of a considerable part of the assimilated matter and energy by respiration and the decay of plant parts and entire individuals. That quantity increases from year to year (Table 32), and in mature forests is tens or hundreds of times as much as the corresponding amounts of matter and energy in other types of plant cover (Fig. 11 and Table 33).

The annually-increasing accumulation of matter and energy in the living mass of trees is one of the most important biogeocoenotic character-istics of forest phytocoenoses.

Very large amounts of organic matter are accumulated in tropical rain forests, but the greatest values have been recorded for countries with temperate climates (on the Pacific coast of America the volume of a single kind of timber reaches 4000 m³ per hectare in the central Chile forest region, to which A. D. Gozhev (1948a) gave special attention). This proves that the biogeocoenotic work of phytocoenoses and their accumulation of organic matter depend not only on hydrothermal and soil conditions but also to a great extent on the biological characteristics of the plants constituting the phytocoenoses. In the Chilean forests, for instance, great amounts of timber accumulate apparently as the result of the very high longevity of the local tree species (up to 2500 years).

DYNAMICS OF PHYTOMASS AND ITS PRODUCTIVITY

As a rule, forest phytocoenoses both accumulate and synthesise organic matter more rapidly than other types of natural terrestrial vegetation, but good factual data on the gross annual production by different types of vegetation are still extremely rare; they are far from being always com-parable; and investigations of this question, in spite of the lively interest taken in it in recent years, have developed both here and abroad in an unjustifiably slow and small-scale manner. Very little is known of the annual growth of underground parts of plant communities. We agree with E. M. Lavrenko (1955) that from both the scientific and the practical points of view it is necessary to press on in every way with planned study of productivity and loss of dead parts by terrestrial plant covers in differ-ent natural zones and provinces.

A priori one might think that the maximum synthesis of organic matter takes place in tropical rain forests, where, because of the hot humid climate and the evergreen nature of the plants, year-round assimilation of carbon dioxide is possible. Fageler (1935) gives the amount of annual increment of organic matter for such forests as 200 tons per hectare, but that scarcely corresponds to the actual figure, at least for primary tropical forests, which include in their composition many tree species of very slow growth (Richards 1961). Production of organic matter reaches a minimum, of course, in forests where climatic and soil conditions are least favourable for photosynthesis and nutrient uptake by roots of trees, e.g., in the conditions of a short growing period, insufficient or excessive moisture, etc. The fall in photosynthetic activity of trees is most extreme at the climatic and edaphic limits of the range of forest growth.

In the Russian plains the maximum average annual productivity of organic matter in forests, measured as the weight of annual increment of dry timber, is attained in seed oak groves in the deciduous forest zone, and is estimated at approximately 5·6 ton/ha (Lavrenko *et al.* 1955). A similar value for average productivity is obtained in coniferous forests on

Table 32. *Accumulation of organic matter in forests with increase in age of stands, total*

Type of forest	Age in years							
	10	20	30	40	50	60	70	80
Pine-bilberry*		39·0		77·0		141·0		213·0
		0·8		2·81		13·09		14·07
Pine-cowberry	—	64·3	108·0	184·6	—	—	274·6	—
Spruce-green-moss	—	—	97·0	151·2	—	260·3	289·3	—
Sedge-goutwort oak	39·2	—	—	—	223·4	—	—	—
Sedge-goutwort oak	18·9	47·8	—	142·5	—	249·3	—	—
	0·07	0·265		0·350		0·475		
Sedge-goutwort poplar	18·0	185·0	—	—	264	—	—	—

* The denominator shows the amount of living soil cover in tons/hectare.

watershed slopes in the southern taiga (5 ton/ha), whereas in the heart of the taiga (in the central taiga) the productivity of coniferous forests is markedly lower than that of seed oak stands (3 ton/ha). In this case, however, it is not the specific features of the producers of the forest that are important, but the less favourable physico-geographical conditions found in the taiga for building up organic matter. At any rate, according

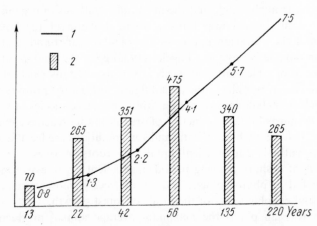

Fig. 12. Dynamics of mass of herbaceous cover and illumination beneath forest canopy at height of 1·3 metres in sedge-goutwort stand, according to age (from data of Matveeva, 1954 and Molchanov 1961a). 1—illumination (in % of light in open space); 2—dry weight of herbaceous cover (kg/ha).

of above-ground and underground parts (dry weight in tons/hectare).

| Age in years | | | | | | | | | | Author |
90	100	120	140	160	180	200	220	240	260	
	248·0	270·0	271·0	267·0	266·0	254·0	252·0	248·0	249·0	A. A. Molchanov
	14·50	15·53	16·81	17·36	18·66	19·41	19·51	19·51	19·51	1961a
248·5	—	—	—	—	—	—	—	—	—	N. P. Remezov, L. N. Bykova, K. M. Smirnova 1959
314·9	—	—	—	—	—	—	—	—	—	N. P. Remezov, L. N. Bykova, K. M. Smirnova 1959
338·6	—	341·5	—	—	—	—	—	—	—	N. P. Remezov, L. N. Bykova, K. M. Smirnova 1959
—	—	—	397·0	—	—	—	413·3	—	—	A. A. Molchanov
			0·340				0·265			1961b
—	—	—	—	—	—	—	—	—	—	N. P. Remezov, L. N. Bykova, K. M. Smirnova 1959

to investigations in England (Ovington and Heitkamp 1961), production in coniferous forests exceeds that of deciduous forests under similar climatic conditions. From the data obtained, productivity of coniferous plantations in England, measured by energy, is from 3 to 7 times as high as that of any broad-leaved forest (birch, alder, oak, chestnut, southern beech) of the same age and in the same growth conditions (Table 33). In evergreen conifers this is probably due to assimilation of CO_2 in the cold early spring and late autumn and to the greater leaf surface. If the same relation, although to a somewhat lesser degree, holds good for continental forests, we must take note of the enormous waste of the energy that falls on the earth's surface over millions of hectares of taiga forest,

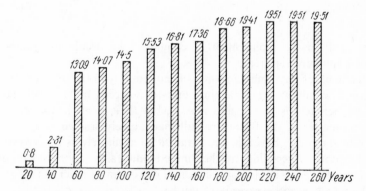

Fig. 13. Dynamics of amount of organic matter in living parts of mosses and berry-plants (in tons/ha) in pine-bilberry stand (from data of Molchanov 1961b).

Table 33. *Gross amounts of energy in tree plantations in England (10^8 cal/ha), including above-ground and underground parts of all living vegetation, litter, and humus (Ovington and Heitkamp 1961).*

Place of observation and species	Age (years)	Total amount of energy in ecosystem	Energy in trunks and crowns
West Tofts			
Pseudotsuga taxifolia	22	9243	8404
Pinus nigra	22	7090	6367
Alnus incana	22	5501	5217
Larix leptolepis	22	4589	4011
Betula alba	22	3121	2863
Treeless areas	—	590	—
Bedgebury			
Picea omorica	21	15,560	14,431
Tsuga heterophylla	23	12,593	12,070
Picea abies	20	11,143	10,407
Pinus nigra	18	9475	8378
Chamaecyparis lawsoniana	21	10,309	9764
Larix eurolepis	23	7734	7167
Pseudotsuga taxifolia	21	5898	5518
Nothofagus obliqua	22	3996	3800
Thuja plicata	22	3782	3241
Quercus petraea	21	2254	1985
Quercus rubra	21	2166	1997
Treeless areas	—	238-295	—
Abbotswood			
Pseudotsuga taxifolia	47	12,577	12,049
Picea abies	47	13,784	12,563
Pinus nigra	46	12,866	11,586
Abies grandis	24	8279	7819
Pinus silvestris	47	8320	7424
Picea abies	47	7595	6618
Larix decidua	46	10,988	9202
Fagus silvatica	39	6565	6131
Quercus robur	47	6157	5929
Castanea sativa	47	5533	5315
Quercus sp.	44	4531	4265

as a result of the replacement of coniferous by deciduous species on burned and cut-over areas.

Increment of phytomass in forests is not entirely in the form of growth of trunk wood; at the same time growth proceeds in branches, roots, shrubs, herbaceous plants, and mosses, but these items in the 'crop' of organic matter in the forest are still inadequately studied and, what is more to the point, have not been studied so accurately as has growth in trunk wood. According to the limited and fairly approximate data available, the gross annual production of dry organic matter in the forests of the U.S.S.R. scarcely exceeds 15 ton/ha.

Phytomass productivity is extremely variable, and in the same forest often fluctuates from year to year because of variations in climatic conditions (drought, excessive rain), attacks by pests, or fires; it also varies with the age of the forest. According to extensive records of the course of growth in different plantations, increment of trunk wood in any forest (without treatment) at first gradually increases with the age of the trees until at a certain point it reaches a maximum, and then it decreases.

In over-mature forests real increment often does not occur, since the effect on total volume of the death of old trees is equal to, or greater than, that of growth of the living trunks. The biomass of soil cover also varies with the age of a forest. In dense young stands soil cover is usually absent or poorly developed. Later, as the stand thins out and the light conditions improve, its biomass begins to increase, and at a certain point (which naturally varies in forests of different species and different environmental conditions) it reaches a maximum (Figs. 12, 13). The development of undergrowth also follows a similar pattern. As a result of these age variations in the phytomass, the biogeocoenetic work of the biogeocoenose varies and is affected by its environment. Often the scale of the biogeocoenotic work of the lower layers of forest phytomass is fairly large and, judging by the annual increment of organic matter and the amount of annual wastage, is sometimes no less (or only slightly less) than that of the tree layer (Melekhov 1957).

The natural rhythm of the biogeocoenotic work of forest communities, and in particular the increment of phytomass and its composition, are to-day subject over wide areas to alteration and regulation by forestry measures, especially improvement cutting, and in boggy districts by drainage, as a result of which the rate of growth of trees is increased and the species-composition of stands is changed for the better.

BREAKDOWN OF ORGANIC MATTER

Coincident with the accumulation of organic matter in forest phytocoenoses, the reverse process takes place on a large scale—the dying of organic matter. It consists of both the death of entire plants and the withering and shedding of separate organs or their parts (leaves, buds, branches, shoots, bark, flowers, pollen, fruits). In the lifetime of one forest generation the amount of dead and shed organic matter is three to four times as large as that retained in the living biomass of the forest (A. A. Molchanov 1961b) (Fig. 14).

The role of dead organic matter in a forest biogeocoenose is no less important and varied than that of living matter. It is a most significant factor of exchange in the phytocoenose-soil and the phytocoenose-atmosphere systems. Its decay adds to the stocks of mineral plant-food in the soil, forms humus in reaction with the soil, makes possible the accumulation of a number of chemical elements in the upper soil layers, largely determines the species-composition and numbers of soil-dwelling animals, warms the lower atmospheric layers and enriches them with CO_2. The accumulation on the soil surface of undecayed or slightly-decayed plant matter is the so-called forest litter, whose role and importance in forest life and biogeocoenotic metabolism are difficult to overestimate. Litter affects the gas-exchange of the soil, penetration by precipitation, surface outflow of water, the temperature regime of the soil, evaporation from the surface, the germination of seeds and development of shoots, and the inter-relationships of plants (by suppressing some species and favouring others). During the historic past forest litter has doubtless played a major role in the formation of a number of living forms of forest plants, e.g., saprophytes, undergrowth plants with roots and rhizomes located in the lower layers of litter, and fungi. We agree with G. F. Morozov (1926) that 'many forest secrets are hidden' in litter, and although at present forest litter receives much attention from sylviculturists, pedologists, and geobotanists, many problems concerning its genesis, structure, and interactions with soil, atmosphere, vegetation, fauna, and micro-organisms remain unstudied.

From the biogeocoenotic point of view it is of great value to have data on litter as a factor in the exchange of matter and energy in the forest

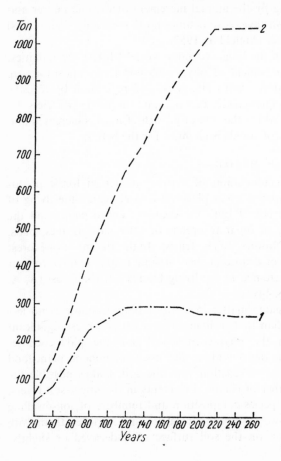

Fig. 14. Productivity of organic matter in pine-bilberry stand (in tons/ha oven-dry) (from data of Molchanov 1961). 1—amount of organic matter in growing above-ground parts of phytocoenose; 2—cumulative total of organic-matter litter.

(especially data on its quantity, etc.) and also as an indicator of that exchange, since such features of litter as its volume (especially in comparison with annual wastage), composition, structure, ash content, and reaction well express the conditions, rate, and direction of the entire biogeocoenotic metabolism. Only a few empiric data on forest litter, its amount, composition, ash content, reaction, etc., have been compiled; they show great variation in these characteristics in nature, and also close dependence on the character of forest biogeocoenoses, habitat conditions, age and density of the forest, and weather conditions.

The dynamics of decay of litter (Shumakov 1941, Molchanov 1961) and its ash content (Zonn and Aleshina 1953) have also been studied. In forests in the temperate zone the amount of litter is at a maximum in autumn (after leaf fall) and spring, and at a minimum at the end of summer. The difference in weight may amount to several tons per hectare (Molchanov 1961). Litter decomposes more rapidly in deciduous forests; more slowly in coniferous forests, especially in very moist conditions and on cold soils. Plant detritus decays most rapidly in tropical forests—there it does not even succeed in forming a more or less distinct layer on the soil surface, and one may speak of litter only academically. Large plant remains are included in the litter—trunks of fallen trees and large boughs as well as the main components of small plant material (leaves, needles, small and medium-sized twigs, and cones, which are scattered on the soil surface more or less in a single layer, although not uniformly in depth and composition).

These larger items stand out well above the general surface of forest litter, decay much more slowly than other litter (Molchanov 1947) and make substantial differences to the micro-environment and distribution of plants. On such decaying trunks special microcommunities of herbaceous plants, mosses, and lichens are often formed, and they are often colonised by tree seedlings, wood-destroying fungi, and certain insects (e.g., ants).

COMPOSITION OF PHYTOCOENOSES

The phytomass of biogeocoenoses is composed of an enormous number of individual plants, which differ greatly in their biogeocoenotic work and consequently in their role in the life of the biogeocoenoses. Even in communities of a single species, of the same age and genetically homogeneous, there are always individuals of greater or less biogeocoenotic activity as a result of differences in competition for light, moisture, nutritive matter in the soil, and living space. This feature is easily seen in arboreal communities, which during the course of competition for the necessities of life separate, according to their individual growth and development, into the five so-called 'Kraft's classes'. Individuals that are much shaded, crowded for space, and generally suppressed (Kraft's classes IV and V) have small biomass, both of leaves and in total; they live in light conditions that are unfavourable for photosynthesis, and produce less organic matter (and therefore less free oxygen in the atmosphere) than trees of classes I, II, and III (Fig. 15). Biogeocoenotic differentiation of plants in a forest is usually favoured by the presence of trees of different ages and ontogenetic maturity. In this connection we may mention the almost universal existence in mature forests of saplings and of under-growth tree species, whose biogeocoenotic work and role in the total metabolism of the biogeocoenose cannot be compared with those of adult trees either in their scale or in their conditions.

In communities composed of many species biogeocoenotic differentiation of plants is, of course, displayed most strongly. Quite apart from the

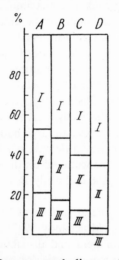

Fig. 15. Percentage relation of weights, leaf-surfaces, total productivity, and annual increment of growth, in trees of three different classes of development in a 12-year-old ash plantation (from Boysen-Jensen 1932). I—dominant trees; II—sub-dominant trees; III—shaded trees. A—weight; B—leaf-surface; C—total productivity; D—annual growth.

different metabolisms of autotrophic and heterotrophic plants living together in biogeocoenoses and prominent in the respective roles of *producers* (creators of primary organic matter) and *consumers* (which consume and transform it), in any multi-species community there are always certain species or groups of species that play a major role in the exchange of matter and energy between the phytocoenose and its en-vironment, and others that have a subordinate or even negligible role. The first type includes the plants with greatest mass in phytocoenoses, which produce the greatest amounts of organic matter, exchange the greatest amounts of mineral matter and water, fix the greatest amounts of solar energy, and simultaneously and consequently make the greatest transformations in the state of the other components of the biogeocoenose. They may be called *macroproducers*, and (with reference to solar energy) *macrotransformers*. With reference to tree and shrub species we may also speak of *macro-accumulators* of matter and energy. In geobotany macro-producers correspond to the concept of plant dominants and determinants. Other autotrophic elements of a phytocoenose may, according to the scale of their biogeocoenotic work, be *meso-* or *micro-producers*. The latter (and partly the former) correspond to the geobotanical concepts of associates or minor species, and the former (in many specific cases) to sub-dominants.

In forest biogeocoenoses the role of macroproducers is generally played by the trees dominating the main forest canopy; they take up the greatest space in the atmosphere and soil, produce most of the organic matter, fix and accumulate the greatest amount of solar energy, draw into circulation the greatest amounts of mineral matter and water, and surpass all other community components in the changes they make in the light regime of the phytocoenose, movement and composition of air, humidity, temperature, and chemical and physical features of the soil. The other autotrophic components of forest biogeocoenoses—other tree species, shrubs, lianas, herbage, mosses, lichens, algae—are usually of much less biogeocoenotic importance than the dominant species of the upper storey of the forest, and are classified as meso- and micro-producers, whose biogeocoenotic activity is heavily suppressed by the macroproducers of the arboreal storey, although it may be on a considerable scale (e.g., bilberries in taiga forests).

The biogeocoenotic role of separate plant species in nature varies greatly in accordance with external conditions and stages of ontogenesis.

Frequently species that act as macroproducers in certain conditions and in certain stages of their ontogenesis appear as meso- and micro-producers in other ecotopes, in different regimes of ecological conditions, or in a different stage of development. We could give many illustrations of this situation, but shall limit ourselves to two. Spruce is a macro-producer and a macrotransformer in the taiga zone on loam or sandy-loam soils with good moisture supply and aeration, where it forms dense shady forests of high productivity. In the same zone on dry and poor soils, or very swampy soils with inadequate oxygen, spruce is either entirely absent or occurs only in small numbers and is heavily suppressed, has a small growth rate, and judged by its metabolic activity is definitely a microproducer. Spruce trees in the first years of their growth in forest clearings are also notable in the same role of microproducers. Another example: haircap moss (*Polytrichum* sp.) and several species of sphagnum occur frequently in pine and spruce forests, but on well-drained substrates their numbers are extremely small and the scale of their biogeocoenotic work is negligible—they are microproducers. In ecotopes with abundant moisture, on the other hand, these plants are represented by numerous individuals, form a dense and luxuriantly-developed ground cover, and in these conditions form and accumulate organic matter (as peat) in quantities not less than—often many times as much as—that of the trees growing there (P'yavchenko 1960). It is clear that in such conditions the mosses are macroproducers, and in their biogeocoenotic work often considerably surpass trees (Table 34). In such cases, one may ask, should not these groups be looked upon as forests?

Table 34. *Increment of organic matter in certain types of swamp forest in Vologda province (in kg/ha per annum, oven-dry) (P'yavchenko 1960).*

Type of forest	Increment of organic matter				
	Tree storey	Under- storey	Herbage-shrub cover	Moss cover	Total
Swamp-herbage spruce stand	850	8	89	959	1906
Shrub-sphagnum pine stand	104	—	15	1661	1780

In forest biogeocoenoses consumers are represented by a considerable variety of lower plants, especially saprophytes and parasitic fungi. Although their biomass is usually small, and is always much less than that of producer plants, their role in the transformation of matter and energy in nature may be very great. This applies mainly to saprophytic fungi, which play a major part in the decay of litter and fallen tree trunks (Chastukhin 1945, Molchanov 1947) and also to parasitic fungi, which often settle on many living trees in a forest and accelerate their deaths. Moreover, many fungi participate in the water and mineral nutrition of the primary producers, as intermediaries between dissolved matter in the soil and root cells. This type of water and mineral nutrition of higher plants (mycotrophy) is widespread in nature and is very characteristic of the biogeocoenotic links in forest phytocoenoses. Among the higher plants that act as consumers in forest phytocoenoses are a number of saprophytes—in our forests, for example, *Monotropa, Neottia, Epipogon, Corallorhiza*—and a very few parasitic forms, e.g., toothwort (*Lathraea squamaria*), which parasitises hazel roots. These plants do not occur in large numbers, however, and the scale of their biogeocoenotic work is therefore practically negligible.

A more substantial role in the life of forest biogeocoenoses is filled by semi-parasitic plants, which obtain only partial nutrition, chiefly mineral matter, from the host-plants. These green plants, which perform independent synthesis of organic matter from CO_2 and water, are represented by a number of herbaceous (*Melampyrum, Euphrasia*, etc.) and shrub (*Viscum, Loranthus*) forms. They occur much more frequently than the real consumers and in incomparably greater numbers. *Melampyrum*, for instance, often forms dense cover in forests. Mistletoe (*Viscum album*) is found in some parts of forests in enormous numbers. In the Tyrol, for example, in the mountain pine forests lying along the spring migration routes of starlings (which eat and distribute mistletoe seeds), these plants exist in thousands of millions and the trees are covered with them from the tops almost to the ground (Beilin 1950). There mistletoe should probably be classified both as a consumer and a producer (mesoproducer), adding a considerable proportion to the total production of primary biomass. As a result of infestation with mistletoe, trees have lower growth rates, wither at the tops, and die prematurely (Beilin 1950).

Unfortunately this question is still little studied. In our forests mistletoe most often infests linden, birch, pear, willow, and aspen, and such conifers as pine and fir.

Plant-species composition is considered to be one of the principal characteristics of a plant community. Different phytocoenoses are distinguished from each other primarily by the species-composition of the plants forming them. The species-composition and inter-relationships of the dominant species are particularly important since the appearance, ecology, and structure of the community depend on them. The species-composition of plant communities is also important from the biogeocoenotic viewpoint, since different species of plants, as a result of the genotypic characteristics of their links with their environment, are fairly specific both in their biogeochemical activity and in the transformation of their habitat. In addition, the genetic composition of the plants may be significant, as has been proved by experiments in raising plants of different origins from seed. For this reason phytocoenoses that differ in the species-composition of their plants differ also in the effects of their biogeocoenotic work. If the species are closely related, however, exceptions can occur. Thus *Picea excelsa* and *P. obovata* are probably very similar in their biogeocoenotic activity, and in similar ecotopes form communities so similar biogeocoenotically that they can be distinguished (either for biogeocoenological or phytocoenological purposes) only on an academic basis—by referring the dominant to one of two minor morphological varieties.

The number of species in forest communities varies, and besides groupings formed by a small aggregation of species there are widely-distributed phytocoenoses formed of hundreds of species of higher plants. The number of species present depends: on the age and density of the stand (the older the stand and the more open the tree canopy, the richer is its floristic composition); on the synecological features of the dominant producer species (the plant-species composition is less rich in forests of shade-tolerant or shade-loving species, capable of forming dense and shady canopies); on environmental conditions (in more favourable climates and on more fertile soils there is greater variety of species in a forest); on the geological history of the earth's surface (there are more species in territories that have over a longer period been unaffected by catastrophic changes in climate, relief, or situation). From the biogeocoenotic viewpoint, the floristic variety of a phytocoenose is essential as a sure indicator of the degree of complexity of biogeocoenotic metabolism: the richer the species-composition of the plants forming a phytocoenose, the more varied and complex are the exchanges taking place in it and the links of the phytomass with the other biogeocoenose components,

STRUCTURE OF PHYTOCOENOSES

The phytomass of plant communities (especially terrestrial ones) is never perfectly uniform either in its structure or in the conditions and effects of the biogeochemical work of its components. Even in single-species stands the phytomass can be divided into above- and below-ground parts. These differ not only in anatomical and morphological structure and in the functions they fulfil but also in that their functions and metabolic relationships are developed in very different environments, with very different materials, under different laws, with results of different kinds. In spite of their interaction and correlation the development of above- and below-ground parts is not synchronised, as is clearly shown, for instance, in the substantial discrepancy in phenology of growth pheno- mena in the roots and in the above-ground organs of trees.

The above- and below-ground parts of the phytomass differ also in their significance in the biogeocoenotic process, as is very evident for trees. Both have active components, distinguished from those that are inactive or only slightly active. In the above-ground biomass the most active parts are leaves and young shoots. They serve as receivers of CO_2 and solar radiation, as the main channels for water-exchange between plants and the atmosphere, and as the chief transformers of the internal environment of the biogeocoenose, since they carry out photosynthesis and transpiration; absorb the greater portion of solar radiation, precipi- tation, and aerosols; break most of the force of the wind; transform moisture and heat; and so on. In the underground biomass the active parts are the young tips of roots. Although their share of the total root mass is very small (especially in trees) they play a major role in the bio- geocoenotic work of the phytocoenose in the soil, since it is through them that plants absorb water and minerals and largely discharge CO_2 and many highly-active organic substances, including a number of enzymes capable of decomposing complex organic matter (Peterburgskii 1959).

These parts also are biogeocoenotically heterogeneous. In the leaf mass, for instance, light and shade leaves differ greatly not only in their structure and the conditions under which photosynthesis occurs (ex- pressed, e.g., in different CO_2 concentration and in different illumination), but also in the results of photosynthesis. Illuminated leaves photosyn- thesise ten times as intensively as do shaded ones (Table 35), and they naturally play a major role in the total production of organic matter and accumulation of energy in the biogeocoenose. Shade leaves often exhibit a negative photosynthetic balance, even those of shade-tolerant plants.

There are also differences in the work done by young, mature, and old leaves, both because of age-differences in their biochemical condition and as a result of the spring-summer-autumn fluctuations in environmental conditions for photosynthesis, transpiration, and respiration.

In multi-species communities biogeocoenotic diversity in the phytomass is greatest, since plants of different species, with their varied genetic charac-

Table 35. *Rate of assimilation of CO_2 by 'light' and 'shade' leaves of trees (in mg per g of dry weight of leaves or needles per hour at a temperature of 18-22°C) (L. A. Ivanov 1946).*

Species	Illumination, in % of full sunlight				
	1	30		100	
CONIFEROUS Light-loving	'Shade' leaves	'Light' leaves			
Pine	− 0·08*	2·1	Average	3·3	Average
Larch	− 0·06	3·1	2·7	4·1	3·8
Shade-tolerant					
Spruce	− 0·06	1·6	2·5	1·7	2·1
Fir	− 0·13	3·4		2·6	
DECIDUOUS Light-loving					
Oak	− 0·12	2·5		4·1	
Willow	0·03	1·2	4·3	8·0	7·2
Birch	0·18	6·0		9·4	
Shade-tolerant					
Maple	0·54	4·9	5·6	5·0	6·1
		6·3		8·3	

* Figures with a minus sign indicate that CO_2 is exhaled, as respiration exceeds photosynthesis.

teristics, interact differently with other biogeocoenose components; they procure their requirements of mineral matter, water, and energy in different ways and in different quantities; their methods of creating organic products are diverse, and often unique; they discharge into the atmosphere and the soil (besides the common life-activity products of CO_2 and O_2) specific metabolites, in their own ways altering the features and characteristics of other biogeocoenose components (see Chapters V and VI). These differences are especially marked in plants of different life forms, so that the more varied the life forms present in the phytomass, the greater is its biogeocoenotic differentiation. Of particular importance in this respect are forest phytocoenoses where the dominant tree species are accompanied by numerous shrubs, lianas, herbaceous plants, mosses, lichens, algae, and fungi. Differing in size, in longevity, in form of growth, in capacity to obtain and transform their requirements in matter and energy, and in relationships with light, moisture, and mineral components of the soil, these forms determine the vertical division of the biomass of forest phytocoenoses (both the above-ground and the underground parts) into a series of layers or storeys, each of which is distinct not only in situation and morphology but also biogeocoenotically, i.e., as a matter-energy system. Thus each layer receives and transforms matter and energy in a specific way, and function in conditions of light, moisture, heat, CO_2 concentration, mineral nutrition, aeration, air movement, and activities of fauna and micro-organisms not duplicated in any other layer.

Unfortunately the plant layers of phytocoenoses have not yet been studied from this point of view, and at present there are only limited data on several features of the phyto-environment in which the biogeocoenotic work of any layer takes place (Sakharov 1940 a, b, 1948).

The number of layers in natural communities varies from one to six, the connection between this variation and environmental conditions being obvious. As a rule the number of layers is greater in communities living in more favourable climatic and soil conditions. Tropical rain forests have the most complex structure, although the layering of the vegetation is not so clearly evident externally as in temperate-zone forests. The history of the earth's surface also plays its part in the organisation of plant communities (the structure of communities is more complex in the more ancient parts of the dry land). A further role is played by the inter-relationships between plants, which restrict the possibilities of association of many species and forms.

The layers of plant communities exist in a state of constant interaction and interdependence among themselves. As observations and experiments indicate, extensive development of the topmost layers seriously impedes the development and metabolism of the lower layers, even to the point of suppressing them, both by severe reduction in light intensity beneath the canopy and deterioration in the physiological composition of the light, and by intensifying competition by roots for moisture and mineral components of the soil. Ya. Ya. Vasil'ev (1935) drew attention to the different roles and fates of different forest layers, and proposed to distinguish: constitutional layers, belonging to a forest of a definite type; demutational, temporary layers taking the place of the main constitutional layer during its rehabilitation (e.g., a layer of young spruce in a pine or birch forest); digressive, arising as a result of destruction (by fire, etc.) of the natural inter-relationships in a forest phytocoenose; and regenerative, rehabilitational (e.g., spruce seedlings in a spruce forest). From the biogeocoenotic viewpoint that classification is worthy of attention, since it emphasises the dynamics of the operative surface of the phytomass and reveals (as far as the decomposition layers are concerned) the past and future types of biogeocoenotic metabolism, and also its stability or disruption.

The layering of the phytomass is the basis of vertical differentiation of all other components of a biogeocoenose, and therefore determines the fundamental features of the total stratification of biogeocoenoses and the biosphere. Unfortunately this subject has not been given proper experimental investigation. It was examined theoretically some time ago by Yu. P. Byallovich (1960), who established a complete system of biogeocoenotic horizons, their laws, and their inter-relationships. Biogeocoenotic horizons are individualised vertically, and the structural parts of a biogeocoenose cannot be further separated vertically. They are of course closely related to plant layers, but are considerably more detailed. For instance, in a three-layer tree-shrub community without herbage or moss cover, Byallovich identified *nine* biogeocoenotic horizons (Fig. 16).

Besides the vertical differentiation of the phytomass of biogeocoenoses, one usually observes considerable diversity (quantitative and qualitative) in its horizontal complexity. That is partly due to: the unequal density of

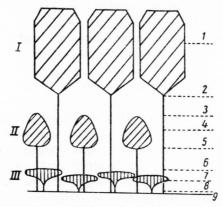

Fig. 16. Relation of plant layers to biogeo-coenotic horizons. I—first layer of tree stand (height 20-22 m); II—second layer of tree stand (height 10-12 m); III—third layer of community, undergrowth (height 2-3 m). 1—upper ('active') photosynthesis biogeo-horizon of first tree layer (thickness about 2 m); 2—lower photosynthesis biogeohorizon of first tree layer (thickness about 4 m); 3—first trunk, or intercanopy, biogeohorizon (thickness 2-6 m); 4—upper ('active') photosynthesis biogeohorizon of second tree layer (thickness about 2 m); 5—lower photosynthesis biogeo-horizon of second tree layer (thickness about 2 m); 6—second trunk, or intercanopy, biogeo-horizon (thickness 4-7 m); 7—photosynthesis biogeohorizon of undergrowth; 8—microdiffer-entiated subcanopy biogeohorizon (thickness about 1 m); 9—nanodifferentiated subcanopy biogeohorizon (thickness about 0·2 m).

plant stands, with a variety of gaps, thinly-colonised areas, and dense plant groups; to the patchwork distribution of different species resulting either from their reproductive peculiarities or from heterogeneity of microconditions (exogenic and endogenic) in a particular area; and the destruction of vegetation by man, animals, or wind. Therefore the bio-geocoenotic work of a phytocoenose in nature may be much diversified horizontally even in homogeneous areas. Horizontal diversity in a phyto-coenose naturally creates diversity also in other biogeocoenose com-ponents, as has been confirmed experimentally with special reference to forest communities (Sakharov 1939, 1950, 1951).

Whereas the vertical layering of the phytomass serves as the basis for horizontal stratification of biogeocoenoses, its horizontal variability leads to diversification of the radial structure of biogeocoenoses. Sakharov (1950) first drew attention to these structures under the name of elements of a biogeocoenose. He regarded them as being recognisable within a phytocoenose by their composition, their structure, their density, the growth and development of their vegetation, and their dynamics; as being intra-coenose groupings of plants, or coeno-elements, corresponding to definite features of soil, phytoclimate, and fauna. Sakharov also gave the first complete descriptions of such elements, applicable to some forests in Poles'e.

More detailed study of the radial structures of forest biogeocoenoses, undertaken in company with A. I. Utkin and I. M. Uspenskaya in complex deciduous-spruce forests in Moscow province, induced us to look upon these structures from another point of view—not as intra-coenose groupings of plants or coeno-elements corresponding to certain features of soil, climate, and fauna, but as complex parts, units, of an equally complex whole (a biogeocoenose), distinguished from each other not so much by the radial arrangement of their components as by specific radially-directed metabolism, and differentiated on that basis throughout the whole radial extent of the biogeocoenose. In this sense the units are apparently close to the radial concept of Yu. P. Byallovich (1960).

The natural diversity of biogeocoenotic units is very great. By origin they may be *radical* or *derivative*, i.e., arising from some human activity; or *basic*, i.e., playing a role in the formation and metabolism of a biogeocoenose. The basic units occupy the greatest area in a certain ecotope and create the principal or, so to speak, 'typical' form of exchange of matter and energy. The *supplementary* units have a minor share in biogeocoenotic metabolism, and are very limited in area. When grouped by developmental tendencies the units may be *moribund* (relict), with regressive boundaries, or *stable* and *progressive*, with boundaries extending laterally.

Study of units is of value from the scientific and scientific-method standpoints and also of practical value, since the various units represent differentiated environments for the appearance, development, and life of young generations of plants, especially tree species. In forests with units one must undertake, for example, such forestry measures as improvement cutting, or selection of trees for felling in selective and gradual cutting systems. For this reason, study of the unit-structure of forest biogeocoenoses is urgent. That study is, to a definite extent, analogous to the synusial analysis of the structures of plant communities made in certain geobotanical works (Minyaev 1963).

INTERACTIONS OF PHYTOCOENOSES WITH OTHER COMPONENTS

A phytocoenose has profound and intricate relationships and interactions with all other components of a biogeocoenose as well as with its environment. They arise during the course of physiological processes (photosynthesis, respiration, transpiration, growth, mineral and water nutrition, excretion) and in most cases are characterised by an exchange of matter and energy.

Some relationships are direct, and others are in various and complex ways indirect, complete understanding of them, especially with respect to matter and energy, being far from always possible.

All of the relationships are unstable, at times being strengthened and at other times being weakened, and occasionally being completely suspended for a longer or shorter period on account of the rhythm in the life-activity of the phytomass or of changes in the state of other components and factors of the biogeocoenose. Changes in the relationships may be oscillatory and of short duration, causing slight reversible reactions in phytocoenoses (e.g., daily and meteorological changes in the interactions between phytocoenoses and atmosphere, soil, and fauna), or they may be made in a single definite direction, producing irreversible changes in the metabolism of phytocoenoses with other biogeocoenose components to the point of complete alteration of the latter (e.g., changes in the amount of flooding of river flats through reduction of their level by erosion, or swamping of dry land through extension of peat beds, etc.).

The links between a phytocoenose and other biogeocoenose components are not universal and essential in all their multiformity for every biogeosphere on the earth. The biogeocoenotic work of phytocoenoses in nature is developed in three spheres, with very different physical and chemical characteristics and consequently differing also in the conditions and relationships of plant life: water, air–water and air–land (terrestrial). In the biogeocoenotic relationships between phytocoenoses and other components there are, strictly speaking, no identical features. Terrestrial and aquatic phytocoenoses are especially differentiated in the interactions linked with the procurement by plants of CO_2, oxygen, mineral matter, and water.

Air–water, and especially air–land (terrestrial) phytocoenoses, which distribute their biomass through environments differing widely in physical features, and which are simultaneously and directly linked with air, water, and soil, obtain their CO_2, water, and mineral nutrition from different sources (gaseous, liquid, and solid), in different forms (gases, solutions), and through surfaces that are functionally differentiated and well adapted to the sources and their functions (leaves, shoots, roots).

The biomass of aquatic (submerged) phytocoenoses interacts directly with only one physically-homogeneous component (liquid water), is functionally undifferentiated (except for reproductive organs), and receives all elements of its nutrition with the entire body surface and in a single general form (solutions of salts). Therefore the interaction of vegetation with environmental factors is here less varied, and interchange of matter is much more simple and direct, than in air–water and terrestrial phytocoenoses.

From the relationships of phytocoenoses with other components there arise local and extremely complex movements of matter and energy (i.e., biogeocoenotic circulation) and processes of expulsion (and repulsion) of matter and energy beyond the limits of the biosphere—upwards, into the atmosphere (and cosmos), and downwards into underground water and to the ocean floor.

The relationships of phytocoenoses with other components are very specific and call for separate analysis.

Relationships of phytocoenoses with the atmosphere

The relationships of a phytocoenose with the atmosphere are peculiar to terrestrial, littoral-aquatic, and semi-aquatic communities, the plants of which exhale into the atmosphere part (albeit small) of their bodies and therefore have direct contact and exchange with it. In submerged aquatic phytocoenoses there are no direct relationships between plants and atmosphere, all exchange being restricted to the aquatic environment.

The basis of interaction between phytocoenoses and atmosphere is gas-exchange, arising from photosynthesis and respiration by plants and involving, on the one hand, the uptake of CO_2 from the air in photo-synthesis by green plants and its return in respiration; and on the other hand movement of oxygen, given off in photosynthesis and absorbed in respiration. The scale of that exchange over the earth's surface reaches astronomical figures and has vast global significance, as it stabilises the gaseous content of the atmosphere and makes animal life possible. Even in regions with temperate climates and short vegetative periods, a single hectare with compact vegetation absorbs annually not less than 5 tons of atmospheric CO_2, and the entire plant cover of dry land takes from the atmosphere 93,600 million tons of CO_2 during a single year's production of living matter (Uspenskii 1956). Altogether, not less than 1/35 of the atmosphere's content of CO_2 takes part in a year's circulation of that gas through photosynthesis by the earth's green plants. Through the activities of a number of micro-organisms that are capable of fixing free molecular nitrogen, a direct nitrogen relationship is added to the CO_2-O_2 relationships between phytocoenoses and the atmosphere. In some phytocoenoses that relationship is on a considerable scale.

The gas-relationships between phytocoenoses and the atmosphere are supplemented by turbulent heat-exchange, moisture-exchange (through transpiration), quantitative and qualitative transformation of light falling on phytocoenoses, and discharge of various organic substances into the atmosphere by phytocoenoses.

All of these interactions take place most intimately between phyto-coenoses and those lower strata of the troposphere that directly penetrate the phytomass and serve as the immediate source of its CO_2 and oxygen supply. And although the relationships of phytocoenoses with the atmosphere, as is well known, are not restricted to the layer of the troposphere adjacent to the earth but extend upwards to the ozone layer, yet the air component of a biogeocoenose, in the strict sense of the term, must be held to be that lowest layer of the troposphere and it alone: the layer that is densest, warmest, most dynamic in its characteristics, most active biologically, and most favourable for the development of fully-balanced life of autotrophic and heterotrophic organisms. Not only does that layer exert the greatest influence on biogeocoenotic exchange and on the life of phytocoenoses, but it is itself subjected to the most profound changes in its gaseous composition, temperature, humidity, and air movement by phytocoenoses, fauna, soil, and micro-organisms. Beyond its limits biogeocoenotic metabolism is replaced by physico-geographic processes, having other origins and subject to other laws.

The interactions of a phytocoenose with the atmosphere are more or less constant in only one limited region on earth, in the humid tropics, since in the remaining vast area they undergo marked fluctuations on account of local thermoperiodism, hydroperiodism, and photoperiodism, and the phases of their active and multiform exchange alternate over longer or shorter periods according to the seasons in which the exchange is less active, or simplified, or even completely suspended.

Clearly-marked daily and meteorological fluctuations are also observed in the relationships within the phytocoenose-atmosphere system, as a result of variations in illumination, spectral composition of light, temperature, atmospheric and soil humidity, and air movement. They are accompanied by wide fluctuations in the CO_2 concentration of the air that penetrates the phytomass of a biogeocoenose; in saturation of the atmosphere with volatile organic substances; in formation of convection currents in the air, which play a major role, for instance, in the distribution of spores and the very small seeds of several forest plants; and in other factors. Daily and meteorological fluctuations in phytocoenose-atmospheric relationships exist in all regions of the earth.

Transformation of atmospheric air by phytocoenoses in nature takes many forms, and depends on the structural features of the phytomass: its radial extent, its density (compactness), the species-composition of its vegetation, and above all on the morphological and biological characteristics of their above-ground parts (colour and structure of leaves, duration of their lives, structure of crowns, body dimensions, form of growth). Therefore a specific atmospheric composition and specific atmospheric characteristics appertain to every phytocoenose, a situation expressed in geobotany in the concept of phytoclimate. The composition and characteristics of the atmosphere are most fundamentally altered by forest communities, especially by evergreens with more or less permanent above-ground organs. Fairly extensive data have been compiled concerning this question in sylviculture; besides much statistical material, formulae have been devised to express the dynamics of changes in various atmospheric features in the forest, linked with the growth of trees from year to year (see, e.g., Molchanov 1961a).

In many cases the natural inter-relationships between phytocoenoses and the atmosphere are altered by pollution of the atmosphere with dust, smoke, or radioactive material from industry, transport, or experiments with nuclear weapons. Smoke and dust, settling on the surface of leaves and young shoots, clog up the stomata and impede gas-exchange by plants, stop light from reaching green cells and retard photosynthesis. SO_2, which is released into the air by many industrial enterprises, is very harmful to plants; evergreen conifers suffer most severely from it. High concentrations of SO_2 cause rapid death of conifers, which are replaced by more tolerant deciduous species. The result is a radical reconstruction of the whole biogeocoenotic metabolism and a lasting change in forest biogeocoenoses, observed on a large scale in the vicinity of many industrial centres. Changes in atmospheric content act on plant communities not only directly, but also indirectly through the soil, since mechanical, gaseous, and radioactive admixtures are washed into the soil by atmospheric precipitation, and in one way or another alter the exchange between roots and dissolved matter in the soil, or are directly absorbed by roots (e.g., radioactive matter).

Of all the manifold relationships between a phytocoenose and the atmosphere, of most interest to forestry is the effect of light on the life of tree species, since the light factor is the only atmospheric one that can be substantially changed and regulated in the forest by the most easily available and the simplest means, tree-cutting. 'Light is a lever whereby the forester can guide the life of the forest in any direction desirable for the national economy' (Beck, quoted by L. A. Ivanov 1946).

In conclusion we may say that, besides its creative functions, the atmosphere has various destructive effects on territorial phytocoenoses. Occasionally these effects are manifested with extreme force and occur over extensive areas. It is sufficient to mention, for instance, the devastation caused in forest biogeocoenoses by violent winds, hail, snowfall, catastrophic rains or (on the contrary) prolonged drought, and excessive frosts.

Relationships of phytocoenoses with the soil

The relationships of phytocoenoses with the soil are inherent in terrestrial plant communities, although not in all of these, since many plants grow on cliffs, rocky scree, new alluvial flats, peat, etc., where soils in the strict sense of the word are almost or wholly non-existent.

In nature the relationships of phytocoenoses with the soil are multiform, reciprocal, very intimate, often very complex, and almost always through the medium of other components or factors. At the foundation of these relationships lies, on the one hand, absorption of mineral matter and water from the soil by plant roots; on the other hand, the introduction of various organic substances into the soil by plants, from their above-ground parts by shedding and decay, and from their root systems, as a result of respiration, secretion (cholines), and dying-off of parts of roots and rhizomes. The scale of these movements of matter is very large. Every year phytocoenoses take quintals and tens of quintals of mineral matter and water out of the soil. At the same time they introduce into the soil (in the form of decayed matter from leaves, branches, flowers, fruits, stems, and roots) tens of tons of dry organic matter and quintals (up to tens of quintals) of salts and nitrogen (Mina 1951, Bazilevich 1955, Remezov *et al.* 1959, Molchanov 1961 a, b, Ovington and Madgwick 1959). In the course of these movements local biological circulation of matter and energy takes place, as well as biogenic accumulation of the principal elements of mineral nutrition of plants (especially phosphorus, sulphur, calcium, potassium, and manganese) and also of a number of micro-elements in the upper soil horizons.

These basic interactions of a phytocoenose and the soil are supplemented by various other contacts between them, affecting both the physical and the chemical aspects of biogeocoenotic metabolism and the various characteristics of the soil and of the plant community. Generally known, for instance, are the interdependent relationships of phytocoenoses (composition, structure, productivity) with such soil characteristics as temperature conditions, acidity, humidity, and aeration.

The mechanical interaction of phytocoenoses and soils is substantial and specific. It makes possible the proper orientation of plants in their aerial environment with reference to the source of energy, and determines the stability of phytocoenoses and soils against the destructive forces of gravitation, wind, and flowing water.

Many of the relationships between a phytocoenose and the soil exist through the medium of, and are complicated by, the activities of fauna (especially of animals inhabiting litter and soil), fungi, and bacteria and other micro-organisms, which transform and break down organic matter that falls on or enters the soil in the litter fall from plants, dead plants, and plant secretions. Especially important in this respect are fungi and bacteria, which by their abundance, rapid multiplication, and high chemical activity strongly affect the processes of chemical exchange between roots and soil. The entire exchange in the phytocoenose-soil system is most closely linked together by micro-organisms in forest communities with wide distribution of mycorrhiza.

All of the relationships between phytocoenoses and soils are very dynamic and depend very closely on climatic conditions, especially on the hydrothermal regime of ecotopes, the ultimate beginning of all exchange interactions. They are weakened or entirely suspended during periods when heat and moisture are unfavourable for the life-activity of the root system, micro-organisms, and soil fauna, and for all the physical and chemical processes that take place within the soil. To some extent variations in the interactions between vegetation and soil are found in the majority of terrestrial plant communities, which experience regular checks in their life-activity with the onset of cold weather, prolonged drought, or floods. Only in tropical rain forests and in some types of subtropical forests are the biogeocoenotic relationships between phytocoenoses and soil apparently uninterrupted and more or less equally strong throughout the whole year. Pauses and declines in soil-vegetation interactions in nature vary greatly in duration and intensity, and the difference between localities in that respect may be very significant in the construction of biogeocoenotic classifications, as a major indicator of characteristics of biogeocoenotic metabolism.

Besides annual periodicity, intra-seasonal, meteorological, and daily fluctuations are also observed in the relationships between phytocoenoses and soil, due to fluctuations in the hydrothermal regime and to the daily and seasonal inequalities in the biogeocoenotic work of separate plants and of the phytocoenose as a whole.

The character of the relationships between phytocoenoses and soils varies with the species-composition of the plants (especially of dominants), the structure (especially underground) of the community, the age of the plants, the density of the stands, etc. The change in relationships observed with change in tree species (e.g., replacement of conifers by deciduous trees) may be given as an example; it includes quantitative and qualitative transformations in organic detritus, in the structure and composition of litter, in the extent of spread of roots, in the composition of their exudates, in soil moisture, acidity, and temperature conditions, in the species-composition of fauna and micro-organisms, and in the degree to which these colonise the various soil horizons.

The thickness of the substrate in which interaction between a phytocoenose and the soil develops varies within wide limits: from a few centimetres to 10 or 20 metres, depending on the morphological features of different plant species and on the physico-chemical features of the soil. As a rule, interaction of phytocoenoses and soil takes place over the greatest depth in forests (in our country, usually in oak forests) and over the least depth in moss and lichen tundras. The roots of many desert and steppe plants penetrate deeply into the soil. With alfalfa and camel's-thorn, for instance, the tap-root goes down 15 to 18 metres into the soil. In forests roots penetrate farthest into deep, friable soils of moderate or good fertility, well watered and aerated. On the other hand, the thickness of the interaction zone is much reduced by bogginess of the soil, permafrost close to the surface, hard mineral strata, underground water, salty layers, or dryness or poverty of the soil.

The relationships between phytocoenoses and soil are not uniform throughout the whole depth of the soil, but change regularly from one genetic horizon to another. The interaction of a phytocoenose with the soil is most intensive and complex in the upper humus layers, where the majority of the most active roots of all plants are located, where there are the greatest masses of fungi, bacteria, insects, and dead organic matter, and where the largest amounts of mineral plant food are concentrated. In the lower depths of the soil interaction is less vigorous and more simple, because root systems are less developed there and fewer species are represented.

Within a homogeneous climatic zone the relationships between phytocoenoses and soil are the principal factors determining the characteristics and productivity of the phytomass of a biogeocoenose. They are also most suitable for regulation and control.

Relationships of phytocoenoses with fauna

The relationships of a phytocoenose with fauna, unlike those with the atmosphere and the soil, are inherent in all plant communities on dry land and in the hydrosphere, although for the existence of phytocoenoses and their normal biogeocoenotic work many of these relationships are not essential and may be superfluous or even injurious.

The relationships are represented by both direct and indirect interactions, involving exchange processes and many other aspects of plant and animal life (pollination of flowers, dispersal of fruits and seeds, and formation of poisons and of various kinds of thorns; and in animals, mimicry, various structural adaptations, etc.).

From the biogeocoenotic point of view the greatest interest attaches to the relationships of phytocoenoses with phytophagous and saprophagous organisms, which transform organic matter created by green plants. The activities of saprophages living in soil and litter accelerate the decomposition of organic matter and its general circulation, enrich the soil with soluble compounds, and improve the physical characteristics of the soil (aeration, structure, water-permeability) and consequently also the conditions of mineral and water nutrition and of the growth and respiration of plants. Phytophages, which eat various parts of living plants or suck their juices, to some extent impede and complicate the life-activity of plants, lessen accumulation of organic matter, increase respiration, and increase the output of various protective substances (gum, resin, etc.), but in nature this activity 'on average' is not on a large destructive scale and does not upset the 'normal' equilibrium of biogeocoenotic work as a whole. Sometimes, however, phytophages concentrate in enormous numbers on certain parts of the earth's surface, and by their destructive activities bring both a phytocoenose and the whole biogeocoenotic exchange into catastrophic confusion (e.g., plagues of locusts in fields and orchards; mass production of larvae of Lepidoptera such as the gypsy moth in oak forests, and the Siberian pine moth in stone-pine forests where they totally destroy vast areas of stone-pine taiga); cockchafer larvae in young pine plantations; etc.). Such phenomena, although their existence is in full normal accordance with the laws of the general scheme of nature (of the biosphere), are quite unnecessary for the development of any specific biogeocoenose, are incidental, and break into the ordered structure of its life and exchange with a harsh dissonance. We do not speak here of the undesirability from the national-economic point of view, much less of the inadmissibility, of such disruptions of biogeo-coenotic exchange in phytocoenoses. In some places great destruction of plant communities is also caused by mammals (e.g., by elk (moose) to young pines, by mouse-like rodents to the seed-crops, shoots, and young saplings of tree species, by hares to young poplars and other trees), as a result of which the species-composition of a forest and the whole course of biogeocoenotic exchange may be substantially altered.

Besides the various direct effects of fauna on a phytocoenose, mainly consisting in the eating and transformation of vegetable matter, there is also a wide range of indirect effects of fauna on a plant community, among which fertilisation of the soil is prominent in some localities (e.g., in forests having bird colonies, and in pastures), also compaction of the soil (on tracks and in lairs of wild mammals, at watering-places, and in pastures) or the opposite process, loosening of the soil (e.g., by wild boars, moles, and earthworms). Occasionally these activities are accompanied by serious changes in the composition and structure of plant communities (e.g., by increase in the role of nitrophylls in forest soil cover, by appearance and spread of weeds). The activities of some kinds of fauna lead to excessive litter fall in communities, to increase of microdifferentiation of environment (e.g., ant-hills and mole-hills), and to creation of a mosaic effect in the horizontal structure of the biomass of phytocoenoses. In tropical forests some insects (termites) play a major role in the mineral nutrition of epiphytes, carrying a considerable amount of soil to a fair height on the trees (Voronov 1960).

The relationships between phytocoenoses and fauna are very dynamic and vary quantitatively and qualitatively in accordance with seasonal and meteorological fluctuations in the food value of the phytomass and its components, and also with seasonal and meteorological fluctuations in other components of the biogeocoenose. In temperate latitudes, the inter-actions of phytocoenoses and fauna as a whole are sharply restricted and simplified during the winter because of the migration or hibernation of a large proportion of the animal and insect population. On the other hand, at that time the contacts between the active fauna and phytocoenoses are profoundly altered, mainly because the animals' summer foods either are buried under snow or lose much of their nutritive matter, and they are obliged to change their diet to the above-ground parts of trees and shrubs that they do not eat in summer.

Like the relationships of phytocoenoses with other biogeocoenose components, those with fauna are not uniform throughout the whole extent of biogeocoenoses, and (not to mention differentiation between above-ground and underground relationships) are clearly differentiated by layers and zones.

The relationships between phytocoenoses and fauna naturally vary greatly from ecotope to ecotope, as was well demonstrated with regard to soil fauna by V. Ya. Shiperovich (1937).

Relationships of phytocoenoses with micro-organisms

The relationships of phytocoenoses with micro-organisms can be observed in any plant stand; they are very diverse, complex, active, and contradictory, and yet have been insufficiently studied. In territorial phytocoenoses they take place through the media of atmosphere and soil, either by contact or in various indirect ways; they may be positive or negative, obligatory or facultative, narrowly specific or universal.

These relationships are based on: the mutual exchange of metabolic products of higher plants and of micro-organisms in the nutrition process, either directly through cell metabolism (in symbionts and parasitic micro-organisms) or through soil solutions, water, and air; the transformation of conditions of life and work (of light regime, temperature, humidity, chemistry of the substrate, solubility of substances, gaseous composition of atmospheric air and air in the soil, etc.) for higher plants by micro-organisms and vice versa. Of great importance to higher plants in a phytocoenose is the activity of saprophytic soil-dwelling micro-organisms, which break down dead organic matter in litter and soil, and set free a variety of mineral substances for fresh use by the higher plants. This relationship is universal, is mutually indispensable for higher plants and for micro-organisms, and occupies a central place in the biological circulation of matter and energy. In different ecotopes, of course, this relationship finds different means of concrete expression, since in different environments life-activity develops in different ways and is performed by different organisms (both higher and lower) on different scales, with different intensities and different results. The external evidence of that diversity may be found in the quantity, quality, or composition of organic matter on the soil surface (litter, peat) or within the soil (thickness, colour, or structure of the humus layer).

From the biogeocoenotic point of view the symbiotic relationships of higher plants and micro-organisms are very important. They greatly intensify the exchange between a phytocoenose and the soil, and assist higher plants to procure from the soil not only water and mineral matter but also a certain amount of organic matter. In forests they are represented by extensive symbiosis of higher plants and fungi. They occur everywhere, but on the largest scale and most effectively on ecotopes where for various reasons (insufficient aeration of the soil, soil temperature, high soil acidity) the water and most of the mineral matter exist in forms unavailable, or available only with difficulty, for direct use by higher plants.

In such cases it is doubtful whether any of the higher members of the coenose could make the transition to a mycotrophic type of mineral-water nutrition. Numerous experiments have demonstrated that such symbiosis has a most beneficial effect on the life of the higher plants. With tree species, for instance, the growth of above-ground and underground organs is stimulated, the size and surface of leaves increase, and the plants become more resistant and fertile.

Also of value are the links between plants and micro-organisms that are able to fix molecular atmospheric nitrogen (nodule bacteria on the roots of leguminous plants, alder, and *Elaeagnus*). They make possible not only improved nutrition of the plants directly involved in that symbiosis, but also accumulation of nitrogenous compounds in the soil with the dying-off of the roots and of colonies of the nitrogen-fixing micro-organisms. In forests, however, this relationship is less widespread and significant than the mycorrhizal one.

The relationships between phytocoenoses and micro-organisms are closest, most varied, and most extensive in the root zones of higher plants, in the immediate vicinity of the source of the plants' various exudates into the soil. The numbers of micro-organisms in that zone are hundreds of times greater than those in soil zones farther from the rhizosphere, and create the impression of a microbial sheath, a dense envelope surrounding the plant roots and controlling the entire exchange between the roots and the soil (Krasil'nikov 1958). It is possible that this description exaggerates the filtering effect of the rhizosphere-sheath of micro-organisms, and that there are actually other direct contacts and exchange between the plant roots and the soil (Peterburgskii 1959). The concentrations of micro-organisms in the immediate vicinity of the receptive and secretory surfaces of the underground organs of higher plants are certainly significant in the exchange of matter between a phytocoenose and the soil. According to the experiments of N. G. Kholodnyi (1951b), 'the roots are actually fed by the cells of microbes existing in a state of lysis, of post-mortem solution, which, as is well known, consists mainly in fermentative breakdown of more complex organic substances into simpler ones by bacterial cells'. On the other hand, many kinds of exudates (including volatile organic matter) of the roots of higher plants are intensively absorbed by various specialised groups of micro-organisms in the rhizosphere.

All the interactions between a phytocoenose and micro-organisms are closely dependent on environmental conditions, and in nature differ in kind in different communities. The scale on which the processes take place, the composition of the interacting groups of higher plants and micro-organisms, and the course and results of the interactions also vary (Krasil'nikov 1958).

The great majority of experimental data on inter-relationships between phytocoenoses and micro-organisms relate to soil microflora and their interactions with the roots of higher plants, whereas the inter-relationships between phytocoenoses and micro-organisms in the atmosphere have as yet been subjected to little investigation.

PRINCIPAL PHYSIOLOGICAL PROCESSES TAKING PLACE IN PHYTO-COENOSES AND THEIR ROLE IN EXCHANGE OF MATTER AND ENERGY

The biogeocoenotic work of plant communities and their interactions with other biogeocoenose components are performed by means of physiological processes taking place in plant organisms. Therefore for better understanding of the place, role, and significance of phytocoenoses in biogeocoenotic metabolism, they must be analysed in detail in the matter-energy aspects of the principal physiological processes. These processes are four in number: photosynthesis, respiration (from these two together the balance of carbon compounds is derived), exchange of water, and exchange of mineral matter. The result of these processes is the growth of plants. The course of the processes of exchange of matter and energy in plant organisms has several peculiarities that distinguish them from corresponding processes in inanimate nature and in animal organisms. The processes of exchange of matter in living organisms require a special environmental characteristic: the presence in living matter of a definite, very labile, spatial structure.

The maintenance of that labile structure requires relative permanence of the internal environment of the organisms. The great ramification of the plant organisms (their large surface in proportion to their volume), however, resulting from the fact that contact of plants with their external environment and their absorption of matter and energy take place on their surface, makes it impossible for them to maintain absolute constancy of internal environment. For instance, plants cannot actively regulate their temperature, and can only relatively regulate the water regime of their tissues. Their metabolic processes therefore depend more on external conditions than do those of animals.

Evaluation of the biogeocoenotic work of physiological processes must take into account two very important aspects: (*i*) qualitative; (*a*) the role of the process in dispersal or accumulation of energy; and (*b*) the qualitative composition of products brought into circulation and of the final products of the reaction; (*ii*) quantitative (the amount of matter and energy brought into biological circulation).

BALANCE OF CARBON COMPOUNDS IN PHYTOCOENOSES

Photosynthesis

Photosynthesis is a fundamental process resulting in creation of organic matter. The total equation representing this process has the following form:

$$6CO_2 + 6H_2O + 674 \text{ kcal} = C_6H_{12}O_6 + 6O_2,$$

or, in other words, creation of carbohydrate molecules requires carbon dioxide gas, water, and external energy. Energy is absorbed by the green pigment of leaves (chlorophyll a and b) in the form of quanta of visible light. The light-energy absorbed by leaves and acting in the photosynthesis process (in the range of wave-lengths from 380 to 700 μ) has been given the name of photosynthetically active radiation, FAR. Gas-exchange takes place mainly through stomata and partly through very small openings in the leaf cuticle. Thus the receptive surface in photosynthesis is the leaf surface. Besides carbohydrates, other organic substances are formed during photosynthesis—proteins and lipoids; as a rule, however, carbohydrates predominate.

Of the energy absorbed by chlorophyll, only one-quarter is converted into the energy of chemical bonds. The remaining three-quarters of absorbed energy is dispersed in the form of heat. The reason lies in the complex, multi-stage reaction of photosynthesis, which is covered by the total equation of photosynthesis.

One of the stages in the transfer and accumulation of energy in photosynthesis is the formation of macro-energetic (i.e., energy-rich) phosphate bonds. That very important stage links photosynthesis with the conditions of phosphate nutrition and also, as will be shown later, with respiration.

Starting with the amount of solar radiation that reaches the earth's surface and is active in photosynthesis, we can calculate the maximum possible production of organic matter by all phytocoenoses on the globe. According to Kleshnin (1954), 70 kcal of total radiation (or 35 kcal of FAR) fall annually on 1 cm^2 of surface at the Equator. Vegetation absorbs 80% of that quantity; with 25% used in photosynthesis, 7×10^8 kcal per ha corresponds to assimilation of 178 ton/ha of cellulose from 275 tons of CO_2.

The average annual fixation of carbon over areas of land and water is considerably less than the calculated amount (Table 36).

Table 36. *Annual productivity of various areas on the earth* (*Rabinovich 1951*).

Plant habitat	Average annual fixation of carbon (ton/ha)	Carbon fixation reckoned as cellulose (ton/ha)*
Oceans	3·75	8·9
All dry land (average)	1·30	3·1
Forest	2·50	5·9
Arable land	1·48	3·5
Steppes	0·35	0·8
Deserts	0·04	0·1

* Calculated by us.

The low figures for annual increment of organic matter for the whole extent of oceans and dry land are partly due to the fact that areas not covered, or thinly covered, by vegetation are not excluded. But even for territory with dense vegetation, such as forests, the annual figure for annual production of organic matter by photosynthesis is approximately one-fifth of the possible figure. It follows that the percentage of use of light-energy (KPD) in photosynthesis is much lower than would be expected theoretically.

Values for KPD experimentally obtained by various authors in natural conditions amount on an average to 0·5-2% (Ivanov 1946, Rabinovich 1951, Nichiporovich 1955, Ovington 1961, Wassink 1959).

In very dim light, values of 15-20% are sometimes obtained for KPD of photosynthesis by terrestrial plants. The highest value obtained experimentally for KPD of photosynthesis by algae is 24%. High values are obtained for KPD of photosynthesis by farm crops with good agrotechnical treatments. Calculations show that during certain periods KPD for use of light in photosynthesis by beet crops reaches 7-9%, by potatoes 5%, and by barley 13·5% (Wassink 1959), but only for a short time.

In order to explain the marked discrepancy between the potentialities of the photosynthetic machinery of phytocoenoses and the figures actually obtained for annual productivity, let us examine the elements composing the total amount of synthesised organic matter. The total amount of organic matter formed by a phytocoenose during a definite period depends on: (*i*) the intensity of the exchange process, i.e., the amount of matter undergoing transformation under the influence of a unit of plant mass (surface) in unit time; (*ii*) the working time of the process; (*iii*) the working mass (surface).

These are general premises, applicable to every process in the metabolism of plants.

If we apply these specifically to the process of synthesis of organic matter, we obtain an equation suggested by L. A. Ivanov (1941a):

$$M + m = fPT - aP_1T_1,$$

where M = plant mass, m = dead parts, f = intensity of photosynthesis, P = working surface of leaves, T = time, a = intensity of respiration, P_1 = mass (living parts), T_1 = time.

In other words, the total organic mass formed by plants is the resultant of two oppositely-directed processes, photosynthesis and respiration. From the photosynthesis equation and the premise that in central latitudes 0·5 cal of physiologically active radiation (FAR) of wave-length 380-700 μ falls on 1 cm^2 of the earth's surface in the middle of a summer day, we may calculate that the highest theoretically-possible intensity of photosynthesis equals approximately 150 mg of CO^2 per 1 dm^2 of leaf surface per hour.

The maximum values for photosynthesis, obtained experimentally by Wielstatter and Stoll in artificial optimal conditions (temperature 25°C, CO_2 content 5%, and lighting 48,000 lux) fluctuate for different species between 17 and 80 mg/dm^2 per hour. In most cases the fluctuations in intensity of photosynthesis, due to species characteristics, were not large (30-45 mg/dm^2 per hour) and the extreme variations quoted above were seldom encountered. The highest values for photosynthetic intensity were obtained from agricultural plants.

In natural conditions the maximum intensity of photosynthesis does not as a rule reach the values given by Wielstatter and Stoll.

In Table 37, from Gabrielsen (1960), we present data for maximum values of photosynthesis, obtained by various authors in natural conditions. The highest figures in this table relate to mountain, Arctic, and farm plants, and the lowest to shade-tolerant forest species and tropical plants. No great differences are observed between maximum values of photosynthetic intensity of trees and herbaceous plants, as was also noted by Wielstatter and Stoll with regard to potential photosynthetic intensity. Among other groups of plants (not flowering plants) the lowest figures relate to mosses (0·5-2·3 mg/dm^2 per hour) and lichens (2·0-8·2), whereas with ferns and marine algae the intensity of photosynthesis fluctuates within the same limits as for plants listed in the table (7·5-11·2 and 1·8-14·9 respectively).

Table 37. *Maximum values of intensity of photosynthesis (in mg/dm^2 per hour) in natural conditions (Gabrielsen 1960).*

Species	Nature of leaves	Temperature (°C)	Intensity of photosynthesis
Sinapis alba	'Light'	20	26·4
Raphanus sativus	'Shade'	29	35·0
Oxalis acetosella	'Shade'	20	1·8
Oxalis acetosella	'Shade'	18	4·8
Statice limonium	'Light'	20	6·4
Chamaenerium latifolium	'Light'	20	15·4
Anemone nemorosa	'Light'	20	17·2
Alchimilla minor	'Light'	20	8·2
Alchimilla sp. and A. alpina	'Light'	14	41-90
Betula verrucosa	'Light'	20	6·8
Fraxinus excelsior	'Light'	20	11·0
	'Shade'		4·6
	'Light'		21·6
	'Shade'		7·4
Corylus maxima	'Light'	20	12·6
Cassia fistula	'Light'	28	12·6
Stelechocarpus burahoe	'Shade'	32	2·9
Phoenix dactylifera	'Light'	29	3·4

Verduin (1953) also summarised published data on intensity of photosynthesis in natural conditions; in many cases he used the same sources as Gabrielsen. According to his data, photosynthesis by trees and shrubs of the temperate zone is approximately 75% of that by herbaceous plants. That refers to wild species, since photosynthesis by orchard trees is as intensive as that by farm field crops. Data collected by both authors agree well with the views of O. V. Zalenskii (1954), who believes that maximum photosynthetic intensity is higher in plants growing in conditions of wide fluctuations in temperatures near the limits of their range, in mountains and in the Far North. In the humid tropics, where temperature conditions are extremely constant, maximum values for photosynthesis are low (Table 38).

Table 38. *Limits of variation of maximum intensity of photosynthesis and respiration in different plant species in different climatic conditions (mg/dm^2 per hour) (Zalenskii 1954).*

Region	No. of species investigated	Photosynthesis		Respiration	
		from	to	from	to
Tropics	28	0·9	11·0	0·1	0·9
Far North	17	3·8	21·3	1·8	10·5
Temperate zone:					
Forest and steppe zones	56	1·6	25·7	2·8	20·0
Subtropics	20	2·8	48·6	1·9	15·3
Deserts	26	1·5	68·5	9·0	26·4
High mountains	63	4·0	100·2	5·1	34·6

While maximum intensity of photosynthesis in natural conditions is below the possible figure, the average value usually lies still further below it. In different climatic zones changes in the average value of intensity of photosynthesis are less marked than changes in the maximum value. That is due to the fact that the highest maxima are found where photosynthesis fluctuates most widely, whereas in more constant and favourable conditions photosynthesis proceeds at a more uniform rate, but does not attain high intensity.

The first reason for low intensity of photosynthesis in natural conditions is shortage of CO_2 in the atmosphere. Studies of the dependence of photosynthetic intensity on CO_2 concentration have shown that in concentrations of from 0 to 0·1-0·3% by volume, intensity of photosynthesis varies directly with the CO_2 content of the air. At higher concentrations photosynthesis does not increase, i.e., the photosynthetic limit of saturation with CO_2 is reached. The average CO_2 content of the air layer nearest the earth in an open space is 0·03%, or from one-third to one-tenth of saturation. Artificial enrichment of the atmosphere with CO_2 produces a considerable rise in photosynthetic intensity. Within a phytocoenose the CO_2 content of the air may be much below the average figure.

On the other hand the soil, in which the processes of decomposition of organic matter proceed intensively, produces large amounts of CO_2. Since there is practically no wind in dense forests (Tkachenko 1955), the CO_2 remains in the lowest layers of the atmosphere, and during the sunlit hours of the day is assimilated only by local plants. Therefore by the end of the night, during which there is no assimilation, the CO_2 content of the lower air layers may reach saturation point.

Some data indicate that CO_2 content is especially high in the air in tropical forests, where the process of decay of organic matter in the soil proceeds most intensively. By day CO_2 is absorbed by leaves, particularly in the air layer containing the crowns of large trees. Therefore by day the CO_2 content of the air at crown level is much below the average. Calculations show that if there is not much air-mass movement leaves may absorb 20-25 % of the CO_2 in the air layers from 0-50 metres above ground level in a day. Consequently in phytocoenoses the CO_2 content of the air displays marked daily fluctuations, and considerable differences according to height above ground level, which are not found in open spaces. That is well illustrated by Huber's (1952, 1960) data, which are presented in Fig. 17.

Light is a factor that exerts strong influence on the intensity of photosynthesis. We have shown above that, with the amounts of solar energy

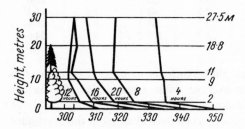

Fig. 17. Dynamics of CO_2 content of air in a forest at different heights above ground level (from Huber 1960).

that reach the earth's surface, photosynthetic intensity could be much higher than the observed figure. Thence we might draw the conclusion that light is not the chief factor regulating photosynthesis in natural conditions. We must not forget, however, that our calculations were based on the direct sunlight falling on plants in the middle of a summer day. If we consider weaker light conditions, in early morning and late evening hours, and also the possibility of plants overshadowing each other, our point of view must be changed.

The curve of dependence of photosynthetic intensity on light is identical in form with the curve of dependence of photosynthesis on CO_2 concentration; with low illumination, photosynthetic intensity is directly proportional to the amount of solar energy, but with high illumination it does not depend on that factor (region of light saturation).

The intensity of light at saturation level is one of two cardinal points on the photosynthesis-light curve. The other is the location of the compensation point, i.e., the point at which absorption of CO_2 in photosynthesis is equal to its exhalation in respiration, and net gas-exchange is zero. The location of the compensation point is mainly determined by the relative shade-tolerance of a species (Lyubimenko 1909). K. A. Timiryazev (1957) and Blackman (1905) examined the form and the causes of the curve of dependence of photosynthesis on external factors. The latter advanced the theory of limiting factors, according to which the region of saturation of the process is created by limitation of the rate of the process by another factor, which then was unknown. For instance, if saturation is reached on the photosynthesis-light curve, further increase may depend on the amount of CO_2 concentration. Gabrielsen (1960) has recently investigated the form of the curve of dependence of photosynthetic intensity on light, and has found that in the region of low light intensity the form of the curve is determined by active (chlorophyll) and inactive (anthocyan, carotinoid) leaf pigments, the structure of the photosynthetic organs, and the spectral composition of the incident light.

With light saturation the magnitude of photosynthesis is determined by the CO_2 concentration in the air, temperature, moisture supply, mineral nutrition, and the condition of the stomata and of the biochemical system of the leaf. The point at which light saturation is reached depends on the plant species (shade-tolerant or light-loving) and the structure of its photosynthetic apparatus: in 'shade' leaves (formed in weak light) it lies at about 4000-8000 lux (approximately one-thirtieth of full midday sunlight), and in 'light' leaves at about 40,000 lux (one-third of full sunlight); in optimal conditions, in only a few plant species is the point of light saturation as high as full sunlight or even higher.

Analysis of the curve of dependence of photosynthetic intensity on the amount of incident radiation and on CO_2 concentration indicates that in the past the earth's plants were adapted to atmospheric conditions different from those of to-day: to weaker light and higher CO_2 content of the air. Apparently these were the conditions of life on earth during the period when arboreal vegetation originated (Komarov 1961).

Therefore lack of CO_2 in the modern atmosphere is one of the principal reasons for the lowering of photosynthetic intensity. Light and CO_2, as is evident from the photosynthesis equation, are the factors taking direct part in the photosynthetic process and therefore directly affecting it.

Besides these, there are several other factors on which photosynthetic intensity depends, but which affect it mainly indirectly, through changes in the internal state of the organism. Among the chief of such factors we must include air temperature, water-supply conditions, and mineral nutrition.

It is a characteristic of the photosynthetic process that it is limited to a fairly narrow range of temperature, from −6 to 38-40°C (Ivanov and Kossovich 1930, Zeller 1951, Larcher 1961). Extremes of temperature not only have a depressing effect on photosynthesis but have a prolonged after-effect, due to damage to the photosynthetic apparatus (Semikhatova 1960). This leads to considerable shortening of photosynthetic working time. Only in the humid tropics can photosynthesis continue during daylight hours throughout the year. In other climatic zones it is interrupted: in hot dry climates by overheating of leaves for several hours every day or, in very arid districts, for the whole hot and dry period; in central latitudes the interruption covers a long frost period. A survey of the effects of temperature on photosynthesis in the temperature range where it is possible shows that with a 10°C rise in temperature the rate of assimilation approximately doubles, i.e., in chemical terms the temperature coefficient (Q_{10}) for photosynthesis equals 2; in purely photochemical reactions it is close to unity. The higher value of Q_{10} for photosynthesis as compared with other photochemical reactions is due to the multi-stage nature of the photosynthetic process and the participation in it of dark as well as light reactions.

The considerable effect of temperature on photosynthesis is the second reason for the actual intensity of photosynthesis being lower than the possible figure, since in most climatic zones of the earth temperature fluctuates widely, and for the greater part of the time is far from the optimum.

The third reason is unfavourable moisture conditions. Only in aquatic organisms is water supply always optimal and favourable, and largely on that account the oceans provide a better environment for synthesis of organic matter than does dry land (Table 36).

Leaves of terrestrial plants are always characterised by a small water deficit, which can be considerably increased either by increases in the drying effect of the atmosphere or by deterioration in water-supply conditions. This leads to lowering, and in extreme cases to complete cessation, of photosynthesis.

Finally, a fourth reason may be unfavourable mineral nutrition. We shall discuss that in the appropriate sections of this chapter.

Internal factors, such as the age of leaves, may have considerable effect on photosynthesis. A young leaf, which contains little chlorophyll, begins photosynthesis 3-9 days after opening (Kursanov et al. 1933, Stålfelt 1960). As it becomes greener and growth of the leaf blade ceases the intensity of photosynthesis rises. It is very probable that the absence of photosynthesis during the first days after opening is because in vigorously-growing leaves the absorbed light-energy goes into synthesis of energy-rich phosphate bonds, photo-formation of nitrates, and other processes, and is insufficient for assimilation of CO_2. The autumn yellowing of leaves lowers the intensity of photosynthesis.

There is a marked difference between the assimilation periods of the two plant groups, evergreen and deciduous. Although in the first group photosynthesis is interrupted by the unfavourable season of the year it is resumed as soon as that season ends, since the assimilating apparatus is retained. Replacement of leaves or needles in evergreens takes place gradually and leaves live for many years.

Perennial leaves and needles have less photosynthetic intensity than annual ones (Ivanov and Kossovich 1930).

The effect of age of pine needles on intensity of assimilation (in mg/g/hour) is seen from the following data:

	One year old		*Two years old*	
Branch layer, counting upwards	1	5	1	5
Intensity of photosynthesis	2·10	2·07	1·65	1·28
Branch layer	1	10	1	10
Intensity of photosynthesis	2·76	2·42	1·92	1·86
Branch layer	1	15	1	15
Intensity of photosynthesis	1·60	1·67	2·00	1·80

The age of the plant has equally strong effect on the intensity of photosynthesis (Table 39).

Table 39. *Effect of age of pines on rate of assimilation (Ivanov and Kossovich 1930).*

Light conditions	Nature of needles	Age of tree (years)	
		12	80
		Intensity of photosynthesis (mg/g/hour)	
S, direct sunlight	'Light'	2·10	1·18
S_0B_1, unbroken clouds	'Light'	1·62	1·15
In shade	'Shade'	0·17	0·24
Screened over	'Shade'	0·07	0·10

According to E. V. Yurina (1957), photosynthesis in 3- to 4-year-old trees is 1·5 times to twice as intensive as in 16- to 20-year-old trees in the same conditions. The intimate interaction of internal and external factors causes a characteristic daily and seasonal rhythm of the photosynthetic process in natural conditions.

Most investigators have noted that in conditions of adequate water supply, even in a cloudless sky with a gradual rise in intensity of solar radiation until noon and a fall towards evening, the daily course of photosynthesis very rarely follows a symmetrical single-peaked curve (Stocker 1956).

Usually a more or less marked depression of photosynthesis is observed at midday, as a result of which the curve has two peaks. If living conditions for the plants are unfavourable, the symmetrical two-peaked curve of photosynthesis is replaced by a single-peaked asymmetrical curve, with relatively high photosynthetic intensity during the first half of the day and very low intensity during the afternoon hours. Such curves are very characteristic of arid regions.

As Zalenskii (1956) has shown, the form of the curve and the intensity of photosynthesis may vary greatly in identical conditions, according to the species of the plants.

Climatic conditions also greatly affect the seasonal course of photosynthesis (Fig. 18).

In the taiga zone, where moisture is adequate (and in many cases excessive), photosynthetic intensity rises until the end of July and then falls towards the end of the season. That is an effect of the age of the leaves, as we have stated above, and also of increases in day-length, intensity of sunlight, and air temperature. But at the same time the intensity of photosynthesis may fluctuate violently and irregularly for short periods of time, because of the prevalence of a dynamic type of weather in that climate (Alekseev 1948) with a great number of cyclones and anticyclones, and sharp fluctuations in air temperature and in light and moisture conditions. In the forest-steppe and especially in the steppe zone maximum values for photosynthesis are observed during the first half of the vegetative period (beginning at the middle of June), as at that time water-supply and temperature conditions are most favourable for normal life-

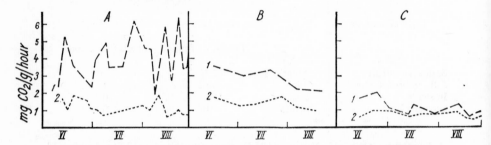

Fig. 18. Seasonal course of photosynthesis and respiration of leaves of tree species in different climatic zones (mg CO_2/g/hour) (from Ivanov, Gulidova, Tsel'niker and Yurina 1963). 1—actual photosynthesis; 2—respiration. A—Kadnikovskoe forest taiga; B—Tellerman forestry station, forest-steppe; C—Derkul' forestry station, steppe.

activity. During the second half of the summer lack of moisture and overheating lead to depression of photosynthesis, and in cases of very severe drought, which occur at intervals of a few years, to death of the photosynthetic apparatus.

In even more arid conditions, e.g., in Central Asia, as Zalenskii (1940) has shown, shedding of leaves during the dry season occurs every year. The same is characteristic of Mediterranean forests (Walter 1951). On the other hand, in tropical rain forests no seasonal rhythm of photosynthesis is apparent: the process continues at a constant rate throughout the whole year (Gessner 1960).

We have discussed the factors that can limit the productivity of phytocoenoses by affecting the rate and working time of photosynthesis. These factors can exert equally powerful influence on another element that determines the rate of growth of biomass in phytocoenoses: the surface of leaves per unit of area occupied by a phytocoenose (leaf-area index, in English terminology). As we shall see later, that index depends almost exclusively on the flow of solar energy to the surface of the earth. To understand the mechanics of the dependence, let us look at the characteristics of the light regime within phytocoenoses.

A single layer of leaves, regardless of plant species, arranged horizontally and covering the whole land surface, absorbs about 80% of the sun's rays between wave-lengths 380 and 700 μ (photosynthetically-active radiation), and approximately equal proportions (about 10% each) are reflected and pass through. Infra-red radiation is practically not absorbed at all, and therefore beneath the absorbing layer of leaves, the so-called infra-red shadow is formed. At the same time absorption of light within the FAR range differs for different wave-lengths, and red rays are absorbed more strongly than green. Consequently, the amount of light passing through a single leaf layer is only 10% of that falling on an open space, and moreover that light is richer in green rays, which are little used by plants in photosynthesis (Kleshnin 1954). As may be seen from the photosythesis–light curves for different plants, in these lighting conditions the organic-matter balance of the second leaf layer, which lies beneath the first, is still positive; the second leaf layer in turn lets through 10% of the light falling on it, i.e., 10% of 10%, or 1%, of full sunlight. After absorption of solar radiation by the third leaf layer, the fourth leaf layer receives only 0·1% of the FAR that falls on an open space.

The light conditions of the third and fourth layers of horizontally-arranged leaves are within the range near the compensation point of photosynthesis. For 'light' leaves, this corresponds to about 0·71% of full sunlight. As the shade deepens, however, the leaves modify their structure from 'light' to 'shade', and consequently the compensation point moves into the region of weaker light (to 0·1% of full sunlight). In addition, the nearer to ground level, the higher is the CO_2 content of the air, and this partly offsets the insufficiency of the light for intensive assimilation. Therefore in the fourth layer of horizontally-arranged leaves the organic-matter production balance may still be slightly positive or close to zero, and so these leaves may still survive and not die off. Thus the maximum area of leaves in a compact phytocoenose is determined by the position of the compensation point of photosynthesis for the lower leaves and by the amount of light passing through the upper canopy. From all these considerations it follows that the area of leaves in a compact phytocoenose, with horizontal arrangement of the leaves, may be four times as large as the area of soil surface on which it grows.

In phytocoenoses, however, horizontal arrangement of leaves is found only in the lowest leaf layer, with the topmost leaves growing at an angle close to the vertical. Consequently the passage of light through the topmost layer is greatly increased, and so there is also an increase in the number of leaves able to use sunlight for photosynthesis with a positive organic-matter balance.

Data on the passage of light, varying according to the angle of inclination of the leaf plane to the horizontal, have been obtained experimentally by Blackman (Blackman and Black 1959), and also by Gulyaev and his colleagues in Nichiporovich's laboratory with regard to farm plant crops. In Fig. 19 we present a diagram taken from Blackman, illustrating his findings. Besides the angle of leaf inclination, as the diagram shows, the relative vertical arrangement of leaves is also significant. It is evident that the leaf-area index for taller plants, even with the leaves arranged horizontally, is somewhat greater than for shorter plants. In addition, naturally, the more FAR a given area receives the greater will the index be, i.e., it increases towards the Equator and decreases towards the Poles.

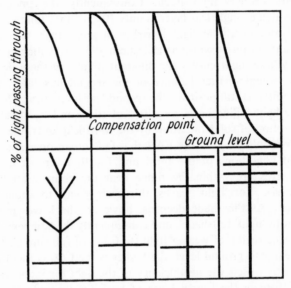

Fig. 19. Illumination at different heights in crops of farm plants, with relation to the spatial arrangement of the leaves (from Blackman and Black 1959). Upper curves—percentage of light passing through the crop at different heights; lower curves—diagram of arrangement of leaves on plants.

Thus as phytocoenoses become more compact these things happen simultaneously: increase of the total surface of the assimilating apparatus; differentiation of the leaves or needles into 'light' and 'shade'; and, finally, lowering of the average intensity of assimilation through the overshadowing of the lower leaves. Since the total productivity of a phytocoenose is the product of the total surface of the assimilating apparatus and the intensity of assimilation, the most productive phytocoenoses are those with some optimal value for total leaf surface. Numerous data compiled by Nichiporovich and his colleagues (1961) indicate that in the most productive farm crops that optimal leaf surface is four to five times the area of the land the crop covers (on 1 hectare of crop it amounts to from 40,000 to 50,000 square metres of leaves).

In a forest the adaptation of plants to making the most profitable use of incident light-energy varies with layer-differentiation. Growth dynamics are clearly shown in the formation of a phytocoenose: in young saplings after formation of the top layer (pole stage) there is only one layer, the others being absent because of lack of light. On the lower branches of the trees 'shade' leaves (or needles) are then formed, possessing different characteristics: lower respiration intensity and a lower illumination level at the compensation point.

N. A. Khlebnikova's (1961) data provide a good illustration of the above point. According to her observations, in compact 17-year-old pure pine stands, at a temperature of 20-25°C, respiration by the needles in the topmost parts of the crowns of trees in growth-class I was 1·1 mg/g/hr, and the compensation point lay at 0·07-0·08 cal/cm^2/min (about 700-800 lux). In needles in the lower parts of the crowns respiration was 0·36 mg/g/hr, and the compensation point was 0·01 cal/cm^2/min.

The proportions of 'light' and 'shade' leaves in a tree stand depends on the degree of its compactness. According to M. Ya. Oskretkov (1953), with a density of 0·4-0·5 'light' needles in a 33-year-old pine constituted 57% of the whole needle surface, and with a density of 1·0-1·1, 49%. A predominance of 'shade' leaves or needles on trees is very unproductive, since intensity of assimilation with saturated illumination (and consequently the maximum possible figure for assimilation) is much lower in them than in 'light' leaves or needles (Table 40).

Table 40. *Rate of photosynthesis in pine needles with different lighting conditions* (*Ivanov and Kossovich 1930*).

Lighting conditions	'Shade' leaves		'Light' leaves	
	Intensity of photosynthesis (mg/hour)			
	per gramme	per 50 cm^2	per gramme	per 50 cm^2
Zone of dying-off	0·49	0·47	0·21	0·31
Diffused light	1·84	1·77	1·18	1·73
Full sunlight	1·42	1·36	1·74	2·55

In poor light conditions where the organic-matter balance for the leaves becomes negative the branches die off (clean-trunk zone). With further growth of the stand, some trees of the same species fall behind in growth and die beneath the canopy. Differentiation of the stand by growth classes takes place. Trees of growth-class III, if they are in bad light conditions, assimilate poorly; their organic-matter balance becomes negative, and as a result they perish.

Data on the productivity of trees of different growth-classes were collected by Boysen-Jensen (1932), one of the first investigators of the organic-matter balance in forests. He provides the following data (Table 41) for a 12-year-old ash forest in Denmark.

Trees of another species, more shade-tolerant than the dominant species, may survive beneath the canopy and form a second storey. As

Table 41. *Productivity of trees of different growth-classes.*

Item	Growth-class			Total
	I	II	III	
Leaf surface, m²/ha × 10³	27·74	17·07	8·07	52·68
Gross productions, tons/ha	6·35	3·33	0·80	10·48
Production per m² × 10³ of leaves*	0·23	0·19	0·09	—

* Calculated by us.

the tree canopy grows older and thins out, shrubs and shade-tolerant herbage become established below these, forming third and fourth layers in the phytocoenose. Thus at different age stages the assimilating surface of the phytocoenose consists of different elements, shown diagrammatically in Fig. 20. Plants of the lower layers live in conditions of somewhat

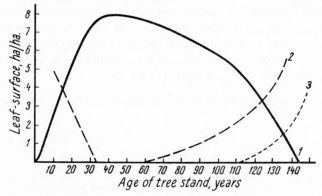

Fig. 20. Diagram of age dynamics of leaf-surface of different layers of forest phytocoenoses. 1—dominant tree layer; 2—herbaceous cover; 3—undergrowth.

higher CO_2 concentration than is found in the upper layers, and in a much-altered light regime. Since the topmost canopy (formed of tree crowns) is nowhere continuous and always has gaps, patches of sunlight fall on the leaves of the lower layers, moving as a result of the sun's changing position during the day and of the swaying of branches. Therefore the location of light in the lower layers of the forest is constantly changing, which doubtless affects the course of life of the phytocoenose. Unfortunately this question has not been studied. Whereas in the patches of sunlight there is no change in spectral composition, but only weaker illumination, in the shade (as we have seen) the light that penetrates is richer in infra-red, and to some extent in green rays. At the same time light that filters through leaves into the shade is blended with light from the blue sky coming through gaps in the canopy, and therefore the light in the shade is enriched with blue rays. The degree of enrichment depends on the area of the gaps, or, in other words, on the compactness of the upper-storey crowns. Consequently, while we may speak of 'infra-red' or 'green' shade light in a compact forest, in a more open forest the shade light is blue. This was first pointed out by Seybold (1936), and investigated in detail by Akulova, Khazanov, Tsel'niker, and Shishov (1964).

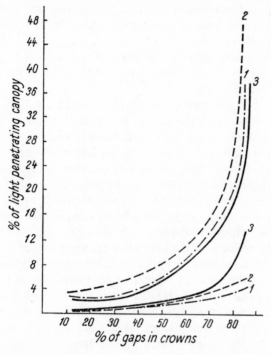

Fig. 21. Amount and spectral composition of light beneath forest canopy, with relation to compactness of canopy (from Akulova, Khazanov, Tsel'niker and Shisov 1964). 1—red rays; 2—green rays; 3—blue rays; upper curves—in diffused light (S_0), lower curves—in direct sunlight (S_4).

In Fig. 21 we give a diagrammatic representation of the results obtained by these authors, which applies to the light regime beneath a forest canopy with direct sunlight. If the sun is covered with clouds, the light under the canopy is unchanged in spectral composition, shadows disappear, and the whole area receives almost uniform lighting; finally, the percentage of incident light passing through the crowns is considerably increased. Fluctuations in light intensity due to the movement of clouds and the obscuring of the sun are much smaller in a forest than in an open space. The percentages of sunlight that pass through a forest canopy in different sunlight conditions are given in Fig. 21.

The quantity and quality of the light penetrating the canopy control the leaf area in the lower layers of the forest. This regulation is accomplished by the dying-off of branches and entire plants and by the direct effect of the quantity and quality of light on growth processes. For instance, data exist showing that direct sunlight with abundant red and yellow rays impedes growth of tree-tops and promotes side-branching, whereby extensive leaf surface is developed; diffused light of low intensity, with predominance of blue rays, has the opposite effect: it favours growth of the tops and restricts appearance of side branches (Leman 1961, Kroker 1951).

This explains the umbrella-like form of tree crowns in tropical rain forests and the pyramidal form of the same species in the lower storey (Richards 1961) (Fig. 22).

Not only the light regime during the vegetative period, but also its seasonal dynamics are important in the regulating role of light. In this respect a forest whose upper storey consists of evergreens differs greatly from a deciduous forest. Whereas in the former absorption of light by the lower storey varies little throughout the year, in the latter two stages are clearly distinguished in the annual cycle; the so-called 'light' stage, when tree leaves are absent and light is absorbed only by the woody parts of trees, and the 'shade' stage, when leaves are present. The understorey plants may exhibit other adaptations to the light conditions, for instance, their growing period is at the time when the upper storey is leafless. This has been well demonstrated by V. A. Kozhevnikov (1950). Similar investigations of the seasonal rhythm of photosynthesis in forest herbage were made by Daxer (1934). He discovered that for herbs with leaves that survive under snow in winter (common in deciduous forests) the maximum synthesis of organic matter takes place when the upper storey is leafless. After the leaves emerge on the trees the intensity of photosynthesis falls sharply, so that the organic-matter balance is almost zero. Tranquillini (1960) has made similar statements.

In deciduous forests of temperate climates most shrubs, as well as herbs, are adapted to existence under canopy in that their leaves open earlier and fall later than those of the trees (Byallovich 1954). Table 42 clearly shows the effect of the state of the upper storey on photosynthesis by plants in the lower layers. After the appearance of leaves in the upper storey photosynthetic intensity in the herbage falls, but the percentage use of light-energy in photosynthesis increases. Therefore shrubs and herbage can survive in light conditions in which an undergrowth of tree species dies off. Survival of the latter beneath the maternal canopy is possible only if the canopy is reduced by thinning or if the undergrowth possesses a much higher degree of shade-tolerance than the parent trees.

In summary, because of the great height of a forest phytocoenose, and the presence of layering and of plants of various species differing in spatial arrangement of leaves, in degree of shade-tolerance, and in rhythm

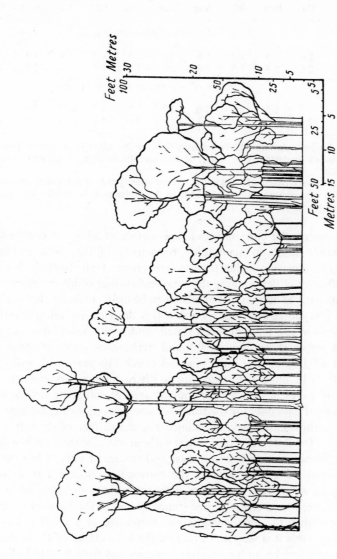

Fig. 22. Profile diagram of virgin tropical rainforest (Richards 1961).

Table 42. *Rate of photosynthesis (mg/g per hour) in plants of different layers, and percentage use of solar energy in photosynthesis (from data of I. S. Malkina) (Serebryano-borskoe forest, Moscow province).*

Layer	1				2					
	May	June	July	Aug	May	June	July	Aug	Sept	Oct
	Intensity of photosynthesis									
Trees and shrubs	2·8	1·9	1·5	1·6	1·2	2·5	1·8	1·3	1·3	—
Herbage	2·6	0·8	1·5	1·8	2·1	1·9	1·0	1·0	2·0	1·6
	% of use of solar energy in photosynthesis									
Trees and shrubs	1·2	1·8	2·1	2·0	1·7	1·7	3·7	3·5	3·0	—
Herbage	2·6	10·2	16·2	18·1	—	2·0	6·1	7·7	6·5	7·6

Notes: 1. Linden-oak-hazel. Crown density 0·5, age 70 years. First storey consists of oak and linden, understorey of hazel; in the herbaceous layer sedge and lungwort predominate.

2. Pine-oak-hazel, 110 years old, crown density 0·4. First storey consists of pine, second of oak, understorey of hazel; in the herbaceous layer bilberry and lily-of-the-valley predominate.

of development, the optimum total leaf-surface of all forest components (which would result in maximum productivity of the work of photosynthesis) is somewhat higher than in farm crops. Unfortunately, because of the difficulty of determining the total leaf-surface of all the plants in a forest, no data of this kind have been published. Only in the work of Pavlova, Davydova, and Tsel'niker (1964) is the total assimilating surface of the plants in two types of forest stated: linden-oak-hazel forest 70 years old, with upper storey density 0·5 and herbaceous cover of sedge and lungwort (70,000 m²/ha), and pine-oak-hazel 110 years old, with pine density 0·4. In most published works where reference is made to the amount of leaves or needles per unit of area of stand the investigators have recorded data only for the tree layer and seldom for the herbage and shrubs. In most cases they determined not the surface of the leaves but their mass. The latter, it is true, is not without value, since there is a fairly close rectilinear relation between mass and surface: 1 dm² of leaf surface weighs about 1·5 g. For needles the conversion coefficient is about 2. In Table 43 we present data from different authors on the amounts of leaves and needles in compact tree stands. Most of the data relate to the pole stage, when the phytocoenose was composed of a single tree layer.

From the table it is seen that as a rule the leaf-surface of the tree layer varies from 4 to 6 ha/ha for deciduous species and from 9 to 12 ha/ha for conifers. The difference in leaf-surface between coniferous and deciduous species is explained by the perennial life of needles and because with deciduous species the calculation of leaf-surface is usually made for only one side of the leaf, whereas with needles, which are not flat leaf blades, the surface of all sides is taken into account. This is correct if one has to calculate the needle-surface that can photosynthesise, but quite wrong if one has to calculate the amount of shade produced by the needles in the lower layers, since the shade is proportional not to the surface of the needles but to their projection on the underlying surface. Calculation of the area of the projection-surface of needles in coniferous forests shows that it is very close to the area of the projection-surface of leaves in deciduous forests.

A second point to be noted in Table 43 is the absence of marked differences in the surface areas of the assimilating apparatus of stands in different zones. This is easily understood, since the limiting factor in

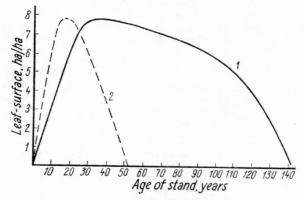

Fig. 23. Age-dynamics of leaf-surface of stands in temperate-climate forests. 1—forest zone; 2—steppe zone.

development of leaves is light, and in the summer months the amount of radiation falling on 1 cm^2 per minute varies little with differences in the geographical latitude of places (Kalitin 1938).

One might expect a considerable increase in leaf-surface in tropical rain forests because of the greater incident radiation and working time of the assimilating apparatus. Apparently that is not so. Although a tropical rain forest is considerably taller than a temperate-zone forest and the entire height is filled with leaves, its upper layer is thinner, light penetrates deeply into the stand, and the amount of foliage is not substantially greater than in a temperate-zone forest (Richards 1961).

Total leaf-surface never reaches optimal amounts in conditions where growth is limited not by light but by other factors, and where formation of a compact plant community is impossible. For instance, the arid regions of Central Asia are characterised by tree stands of low density.

Examples are also available where soil and climatic conditions favour the formation of plant (especially forest) coenoses, which during their early years grow well and may become compact; but as a result of the life-activity of the phytocoenose its living conditions deteriorate. In such circumstances the life of the closed community is brief, and the maximum

Table 43. *Leaf mass (tons/ha) and leaf surface (ha/ha) in compact stands in different forest zones.*

Zone	Stand	Density	Age (years)	Place of investigation	Leaf mass	Leaf surface	Author
				Deciduous species			
Forest zone	Birch-bil-berry		30	Germany	6·6	5·0	A. A. Molchanov 1952 Polster 1950
	Birch	1·0	22	Near Moscow	4·9	—	V. V. Smirnov 1961
	Birch	1·0	24	Chuvash A.S.S.R.	12·3	8·5	M. D. Danilov 1953
	Birch	—	—	Denmark		4·6	Ellenberg (quoted by Walter 1951)
	Poplar			N. America, Cordilleras, 2700 metres above sea level		8·2	Vareschi (quoted by Gessner 1960)
	Poplar	1·0	25	Chuvash A.S.S.R.	6·9	4·1	M. D. Danilov 1953
	Oak		20	Chuvash A.S.S.R.	8·3	5·5	M. D. Danilov 1953
	Oak-horn-beam			Denmark		8·9	Ellenberg (quoted by Walter 1951)
	Beech		65	Denmark	7·9	5·5	Møller 1945
	Linden		40	Chuvash A.S.S.R.	5·7	4·1	M. D. Danilov 1953
	Ash	1·0	12	Denmark		5·4	Boysen-Jensen 1932
	Oak with ash		25	Voronezh province	8·0	4·5	P. B. Raskatov 1940

Average for forest zone, 5·8 ± 0·5

Zone	Stand	Density	Age (years)	Place of investigation	Leaf mass	Leaf surface	Author
Forest-steppe	Oak		40	Voronezh province, Tellerman forest	8·2	5·5	N. F. Polyakova 1957
	Oak		15	Voronezh province, Tellerman forest	4·4	4·1	T. A. Alekseeva 1957 a, b
	Oak with maple		15	Voronezh province, Tellerman forest		4·7	T. A. Alekseeva 1957 a, b
	Poplar		25	Voronezh province, Tellerman forest	10·0	5·1	V. V. Smirnov 1957

Average for forest-steppe zone, 4·9 ± 0·1

Zone	Stand	Density	Age (years)	Place of investigation	Leaf mass	Leaf surface	Author
Steppe	Oak with maple		30	Belye Prudy	8·0	5·0	I. V. Gulidova 1955
	Oak with ash	1·0	17-20	Lugansk province, Derkul' station	8·0	5·0	Yu. L. Tsel'niker 1958a
	Oak		20	Volgograd province		3·8	L. K. Serebryakova 1958
	Maple		40	Volgograd province	8·0	4·5	L. K. Serebryakova 1958
	Oak with ash		14	Nikolaevsk province	6·7	6·3	E. G. Kucheryavykh 1954
	Oak with ash		24	Nikolaevsk province	12·9	10·5	E. G. Kucheryavykh
	Oak with ash		23	Lugansk province, Derkul' station	6·4	—	N. F. Polyakova-Minchenko 1961
	Oak with shrubs	0·9-1·0	15-16	Rostov province	10·5	—	V. F. Kol'tsov 1954
	Oak with ash and shrubs		4	Rostov province	5·88	—	V. F. Kol'tsov 1954
	Oak with ash	0·8-0·9	56-57	Rostov province	12·2	—	V. F. Kol'tsov 1954

Average for steppe zone, 5·8 ± 0·9

Table 43 (continued)

Zone	Stand	Density	Age (years)	Place of investigation	Leaf mass	Leaf surface	Author
				Coniferous species			
Taiga	Pine		33	Moscow province, Prokudin forest		9·3	A. A. Molchanov 1949
	Pine		27	Chuvash A.S.S.R.		9·7	M. D. Danilov 1953
	Pine		12	Moscow province	12·6	6·0	A. V. Savina 1949
	Pine		24	Bryansk province		9·0	Oskretkov 1954
	Larch with stone pine			Germany	12·5		Polster 1950 (quoted by Gessner 1960)
	Larch with stone pine			Alps, 1800 metres above sea level		11·7	
	Spruce		37	Vologda province		11·3	V. V. Smirnov 1961
	Spruce		37	Leningrad province	20·8		A. L. Koshcheev 1955
	Spruce			Germany	31·0		Polster 1950
	Douglas fir plantation			Germany	40·0		Polster 1950
	Larch			Germany	13·9		Polster 1950

Average for taiga zone, $9·5 \pm 0·4$

possible surface of the photosynthetic apparatus lasts only a short time. Such leaf-surface dynamics are observed in artificial tree plantations on the steppes. We present in Fig. 23 a diagram of the age-dynamics of leaf-surface of tree stands in different zones.

To summarise this section devoted to photosynthesis, we may remark that:

1. The biogeocoenotic role of the photosynthetic process results in the creation of organic matter and in the accumulation of solar energy in chemical compounds.

2. The amount of energy accumulated by photosynthesis is, on an average, from 0·5 to 2% of that reaching the earth's surface. Only in the centres of oceans and forests is the percentage of accumulated energy substantially higher (about 9% for oceans and 5% for forests).

3. The amounts of matter and energy assimilated during the photosynthetic process are regulated in more favourable conditions by the intensity of photosynthesis, and in marginal living conditions by the area of the receiving surface.

4. The principal factor that limits the rate of photosynthesis in terrestrial plants is shortage of CO_2 in the atmosphere.

Respiration

Whereas from the biogeocoenological point of view photosynthesis is a positive item in the organic-matter balance, respiration is a negative item. The total equation of the respiration process, like that of the combustion process, is:

$$C_6H_{12}O_6 + 6O_2 = 6CO_2 + 6H_2O + 674 \text{ kcal}$$

The freed energy is released in the form of heat. Biological oxidation, i.e., respiration, is summarised by the same equation as combustion. Analogy between the two processes, however, is far from complete. Above all, respiration is slow oxidation. The multi-stage nature of the process causes the release of energy in respiration to take place gradually, and the energy may be used in biosynthesis accompanying respiration. As a result substances with a higher level of energy than carbohydrates, e.g., proteins and fats, are formed. But the most important function of respiration is maintenance of the structure of living tissue by means of the energy released in respiration. Only part of the energy released in respiration is converted into heat, causing a loss of energy. The proportions of 'useful' and 'useless' respirational energy depend on the state of the plant (Zholkevich 1961). The chief role in energy transfer during respiration involves macro-energetic phosphate compounds, as in photosynthesis. Respirational energy is not required for active absorption of water and mineral matter by roots, which we shall discuss below.

The connection between respiration and all other processes involved in the conversion of matter within the organism is not only through transfer of energy but also through creation of intermediate products, which in most cases are products of the incomplete oxidation of carbohydrates. They include organic acids, converted into each other by a number of reversible reactions (the so-called Krebs cycle). Amination of these organic acids converts them into amino acids, thus linking carbohydrate and protein metabolism. A number of indirect secondary conversions, on the other hand, link carbohydrate and lipoid metabolism. The respiration process, unlike photosynthesis, possesses no special organ. All living tissues of a plant breathe, although with different intensities. The relation between the processes of photosynthesis and respiration determines the productivity of a phytocoenose.

We have discussed the significance of respiration by leaves in determining the degree of shade-tolerance. Here we need only say that the dependence of the rate of respiration on external and internal factors is different from that for photosynthesis. Therefore in different external conditions there may be variations in the comparative intensity of the two processes, and therefore in the degree of shade-tolerance of a species. Plant respiration depends on temperature much more than does photosynthesis. The temperature coefficient for respiration varies mainly between the limits 3-4 (James 1956). The temperature range is wider for respiration than for photosynthesis, and perceptible respiration is observed at temperatures of $-12°C$ and $-15°C$, and also above $40°C$. This is shown in the location of the compensation point: at high temperatures it lies at a higher level of light intensity than at low temperatures (Pisek and Knapp 1959). The water supply of plants greatly affects their respiration: excessive water supply (which often produces anaerobic conditions, e.g., in flooded soil) depresses it, and insufficient water supply in its early stages of action increases the rate of respiration. But in that case the KPD of respiration changes abruptly: the energy produced by respiration ceases to be used in biosynthesis and is uselessly dispersed in the form of heat (Zholkevich 1961). Data exist to show that lighting conditions affect the rate of respiration by leaves (Zalenskii 1957, Voskresenskaya 1961), but opinions differ.

As a result of the simultaneous action of many external and internal factors on photosynthesis and respiration there is a regular variation in the relation between photosynthesis and respiration in different climatic zones (Table 44).

As conditions become hotter and drier the rate of photosynthesis falls and the share of respiration in gas-exchange rises, producing a negative effect on the organic-matter balance. Such a relationship is characteristic of continental climates, but in tropical rain forests respiration is much less than photosynthesis.

The rate of respiration, like that of photosynthesis, fluctuates regularly during the course of the year, the fluctuations being partly due to the annual rhythm of meteorological conditions and partly to the state of the plant. For instance, Pollock (1953) discovered that maple buds and trunk wood breathe much more intensively in April than in October and November even though the temperature is the same.

The ending of the quiescent period and the beginning of growth processes produce an increase in respiration. The connection between rate of respiration in buds and in the points of growth of shoots and the age of the shoots is shown in Fig. 24.

A considerable increase in respiration by evergreens in spring or early summer, during the period of growth of new shoots, leads to a decrease in the amount of photosynthesis observed (Neuwirth 1959) (Fig. 25). The average rate of respiration by leaves and needles during the growing

Table 44. *Average rate of actual photosynthesis and respiration in different forest-growing zones (mg/hour) (Ivanov et al. 1963).*

Zone	Photosynthesis				Respiration				Share of respiration in gas-exchange (%)			
	White birch	English oak	Red-leaved ash	Norway maple	White birch	English oak	Red-leaved ash	Norway maple	White birch	English oak	Red-leaved ash	Norway maple
Central taiga	4·8	—	—	—	0·5	—	—	—	10	—	—	—
Coniferous-deciduous forest	—	3·1	3·2	2·7	—	0·7	0·5	0·5	—	23	16	19
Forest-steppe	—	2·6	—	1·4	—	1·0	—	0·6	—	39	—	43
Steppe	—	1·2	1·5	1·1	—	0·6	0·5	0·6	—	50	30	55

period depends largely on the specific characteristics of plants, as is shown
in data obtained by Polster (1950):

Species	Intensity of respiration, mg/g/hr
Birch	2·01
Oak	1·64
Beech	1·00
Pine	0·78
Larch	0·73
Douglas fir	0·64
Fir	0·46

It is characteristic that in the above list of species, arranged by rate of
respiration, broad-leaved species occupy the first places and conifers the
last. This is also true for other physiological processes: photosynthesis
and transpiration.

Organic matter is used up in respiration by other living parts of plants
as well as by leaves: roots, branches, and trunks or stems. Data on rate
of respiration by trunks and branches, however, are very scanty in the
literature. The intensity of respiration by roots of tree species was investigated by Eidmann (1943, 1950). We present data from his work in
Table 45.

Table 45. *Rate of respiration by roots (Eidmann 1950).*

Species	Intensity of respiration	
	mg/g of dry weight per day	mg/g of fresh weight per hour*
Coniferous species		
Larix europaea	63·4	0·26
Pinus silvestris		
from East Prussia	62·3	0·26
from Rhine valley	48·4	0·20
Pseudotsuga douglassi	39·9	0·17
Picea excelsa	29·0	0·12
Abies pectinata	17·8	0·08
Deciduous species		
Populus canadensis	381·8	1·59
Betula verrucosa	108·4	0·46
Populus tremula	83·4	0·35
Alnus glutinosa	74·5	0·31
Tilia parvifolia	53·7	0·22
Carpinus betulus	43·6	0·18
Acer platanoides	30·8	0·13
Fagus silvatica	30·8	0·13
Quercus pedunculata	26·6	0·11
Quercus sessiliflora	20·9	0·09

* Calculated by us, assuming that dry weight of roots is 10% of fresh
weight.

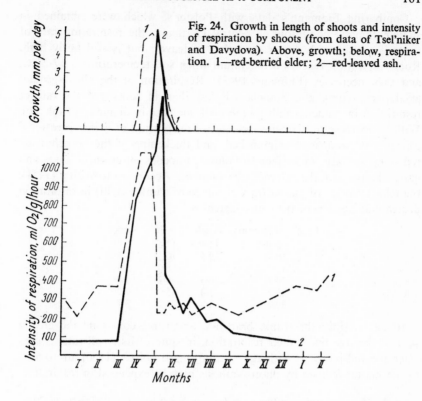

Fig. 24. Growth in length of shoots and intensity of respiration by shoots (from data of Tsel'niker and Davydova). Above, growth; below, respiration. 1—red-berried elder; 2—red-leaved ash.

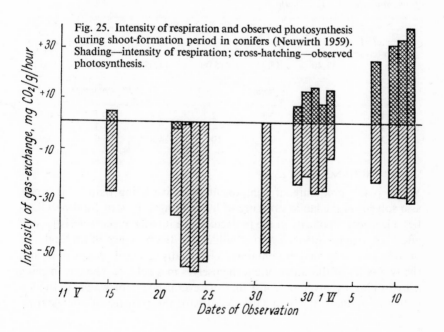

Fig. 25. Intensity of respiration and observed photosynthesis during shoot-formation period in conifers (Neuwirth 1959). Shading—intensity of respiration; cross-hatching—observed photosynthesis.

Comparing Eidmann's data with Polster's, which were obtained in similar climatic conditions in central Germany, the respiration rate of roots is considerably lower than that of leaves, but is still fairly high. Root respiration reaches a maximum when soil temperature is 17-22°C, and then decreases (Eidmann 1943). Respiration in the above-ground axial parts (trunk and branches) is less than in roots and the rate of respiration in branches falls steeply with increase in diameter (Table 46). With an increase in diameter the relative proportion of living 'breathing' elements in branches is diminished, and thickening of the branches, by reducing the ratio of surface to volume, impedes diffusion of gases and gas-exchange with the external environment. According to Walter (1951), the relative losses of matter (in percentage of trunk weight) in respiration decrease as the size of the trunk increases:

Height of tree (m)	Beech	Ash	Spruce
5·9	11·0	12·5	8·0
10·9	9·1	10·1	6·6
15·9	7·6	8·0	5·6
20·9	6·3	6·2	4·8
25·9	5·4	5·0	4·2
30·9	4·8	4·2	3·7

However, since the trunk, branches, and roots constitute the greater part of the tree mass, their respiration, in spite of its small rate, is considerable and increases with age. According to Polster (1950), up to 60% of the matter formed by photosynthesis may be expended in respiration.

Table 46. *Rate of respiration (mg/g per hour) in linden branches of different diameters (from data of A. M. Yakshina).*

Diameter of branch (cm)	Time of day					
	9.40-11.40	11.54-13.54	14.09-15.09	9.35-11.05	11.18-12.49	13.02-14.32
	9 August			13 September		
8·4	3·3	2·3	2·7	0·8	0·4	0·5
5·3	—	—	—	2·2	1·3	1·5
3·9	10·0	7·0	10·5	4·0	6·0	2·9

WATER-EXCHANGE PROCESSES

Water is the predominant component of active living tissues. It serves as a solvent in the labile structure of living things. Its structural peculiarities, electrical bipolarity in the molecule, give it specific characteristics and affect its specific role as a tissue-builder. In the presence of molecules of protein and other matter containing electrically-charged groups of atoms, the two poles of the water orient themselves in a definite manner, forming a hydrophilous envelope consisting of several layers of water molecules.

In the opinion of Czent Györgyi (1960), the transfer of energy from activated molecules of chlorophyll to macro-energetic phosphate compounds and thence to other substances can take place only because it is possible for molecules that are acceptors and donors of energy, in a slowly-stimulated (triplet) state, to pass over in an environment of water with appropriate structure.

Hence the rate and direction of exchange processes are to a great degree regulated by the amount and state of water in the cell. Normal life-activity of an organism and a positive organic-matter balance in it are possible only with a high level of water-saturation of the sub-microscopic structure of the cell. Because of the drying action of the atmosphere, a constant flow of water through the plant, with an equal supply of water to the roots from the soil and output through the leaves, is necessary in order to maintain the requisite water level in a terrestrial plant. In this way plant communities serve as intermediaries between the soil and the atmosphere, pumping great masses of water from the soil and converting it into vapour. Because of the deep penetration of roots into the soil, the soil layers participating in moisture-exchange with the atmosphere extend deeper in a plant community than in areas without vegetation. Relative constancy of the water regime in plant tissues is an adaptation developed in the course of evolution by the higher groups of the plant kingdom.

The amount of water required by plants may be expressed in two equations. Using transpiration as the basis, the equation is:

(1) $W_1 = iPT_2 + \Delta Mw + M\Delta w,$

where W_1 = amount of water;

P = working surface (the same as in photosynthesis);

i = average intensity of transpiration;

T = time;

M = mass of the whole plant;

w = its moisture content;

ΔM and Δw = change in the plant's mass and moisture content during time T.

Thus the equation by which we may express the amount of water required by a plant includes the amount of evaporated water and the amount enclosed in the plant's tissues. If we take the absorption of water by the roots as the basis of the equation, the amount of absorbed water is given by:

(2) $W_2 = lP_2T_3,$

where P_2 = absorbing surface (of the roots);

l = average intensity of absorption;

T_3 = time.

In all cases $W_1 = W_2$.

We shall now pass to a more detailed examination of these equations.

Water supply

As is seen from the second equation, the amount of water absorbed by the roots is determined by the rate of absorption, the absorbing surface, and the time interval when absorption occurs.

Let us first consider the dimensions of the absorbing surface.

In spite of the great biological significance of the dimensions of the absorbing surface of roots, almost no data are to be found in the literature. It is not known whether there is an upper limit to the density of occupation of soil by roots, or a maximum possible root surface in a given soil volume.

The reason doubtless lies in the great methodological difficulty in investigations of this kind, the chief problem being that the absorbing function of a plant is performed not by the entire root system but only by the smallest root tips with little mass but great surface and length, which are very delicate and easily broken off in digging. According to the data of I. N. Elagin and V. N. Mina (1953), one gramme of air-dry root less than 0·1 mm in diameter has a length of about 200 metres, while one gramme of root more than 5 mm in diameter is 0·025 metres in length. Therefore even the weight of roots, not to mention their surface, can be closely estimated only for plants grown in water. But even in this case, determination of the size of the absorbing surface remains a difficult problem, since the size varies in accordance with external conditions, age of the root, etc.

Most botanists and pedologists who study the root systems of plants in natural conditions limit themselves to discovering only the weight of the roots, provisionally dividing them into two groups, conducting and absorbing. Unfortunately these studies refer only to roots more than 1 mm in diameter.

Painstaking anatomical studies made by L. A. Ivanov (1916b) on seedlings growing in sandy soil have shown that absorption is performed mainly by light-coloured root tips, of diameter less than 0·4 mm. After they become dark-brown the faculty of absorption is partly lost.

Differences in the physiological functions of light-coloured and dark-brown roots were later confirmed by I. A. Muromtsev (1940), A. Ya. Orlov (1955), R. K. Salyaev (1961), and others.

An estimate of the absorbing surface and the number of root tips enabled L. A. Ivanov (1953) to calculate the total absorbing surface of the roots of a single seedling of different species, and its relation to leaf-surface. That author, however, did not take into account the surface of root filaments, which considerably increases the root surface.

From Table 47 it is seen that the ratio of root-surface to leaf-surface varies within very wide limits, from $0·07 \times 10^2$ to $237·9 \times 10^2$.

L. A. Ivanov's investigations were pursued in further detail by L. N. Zgurovskaya (1958, 1961, 1962). She made detailed studies of the structure of root tips, including root filaments, in a number of species in different growing conditions. She showed that roots having primary structure (retaining their original cortex with root filaments) possessed the largest absorbing surface. But as the roots grow older, or because of unfavourable external conditions (such as drought or overheating), their filaments and original cortex die off, some of the endodermal cells become subereous and so lose ability to absorb, and the absorbing surface is much diminished.

It is clear that roots of the same diameter may vary greatly in their functions, and also that the ratio of absorbing surface to root mass is not constant. Therefore determination of the weights of conducting and absorbing roots does not enable us to state even approximately what amount of absorbing surface corresponds to a given weight of 'absorbing' roots.

The problem is further complicated by the presence on many plant roots of mycorrhizae: hyphae of fungi, whose surface area it is almost impossible to estimate. The functions of mycorrhizae, however, especially with regard to the water regime of higher plants, have not been clarified.

Although studies of the weight of roots do not enable us to calculate their absorbing surface, we can judge how densely the various soil horizons are permeated with roots and consequently how completely the water supplies in these layers can be used. There are numerous data in the literature on the morphology and mass of roots (Kuz'min 1929-30, Petrov 1935, Eremeev 1938, Gurskii 1939 a, b, Shumakov 1949, Karandina 1950, Labunskii 1951, Guseinov 1952, Kucheryavykh 1954, Grudzinskaya 1956, Orlov 1959).

Usually the main mass of absorbing root tips is located in the humus

Table 47. *Dimensions of absorbing surface of roots of seedlings of tree species* (Ivanov 1953).*

Species	Age (years)	Diameter (mm)	Length (mm)	Absorbing surface of tips (mm)	No. of tips on one seedling	Surface of all tips (mm²)	Surface of leaves (cm²)	'Coefficient of service' (ratio of root-surface to leaf-surface × 100)
Spruce	1	0·26	0·70	0·57	50	29	29·6	0·98
	2	0·20	0·25	0·16	780	125	1628·0	0·07
Pine	1	0·32	1·00	1·00	250	250	30·1	8·31
	2	0·26	0·60	0·50	3800	1900	4084·0	0·46
English oak	1	0·11	2·00	0·69	3000	2070	355·0	5·80
	2	0·11	1·40	0·48	6365	3055	—	—
Norway maple	1	0·33	3·00	3·11	407	1266	19·8	64·00
	2	0·31	2·40	2·34	1415	3311	—	—
Elm	1	0·20	3·00	1·88	420	790	10·9	72·50
White birch	1	0·80	1·00	0·94	445	418	2·3	181·50
	2	0·29	1·36	1·24	1600	1984	29·0	68·50
Small-leaf linden	1	0·22	1·50	1·04	912	950	4·8	198·00
Common ash	1	0·33	19·00	19·69	678	13,349	55·9	137·90

* Surface of root filaments not included.

soil layer near the surface. That layer warms up more rapidly in spring, is more rapidly soaked by precipitation, has a looser structure, is better aerated, and is rich in nutrients. The large absorbing surface of the roots permits rapid entry of water and nutrients into the roots in that layer. In addition to these factors that positively affect the life-activity of roots, there are a number of negative factors; in the surface soil layers, for instance, variations in the water and temperature regimes are more marked than in deeper soil layers.

In soils that are cold, poor in nutrients, or boggy, the spreading of roots is restricted to the surface soil layers alone; but in richer soils of chernozem type, grey forest soils of the forest-steppe, or chestnut soils, a comparatively small part of the root system may penetrate to a great depth (to 8 or 10 metres). The water regime there is more constant than in surface soil layers (Table 48).

The area of contact of roots with soil is one of the factors that determine the rate of water absorption from a given soil layer. Therefore we may expect roots to absorb water from the upper soil layers more rapidly than from the lower, other things being equal. In natural conditions such equality does not exist, and therefore the ratio of rates of water absorption from different soil horizons may vary with the actual circumstances and with the state of the plant.

We know of two different physiological mechanisms, active and passive, within plants, providing for entry of water into the roots. The first is connected with the life-activity of the roots and especially with their osmotic characteristics. The moving force of active water absorption by root systems is the difference between the water-retaining power of soil and that of living root cells. On account of the gradual increase of suction power from the periphery to the centre of a root and thence to the contents of the xylem vessels, water absorbed by the root filaments travels to the

Table 48. *Distribution of root mass of trees by soil horizons in ordinary chernozems in steppe zone (Afanas'eva et al. 1955).*

Depth below surface (cm)	Soil horizon	Total weight of roots		Roots over 1 mm in diameter		% of thin roots in total weight of roots*
		Air-dry weight (g)	%	Air-dry weight (g)	%	
0–50		8100	73	1534	65	19
50–100	Humus	1353	12	110	5	8
100–150	Transitional carbonate	842	8	168	7	20
150–200		346	3	141	6	41
200–250		142	1	90	4	64
250–300		109	1	84	3	78
300–350	Bedrock	127	1	107	5	84
350–400		120	1	108	5	90

* Calculated by us.

vessels in the centre of the root and rises through them to the trunk (Sabinin 1925, 1949). Active water absorption is a comparatively slow process. Much more rapid is the passive absorption of water by roots, due to the sucking action of the transpiration organs. When output of water by the above-ground organs is much reduced, the roots absorb water only slowly.

In the latter case, in equation (1), $W_1 = iPT_2 + \Delta Mw + M\Delta w$, the item iPT_2 is considerably less than the other two. Moisture absorbed by the roots ($W_2 = W_1$) goes mainly to increasing the moisture-content of the existing mass $M\Delta w$ or to moistening the newly-formed mass ΔMw.

As the moisture output by transpiration, and consequently the first item in equation (1), increase, passive water absorption first commences. After a rapid passive flow of water through the roots has been established the root cells may provide resistance, checking the flow of water (Maksimov 1952, Sabinin 1955).

Active and passive water absorption react differently to changes in environmental temperature. The former, being a process depending on chemical reactions within the plant, has the temperature coefficient Q_{10} within the limits 2 to 7; but in the latter, which is more a result of physical processes, Q_{10} is much smaller (Trubetskova 1962). Water absorption by roots from the soil in perceptible amounts proceeds at temperatures above -3 or $-5°C$, at which temperatures water in soil solutions and in plants is not yet frozen. Therefore in temperate latitudes the period of absorbent action by roots is limited to the warmer half of the year.

The rate of water absorption per unit of absorbing root surface depends on four factors: (i) environmental temperature; (ii) differences in the water-retaining power of soil and of roots; (iii) resistance by cells of living protoplasm in the roots to passage of water; and (iv) rate of movement of water in the soil to the points of water uptake, the root tips.

As we have stated, active water absorption by roots is a comparatively slow process. Therefore when active absorption alone is taking place, its rate is determined solely by soil temperature and the difference in the water-retaining power of the soil and of the roots. The latter item is functionally linked with the amount of soil moisture, and therefore fluctuates directly with changes in it. As is well known, when moisture in the soil decreases the soil's water-retaining power (soil-moisture tension H, in pedologists' terminology) increases. The corresponding relationship may be expressed in a hyperbolic curve. At first, as soil moisture decreases, H increases very slowly, but in very dry conditions it rises very quickly. Investigations by the school of D. A. Sabinin (1925, 1949) have shown that, as a rule, when active water absorption is taking place the suction power (S) of roots exceeds by 1-2 atmospheres the water-retaining power of the surrounding soil. As the latter increases, a corresponding increase in S occurs in the roots, so that the difference ($S-H$) remains almost constant,

With rapid increase in H, however, further increase in S leads to a breakdown of the normal life-activity of the roots. Therefore the value of S, and consequently the possibility of active water absorption by the roots, has definitive limits. Unfortunately the literature contains almost no data on the value of S in roots when only active water absorption is taking place. To judge, however, from data given for the suction power of roots at other periods, the upper limit of availability of water for active absorption is about 8-10 atmospheres (Zgurovskaya 1958). Dependence of the rate of passive water absorption by roots on external factors differs from that of active water absorption. In the first place, environmental temperature has little effect on the rate of the passive process, but on the other hand the effects of the moisture content of the soil, and of the rate of moisture movement to the absorption foci, are greater. Let us first examine the dynamics of the movement of moisture in the soil while it is being removed by plants.

The rate of movement of moisture in the soil depends on the soil's capacity to conduct moisture and the gradient of its water-retaining power. It is known that the water-conducting capacity of the soil is a function of its moisture content. As soil moisture gradually decreases, the water-conducting capacity is high up to a definite limit of desiccation, but later falls rapidly on account of the breaking of the water filaments in the soil capillaries (moisture-breakage-point in the capillaries, abbreviation VRK, according to Rode 1952). When soil moisture is lower than VRK, water moves in the soil mainly in the form of vapour. That is accompanied by a rise in the gradient of water-retaining power, since local increases in H near the roots persist because of the slow movement of water towards the absorption foci. The rise in the gradient of the soil's water-retaining power near the roots, however, cannot offset the sharp fall in its water-conducting power. Therefore with a moisture lower than VRK, the water supply to the roots is limited by the rate of water movement in the soil and VRK represents the limit of available water supply to plants (Sudnitsyn 1961). For supply of water to plants when the moisture level falls below VRK, the surface of contact between roots and soil is of primary importance. That is why marked extension of root systems is always observed in dry localities. The underground parts of plants become densely aggregated, while the above-ground parts remain dispersed; and open tree stands, such as are characteristic, for instance, of Central Asia, result.

A special and urgent problem is to elucidate the significance of the resistance by living root cells to passage of water in passive absorption, and also of the difference between H and S in different water-supply conditions. Comparatively little study is being given to the resistance by living protoplasm to the passage of water.

There are far more data on the amount of the water-retaining power of soil and roots but unfortunately these two quantities have been studied separately. Therefore we have to evaluate changes in the value of the moving force of water flow in passive absorption only on the basis of suppositions and estimates. In world literature there are only a few works in which the dynamics of soil-moisture tension H and suction power of roots S have been studied simultaneously. In particular, a series of figures obtained by Sudnitsyn and Tsel'niker for trees in steppe and semi-desert conditions indicate that in passive absorption the difference between the water-retaining powers of soil and of underground organs (but not roots)

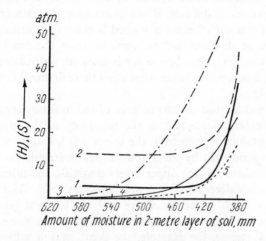

Fig. 26. Soil-moisture tension (H) and suction power of leaves (S) with relation to amount of water in the soil (Sudnitsyn and Tsel'niker). 1—S at sunrise; 2—S at 2 p.m.; 3—H in 10-cm layer; 4—H, average for layers 10-50 cm; 5—H, average for layers 50-200 cm.

is significant, and water travels through roots as through a candle-wick. In that case absorption of water from the corresponding soil layer takes place if the difference between S (of the leaves) and H (of the soil) is greater than zero, i.e., S (of the leaves) is greater than H. If this condition does not exist, the soil layer concerned takes no part in water supply. It is clearly seen from Fig. 26 how, with drying of the soil, successive soil layers from the uppermost downwards gradually cease to supply water. The value of H at which moisture becomes thermodynamically inaccessible to plants (the moisture-level of wilting, VZ according to Rode) is placed by a number of American authors at 15 atmospheres (Richards and Weaver 1943, Furr and Reeve 1945). It is possible, however, that this value is too high, especially for trees, and also may vary with weather conditions (Rode 1961).

Let us examine the question of the significance of different soil horizons in the supply of water to plants. In conditions of adequate or excessive moisture, where there is no competition for water in a phytocoenose, this question has little significance. Roots can easily take water from the upper soil horizons, which are better provided with nutrients, less compact, and better aerated.

In dry conditions (in the forest-steppe, steppe, and semi-desert zones) competition for moisture becomes of primary importance. As we have already stated, in these zones root systems in deep fertile soils (chernozem, grey forest soil, and chestnut soil) are fairly sharply divided into two parts: vertical and horizontal.

Direct determination of the amount of water absorbed by different parts of the root system is difficult, if not impossible, in natural conditions. Therefore the amount of water absorbed is generally estimated from the water balance in different soil horizons or from the amount of water expended in transpiration when water is unevenly distributed in the soil layers, on the assumption that in most cases these figures will be very close to the actual ones.

Data obtained by studying the balance of soil moisture are fairly plentiful. They clearly indicate that at the beginning of the growing period moisture is obtained mainly from the upper soil horizons, which are the most densely permeated by roots and the best warmed. As the supplies of available moisture in the upper layers diminish, the principal sources of moisture move deeper into the soil (Shockley 1955, Afanas'eva et al. 1955), and so much larger masses of soil are brought into the water-exchange process than is the case in soils without vegetation. In a forest in the Derkul' steppe, for example, the depth of soil subject to annual desiccation to the VZ level varies from 3·5 to 4 metres. Since the depth of autumn soil wetting does not, as a rule, exceed 2 to 2·5 metres, in these conditions there is formed beneath forest growth the so-called 'dead horizon' (Vysotskii 1912), which is always dry. This phenomenon is not found over the entire steppe or in farm crop-lands, where the depth of soil subject to desiccation is only 1 to 1·5 metres, corresponding roughly to the depth of moistening (Zonn 1959). With physical evaporation from the soil surface, serious desiccation affects only the top (5-10 cm) layer of the soil (Abramova 1953, Abramova et al. 1956). The role of surface and deep roots in supplying water in the steppe zone has been determined by Tsel'niker (1956) by a second method. That author estimated the intensity of transpiration and consumption of water by a plantation, by cutting through horizontal roots at a depth of 40 cm or by cutting through vertical roots. The trees survived for several years after that treatment, indicating that they did not suffer fatal injury.

As is seen from Table 49, cutting through horizontal roots more strongly affects uptake of water from the soil, and therefore they play a greater part in supplying water than do vertical roots. This was confirmed by another of Tsel'niker's experiments, in which transpiration intensity was determined after artificial moistening of the upper (humus) layer or the deep soil layers during a severe drought (Tsel'niker 1957). In the first case transpiration rose rapidly, and in the second it remained low. Therefore deep roots cannot absorb water quickly, and plants obtain most of their water requirements from surface roots.

Table 49. *Consumption of water from the soil after undercutting of roots (Tsel'niker 1956).*

Stand composition	Control (mm)	Horizontal roots undercut		Vertical roots undercut	
		mm	%	mm	%
Oak (unmixed)	194	155	80	191	98
Oak with Norway maple	129	81	63	114	88
Red-leaved ash	150	77	52	125	83

Nevertheless, deep roots play an important role in the life-activity of plants, as they supply them with water when there is no available water in the humus layers. The role of deep roots is particularly important when underground water is present. In that case, even in very dry places, plants with roots that reach underground water do not suffer from lack of moisture (Ivanov *et al.* 1953, Tsel'niker and Markova 1955, Erpert 1962).

Water output

When water supply is adequate, water output by leaves is mainly subject to the same laws as physical evaporation: the rate of evaporation is proportional to the deficit in saturation of the atmosphere with water vapour and to the diffusion coefficient. This makes evaporation also dependent on air temperature, since the vapour pressure of water saturating an area and also the diffusion coefficient increase with rising temperature. Moreover, temperature affects transpiration by changing the permeability of protoplasm by water and by altering other life processes linked directly or indirectly with water-exchange. As has been shown by L. A. Ivanov *et al.* (1951), for a wide selection of tree species the rate of transpiration is so closely correlated with air temperature that it can be estimated from air temperature without measuring it directly (Table 50).

Light also has a strong effect on transpiration. Direct light raises the rate of transpiration as compared with diffused light by 30-150% (Ivanov 1956a, Gulidova 1958). At night transpiration falls markedly, but zero transpiration is observed only for a very short time (Fig. 27). The effect of light is much stronger on transpiration intensity than on physical evaporation, because of the specific action of light on the permeability of the protoplasm in transpiration cells. Differences are similarly observed in the effect of light of different wave-lengths on transpiration (Ivanov and Yurina 1961).

The maximum water requirements of vegetation over a period of time are calculated by means of formulae suggested by Zennman and Tornveit, and also by Ivanov (1948) and others. The chief component in these formulae is the amount of solar radiation providing energy for vapour-

Table 50. *Increase in transpiration rate in trees and shrubs with rise of 1° in air temperature (Ivanov et al. 1951).*

Species	Increase in transpiration rate (mg/g per hour)	Coefficient of correlation between transpiration rate and air temperature
Berlin poplar	66·5	0·96
Mongolian poplar	54·6	0·94
European ash	37·6	0·85
Aspen	35·2	0·98
Small-leaf linden	33·1	0·94
European mountain ash	24·9	0·94
English oak	22·6	0·92
White birch	20·6	0·79
Hazel	19·0	0·87
Red-berried elder	18·9	0·67
Honeysuckle	17·8	0·86
Norway maple	14·7	0·77
Russian elm	13·0	0·76
Scots pine	10·0	0·96

ising water. A simplified calculation of water requirements may be made from the fact that 585 calories are required to convert one gramme of water into vapour. If we assume that leaves in a dense forest absorb

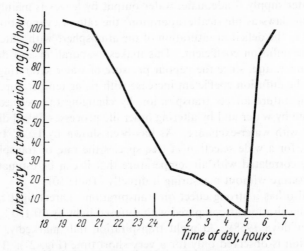

Fig. 27. Intensity of transpiration by Chinese elm during night hours (from data of Tsyganova and Tsel'niker).

about 60% of the total radiation (physiological and infra-red), then, neglecting the amount of radiation used in photosynthesis, which is a very small quantity, we may write the equation:

$$W = 0.6 \, Q/585,$$

where W = amount of water evaporated (in mm), and Q = amount of solar energy in kcal/cm².

In conditions of adequate moisture, data obtained by this formula agree well with data obtained by direct determination of the transpiration intensity of the leaf masses during the time when transpiration occurs (Ivanov and Silina 1951).

Transpiration is largely limited to daylight hours and to the warmer part of the year; in winter losses by transpiration are reduced so greatly that they need not be taken into account in reckoning water consumption by trees (Ivanov 1941b, Girnik 1955). They are important only in that during winter transpiration may cause severe desiccation or death of shoots, since at that time water supply from the soil is less than transpiration.

An empiric determination of maximum water requirement may be made from evaporation potential, i.e., the amount of water evaporated from an open water surface. Generally the evaporation potential is the upper limit of the water requirement by plants.

An excess of water output by vegetation above the evaporation potential may result from the possible absorption of additional heat energy from the surrounding air by leaves (Alpat'ev 1954). That source of energy cannot have much effect on the value of the evaporation potential, because the area of the evaporating surface in contact with the atmosphere is considerably less than that of the surface of the leaves.

All of these factors determine the amount of moisture expended by plants only when there is unrestricted water supply. Otherwise physiological regulation of the transpiration process comes into play. In cases of moisture shortage, rises in temperature and in saturation deficit increase potential evaporation but lead to a reduction in water consumption by plants. The moisture content of the soil becomes the major factor in the amount of transpiration in these circumstances (Kokina 1926, Tsel'niker 1957). The excess of radiant energy absorbed by the leaves goes to heating the leaves above air temperature, and the more transpiration is reduced, the greater is the proportion of energy used in heating (Kleshnin 1954, Akulova 1962). Since gas-exchange occurs mainly through the stomata, reduction of water loss is achieved primarily by closing the stomata but in addition an increase in the leaves' water-retaining power occurs through the chemical combination of water.

L. A. Ivanov gave the name 'transpiration resistance' to the whole assemblage of factors reducing the output of water by leaves. Because of it, when there is a sharp fall in the supply of water from the root system the leaves not only do not dry up but often increase the amount of water content per unit of leaf area, or per leaf (Tsel'niker 1958). Transpiration resistance may be caused both by drought and by extreme (high or low) temperatures. Thus plants are able, through transpiration resistance, to lower water loss by transpiration to a minimum, while avoiding lethal effects of a decrease in transpiration. Determination of the minimum value for loss by transpiration is extremely important, since it is this value that determines the survival of compact tree stands in drought conditions. Since transpiration depends to a great degree on external conditions, the minimum transpiration loss probably differs in different zones but lack of data prevents us from making a definite statement on this point.

The minimum moisture requirements by dense tree stands may be found by three methods: (*i*) empirical, determining the water output by stands at the extreme southern limits of forest growth in arid regions and in particularly dry years; (*ii*) experimental, imitating drought conditions for a tree stand and recording the water loss by transpiration; and (*iii*) by calculations from data of minimum transpiration intensity, leaf mass, and the period in which transpiration occurs (Table 51).

Table 51. *Minimum rate of transpiration in different tree species. Steppe zone, southeast of European part of U.S.S.R. (Tsel'niker 1956).*

Species	No. of experiments	Minimum transpiration	
		mg/g per hour	mg/dm² per hour
English oak	59	101	141
Red-leaved ash	39	118	153
Norway maple	14	75	60
Elm	15	101	202
Apple	32	191	286
Pear	15	114	137
Tatarian maple	31	112	112
Caragana	35	198	—
Average		125±16	156±27

Empirical determination of minimum loss by transpiration during the vegetative period gave the following results: for an oak-maple plantation near Volgograd (Serebryakova 1951) 110-115 mm; for an oak-ash plantation at Derkul' (Lugansk province) (Tsel'niker 1958) 110-120 mm.

An experiment in artificially creating drought conditions by cutting horizontal roots of trees or by sheltering the soil from summer rain gave a loss of 100-110 mm (Tsel'niker 1956).

From calculations, the following data were obtained: minimum intensity of transpiration during drought (short of dying-off of leaves) about 125 mg per hour; leaf mass per hectare of compact stand 7·5 tons; number of working hours of transpiration during vegetative period (4 months) 120. These values gave a transpiration loss of 113 mm. Thus all three methods of estimating minimum water requirements produce closely similar results.

It is to be noted that both maximum and minimum values for loss of moisture by compact stands depend very little on the species-composition. In the first case the value determined is the evaporation potential for a given place, and in the second case it is the amount of available moisture in the soil (Fig. 28). The rate of desiccation of the soil may vary in accord-

ance with the structure of the root system and the physiological characteristics of the species, but the degree of desiccation at the end of the vegetative period will be almost identical for different species. The depth of
the root system does not play a major role in this case, since the amount
of moisture accessible to deep roots is relatively small.

The loss of moisture by transpiration in a compact forest coenose in a
temperate climate, in districts with different hydrothermal regimes, may
vary from 100 mm (in arid districts) to a figure near the evaporation
potential for a given locality. If the amount of moisture lost by transpira-

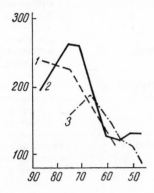

Fig. 28. Intensity of transpiration by English oak
with relation to amount of moisture in the soil
(Tsel'niker 1957); abscissa axis—amount of water
in soil layer 0-50 cm as % of minimum water
capacity; ordinate axis—intensity of transpiration
in mg/g/hour. Soil temperature: 1—13-15°C;
2—16-19°C; 3—20-24°C.

tion in a hot dry climate is less than 100 mm, compact stands cannot
survive.

In Table 52 we summarise data on the amounts of moisture lost by
tree stands by transpiration, obtained in different climatic zones of the earth.

From Table 52 it is clear that only in tropical rain forests is loss of
water by transpiration equal to or greater than the evaporation potential.
In other cases loss by transpiration is less than the evaporation potential,
and is one-third or one-half of the annual amount of precipitation.
Apparently the reason is that, even with a high average coefficient of
wetting, there may be periods during the height of the growing season
when moisture is inadequate; also the periods with high wetting coefficient, as a rule, occur at times when transpiration by plants is low
(Ivanov 1948).

Loss by transpiration is very high with artificial irrigation or in places
where underground water is near the surface. One may divide the earth
into three different regions according to the amounts of water lost by
transpiration: the humid tropics, where loss of water by forests varies
from 800 to 3000 mm per annum; temperate-climate forests with adequate
moisture, from 200 to 600 mm; and temperate-climate forests with inadequate moisture, from 100 to 200 mm. The division between the two
latter regions is not distinct, since fluctuations in the amount of water lost
by transpiration in the same place in different years may be greater than
the difference between separate regions in a temperate-climate zone.

Table 52. *Annual removal of water by transpiration by plant communities in different climatic zones of the earth.*

Climate	Place of investigation	Vegetation	Transpiration loss (mm)	Precipitation (mm)	Evaporation potential (mm)	Ratio of precipitation to evaporation potential	Author
Equatorial	Indonesia, Java	Rain forest	2300-3000	4200	—	—	Coster (from Kramer and Kozlowsky 1960)
Equatorial	Indonesia, Java	Mountain forest: Trees Herbage	740 130 }870	3600			Coster (from Kramer and Kozlowsky 1960)
Tropical, with slight drought period	Brazil	Coffee plantation	600	1390	1073	1·3	Franco and Inforzato (from Kramer and Kozlowsky 1960)
	Africa	Plantation of *Acacia mollissima*	2500	760			Franco and Inforzato (from Kramer and Kozlowsky 1960)
	Africa	Plantation of *Pinus insignis*	760	760			Franco and Inforzato (from Kramer and Kozlowsky 1960)
	Africa	Eucalyptus plantation	1200	760			Franco and Inforzato (from Kramer and Kozlowsky 1960)
Tropical savannah	Africa	Shrubs	48	200			Henrici (from Kramer and Kozlowsky 1960)
Temperate	U.S.A., Appalachian Mountains	Hardwood forest	425-500				Hoover (from Kramer and Kozlowsky 1960)
	U.S.A.	Average for U.S. forests	126-382				Kittredge 1951
	Germany	Mixed forest	290	771			Troll 1956
	Germany, Tarandt	Birch	563	670	698	0·96	Polster 1950
		Beech	448				
		Larch	555				
		Douglas fir	626				

Table 52 (*continued*)

Climate	Place of investigation	Vegetation	Transpiration loss (mm)	Precipitation (mm)	Evaporation potential (mm)	Ratio of precipitation to evaporation potential	Author
Temperate	Austria, Innsbruck	Spruce-beech	250	861	558	1·53	Pisek and Cartellieri 1939
		Pine-larch	300				
		Birch-beech	360				
	Southern Sweden	Spruce	211	545	472	1·16	Stålfelt 1956
	U.S.S.R., Archangel province	Pine	260	500			Molchanov 1952
		Spruce-birch	197	500			Molchanov 1952
	U.S.S.R., Leningrad province	Spruce-deciduous	374-610	620	406	1·54	Koshcheev 1955
	U.S.S.R., Vologda province	Spruce-deciduous	193-329	570	457	1·25	Gulidova 1958
	U.S.S.R., Velikoluksk province	Spruce	226	550			Koshcheev 1955
	U.S.S.R., Moscow province	Pine	181-371	525			Molchanov 1952
	U.S.S.R., Moscow province	Oak	250	525			Davydova 1964
		Birch	286-350	600			Molchanov 1952
		Pine	260	600			Molchanov 1952
Temperate, forest-steppe zone	U.S.S.R., Vinnitsk province	Ash-hornbeam	354	615	700	0·88	Rats 1938
	U.S.S.R., Voronezh province	Oak-ash	217	400	626	0·77	Silina 1955

Table 52 (*concluded*)

Climate	Place of investigation	Vegetation	Transpiration loss (mm)	Precipitation (mm)	Evaporation potential (mm)	Ratio of precipitation to evaporation potential	Author
Temperate, steppe zone	U.S.S.R., Lugansk province	Unmixed oak plantation	148	420	772	0·59	Ivanov *et al.* 1953
		Oak-ash plantation	135				
		Oak-maple plantation	158				
	U.S.S.R., Rostov province	Oak plantation	319	465	916	0·51	Kol'tsov 1954
		Oak-ash plantation	271				
Dry steppe and semi-desert zone	U.S.S.R., Volgograd province	Oak-maple plantation	115	300	887	0·38	Serebryakova 1951
		Maple plantation	110				
		Pine	146	325	986	0·33	Molchanov 1952
		Mixed deciduous plantation on black soil, near underground water	1000	280	1000	0·28	Khlebnikova and Markova 1955
		Chinese elm plantation irrigated	1655	280	1000	0·28	Erpert 1962

As may be seen from Table 53, loss of moisture by transpiration is correlated with the amount of leaf mass, and therefore annual changes in the amount of moisture lost by transpiration result from changes in leaf mass as well as in transpiration rate.

The age of a stand, especially before it becomes compact, has a strong effect on loss of moisture by transpiration; the age dynamics of moisture loss by tree stands in different zones are different, and generally depend on leaf-mass dynamics.

A. L. Koshcheev (1955), who studied moisture loss by transpiration in the taiga zone (Lisinsk leskhoz, Leningrad province), found that moisture loss by transpiration gradually rises after tree-felling until the new growth reaches the pole stage (20-30 years), and afterwards falls gradually (Fig. 29).

Observations made at Derkul' (steppe zone) have shown that initially plantations have a very high transpiration rate. Because of their small leaf mass transpiration is low in seedlings in nurseries. In the second year transpiration loss and leaf mass increase, and in the third year transpiration loss is already higher than in a mature stand. The possibility of such a high loss (regardless of the exceptionally dry year of the investigation) is due to water reserves accumulated during preceding years when the soil was lying fallow and during the first two years after planting. Reserves from deeper soil layers are gradually brought into use as the root systems of the trees go deeper.

From the fourth to the sixth year loss of moisture by transpiration falls, and then in later years fluctuates around an average level of 153-157 mm. The constant level of water loss between the ages of 5 and 15 years results from the increasing leaf mass and the gradual decrease in rate of transpiration.

We may assume that that constant level of moisture loss by transpiration will be maintained for a very long time, even when natural old age begins and the stand is less dense, since the available moisture will be consumed by the remaining trees and later by the herb layer.

Similar relationships have been observed with artificial thinning of a forest area in the steppe zone (Derkul'). With relatively light thinning, all of the moisture was consumed by trees and shrubs, and with heavy thinning, by the herb layer. Loss of water by the herb layer amounted to

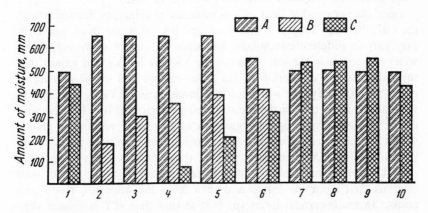

Fig. 29. Moisture balance beneath forest stands of different ages and on clearings in taiga zone (from Koshcheev, 1955). A—annual precipitation reaching the soil; B—total moisture consumption; C—moisture consumption by tree stands. 1—100-year-old stand of density 0·8; 2—clearing, first year after cutting, in bilberry cover; 3—clearing with continuous cover of haircap moss (*Polytrichum*); 4—clearing with continuous sphagnum cover and sparse regrowth of tree species; 5—clearing with 5- to 8-year-old regrowth of conifers and deciduous species; 6—12- to 15-year-old coniferous-deciduous saplings with crown density 0·9 to 1·0; 7—25- to 30-year-old tree stand; 8—50-year-old stand; 9—submature stand, age 70-80 years; 10—85- to 90-year-old stand.

Table 53. *Water loss through transpiration by tree plantations in steppe zone (L. A. Ivanov 1956, Tsel'niker 1958a).*

Item	Nursery			60-metre strip		
	Age of plantation (years)					
	1	2	3	4	5	6
Year of investigation	1952	1953	1954	1954	1955	1956
Moisture conditions of the year	Wet	Dry	Very dry	Very dry	Very wet	Wet
Leaf mass (ton/ha)	0·90	3·98	3·78	2·44	4·03	4·58
Transpiration loss (mm)	26·7	89·5	275·0	155·9	164·0	141·8

half of the total loss; the total loss, in spite of heavy thinning, was not substantially altered (Tsel'niker 1959).

Effect of water-supply conditions on growth processes and synthesis of organic matter

We have shown how curtailment of losses by transpiration during desiccation of the soil is possible within certain limits without serious damage to the water regime of plants.

What is the limit of soil desiccation at which breakdown of normal life-activity in plants begins? This question has long been the subject of lively discussion (Litvinov 1932, Furr and Reeve 1945).

Since the permissible limit of soil moisture is primarily determined by the ratio between supply of water to, and loss of water from, plants, it may vary in different conditions; and since an abrupt slowing-down of water movement in the soil occurs at the VRK point, we may expect that the most marked disruption of the water regime and of life-activity of plants would take place with moisture levels close to VRK. In fact, investigations show that in the moisture range from HB[1] to VRK, because of the efficient physiological regulation of the water regime by plants, values for the water regime of leaves remain almost constant during a fall in soil moisture. As drought intensifies, that method of regulating water output becomes inadequate, and moisture is used from the plants' own reserves, first from the leaves and then from the branches, roots, and stems. In mathematical terms (p. 163) at that time iPT is greater than W_2, and ΔwM has a minus sign. The quantity ΔwM, however, is small. The water reserves of plants play a large part in the regulation of daily variations in the water regime, but when they are the sole source of water supply these reserves, even with greatly reduced water output, last for only 2-3 days (Tsel'niker 1955b).

[1] HB is not defined in the original Russian edition. It is presumed to mean that the soil is saturated with water. (*Tr.*)

Table 53 (*continued*)

Item	60-metre strip		Forest area				
	Age of plantation (years)						
	13	14	15	16	17	18	19
Year of investigation	1950	1951	1952	1953	1954	1955	1956
Moisture conditions of the year	Average	Average	Wet	Dry	Very dry	Very wet	Wet
Leaf mass (ton/ha)	5·86	7·31	7·68	5·10	6·69	10·70	8·89
Transpiration loss (mm)	120·1	135·3	177·2	134·7	115·6	276·0	144·1

Loss of water in the leaves through transpiration increases the water deficit, the suction power (S), and the osmotic pressure (P) of the cell sap. In different species of plants the increases in osmotic pressure and in suction power proceed at different rates from the increase in water deficit; in some (e.g., oak) a small water deficit produces a steep rise in P and S, and in others (e.g., ash) the rise in these quantities is slow and gradual (Tsel'niker 1955a).

Increase in the water-retaining power of leaves as a result of desiccation plays a negative role in the life-activity processes of plants, since water in combination has a low chemical potential and cannot take part in chemical reactions. Therefore an increase in combined water leads to a decrease of photosynthetic intensity, and to a change in direction of biochemical processes from synthesis to hydrolysis (Sisakyan 1940).

Hydrolysis of compounds so important to life as proteins causes irreversible harm to the life-activity of an organism (Mothes 1931). This applies especially to the photosynthetic apparatus of plants, which is the first to suffer from desiccation.

According to E. V. Yurina (1957) and Yu. L. Tsel'niker (1958b), an increase in the osmotic pressure in tree leaves above 20 atmospheres depresses photosynthesis. According to Volk (1937), in several plant species stoppage of photosynthesis begins at osmotic pressure of 35 atmospheres. Rapid increase in suction power and osmotic pressure in leaves takes place, according to Tsel'niker (1958b), with soil moisture below VRK. Thus values for soil moisture corresponding to VRK may be considered to be the lower limit of permissible soil desiccation. With further desiccation normal functioning of the photosynthetic apparatus is impaired, and ultimately the leaves die off. The limit of water deficit that leaves can endure without dying off is only 10-20% of the dry weight of the leaves, or 20-30% of the weight of water in leaves in a saturated condition (Tsel'niker 1955a, Pisek 1960). In herbs this limit is usually higher, up to 80% of the water content (Höfler *et al.* 1941).[1] The limiting values of

[1] In light-loving plants with xeromorphic structure of leaves the limit is much higher than in shade-tolerant plants with mesomorphic leaves.

water deficit in different plant species are well correlated with the rate of increase in P and S.

Apparently dying-off of leaves occurs in different species of plants with identical values of P and S. Herbs are not necessarily more drought-resistant than trees, since resistance when certain organs are deprived of water is only one of the items contributing to the drought-resistance of a plant as a whole.

Damage resulting from loss of the assimilating apparatus differs in plants with different rhythms of development. It is greatest in annual mesophyte-herbs, in which growth of leaf-surface is comparatively slow and seeds ripen only in the second half of the growing period. In perennial plants, as a rule, the leaf mass develops in a short time, from the reserves of preceding years. Formation of these reserves during the growing period and also formation of the next year's shoots as buds take place differently in different species. For instance, in favourable conditions many trees and shrubs possessing a high potential for photosynthesis rapidly accumulate reserves of organic matter in trunks and branches, and stop further development of the next year's shoots. In such plants the dying-off of leaves during drought is a useful adaptation, since it prevents further desiccation of the plant and soil as well as loss of matter in respiration. In such plants transpiration is usually high when the soil is well watered, but the water regime of the leaves is variable, and resistance of their leaves to desiccation is higher than the average for trees and shrubs.

In trees not possessing the faculty of rapid building-up of organic reserves and rapid completion of the embryonic growth of buds, the water regime of the leaves is less variable because of better regulation of the transpiration process and use of moisture from deep soil horizons. Therefore in spite of the low resistance of leaves to desiccation during drought, their natural water deficit remains below the lethal point and they do not die off (Stocker 1929, Tsel'niker 1960a).

The ratio between the amount of water lost and the amount of organic matter synthesised (the transpiration coefficient of assimilation) is of great interest.

There are many data for farm plants on the subject (Maksimov 1952) but few for tree species. The little material available indicates that the transpiration coefficient of assimilation for forest trees varies within narrow limits, from 200 to 250 (Ivanov *et al.* 1963). Polster (1950) gives the following series of transpiration coefficients of assimilation:

Oak	344	Spruce	231
Birch	317	Douglas fir	173
Pine	300	Beech	169
Larch	257		

Shade-tolerant plants use less water to create a unit of organic matter than do light-loving plants.

Although the limits of variation of the transpiration coefficient of assimilation, according to Polster's data, are much wider than those given by the authors quoted above, the average values for all authors are very close. With a small moisture shortage transpiration falls more quickly than does photosynthesis, and therefore the transpiration coefficient of assimilation falls. But with severe drought, when the normal life-activity of a plant is disrupted, the coefficient rises sharply (e.g., in the steppe zone during severe drought (Ivanov *et al.* 1963); in the semi-desert zone (Khlebnikova 1958)).

MINERAL NUTRITION

The elements of mineral nutrition may be divided into four groups according to their biogeocoenotic significance:

(*i*) those taking part directly in energy conversion (phosphorus);

(*ii*) those entering into the composition of the chief products of life-activity (phosphorus, nitrogen, sulphur, potassium);

(*iii*) those entering into the composition of physiologically-active substances, such as chlorophyll, and also those acting in small quantities, such as enzymes and hormones (iron, sulphur, and micro-elements—magnesium, copper, boron, zinc, cobalt, etc.);

(*iv*) those regulating the internal environment of an organism—the arrangement of protoplasmic structures, the hydrophilous condition of biocolloids, etc. (mainly univalent or bivalent cations—potassium, sodium, calcium, etc.).

The role of each element in exchange of matter and energy depends on the nature of its transformations after entering an organism. For example, phosphorus characteristically forms labile bonds possessing different amounts of chemical energy, about 2 kcal/mole for simple bonds and about 8-10 kcal/mole for macro-energetic bonds. In such cases the valency of phosphorus does not change.

Nitrogen enters into the composition of amino acids and proteins, always in a reduced state, whereas it is absorbed from the atmosphere in both reduced and oxidised states. In the latter case, energy (possibly light-energy) is required for reduction of oxidised nitrogen compounds.

Elements of the third group, as a rule, form compounds with reversible changes of valency within living organisms, and possibly it is this circumstance that determines their special physiological role.

Finally, elements of the fourth group do not change their valency within plants.

Mineral substances are taken up by the absorbing root tips, and reciprocal transformation of matter takes place between all living tissues. There is, however, specificity in the direction of the exchange process, e.g., a number of amino acids and alkaloids are synthesised only in roots, whereas other forms of biosynthesis take place only in leaves or in the growing points of stems,

The quantity of matter absorbed from the soil by a phytocoenose may be expressed by the equation:

$$N = kP_2T_4,$$

where N is the amount of matter absorbed; k is the rate of absorption; P_2 is the absorbing surface of the roots; T_4 is the time.

The absorbing surface for mineral matter is the same as for water. Suberised root cells, impermeable to water, are also impermeable to mineral elements. The rate of absorption of mineral elements depends, firstly, on the state of the roots themselves, and secondly on the characteristics of the external environment (soil) and especially on the mineral nutrient content of the soil in a form available to plants. In many soils, for instance, large amounts of nutrients are firmly linked to the absorbent complex of the soil and so cannot enter into biological circulation. Such elements are principally phosphorus and potassium (Ratner 1950). The availability of nutrients depends on the compounds in which they exist in the soil. It has been shown beyond doubt that mineral matter in solution is easily available to plants. Data exist showing that low-molecular organic compounds (e.g., amino acids and amides) may also enter plants (Kolosov 1962). The degree of availability to plants of different compounds, however, has not yet been clarified. This question is extremely important, since the composition of the soil varies greatly in the course of the year as the result of microbial activity, and apparently so does the degree of availability of nutrients to plants, even when the total nutrient content remains unchanged.

Table 54. *Amount of phosphorus in leaves of oak, without and with mycorrhiza (Tsel'niker 1960).*

Item	Amount of phosphorus					
	Without mycorrhiza			With mycorrhiza		
	% of dry weight	per plant (mg)	per g of small roots (mg)	% of dry weight	per plant (mg)	per g of small roots (mg)
Control	0·073	29·6	313	0·075	26·1	311
Irrigation	0·070	30·1	440	0·107	62·7	646
Fertiliser	0·075	20·0	191	0·091	44·4	217
Fertiliser + irrigation	0·105	76·2	978	0·117	106·0	559

Mycorrhizae may have considerable effect on absorption of matter by roots (Table 54). Their role in the absorption of phosphorus has been particularly well studied (Trubetskova et al. 1955, Tarabrin 1957, Tsel'niker 1960b). In that process the mechanism of mycorrhiza action may consist in the fungus first absorbing phosphorus and then passing it on to the plant, and also in stimulation of the plant by hormones and vitamins secreted by the fungus (Shemakhanova 1955a).

It has long been held that there is a direct link between ingestion of water and ingestion of mineral nutrients, or, in other words, that mineral nutrients enter roots in soil solutions. Sabinin (1925) and his disciples (Trubetskova 1927, 1935) have shown that the mechanisms of ingestion of water and of mineral elements are quite different. Nevertheless there is an indirect connection between these two processes. It is based on the facts that (i) the absorbing surface for water and for mineral nutrients is the same; (ii) the life-activity of roots depends on the water content of the soil: when there is shortage of moisture, roots become less active, and when there is excess, aeration deteriorates and poisoning from excess of CO_2 may follow; (iii) when moisture is inadequate, the elements of mineral nutrition become more firmly linked with the absorbent complex of the soil, and when it is excessive, substances injurious to plants (e.g., acid iron compounds) may be formed; (iv) mineral nutrients in solution migrate in the soil, and the rate of movement increases with increase in the rate of movement of soil moisture.

As the investigations of D. A. Sabinin (1940, 1955), D. A. Sabinin and I. I. Kolosov (1935), and I. I. Kolosov (1962) have shown, the first stage of absorption of matter from the external environment is an exchange-absorption, depending on the quantity and the type of electrical charge of the matter entering the plant in an ionised state. To complete this stage of absorption, the roots should possess an exchange-reserve of ions. One of the chief sources of supply for the exchange-reserve is the ions of CO_2 released by respiration. This is one reason for the close dependence of the rate of mineral-element absorption on respiration. Since the charge of protoplasmic structures that accept ions in the root depends closely on the pH of the environment, the latter factor also has considerable influence on the rate of absorption.

After entering a plant most elements of mineral nutrition undergo further transformations. The rate and nature of these transformations depend on the presence of acceptors of the element, especially carbon chains (e.g., organic acids), and also on the oxidation-reduction potentials of the tissues where changes in the degree of reduction of the element are required for further transformation. This links mineral nutrition of plants with photosynthesis and with the respiration cycle. A stoppage in the transformation of an element within the plant prevents further entrance from the external environment. This explains the dependence of the rate of absorption of mineral matter on the phase of growth, and also the seasonal periodicity of absorption, especially in temperate climates.

For instance, young growing shoots and leaves of trees are distinguished by a high content of all mineral elements: nitrogen (up to 4-5% of dry weight), phosphorus (up to 1%), potassium, and others. During growth, proteins and nucleoproteins are intensively synthesised in these young leaves and shoots.

Nitrogen and phosphorus absorbed by plants enter newly-synthesised molecules. The amount of mineral elements absorbed by a plant is then at a maximum. After elongation ceases and differentiation of plant tissues begins, synthesis of carbon compounds proceeds rapidly. The rate of uptake of mineral matter falls almost to zero, and its content in the tissues becomes low (nitrogen 1-2%, phosphorus 0·2-0·4%) (Kramer and Kozlowsky 1960). Thus the upper limit of content of mineral nutritive elements in the tissues is determined by the possibility of their use in forming compounds. The more intensive the growth processes and bio-synthesis, the greater is the uptake of mineral elements by the plant. When biosynthesis ceases, accumulation and a small rise in the content of mineral matter take place, and then uptake ceases. The lower limit for the content of a given element is determined by its availability from the soil. If the characteristics of the soil are such that uptake of a given element is not intensive enough, growth and biosynthesis cease after the content of the element falls below a certain threshold, and further re-duction in the content of the element does not take place. When plant tissues grow old and die off most of the elements of mineral nutrition can be re-utilised, i.e., they pass over into younger growing parts and are used again in different biosyntheses. Elements suitable for re-use are nitrogen, phosphorus, and potassium, but calcium is not re-used.

There are several methods of determining a plant's requirements of nutrient elements in different habitats. In the first place the amount of an element in tree leaves or needles is determined (foliar analysis), since the mineral element level is highest in these organs (Lundergårdh 1951). This method is based on the fact that for normal life-activity of leaves the content of mineral nutrient elements cannot fall below specific levels. For instance, Kramer and Kozlowsky present data on the nitrogen, potassium, and phosphorus content of normal leaves and of leaves suffering from shortage of these elements (Table 55).

Table 55. *Variation in content of mineral nutrient elements in leaves during shortage of these elements, and optimum levels (% of dry weight) (Kramer and Kozlowsky 1960).*

Species	Shortage			Optimum		
	N	P	K	N	P	K
Pinus silvestris	1·2-1·3	0·08	—	3·0*	—	—
Pinus strobus	0·70-1·33	0·10-0·28	0·82-1·02	3·26	0·67	1·72
Picea abies	0·80-1·00	0·06	0·13-0·21	—	—	—
Betula sp.	1·8-2·2	0·08-0·10	0·29-0·34	—	—	—
Populus tremuloides	2·0	—	—	2·6-2·8	—	—
Acer saccharinum	1·75	—	—	2·8-2·9	—	—
Fraxinus americana	2·01	—	—	2·8-2·9	—	—
Tilia americana	2·32	—	—	3·1-3·2	—	—

* Seedlings. All other data refer to mature trees.

Data on the content of substances in leaves, however, are often very difficult to obtain, since one must know what is the norm for a given plant species and a given phase of shoot development. For forest tree species such data are far from adequate. To ease this task it is expedient to study simultaneously shoots of slowly- and quickly-growing trees or shoots from different locations on the crown of the same tree, and therefore possessing different growth-energy. The content of an element that impedes growth should be lower in slowly-growing shoots than in quickly-growing shoots, and the content of other elements should be higher, since they are not expended in growth processes and are accumulated. According to the data of Gulidova and Tsel'niker (1962), during the period of intensive growth in birch shoots of growth classes I and III the nitrogen content was higher in quickly-growing, and the phosphorus content higher in slowly-growing, shoots (Table 56).

The literature contains a number of works in which studies are made of the total amount of mineral matter used annually by tree stands. Part of the mineral uptake is returned to the soil each year in shed leaves and branches and dead roots. The return is small in young growth, but increases with age (Remezov and Bykova 1952, Remezov 1956) (Table 57). At the same time some of the nutrients are re-used within the plant itself, and consequently consumption of mineral elements falls as age increases.

An interesting summary of consumption of elements of mineral nutrition in the forest is given by Ehwald (1957) (Table 58).

Table 56. *Total nitrogen and phosphorus content in birch shoots at the ends of branches (% of dry weight) (Gulidova and Tsel'niker 1962).*

Element	Growth class	Date of taking specimens						
		13.V*	18.V	23.V†	9.VI	21.VI	14.VIII	17.IX
Total nitrogen	I	3·0	3·0	2·7	1·8	2·1	1·8	2·1
	III	2·3	2·4	2·1	1·4	2·1	1·5	1·8
Total phosphorus	I	0·54	0·39	0·39	0·30	0·20	0·26	0·21
	III	0·64	0·52	0·46	0·30	0·25	0·22	0·38

* Opening of buds.
† Intensive growth.

Table 57. *Return to soil of nutrient elements in trees (from Remezov 1956).*

Tree stand	Stock of organic matter, ton/ha (dry weight)	Returned to soil, kg/ha									
		Na	Si	Fe	Al	Mn	Ca	Mg	K	P	*
Pine, 100 years	222	424	35	10	59	43	352	77	307	55	70
Spruce, 100 years	315	477	71	15	95	60	557	65	294	86	191
Oak, 100 years	366	2924	107	40	39	52	3503	231	944	150	282
Linden, 75 years	222	1052	59	13	105	13	1489	181	647	78	182
Poplar, 50 years	251	953	68	8	39	3	966	880	382	115	72
Birch, 60 years	275	914	43	6	104	49	774	149	599	107	106

* This column heading is left blank in the original Russian book. (*Tr.*)

Comparing annual consumption of mineral nutrients in the field and in the forest, the author concludes that to create dry matter the forest consumes (in kg) considerably less mineral matter than do field crops:

	Forest	Field crop	Ratio of field crop to forest
Nitrogen	4-7	10-17	2·5
Phosphorus	0·3-0·6	2-3	5·6
Calcium	3-9	3-8	1·0
Potassium	1-5	8-26	5-8

The reason doubtless lies in the re-use by trees of the majority of elements, an exception being calcium. Characteristically a considerable proportion of the nutrients in forest phytocoenoses is consumed annually by the ground cover (herbs, ferns, and shrubs) (Table 59).

Ehwald also presents a diagram of circulation of nutrients in forest biogeocoenoses, well systematising and generalising the available data on this subject (Fig. 30). A similar diagram, but with quantitative data, is given by Weetman (1961) for nitrogen circulation (Fig. 31).

The material we have examined shows that, although in forest biogeocoenoses mineral elements are taken out of circulation by being incorporated in trees, the balance of these elements in a forest is much more favourable than in herbaceous biogeocoenoses, for the following reasons: (*i*) every year a considerable amount of these elements returns to the soil, and to the soil layers most permeated by roots (Table 60); (*ii*) part of the matter incorporated in trees is re-used, and therefore is not entirely excluded from circulation; when a forest is cut down the part of the organic matter removed is that containing the least mineral matter, whereas in hay-mowing and harvesting of field crops the parts richest in mineral elements are removed every year.

Table 58. *Consumption of nutrients by forest plant communities (kg/ha per year) (Ehwald 1957).*

	Consumption of nutrients														
	Quality class I					Quality class II					Quality class III				
	N	P	Ca	Mg	K	N	P	Ca	Mg	K	N	P	Ca	Mg	K
Pine															
Total	56·1	4·5	44·6	4·2	16·5	33·2	2·9	19·9	2·7	10·7	16·9	1·6	10·0	1·5	5·6
Needles	39·0	3·6	13·8	3·0	14·2	26·0	2·4	9·2	2·0	9·4	13·0	1·2	4·6	1·0	4·7
Roots	2·5	0·2	1·0	0·2	0·5	1·6	0·6	0·1	0·3	0·9	0·9	0·1	0·3	0·1	0·2
Spruce															
Total	61·8	5·5	86·6	6·7	21·7	43·6	4·0	63·2	4·7	15·6	—	—	—	—	—
Needles	41·2	3·7	32·6	4·5	15·0	33·0	3·0	26·1	3·6	12·0	—	—	—	—	—
Roots	3·5	0·3	3·7	0·3	1·3	1·8	0·2	1·9	0·2	0·7	—	—	—	—	—
Beech															
Total	57·2	4·0	59·0	8·5	47·6	39·0	2·8	39·0	5·6	32·9	—	—	—	—	—
Leaves	4·3	2·8	32·9	5·9	38·8	29·5	2·0	23·5	4·2	27·8	—	—	—	—	—
Roots	3·4	0·3	5·3	0·5	1·9	2·2	0·2	3·5	0·3	1·3	—	—	—	—	—

Table 59. *Annual production of lower vegetation in a forest.*

Locality	Character of stand	Predominant plants in lower vegetation	Dry matter produced annually (ton/ha)	Elements included (kg/ha)				Author
				N	P	Ca	K	
Eberswald	Pine, 70 yr Density:	Raspberry, mayflower						Ehwald 1957
	0·6		2·6	25·5	—	5·5	—	
	0·7		2·5	36·5	—	10·5	—	
	0·3		3·3	23·0	—	5	—	
	Pine, 60 yr Density: 0·7	*Calamagrostis*	4·1	37·5	—	8	—	
Northern Sweden	Pine, 200-250 yr	Bilberry and moss	1·1	13·7	2·2	6·6	8·2	
Northern Sweden	Pine with spruce, 100 yr	*Erica arborea*, bilberry	1·5	17·3	—	—	—	
Norway	Pine with spruce, 100 yr	Ferns	2·6	—	2·8	12·8	14·1	
		Mayflower, herbage	0·8	—	1·2	7·3	7·4	
U.S.S.R., Velikie Luki	Pine, 83 yr Density: 0·8		—	27	—	16	33	K. M. Smirnova 1951a
U.S.S.R., Voronezh province	Oak, 25-212 yr	*Carex hirta*, goutwort	0·2-0·4	3·5-5·3	0·5	2·6	8·1	V. N. Mina 1954a

For these reasons Walter (1936, 1951) concludes that the nutrient content of the soil is not, as a rule, a factor limiting forest productivity. Even in tropical forests, where all available mineral nutrients in the soil are absorbed by the trees, no shortage of such elements occurs; shed parts and dead trees quickly decay and their contained minerals are quickly reabsorbed. Thus a high rate of circulation of mineral nutrients is achieved.

A number of other authors (Ivanov 1948, Armand 1950, Volobuev 1953, Troll 1956), discussing the effect of edaphic factors on the nature of vegetation, also conclude that the hydrothermal regime of the soil is pre-eminent among soil factors, since it directly or indirectly affects all processes of the life-activity of phytocoenoses, including mineral nutrition. Within each soil-climate zone, however, the productivity of vegetation is often determined by the level of mineral nutrition. Shortage of mineral elements is most pronounced on young, eroded, sandy, or gravelly soils, which are almost devoid of humus. This is well seen in East German forests, where forest productivity has been substantially increased by the use of mineral fertilisers. Low productivity of a forest may be due to unfavourable pH of the soil, or to heavy concentration of injurious salts as well as to shortage of mineral matter.

POSSIBILITY OF EXISTENCE OF FOREST PHYTOCOENOSES FROM THE
PHYSIOLOGICAL POINT OF VIEW

The various organs of a tree (roots, leaves, axial parts) are specialised to
fulfil different functions. The active role of exchange of matter is played
mainly by roots and leaves, since the axial parts serve as supports and
conductors of matter and are conceptacles of reserves.

The presence of a massive supporting structure is the outstanding
feature of trees as a form of life (Serebryakov 1954). This feature was
developed through competition for light with plants whose above-ground

Table 60. *Consumption of nutrients by different types of forest (kg/ha per year) and percentage of
annual consumption returned to soil (Remezov 1956).*

Age (yr)	Consumption								Return to soil							
	N	Ca	K	P	S	Si	Al + Fe	Mg	N	Ca	K	P	S	Si	Al + Fe	Mg
Pine-cowberry																
14	37	22	17	4	6	3	6	5	48	64	27	41	50	95	38	54
30	47	44	19	6	6	5	9	8	45	54	32	44	38	76	50	47
45	57	36	20	5	9	6	9	8	55	53	26	38	49	46	27	42
70	25	22	9	3	3	2	3	3	60	80	64	74	58	86	78	80
95	13	13	5	1	2	2	2	2	84	87	68	80	80	92	85	83
Spruce with green moss and Oxalis																
24	16	15	8	3	2	4	3	2	17	16	6	14	7	39	16	20
38	62	52	38	12	9	12	17	10	39	34	20	37	22	43	30	36
60	40	34	19	6	6	10	9	6	62	60	42	43	28	73	52	52
72	33	31	15	4	5	10	7	5	69	62	48	52	36	81	56	63
93	28	25	9	3	5	10	5	3	75	71	60	63	46	89	72	76
Linden with goutwort and sedge																
13	59	77	30	6	7	6	4	10	58	66	53	61	50	66	54	62
25	85	111	50	9	13	7	6	15	56	57	51	64	40	58	44	61
40	73	96	42	9	11	6	5	13	63	64	58	71	48	65	51	68
74	87	115	46	11	9	8	6	17	82	82	79	89	70	85	71	84
Poplar with sedge and goutwort																
10	68	119	42	12	8	26	7	17	37	43	42	25	26	71	50	55
25	107	163	83	12	13	32	11	15	45	43	58	35	28	64	62	52
30	120	183	90	12	15	33	12	17	43	41	50	34	30	83	50	46
50	85	124	78	9	7	23	7	11	68	79	85	67	72	92	77	78
Birch with mixed herbage																
10	199	109	125	24	20	10	6	24	17	25	21	36	25	48	22	12
25	160	129	112	23	19	10	5	14	24	39	25	43	32	55	51	23
37	67	63	45	12	9	7	7	8	63	79	68	84	70	51	53	56
62	42	55	32	11	8	6	4	4	91	94	91	90	93	93	82	83
Oak with sedge and goutwort																
12	36	72	20	12	8	24	3	17	72	75	67	83	70	80	67	73
25	56	94	26	12	6	24	4	13	36	42	38	63	36	85	45	48
48	40	75	19	12	7	24	3	11	53	56	46	66	63	88	55	62
93	36	67	17	11	6	24	3	10	73	80	83	92	63	98	86	78
130	41	80	19	9	6	18	6	9								

parts die off every year. The annual growth in height of the trunk enables a tree to raise its crown much higher than the assimilating organs of all other plants and to occupy a dominant position in a phytocoenose.

The axial parts of trees (trunks, branches, and conducting roots) are mainly composed of carbohydrates, cellulose and lignin. These organs consist basically of dead wood vessels with a large quantity of mechanical elements. Cellulose and lignin take no part in the further circulation of matter during the life of a tree, and only after the death of parts or of a whole tree are they utilised by micro-organisms and fungi. Every year, therefore, part of the organic matter formed by a forest phytocoenose is removed from circulation and with increasing age the relative proportion of inactive parts in the total mass of a tree becomes larger (Table 61). Therein is the radical difference between forest and herbaceous phytocoenoses.

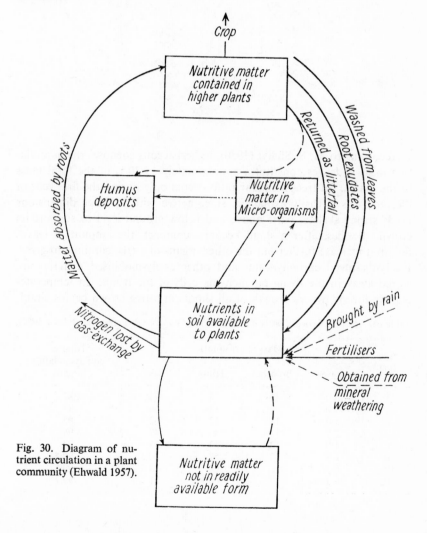

Fig. 30. Diagram of nutrient circulation in a plant community (Ehwald 1957).

Fig. 31. Diagram of nitrogen circulation in forest biogeocoenoses (Weetman 1961).

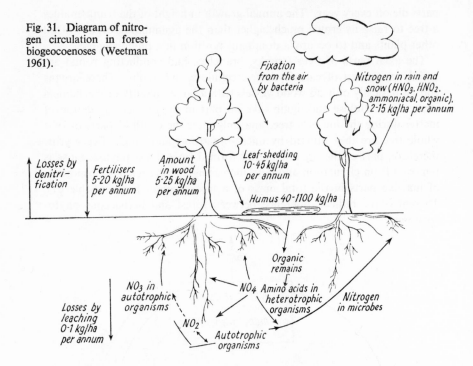

According to M. S. Shalyt (1950), in herbaceous coenoses the assimilating mass of leaves constitutes approximately 15-30% of the total mass of the plants. In trees a similar ratio occurs only during the first year of life, and this is easily explained, since a tree with its large dimensions needs strong skeletal parts above and below ground level to support its crown. Besides, after a stand becomes compact, the amount of leaves does not increase further but the other organs of a tree continue to grow. Easily-dissolved carbohydrates and other easily-mobilised material are stored away in the living parenchyma cells of the trunk. In temperate-climate forests, where growth of all shoots in a tree crown begins simul-

Table 61. *Relation between parts during growth of an oak (Remezov and Bykova 1952).*

Age (yr)	Mass (% of total)				Total of non-assimilating organs
	Leaves	Branches	Trunk	Roots	
1	13·0	5	8	74	87
2	7·0	9	18	66	93
3	5·0	11	22	62	95
10	2·0	12	43	49	98
20	1·0	12	52	35	99
30	0·5	12	57	30·5	99·5
40	0·4	12	64	23·6	99·6
50	0·3	12	66·7	21	99·7

taneously, the stored material is used to form leaves (while little or no photosynthesis is going on). This enables trees to develop leaves quickly and thus, when the maximum possible amount of leaves has been attained, to prevent development of plants in the lower storeys of the forest.

Reserves of non-nitrogenous matter and also of compounds of nitrogen, phosphorus, etc., which during the shoot-growth period were mobilised by the plant to construct new parts, are also held in large amounts in the bark and cambium.

A large amount of reserve material from which assimilating organs are produced is a distinguishing feature of all perennial plants, not only of trees; but in trees the predominance of non-assimilating parts over assimilating parts is most strongly developed.

The consumption of matter by trees in respiration is therefore high and it is very probable that the short growth period for tree shoots is an adaptation to reduce organic consumption in respiration so as to make the organic-matter balance more favourable. During the growth period consumption of organic matter by respiration increases greatly, and if growth were much prolonged that would prevent storage of organic matter to provide for growth in the following year.

Because of the increasing proportion of non-assimilating parts in the total mass of a tree stand with advancing age, consumption in respiration also increases. Therefore, as G. F. Khil'mi (1957) points out, the amount of energy used in maintaining the existing biomass of a tree increases with age. Furthermore the photosynthetic capacity is falling at the same time so that the organic-matter balance may finally be negative.

That is one reason for the death of the crown, and later of the whole tree.

The ratio between the amount of matter contained in leaves and in other organs depends not only on the age of the tree but also on its growing conditions. As Kramer and Kozlowsky (1960) point out, when trees grow in insufficient light the relative weight of the roots decreases. With insufficient water supply, on the other hand, salinisation of the soil lessens the proportion of the total organic matter in the leaves, while the proportion in roots and branches increases (Table 62). The warmer and drier the climate, the more organic matter is consumed in respiration.

It can be seen from Table 62 that even in compact stands with masses of leaves close together and consequently with hectares of leaf surface, productivity varies, because of different rates of photosynthesis, different ratios between photosynthesis and respiration, and different distribution of organic matter among the organs of the trees. The highest ratio of trunk wood to the total increment of organic matter is found in coniferous-broad-leaved and broad-leaved forests. In severe drought conditions the balance may be negative, because of use of previous years' reserves in the formation of leaves and partly also in increment of trunk wood.

If a tree has a negative balance over several years, the result is withering and death of the tree and thinning-out of the canopy. That is the reason for the short life of forest plantations in the steppe zone.

It follows that the existence of dense forests is possible only where there is an annual positive balance of organic matter and energy. For that a favourable local hydrothermal regime is necessary, since assimilation is limited to comparatively narrow ranges of temperature (narrower than the temperature range for respiration) and of moisture conditions. Analysis of the physiological curves of the photosynthetic process (its dependence on light and on CO_2 concentration), and also data on the earth's history, indicate that the arboreal flora developed during those epochs when the earth's climate was equable, warm, and humid, when the flow of solar radiation through the atmosphere was smaller, and when there was more CO_2 in the air.

Analysis of the present distribution of forests on the earth, made by various authors (Armand 1950, Weck 1957), shows that nowadays forests are adapted to humid and comparatively warm climatic zones, where prolonged periods of assimilation are possible. According to Richards (1961), in the humid equatorial zone trees constitute by far the greater part of natural vegetation. Where prolonged assimilation is impossible because of unfavourable conditions (on the borders of steppes and deserts where moisture is lacking, or in boggy places where it is excessive) the annual organic-matter balance of trees becomes negative at a com-

Table 62. *Balance of organic matter, reckoned as cellulose, in tree stands in different zones* (*Ivanov et al. 1963*).

Forest-growing zone	Composition of stand	Assimilated (ton/ha)	Consumed							
			In respiration		In growth				Total	
					of leaves		of trunk wood			
			(ton/ha)	%	(ton/ha)	%	(ton/ha)	%	(ton/ha)	%
Central taiga	Birch with 2nd-storey spruce	19·0	4·6	24	2·6	14	—	—	—	—
Coniferous broad-leaved forest	Poplar with some oak	23·8	6·7	28	2·6	11	8·8	37	18·1	76
Forest-steppe	Oak with some ash	12·2	4·3	43	2·6	21	3·9	32	11·8	96
Steppe	Oak-ash stand	9·5	6·3	66	2·6	27	2·8	30	11·7	123
Broad-leaved forest:										
Denmark*	Ash	9·9	3·1	31	2·7	27	4·1	41	9·9	100
Germany†	Oak	24·1	14·5	60	2·9	12	6·5	27	23·6	99

* Boysen-Jensen 1932.
† Polster 1950.

paratively early age, i.e., reserves of earlier years are drawn upon for growth, and are not replenished by current photosynthesis. The life period and productivity of such forests are reduced.

Partial improvement of the water regime is achieved by more vigorous development of the root system. The roots, but not the crowns, of the trees become denser, and the forest becomes sparser above ground. Increased consumption of matter in the formation of roots, however, also has a negative effect on the organic-matter balance, and therefore does not substantially improve the situation. In such conditions trees are replaced first by perennial and later by annual herbs (Varming 1901).

Forest distribution is also restricted by shortage of heat. Where the vegetative period is short and a sufficient amount of organic matter cannot be accumulated in summer (in the Far North and on mountains) forests are replaced first by sparse forest and then by tundra. The existence of forests in such conditions is also hindered because soil formation proceeds extremely slowly in deserts and tundras, both because little organic matter enters the soil on account of the scarcity of vegetation, and because of the unfavourable physical conditions (Pisek 1960). Therefore such soils are poor in nutritive elements, especially nitrogen. This is well confirmed by available data on forest productivity in different climatic zones of the earth. According to Peterson's data (quoted by Gessner 1960), annual production of organic matter in different climatic zones is determined by the hydrothermal regime of the territory. The latter term was called CvP by that author, and calculated as follows:

$$CvP = \frac{T_v.P.G.E}{T_a.12.10^2},$$

where T_v is the average temperature of the warmest month; T_a is the difference in temperature between the warmest and the coldest month; G is the length of the vegetative period in months; P is the annual total of precipitation; and E is the evaporation index (calculated as the amount of incident solar energy).

According to Weck's (1957) calculations, tropical forests with a CvP index from 5800 to 125,000 form on average 50-80 tons of cellulose per hectare per annum, i.e., close to the maximum possible, whereas in European forests with an index of 67 to 770, 6-30 tons of organic matter are formed per hectare (Fig. 32). The calculated and empirical figures are in good agreement.

Forests of different climatic zones are characterised by specific rhythms of conversion of matter and energy.

1. In tropical rain forests, conversion of matter proceeds regularly throughout the year at a comparatively low rate and the amount of matter brought into circulation is greatest there. Duration of life, and total mass of matter composing tree stands, reach very high figures.

2. With location nearer the Poles and a drier climate, the duration of

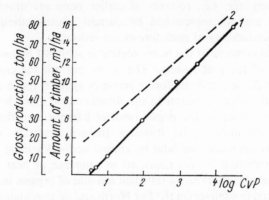

Fig. 32. Annual productivity and increment of wood in forests, with relation to hydrothermal regime of the territory (from data of Weck 1957). 1—annual increment of organic matter; 2—increment of trunk wood.

the annual period of life-activity gradually diminishes. Simultaneously the rate of matter and energy conversion per unit of time increases and the rhythm of matter conversion becomes pulsatory. The total amount of matter brought into circulation falls, as also do the general duration of life and the mass of matter composing tree stands.

3. Where conditions approach the limits for forest existence the active period is even shorter. High intensity of metabolic processes is possible for only a very brief period, and only in youth (Fig. 33). Afterwards the balance of conversion of organic matter and energy becomes negative. Productivity and duration of life are low.

Differences in matter conversion due to climate also affect the individual organs of trees. Storage of reserves in wood and bark is very characteristic of forests with periodic growth activity, the stored material being

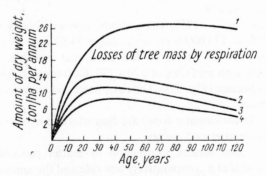

Fig. 33. Age-dynamics of organic-matter balance in a forest (from Polster 1950). 1—observed assimilation; 2—losses by leaves; 3—losses by branches and roots; 4—increment of above-ground mass.

starch, or soluble carbohydrates, or oils (Gulisashvili 1962). When starch is converted into oil the energy level is raised considerably. Consequently in that case loss of organic matter by respiration does not imply great loss of energy. In tropical climates, where even in deciduous tree species growth of shoots takes place at different times throughout the year, storage of reserves is usually not observed and the growth of new shoots is covered by current photosynthesis.

Let us examine further the chemical composition of the different organs of trees.

Since roots are organs of absorption of mineral matter, they contain much nitrogen, phosphorus, and mineral elements. Feeding roots are poor in mechanical elements, but there are many of these in conducting roots. The content of mechanical elements (cellulose) is small in leaves only in the young stage. The relative content of carbon compounds (mainly carbohydrates) in leaves is greater than in roots, but they contain less mineral elements and nitrogen. Large amounts of soluble carbohydrates and proteins (especially in young leaves) make possible their use as food by herbivorous animals.

Besides carbohydrates, non-nitrogenous matter of non-carbohydrate types may be formed in leaves, e.g., organic acids, lipoids, resins, and tanning substances. These and other substances of specialised metabolism may be contained in trunk bark in large quantities (e.g., rubber, gutta-percha). Often they have a higher energy level than carbohydrates. Storage of high-calorie reserve matter (oils) is very characteristic of seeds and fruits, which are therefore favourite foods of animals.

Differences in chemical composition create differences in calorific value per gramme of dry matter in trees. Ovington (1961) has given the calorific value of different parts of a pine tree in the following figures (in calories per gramme):

Needles:		Living branches	4892
1 year old	5077	Dead branches	4937
2 years old	5093	Trunk	4781
3 years old	5157	Average	4870
Average	5109	Thick roots (>0.5 cm)	3581
		Thin roots (<0.5 cm)	4612

These data show that needles possess the highest calorific value, axial parts occupy an intermediate place, and thick roots have the lowest calorific value.

The calorific value of thick roots is lower than that of carbohydrates (about 4000 calories), and that of axial parts and needles is higher. This may be because thick roots contain a large amount of inorganic matter.

As we approach the end of this section of the chapter, we may dwell on two items: (i) the biogeocoenotic role of the separate physiological processes that we have discussed, and (ii) the amount of matter brought into biological circulation on the earth.

All that has been said above indicates that the life-activity processes of plants are closely inter-related among themselves. We have tried to evaluate the role of each of them, from the point of view of accumulation of energy, in Fig. 34. In it the processes resulting in dispersal and deterioration in value of energy are enclosed in ellipses, while stocks of energy are enclosed in rectangles.

As we see, only one process, photosynthesis, leads to creation of stocks of matter and energy. Respiration, which provides energy for various

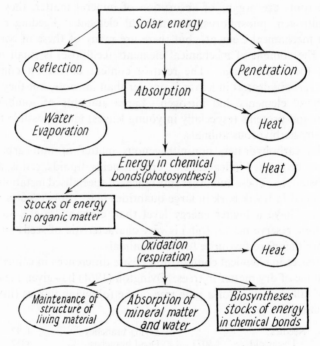

Fig. 34. Diagram of use of solar energy by plants.

biosyntheses, also partly helps in storage of energy, but the latter is actually the transformed energy of products of photosynthesis. Other aspects of life-activity carried on with energy derived from respiration (for instance, maintenance of the structure of living matter and absorption of minerals and water), although necessary for normal functioning of the whole system, lead to dispersal and depreciation of energy.

We may summarise the role of each process, from the viewpoint of its participation in the circulation of matter, in a simplified way as follows:

photosynthesis is the income side of the organic-matter balance;

respiration is the expenditure side of the balance;

water-exchange involves two processes: a very small part of the water enters into the composition of living tissues of plants, but most of it takes part in the biogeocoenotic circulation of matter (inevitably, because of the physical features of the environment of terrestrial plants);

absorption and conversion of mineral elements are necessary for regulation of the processes of organic-matter synthesis and plant growth.

The upper limit of production of organic matter on earth is determined: (*i*) by the amount of solar energy required to produce the maximum possible development of leaf-area of phytocoenoses per unit of area occupied by vegetation; (*ii*) by the amount of CO_2 in the air, on which the intensity of photosynthesis largely depends. A considerable fall in the actual productivity of a phytocoenose below the maximum possible is due to the effect of unfavourable external conditions: air temperature, water supply, and mineral nutrition. Under the influence of these factors the photosynthetic period is reduced, the rate of photosynthesis falls, leaf mass decreases, and losses by respiration increase because of increase in its rate and increase in the amount of the respiring mass.

The existence of forest phytocoenoses on earth is possible only in the most favourable hydrothermal regime.

INTER-RELATIONSHIPS BETWEEN PLANTS AND THEIR SIGNIFICANCE IN THE LIFE OF FOREST BIOGEOCOENOSES

When growing in the same locality and using the matter and energy resources of their habitat (ecotope), plants enter into relationships with each other that are complex and multiform in their mechanism. These inter-relationships represent one of the most important and characteristic forms of interactions in a forest biogeocoenose. Moreover, they are an indispensable condition of the existence of forest biogeocoenoses as complex matter-and-energy systems. Every biogeocoenotic process in a forest takes place with the direct participation of plants, and consequently it is closely linked with the complex and varied interaction among them, which in many ways determine the direction, nature, and rate of conversion of matter and energy. This is easily understood when one considers that one plant affects another mainly through the profound and many-faceted transformation of other components of a forest biogeocoenose (atmosphere, soil, microbiocoenose, animal population) and is based on the alteration of a large number of diverse factors and conditions that strongly affect the metabolism of plants.

A detailed historical survey of the development of views on mutual relationships between plants from the phytocoenological standpoint does not come within the scope of the present work. That has already been done in a number of summaries and reviews (Clements *et al.* 1929, Shennikov 1939, 1948, Sukachev 1946, 1953, 1955b, Knapp 1954, Grümmer 1955, Börner 1958, Rademacher 1959, Milthorpe 1961).

The present section undertakes to clarify the mutual relationships between plants as one of the forms of relationships and interactions in forest biogeocoenoses, and to explain the significance of these relationships in the processes of metabolism and conversion of matter and energy. This task is complex and difficult, as the biogeocoenotic approach to the problem of mutual relationships between organisms, described in the works of V. I. Vernadskii (1926, 1934), V. N. Sukachev (1946, 1956a, 1957), B. B. Polynov (1948, 1953), E. M. Lavrenko (1945, 1959), and N. V. Timofeev-Resovskii (1962), has not yet been adequately developed.

Works published to date on the problem of mutual relationships between plants have usually discussed only certain ecological, phyto-coenotic, and evolutionary aspects of the problem, without attempting to determine and explain the biogeocoenotic content of different forms of interaction among higher plants in natural biogeocoenoses. At present it is very difficult to draw a line between the phytocoenotic and the bio-geocoenotic studies of the problem, as both these aspects are closely bound together. Before we can discover the biogeocoenotic content of any relationship between plants we must know precisely: what role the relationship between them plays in the species-composition and structure of a forest biogeocoenose; what is the nature of the mechanism whereby one plant affects another; and what factors affect the nature and intensity of interactions between plants. Only after study of all these problems and determination of quantitative indexes characterising the intensity and mechanism of interactions between plants can we explain the significance of the relationship in the processes of accumulation and transformation of matter and energy in a forest biogeocoenose.

Fulfilment of the latter task, however, is impeded because many forms of mutual relationships between plants in natural biogeocoenoses are little studied and analysed from the standpoint of their significance in the life of a forest phytocoenose and of the internal mechanisms of these relationships. We therefore restrict ourselves here to the most general outlines in evaluating the biogeocoenotic significance of the mutual re-lationships between plants. All these difficulties do not discredit, or cause rejection of, the biogeocoenotic approach to the problem of mutual rela-tionships between plants, but merely show the necessity of broader and deeper analysis of the problem both in geobotany and in forest biogeo-coenology.

The present section falls into two parts. In the first part we discuss the general principles of the problem of mutual relationships between plants in forest biogeocoenology; the second part contains a systematic review of the chief forms of relationships between plants in a forest, with an attempt to evaluate their significance in the life and dynamics of forest biogeocoenoses.

INTER-RELATIONSHIPS BETWEEN HIGHER PLANTS, AND BIOGEOCOENOTIC PROCESSES IN A FOREST

Any biogeochemical manifestation of the life-activity of a forest plant depends on, and is determined by, its inherited characteristics and its physical environment in the ecotope as well as its relationships with other plants, which in many ways determines the level of the biogeochemical activity of a species and the share taken by it in the chemical work and synthesis of organic matter in a forest biogeocoenose.

The significance of mutual relationships between plants for biogeocoenotic processes in a forest is essentially that these relationships are a principal factor that determines and regulates species-composition, population numbers, and the structure and productivity of forest phytocoenoses. By changing and controlling the composition and structure of forest phytocoenoses, mutual relationships between plants substantially affect the whole course of biogeocoenotic processes.

Without going into detailed consideration of this important premise of forest biogeocoenology, we would point out that modern progress in the field of experimental study of plant communities clearly demonstrates the leading role played by mutual relationships between plants in developing phytocoenoses of definite species-composition and structure (Shennikov 1942, Sukachev 1946, 1956a, Karpov 1958, 1962, Ellenberg 1950, 1953, Knapp 1954). It has been shown that the simplified interpretation of composition and structure of a phytocoenose as merely the direct product of a complex of physical factors of the ecotope is completely erroneous and is unsupported by the latest experimental data.

The relationship of a plant to the complex of physical environmental factors in the ecotope is altered and regulated by other species of higher plants, with the biocoenotic relationships between different species of higher plants constantly providing the close correlative links existing between different types of ecotope and types of phytocoenose (Shennikov 1942, Ellenberg 1950, 1953).

This premise has been developed and demonstrated in studies of the ecological and biocoenotic ranges of higher plants, such studies having a very important and direct relation to forest biogeocoenology.

From observations in nature and direct experimental data, A. P. Shennikov (1942) and Ellenberg (1953) concluded that the role of individual species in the formation of phytocoenoses, and consequently their importance in the creation of organic matter and accumulation of energy, should be studied by making a sharp distinction between the relationship of species to the complex of physical factors in the ecotope and their relationship to the complex of biocoenotic factors in the phytocoenose. The distribution range of a species in natural conditions that determines its relationship to the first group of factors was called by Shennikov the 'ecological range' of the species. The distribution range of the same species that results from its relationship to other groups of higher plants is its 'biogeocoenotic range'.[1] It has been discovered that the ecological and biogeocoenotic optima of distribution of higher plants do not coincide.

[1] T. A. Rabotnov (1959) maintains that it is better to speak of 'synecological' and 'autecological' optima of distribution of species.

Many species of higher plants in monoculture plantations may thrive and (what is more to the point) may produce the largest amounts of organic matter in those very ecotopes where they are scarce in natural biogeocoenoses or where they are represented by severely suppressed specimens. The role of several species of higher plants in the production of organic matter and accumulation of energy may be reduced to a minimum in natural biogeocoenoses through restriction of their growth and development by the competition of other higher plants, although the potentialities of these species for biochemical work in the ecotopes concerned may be extremely high.

To illustrate these statements we quote the results of experimental investigations by Ellenberg (1953) with meadow grasses.

Foxtails (*Alopecurus pratensis*), French ryegrass (*Arrhenatherum elatius*), and brome grass (*Bromus erectus*) are prominent in Western Europe as the dominant producers of meadow communities in different habitats; the first dominates the herbage in damp meadows, the second on soils with average moisture, and the third in dry habitats.

By cultivating these grasses in special buildings, enabling him to control the underground water table within the range of 0 to 150 cm, Ellenberg discovered that all three species grew best and produced the greatest amounts of dry matter with an intermediate level of underground water (Fig. 35, A). By mixing these species experimentally, however, a quite different relationship between the grasses and the level of underground water was observed, more in accord with their behaviour in natural conditions. In that case also French ryegrass develops best with an intermediate level of underground water, and because of its higher competitive ability it forces out foxtails to damp habitats and brome grass to dry habitats, where they produce the greatest phytomass (Fig. 35, B). This experiment well illustrates the fact that such a valuable biogeocoenotic indicator as plant productivity depends on competition by other species and may alter according to the species-composition of the competing neighbours.

A. P. Shennikov has suggested that the nearer a plant is to the ecological optimum for its growth, the better and the more quickly it grows, as shown by more vigorous development of above-ground and underground parts and by more normal seasonal and age development cycles. An indication that a given species is existing in biocoenotically optimum conditions is provided by high frequency and numbers of the species and the intensity of its environment-forming effect on the ecotope in natural biogeocoenoses. Knapp (1954) and Walter (1962) define a number of types of higher plants by the relation between their ecological and biocoenotic optima. We shall restrict ourselves here to descriptions of only those of exceptional interest from the point of view of forest biogeocoenology (Fig. 36).

The first and most widespread group includes plant species whose

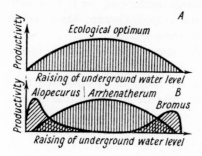

Fig. 35. Productivity of meadow grasses with relation to the static height of underground water. A—in unmixed cultures, all three grasses (*Alopecurus, Arrhenatherum, Bromus*) grow and produce the greatest mass with an intermediate level of underground water; B—in mixed sowings, French ryegrass produces the greatest mass with an intermediate level, foxtails with a low level, and brome grass with a high level of underground water (Walter 1962).

ecological and biocoenotic optima coincide (Fig. 36, A and B). It includes species that produce the greatest phytomass in ecotopes with an exceptionally favourable combination of physiologically-active environmental regimes. The plants of this group are further divided into two subgroups: (*i*) plants with very wide and almost co-extensive ecological and biocoenotic ranges; (*ii*) plants with a comparatively wide ecological range but a restricted biocoenotic range.

As an example of a species in the first subgroup we may cite spruce (*Picea excelsa*). Comparatively demanding with regard to soil and subsoil conditions, spruce produces most phytomass on the richest well-drained soils of the taiga zone. Its phytocoenotic optimum is also found there.

Examples of species in the second group are many broad-leaved tree species and associated grasses (Fig. 36, C). In taiga forests such broad-leaved species as linden (*Tilia cordata*), Norway maple (*Acer platanoides*), and European ash (*Fraxinus excelsior*) form part of forest biogeocoenoses

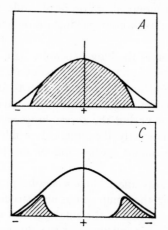

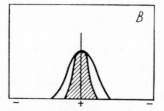

Fig. 36. Main types of relationships between ecological and biocoenotic optima for growth of higher plants. The perpendicular line with a plus sign marks the ecotope with the most favourable regime of physiologically-acting factors, and the minus signs mark extreme conditions for plant growth. The area within the curve denotes the ecological range of distribution of a species (in the absence of competition by other species); the shaded part denotes the biocoenotic range of the species. The higher the curve, the greater the productivity of the species. (From Knapp 1954.) Detailed explanation in text.

and have maximum production of organic matter only on rich or well-drained soils, typical of forest types of the *composita* group. Many species of 'oak wood' plants behave similarly to these species.

It has been proved experimentally that many ecotopes in the southern taiga zone (in particular, shallow podzols in strongly-podzolised soils of spruce-bilberry stands) are still within the range of broad-leaved trees and their associated species of grass, but these are absent because of competition by the dominant primary producers of taiga, i.e., spruce (Karpov 1960a, 1962). Many dominants of meadow phytocoenoses behave similarly (e.g., *Festuca pratensis*, Knapp, 1954).

Another important group of forest plants consists of species whose ecological and biocoenotic optima do not coincide. As a rule these species have a very wide ecological range and in the absence of competition from other species may grow in the most varied habitats, but produce the greatest amounts of organic matter in fertile soils with the most favourable regimes of physiologically-active environmental factors. These species have very low competitive ability and so are forced from the better soils by other species. In nature these species form a forest only in the less favourable habitat conditions, beyond the ecological range of their competitors. Consequently the productivity and biogeochemical work of such species in natural biogeocoenoses are small. These species include pine, which is dominant in biogeocoenoses in the taiga zone only on poor, dry, sandy soils (*Pineta cladinosa*) or on very wet sphagnum peat beds (*Pineta sphagnosa*). As seen from Fig. 36, pine has two biocoenotic growth optima, neither of which coincides with its ecological optimum.

Among meadow grasses, matgrass (*Nardus stricta*) behaves like pine, producing maximum phytomass in a monoculture on fertile soils, but in natural conditions only forms a sward with poor production on acid and poor soils (Knapp 1954).

It is quite evident that an evaluation of the biogeocoenotic significance of any species or group of similar species in a forest biogeocoenose should include examination of their ecological and biocoenotic growth optima as a basis for discovering the factors limiting the productivity and biogeochemical work of the species composing the forest biogeocoenose. Such an approach aids in solving correctly the practical problems of raising the productivity of valuable timber species and of the biogeocoenose as a whole.

The inter-relationships between plants affect biogeocoenotic processes in a forest both indirectly by altering the composition and structure of forest phytocoenoses and directly by affecting the total productivity of coenoses, the photosynthetic processes in plants, and the dynamics of organic matter accumulation. Unfortunately there are comparatively few data throwing light on this problem, which is so important for forest biogeocoenology, and these data refer mainly to monoculture tree plantations and fields of perennial grasses (Eitingen 1918, 1925, Uspenskaya 1929,

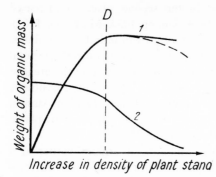

Fig. 37. The relationships usually observed among density of plant stand, total accumulation of organic mass (1), and average weight of an individual plant (2) (Nichiporovich, 1955). D—point of optimal density of plant stand. Scales of curves (1) and (2) are different.

Kondrat'ev 1935, Tol'skii 1938, Sukachev 1953, Knapp 1954, Rubtsov 1960, Milthorpe 1961).

We may conclude that, within certain limits, increasing numbers of plants per unit area gives a directly proportional increase in the weight of the phytomass (Figs. 37 and 38). With further increase in the density of plants there is an abrupt check to phytomass accumulation (point D), and in some cases of very high density a decrease occurs in the accumulation of phytomass by plantations and crops (Knapp 1954).

The total phytomass produced by very dense tree and grass stands seems little different from the mass they would produce with the lower density at a certain optimal number of plants per unit of area. This is because, as a plant stand increases in density and the feeding area of each single plant diminishes, the intensity of competition among the plants for light, moisture, and nutrients rises quickly, leading in the first place to

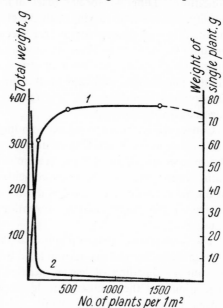

Fig. 38. Effect of density of grass stand on total productivity and individual plant weight of ryegrass. 1—relationship of total crop of aboveground mass to density of stand; 2—relationship of individual weight to density of stand (Shennikov and Serafimovich 1963).

reduction of individual weight and in the second place to increased dying-off of plant parts (Sukachev 1953, Knapp 1954, Milthorpe 1961).

Thus by decreasing weight and intensity of growth in individual plants, heavy competition among plants in dense stands also controls the total productivity of the coenoses. It is important to point out that the greatest efficiency in use of radiation energy and the highest photosynthetic productivity are observed at a certain optimal density of plant stands, but with excessive stand density the indices of energy utilisation by plants decrease (Ivanov 1946, Nichiporovich 1955).

Phytocoenoses produce the greatest amounts of organic matter at a density where mutual suppression of plants and a decrease in their individual weights can already be clearly traced. Within definite limits of increase in density, however, the weight decrease for individual plants is offset by increase in their numbers, so that the total production of phytomass of the entire phytocoenose is increased.

From analysis of numerous data on the connection between stand density and productivity of coenoses, we conclude that there is an optimum density at which forest biogeocoenoses possess the greatest total productivity and the highest rate of accumulation of phytomass by the separate plants composing them. Determination of that optimum density is one of the urgent tasks of forest biogeocoenology. We must keep in mind, however, that, as many investigations have shown, the plant-stand density (D) at which a sharp fall in individual productivity is observed and the curve of total productivity of the stand flattens out depends not only on the biotype and the plant species but also on the environmental conditions. Different species react differently to the same plant-stand density, the reaction varying with the climatic and soil conditions of the stand. Thus in shade-tolerant tree species the optimum density corresponding to point D is observed at a much higher stand density than in light-loving species. On poor or dry soils the intensity of competition among plants rises, and they need a greater feeding area, which lowers the optimum stand density. The relationship of plants to stand density varies also with their age.

Consequently, in analysing such an important function of a forest biogeocoenose as its production of organic matter, we must look at it from the biocoenotic point of view, remembering that the productivity of an entire forest phytocoenose never coincides with the potential productivity of all the individuals composing it, and that competition between plants is one of the major factors preventing attainment of these potential figures.

The interactions between plants also play a large part in the conversion of organic matter already accumulated by a phytocoenose, and in causing the annual shedding of leaves, needles, branches, and stems of forest vegetation and their subsequent decomposition and mineralisation. Intensive competition among plants and severe suppression of some by others is always accompanied by an increase in annual dying-off of above-ground and underground parts of plants (Knapp 1954), and also death of a large number of trees, herbaceous plants, and shrubs, thus causing considerable amounts of new organic matter to enter the soil. That organic matter is a source of food and energy for various hetero-trophic organisms, as a result of whose activity it is broken down and undergoes a complex cycle of chemical transformation. The final products of its decay are again used by higher plants and are included in the continuous circulation of matter and energy in a forest biogeocoenose.

There are abundant data on the scale of tree mortality during the natural thinning-out of a stand with age; these data have often been discussed in the literature in connection with the problem of over-population and mutual relationships in forest phytocoenoses (Morozov 1926, Sukachev 1946, 1953). From experimental records of the course of growth of various stands it is known that out of hundreds of thousands and even millions of tree seedlings springing up in burned and cut-over areas, usually not more than 5000 to 10,000 trees survive to the age of 20 years, and at the age of 120 years the number falls to 400 or 500 per hectare. It is quite obvious that great masses of organic matter enter the soil as a result of the death of these huge numbers of trees in the course of competition and development of stands.

The quantitative aspect of that process is less well studied for lower-storey plants, although we may suppose that in these layers also there is colossal mortality of individuals through intensive competition for the means of living. For instance, herbaceous communities whose productivity reaches 3·5-4·5 ton/ha in an open cut-over area die off almost completely as the forest becomes re-established, causing the entry into the soil of large quantities of organic matter differing somewhat in chemical composition from tree litter. The whole future course of syngenetic succession of forest biogeocoenoses, whose activating force is undoubtedly the relationships between plants (especially between forest-forming species of original and derivative forest phytocoenoses), is accompanied by continual and colossal mortality of individuals of many species, to the point of their almost total disappearance from the composition of the forest phytocoenoses. For instance, in the process of replacement of birch stands by original spruce phytocoenoses the whole stock of birch wood and the solar energy accumulated in it enters the soil and is included in a new cycle of matter and energy transformation, unless, of course, the stock is removed and used for human needs. Specific calculations show that throughout the life of the birch stand about 150-200 tons of organic matter per hectare may thus enter the soil. In the course of syngenetic successions forest biogeocoenoses are replaced by others differing substantially in the intensity and nature of the processes of exchange of matter and energy (e.g., birch and spruce biogeo-coenoses), these alternations in types of matter circulation are again determined by the inter-relationships among plants. Unfortunately the biogeocoenotic significance of all these phenomena resulting from the inter-relationships among plants still remains unstudied and undiscovered. Recently some data have appeared throwing light on the biogeocoenotic relevance of the natural thinning-out of tree stands, e.g., in studies of the circulation of nitrogen and mineral elements in forest biogeocoenoses, figures showing the amount of organic matter, and its content of nitrogen and salts, entering the soil from dead trees (K. M. Smirnova 1951 a, b, 1952, Mina 1954, Remezov et al. 1956, Rodin 1961, Parshevnikov 1962). According to these data, the return of nitrogen and salts from trees dying during the struggle for existence constitutes a large part of the total balance of organic matter entering the soil. No less important from the biogeocoenotic viewpoint is competition among plants, which is accompanied by changes in the chemical composition of plant detritus entering the soil. The chemical composition of organic matter entering the soil from dead trees is quite different from that of litter fall (Remezov et al. 1959), and competition between plants affects the chemical composition of the competing plants. For instance, it has been shown that spruce seedlings and forest grasses growing competitively with tree roots in taiga biogeocoenoses have low contents of total nitrogen, calcium, and other elements as compared with the same plants when protected from such competition (Karpov 1959a). Fragmentary data exist to show that severely-suppressed and stunted trees differ from dominant trees of Kraft's classes I to III in the chemical composition of leaves and needles (Aaltonen 1942, Gulidova and Tsel'niker 1962). Unfortunately the extreme scarcity of such data at present prevents us from drawing a conclusion regarding the significance of such changes in biochemical composition of plants during the exchange of matter in a forest biogeocoenose. It is quite evident, however, that they play some lasting role in the life of forest biogeocoenoses.

Finally, the formation and evolution of plant species are closely linked with constant competition for living requirements between plants in a phytocoenose and with natural selection, and are accompanied by the appearance of unique and sometimes surprisingly precise adaptations to community life and to the fullest use of the matter and energy resources of forest biogeocoenoses (e.g., epiphytism, liana growth, symbiosis). Some scientists are now inclined to regard a biogeocoenose as a system basically controlling and regulating the process of evolution of species through evolutionary transformation and natural selection (Timofeev-Resovskii 1958, Shmal'gauzen 1958).

Without touching on the genetic and evolutionary aspects of this problem, we merely state that in this case the forest biogeocoenose itself stands out as a factor in the directed and regular evolutionary transformations by which ecologo-physiological, morphological, and biological features of higher plants, and consequently their competitive ability and biogeochemical activity, are altered. Linked with these evolutionary transformations are the origin and development of new plant coenotypes adapted to fuller utilisation of the matter and energy resources of the currently unoccupied ecological niches and units in a forest biogeocoenose. In turn this leads to complication of the composition and structural organisation of the biogeocoenose, and consequently increases the efficiency of its chemical work.

We have not discussed all the many-faceted complexities of the biogeocoenotic problem of inter-relationships among plants. We have restricted ourselves to a few statements and factual data illustrating the significance of the interactions among higher plants in the life of forest biogeocoenoses. Nevertheless, it is clear that study of the many kinds of interactions among plants in a forest is one of the most urgent tasks of forest biogeocoenology, since it will assist in more profound analysis of all the complex processes of transformation of matter and energy that take place in forest biogeocoenoses.

FACTORS REGULATING THE INTENSITY AND MECHANISMS OF INTERACTIONS BETWEEN PLANTS

The intensity of interactions between plants, their mechanisms, and their biogeocoenotic significance vary within wide limits in natural biogeocoenoses, depending on the complex assemblage of factors. This fact, unfortunately often overlooked, sometimes leads to serious disagreement by different investigators in the evaluation of the role of the various forms of inter-relationships between plants in the life of forest biogeocoenoses.

In this respect we remember the very lively discussion that arose after the appearance of the work of K. Fricke (1904), who was the first to demonstrate experimentally the depressing effect of competition by mature tree roots on the growth and development of undergrowth beneath forest canopy. Some investigators, repeating Fricke's experiments, agreed with him and considered competition by tree roots to be the chief factor impeding restoration processes beneath a forest canopy (Toumey and Kienholz 1931, Korstian and Coile 1938). Other investigators at first stressed competition for light as the main cause of death of undergrowth in dense forests (Fabricius 1927, Olmstead 1941, Shirley 1945, Lutz 1945, Oosting and Kramer 1946). In recent years it has been clearly shown that this problem, so important to sylviculture, has not and cannot have a universal solution. The mechanism and intensity of the depressing effects of mature trees on undergrowth in forest biogeocoenoses vary considerably with the physico-geographical conditions of the forest biogeocoenoses, with the specific soil conditions of the ecotope, with the ecologo-biological features of the species composing the forest and the undergrowth, and with the nature and the intensity of the environment-creating influence of mature trees on phytoclimate and soil (Shirley 1945, Karpov 1955a, 1960b).

The discovery of the mechanism controlling different relationships between plants and comparative evaluation of their significance in forest biogeocoenoses is possible only on the basis of differential analysis of the effect of different factors on the relationships between forest phyto-coenose components. Such analysis is absolutely necessary for the discovery of effective and active methods of regulation of the mutual relationships between plants in forest biogeocoenoses and for control of them. By modifying these factors we can control the course of competition between plants for the food and energy resources of the habitat, and consequently regulate the biogeocoenotic processes.

Among the outstanding factors affecting the intensity and mechanism of interactions between plants in forest biogeocoenoses are the following:

1. *Ecologo-physiological and biological features of species* of higher plants, which in the aggregate constitute 'the biological equipment of the species' (Salisbury 1929). Charles Darwin showed that hereditarily-fixed biological differences between species and between individuals of the same species have profound effects on the outcome of competition between plants for space and living requirements. Since Darwin an enormous number of facts, observations, and experimental data have accumulated in forestry and phytocoenology, confirming his statements.

Valuable factual data on this subject may be found in numerous works on geobotany and forestry (Morozov 1912, 1926, Salisbury 1929, Clements *et al.* 1929, Toumey and Kienholz 1931, Shennikov 1938, 1946, 1950, Sukachev 1941, 1946, 1953, 1959a, Shirley 1945, Ellenberg 1950, 1953, Knapp 1954, Rabotnov 1950, 1959). Investigations devoted to the discovery of the significance of small individual biotic differences within a species in competition between plants and in natural selection are of exceptional interest (Morozov 1912, Sukachev 1933, 1959b).

Analysing all these data, we conclude that a decisive role is played in the inter-relationships between plants by the following ecologo-physiological features: (*i*) capacity to carry on photosynthesis in light of low intensity; (*ii*) high efficiency of root systems in absorbing moisture and nutrients when these are in short supply; (*iii*) resistance to unfavourable factors in the physical environment (e.g., low and high air temperatures, poor aeration of the soil, presence of harmful salts in the soil); (*iv*) general high level of metabolism.

These characteristics of plants are the most important biologically: (*i*) life form; (*ii*) abundance and periodicity of seed production; (*iii*) adaptation to seed distribution; (*iv*) vigour and means of vegetative reproduction; (*v*) rate of growth and amount of development of above- and under-ground parts; (*vi*) resistance to pests and fungal diseases; (*vii*) longevity. The range of differences among species of forest plants in ecologo-physiological and biological features is so great that within forest biogeocoenoses interactions occur differing greatly in intensity and mechanism. The situation is complicated because, while certain characteristics of a species give it a competitive advantage over others, other characteristics make it less able to compete, so that it disappears from a forest biogeocoenose.

Many plant species can carry on photosynthesis with a large positive balance only in high light intensity, which is not as a rule found within forest biogeocoenoses. That characteristic is found in Scots pine and birch, and not only jeopardises these forest-forming species in the taiga-zone forests but also determines the mechanism of the competitive effect upon them of shade-tolerant species such as spruce, mainly through interception of light by the crowns of spruce. On the other hand, ability to carry on photosynthesis in the dim light and low temperature of the air layer nearest the earth enables spruce to compete successfully with pine and small-leaved species and to force them out of forest biogeocoenoses. Spruce, however, is more sensitive to nutrient shortage in the soil, and under a canopy of deciduous species suffers severely from the interception of moisture and nutrients by their roots, which greatly impedes replacement of deciduous biogeocoenoses by spruce (Karpov 1960b). Generally, where competition is taking place between plants of the same life form and with the same development of above- and under-ground parts, the species with greater requirements of a factor existing at the ecological minimum cannot compete with less demanding species, and either drops out of the forest biogeocoenose or remains in a suppressed state.

In many cases, however, the deciding factors are rate of growth and extent of development of above-ground and underground parts, vigour of sexual and vegetative reproduction, duration of life, and resistance to unfavourable action by climatic and edaphic conditions and to pests and diseases.

Light-loving species that reproduce and disseminate quickly and have high rates of growth may hold their territory for a long time and prevent shade-tolerant species, coenotically stronger, from invading a forest biogeocoenose. Sometimes very small differences in individual growth rates may have a decisive effect on the final outcome of competition between plants for space and living requirements. Striking data in this respect are presented by Salisbury (1929), and Weaver and Clements (1938), and show that differences of only a few millimetres in the height of grasses may decide the outcome of competition between them.

The rates of growth of root systems, their structure, and their efficiency in absorbing nutrients are of outstanding importance. Such species as ash and birch possess higher rates of growth and regeneration of root systems than oak and spruce, and can suppress the latter by occupying soil space more rapidly and lowering its content of moisture and nutrients (Karpov 1952, Orlov 1960, Rakhteenko 1958a, 1961).

Ecologo-physiological and biological features not only determine the mechanism of interactions between plant species but also strongly affect the course of competition between individuals of the same species. For instance, pure stands of light-loving species have a higher rate of thinning-out and of differentiation of trunks into Kraft's classes than stands of shade-tolerant species (Morozov 1926, Sukachev 1953). The ecologo-physiological and biological features of forest plants that determine the dominance of one species over another when growing together have not yet been studied in many individual cases. This greatly impedes analysis of the inter-relationships between plants and discovery of their mechanism. As the experimental studies of V. N. Sukachev (1959b) have shown, biotypes that are only slightly different morphologically sometimes exhibit considerable differences in competitive ability. Consequently one cannot always judge the competitive advantage of a given race, or of a hereditary characteristic, by its morphological features.

The intensity and mechanism of the interactions between plants can be studied in a forest biogeocoenose only by comparative study of the ecologo-physiological and biological features that enter into its species-composition (Shennikov 1946, Sukachev 1953). Account must be taken, however, of the variability of these features within the same plant species according to the conditions of formation of a forest biogeocoenose. These features are best developed in optimal physiologically-active regimes of the ecotope and are weakened or lost in extreme environmental conditions. This situation is reflected in the competitive ability of a given plant species, and clearly each item in the physiological and biological charac-teristics of a species should always be studied at the borders of the eco-logical range of the species.

Only by discovering the features of economically-valuable species of forest plants that impair their position in competition with weed species can we know how to protect them.

2. *The environment-forming capacity of species* determines in many ways the means and form of the influence of certain plants on others when growing together in a forest biogeocoenose. Plants have extremely varied effects on their environment, these effects extending over all the chief components of a forest biogeocoenose: atmosphere, soil, microbial popu-lation, and animal life. All plant species in a forest biogeocoenose take part in altering these components. The various species, however, differ widely in both the intensity and the nature of their transforming actions. Some cause marked alteration in the physiologically-active regimes of the ecotope, while other species have very little effect in this respect. Undoubtedly, the former group includes the dominants of the forest biogeocoenose, which make great changes in the phytoclimate and the soil and, by creating beneath their canopy a specific phyto-environment, determine the composition and structural organisation of species with little environment-forming capacity (acceptors) growing in the lower storeys of the forest biogeocoenose.

The quantitative value and the nature of the environment-forming effects of the various species in a forest biogeocoenose are determined by many factors, mainly: (*i*) the life form of the plant; (*ii*) the number of individuals of the species; (*iii*) the mass, structural features, and dimensions of the above-ground and underground organs; (*iv*) the biological characteristics of the plant, e.g., its longevity and the features of the seasonal development of its above-ground and underground organs, which in many ways determine the duration and regularity of its environment-forming actions; (*v*) the level of the general metabolic activity of the species, which determines the efficiency and rate of its use of the food and energy resources of its habitat.

Depending on these factors, and according to their ecologo-biological characteristics, different plant species have different quantitative and qualitative effects on the physical aspects of their environment. Some species affect others by modifying the phytoclimate by means of their above-ground parts, mostly through interception of radiant energy and change of its qualitative composition. This type of modification of some plants by others has been described in numerous studies of the spectral features of the above-ground parts of plants and the light regime in forest biogeocoenoses. The ability of species to alter the light regime is in many ways determined by the specific optical properties of their leaves, by features of crown structure, and by stand density (Ivanov 1932, 1946, Sakharov 1948, Seybold 1932, Shirley 1945, Kittredge 1948, Walter 1951, Gar 1954, Molchanov 1954a, Kleshnin 1954, Karpov 1955b). Other plant species affect the ecotope most strongly through change of edaphic factors. In some types of forest biogeocoenoses this is shown by acute drying-out of the soil by the roots of dominant synusiae; in others, in a lowering of the concentration of various nutrient solutions in the soil, in discharge of useful or harmful organic compounds by roots into the soil, or in the accumulation of forest litter with varying physical and chemical properties.

Higher plants may strongly affect each other by changing the composition and activity of various groups of micro-organisms in their rhizospheres and by changing soil fauna (Krasil'nikov and Garkina 1946, Bonner 1950, Becker and Guillemat 1951, Winter and Willeke 1951, Franz 1950, 1955, Wittich 1953, Runov and Kudrina 1954, Runov and Enikeeva 1959, Rademacher 1959, Börner 1959).

T. A. Rabotnov (1962) points out that change of environment by plants may be current (e.g., absorption of water and nutrients, alteration of lighting) or cumulative, i.e., increasing from year to year and being expressed in such changes as accumulation of litter, gradual alteration of hydro-physical and chemical features of the soil (accumulation of humus, change in the reaction of the soil, etc.), general alteration of the soil-forming process, etc. The fact that cumulative environment-forming action plays a decisive role in the inter-relationships between plants in forest biogeocoenoses is proved by numerous studies of the effect of litter-accumulation on the growth and development of tree shoots and herbage, the effect of soil-leaching on the growth and development of trees in the steppe zone, and the changes in hydro-physical and chemical properties of the soil with succession of species (Bautz 1953, Bublitz 1953, Winter and Bublitz 1953a, Runov and Kudrina 1954, Karpov 1955 a, b, Parshevnikov 1957, 1962). Knowledge of the complex mechanisms of inter-relationships between plants in forest biogeocoenoses can only be obtained by careful and thorough study of the various kinds of action by higher plants on their environment, and by determination of the reaction of other components of a forest biogeocoenose to these changes. The quantitative and qualitative results of the environment-forming action of a plant species also vary within wide limits according to factors of the physico-geographical environment and biocoenotic conditions.

3. *The physical factors of the ecotope and general physico-geographical conditions* of growth of a forest biogeocoenose greatly affect the intensity and mechanisms of interactions between plants. Such factors directly constitute and determine the matter and energy resources of the habitat, or, in other words, the level of supply of solar energy, atmospheric CO_2, soil moisture and nutrients, etc., to the plants; they also determine the parameters of the chief physiological processes in plants, including metabolism and use of the various resources of the habitat for carrying out life functions (e.g., absorption of water from the soil for transpiration).

The complex and varied processes of carbon and mineral nutrition of plants depend primarily on the presence, in both the atmosphere and the soil of a given ecotope, of adequate amounts of the materials composing the plant body and involved in its metabolism. Generally, in natural ecotopes shortages occur in the matter and energy resources in habitats (on account of high overpopulation of forest phytocoenoses); this may be true for different nutrient elements. In natural biogeocoenoses the quantitative and qualitative inter-relationships between factors determining the energy and food resources of a habitat vary within wide limits, some inevitably being at ecological minima. Competition between plants is especially keen for the nutrients at ecological minima, and since these differ in different ecotopes, we find different mechanisms of dominance by certain plants over others. This is proved by the great number of observations and experimental data demonstrating the close connection between the intensity of competition between plants in a forest biogeocoenose and the amounts of various basic requirements: soil moisture, different kinds of nutrients in the soil, CO_2, light, etc. (Toumey and Keinholz 1931, Korstian and Coile 1938, Kramer and Decker, 1944, Oosting and Kramer 1946, Sukachev 1946, 1953).

We can often deduce, simply from studying the dynamics of the basic requirements in natural biogeocoenoses, for what items and with what intensity competition is taking place between plants. For instance, G. F. Morozov (1899), from observations on seasonal soil humidity in the Khrenovsk pine forest, concluded (even before the appearance of Fricke's work in 1904) that suppression of pine undergrowth by mature trees in lichen-pine forests results mainly from drying-out of the soil by the roots of the parent trees.

Many physical features in the ecotope exert considerable indirect influence on the use of nutrients and solar energy by plants in a forest biogeocoenose. These factors include the temperature of air and soil, and the hydro-physical and some chemical features of the soil (acidity, humus content, etc.). By modifying the course of physiological processes in plants, these factors control the growth and extent of development of the above-ground and underground organs and their rates of seasonal development, and alter the light, moisture, and nutritional requirements of plants and so indirectly affect the plants' interactions in forest biogeocoenoses.

The significance of these factors is illustrated by the dependence of the rate of thinning-out of tree stands on the climatic and soil conditions (Morozov 1926, Sukachev 1953). As climatic and soil conditions for tree growth deteriorate, the natural thinning-out of tree stands with age decreases notably. According to records of the growth of pine forests in Leningrad province, for example, the numbers of 100-year-old pine trees according to soil fertility are: quality class I, 532; class II, 668; class III, 759; class IV, 933; and class V, 1189 per hectare.

The poor growth and low quality of spruce stands on peat-humus soils in the central taiga zone is due to prolonged high levels of underground water with very low oxygen content and to inadequate warming of these soils (Orlov 1962). On such soils spruce stands of quality classes IV and V are formed, with the number of 100-year-old trees being from 1400 to 1800 per hectare. In the same climatic conditions, on better-drained and warmer, weakly-podzolic soils, highly-productive spruce stands of quality classes I and II grow, with 800-950 trees per hectare at 100 years of age.

Excessive saturation of peat-humus soils with water of low oxygen content, combined with their other unfavourable features, impedes the growth of spruce trees, reduces the vigour of development of their above-ground and underground parts, and so lessens the intensity of interaction among them and consequently the rate at which the number of trees in stands decreases.

Similar relationships can be identified in analysis of the effects of climatic conditions on the growth of tree stands and their stocking density (Morozov 1926, Tkachenko 1939, Rubner 1953, Aaltonen 1942).

These considerations indicate that interactions between plants are controlled and directed by a group of factors of the physical environment of an ecotope, which are very complex, act together simultaneously, and are extremely variable in time and space. Therefore field and experimental studies of the various biotic factors and their combinations must be made in order to discover the mechanism and significance of the various kinds of relationships between plants in the life of forest biogeocoenoses. By adjusting various factors of the physical environment we can regulate not only the productivity of stands but also the relationships between different components of a forest phytocoenose, stimulating or checking the intensity of competition among them. Meanwhile we must remember that a forest phytocoenose reacts to change in any single factor of physical environment with a very complex reconstruction of the inter-relationships between plants. If by agrotechnical and forest-improvement measures we improve the soil-nutrition conditions of forest plants, and raise the rate of growth and the vigour of development of trees, we may simultaneously stimulate competition between them for light and thus create the prerequisites for valuable slow-growing species to be forced out of the composition of forest biogeocoenoses by weed species having little economic value, but responding quickly to these measures.

4. *Structural features* of a forest phytocoenose (e.g., density of tree or herb layers, and consequently the degree of proximity of their above- and under-ground organs; or the arrangement of plants in the coenose) may exert the strongest and most direct influence on competition between plants and the mechanism of interactions among them in a forest biogeocoenose.

The significance of these factors in inter-relationships between higher plants has been studied by many authors and discussed in great detail in a number of works devoted to the study of the effects produced by sowing density and by the placement of components of a forest phytocoenose on the growth and development of forest plants, on their resistance, on their ability to survive, on their productivity, etc. (Eitingen 1918, 1925, Sochava 1926, Smirnova 1928, Krasovskaya 1931, Sukachev 1941, 1959b, Timofeev 1947, Karandina 1953, Rubtsov 1960, Sidel'nik 1960). These studies have shown that every plant species requires a definite feeding area for its normal development, the dimensions of the area varying in accordance with the age of the plant, the climatic and soil conditions of its growth, and the species of its competitors. Excessive density of sowing and planting induces more intense competition between plants for living requirements, which is manifested in general severe stunting of plants and high mortality in dense stands, and in weakening and sometimes loss of the power to reproduce by seed or vegetatively (Shennikov 1939, Sukachev 1953, Knapp 1954). The ultimate effect of competition on plant species is alteration of many anatomo-morphological features and breakdown of their principal physiological functions and rhythm of development, which in one way or another affects their competitive ability (Sukachev 1953, Rabotnov 1950, Karpov 1955a, 1960b).

Numerous observations and experimental data also show that the intensity of interactions among plants depends on the distribution of species in a forest phytocoenose (Vysotskii 1912, Kharitonovich 1951, Pyatnitskii 1951, Karpov 1952, Sidel'nik 1953, 1960). The resistance of many tree species, like that of entire artificial plantations, depends not only on the biological features of the species and the forest-growth conditions but also on their combination and situation in the plantation. Some tree species when placed in direct proximity to each other mutually or unilaterally suppress each other (e.g., ash and elm severely suppress oak), making it necessary to separate them by means of rows of trees of other species or shrubs. Many examples could be quoted from experience gained in the culture of meadows, where rational placement of components in artificial seeding of grasses has proved of great value (Shennikov 1939, 1941, 1950, Ponyatovskaya 1941).

The intensity of interactions between plants and the methods and forms of action by some plants on others control and regulate a complex assemblage of factors, analysis of which is absolutely necessary for understanding the biogeochemical role of any species of higher plant in a forest biogeocoenose. The quantitative and qualitative expression of these factors depends on the nature and ecological characteristics of a given plant and its competing species, on how different factors of the physical environment are combined together, and on how they affect the basic physiological processes and the condition of the plants and their competing species.

BRIEF REVIEW OF THE CHIEF FORMS OF INTERACTIONS BETWEEN PLANTS

Attempts to classify the mutual relationships between higher plants have been made by many investigators (Clements *et al.* 1939, Sukachev 1946, 1953, 1956b, Shennikov 1950, Braun-Blanquet 1951, Knapp 1954, Grümmer 1955, Sokolov 1956, Korchagin 1956, Lavrenko 1959, Rademacher 1959). Perhaps the most successful is the classification by V. N. Sukachev (1956b). In that classification mutual relationships between plants are grouped according to the methods whereby they affect each other; as a basis for separating categories such as 'transabiotic' and 'transbiotic' interactions between plants, Sukachev took the chief forest-biogeocoenose components through alteration of which one plant usually affects another. At the present level of our knowledge of the interactions between higher plants in forest biogeocoenoses such an approach best fulfils the most important task of forest biogeocoenology: discovery of the principal forms of the relationships and interactions between the chief components of a forest biogeocoenose, and clarification of the mechanism of these interactions. Study of the inter-relationships between plants is closely connected with that of the other diverse kinds of interaction between the components of forest biogeocoenoses, and of the processes taking place in them.

Before discussing the characteristics of different forms of relationships between plants, we must examine some of the more general and universal forms of interactions between them, not mentioned previously.

Interactions between plants can be favourable or unfavourable for species or individuals growing together in a forest biogeocoenose. The unfavourable influence of one plant on another may be manifested in mutual or unilateral suppression of one species or individual by another. The mechanism of the unfavourable action of certain plants on others in a forest biogeocoenose is very complex, and includes widely differing methods of dominance and suppression, which will be discussed in detail later. Here we merely remark that the unfavourable influence is expressed largely in competition between plants for living requirements, and results from their simultaneous use of food and energy resources of the same habitat. Favourable influence ('mutual aid') also may take the form of unilateral or reciprocal improvement or stimulation of growth of plants growing together. Usually favourable interactions between plants are based on such environment-forming effects as moderation of violent fluctuations in temperature, reduction of insolation, breaking of wind force, and raising of air humidity when plants grow together, i.e., alteration of factors affecting the general physiological state and living conditions of plants. Sometimes different types of favourable influence result from simultaneous use of food and energy resources of the same habitat, usually taking the form of unilateral benefit (host-tree and plant-parasite; tree and epiphyte; etc.).

According to V. N. Sukachev (1946), mutual aid between plants in a forest phytocoenose (both intra- and inter-specific) is either a simple result of close proximity due to overpopulation, or has been developed through natural selection in the process of competition with other species.

Other scientists have come to the same conclusion, attributing the origin of such relationships as parasitism, symbiosis, and epiphytism to intensive competition between plants for space and living requirements (Schimper 1935, Schmucker 1959 a, b).

Favourable and unfavourable effects of certain plants on others may occur between individuals of the same species or of different species. In the first case we are dealing with intraspecific, in the second with interspecific relationships between plants. These two major categories of inter-relationships between plants have much in common phytocoenotically and biogeocoenotically, but at the same time have their own specific characteristics. The fact that all higher plants in forest biogeocoenoses are interconnected by a definite system of interspecific relationships, varying in their duration and mechanisms, now requires no special proof. But it must be stressed that from the biogeocoenotic point of view it is most important to study and understand the mechanism of interactions between the plant species having the greatest phytocoenotic significance and strongly affecting the environment and the biogeocoenose as a whole. Such relationships include interactions between tree species forming a stand, and between the dominant species in the stand and the dominant species in the undergrowth, the shrub-herbaceous layer, and the moss layer of a forest community.

The biogeocoenotic significance of interspecific relationships is mainly that they enable plant species, ecologically and biologically heterogeneous, to combine in a forest phytocoenose to form different structural parts of the coenose and use matter and energy from different layers of atmosphere and soil. Not only separate plant species but also entire structural combinations of them in the form of layers, synusiae, or parcels exist in a forest biogeocoenose in a state of very complex dependence on one another. That has been proved by many observations and experimental data, which show that destruction of the compactness and density of the tree stand in the upper layers or removal of the influence of their root systems is accompanied by alteration in the species-composition and density of the undergrowth and of the herbaceous-shrub and moss layers. As the competitive effect of trees weakens with age, shrubs in the underbrush and shrubs and herbs in the lower layers begin to flourish and in their turn suppress and kill the moss cover, and impede the germination of seeds and the growth and development of shoots of the tree species (Toumey 1929, Toumey and Kienholz 1931, Snigireva 1936, Shirley 1945, Sukachev 1946, Karpov 1958, 1960). The lower layers in a forest biogeocoenose may in turn exert no less influence on the composition and structure of the tree stands. When the effects of the undergrowth or of the synusiae of herbaceous-shrub and moss layers weaken with age, generally regrowth of tree species is stimulated, and the growth of saplings and, apparently, of mature trees is improved (Tkachenko 1939, Sukachev 1946).

The influence of the lower plant layers on tree stands is very clearly seen in artificial plantations of the steppe zone, where layers of underbrush and weedy herbage require and consume a considerable part of the soil moisture and so may impair and restrict both the growth and the restoration of tree species (Vysotskii 1915, Raskatov 1940, Al'bitskaya and Bel'gard 1950, Afanas'eva et al. 1952, Ivanov et al. 1952, 1953, Olovyannikova 1953, 1962, Karpov 1955b, Al'bitskaya 1960). In practical sylviculture very often one has to regulate and direct the inter-relationships between entire structural parts of a forest phytocoenose. Clearly, only an explanation covering the mechanism of these interactions can show us how to regulate them for the purpose of intensifying regeneration processes and improving the composition and productivity of tree stands.

With regard to intraspecific relationships between plants, we must point out that the entire assemblage of individuals of a single plant species growing in a forest phytocoenose is combined in the concept of a species-population, regarded as an important element of the structural organisation of coenoses (Rabotnov 1950, Lavrenko 1959). All individuals in a species-population in a forest phytocoenose not only interact with individuals of other populations but exist in complex and varied relationships with each other. These intraspecific relationships usually result in competition between individuals of the same species for space and living requirements, but also may involve mutual benefit. Excellent proof of the intensity of competition between individuals of the same species is provided by the phenomenon, frequently described and well studied by sylviculturists, of the natural thinning-out of monoculture tree stands with increasing age and their differentiation into Kraft's tree classes (Eitingen 1918, Morozov 1926, Cajander 1925, Tkachenko 1939, Sukachev 1946, 1953, Rubtsov 1960). The competitive relationships between individuals of the same species are shown no less clearly in the severe suppression of saplings by mature parent trees. That form of intraspecific competitive relationship between trees in a forest has for a long time been studied very thoroughly by many sylviculturists and geobotanists, who have also applied experimental methods of investigation. Mature trees may suppress and kill undergrowth of their own species in various ways, the chief of these being interception of radiant energy by the crowns of mature trees, and of soil moisture and mineral nutrients by their roots (Morozov 1899, 1926, Fricke 1904, Aaltonen 1926, 1942, 1948, Fabricius 1927, 1929, Shirley 1945, Oosting and Kramer 1945, Romell and Malmström 1945, Sukachev 1946, 1953, Karpov 1955a, 1959b, 1960a, Karandina and Erpert 1961).

As G. F. Morozov (1926) and V. N. Sukachev (1946) justly remark, suppression of undergrowth by trees in the process of competition is practically inseparable from the beneficial influence of parent trees on young trees in a forest. An example of such beneficial influence is the well-known phenomenon of shelter of undergrowth by mature trees from severe insolation, from frost, from harmful drying by wind, etc. (Morozov 1926, Tkachenko 1939). The extent to which these favourable effects are often closely linked with competition between plants may be judged from the fact that the high resistance of spruce saplings beneath the forest canopy to frost is due not only to the warming effect of trees on the lowest air layer but also to the fact that competition by tree roots slows down the rate of seasonal development of the saplings and enables them to escape damage by late frosts (Karpov 1960b).

Thus both inter- and intra-specific relationships between plants, differing widely in their nature and mechanism, including and linking together all individuals and species in a single whole, the forest phytocoenose, are very clearly demonstrated in forest biogeocoenoses. These two forms of inter-relationships are closely connected in their formative and organising effects on the phytocoenose, and constitute a necessary condition for the existence of forest coenoses and also for progressive evolution of the plants composing them. The evolutionary significance and role of inter- and intra-specific relationships in the life of forest phytocoenoses have been discussed in greater detail in a number of works by V. N. Sukachev (1946, 1953), I. I. Shmal'gauzen (1946, 1958), and N. V. Timofeev-

Resovskii (1958). In particular, Sukachev remarks that both inter- and intra-specific relationships between plants, together with the struggle for existence in a very unfavourable physical environment, are a necessary prerequisite for further development and perfecting of species in their progressive evolution. While interspecific competition may be accompanied by the death of all individuals of the competing species, that is not observed in intraspecific competition, and is a major difference between these two principal forms of inter-relationships between plants in forest biogeocoenoses. It appears that Charles Darwin's theory that intra-specific competition between plants is in all cases keener than interspecific is not always true. This does not, however, lessen the tremendous significance of intraspecific competitive relationships between plants. These are the most important factors determining the numbers of individuals, their living conditions, and their biogeochemical work.

Interspecific and intraspecific relationships between plants may be divided, according to the methods and mechanisms whereby one plant affects another, into the following three main groups:

(i) direct or indirect effects of one plant on another;
(ii) indirect transabiotic inter-relationships between plants;
(iii) indirect transbiotic effects of one plant on another.

Inter-relationships between plants may be unfavourable or they may be reciprocally or unilaterally beneficial. In natural biogeocoenoses, as a rule, it is very difficult to draw a line between these main groups of interactions between plants, since they are all based on alterations in very complex, mutually-conditioned, and closely-linked aggregates of factors and processes occurring in forest biogeocoenoses.

Direct or contact interactions

These include inter-relationships between plants differing in their complexity, duration, and mechanisms, an indispensable condition of which is direct contact of one plant with another.

Based on the duration and mechanism of relationships between plants in contact, this group of inter-relationships may be divided into two sub-groups: (i) contact interactions of mechanical type; (ii) contact interactions of physiological type.

Contact interactions of mechanical type include simple, purely physical effects of one plant on another when they grow together. Examples of such effects are the breaking-off of branches and damage to crowns of coniferous species by branches of broad-leaved species, mechanical pressure and cohesion of underground parts and root systems of plants, use of some plants by others (lianas) as mechanical supports, or attachment (epiphytism), etc.

Knocking-off of branches and damage to crowns of coniferous species when they grow beside broad-leaved species is observed frequently, and has been well described in sylvicultural literature (Morozov 1926, Tkachenko 1939). Deciduous species with very long and flexible branches (e.g. species of birch) may seriously damage needles, young shoots, and terminal buds of spruce and pine and so hinder their growth and sometimes cause dying-off. The branches of deciduous species are moved even by light winds, and consequently damage to conifers is observed not only on the outskirts of forest areas but also in the centre of forests. Damage by branches of deciduous trees to terminal shoots and buds of spruce at the period when the latter are beginning to occupy a deciduous canopy is on a wide scale, and is one of the main reasons for delay in the replacement of deciduous by coniferous forest (Morozov 1926).

Mechanical pressure and cohesion of trunks and root systems is one of the simplest physical methods whereby one plant can affect another in a forest biogeocoenose. Direct pressure and forcing-out of one plant by another, apparently, is extremely rare in the life of forest biogeocoenoses, although such happenings are fairly often observed in vegetable cultivation (Edel'shtein 1946, Sukachev 1953). In forest phytocoenoses, however, one may often observe very close contact between the trunks of two trees, accompanied by friction and damage to the cambium of both, sometimes culminating in their mechanical cohesion and finally in their coalescence. This may be observed both in trees of the same species and in trees of different species.

Generally mechanical pressure by one plant on another with consequent cohesion is observed in underground parts of forest phytocoenoses where, because of extreme over-population, small soil depth, and special environmental conditions, the most favourable conditions for mutual contact of plants are created. N. I. Rubtsov (1950) has shown that, when such contact occurs and the roots grow in thickness, the mechanical pressure by one plant on another is accompanied by different types of union, beginning with simple mechanical cohesion and culminating in complete and permanent fusion of the tissues of the two root systems and physiological exchange of matter between them. The latter form of interaction between plants belongs to a type of contact-physiological interaction, which we shall discuss in more detail later.

On the whole this form of interaction plays no substantial role in the life of forest biogeocoenoses.

Epiphytic interaction consists in the settling of many species of mosses, lichens, pteridophytes, and flowering plants on trunks and large branches, and sometimes on leaves, of trees, where they lead a highly specialised mode of life. Epiphytes have no physiological contact with the plants on whose surface they grow, and are fully-independent autotrophic organ-

isms, capable of independent synthesis of organic matter. Development
of an epiphytic mode of life in plants is due to competition for light; by
settling on tree-trunks and branches epiphytes, being light-loving plants,
improve their light supply, although their conditions of supply of moisture
and nutriment are fairly difficult (Schimper 1935, Went 1940, Walter
1962). With a few exceptions, epiphytes have very little effect on the life-
activity of their hosts.

In temperate-zone forests epiphytes are represented only by mosses,
lichens, and algae, the proportional weight of which in the total organic
mass produced by forest biogeocoenoses sometimes reaches considerable
figures. According to Scotter (1960), in spruce (*Picea mariana*) forests
the air-dry weight of lichens growing on tree-trunks is 1212 kg/ha, and
in pine (*Pinus banksiana*) forests, 2075 kg/ha. In tropical forests epiphytes are represented by a large number of different forms of flowering
plants and pteridophytes, sometimes possessing very complex and finely-
adjusted adaptations to the epiphytic mode of life. Some types of tropical
forests are heavily laden with epiphytes, but there are no precise data in
the literature on the mass of organic matter produced by them (Richards
1961). Went (1940) remarks that the abundance and distribution of
epiphytes in tropical forests depends on light conditions, the structure of
tree bark, and the possibility of accumulation in tree forks of decaying
litter brought there by water flowing from trunks and leaves. Until
recently it had been thought that epiphytic synusiae lived mainly on
nutrients derived from atmospheric dust and plant litter accumulating in
cracks in the bark and in forks of trunks and branch intersections. In
recent years, however, many works have appeared showing that precipitation washes a large amount of nutrients out of leaves and needles,
which probably also assists epiphytes.

With regard to the relative biogeocoenotic significance of this form of
relationship between plants, the existence of epiphytic synusiae complicates the structure of forest phytocoenoses and so to some extent increases
the efficiency of use of radiant energy and the production of phytomass.
It must be stressed that the chemical composition of epiphytes has its own
specific features, which, of course, may have definite significance in the
general metabolism of forest biogeocoenoses. Finally, the use by epiphytic synusiae of matter washed from the above-ground parts of host
plants complicates the cycle of its transformation and circulation in a
forest biogeocoenose.

Relationships between trees and lianas are also simple contact interactions: creeping plants use trees as mechanical supports to carry their
above-ground parts to the light. There is no doubt that lianas as a unique
coenotype arose during prolonged evolution and natural selection in
conditions of keen competition among plants for light. This view of the
origin of lianas was developed in the classic works of Darwin (1941) and

of many other investigators who have studied this unique plant form (Schenk 1892-93, Schimper 1935, Baranov 1960, Walter 1962).

In temperate-zone forests this type of relationship between plants sometimes plays a substantial role in biogeocoenotic processes. In the forests of the Far East and the Caucasus, for instance, lianas are represented by a comparatively large number of species, sometimes forming considerable aggregates on trees and bushes. Among the common lianas of broad-leaved forests in the temperate zone we may mention hops (*Humulus lupulus*), species of ivy (*Hedera helix*, *Hedera colchica*), wild grapes (*Vitis amurensis*), actinidia (*Actinidia kolomikta*), etc. Over vast areas of the taiga zone lianas are represented by a very small number of species, small in size and infrequently occurring (e.g., *Atragene sibirica*), so that their share in the accumulation of organic matter is negligible.

An entirely different situation occurs in tropical forests, where lianas are found in large numbers everywhere and are represented by a great variety of species and forms, constituting one of the notable features of tropical biogeocoenoses. There lianas may reach a large size (about 70 metres, and sometimes much more) and, penetrating into the upper storey, sometimes grow to such an extent that they seriously impair light conditions for the lower layers. In the upper storeys of tropical biogeocoenoses lianas compete with trees for light; they shade and deform the crowns of trees, and by their weight break the trunks and branches of the trees supporting them (Schimper 1935, Richards 1961, Walter 1962).

An example of this type of relationship is that between trees and 'smothering' plants of the genera *Ficus* and *Clusia*. Members of these genera, settling on the upper parts of the crowns of their hosts, at first lead an epiphytic mode of life. Later they root in the soil and begin to grow luxuriantly, ultimately causing the death of their hosts. As Richards (1961) writes, the immediate cause of the hosts' death in such cases has not yet been discovered, but apparently death results from overshadowing, mechanical pressure, and root competition by the fig.

Unfortunately no data are available on the role played by lianas in the accumulation of organic matter in forests, which naturally makes it difficult to obtain a biogeocoenotic evaluation of that form of relationship between plants. It is clear, however, that heavy growth of lianas in forest biogeocoenoses must raise their productivity, providing, of course, that it does not lower the productivity of other species of higher plants.

The above-mentioned types of interactions are far from exhausting the list of multiform, contact-mechanical interactions between plants in forest biogeocoenoses, which are still very incompletely studied and evaluated from the phytocoenotic and biogeocoenotic points of view.

Contact-physiological types of relationships between plants are characterised by more intimate and finely-adjusted, or sometimes extremely specialised, physiological bonds, consisting generally in mutual or unilateral exchange of photosynthetic products, mineral compounds, and water between the partners. These interactions are always accompanied by close union and fusion of the organs and tissues of two plant organisms, sometimes so intimate that one plant is physiologically part of the other (e.g., parasites and host plants). This includes the fusion of root systems, semi-parasitism, parasitism, saprophytism, and symbiosis.

Fusion of root systems of trees in forests has recently been the subject of many investigations, sometimes leading to very contradictory con- clusions about the frequency and significance of this phenomenon in the life of forest biogeocoenoses (Rubtsov 1950, Yunovidov 1951, Shishkov 1953, Ogievskii 1954, Beskarabainyi 1955, Orlenko 1955, Koldanov 1958).

It has been proved that the roots of trees of the same species or of different species can fuse together; this is most often seen in individuals of the same species or of closely-related species.

The frequency of fusion of root systems depends on the biological features of tree species, the density of tree stands, and the soil conditions in which they grow.

According to N. I. Rubtsov (1950), in protected plantations in the steppe zone very many species of trees fuse their root systems; all stages of such fusion can be traced, from simple mechanical interlacing to thorough penetration of the tissues of one root by another. I. I. Shishkov (1953) states that in dense Norway spruce stands not fewer than 30% of the roots are joined. D. E. Kunts and A. D. Raiker (1956) have discovered that in Wisconsin forests all oak trees (*Quercus ellipsoidalis*) show inter- connected root systems; but, according to these authors, in such species as aspen (*Populus tremuloides*), white birch (*Betula alba*), and Rocky Mountain ponderosa pine (*Pinus ponderosa*) fusion of root systems is extremely rare, and in white spruce (*Picea alba*) and black spruce (*Picea mariana*), and also in balsam fir, fusion is never seen.

All authors state that fusion of roots of trees of different species and genera is comparatively seldom observed. The frequency of fusion in- creases with increase in density of sowings and plantations of trees (Ogievskii 1954, Orlenko 1955); this, however, is disputed by some in- vestigators (Lisenkov 1957, Koldanov 1958).

At a certain stage of root fusion, exchange of nutrients and water occurs, as is well illustrated by the results of experiments by Kunts and Raiker. Using isotopes (radioactive iodine and rubidium) these authors found an exchange of matter between trees whose roots had fused. Radioactive fungal spores were also carried with the flow of materials, indicating the possibility of very rapid spread of fungal diseases among trees connected to each other by fused roots.

As I. N. Rakhteenko (1958a) has pointed out, however, exchange of matter between tree roots may take place not only through root fusion but also through simple contact or solution of matter in the soil.

There are wide divergences in the evaluation of the phytocoenotic and biogeocoenotic significance of root fusion. Some authors try to prove that, in the first place, fusion of roots of individuals of a single species shows the absence of intraspecific competition in a forest, and in the second place it benefits tree growth, raising the total productivity of forest bio-geocoenoses (Beskaravainyi 1955). That view, however, is not based on any reliable factual data.

As stated above, fusion of root systems is one result of overpopulation in a forest, which is in no way reduced by the heavy tree mortality observed during the process of thinning-out with age.

Some authors state that fusion of root systems increases resistance to wind, and that 'adoption' of external root systems enables trees to grow more vigorously; but that does not exclude competitive relationships between them.

Stronger trees may intercept their partners' supply of moisture and nutrients, and thus weaken and hasten the death of trees suffering from such deprivation.

The inter-relationships between trees with fused roots are still inadequately studied from the physiological point of view, which makes biogeocoenotic evaluation difficult. In any case, what is occurring is merely redistribution of organic and mineral matter already synthesised and incorporated from the soil by higher plants.

Semi-parasitism and parasitism, as types of relationships between higher plants, do not play a substantial role in the life of forest biogeo-coenoses. In forests only a few higher plant species are parasitic or semi-parasitic, and relatively seldom are they found in large numbers. The most common semi-parasites in temperate-zone forests are species of the herbaceous genera *Melampyrum* and *Pedicularis*. Semi-parasites and parasites are comparatively scarce in tropical forests, where they mostly lead epiphytic modes of life (species of the families Loranthaceae, Balano-phoreae, Rafflesiaceae, etc.). The biogeocoenotic functions of these groups of higher plants are very peculiar. Semi-parasites have not yet lost the capacity for autotrophic nutrition, i.e., independent creation of organic matter by photosynthesis, and they possess a fairly-well-developed assimilating system and chlorophyll in the leaves. For their supply of water and mineral compounds, however, they are entirely dependent on their higher-plant hosts, on the root systems of which they are parasitic. In the process of evolution they have lost the ability to absorb moisture and nutrients from the soil, replacing absorbing root tips with special organs, suckers, to absorb the necessary moisture and nutrients from the roots of their hosts.

Still more intimate physiological contact is that established between autotrophic plants and true parasites. The latter have passed completely to a heterotrophic mode of life and depend altogether on their hosts for photo-synthetic products, moisture, and mineral compounds. The biogeocoenotic functions of parasites, of course, consist solely in the re-use of organic matter synthesised by autotrophic plants and of the energy contained therein.

Among fairly common semi-parasites found on trees in temperate-climate forests are mistletoe (*Viscum album*) and loranth (*Loranthus europaeus*). When strongly developed they may cause withering of the branches and tops of coniferous and deciduous trees.

Three species of dodder (*Cuscuta europaea, C. lupuliformis,* and *C. monogyna*) parasitise willows, alders, and poplars, and sometimes may cause considerable injury to young saplings, especially of willow. Inter-relationships of this type have a very clear unilaterally-beneficial character.

Cases of serious injury or death of trees caused by semi-parasitic or parasitic plants are actually very rare, and occur mainly in tropical forests.

The reasons for the origin and development of parasitism and semi-parasitism among higher plants are not yet clear. Some scientists are inclined to attribute the origin of this form of relationship between plants mainly to competition between roots for water and nutrients. By using another plant's root system, semi-parasites escape the necessity of competing with other plants in the soil, and parasites also escape from competition for light (Härtel 1959). Very little is known of the mechanism of the inter-relationships between host plants and semi-parasites or parasites. In recent articles by Härtel (1959) and Schmucker (1959a) devoted to these groups of higher plants, it is stated that the biological and physiological features of semi-parasites and parasites indicate that they obtain many substances necessary for existence directly from the tissues of the host plant.

Carbohydrate and protein exchanges between host plants and parasites have not yet been studied, although there is every reason to believe that they occur.

It is of interest to note that both parasites and semi-parasites exert considerable influence on the course of physiological processes and metabolism in host plants, that being displayed, for instance, in increased transpiration by whole plants or branches occupied by parasites (Härtel 1959).

Parasitism by the lower fungi (especially rusts) on higher plants is very widespread in forests, sometimes playing an important part in the life of forest biogeocoenoses.

Lower fungi that parasitise the leaves and branches of trees produce such diseases as powdery mildew of leaves, leaf-spot, yellowing and wilting of leaves and needles, death of whole branches, or abnormal branching of shoots ('witch's broom'). These diseases may seriously diminish the photosynthetic work of the infested plants, at the same time causing entry into the soil of increased amounts of plant litter.

The severe injury done to young pine seedlings by the fungus *Lophodermium pinastri* (Schütte disease); by species of the genera *Fusarium,*

Alternaria, and *Rhidoctonia*, which cause the seedlings to fall; and also by fungi that cause rotting of above-ground parts (*Phytophtora omnivora*) or of root systems (*Rosellinia quercina*) is well known. This type of parasitism is described in more detail in special works and handbooks on forest phytopathology (Vanin 1948, Kern 1959).

Saprophytism. Saprophytic higher plants form a unique group of mycotrophic plants, almost or quite lacking in chlorophyll and therefore having lost the faculty of independent synthesis of organic matter. These plants obtain organic matter from the soil through fungi, on whose nutritive substances they mainly subsist. Consequently one may speak of their saprophytism somewhat conditionally, since in this case only the fungus is a true saprophyte, and the flowering plant is parasitic on the fungus. The name of saprophyte, however (often applied to the flowering plant alone), is still retained for the complete partnership of flowering plant and fungus. In temperate-zone forests such saprophytes include Indian pipe (*Monotropa hypopithys*), bird's-nest orchis (*Neottia nidus-avis*), coral-root (*Corallorhiza trifida*), *Epipogon aphyllum*, and *Calypso bulbosa*. These species play a very small role in the life of forest phytocoenoses, mainly because of their extreme scarcity and small size. They are all low-growing plants with reduced scaly leaves of yellow-brown or brown colour.

According to Björkman (1956), the fungi parasitised by Indian pipe, which is fairly common in our forests, are those that live on tree roots. Consequently that plant may parasitise tree roots by means of mycorrhiza fungi. Melin (1953) has expressed the opinion that Indian pipe is a typical mycotrophic plant, able to incorporate organic matter from forest humus by means of fungi.

Valuable information on the ecology, biology, and structure of saprophytic higher plants is contained in the work of Schmucker (1959), from which it is evident that many physiological aspects of the partnership bonds between fungi and saprophytic higher plants have not been studied.

There is no doubt that these heterotrophic organisms take part in the re-use of decaying organic matter and that their role consists mainly in bringing complex organic compounds into a new supplementary cycle of transformation of matter and energy.

Typical saprophytes are fungi and bacteria, the number of them in forest soils being extremely high. These micro-organisms transform enormous masses of organic matter in their life-activities, thus providing an important and specific link in the total exchange of matter and energy in forest biogeocoenoses.

Symbiosis as a contact-inter-relationship between organisms is characterised by a high level of specialisation in the mutual exchange of metabolic products, beneficial in some degree to both symbionts.

A classic example of symbiosis is the partnership of mycorrhiza fungi and nodule bacteria with higher plants. This type of symbiosis is typical of forest biogeocoenoses, and plays an extremely important role in the exchange of matter and energy in the phytocoenose-soil system. The hyphae of mycorrhiza fungi, which live on roots of trees, shrubs, and herbs, substantially increase and extend the ability of root systems of higher plants to absorb and incorporate mineral compounds from the soil and consequently to bring them into new cycles of circulation of matter and energy in a forest biogeocoenose. By means of mycorrhiza fungi higher plants can obtain some complex organic compounds directly from the soil.

The hyphae of symbiotic fungi secrete growth substances (including auxin), which stimulate development and branching of root systems and increase their absorbing surface (Melin 1959). In return the hyphae receive from the roots mostly carbohydrates and some other substances necessary for their life. Björkman (1942, 1944) discovered, for instance, a direct correlation between the frequency of formation of mycorrhiza on tree seedlings and the content of soluble carbohydrates in their roots, the most favourable conditions for mycorrhiza formation being observed when soluble carbohydrates were abundant in the tree roots. Melin (1955), using tagged C^{14}, showed that a considerable proportion of the products of photosynthesis passes comparatively rapidly from roots into fungal hyphae, much of it being used by the symbiotic fungus as a source of energy. The complex questions of symbiosis of fungi with higher plants are discussed in more detail in another chapter of this book and in special works devoted to mycotrophy in plants (Melin 1959, Shemakhanova 1962).

Another classic example of symbiosis is provided by lichens, which in some types of forest play an important role in the formation of soil cover and epiphytic synusiae. In this case, intimate symbiosis between fungi and algae serves as the foundation for the formation of a complete individual organism, which carries out specific biogeochemical work in forest biogeocoenoses. In the morphogenesis of lichens as combined organisms, the leading role is apparently taken by the algae; the same species of fungus may form thalli of different morphology in partnership with different species of algae (Quispel 1959). The fundamental partnership bonds of the fungus-algae association are exceedingly complex and include elements both of mutual benefit and of suppression, which to a great extent are determined by the ecological conditions in which the lichens live. In photosynthesis the algae produce carbohydrates and other organic compounds, which are used by the fungi as sources of matter and energy. In addition, the fungi receive growth substances from the algae, and possibly also proteins and nucleic acids, which take part in thallus-formation. The algae in turn receive from the fungi mineral compounds and moisture, and in dim light where normal photosynthesis is impossible they also receive organic compounds absorbed by the fungi from the substrate. During the mutual physiological exchange each partner may have a suppressing effect on the other to such a degree, in sharp fluctuations of ecological conditions, that the symbiosis is converted into uni-lateral parasitism, generally of the fungi on the algae (Quispel 1959). On the whole, the symbiosis of fungi and algae increases the ability of some forest biogeocoenoses to accumulate organic matter, since it introduces a new plant form, possessing specific biochemical features for the exchange of matter and energy.

Another form of symbiosis is that of higher plants and nodule bacteria. It plays a small part in the life of forest biogeocoenoses, since the bacteria colonise only a relatively small number of forest plants on 'club-shaped' roots. The nitrogen fixed from the air penetrates the intercellular spaces of the roots. Such plants include acacias, alders, oleasters, and sea-buckthorns, and also some species of leguminous plants. The bacteria use carbohydrates synthesised by higher plants, and the latter use nitrogenous compounds created by the bacteria.

It is quite evident that this form of relationship between plants and bacteria makes possible the inclusion of atmospheric nitrogen in the exchange of matter in a forest biogeocoenose. The leading role in this respect is played by free-living micro-organisms in the soil, including bacteria. The latter (saprophytes) bring nitrogen into the circulation of matter in a forest biogeocoenose only by means of energy derived from organic matter entering the soil in the form of plant litter and dead higher plants.

Indirect transabiotic interactions

Indirect transabiotic interactions between plants play an extremely im-portant role in the life of forest biogeocoenoses, and are universal in character. In any forest biogeocoenose some plants affect others through alteration of the complex aggregate of physical factors of the ecotope. Some plant species possess considerable ability to alter the physical factors of the ecotope and thus make it difficult for others to obtain light, moisture, nutritive matter, etc.

Indirect transabiotic interactions may be divided into the following three main groups:

(*i*) competitive relationships between plants;
(*ii*) environment-forming effects of some plants on others;
(*iii*) allelopathic (biochemical) forms of interactions between plants.

Competitive relationships between plants are usually treated very broadly in the literature and include the most varied forms of injurious effects of one plant on another growing beside it. We think it is useful to narrow that meaning, and we propose to consider competition between plants as including only those mutually or unilaterally harmful interactions that originate from use of the food and energy resources of the habitat. When we remember that interactions between plants through the medium of the environment are extremely varied in their methods and results both for the plants and for the forest biogeocoenose as a whole, we see that it is very important to discover the principal link in the complex chain of mutual influences. Without doubt the decisive and leading role in the life of phytocoenoses is taken by nutrition, including both carbon nutrition and the absorption of water and mineral elements from the soil. In nutrition the closest relationships are established between plants in a forest biogeocoenose, these being competitive for the principal means of existence. This is proved by numerous observations, field investigations, and direct experiments that show that shortages of radiant energy, of moisture, and of nutrients in the soil are the main reasons for the very keen and intense competition between plants and play a leading role in the formation of forest phytocoenoses of definite species-composition and structure. From this point of view we cannot speak, for instance, of competition between plants for the moisture or the warmth contained in the lowest layer of the air; nor can we include here the effects produced by some plants on others through alteration of the hydro-physical and some chemical properties of the soil (e.g., acidity). Such an effect is best regarded as an environment-forming interaction between plants, since in this case we are concerned with factors and conditions that affect the nutrition system of plants in a forest biogeocoenose, and not with the direct requirements of plants for matter and energy to carry out their living and biogeochemical functions.

It is sometimes difficult, of course, to separate competitive and environment-forming relationships, since any essential factor (e.g., water) may be absorbed and consumed by a plant in body-building and transpiration and at the same time have an environment-forming effect on it (e.g., excessive watering).

We shall now briefly discuss the principal kinds of competitive relationships between plants in forest biogeocoenoses.

Competition for light. As a result of high stand density and extremely close proximity of the above-ground parts of plants in forest biogeocoenoses, the plants intercept each other's light and change its qualitative composition, so producing notable effects on all the layers and synusiae that form a forest phytocoenose. In such constant and intense competition for light the greater height of trees gives them an advantage over low-growing shrubs, bushes, and herbage, which as a rule are much shaded by the crowns. But in some types of forest biogeocoenoses these suppressed layers grow vigorously and in turn impair the light-supply conditions of tree seedlings, so affecting the composition and structure of the tree layer. That is seen, for instance, in dense pine and spruce stands, where a compact shrub layer reduces the light on the soil surface to a level very close to the compensation point for many trees and herbaceous plants (Yurina and Zhmur 1962).

The existence of keen competition for light between plants in a forest is confirmed by numerous studies of the light regime, in the course of which it has been discovered that illumination within forest biogeocoenoses is as a rule below the level required for normal assimilation by plants. According to many authors, the light within forest biogeocoenoses varies generally within the limits of 2-3% to 9-10% of full sunlight, depending on the composition and structure of forest phytocoenoses and the weather conditions; sometimes, in open ash and larch stands, it reaches 20-25% of the full radiation in an open space (Ivanov 1932, 1946, Sakharov 1940b, 1948, Kossovich 1940, 1945, 1952, Shirley 1945, Pozdnyakov 1953, Karpov 1955b, Siren 1955, Oskretkov 1957, Lundegårdh 1957, Yurina and Zhmur 1962).

Direct experiments have proved that in many types of forest biogeo-coenoses the interception of light by crowns of mature trees is a main reason for poor regeneration and growth of seedlings of both light-loving and shade-tolerant species. This is illustrated by the results of experiments by Fabricius (1927, 1929) and V. G. Karpov (1962), which showed that seedlings of spruce and many other tree species responded very weakly to elimination of competition by tree roots in spruce forests, since light was the chief limiting factor at an ecological minimum.

Still more decisive results were obtained in experiments with light-loving species, showing that in forest biogeocoenoses seedlings of these species (e.g., pine) usually perished on account of interception of light by tree crowns, especially when there was a dense layer of undergrowth or tall herbage (Yakhontov 1909, Lutz 1945, Shirley 1945, Oosting and Kramer 1946, Karpov 1954, Oskretkov 1957).

Competition by tree species for light coming through the crowns is very intense; the overshadowing of some crowns by others is a regular and common occurrence. The lower parts of tree crowns, as a rule, exist in less favourable light conditions, and therefore leaves and needles there work with a small positive balance and sometimes do not compensate for their expenditure in respiration (Ivanov and Kossovich 1930, 1932).

The seasonal rhythm of development of forest-forming species strongly affects the outcome of competition for light in forest biogeocoenoses; and closely linked with it is the annual light rhythm beneath the forest canopy. In this respect great differences are observed between evergreen dark-conifer and deciduous broad-leaved forests (Ivanov and Orlova 1931, Tolmachev 1954).

In biogeocoenoses of dark-conifer taiga a constant low level of light exists throughout the year, which dooms the seedlings and plants in the lower layers to a low level of assimilating activity throughout the whole cycle of their seasonal development. On the other hand, in deciduous forests of the temperate zone the intensity of competition for light is somewhat modified by the leafless periods for the tree stands, when the light beneath the canopy increases considerably and reaches 45-60% of full sunlight. In small-leaved forests of the taiga zone, spruce seedlings and herbs make good use of these favourable ecological circumstances for intensifying photosynthesis, which in many ways determines the success of spruce in competition with small-leaved species (Ivanov and Orlova 1931, Karpov 1960b). With regard to broad-leaved forests, the very interesting ecologo-physiological investigations of Daxer show that maximum intensity of photosynthesis in the understorey plants occurs early in spring, before the leaves of broad-leaved tree species come out.

All of the above data indicate that competition between plants for light plays a very important role in the formation, composition, structure, and life of biogeocoenoses. We must warn against the overvaluation of this factor that is typical of several geobotanical and sylvicultural works, since competition for light is not always the sole and decisive factor determining the composition and structure of forest biogeocoenoses and the biogeochemical work of their components (Fricke 1904, Toumey and Kienholz 1931, Morozov 1926, Romell and Malmström 1944, 1945, Sukachev 1946, Karpov 1955a, 1962). In the life of many types of biogeocoenoses it is of secondary importance.

Before planning forestry measures for regulation of light to obtain definite sylvicultural effects, one must determine the real significance of competition for light in the structure, productivity, and regeneration processes of specific biogeocoenoses.

Competition for soil moisture. The resources of soil moisture in forest ecotopes are often limited, which is a prerequisite for competition for it in forest biogeocoenoses. The fact that in forest biogeocoenoses soil moisture is not always adequate is well illustrated by the large number of works on the water regime in forest soils and on the effects of that regime on forest phytocoenoses (Vasil'ev 1950, Bol'shakov 1950, Rode 1950, 1952, Zonn 1951, Afanes'eva *et al.* 1952, 1955, Molchanov 1952, 1954a, Mina 1954). The stocks of available water in forest soils are often exhausted by tree stands by the end of the first half of the vegetative period, the soil being desiccated by tree roots to a point near the coefficient of wilting. This usually occurs in forest biogeocoenoses in plakor habitats in arid regions (Bol'shakov 1950, Rode 1950, 1952, Zonn 1951, Afanas'eva *et al.* 1952). In forests within the zone of adequate moisture, however, one can sometimes observe very clearly a water shortage in the soil of forest biogeocoenoses, including spruce forests (Vasil'ev 1950, Orlov 1960b, Orlov and Mina 1962). The reasons for deterioration of the water regime in forest soils have become clear from recent investigations: it has been discovered that colossal stocks of soil moisture are consumed by forest biogeocoenoses in transpiration (Ivanov 1946, Molchanov 1952, Rode 1952, Ivanov *et al.* 1953).

These data in the aggregate show that competition for soil moisture by the root systems of plants develops in forest biogeocoenoses, and in many

cases plays a leading role in the formation and life of a forest. This conclusion is supported by direct experimental studies of the mechanism of
the effects of mature trees on seedlings and plants in lower layers (Sukachev
1946, Karpov 1955a, 1956, Fricke 1904, Morozov and Okhlyabinin 1911,
Morozov 1926, Toumey 1926, Craib 1929, Toumey and Kienholz 1931,
Snigireva 1936, Korstian and Coile 1938, Romell and Malmström 1944,
1945, Karandina and Erpert 1961). In one of the earliest works on this
subject, written by Fricke in 1904, it was reported that elimination of the
influence of pine-tree roots on seedlings by tree-felling around the periphery of test areas was followed by a rapid improvement in water-supply
conditions for plants in the lower layers; soil moisture was doubled or

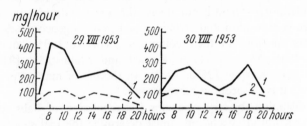

Fig. 39. Intensity of transpiration by oak seedlings in conditions of competition with roots of mature trees (1), and when
that factor is eliminated (2).

trebled, reaching 13·8% in the top soil layer as compared with 4·5% in
the zone of influence of the roots of mature trees. Similar results have
been obtained by many investigators who made experiments in the isolation of mature tree roots in forest biogeocoenoses differing in structure,
composition, and physico-geographical conditions (Fig. 39). From these
data it has been concluded that competition for moisture between tree
roots is much more important in the life of a forest than competition for
light (Fricke 1904, Toumey and Kienholz 1931, Korstian and Coile 1938).

Clements and his colleagues came to the same conclusion on the basis
of numerous experiments, holding that competition between plants growing close together is primarily for water, and then for light and mineral
nutrients (Clements et al. 1929). That conclusion, however, is justified
only in application to plakor forests in arid regions and in the zone of
inadequate moisture, and to forests of some ecotopes in the taiga zone
(e.g., pine-lichen forests), where soil moisture is at the ecological minimum.
The intensity of competition for water between tree root systems varies
substantially according to physico-geographical conditions. This is
illustrated in Table 63, where we present the results of our studies of the
intensity of competition between plants for moisture in artificial forests in
the arid steppe zone and in forests in the taiga zone.

From these data it is clearly seen that soil moisture in arid-zone planta-

Table 63. *Effect of tree roots on stocks of soil moisture (in mm above the coefficient of wilting) in the 0-50 cm layer in biogeocoenoses in different zones.*

Zone and type of forest biogeocoenose	With tree-root competition		Without tree-root competition	
	30 June	30 Aug	30 June	30 Aug
Arid steppe subzone				
Oak-maple plantation	65·3	4·0	70·1	33·8
Ash-elm plantation	59·2	2·0	75·3	50·6
Southern taiga zone				
Oxalis-fern spruce stand	89·4	56·9	120·1	98·2
Goutwort-mixed-herb birch stand	78·7	28·3	115·3	87·8

tions is almost entirely consumed by the trees in the second half of the growing period, which is the chief reason for poor growth and death of seedlings in biogeocoenoses in that zone (Karpov 1955a).

In southern taiga forests competition between plants for water is less intense, for a number of reasons which we shall not discuss because of lack of space. We may state, however, that even in forests in the zone of adequate moisture that form of relationship between plants sometimes becomes important, e.g., in dry years with little summer precipitation and in ecotopes with obvious signs of moisture deficit in the soil. In the latter, competition between plants for water can be traced all the way to the forest-tundra zone (Norin 1956).

Interception of moisture by tree roots is the basis of the suppressing action of tree roots on the growth and development of plants in the lower subordinate layers of forest biogeocoenoses. This is well illustrated by the results of experimental studies by A. V. Snigireva (1936), I. N. Olovyannikova (1953), and a number of foreign authors (Toumey 1926, Toumey and Kienholz 1931, Korstian and Coile 1938). On the other hand data exist showing that synusiae of shrubs and herbage may seriously harm trees by drying up the soil layer occupied by their roots, which is observed mostly in arid regions (Walter 1951, 1962, Afanas'eva *et al.* 1952, Olovyannikova 1953, 1962, Karpov 1955a, Al'bitskaya 1960).

Special studies devoted to the physiological reaction of tree seedlings to competition for moisture in the soils of forest biogeocoenoses have shown that severe competition for water is accompanied by changes in morphological structure and marked injury to growth of seedlings; their rate of transpiration and the water content of their tissues are lowered and the suppression of growth processes causes serious disruption of the carbohydrate and phosphorus metabolism of the plants (Karpov 1956, 1959b, Levitskaya 1961).

Consequently, in certain conditions of the physico-geographical environment and in certain types of forest biogeocoenoses competition between plants for soil moisture is of primary significance in the life of forest biogeocoenoses. Naturally in such circumstances the whole

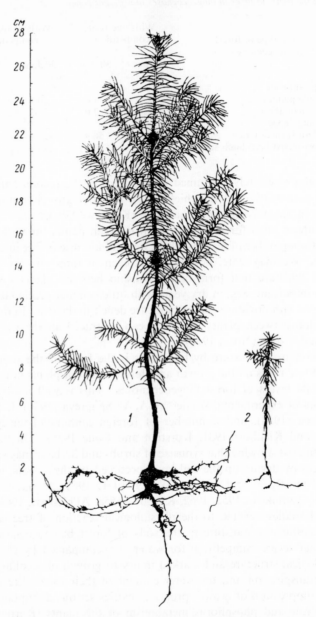

Fig. 40. Competition for nutrients in relation to growth of spruce seedlings: 1—when competition is eliminated; 2—in conditions of competition with roots of mature trees for nutritive matter in the soil (mainly soluble forms of nitrogen).

system of forestry and sylvicultural measures should be directed towards regulation of the relationships concerning competition for soil moisture.

Competition for nutrients in the soil. The most complex and least studied form of competitive relationships between plants in forest bio-geocoenoses is competition for nutrients in the soil. Adequate supply of minerals is one of the most important requisites for the normal growth and general life-activity of plants. If there is a shortage of several elements of mineral nutrition in forest soils, competition between plants for soil nutrients inevitably occurs. The roles of different nutrient elements in the life of plants vary, and different plant species vary considerably in their requirements, which also change with plant age. On the other hand, different soil types differ substantially as providers of mineral nutrients; this, in combination with the varying requirements of higher plants, causes great variation in the types of competition for soil nutrients.

Unfortunately the nutrient regime of most forest soils has as yet been little studied, which makes it difficult to evaluate this form of relationship between plants in the life of forest biogeocoenoses. Valuable data on this subject can be found in the works of S. A. Kovrigin (1952), N. P. Remezov (1953) and Remezov and his colleagues (1959), K. M. Smirnova (1951 a, b, 1952, 1956), A. Ya. Orlov (1960), Orlov and Mina (1962), and several others. The first attempts at direct determination of the intensity of competition between root systems in forest biogeocoenoses were undertaken by the American sylviculturists Korstian and Coile, who analysed the soil in areas isolated from the influence of tree roots and did not find any substantial difference in the nutrient content there as compared with adjacent parts of the forest. The reason was the unsuitability of agro-chemical methods of analysis of soils, which are not always reliable for evaluating the nutrient regime of forest soils (Lutz and Chandler 1947, Wittich 1958). Much more precise and reliable results have been obtained by other investigators, who, in their studies of competition between plants in forest biogeocoenoses for nutrients, used liquid fertilisers and methods of leaf analysis and leaf diagnostics, in combination with trans-planting of indicator plants and use of radioactive isotopes (Romell 1938, Wallihan 1940, Karpov 1959 a, b, 1960b, 1962).

It appears that in plakor biogeocoenoses of the dark-conifer central and southern taiga zones competition by plants through their roots is mainly for absorbable nitrogen. Such relationships between plants play a very important role in regeneration processes and succession of species, and also in the composition and structure of the lower layers of taiga biogeocoenoses. For instance, Romell (1938) demonstrated the great beneficial effect of supplying nitrogenous fertilisers to spruce seedlings and plants in the lower layers of spruce forests in Sweden. V. G. Karpov has shown that spruce seedlings in spruce-deciduous forests suffer badly from deprivation of available forms of nitrogen by birch roots, and that

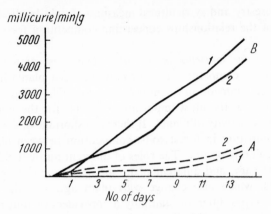

Fig. 41. Dynamics of accumulation of radio-active phosphorus in needles of 2-year-old spruce seedlings. 1—in
reedgrass-bilberry birch stand on strongly-podzolic soil;
2—in goutwort-mixed-herb birch stand on soddy weakly-
podzolised soil; A—in competition with tree roots; B—
with tree-root competition eliminated.

the numbers, growth, and development of many taiga shrubs and herbaceous plants in the lower layers of taiga biogeocoenoses are dominated
by tree roots mainly through decrease of the concentration of soluble
nitrogen compounds in the soil. Table 64 shows how intense the competition is between tree roots for nutrients in the soil; from it one may
see that seedlings of spruce and many taiga herbaceous plants growing in
competition with tree roots have much lower nitrogen concentration in
their leaf tissues than those growing in areas where tree roots had been
cut away. The effect on the growth of spruce seedlings due to deprivation
of nitrogen by roots of mature trees is seen in Fig. 40. Competition
between root systems of plants for soluble phosphorus compounds is no
less intense in forests in the taiga zone, as is well illustrated by the results
of experiments using tagged phosphorus (Fig. 41) (Karpov 1962).

Table 64. *Content of certain substances in needles of spruce seedlings and leaves of cover plants,
in experiments with trench-cutting of root systems of birch in bilberry communities* (% *of oven-
dry weight*).

Species	In contact with tree roots				Out of contact with tree roots			
	Ash content	N	P	Ca	Ash content	N	P	Ca
Picea excelsa seedlings	3·69	1·38	0·23	0·11	3·73	1·70	0·26	0·19
Oxalis acetosella	9·89	2·06	0·57	0·89	9·71	3·18	0·43	1·07
Vaccinium myrtillus	4·85	2·19	0·18	0·94	4·84	2·70	0·24	0·95
Calamagrostis langsdorfii	14·26	1·85	0·31	0·11	13·87	2·82	0·31	0·16
Aegopodium podagraria	12·52	1·49	0·50	1·40	12·00	3·11	0·58	1·60

In forest biogeocoenoses formed on rich soils, competition between plants for nutrients in the soil is much less active. This has been proved by experiments made in plantations in the southern chernozem of the steppe zone, which is comparatively rich in nutrients. In such southern chernozem plantations oak seedlings show almost no reaction to elimination of competition by roots of mature trees, measured as increase in the nutrient content of their tissues (Karpov 1959a).

Recent data indicate that roots of trees growing close together not only absorb, but may also discharge into the soil, various kinds of nutritive matter (Akhromeiko 1936, Rakhteenko 1958b). From these facts some investigators have tried to prove that with definite combinations of tree species conditions are created whereby some species may use the root excretions of others, so increasing the stability and productivity of mixed stands. While in no way denying the possibility of such forms of beneficial interaction between components of forest phytocoenoses, we should point out that they do not reduce competition between plant roots for nutrients in the soil. To reduce competition for nutrients it is more important to select tree species that use mineral elements from different soil horizons, and do so at different times during the growing period.

On the whole much field and experimental work still has to be done to discover how intensely, and for what nutrients, competition takes place between plants in the principal biogeocoenotic formations and in the forest types composed of them in various physico-geographical districts of our country.

Competition for carbon dioxide in the air. To ensure normal photosynthesis it is necessary to have sufficiently high concentrations of CO_2 in the surrounding air and a constant flow of CO_2 to the assimilating organs of plants. Lowering the CO_2 content of the air by 10-20% of its normal concentration produces a marked fall in the rate of photosynthesis (Nichiporovich 1955). Precise investigations of the dynamics of the CO_2 content of the air made in recent years show that during the daytime hours, as a result of the assimilating activity of plants, the CO_2 content of the air within forest biogeocoenoses may decrease considerably, this being observed in the zone of the lower parts of the crowns (Huber 1960, Miller and Risch 1962) and in particular weather conditions (clear windless days). In the layer of air adjacent to the soil (up to 0·5 metres), on the other hand, an increase in CO_2 concentration is observed, because of release of the gas from the soil. Some investigators believe that the latter circumstance improves carbon-nutrition conditions for plants in the lower layers. There are, however, no reliable data on whether some plants affect others by reducing the CO_2 content in the lower layers of air in a forest. Even slight turbulent movement of air and wind rapidly equalise the CO_2 concentration in the lower layers of air, so that we may conclude that if there is competition between plants for CO_2 in the air it is for a very short time and its role in the life of forest biogeocoenoses is very small.

While studying and analysing the various kinds of competitive relationships between plants, we must bear in mind that in natural biogeocoenoses all of these act on a plant and on the whole phytocoenose not singly but in combination and in close interaction. A number of experimental data have shown that the result of competition between plants for light is in many ways determined by the competition of their underground parts for moisture and mineral nutrients, and vice versa (Karpov 1959a).

Environment-forming effects of some plants on others take place through modification of the multiform physical factors of the ecotope, as a rule not those directly entering into the composition of the food and energy resources of the habitat, but those with considerable influence on the course of many of the physiological processes of plants. These effects may be subdivided into two main groups: (*i*) environment-forming effects of some plants on others through alteration of microclimatic factors: (*ii*) environment-forming interactions between plants based on alteration of soil factors.

Environment-forming microclimatic effects are produced mainly by changes in wind velocity, temperature and humidity of the lower air layers, intensity of solar radiation, etc.

For instance, in forest biogeocoenoses the beneficial effects of trees on each other are clearly shown in reduction of wind velocity. Thus the wind-resistance of stands is raised; the harmful physiological and mechanical effects of strong air currents on trees growing close together and on plants in the lower layers are reduced; the harmful drying action of wind on the assimilating organs of plants is modified; and the frequency and extent of mechanical damage (breaking of branches, crown tips, and trunks, lashing of trees by each other's branches, breaking of roots) are decreased.

Some plants affect others in a forest biogeocoenose by changing the temperature of the lower air layers. The upper plant layers decrease the inflow of heat during the hours of maximum solar radiation, and also protect the soil and the air layer adjacent to it from loss of heat by radiation at night. Similarly they smooth out sharp temperature fluctuations in the air surrounding the above-ground parts of lower-storey plants. Consequently plants in the lower layers, including tree seedlings, grow in more favourable temperature conditions and escape damage from both high and low temperatures during the growing period (Morozov 1926, Tkachenko 1939, Geiger 1942, Shennikov 1950, Chugai 1960). Many forestry measures are based on such protective effects of some plants on others, including, for instance, measures for the first stages of natural or artificial forest regeneration (Morozov 1926, Tkachenko 1939).

The lower atmospheric layers in forest biogeocoenoses are characterised by higher amounts, and a more regular daily and seasonal rhythm, of air humidity, which also is a result of higher plants growing close together and exerts a definite influence on their life-activity in a forest (Molchanov 1961). Some mesophyllous species are so well adapted to the phytoclimate in forest biogeocoenoses that they are unable to grow in open spaces (e.g., *Oxalis*, European starflower (*Trientalis* sp.), and several other species of small herbaceous taiga plants) (Sukachev 1934, Tolmachev 1954). Seedlings of many tree species, including spruce, suffer greatly in clear-cut areas from critically low and high temperatures, which impede reafforestation processes and call for special forestry measures to protect the new generations of trees from frost damage (Morozov 1926, Tkachenko 1939, Molchanov 1961c).

Environment-forming edaphic effects of some plants on others are due to alteration by plants of the very complex physical and chemical properties of the soil, and on subsequent varied reactions to such alteration by various groups of higher plants. Different forest-forming species, and species of shrubs and herbage, affect the soil in different ways and by different means, depending on the climatic and soil conditions in which they grow. It is now known that some tree species and their accompanying forest plants can improve the forest-growing properties of the soil and thus raise the growth rates and productivity of the trees that grow with them or after them. Other species, on the contrary, impair the hydrophysical and chemical properties of the soil, and thus unfavourably affect other components of a forest phytocoenose (Morozov 1926, Tkachenko 1939, Zonn 1954a). Excellent proof of this type of interaction between plants is provided by many sylvicultural observations and records, on which the classification of tree species into soil-improvers and soil-degraders is based (Tkachenko 1938, Zonn 1954). Soil-degraders include, for example, spruce and pine, which cause raw humus to form and impair the hydro-physical properties of the soil. Soil-improvers include most deciduous trees, especially birch, and larch among conifers. These species draw mineral nutrients from deep soil horizons and enrich the upper layers of the soil with them, and because of the morphological and chemical characteristics of their litter they raise the rate of mineralisation of plant refuse, preventing formation of thick layers of raw humus (Tkachenko 1939, Aaltonen 1942, Zonn 1954a, Parshevnikov 1957).

The environment-forming edaphic effects of some plants on others consist mainly in prolonged and annually-increasing alteration of the characteristics of forest soils as a result of the life-activity of the plants themselves. Such alterations may differ widely in nature and in their consequences for plant life-activity; they are still far from being fully studied and analysed from the point of view of their significance for biogeocoenotic processes in the forest.

In their most general and universal form these interactions are displayed primarily by dominant tree species, which affect soil development so that definite genetic varieties of soil are formed with specific features and properties. Different plant species react differently to the same genetic variety of soil, which in many ways determines their part in the formation of different types of forest biogeocoenoses. The question of the significance of higher plants in soil-formation processes is very complex, and will be discussed in detail in another part of this book. Here we shall merely list some particular forms of interactions of this type between plants. In the first place, some plants may exert definite influence on others by changing the structure and other important physical properties of soils (e.g., porosity, water-permeability, and bulk-density). This has been proved by many experiments dealing with the water-retaining properties of forests, which showed that in their life-activity processes many tree species improve the hydro-physical properties of soils, thereby favourably affecting the growth of plants in their immediate vicinity (Tkachenko 1939, Sozykin 1939, Molchanov 1952, Remezov 1953, Zonn 1954a, Stadnichenko 1960). In sylvicultural literature attention is often drawn to the interesting fact that roots of several tree species (e.g., birch) penetrate deeply into the soil and, after their death, enable roots of conifers (spruce) to reach lower horizons, thus permitting the latter to use stocks of water and nutrients from deeper soil layers (Tkachenko 1939). This helps to explain the soil-improving role of several tree species and the beneficial effects of succession of species in the taiga zone.

On the other hand, one has constantly to remember that the development of a dense herbaceous cover in thinned-out forests of light-loving species is accompanied by deterioration of the hydro-physical properties of the soil, with consequent lowering of the productivity of stands and of the rates of their regeneration from seed or vegetatively.

The role of forest litter in interactions between plants in forest biogeocoenoses is a major one. In this connection the differences between forest-forming species in the morphological and chemical characteristics of the leaves, needles, and twigs shed every year, and in the rates of mineralisation of these, are very significant. Some species (spruce, pine) form coarse-humus forest litter, whose hydro-physical and chemical properties are unfavourable for seed germination and for the growth and development of many species of higher plants, including several tree species. Other species form litter of an intermediate character, possessing more favourable hydro-physical properties and creating a better mineral-nutrition regime for plants growing in a forest phytocoenose. Sylvicultural control over the condition and fertility of forest soils (and consequently over the growth and productivity of tree stands) often leads to changes in the characteristics and properties of forest litter. The same result may be achieved by regulating the mutual relationships between tree species, including the introduction into forest biogeocoenoses of trees and shrubs that, by the morphological and chemical properties of their litter fall, change the structure and the hydro-physical and chemical properties of the litter, and the rates of its mineralisation and of the liberation of nutrients from it (Tkachenko 1939, Aaltonen 1942).

Litter has a marked effect on the formation of herbaceous cover in forest biogeocoenoses, the effects partly consisting in such purely mechanical actions as the burying and mulching of seeds and delicate seedlings, creating unfavourable conditions for seed germination in loose and quickly-drying layers of litter, etc.

Finally, the redistribution of matter from the soil by forest plants is very important in the life of forest biogeocoenoses. Plants with deep root systems draw mineral elements from lower soil horizons and thereby enrich the upper layers of the soil profile with litter of leaves, bark, branches, etc. In this way these plants bring into use and into new biological cycles material from deep soil layers inaccessible to other plants.

We could give many more examples of the environment-forming effects of some plants on others in forest biogeocoenoses, but it is enough to sum up by saying that they include many harmful and beneficial interactions between plants, differing greatly in their direction and nature, and not included in the narrow meaning of competition between plants for light, moisture, and mineral nutrients, but indirectly affecting these important kinds of inter-relationships between plants.

Allelopathic (*biochemical*) *forms of relationships between plants* consist in mutual or unilateral effects of one plant on another through secretion of specific products by the leaves and roots of living plants, or through organic compounds produced by decay of dead plant remains. The effect of one plant's secretions on another may be detrimental (toxic) or beneficial, stimulating plant growth. This form of relationship between plants has been studied most intensively (Molisch 1937, Bonner 1950, Grümmer 1955, 1961, Chernobrivenko 1956, Börner 1958, 1960 a, b, 1961, Chasovennaya 1961, Rademacher 1959, Sanadze 1961, Winter 1961, Grodzins'kii 1962). But up to the present the role of biochemical interactions between plants in the life of forest biogeocoenoses has not been adequately described.

In the study of biochemical forms of interactions between plants in natural biogeocoenoses the following main problems have to be solved: (*i*) do plants in natural biogeocoenoses actually affect each other by means of products of their life-activity, and what is the extent and coenotic significance of these effects; (*ii*) what is the nature of these active chemicals, and what constitutes the physiological mechanism of their action on other plants; (*iii*) what are the concentrations of these substances in natural biogeocoenoses, and the dynamics of their accumulation and breakdown. Unfortunately all these questions as they apply to forest, and also to other, biogeocoenoses are practically unstudied and unclarified in the literature now existing on the subject of biochemical interactions between plants. Actually most studies of these problems have been made in laboratory conditions, where the concentrations of active chemicals have been raised to artificial levels in the experiments. In order to extract these regulatory substances, 'cholines', from plants, macroscopic methods of dissection of tissues of living plants have been used, and the plants have been, as a rule, species not playing a large part in the formation of natural plant communities. Obviously the results of such investigations cannot be applied to natural coenoses, in which biochemical forms of inter-relationships between plants are practically uninvestigated.

The problem is still further complicated by the fact that in investigating biochemical kinds of relationships between plants in natural biogeo-coenoses it is extremely difficult to isolate the reactions of plants to metabolic products of other plants, and to investigate them separately from the reactions to the competitive and environment-forming effects of the other plants. This problem, however, is constantly overlooked and several phenomena in the life of plant communities are attributed to the biochemical action of one plant on another without any attempt being made to discover whether competition between the plants is not respon-sible for the phenomena.

Consequently, we shall discuss some factual data on biochemical rela-tionships between plants growing close together. Biochemical relation-ships between plants are usually divided into the following principal types (Rademacher 1959, Börner 1962):

1. Effects of certain plants on others through secretion of organic compounds by living roots.

2. Effects of certain plants on others through secretion of substances by living above-ground parts of plants.

3. Effects of certain plants on others through products of decay of dead plant remains.

Effects of certain plants on others through secretion of organic compounds by living roots. De Candolle (1832) was the first to propound the theory that plant roots discharge into the soil organic compounds that 'poison' it and hinder and suppress the growth of following generations of plants. That view was further developed in the theory of 'exhaustion' and soil toxicosis with reference to the accumulation in the soil of substances harm-ful to plant growth and secreted by the roots of the plants themselves.

The fact that plant roots secrete organic and mineral compounds differing widely in composition has been confirmed by many investigations, in which the chemical composition of these substances has also been more or less precisely defined (Akhromeiko 1938, Knapp 1954, Krasil'nikov 1958, Rakhteenko 1958b, Rademacher 1959, Börner 1960a, 1962).

The discharge of root secretions is also indicated by the concentration of micro-organisms, parasitic nematodes, and other pathogenic organisms in the rhizosphere zone. At the same time that the substances are being excreted by living roots, root excretions are enriched by compounds formed as a result of the dying cells and root tissues, these compounds being found in complex combinations with products of the life-activity of micro-organisms living in the rhizosphere of the roots. The products of microbial life-activity may also be toxic to plants (Krasil'nikov 1958). It has proved much more difficult to establish the significance of root secretions in the formation and life of plant communities. The great majority of these secretions consist of different kinds of amino acids and sugars, which have practically no effect on plant growth (Rademacher 1959, Börner 1962).

In the course of very precise and careful studies of plant root secretions a number of substances have been isolated, possessing high toxicity for germinating seeds and juvenile plants; these organic compounds retard or suppress plant growth in very low concentrations. For instance, Bonner and Galston (1944) and their colleagues isolated from the roots of guayule shrubs trans-cinnamonic acid, which has an inhibiting effect on the growth of seedlings of the shrub in a concentration of 50 mg in 1 litre of nutrient solution in water. From oat root secretions scopoletin (a substance derived from coumarin) has been isolated; it seriously retards the growth and development of weed grasses (Eberhardt 1954, 1955, Martin 1956, 1957, 1958). With regard to secretion of toxic substances by tree roots, data exist concerning only a few species of fruit trees. For instance, in a study of the problem of 'exhaustion' of the soil in peach orchards, amygdalin was found in the cortex of peach tree roots; it gives rise to benzaldehyde (which is highly toxic to peach seedlings) when it decomposes in the soil. A similar type of substance, fluricin, has been found in the cortex of apple-tree roots (Börner 1959, 1960b). It is still an open question whether the roots of the principal forest-forming species secrete strongly-acting substances of such types and, if so, what is the chemical nature of these substances; this naturally makes it difficult to evaluate the role of root secretions in the life of forest biogeocoenoses. Even with regard to the toxic secretions by living roots that have already been discovered, there are no reliable data proving that they have much significance in relationships between plants. In the classic work by Bonner and his colleagues, often quoted by supporters of allelopathy, an attempt to discover the toxin isolated by them (trans-cinnamonic acid) in the soil of guayule plantations was unsuccessful. It appears that that toxin is very unstable; in natural conditions it is rapidly broken down by micro-organisms and does not accumulate in the soil. This entirely agrees with Martin's quantitative investigations (Martin 1957, 1958, Börner 1962) on secretion of another toxin, scopolin. In natural conditions oat roots secrete scopolin in very small amounts, quite insufficient to produce any retardation of plant growth. According to Martin's investigations (1958), the process of secretion of organic compounds by roots has a passive character and is due to the inability of root cells to retain these compounds when conditions are unfavourable for their life-activity and when the general metabolism of the plant is disrupted. This means that in natural conditions one cannot expect high concentrations of these substances in the soil, or prolonged effects of them on plant roots (Börner 1962).

The problem of toxic root secretions in the soil is usually considered to be closely connected with the problem of toxicosis, or 'exhaustion', of the soil under some agricultural or fruit crops. It has been found, however, by numerous investigations, that 'exhaustion' or toxicosis of soils is a very complex phenomenon, in most cases due to accumulation in the soil of pathogenic soil organisms (nematodes) or disruption of the microbiological equilibrium in the soil, and not to the accumulation of toxic substances of vegetable origin (Rademacher 1959, Börner 1960a, 1962). This does not, of course, exclude the possibility of development of toxicosis in the soil as a result of accumulation of toxic plant-root secretions in it, but such phenomena are apparently observed only rarely, and then mostly after prolonged artificial cultivation of certain species of agricultural crops and fruit trees (Patrick 1955, Börner 1959, 1962).

Effects of certain plants on others through secretion of organic compounds by above-ground parts of higher plants have been investigated by many authors, mostly in laboratory conditions (Chasovennaya 1954, 1961, Knapp 1954, Grümmer 1955, Bode 1958, Rademacher 1959, Kolesnichenko 1960). We may consider it as proved that the fruits and leaves of several species of higher plants (especially of the families Umbelliferae, Compositae, and Labiatae) may secrete substances that are very active physiologically, inhibiting the growth and development of other plants. Molisch (1937, Grümmer 1955) demonstrated the effect of ethylene, which is secreted by apples, on the growth and development of a large number of higher plants. A. A. Chasovennaya (1954, 1961), in a number of investigations, has shown the effects of organic substances found in the leaves of meadow herbs and weeds on the growth of many species of cultivated plants. Bode discovered that the leaves of black walnut (*Juglans nigra*) and wormwood (*Artemisia absinthium*) secrete substances inhibiting growth in other plants. Recently Kolesnichenko (1960, 1962) has demonstrated the existence of physiologically-active substances in the leaves and needles of many tree species; he believes that he has succeeded in proving that these substances have toxic effects on the processes of photosynthesis in tree species. All of these observations have been described in detail in comprehensive reviews by Knapp (1954), Grümmer (1955), and Rademacher (1959), where information is also given on the chemical nature and physiological action of substances of various kinds extracted from the fruits and above-ground parts of many plant species. The real significance of volatile secretions by above-ground parts of plants in the formation and life of plant communities is still unknown, since it is not known in what concentrations these substances remain in the atmosphere of forest biogeocoenoses, and the effects of these substances on plants have not been studied in a natural environment, where the conditions of their accumulation and physiological action on plants are entirely different from those in the laboratory.

Only in connection with herbaceous communities in steppe districts, where species producing highly-active volatile organic compounds are dominant and occur in great abundance, can we speak with some confidence of the effects produced by some plants on others through secretion of cholines by their above-ground parts.

At the same time that organic compounds are being actively secreted by leaves and fruits, mineral and complex organic compounds are washed from the above-ground parts of plants by precipitation, sometimes in considerable quantities (Tamm 1950, 1951, Pozdnyakov 1956, Rademacher 1959). Unfortunately the phytocoenotic and biogeocoenotic significance of this in the life of forest biogeocoenoses is entirely unstudied. On the one hand it may be assumed that in this way plants in the lower layers receive additional amounts of nutrients (e.g., available compounds of nitrogen, phosphorus, potassium, and calcium) to improve their living conditions (Tamm 1951). On the other hand it may be that some of the washed-off organic compounds have a definite toxicity for plants growing in the lower layers of biogeocoenoses, and to some extent suppress them. The problem requires further study and presents great interest for forest biogeocoenology, since a fairly important form of exchange of matter is concerned.

Effects of certain plants on others through products of decay of dead plant remains are much more important in the life of forest biogeocoenoses than those through secretions by living plants. Huge amounts of plant remains reach the soil every year as the result of the shedding of above-ground plant parts of different kinds, of annual dying-off of parts of root systems, and of death and breakdown of individual plants. A great variety of organic compounds are leached by rain out of fresh plant remains and from those already undergoing mineralisation, and enter the soil; these include compounds that either inhibit or stimulate plant growth. This has been proved by many experiments (Bautz 1953, Winter and Schönbeck 1953, Bublitz 1953, Winter and Bublitz 1953, Knapp 1954). For instance, Bautz (1953) and Bublitz (1953) discovered that fresh litter in spruce and beech forests contains substances that inhibit seed germination and seedling growth of spruce and pine. The concentrations of these substances in fresh litter depends on the time of year and on the climatic growth conditions of forest biogeocoenoses; in arid districts with little precipitation small amounts of these substances are leached out of litter, so that they accumulate and may inhibit natural regeneration of such tree species as spruce and beech (Knapp 1954).

The experiments of Runov (1954) have proved that water-soluble extracts from forest litter in artificial plantations inhibit germination of the seeds of many steppe grasses.

Finally, in a series of works by Winter and his colleagues (Winter and Schönbeck 1953) it has been demonstrated that straw and stubble of grain crops contain substances that inhibit the growth and development of several cultivated plants and weeds; these authors have traced the accumulation of these substances in the soil and determined the duration of their physiological activity.

Clearly certain plant species may affect others growing beside them in a phytocoenose through the products of litter decay. Since the roots of higher plants can absorb complex organic molecules, the probability of some plants affecting others by means of organic compounds differing widely in their composition and properties becomes very much greater. Moreover, these products are constantly entering the soil, and conditions for their accumulation there are much more favourable than in the atmosphere.

Unfortunately little study has been given to the effects on plant growth of organic substances released by decomposition of plant remains, which makes evaluation of this type of interaction between plants in forest biogeocoenoses extremely difficult.

Summarising, we may state that allelopathic relationships between plants have incomparably less significance in the formation and life of natural biogeocoenotic systems than has competition between plants for the matter and energy resources of their habitat. This view is held by the most eminent specialists in the field of plant allelopathy, Grümmer (1955), Rademacher (1959), and Börner (1960a, 1961), who point out the speculative and hypothetical character of many works on the biochemical relationships between plants.

Indirect transbiotic interactions

The effects of certain plants on others are produced not only through alteration of factors of the physical environment in the ecotope but also through the medium of animals. Animals exert tremendous indirect influence on the system of established relationships between plants in a phytocoenose, sometimes changing the system so much that radical modification of a forest biogeocoenose takes place.

The role of insect pests and rodents in destroying seeds, seedlings, and mature plants, and also the results of pasturing domestic livestock in a forest, are well known. The activities of these animals, however, are not always given adequate biogeocoenotic interpretation, and people do not always realise the serious damage to the relationships and interactions between plants that is caused directly by animals and leads to profound changes in the composition and structure of forest biogeocoenoses.

Plants are the primary food of most animals, and constitute the beginning of all the complex 'food chains' and 'food cycles' that develop among forest animals.

By eating or damaging certain plant species animals partly or completely destroy the competitive effects of certain plants on others, improving and strengthening the position of plants not damaged by them or more resistant to their action, in comparison with the position of other components of a forest phytocoenose. This form of transbiotic relationship between plants is universal, and the most diverse animal groups take part in it. Different animal species have their own peculiar ways of using or damaging different species of plants, and their methods of action differ greatly. On the other hand, different plant species react differently to animal damage. These facts give a very complex character to transbiotic forms of relationships between plants, which usually develop thus: action of animals on plants–change in the environment-forming and competitive ability of plants–reaction of other plants in the neighbourhood to that change in their ability.

Animals may not only break down existing systems of relationships between plants by damaging and destroying certain species, but also strengthen certain links and interactions between plants. For instance, birds destroy enormous numbers of insect pests and mouse-like rodents, and form one of the most important factors restricting the numbers and multiplication of these animal groups in a forest. Thus they may benefit the regeneration and growth of several tree species and raise their numbers appreciably. The value of many birds in distributing seeds and in extending the range of various tree species is well known; their actions lead to the introduction of new species into existing phytocoenoses, sometimes of species with more competitive ability than the native ones. Finally, many insectivorous species play a valuable role in the pollination of forest plants.

Some very complex interactions between plants take place through the medium of soil fauna, especially invertebrates: insects, nematodes, millipedes, crustaceans, arachnids, etc. These animals take an important part in the conversion of plant remains, in enriching soils with various compounds, and in changing the organic and hydro-physical properties of soils. The species-composition, numbers, and (in many ways) the activities of these animals depend on the presence in a forest biogeocoenose of certain species of higher plants that provide them with food. The presence of these plants is necessary for the participation of animals in the biocoenose, and stimulates the activity of soil fauna in altering the chemical and physical properties of the soil, so leading to corresponding reactions from other plants. The outstanding role of earthworms, for instance, in improving the structure and chemical properties of the soil is well known, but is best displayed only with a definite combination of tree species in a forest phytocoenose. Some animals, on the other hand, impair the hydrophysical properties of the soil by compacting it, which may weaken the

FFB R

competitive ability of certain plant species to contend with others more resistant to that soil condition.

As a unique and specific type of transbiotic interactions between plants, we must examine the effect of certain plants on others through alteration in the abundance, composition, and activities of micro-organisms in the soil.

Extremely intimate and multiform relationships have been established between higher plants and micro-organisms in forest biogeocoenoses. Micro-organisms consume and transform various root exudates, some of these substances stimulating their reproduction and others depressing it. By altering the conditions of root nutrition for plants and excreting a wide variety of products in their life-activity processes, micro-organisms may in turn stimulate or depress higher plant growth.

The composition and abundance of micro-organisms in rhizospheres and soils depends on the species-characteristics of higher plants, whose roots are a factor not only in accumulation of micro-organisms in the soil but also in deciding their species-composition (Krasil'nikov 1954, 1958, Krasil'nikov and Gorkina 1954, Runov and Kudrina 1954, Runov and Enikeeva 1959). Different species of higher plants apparently favour the development of specific groups of micro-organisms, and by changing the composition of the microflora may affect other plants. The possibility of such indirect effects of plants on each other is confirmed by a number of observations and direct experimental data, obtained so far only in laboratory conditions (Runov and Zhdannikova 1956, Runov and Egorova 1958, Runov and Enikeeva 1959, Egorova 1962).

Soil toxicity is due to the breakdown of relationships between the most important groups of micro-organisms, the direct cause of change in microflora composition often being the plants themselves (Börner 1960a, 1962). On the whole this type of transbiotic relationship between plants has been studied very little in natural biogeocoenoses, although undoubtedly it plays a very important role in the life of the forest.

PRACTICAL SIGNIFICANCE OF THE PROBLEM OF MUTUAL RELATIONSHIPS BETWEEN PLANTS IN FOREST BIOGEOCOENOSES

It has been shown previously that relationships between tree species and between them and lower-storey plants are one of the main factors controlling and regulating the composition, structure, and dynamics of forest biogeocoenoses and of the biogeocoenotic processes. Since many sylvicultural measures are aimed at altering the composition and structure of forest biogeocoenoses and at raising their productivity and their valuable economic properties, the intelligent and purposeful guidance of mutual relationships between plants is a vital link between theory and prectice in forest economy.

In fact, the whole activity of a sylviculturist in the forest is oriented towards the management of mutual relationships between plants, in eliminating the harmful effects of weeds and grasses on the growth and regeneration of valuable species, in regulating the relationships between trees in valuable stands in order to produce trunks with the required technical features, in eliminating the harmful effects of mature trees on young growth, and so on.

The success of any sylvicultural or forestry measures in a forest depends in many ways on having a clear picture of the character and mechanisms of those interactions between plants that can be artificially altered or eliminated by a sylviculturist in order to obtain a definite sylvicultural effect. Many failures and errors in solving practical sylvicultural problems occur through insufficient knowledge, or ignorance, of the mechanism and factors of regulation of interactions between plants in forest biogeocoenoses. This statement could be illustrated by many examples from widely-differing branches of sylviculture. Here we shall restrict ourselves to two that demonstrate very clearly the importance of sound knowledge of the character and mechanism of interactions between plants in a forest.

We should bear in mind the errors in selection and placement of tree species in plantations that were permitted by sylviculturists during the first stages of afforestation of our steppes. Such unstable combinations of tree species as the ash-elm 'Don' and 'normal' types of plantations were then created. In conditions of inadequate soil moisture these plantations either did not survive competition with steppe weeds more adapted to the steppe-zone climate, or their best varieties were displaced by second-rate ones because the latter proved to be more adaptable to competition for moisture (e.g., in the case of ash and elm).

There is no doubt that one of the most urgent tasks of forest management in the taiga zone is the restoration of valuable coniferous forests and the prevention of deterioration of the species-composition of forests during the process of succession of species. Successful solution of these problems is possible only if we find effective and economically-profitable means of combating the harmful effects of weed vegetation in the process of regeneration and growth of valuable coniferous species on clear-cut areas, and if we ensure the protection of new conifer populations against the inhibiting and destructive action of small-leaved species. All forestry experience in the taiga zone goes to show that without sylvicultural control over the mutual relationships between plants, without active measures of control of weed grasses and other species during the most important periods of stand formation, we cannot achieve success in the rapid restoration of valuable forests.

Unfortunately the tremendous national-economic importance of timely and correct intervention in the system of established inter-relationships between plants in forest biogeocoenoses is not always recognised. At present our forest economy annually suffers colossal losses as a result of the suppression and annihilation of plantations by the natural self-sowing of weeds in the first phases of replanting clear-cut areas. The losses in annual increment of valuable coniferous species through suppression by less-valuable deciduous species, within the European part of the U.S.S.R. alone, are estimated in tens and hundreds of thousands of cubic metres of wood, which ultimately means prolongation of the period required to produce timber.

Up to the present many sylvicultural measures are based on very primitive and sometimes erroneous ideas about the inter-relationships between forest-forming species and between them and seedlings, herbs, and shrubs in the lower layers. Recent investigations have shown that the growth and regeneration of tree stands are controlled by a very complex aggregate of biogeocoenotic factors, among which competition for light is not the leading and basic factor in all biogeocoenotic formations and forest types. In many cases, to obtain the best sylvicultural results, it is necessary to regulate the intensity of competition between trees for soil moisture and mineral nutrients, and sometimes also to regulate other forms of inter-relationships between plants.

Meanwhile, over vast territories and in the most diverse types of habitat and of forest, identical equipment and technical measures for regulating mutual relationships between plants are being used. The situation is complicated by the fact that our information about the character, and especially about the mechanism, of relationships between the chief species of forest plants is still fragmentary and inadequately covers all the varied conditions in which biogeocoenotic formations and the forest types composing them have arisen. This certainly handicaps forest science in its control of the very complex processes that take place in the vast forest areas of our country.

Therefore the devising and application of all sylvicultural measures for improving the composition and raising the productivity of forests should be based on thorough knowledge of the conditions and factors that regulate inter-relationships between plants in forest biogeocoenoses, and on intelligent and purposeful management of these inter-relationships with the aim of obtaining the best possible growth and regeneration of valuable species in the forests of our country.

CHAPTER IV

Animal life as a component of a forest biogeocoenose

'The forest,' wrote G. F. Morozov (1931, p. 239), 'like terrestrial surface relief . . . increases the surface on which life may develop, but it also adds complexity and multiformity to living conditions themselves.' Later on, Morozov (pp. 300-302) stresses the interactions of the environment with the biocoenose related to it, and especially with the animal communities formed in it.

The specificity of animal life in different types of forest, and the close connection between animals and the many phenomena occurring in the forests, are outstanding characteristics of forests as biogeocoenoses.

The process of development of a biogeocoenose is usually accompanied by colonisation of organisms, leading to the creation of specific food chains in which animals participate as a distinctive link. Thus the establishment of vegetation leads to the arrival of herbivores that live upon it; the presence of these herbivores attracts specific predators and parasites. The mutual adaptation of species in the particular conditions of a developing biogeocoenose creates animal populations representing definite living forms of their species (Gilyarov 1954, 1959b), and as such they take their place in the food chains.

Adaptation of animal species to the climate of a given locality also occurs during the formation of a local population. The population is characterised, in the first place, by the fact that the range of tolerance of the organisms composing it corresponds to the range of climatic fluctuations; and in the second place, by ability to increase population density with favourable variations in the environment, i.e., with decrease in the strength of unfavourable factors regulating the environment. Such increases in population should be offset by decreases in unfavourable years; if there is no such compensation, or if the action of an unfavourable factor (e.g., frost) is too powerful, the result may be extermination of the population.

Naturally, in every generation the part of the population unable to use favourable conditions to protect it against unfavourable ones dies off. Sufficient numbers of individuals must remain to perpetuate the population.

K. Frideriks (1932) maintains that in ecological investigations it is necessary to consider the following climatic factors: incident solar radiation, heat emanation, air temperature, air humidity and evaporation,

253

precipitation and cloudiness, wind, and atmospheric pressure. Clearly, the first two of these determine the third, and all three together constitute the thermal regime. The rest create the humidity regime, which for organisms is inseparable from the thermal regime. In fact, the optimum and the extreme limits of environmental temperature vary with the humidity regime, and moisture exchange varies with the temperature. With increased humidity, for example, animals can endure higher temperatures; with comparatively high temperatures, shortage of moisture becomes lethal.

Because of its gaseous composition, the atmosphere is an essential component of every biogeocoenose, and as the vector of climate it makes the effects of the latter inescapable, and almost independent of population density. The postulate of independence of action of climatic factors has been stressed by a number of investigators in cases of mass insect reproduction, in spite of the action of biotic factors whose effects are always dependent on population density. B. P. Uvarov (1931) wrote that in nature it rarely happens that competition for food, shelter, etc., is inoperative, but independence from climatic effects is impossible. Only within a restricted area can a change in population density affect temperature and humidity and thereby modify the effect of meteorological conditions. In all other cases the action of the latter is independent.

Determination of the dependence of animals on the biogeocoenose in general and on other components in particular, and also the animals' effects on the other components, implies determination of the role and the potentiality of the animals in maintaining or disrupting the biotic balance. The biotic balance, according to G. G. Vinberg (1956, 1962 a, b), is the equation representing, on the one hand, the difference between primary production and its destruction (i.e., its utilisation by all species), and on the other hand the augmentation of energy, and consequently of biomass, during a certain period. Investigation of the role of animals in maintaining or disrupting a particular biotic balance is a major part of the study of animals as components of given biogeocoenoses.

To solve that problem it is first necessary to discover the species-composition of the animal life and its numbers (recording its distribution in space and time), and, secondly, to discover the relationships of the animals with plants, with micro-organisms, with one another, and with abiotic components. Some authors (Doppel'mair 1915, Shiperovich 1936) have dealt with the characteristics of animal life in the forest; they have, however, described forest fauna and the ecological characteristics of separate species of forest animals, but touched only briefly, or not at all, on the questions discussed here.

During the last two or three decades investigations have begun on the animal groups typical of different forest types and on their role in the circulation of matter and energy (Kühnelt 1944, Shelford 1951, Rabeler 1957, 1962, Trautmann 1957 a, b).

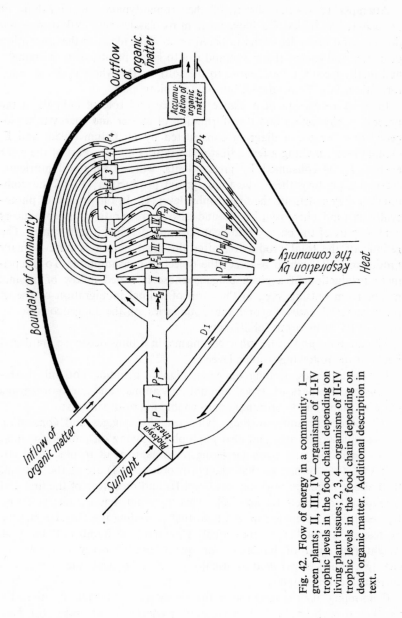

Fig. 42. Flow of energy in a community. I—green plants; II, III, IV—organisms of II-IV trophic levels in the food chain depending on living plant tissues; 2, 3, 4—organisms of II-IV trophic levels in the food chain depending on dead organic matter. Additional description in text.

The study of biogeocoenoses (particularly biohydrology) transcends the boundaries of biology, as it deals with the participation of abiotic as well as biotic components in the circulation of matter and energy.

Attempts to restrict survey of the trophodynamic relationships of animals to the limits of a biocoenose in its classic sense (Viktorov 1960) do not stand up under criticism based on modern ideas on the circulation of matter and energy (from sun and soil to green plants, then to animals, and finally from the biocoenose to the soil), as the whole cycle of transformations is not considered, but only its biotic part.

In order to present at a glance the place and role of animals in the process of circulation and transformation of matter and energy in a biogeocoenose we use a diagram constructed by H. Odum (1956) and E. Odum (1959), making a few adjustments in it (Fig. 42). Part of the solar energy (E_1) is consumed by green plants, which (using soil nutrients) create by photosynthesis the total (gross) production (P) of the community; they constitute the first trophic level (I). In the process of photosynthesis (and chemosynthesis), and also in respiration, the producers convert part of the energy into heat (D_1) and disperse it into space. The residue of the total production after this heat output is pure primary production (P_1), which becomes the object of consumption. To the primary production of a given biogeocoenose is added inflow of organic matter from outside, e.g., in the form of animal immigration or use of plant material eaten beyond the boundaries of the biogeocoenose by some of its herbivorous inhabitants.

The animals, and some other organisms, in a biogeocoenose are distributed at the following trophic levels:

II   phytophages, feeding on green plants (including not only herbivorous but also parasitic types, and not only animals and micro-organisms but also plants) and consuming autotrophic micro-organisms;

III   predators and parasites, living on phytophages and consuming autotrophic organisms, and also phytophages living on parasitic plants;

IV   predators and parasites living on organisms at trophic level III.

If Fig. 42 is turned so that plants (trophic level I) are at the bottom, then above them are phytophages (level II) and predators of the first and second orders (levels III and IV). This 'pyramid' has the characteristic that each level is smaller in size than that preceding it. The size may be expressed in number of individuals ('pyramid of numbers'), mass or weight ('pyramid of biomass'), or quantities of energy ('pyramid of energy'). Such a pyramid is usually called ecological, but it is more correct to call it trophic.

Continuing our examination of the diagram in its normal position, we see that at each level a certain amount of energy is consumed (E_n), part is used in respiration (D_n), i.e., converted into heat, and the remainder is pure production at the given level (P_n).

Besides the main flow of production along the central channel from left to right in the diagram, a certain part flows out in the 'lower channel', which lies beneath the central horizontal channel and shows accumulation of organic matter. This includes the matter not consumed by representatives of the following level and not yet decomposed (in particular, the stock of wood that performs skeletal, not metabolic, functions). It also includes the matter within living animals and in carcases that have not yet begun to decay. In other words, here we have all the stock of living and dead organic matter accumulated in a biogeocoenose. It is the source of all the output of organic matter in such forms as animal emigration, removal of litter by outflowing water or emigrating animals, etc.

Another part enters the 'upper channel' above the central horizontal channel, leading to decaying organisms. All types of plant litter and animal excretions, as well as accumulated matter when it dies off and decays, enter that channel. Dead wood, sometimes even without losing connection with the living parts of plants, may be acted upon by the organisms of decay; other forms of dead organic matter decompose, as a rule, after falling on the surface of the soil.

Decomposition of the organic matter in litter and other soil horizons is aided by soil-dwelling saprophages and saprophytes, among which animals occupy a prominent place. Decaying above-ground parts of the dead wood of standing trees constitute the sphere of action of saprophages and saprophytes (including micro-organisms) that live on trees; after these parts of a tree fall off, or after such trees fall, a change-over occurs, the 'timber-eating' saprophages and saprophytes gradually give way to soil-dwelling ones that live on decaying organic remains. The period occupied by this change-over depends on the conditions of the decaying wood.

The above description shows that when certain parts of plants die (independently of whether they maintain or break off connection with the organism as a whole) they still have a place in the community's primary production. The matter produced by autotrophic organisms serves as the base of the trophic pyramid, i.e., as a source of food energy for organisms on the next level. We must remember that matter of animal origin and the remains of micro-organisms combine with plant litter. Thus the 'upper channel', parallel to the central horizontal channel, also has its trophic levels:

1. organic matter of plant and animal origin and remains of micro-organisms;

2. saprophages and saprophytes from meso- and micro-fauna and -flora;

3. phytophages, predators, and parasites (animals, plants, and micro-organisms) that live on organisms of the second trophic level;

4. predators and parasites living on organisms of the third trophic level.

Organisms that live on decaying matter (as well as those deriving energy from living autotrophic organisms) form a second trophic pyramid, from which part of the energy is dissipated by respiration and in the form of heat, part (not consumed by succeeding levels) returns into the same channel, and part enters into accumulated matter.

It is of great significance that a large part of the organic matter in the 'upper channel' is mineralised by micro-organisms, i.e., decomposed into elements that, together with humins and other soil material, become plant food. Thus the cycle of matter and energy is completed.

In some cases the movement of a given organism to a certain level is complex. Thus predators and parasites may attack not merely organisms in the trophic levels directly preceding them, but also those in other levels or not even in 'their own' pyramid. For example, a number of seed-eating birds feed their young on insects; a number of entomophagous insects feed also on phytophages and saprophages, and so on. Some species change their diet with the season of the year, and then distribution by levels may vary with the seasons. In other species different populations have different diets and their level-distribution is zonal (by landscape or habitat). But diet complexity may be constant in time or space, and then a species should be referred to certain levels in accordance with its consumption of certain items. From the biogeocoenotic point of view this forms one of the most interesting problems, since the biochemical composition of matter varies, of course, with the nature of the food consumed.

The chief deductions to be made from the diagram of energy flow (Fig. 42) are as follows:

Primary production, formed from a single source of energy (solar radiation), is a stock of food or chemical energy that can: (i) enter living matter in the food of second-level organisms (phytophages); (ii) enter the food of second-level organisms (saprophages and saprophytes) after death; (iii) be stored for some time in living or dead biomass, and after death decompose and become food for saprophages and saprophytes.

In every biogeocoenose the specific features and the quantitative ratios of these three channels determine its trophic structure and the nature of its biogeocoenotic processes, which remain, with slight variations, close to the long-term averages; that is what determines the specific nature of a given biogeocoenose.

Fluctuations in the density of animal populations, and especially mass reproduction, cause changes in the relationships between the channels. An increase in the proportion of matter and energy consumed by organisms of the second trophic level, phytophages (e.g., leaf and needle pests) and saprophages (e.g., timber-eaters), with a corresponding fall in accumulation of organic matter (mainly the wood stock), may in forest communities lead to the death of a tree stand and to the development of another biogeocoenose (Rafes 1964).

In any biogeocoenose, animals are merely consumers of organic matter; but while they possess that general characteristic, their role in the transformation of matter and energy may be extremely varied. Some eat plant matter, living or dead; others eat animals, vegetarian or carnivorous.

Although plant matter is the basic resource for all animals, animals must not be regarded as organisms merely parasitic on green plants. From the geochemical and biogeochemical points of view, one of the chief functions of animals in nature consists in accelerating the circulation of matter and energy and, in the course of time, the accumulated plant matter passes into the soil. A considerable proportion of the plant mass, especially in forest communities (in the form of wood), remains out of circulation for a long time. Consumption of wood by xylophages brings the eaten wood back into circulation and hastens its mineralisation. Green matter, which in temperate-zone forests annually falls in large amounts and becomes part of forest litter, is mineralised more rapidly than wood, but not in the same year that it falls. The consumption of green matter by phytophages accelerates the mineralisation of that portion of plant matter also. Thus in a forest biogeocoenose (as also in others), besides the portion of plant matter that enters the soil directly, a portion is used to form animal matter and reaches the soil as excreta or in the form of carcases.

The role of animals in biogeocoenotic processes consists not only in accelerating mineralisation but also in redistributing organic matter. If plant matter entered the soil only through death, i.e., without the participation of heterotrophic organisms, the process of mineralisation would be very prolonged and the distribution of matter would be patchy, reflecting plant distribution.

By accelerating mineralisation and redistributing organic matter within a biogeocoenose and outside it (on account of their mobility), animals affect the evolution of a biogeocoenose. They act upon soil processes, upon the physiological state of the vegetation, and consequently upon natural regeneration and succession of species, and thereby produce a succession of biogeocoenoses, as stated previously and to be discussed later.

The links between animals and other components of a biogeocoenose lead to the conclusion that quantitative increase in the flow of matter and energy through a food chain from plants to phytophages, e.g., through mass reproduction of pests, produces an increase in numbers of consumers ('enemies') of the latter, on account of the increase in their food supply. On the other hand, the growth of phytophage populations is sharply checked by exhaustion of their food resources. Although these regulators suppress mass reproduction of phytophages, the recuperative powers of the food plants may be exhausted, and the forest biogeocoenose perishes and succession occurs.

Trophic relationships are always decisive in the inter-relationships of animals among themselves and with other biogeocoenose components. The presence of food determines the entrance of an organism into a given biogeocoenose, and its diet determines the exchange of matter and energy in that link of the chain. When food is present the organism adapts itself to the hydrothermal living conditions and to the biotic factors of the environment, as is demonstrated by its choice of shelter, formation of daily and seasonal rhythms of living, establishment of phenological periods of development, form of life-activity, etc., to the point of acquirement of morphological characteristics ('morphs'). From study of food relationships we must advance to the study of all other possible relationships and their systems. Therein lies the task of biogeocoenological investigations.

The fact that a particular fauna is typical of every forest type is due primarily to the specific composition of the vegetation, which results in specific composition of the phytophagous animals and micro-organisms. The biocoenotic complex of a given forest type determines the distinctive features of its organic components, and consequently the species-composition of its saprophages, coprophages, and necrophages. The species-composition of the consumers of vegetation and organic remains determines the predatory and parasitic fauna.

Food relationships create specific food chains for each forest type, and ultimately the geochemical characteristics of the biogeocoenose are created, i.e., the specific characteristics of migration of chemical elements.

The entrance of animals into a biogeocoenose and also their spatial distribution in it are determined by food chains. Most invertebrate phytophages feed directly on their substrate. Animals whose shelter is elsewhere than on their food substrate are also linked with definite places in the biogeocoenose through their life-activity. These places are determined for feeding by the location of food, and for shelter by the possibilities of shelter-construction and by the specific requirements of an animal in relation to the hygrothermal regime in these microstations. The diet of phytophages is determined directly by the nature of the vegetation. The hygrothermal regime in any part of a biogeocoenose, and especially of a forest biogeocoenose, is determined by the vegetation, which creates the microclimate (phytoclimate).

It is worthy of note that a tree stand (the foundation of a forest biogeocoenose) creates, as stated above, a special form of terrestrial surface. The complex differentiation of living space beneath forest canopy, caused by tree stands, is an important feature in the life of animals in a forest (Doppel'mair 1915).

Every forest type has its own distinctive potentiality for animal distribution, determined by layering, form of crowns, character of underbrush and herbaceous cover, and nature of the soil; these create much-varied microclimatic features. All of these environmental conditions produce ecological niches for the animals dwelling in the forest, and determine the carrying capacity of the forest type for them. In this way the spatial distribution of animal life in the forest is determined.

The distribution of animals by layers and horizons is not fixed: the woodpecker nesting in a hole in a trunk feeds in the crown; rodents living in burrows feed on the soil surface; many larvae that develop in the crowns go into the soil to pupate; and there are insects that develop in the soil and spend a large part of their adult life on the wing. These phenomena show how animals are more or less linked with certain layers, and how the layer-grouping varies periodically (daily, seasonally, etc.). Since such movements are phenological phenomena, i.e., connected with specific periods, it is correct to speak of seasonal aspects of animal grouping.

Distribution in time may be displayed not only in regrouping by layers but also in changes in the composition of animal life in a forest biogeo-coenose as a whole. Thus some animals are linked with a certain layer or forest biogeocoenose throughout their lives, some may have activities beyond the forest limits (e.g., nesting in the forest and hunting in surrounding open spaces) or covering several types of forest biogeocoenoses (e.g., adjacent parts of the forest of different species-composition), and the connection of some with a forest biogeocoenose may be seasonal. Examples of the last type are birds that migrate elsewhere for the winter, or chinch-bugs that winter in the forest.

Animals may be divided into three groups according to their radii of activity: (*i*) the radius of activity does not extend beyond a given biogeo-coenose or part of it; (*ii*) the territory of activity is a given biogeocoenose as a whole, but from time to time the animals go outside it; (*iii*) the radius of activity extends over several biogeocoenoses. Invertebrate forest-dwellers belong mostly to the first and partly to the second group, and vertebrates mainly to the third group. Inflow of matter and energy from outside, and also its outflow from a biogeocoenose, take place partly through the activities of animals in the second and third groups (see Fig. 41). It is appropriate to remark here that 'a biogeocoenose without the necessary introduction from outside of organic matter, soil, plant seeds, animals, etc.' (K. V. and L. V. Arnol'di 1963) is unlikely, as exchange between biogeocoenoses would thereby be eliminated. In nature inflow and outflow take place both through abiotic factors (wind, water) and directly through 'exchange of forms' (the phrase of Beklemishev 1951).

Depending on the size of its radius of activity, a single animal population may be linked with one or several biogeocoenoses. The loyalty (Beklemishev's term, 1931) of a species to a certain biogeocoenose is determined by the abundance and the nature of its links with it (nesting, feeding, movement through the territory). Therefore the extent of animal relationships with different types of biogeocoenoses may vary, but the species of animals present (faunistic complex) and the amplitude of fluctuation in their numbers are characteristic.

The biogeocoenotic significance of animals is largely determined by the nature of their diet, and consequently the abundance of individuals of separate species in the groups must be considered.

In discussing species populations, we give the name 'dominants' ('dominant species', 'predominants') to those species that surpass others in population density. It is not only the physiognomy of grouping that gives dominance to a species. Relatively permanent abundance of an animal species in a certain place suggests a favourable environment there. To determine the mutual suitability of a species and its habitat we must discover the nature of their interactions and compare them with those in other habitats, also comparing the relative numbers of animals. It will then appear that in one biogeocoenose that species is numerically dominant, and in another it is suppressed.

Dominants, which constitute a major part of the biomass, have great quantitative significance in the circulation of matter, but they are not exclusively characteristic of a biogeocoenose. Fluctuations in their abundance may result from various causes and combinations of them. Qualitative differences are illustrated by indicator species (specific species), present in one biogeocoenose and absent in others. For instance, the group of oak-gall wasps (typical monophages) is an indicator of forests containing oaks as compared with forests where oaks are absent, and for the tree storey in an oak forest as compared with other storeys in the same forest.

The multiform relationships of animals among themselves and of animals with other biogeocoenose components are classified by V. N. Beklemishev (1951) in the following four types: (*i*) topic (change in habitat conditions); (*ii*) trophic (nutrition); (*iii*) fabric (use for construction); and (*iv*) phoric (phoresia, i.e., displacement). He also concludes that the relationship between organisms is an interaction of populations, the result of which can be expressed quantitatively. In the classification of relationships devised by E. Odum (1954, 1959) account is taken of the effect of the process of interaction on the populations taking part in it, expressed in either increase or decrease in their total of living matter (without quantitative determination).

To evaluate the significance of interactions between the components of a biogeocoenose, quantitative determination is absolutely necessary. Trophic relationships between biotic components are comparatively easy to evaluate, especially those between plants (in a state of vegetation or decay) and their consumers (phytophages, saprophytes, saprophages) and between different species of animals (predators and prey). At present it it possible in practice to measure the transfer of matter and energy from one trophic level to another (Vinberg 1962 a, b).

Fabric relationships, as defined by Beklemishev, occur on the rare occasions when parts of living organisms are used as material for building shelters, e.g., leaves for the nests of leaf-cutting bees (*Megachile*). We are not convinced that separation of fabric relationships as an independent type can be justified: they have a dual characterisation. Regarding them from the victim's standpoint they might be classified as predation, since a plant that loses a leaf suffers the same damage regardless of the use to which it is put. In this case, however, the insect uses it for nest-building, i.e., it obtains a positive effect of 'topic' type, improvement of environmental conditions. The quantitative analysis of such phenomena has not yet been worked out.

Beklemishev included in topic relationships effects produced on air and water movement, on light, on temperature, on humidity, on the chemical properties of air and soil, etc. Here, however, there is no direct interaction between two organisms, but one organism accepts a set of environmental conditions produced or altered by another. These, according to V. N. Sukachev (1956), are transabiotic relationships, i.e., interactions through the medium of abiotic biogeocoenose components. With direct ('contact', according to V. N. Sukachev) interactions between two partners indirect relationships arise, i.e., interactions with third (there may be several) populations. In other words, every interaction has indirect effects on the environment, i.e., also on biogeocoenose components and on processes taking place in them.

The phenomenon called 'metabiosis' is associated with indirect actions. It is the use by organisms of environmental conditions created by a group of other organisms that has ceased to exist. Here we see a succession of groups in a habitat, the resources of which at a certain period meet the requirements of a certain group of organisms; these organisms fully use and alter the resources, which then cease to satisfy the first group but become suitable for a new group.

It follows that topic relationships are not displayed in direct exchange of matter and energy, but affect the distribution of populations in a biogeocoenose and may act to change their density. Investigation of their quantitative significance is necessary.

Examples of phoric relationships are the transfer of seeds from thornbushes to mammal wool, the transfer of triungulin-larvae of oil-beetles (Meloidae) by bees of the genera *Halictus* and *Andrena*, the transfer of nematodes (Anguillulidae) on dung-beetles, earthworms, ticks, etc. Interactions between entomophilous plants and pollinating insects consist of 'predation' (the animal feeds on the plant products, pollen and nectar) and phoresia (the plant, as pollen, is transported by the animal). The feeding of vertebrates on succulent fruits and dispersal of their seeds is an analogous situation.

Vertebrates are less closely linked to specific vegetation than are phytophagous insects, which are often intimately linked to definite plant species (the insects being accompanied by their predators, parasites and commensals, forming a consortium), and soil-dwellers are often linked to definite varieties of soil occupied by specific microbiocoenoses.

Differences are displayed in other ways. The mass of vertebrates in a biogeocoenose is represented by comparatively few but comparatively large individuals, frequently connected with several types of biogeo-

coenoses. Such megafauna is contrasted with macrofauna (small vertebrates) and meso- and micro-fauna (invertebrates). The mass of the
latter consists of a large number of comparatively small individuals, often
attached to a definite type of biogeocoenose.

Vertebrates, because of the wide area of their activity, are seldom 'loyal'
to one biogeocoenose or, more precisely speaking, to a single type of
biogeocoenose; their lives are frequently associated with several.

It is incorrect to assess the role of an animal in the life of a biogeocoenose merely by its mass. B. B. Polynov has pointed out that the error of
ignoring the significance of geological action by organisms is due to the
habit of judging the importance of an object by its mass (Perel'man 1961).
It would be equally erroneous to estimate the biogeocoenotic significance
of any animal on the basis of its mass. We must take into account all
forms of interaction between the animal and all other components of the
biogeocoenose.

PHYTOPHAGOUS INVERTEBRATES IN A FOREST BIOGEOCOENOSE

Invertebrate phytophages are of very great importance in the life of a
forest biogeocoenose, especially in the circulation of matter and energy
within it. Invertebrates consume almost 40% of the matter synthesised
by plants, and create new animal protoplasm (the share of vertebrates in
that process is much smaller). Invertebrate excretions, and later their
carcases, form a considerable part of the organic matter that enters the
soil.

An assessment of the role filled by invertebrate phytophages in a forest
biogeocoenose must naturally be accompanied by an estimate of the
action upon them of phytophage-eating carnivores and also of second-
order predators, i.e., those that eat carnivores, as well as of the invertebrates that eat organic remains (necrophages, saprophages, coprophages),
and finally of those carnivores that feed upon the latter group of invertebrates. A special section of this book is devoted to description of the
groups of invertebrates living on organic remains in the soil and actively
affecting soil processes.

The variety of questions relating to this problem, the varying degrees
to which they have been studied, and also shortage of space, mean that
we can present only a few illustrations or quotations from the literature.
Most of the examples are entomological, insects occupying the dominant
position among invertebrate phytophages in a biogeocoenose.

We shall discuss the activity of only those phytophages that feed on
trees and shrubs. Compared with that group, consumers of herbaceous
and moss cover play a quantitatively insignificant role in biogeocoenotic
processes, and they have been studied much less. Nevertheless, investigation of the activities of dwellers in low-growing vegetation is very important, and should find a place in studies of forest biogeocoenoses.

The distribution of phytophages in a biogeocoenose follows that of their food plants, and they in turn are followed by their own predators, parasites, and commensals. As the result a 'faunula'[1] is created, consisting of separate plants and plant species. The plant is the basis for a grouping of consortia, i.e., a combination of organisms of different kinds closely linked in their life-activities, e.g., an oak with its own parasites from the worlds of micro-organisms, plants, and animals—epiphytes (lichens, mosses), symbionts (mycorrhiza, microbes of the rhizosphere), etc. (Ramenskii 1952). The plant is the foundation of the consortial group; for each member of the consortium it is more or less a substrate, and in a series of interactions it is a living partner. Together with the word 'consortium' it is appropriate to use the word, accepted in foreign literature, 'consorts' (Latin *consortes*, singular *consors*), i.e., partner-organisms. Data from investigations in England, the U.S.S.R., Sweden, and Cyprus indicate that consort-relationships developed during the Quaternary period (Southwood 1961). The composition of consortia, in combination with the corresponding area of soil and the surrounding atmosphere, creates the distinctive characteristics of the exchange of matter and energy in the biogeocoenose.

Since consortial relationships are phylogenetically linked (they are specific for each species), we may conclude that on each plant of a certain species, in the same environmental conditions, there is created (more or less completely) the same consortium. In different biogeocoenotic conditions, different consortia may be formed on other plants of the species.

The majority of phytophagous invertebrates are closely associated with specific plants: sometimes (monophages) with a single species, sometimes (oligophages) with a group of related species, and least often (polyphages) with many species. Since each plant species in a forest takes its place in a definite synusia, phytophagous animals (like the micro-organisms associated with the plants) enter into a corresponding synusial consortium.

Thus the complex of living organisms in a biogeocoenose, i.e., the biocoenose, contains synusial consortia consisting of population consortia, composed in turn of consortia of separate plants. In such individual consortia individualised elements may be discerned. They appear in the form of individualised groupings linked with definite locations on the plant or on its organs. Thus Winston (1956), listing the inhabitants of an acorn, includes the insects and moulds that feed on its tissues, the entomophages and mycetophages linked with them, secondary entomophages, and so on. In this type of microgrouping we include cavity-dwellers living in tree-trunks (Hirschmann 1954). Special groups of invertebrates (predators, commensals, parasites) and micro-organisms are formed in the burrows of bark-beetles (Zinov'ev 1958) and other tree-trunk pests.

In investigations of consortial relationships (Arnol'di 1960, Arnol'di and Lavrenko 1960, Arnol'di and Borisova 1962), monophagy (feeding

---

[1] 'Faunula': a small fauna, i.e., a complex of species inhabiting a small individual area.

on a single plant species) is considered to be a first-degree relationship; oligophagy (feeding on several species related systematically, biochemically, or morphologically) second-degree; and wider relationships, third-degree. As a rule, monophages are associated with individual populations of a species and oligophages with synusiae, and polyphages may be associated with several synusiae. As for non-phytophagous species, those with restricted diets gravitate to population-groupings. The polyvalent species (those with varied diets, e.g., predators) may be included in several groups.

The oligophagy of a predator may be based, not on its prey species being close to each other systematically, but on proximity or even communal places of shelter, or on similarity in the forms of their life-activity. For instance, the saxaul ladybird (*Brumus jacobsoni* Bar.), which is adapted to dwelling on plant-lice galls on saxaul, eats both aphids and thrips (Savoiskaya 1955). Pschorn-Walcher and Zwölfer (1956) have reported the formation of groups of organisms that live on fir-bark lice. The prey of polyphagous predators may consist of species systematically very far apart. Thus among predators consortial relationships of different degrees are observed. As a consequence, with first-degree relationships the radius of activity of a predator is relatively small, and its shelter is close to, or coincides with, that of its prey. With third-degree relationships a predator's radius of activity is large, and as a rule its shelter is distant from that of its prey.

The groupings of animals (like those of the plants themselves) in the herbage-low-shrub and moss-lichen layers play a relatively small role in a forest biogeocoenose. The leading role in forest-biogeocoenotic processes belongs to consortia of the tree and shrub layers, especially of the former. Invertebrates feeding on trees (especially on those of the dominant species) can exert a decisive influence on the exchange of matter and energy in a forest biogeocoenose.

Dietary characteristics link phytophagous invertebrates not only with their specific food plants but also with a definite place on them. We have given examples above of the formation of individualised groupings within arboreal consortia. If we examine the separate horizons of a forest biogeocoenose, we may divide dendrophilous invertebrates (or, more correctly, invertebrate dendrobionts) into crown, trunk, and root-system dwellers. The activities of these phytophages may cause plant damage: (*i*) to the generative organs (impeding reproduction), (*ii*) to the green-mass tissues (impairing aerial nutrition and transpiration), (*iii*) to cortical tissue, periderm, and wood (impairing the movement of solutions and substances, and also the activity of the cambium), and (*iv*) to the root system (impairing water and mineral nutrition).

It is clear that the exchange of matter and energy between trees and their inhabitants is also adapted to separate horizons and layers. The intensity of insect participation may vary greatly and in many cases leads to the death of the plant. For instance, the stripping of tree crowns by masses of leaf-eating pests leads to the withering and death of trees; attacks on root systems by cockchafers or mass infestation of vessels by tree-trunk pests may produce the same results.

The disruption of normal relationships in any layer of a biogeocoenose does not remain localised. Mass reproduction of leaf-eating insects creates a large concentration of them in the tree crown, impairs the normal state of their food plant, produces changes in its physiological processes and in its inter-relationships with the soil, so disrupting the regime of the latter, and so on; moreover, destruction of foliage thins the crown and changes the microclimate; at the same time, abundance of food attracts predators and parasites of the abundant insects to the crown, epizootic diseases develop among them, and consequently disease-bearing organisms multiply. Thus the crowns are the focus of variations from the norm in many processes, but all biogeocoenose components take part to varying degrees in compensatory processes.

The exchange of matter and energy in any biogeocoenose fluctuates qualitatively and quantitatively through both the day and the year. During the day the form and degree of participation by invertebrate phytophages depends on species-characteristics. In the temperate zone, however, the importance of their role usually decreases during the night hours, since the activity of poikilothermal animals decreases during that period.

The qualitative and quantitative composition of groupings of invertebrates in consortia, like the composition of their populations, fluctuates during the course of the year. Stages of development succeed each other, and consequently the character of interactions changes; at particular stages several species migrate from one horizon to another (e.g., from crown to soil), to other consortia, or even to other biogeocoenoses (migration from steppe to forest for wintering). Individual animals perish from various causes but new generations emerge from eggs, so that consortia have seasonal aspects, and the nature of the exchange of matter and energy varies from period to period. Consortial relationships determine the adaptation of the life-cycle stages of phytophages to the phenology of their food plants in given environmental conditions (Čapek 1962). For instance, the feeding of certain organisms on green matter may be an adaptation to spring conditions, when the growing leaf is rich in protein and water, whereas other organisms are adapted to the summer-autumn period, when the leaf is richer in carbohydrates and other substances (Kozhanchikov 1960). Phenological dates for these changes may be established in each individual food chain. The exchange of matter and energy in each food chain may, in theory, be measured, but as yet only a negligible amount of factual material has been collected. In temperate-zone forests the most intensive interaction between plants and phytophages takes place in spring and early summer, when most of the phytophages (especially serious pests) are in the feeding stages.

The seasonality of many phytophages (more precisely, of certain phases of their development) is affected by adverse phenological periods. The timing of the feeding stage of an animal must coincide with the appearance of the food that it requires. Such synchronisation of timing is decisive in the development or suppression of mass reproduction of many pests (Thalenhorst 1951). Several authors (Schütte 1957, 1959, Thalenhorst 1960) have confirmed the necessity of correspondence of the hatching dates of larvae of the green oak tortrix moth (*Tortrix viridana* L.) with the opening of oak buds. A. S. Moravskaya (1957) has proved by investigations extending over many years that the species-composition of the chief oak-leaf consumers in the Tellerman forest is the same on oaks of the types with early and with late bud opening, but the leaves of the late type open 20-30 days later than those of the early type and the number of pests on them is 73% less. The fact that the development of these insects is adapted to the early type often leads to the death by starvation of larvae hatching on trees of the later type. Confirming that observation with regard to the green oak tortrix moth, A. S. Danilevskii and I. G. Bei-Bienko (1958) considered that it explains the recent invasion of the forest-steppe oak forests by the late type of oak. The female of the little spruce sawfly (*Pristiphora abietina* Christ.) can lay eggs only in buds that have begun to open and have already shed their scales, but still have their needles in a compact mass (Ohnesorge 1958). If the spruce buds are not at the correct stage many females of the spruce sawfly find it impossible to lay eggs, and the numbers of the next generation are considerably decreased.

Changes in the grouping of invertebrates in consortia on separate trees may be due to the succession of seasons but also to other factors. For instance, the populations of bark and wood dwellers depend directly on the physiological condition of the tree. The species-composition of associations of tree-trunk pests and organisms feeding on them becomes richer, and to a certain degree fluctuates, as the tree weakens with the passage of years. These relationships were discovered by Z. S. Golovyanko (1926), and were later analysed by several investigators (Il'inskii 1928, 1931, 1958, Zinov'ev 1958, Jamnicky 1958). The species composing such groupings are included in the fauna of a given biogeocoenose and in corresponding synusial and population consortia, and can enter the consortia of individual trees only when the latter lose their power of resistance.

INTERACTIONS OF PHYTOPHAGOUS INVERTEBRATES WITH TREES

Even though phytophages live at the expense of vegetation, they need not be classified as parasites. The consumption by phytophages of organic matter created by plant-producers accelerates its circulation and facilitates its mineralisation, and consequently the creation of new organic matter.

To determine the role of phytophagous invertebrates in a biogeocoenose, one must first discover the nature of their consumption of plant matter, and then the changes that take place in the soil when animal matter replaces vegetable matter in it. At the same time the ability of plants to defend themselves against phytophages should be considered. To evaluate the significance of phytophagous invertebrates in a forest biogeocoenose one must assess quantitatively the processes of transfer of matter and energy from plant to phytophage and from the latter to the soil.

These inter-relationships cannot exist independently: they indirectly affect other processes, and are themselves affected by these; in particular, interactions of phytophages with the soil take place largely through other organisms. Phytophagous invertebrates exist in a state of interaction with other animals and micro-organisms, as well as with plants and soil, and the entire life-activity of phytophages is affected by the atmosphere. This creates a very complex network of relationships, producing a chain of phenomena in a biogeocoenose.

Invertebrate phytophages are specialised for eating either green matter (chlorophyll-bearing tissues) or generative organs, and those living on tree and shrub vegetation are specialised for eating cambium and wood, which may be alive, dying, or dead. Damage to the wood weakens the plant (mechanically and not physiologically) by affecting the skeletal mechanical functions of wood.

The damage done by phytophagous animals occurs in three ways: (*i*) wounding, i.e., disrupting the integrity of the plant organism by the destruction of a very small amount of tissue, but providing access for the external atmosphere (breaking down the internal gas and hygrothermal regime) and for disease-bearing micro-organisms; (*ii*) more extensive destruction of tissues, and sometimes even of organs, which increases the disruption of the plant organism's integrity and adds to the loss of matter by reducing (sometimes terminating) the physiological activities of the tissues and organs eaten; (*iii*) disruption of the normal physiological state, i.e., onset of disease.

Wounding, including that done by animals, is not necessarily followed by disease, since not every injury disrupts the normal physiological state of the plant, which has been developed 'as the result of adaptation to a definite ecological environment' (Taliev 1930). Local damage to a plant induces a pathological process that cannot be considered to be a disease if it does not affect the organism as a whole and does not involve all its elements and activities (Slepyan 1962).

For instance, where mechanical injury occurs to bark due to a blow, conditions may be created for infestation by jewel (Buprestid) beetles; increased flow of nutrients as a response-reaction creates a cambial excrescence there (a callus), and the consequent impediment to the vascular system in one place is compensated for by increased flow in the remaining vascular bundles, so that the tree continues to grow. Further attacks by jewel beetles (a number of species of which are attracted by the cambial excrescence) leads to ringing of the trunk by their burrows, disruption of the vascular system, and interruption of the movement of substances, thus interfering with the work of the cellular system as a whole and producing a disease of the plant.

The normal work of leaves, root and stem induces normal activity in all cells, these being in a state of turgor due to accumulation of cell sap. That is why values of the soil-water tension of a tree or of its osmotic pressure may serve to indicate its state of health or disease. Loss of leaves due to eating by insects is often made good in the same season, but loss of needles may cause serious weakening, enabling tree-trunk pests to infest the tree.

Damage by the so-called sucking pests (mites, aphids, scale insects, etc.) does not injure the entire organism: the pest makes a small puncture wound, and so access of the surrounding atmosphere to the interior of the plant is avoided. In such attacks, however, the plant is poisoned by the excretions of the phytophage.

An important question of this relationship is whether the pest leads an exposed mode of life (i.e., remaining outside the plant) or a cryptic one (developing within its tissues); also whether it moves freely or is attached (as in the case of scale insects). Cryptic and attached pests are essentially parasitic on plants.

Phytophages may consume either living or dead parts of plants. In the first case there is not only quantitative flow of matter, with the primary production of the biogeocoenose (plant matter) being consumed in the formation of secondary (animal) products, but in addition the physiological state of the plant is disrupted and its productivity is thereby diminished. In the second case, i.e., when dead parts of plants are consumed, the condition of the plants does not deteriorate, but the decomposition process is accelerated, thus increasing the intensity of circulation of matter and energy.

Defensive reactions of plants are of six main types: (*i*) restoration of loss by production; (*ii*) transformation of matter, making tissues biochemically or physiologically unsuitable for eating; (*iii*) positive action against the phytophages; (*iv*) formation of protective tissues; (*v*) localisation of damage; and (*vi*) mechanical defence. These forms of defence may occur in combination.

Every living organism tends to compensate for possible failure of regeneration by maximum fertility: seed production is many times greater than the minimum required for regeneration. Similarly losses of green matter may prove to be insignificant, since trees generally form more leaves than are needed for normal growth and development. Losses of assimilating apparatus, however, are reflected in reduced increment of wood and changes in the quantity and quality of annual litter fall, so affecting soil processes. Restoration of the physiological condition of the trees may be delayed or may not take place at all.

The biochemical unsuitability of a plant as food, and sometimes even its toxicity for a certain animal species, is not an absolute defence against pest attacks. After the principal (preferred) plant food of mass pests has been exhausted, or when it is absent from a given locality, animals move to other plants that normally are almost untouched by them (Kovacevic 1956, Jancovic 1958, Zagaikevich 1959, Klepac 1959, Gauss 1960). 'Unsuitability' may be specific: because of its biochemical nature a certain species of plant may not produce normal development of a certain pest (in normal periods, to normal dimensions, and with normal productivity). Unsuitability of a plant for animals is usually due to conditions in a certain part of the plant range or to conditions (especially soil conditions) in a certain habitat. The freedom from pest damage enjoyed by such plants, and in fact their unsuitability for eating, are the result of the environmental conditions surrounding the phytophage population of the biogeocoenose concerned.

Feeding on unsuitable food reduces accumulation of fatty matter and resistance to diseases, and produces general weakening and depression of a population (Kurir 1952, Gur'yanova 1954, Bel'govskii 1955, Burnasheva 1955, Basurmanova 1958, Steger 1962). That has been well demonstrated with regard to the gypsy moth, for instance, by D. F. Rudnef (1936, 1962), V. L. Tsiopkalo (1940), Kurir (1953), and A. V. Likventov (1954). I. V. Kozhanchikov (1951) has shown that the diet of dendrophilous Lepidoptera depends on seasonal changes in the chemistry of food plants, and consequently the metabolic chemistry of the feeding stages of the insects has had to be adapted to the natural dynamics of the chemistry of the food substrate. That explains P. M. Rafes's observation (1960) that in the Narynsk sands aspen species, usually considered to be preferred foods, are unsuitable for the development of the gypsy moth, and it agrees with N. M. Edel'man's deduction (1956) that in certain zones the range of tree species preferred by the gypsy moth is restricted, in spite of the evidence of its polyphagy.

M. M. Padii (1959) has demonstrated that many tree-trunk pests of larch in Siberia not only do not attack larch in the Ukraine but do not even live on it there. Yet pine stands in the Amur basin, where larch is dominant, are infested with pests usually specific to larch (Tarasova 1962). Variations have been discovered in the distribution and harmfulness of the larch thrips (*Taeniothrips laricivorus* Krat.) on larch in northern Germany, France, and Switzerland, and there is evidence that these variations are related to height above sea level and also to the conditions of planting (Vietinghoff-Riesch 1957). Investigations in Germany and Denmark have revealed variations in infestation of Douglas fir by the aphid *Giletella cooleyi* (Gill.), depending on the origin of the trees in North America (Teucher 1955, Petersen and Soegaard 1958).

Two races of larch tortrix (*Zeiraphera griseana* Hb.) have been recorded which are genetically similar, sexually isolated, living in the same areas but one developing on larch and one on pine (Bovey and Maksymov 1959). Substances have been discovered that attract or repel phytophages and so cause them to settle on a preferred substrate (Hesse *et al.* 1955, Fisher 1956, Merker 1956a, Adlung 1957, Chararas 1960a). The occupation of illuminated or shaded parts of plants by different species of dendrophilous aphids is determined mainly by their nutritional, not their hygrothermal, requirements, since differences in lighting change the composition of plant proteins (Fomicheva 1962). The higher survival rate of needle-eating pests on weakened pines is due to the lowered content of volatile oils, toxic to the larvae, in the needles (Grimal'skii 1959, 1961 a, b). Rudnev (1962), after evaluating many investigations, concluded that plants have a natural resistance to leaf- and needle-eating insects; the resistance varies with conditions of growth, weather, age of trees, etc. Bevan (1958) suggested that the leaves of young pines are biochemically unsuitable as an insect diet. For instance, pine groves planted in England in 1921 grew without pest damage until 1953, when a considerable area was stripped by pine-loopers, which prefer 30-year-old stands.

Investigations by Edel'man (1963) have demonstrated the dependence of the physiological state of phytophages on the biochemical composition of their food. In particular, prolonged feeding on vegetation of a definite biochemical composition determines the dynamics of accumulation and expenditure of reserve matter in insects, and also the kind of food they require. Changes in the sugar and nitrogen content of food plants produce corresponding changes in the fat and nitrogenous-matter content of the insects' bodies. The acidity of their excrement varies with the acidity of plant cell sap. When the seasonal variations in plants' biochemical composition do not coincide with insects' requirements, that may depress reproduction of the insects and even cause them to die off.

The fluctuations (demonstrated by V. I. Grimal'skii 1961 a, b) in the volatile oil content of pine needles sometimes improve the diet of larvae but when unfavourable make the needles toxic to them. Another example of the positive action of plants on phytophages is the toxicity of the wood of certain species for termites (Sen-Sarma 1963). In vigorously-growing aspen plantations the holes made by the small poplar borer (*Saperda populnea* L.) are grown over, with the consequent death of from 75 to 100% of the larvae (Kudler 1961).

The ability of plants to form calluses covering wounds is well known. Another example of defence is the ability of a tree to form scarperiderm where aphids have been sucking, protecting it from further harm (Oechssler 1962). Callus formation has become the chief defence reaction against several phytophages, causing encapsulation of them in a burr or a gall.

An example of mechanical defence by a tree against pest attacks is thick bark (Kriebel 1954).

The above descriptions are far from exhausting the immense variety of defence reactions by plants, but they illustrate some types and permit the following conclusions to be drawn.

First, the nature of the defence reaction is determined by the phylogenesis of the reciprocal adaptations of the phytophage and the plant. Thus many tree-trunk pests have developed the instinct to attack weakened trees. Gall-formers are adapted to development in newly-formed structures, and their food plants to continued growth and development. Secondly, all types of defence reaction are not caused simply by interaction between plant and phytophage, but depend on environmental conditions. In other words, plant and phytophage interact on each other like two populations participating in the metabolism of a biogeocoenose; they act upon one another and are affected by all changes taking place in the life of the biogeocoenose (Weingärtner 1962).

The highest degree of association, involving preference for a single plant species, occurs only with monophages, which feed on only one species. Polyphages have several preferred species, often, but not always, closely related systematically (Kondakov 1963, Prozorov et al. 1963).

In a biogeocoenose a pest shows greatest preference for the species with the highest biochemical value to it and with no repellent (certainly no toxic) properties (Friend 1958, Thorsteinson 1960).

Macfadyen (1963) has suggested a method of determining the efficiency of diet and the intensity of metabolism in relation to different environmental factors. Comparison of the caloric values of food, body tissues, and excrement enables one to demonstrate preferences and to make an objective evaluation of conditions for growth and development of an animal, particularly a phytophage.

Resistance, according to R. Painter (1953), is defined as the presence in a plant species of hereditary features affecting the ultimate extent of the damage that might be caused by a particular phytophagous species. There may be 'false' resistance in a plant, because the vulnerable stage of a plant's development passes fairly quickly or declines during a period of minimum activity of a phytophage, or else because of an artificial or accidental coincidence of favourable conditions for growth and development.

In forestry practice 'false' resistance is of great importance: the creation of favourable conditions, especially the arranging of periods of development so that a plant evades pest damage, is one of the foundations of forestry measures for protecting tree stands from pests.

The investigations of Grimal'skii have demonstrated the connection between the biochemical nutritive properties of needles and the varieties of forest soils. A number of investigations have shown that the vulnerability of various tree species to pests varies according to the way they are mixed in tree stands (Turner 1952, Kinghorn 1954, Westveld 1954, Vietinghoff-Riesch 1957, Courtois et al. 1960), to the density of planting

(Lozovoi 1956, Stebaev and Polivanova 1959), and to other forestry-management conditions, especially the kind of tree-felling.

The examples quoted undoubtedly show that the vigour of the defence reaction depends directly and primarily on the physiological condition of the tree. Since the physiological state of the tree depends on the conditions in which it grows, the vigour of its defence reactions is determined by habitat (forest type, character of the forest territory, etc.) and time (age of the tree and of the whole stand, conditions of a given year, etc.). It follows that 'true' (hereditary) resistance may vary with environmental conditions, and to make a precise distinction between it and 'acquired' resistance is difficult.

The ability of a tree to retain viability, and also to develop and add increment depends upon its resistance, which is determined by its capacity for defence reactions and for creation of new cells and tissues. The resistance of a tree stand, like that of a forest biogeocoenose as a whole, seems to depend on the resistance of the trees composing it. This is a reciprocal interdependence and conservation of a tree stand is achieved not by retention of all the trees but by constant differentiation, whereby the most resistant individuals are preserved at the cost of the death of the weakest.

The specificity of a biogeocoenose and of its types is again shown to be important, since it decides the nature of the adaptation by plants to defence and of the adaptation by phytophages to overcome the defence. Some authors (De Leon 1954) doubt whether 'real' resistance actually exists, since sooner or later pests overcome it.

Schwerdtfeger, in *Forest Diseases* (1957), has described in great detail the variations in a forest's resistance to the harmful action of abiotic factors and of organisms; his work contains special chapters on infestation of trees and tree stands. V. N. Stark (1961) has made a brief survey of data on the resistance of forest plantations to pests. Both authors mention ways of increasing the resistance of a forest; other reports (Francke-Grossmann 1953, Pschorn-Walcher 1958) are also devoted to recommendations for forestry measures for the protection of forests, as is the special report by A. Voûte (1960) to the Fifth World Forestry Congress.

The vulnerability of wood to infestation by trunk pests literally means lack of resistance. That is true, however, only with regard to those phytophage species that attack weakened but still living trees. A. I. Il'inskii (1928, 1931, 1958) has demonstrated that as wood dries up and dies the species-composition of the pests infesting it changes in a regular pattern. Later studies have revealed successions of different species (Schimitschek 1952-53, Mogren 1955, Valenta 1960, Mamaev 1961, Rafes 1962, Jahn 1962, Lindeman 1964), clearly showing that some dwellers in wood behave like parasites and some like saprophages. The former type need to inhabit living wood at the stage of weakening when there is sufficient time before death of the tree for the life cycle of the pests to be completed. It is essential for them that the movement of sap in sapwood should maintain its vital nutritive properties, and therefore these species never infest fallen trees. The second type inhabit only dead wood, and some of them need a certain degree of wood decay. Intermediate forms inhabit a tree for some time before its death, but development of their descendants continues in dead wood.

The succession of groups of xylophages is determined by metabiosis, i.e., by the circumstance that each group uses environmental conditions until they become unsuitable for it but are fully acceptable for the next group.

The need for food plants that serve as a substrate for breeding purposes has led to the development of a number of adaptations in wood-dwellers for their protection against the plants' defence reactions, and in the hosts for maintenance of viability during infestation.

The long-known ability of conifers to fill up bark-beetle holes with resin or gum provides evidence (by a decrease in the secretion) not only that the trees' defence reaction is weakening but also that the bark beetles have ensured conditions for their development. Experimental study of the infestation has shown that the first outflow of resin from a wound covers up not more than 12-14% of the attacking individuals, and the second outflow (which occurs at different times in different species) has no effect, as the insects have already penetrated some distances into the wood (Chararas 1959a). The studies have shown that ability to secrete resin is a less reliable sign of infestation of a tree than osmotic pressure. According to many data (Callaham 1955, Merker 1956b, 1960, Zwölfer 1957, Chararas 1959 a, b, 1961, 1962, Chararas et al. 1960, 1962, Vité and Rudinsky 1962, Schimitschek and Wienke 1963), osmotic pressure changes because the breakdown of the water-supplying capacity of the tree leads to a rise in the concentration of cell sap, followed by an increase in suction pressure: the normal semi-permeability of the cells gives place to irreversible permeability, and a rapid rise in osmotic pressure and death of the cell takes place. Bark beetles depend on that condition, and the subsequent withering results from their activities.

Fluctuations in osmotic pressure in a tree determine its vulnerability to bark beetles (Merker 1956b, 1960, Zwölfer 1957, Chararas 1959b, 1960 a, b, c, 1961, Courtois et al. 1960). Osmotic pressure depends, in particular, on the water-metabolism of the tree; based on these premises, studies have been made comparing the vulnerability of trees to bark beetles with the rate of water flow in their vessels (Kraemer 1953, Georgescu et al. 1960). A decrease in water-supply also creates conditions for infestation by poplar borers (Rafes 1956).

Vulnerability of trees to bark beetles increases with age, and also with weakening of a tree stand by improper felling, industrial gases, and other harmful activities (Pogorilyak 1962 a, b, Pfeffer 1963).

Disruptions of normal osmotic pressure, which cause the weakening of a tree and therefore a fall in its resistance, are accompanied by formation of chemical compounds that attract bark beetles, and also by a decrease in the food value of the bast and sapwood tissues (fall in carbohydrate content). The bark beetles are adapted to a diet of tissues of poorer food value. In this respect chemotropism has both qualitative and quantitative characteristics. Thus the transpiration of terpenes (mainly α-pinene) attracts the larger pine-shoot beetle, but only with some lowering of concentration below the norm; maximum secretion serves as a defence against the beetles, but a dying tree with exhausted terpenes does not attract them (Chararas and Berton 1961).

In many cases xylophages require specific developmental conditions. Ambrosia fungi, on which the larvae of striated ambrosia beetles (*Xyloterus lineatus* Ol.) feed, are able to raise the nitrogen content (necessary for their diet) of sapwood to the requisite level; but the fungi themselves require 42-62% of moisture in the sapwood (Chararas 1961). Thus inclusion of a symbiotic fungus in the trophic relationship of the bark beetles with the tree makes the nitrogen content and the water content interdependent.

The capacity of a plant with regard to a phytophage may be defined as the point at which loss of tissues and damage to the normal physiological condition of a plant takes place while the plant continues to grow and develop to a minimum degree. The previous comments show that the use of plant material by phytophages is important not only in the transformation of biomass, since phytophages, by diminishing the mass of plants and increasing their own, stimulate their own life-activity and depress that of plants.

Data on the capacity of plants in relation to pest population density are the basis for forecasting mass multiplication of pests and deciding on the need for control measures.

From the biological and economic points of view, of course, it is important to discover the capacity of a tree stand and not that of the separate trees, since the relationships between trees and phytophages amount to interactions between populations existing in definite conditions, i.e., within a biogeocoenose.

In every forest biogeocoenose there are many invertebrate species characterised by small numbers of individuals and negligible fluctuations in these numbers. At the same time there are some (comparatively few) species characterised by wide fluctuations in the numbers of populations, whose density sometimes reaches colossal figures.

Consequently there is a widely-used division of phytophages into two groups: (*i*) indifferent species, and (*ii*) pests; among the latter 'mass pests' are distinguished. Indifferent species are those that eat plant matter, but practically to an insignificant extent, as their numbers are

small, and the physiological condition of the plants is unharmed. Therefore, these species are not even regarded as pests, although in certain
circumstances the harm done by them is economically noticeable and
requires attention (Eidmann 1949, Neugebauer 1951, Brauns 1953,
Vorontsov 1955, Rafes 1960, Gauss 1960, Postner 1961). Mass pests, by
building up to enormous numbers, do colossal damage and at various
periods markedly change the ratio between animal and vegetable substances entering the soil. Auer (1961), for instance, reports that the
population density of the larch tortrix moth (*Zeiraphera griseana* Hb.) at
the peak of its reproductive period increased 18,500-fold! Other (non-
mass) pests consume plant matter to varying extents, but annual figures
for their activities may be considered to fluctuate little from long-term
averages.

Thus consumers of plant matter, from the point of view of plant
capacity, may be placed in the following groups corresponding to their
numbers: (*i*) doing insignificant damage (indifferent species); (*ii*) doing
damage not exceeding the plant capacity (pests whose damage does not
fluctuate much from year to year and does not, as a rule, reach capacity
limits); (*iii*) pests whose consumption during periods of high numbers
exceeds the capacity limits of the plants. This classification shows that the
first two groups of invertebrates do not disrupt the normal circulation of
matter in a biogeocoenose, whereby trees accumulate their annual increment of wood and shed litter annually, with the usual addition of organic
matter to the soil. We call amounts normal or usual when they are near
the long-term average, i.e., varying from it to an extent that does not upset
the normal state of the biogeocoenose. Only members of the third group
(mass pests) are able to disrupt both normal accumulation of plant
matter and normal movement of part of it into the soil.

In cases where pests of the third group do not reach critical (i.e.,
dangerous for tree stands) numbers, we may agree with the view that
'animals as a whole are broadly symbiotic in their relationships with
vegetation . . . eating its surplus and not doing damage to it that is
dangerous to life' (Allee *et al.* 1949).

Henson and Stark (1959) propose defining three degrees of population
numbers of mass pests. The numbers are called 'tolerable' when they do
not use up all the surplus biological production of the food plant; 'critical'
when they use up more than the surplus, but less than the total production
(leading to loss of annual increment): 'intolerable' when they consume the
production not only of the year but also of preceding years, leading to the
death of the food plants. Stark (1961) suggests the term 'population intensity', expressing the ratio between pest numbers and food stocks; that
term corresponds to the 'relative density' of some other investigators.

The capacity of vegetation as a source of food has a limiting effect on
phytophages. That capacity, as primary production of the biogeocoenose,
is determined by the quantity and quality of matter and therefore by its

energy significance. It limits the amount of secondary production, i.e., the biomass of phytophages; in other words, it limits the capacity of the latter for the third trophic level—carnivorous organisms.

Annual fluctuations in the production of the various levels, depending on fluctuations in weather and soil conditions, alter the capacity limits. In these fluctuations a considerable part may be played by pests when their population densities become, in Henson and Stark's terms, intolerable or perhaps critical.

Losses by trees during mass pest reproduction periods are not restricted to destruction of leaves, which are regenerated in deciduous species to some extent (especially in the earlier years). Losses of leaves lead to a temporary suspension of assimilation and a sharp fall in transpiration. Regeneration of leaves takes place at the expense of reserves of nutrient material; the rate of regeneration depends directly on growth and weather conditions. According to A. I. Il'inskii (1959), a new leaf is often infested with powdery mildew, and shoots that do not manage to lignify are killed by frost. Trees are usually noticeably weakened during maximum mass reproduction of pests, when annual increment may fall sharply (by 40% or more) in the best developed and most vigorously growing, outwardly healthy-looking trees. In trees of average growth and development weakening is much more evident: leaves are not so well regenerated, and soil solutions are drawn to the crowns at the expense of suckers. In the trees of poorest growth and development the tops wither, and the trees are infested with trunk pests and suffer from tracheomycosis. When foci of these pests and diseases are present in tree stands infested with mass reproduction of gypsy moth, the withering of the trees stimulates the activity of gypsy moth, which may kill the trees. If that does not happen, the stand begins to recover after the peak of gypsy moth reproduction. The cycle from the beginning of weakening to the recovery of a stand infested by gypsy moth lasts from 10 to 12 years, and losses in increment reach 25-50%. These increment losses, due to the decrease in the physiological activity of the trees and in their shoot-producing ability, indicate the inevitable decrease in leaf-formation that occurs during plagues of pests.

Coniferous species in the temperate zones (excluding larch) suffer more from loss of green matter than do deciduous trees, since they are able to regenerate only part of the lost needles each year.

According to Schwerdtfeger (1957), pines have greater reserves than spruce. In pines losses are made good by additional growth of the remaining needles (even by growth of the needle bases), by formation of secondary short shoots from dormant shoots between the ephedrae, and by rosette shoots (bearing abnormal needles, formed from the last resources of a dying tree). Spruce not only possesses less regenerative power but also suffers from scald of the cambium, which is usually protected by needles;

therefore periods of hot weather after leaf-stripping hinder the regeneration of spruce stands.

Needles regenerated after being eaten off differ in quality from the original ones. According to A. S. Rozhkov, B. P. Sendarovich, and K. N. Danovich (1962), regenerated larch needles are distinguished from the original ones by larger size and contain less carbohydrates, fats, and protein. The shortage of nutrients injures the physiological state of the Siberian pine moth, and the increased amount of moisture changes the course of physico-chemical processes in its organism, lowering its cold-resistance.

Observations on trees part of whose assimilating apparatus has been destroyed by pests reveal a fall in wood increments and withering of tips; the dependence of changes in productivity on such losses, however, is often extremely complex (Moiseenko and Kozhevnikov 1963, Brown 1963, Kulman et al. 1963, O'Neil 1963, et al.). Some loss of needles is occasionally accompanied by increased wood growth during the same year, and a fall in wood increment takes place later; the dependence of fertility on needle loss is also complex (Čapek 1962a).

Consequently mass attacks by phytophages on any (especially a forest) biogeocoenose do not merely lead to an increase in the conversion of plant into animal matter and thus to a change in the composition of organic remains entering the soil. Besides the disruption of geochemical circulation there is a fall in the production of green vegetation, which alters the very foundation of biogeocoenotic processes in a forest. The resistance of a tree stand also declines, which, to use Schwerdtfeger's expression (1957), produces 'chain sickness' (Kettenkrankheit), i.e., a chain of illnesses that can have very serious results, even death of a stand and replacement by another biogeocoenose.

One frequent result of the eating of green matter and the consequent weakening of a stand is infestation by tree-trunk pests.

DEPENDENCE OF PHYTOPHAGOUS INVERTEBRATES ON SOIL

The soil is the habitat of many invertebrates. These animals depend on the soil as the substrate in which they dig burrows or construct breeding chambers. Their hygrothermal and gas regimes depend on the soil. As an environment, of course, the soil acts on the life-activity of its inhabitants, and changes in its state are reflected in the nature of their metabolic processes. These problems are examined here only in relation to organisms living on organic remains and taking part in soil processes.

The relationships between soil and phytophages are regarded from two points of view: (i) dependence of phytophages on soil, as manifested through plants, and (ii) dependence of soil on phytophages on account of the return of the products of their metabolism and of their carcases after death. The second group of relationships, consisting of the action of animals on the soil, and in particular the entrance of organic matter of animal origin into the soil, will be discussed later.

The dependence of phytophages on the soil, although it exists through the medium of food plants, is very clearly displayed. As early as 1926 V. N. Stark published his observation that poplars growing on marly, humus-calcareous, and boggy soils suffer less from pests (25 species of pests were observed) than those on pine sands (where they were infested with cordiform rot and 67 species of pests). He also states (Stark 1931) that pine stands whose roots were in a podzolic layer or on the surface of an ortstein layer suffered more heavily from insects than those whose roots either did not reach the podzolic layer or were between the podzolic and the ortstein layers. Among pines growing in black alder swamps and at the margins of bogs, the most heavily infested by pests were those whose roots adjoined a vivianite horizon. Stark had already (1926) made experiments with addition of fertilisers and discovered their depressing effect on the development of several pests. Increased pest activity as the result of bad soil conditions has been observed more and more often in recent years (Golovyanko 1952, Zwölfer 1953, Zinecker 1957, Niechziol 1958 a, b, Rudnev 1958, 1959, 1961). Experimental work has been done in this field, in the course of which the essential features of the effects of soil on phytophages have been discovered and at the same time methods of fertilisation have been devised, which not only improve the condition of the plants but also suppress pest activity (Büttner 1956, 1959, Zwölfer 1957, Oldiges 1958, 1959, 1960, Merker 1958, 1961, 1962a, Schwenke 1960, 1961).

We may consider the connection between nitrogen and water in the soil and in trees from the point of view of interactions between invertebrates and soil. Schwenke (1961, 1962b, 1963) remarks that (*i*) soil moisture is necessary for dissolving nutrients required by roots, for the activation of humus (conversion of raw humus) by micro-organisms, and for introduction by earthworms of parts of the humus into the mineral layer of the soil; (*ii*) prolonged water shortage lowers plant respiration and thus increases the unoxidised sugar content, increases assimilation and thereby the formation of carbohydrates (including sugars), and raises osmotic pressure; (*iii*) good supply of water assures its passage from colloids into cell sap and also better nitrogen supply, which stimulates root growth and increases assimilation, transpiration, and protein content. Adequate soil moisture restricts the relative proportion of sugar; as a result either of sugar shortage or of protein excess, the food value of leaves is diminished, which depresses phytophage reproduction. In the section on phytophages as consumers of wood, the dependence of trees' defence reactions on osmotic pressure was also noted. Growing conditions that favour plant growth and depress reproduction of leaf-eating pests are typical of biogeocoenoses not heavily infested by these pests; they may be created by artificial fertilisation, mainly with nitrogenous fertilisers. Water shortage in the soil produces shortages of both water and nitrogen (as a result of limited availability of the latter to micro-organisms) in trees, and the protein-carbohydrate ratio is displaced in favour of the latter (sugar). That improves the food value of the leaves for phytophagous insects and reduces insect mortality; the percentage of females, and sexual productivity, are raised, leading to mass reproduction.

Merker (1962 a, b) makes the significant statement that change in food quality, arising out of soil conditions, has decisive effects on phytophage populations only when there are supplementary environmental conditions. For instance, increased concentration of nutrient matter is not utilised in all cases; in particular, bark beetles cause their chief damage in forests of quality classes II and III, where an increase of 50% in sugar concentration is rarer than in forests of quality classes IV and V, which the beetles avoid. The effect of the inadequacy of the food in turgescent plants on the condition and numbers of phytophages is seen only in combination with rainy weather. Moreover, Merkel has discovered two ways in which fertilisation acts on phytophages: indirectly through change in the physiological state of plants, and directly (shown by radioactive tagging) through penetration of the body of the phytophage by particles of nitrogen, calcium, phosphorus, caesium, etc., contained in the fertiliser, so upsetting their metabolism. It follows that materials introduced into the soil by other than artificial means may also have direct toxic effects.

Apart from the discovery that after applying nutrients forests growing on poor soils have increased resistance, the causes of suppression of insects have not been adequately explained (Francke-Grossmann 1963). There is a danger that increase in nitrogen, which checks the activity of needle and leaf pests, will stimulate the activity of sucking pests, especially spider-mites; detailed study of the action of fertilisers on plant and animal organisms is necessary, also study of possible changes in biogeocoenoses resulting from the fact that increase in nitrogen suppresses some animals and stimulates the life-activity of others (Schwerdtfeger 1962, Thalenhorst 1963). Thalenhorst (1963) says that one of the tasks involved in experimental fertilisation of forest soils is discovery of the significance of the 'nutrient-plant-animal' trophic links in the complex system of biotic and abiotic factors governing pest population density, in other words, study of the biogeocoenotic significance of fertilisation and its place in the circulation of matter.

The importance of the physiological state of a tree stand in the development of mass multiplication of pests has been pointed out in another of Merker's works (1960), and has been analysed in great detail by D. F. Rudnev (1962), who shows that growth conditions regulate not only the nutritional properties of plant tissues but also their defensive properties (see the section dealing with feeding of invertebrates on plants).

INTERACTIONS OF PHYTOPHAGOUS INVERTEBRATES AMONG THEMSELVES AND WITH OTHER ANIMALS

Animal phytophages, mainly invertebrates, constitute the majority of

FFB T

consumers of plant matter, but animals that feed on phytophages regulate the numbers of the latter. Most parasites and predators are closely linked consortially with phytophages, and the numbers of carnivores are directly dependent on the numbers of phytophages. The interactions of many species in biogeocoenoses are complex, and their entry into any consortium is sometimes manifested not in a single food chain but in two or three. Cockchafers feed on roots as larvae and on leaves as adults. Several parasites possess hosts both among phytophages and among other parasites on phytophages, in this way displaying parasitism of both the first and the second orders. Moreover, multiformity of the links between a species and its environment strengthens its position in a biogeocoenose. That explains the patchy, irregular distribution of a species within its range ('range lacework'). Consequently, the ecological optimum for a species is relative and depends, in particular, on conservatism in the distribution of the organism and on the coenotic relationships within a community (Arnol'di 1957).

Trophic relationships determine the position of a phytophage in two links in the flow of matter and energy: plant-phytophage and phytophage-soil (often the second link becomes triple: phytophage-carnivore-soil). The activity of each phytophage population is determined by: (*i*) its characteristics, both hereditary and peculiar to a particular generation; (*ii*) its interactions with food plants; (*iii*) its interactions with predators and parasites; and (*iv*) its interactions with other components of the environment. In the present section we shall discuss the characteristics of a phytophage population and the interactions of phytophages with animal predators and parasites, i.e., within a single environmental component.

The structure of a population of invertebrate phytophages may vary from generation to generation. It is determined by (*i*) morphological features, (*ii*) age composition, (*iii*) constitution (physiological state) of individuals, (*iv*) sexual index, or numerical ratio of females to males, (*v*) egg-production by females, (*vi*) morbidity (percentage of parasitised and diseased, but living, individuals), and (*vii*) behaviour (capacity for accumulation, migration, etc.).

The living conditions in a given biogeocoenose govern, by natural selection, the adaptive features of individuals that morphologically and ecologically characterise a local population. Conditions in any climatic period or period in the life of a population, and even conditions in a single year, may divert the ecological characteristics (again under the influence of selection in given conditions) in one direction or another. Thus a period of wet weather or of drought gives preferential selection to hygrophilous or to xerophilous individuals respectively. Food shortage, abundance of enemies, and epizootics during the critical period of mass reproduction select the most resistant individuals, usually the smallest of them, i.e., those able to complete their development during a period of food shortage; in other words, those with a high coefficient of food utilisation for growth and development; but at the same time smaller size of individuals means lower egg-production.

The biochemical composition of their food affects the physiological state of phytophages (Edel'man 1963). If it does not correspond to their requirements, not only is the reproduction of several species of wood-eating insects depressed, but their larvae perish.

The age-composition of a population varies with the stages of the life cycle. For example, synchronisation of the appearance of the feeding stages with the period of abundance and high nutritional value of their food, or coincidence of the most defenceless stages with the periods of minimum predatory activity, is very important in the development of a population. The viability of a population fluctuates with the presence or absence of these coincidences, and is expressed in its density dynamics and the constitution of its individuals.

Egg-production, which with a certain sexual index decides the fertility of a population, depends on that constitution.

The presence of parasitised or diseased individuals indicates the action of animal or microbic components of the environment on the population; the individuals are still alive, but either do not complete development, or do not attain normal size and weight, or do not produce offspring.

Individual behaviour alters population density either through emigration of part of the population beyond the boundaries of the biogeocoenose or through concentration in certain parts with depletion in others.

Analysis of the factors affecting population structure discloses its internal potentialities for fluctuation in biomass, and also the elements in it that may vary under the influence of external environmental factors.

Thus in many species unfavourable conditions for development often reduce the relative number of females, whose development is more prolonged than that of males. Several predators select their prey at definite stages of development. Some parasites sterilise their hosts. Various parts of a biogeocoenose may, because of their exposure, depth of underground water, or other conditions, attract or repel individuals, thus altering the spatial distribution (dispersal) of the population.

The action of enemies on phytophages may be substantial. According to McCormick (1959), carnivorous organisms of the first order consume an average of 42% of the mass of phytophages. The efficacy of that regulator may, of course, vary considerably in different localities and in different years. Apart from the effects of diet and weather, the regulating efficacy of carnivores is largely dependent on the action of their own enemies, carnivores of the second order (large predators, parasites on vertebrates, blood-sucking organisms, etc.). Second-order carnivores (according to the same author) consume, on an average, 45% of the mass of first-order carnivores. We have already pointed out the complexity of analysis of the situation where some parasites attack both phytophages (depressing effect) and their parasites (beneficial effect on the phytophages); some predators behave similarly. When surveying the complex of dependences (regulating relationships) we must consider the processes of interaction and not the species, since the latter may take part in several, sometimes antagonistic, interactions.

When discussing the significance of animals that feed on phytophages and consequently are dependent on their presence, the name of 'parasites' is given to those whose phytophage host serves as a source of food and as a habitat. At present a large number of organisms (including both insects and nematodes—Polozhentsev 1952) are known that parasitise forest pests, but members of other groups parasitise the eggs and all other stages of development of phytophages. The name of 'predators' is given where the phytophage-feeders live in the same place as their prey populations; this group, of course, includes commensals. G. A. Zinov'ev (1958, 1959) has compiled, from personal observations and published sources, extensive material about the regulating effect of the complex of entomophages inhabiting the tunnels of bark beetles. In monographs about pests information is usually given about their parasites and predators; most pests are consortially linked with their prey. Studies of host-parasite relationships have shown, for instance, that the specific egg-parasite *Anastatus disparis* may be considered as an indicator of the presence of the gypsy moth (Vaclav *et al.* 1959).

At the same time, the complexity of the inter-relationships between host and parasite and between predator and prey has been discovered. Since host and parasite react differently to environmental action, their relationships are dynamic and are often separated in time and space (Thalenhorst 1951, Clark and Brown 1962). Variations in the effectiveness of a parasite and resistance of the host have been observed (Muldrew 1953, Lejeune and Hildahl 1954, Malysheva 1962). Thus the ichneumonid *Mesoleius tenthredinis* Morl. is an effective parasite of the larch sawfly *Pristiphora erichsonii* Htg. in Pennsylvania, but in Michigan the latter is almost immune to that ichneumonid (Drooz 1961). Another ichneumonid, *Itoplectis conquisitor* (Say), infested pupae of the European pine-shoot moth (*Evetria buoliana* Schiff) less on Norway pine than on Scots pine: the latter had smaller (and therefore more accessible to the parasite's ovipositor) buds (Arthur 1962).[1] According to other data (Haynes and Butcher 1962) the European pine-shoot moth lays fewer eggs per unit of shoot-length on Scots pine, Banks pine, and Weymouth pine than on Norway pine, whereas on Scots pine the egg-parasite *Trichogramma minutum* Riley is relatively more active.

It has also been discovered that parasitic infestation of the eggs of the

---

[1] The length of the ovipositor is of special significance for parasites of bark beetles (Ryan and Rudinsky 1962).

sawfly *Neodiprion swainei* Midd. depends directly on the number of eggs in a clutch (Lyons 1962). Competition between parasites is important (Graham 1949, Telenga 1953, Bjegovic 1963), being complicated by the fact that many parasites are linked with several hosts on different food plant species (Zwölfer and Kraus 1957). The above examples are enough to show that the interactions are variable and that the whole network of relationships in a biogeocoenose acts on each one of them; therefore more reports on parasitism are constantly appearing, dealing with the complex of interactions in a community (Balch 1958, Simmonds 1959).

Some polyphagous predators, as stated above, may be linked with several consortia, and the links with each of them may be weak (third-order). Their regulatory work varies little in intensity. They include, for instance, arthropod predators living on the soil surface and feeding on insect phytophages that enter the soil to pupate (Drift 1959). Ants play a large part in maintaining 'forest hygiene' in Europe, chiefly the wood ant *Formica rufa* L.

As the result of 25-year studies Wellenstein (1953) reported that the population of a single nest of wood ants destroyed from 3 to 5 million insects per annum, including from 150,000 to 360,000 pests (in the absence of outbreaks of mass reproduction), in an area of 0·2 to 0·5 hectares; many insects avoid settling near ant-hills. According to Pavan's investigations (1960), on an area of 1,000,000 hectares of forest (fir, spruce, larch) in the Italian Alps ants consume 24,000 tons of food; if we assume that 60% of the food is live prey (most of it harmful insects) we may conclude that the ants consume 14,400 tons of pests per annum. Many studies have been devoted to the role of ants and their practical utilisation (see the reviews by Thalenhorst 1956, and Khalifman 1961). It has also been noted that only certain races of wood ants play a useful role in the forest; at the same time ants are beneficial to aphids, which do considerable harm in some forests (Kautsis 1956, Schwenke 1957, Müller 1958).

Another important group of carnivores is insectivorous birds, which are usually attracted by mass multiplication of pests. The value of birds as regulators of insect numbers, and therefore of their effects on forest bio-geocoenoses as a whole, has been described in many publications. The work of G. E. Korol'kova (1963) contains not only a wealth of material but also a long list of publications on this subject. Forest invertebrates also have enemies among other vertebrate classes (amphibians, reptiles, and, among mammals, insectivores and bats).

We have reviewed the chief relationships between the activities of phytophagous invertebrates and other animal species. These relationships are included among the direct trophic links with regulatory organisms. The regulators themselves, however, depend not only on the presence of food but also to a great extent on other environmental conditions, especially on their predators and parasites. The 'energy principle' of study of relationships between populations included in a single food cycle

(Vinberg 1962 a, b) enables us to determine the biomass of populations at each stage. The mass of plant food, the biomass of phytophages, and the biomass of first-order and later-order carnivores feeding on them, can be calculated.

Indirect effects of animals on phytophage activities are displayed in different forms of variation in regulator activities. For instance, some insectivorous birds distribute plant seeds, thus leading indirectly to the development of insects that parasitise phytophages (Turček 1961, 1962, 1963).

Pest attacks have very great indirect effects on forest biogeocoenoses. Each attack weakens the resistance of the food plant and facilitates access by other phytophages typical of different tree species and of different stages of the weakening and dying of trees (Golovyanko 1926, Il'inskii 1931, 1958, Schimitschek 1952-53, Mamaev 1960, 1961).

The effects of certain phenomena, which at first glance are apparently quite unconnected, may be related to each other. For instance, excessive livestock pasturing first kills off flowering plants (which feed Larvae-voridae and Chalcididae), and afterwards underbrush (required for bird nests and for retention of moisture in the upper soil horizons to permit disease-bearing organisms to survive), and finally leads to the destruction of roots and death of trees in the upper storeys; in this way a regime is created favouring regular breeding of the gypsy moth, since unfavourable factors for the pest are almost absent (Il'inskii 1959).

Among the interactions of animals with their environment we must mention also secretion of the biologically-active substances 'telergons' (Kirshenblatt 1962), which are varied in both chemical structure and physiological action. Some telergons act upon individuals of the same species (e.g., attracting individuals of the opposite sex, or giving a distinctive smell to the nest), others on animals of other species (defensive or repulsive substances, substances that paralyse or kill prey, etc.).

The variety of indirect interactions between animals has been well described by V. N. Beklemishev (1951) and there is no need to cite a large number of examples here. We must again state, however, that any relationships affect both the interacting organisms and the environment, and give rise to many different indirect effects on other organisms.

ACTION OF PHYTOPHAGOUS INVERTEBRATES ON THE SOIL

Phytophagous invertebrates consume plant matter, transform it, and transfer it to the soil in the form of excretions and (at the conclusion of their life-activity) carcases. In many cases individual phytophages are eaten by predators and parasites and so increase the amount of organic matter reaching the soil from carnivores. Being mobile, animals transport matter both within and beyond the boundaries of a biogeocoenose (the latter is offset by introductions from other biogeocoenoses). In their relationships with the soil, phytophagous animals take part in the circulation of matter in only one direction, to the soil. They themselves, although they depend on the soil (through food plants), take nothing directly from it.

The proportion of organic matter entering the soil from phytophages is usually small in comparison with that of matter of plant origin, and in the process of formation and development of a biogeocoenose it attains a certain constant figure.

Phytophages may exert considerable influence on the soil in years of their mass reproduction. Rafes (1964) attempted to make a close estimate

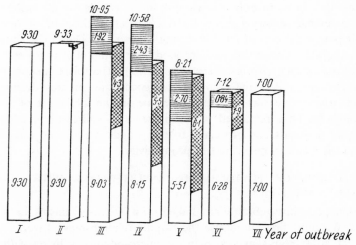

Fig. 43. Weight of organic matter reaching one hectare of soil during the vegetative period of each year of a 7-year cycle of mass gypsy moth reproduction. Explanation in text.

(as dry weight) of the changes in matter circulation in an oak grove caused by gypsy moths in a 7-year outbreak (Fig. 43). At the foot of the columns are data on the amount of tree leaves in spring, 9·3 tons per hectare (including oak leaves, 4 ton/ha), and the dynamics of organic matter reaching the soil each year are shown. During the first year of increase in the population of gypsy moths their activity was negligible, but in the second they consumed 0·07 tons of leaves; the amount of organic matter of animal origin (carcases of pests and of entomophagous insects, excrement of pests and of vertebrates eating them) was 0·03 ton/ha. In later years there were increases in both the amount of leaves consumed (cross-hatching) and the amount of matter of animal origin reaching the soil (shaded). In the fourth year of the outbreak the total amount of litter fall was very high, since it included remains of spring leaves not touched by the pests, gnawed pieces of leaves dropped by the larvae, leaves regenerated during the second half of the summer, and matter of animal origin. Beginning in the fourth year of the outbreak, the spring supply of leaves decreased because of the weakening of the infested trees. After the peak of the outbreak in the fifth year the proportion of animal matter in the circulation of organic matter fell sharply, and in the seventh year was negligible. When large numbers of pests are absent, the decrease in production of green matter by the trees does not occur.

It follows that multiplication of insects affects metabolic processes mainly by accelerating the mineralisation of organic matter entering the soil and by changing its composition. Its significance in soil-formation processes and mineral nutrition of plants, and consequently in the entire biogeocoenose, calls for thorough investigation.

These data concern only a single link in the circulation of matter: the organic remains entering forest litter. A positive effect of the animal matter is increase in soil fertility during the period. Data exist showing that mass pest reproduction has increased the amount of nitrogen entering forest litter by 50% (*Nature Conservancy* 1962). That phenomenon, however, does not reflect the complete biogeocoenotic process during the period, when there was a marked decrease in the increment of wood. Although we have remarked on the increase in soil fertility, which is beneficial to plants and especially to tree stands, we still do not know how much it aids in restoring tree productivity.

INTERACTIONS OF PHYTOPHAGOUS INVERTEBRATES WITH MICRO-ORGANISMS AND FUNGI

Phytophagous invertebrates are linked with micro-organisms in many ways. We must first emphasise the trophic relationships. Invertebrates that eat green plants may also ingest micro-organisms living with the latter as symbionts or parasites. There are also cases of direct eating of micro-organisms. The weevil *Sitona*, for example, eats nodules on the roots of leguminous plants, i.e., consumes nitrogen-fixing bacteria (*Bacterium radicicola*). The outcome of that diet may be classified as relatively simple conversion of the substance of an autotrophic plant (and with it the substance of micro-organisms) into the substance of a phytophagous insect. In this case the amount of nitrogen fixed by bacteria is reduced, and the nitrogen exchange in that element of the biogeocoenose is disrupted. This provides additional grounds for regarding the elements of a biogeocoenose as consortia together with the corresponding area of soil and volume of atmosphere.

Bark beetles, termites, and some ants (Bennett 1958) tend 'fungus gardens' for food. According to Hadorn (1933), in 1836 Schmidtberger first described the presence of a white substance, which he called ambrosia, in the tunnels of some bark beetles; the fungal origin of ambrosia was described by T. Hartig in 1844. Bark beetles (called 'ambrosia beetles' in English literature) have special adaptations, both morphologically and in their mode of life, for transporting spores of the fungi whose mycelia serve as food for their larvae. The fungi are adapted, firstly, for transportation of their spores by the beetles; secondly, for penetration (facilitated by the beetles) into the wood on which they live and in which they cause diseases (e.g., Dutch elm disease); thirdly, to loss of part of their mycelia. It is known (Escherich 1940-41) that some wood-wasps are linked by a similar symbiosis with certain fungi, since the wood-wasp larva gnaws its way through the wood after the wood has been permeated by the fungus. Several other wood-dwelling insects similarly follow the spread of fungi.

Termites, which feed on decayed wood, are guided in their search for it by changes in the composition of substances produced by the action of wood-decomposing fungi (Esenter *et al.* 1961).

Symbiosis of insects and fungi does not occur without participation by the other components of a biogeocoenose. The investigations of B. V. Lindeman (1963) have shown that the spread of Dutch elm disease is not an automatic consequence of the 'bite' of the elm bark beetle. Although many wounds may be made by the beetles, which carry spores of the disease agent *Ophiostoma* (*Ceratostomella*) *ulmi*, not all trees become diseased, but only some of the less resistant; resistance to infection is determined by a number of the conditions of growth, which also depend on the type of forest biogeocoenose, on weather conditions of the year or other period, and on several other factors. Experiments (Al-Asawi and Norris 1959) have revealed that infestation with Dutch elm disease occurs only if the bark beetles carry spores to a depth of over 3 mm in the tree.

Interactions between invertebrates and fungi may be competitive. Studies in the Serebryanoborsk forest (Moscow province) by T. M. Turundaevskaya (1963) have shown that growth of mycelia of butt-rot fungus (*Fomes annosus* Cooke) in the root collar of pine shoots leads to death of banded pine-weevil larvae (*Pissodes notatis* F.).

There are many cases of parasitic feeding by micro-organisms on phytophagous invertebrates. This form of parasitism is very different from that in which the parasite of an invertebrate is also an invertebrate; in the latter type, for instance, one or more larvae of a nematode or of one of the Larvaevoridae live inside the body of a phytophage, and at the conclusion of the parasite's development the death of the individual host (or of several, if the parasite changes hosts) takes place. In the great majority of cases parasitism by micro-organisms causes epizootics among the host-phytophages; some epizootics spread with extreme rapidity and are powerful regulating factors, suppressing mass reproduction of phytophages. Especially important are virus diseases (Bucher 1953, Bergold 1953, Husson 1954, Weiser 1956, Vasiljevic 1959, Atger and Chastang 1961, Smirnoff *et al.* 1962, Benz 1962, Orlovskaya 1962), diseases caused by microsporidia (Weiser 1957), and diseases caused by bacteria (Talalaev 1956, Gukasyan 1960, Gukasyan and Domb 1961). The importance of the activity of micro-organisms as a cause of disease in phytophages is so great that study of it has given rise to a special discipline, insect pathology (Shteinkhauz 1952). Use of such epizootics as a means of practical forest protection has begun (Gukasyan 1963 a, b, Kudler and Lusenko 1963).

It is of interest to mention here that interactions between two organisms can never take place independently of the environment. A. B. Gukasyan and N. S. Domb (1961) have observed that the needles of fir, stone pine, and pine (but not those of spruce and larch) have a bacteriostatic effect on *Bacillus dendrolimus* var. *sibirica*, an agent of disease in the Siberian pine moth. That phenomenon, no doubt, affects both the population dynamics of the pest and the development of preference shown for a food plant species.

Some micro-organisms settle in the shelters of tree-trunk pests, not as parasites on them but as commensals. The interactions of the bacteria that break down cellulose in the alimentary tract of xylophagous insects is mutualism arising from two processes of the commensal type: of the bacteria with relation to the insects and of the insects with relation to the bacteria (digestion of broken-down cellulose, which would be impossible without the bacterial action).

The transmission of tree diseases caused by micro-organisms (Carter 1961) is an example of phoresia. The combined feeding of invertebrates and micro-organisms on a single plant gives rise to inter-relationships of the two species connected with that situation, which may be symbiotic, competitive, or neutral.

PHYTOPHAGOUS INVERTEBRATES AND THE ATMOSPHERE

To discover the interactions of invertebrates with the atmosphere we must first take into account the fact that the atmosphere of a biogeocoenose is in constant connection with the atmosphere of the surrounding area, and therefore its properties as a source of oxygen do not vary, even if the oxygen-forming capacity of the biogeocoenose is seriously impaired. A shortage in the oxygen formed by green plants in the biogeocoenose, with consequent increase in CO_2 content, is levelled out by air currents. An exception exists in the case of small compact spaces in the retreats of animals with a cryptic mode of life, where the connection with the outer atmosphere may be very tenuous. These animals, however, are phylogenetically adapted to such retreats selected or created by them, and therefore gas-exchange takes place normally there. It must be pointed out that in a forest, as compared with other biogeocoenoses, atmospheric movement is very slow, and therefore in certain parts of a forest gases occasionally accumulate (e.g., from decomposition) and may linger for a relatively long time. We still consider it beyond question that in natural conditions animals, in particular invertebrates, find no difficulty in respiration or in other forms of gas-exchange. The chemical composition of the atmosphere as affected by local conditions, however, plays an important role. The effect on gas-exchange caused by animals, especially invertebrates, is in ordinary biogeocoenose conditions practically imperceptible. Even during a period of mass reproduction the gas-exchange processes encounter no difficulty, on account of the constant movement of air currents. Only in certain circumstances (piles of carcases, excrement, etc.) there may be pollution of the atmosphere, but it does not involve successional changes in the biogeocoenose.

The atmosphere, as the environment of life, fulfils another very important function for animals: it determines weather and the climate in a biogeocoenose. In that function also it interacts with all the components of the biogeocoenose.

Animals, including invertebrates, exert no influence on weather or climate, but exist in a state of perpetual dependence on them.

MICROCLIMATIC CONDITIONS OF EXISTENCE FOR PHYTOPHAGOUS INVERTEBRATES

The relatively small radius of life-activity of phytophagous invertebrates, their specific requirements of hygrothermal conditions for production of offspring, and the extreme limitation of their ability to regulate the temperature and moisture regimes of their own bodies prevent us from restricting ourselves to consideration of the phytoclimate of a forest biogeocoenose as a whole when we are studying the dependence of insects on climate; we must take into account that every tree, every population of every kind of tree species, and every layer of vegetation creates its own microclimatic conditions.

Therefore we must always remember that the invertebrate inhabitants of a forest biogeocoenose are acted upon not by the climate of the locality as recorded by the national meteorological network but by specific microclimatic conditions. In general the effect of weather is much more pronounced on invertebrates, because of their poikilothermal nature, than on mammals and birds, which are capable of thermoregulation.

The zonality of climate determines the complexes of plant and animal organisms that occupy different areas; the ecological characteristics of invertebrates determine their distribution within zones and in biogeocoenoses. For instance, the larger pine-shoot beetle (*Blastophagus piniperda* L.), which is typical of the taiga zone, lives in pine forests and settles under the bark of the lower part of the trunk of mature and submature pines; thus it selects definite types of biogeocoenoses in the zone and occupies ecological niches in them. In that process two factors play leading roles: (*i*) the presence of specific food, and (*ii*) the climate of the zone, of the biogeocoenose, and of the microhabitat. It frequently happens that, when tree species are introduced into a zone new to them (e.g., pine groves planted on the sands of semi-deserts), animals that are also foreign to the zone enter it, following their food plants. For instance, the larger pine-shoot beetle has entered the pine groves just mentioned. There, however, it settles not only under the bark of the lower part of the trunk, but mainly under the bark of the butt end and even under that of root spurs; that may be due merely to an attempt to escape from the danger of overheating and to get as near as possible to moister conditions (Rafes 1957). According to Stark (1952), a similar lower position on trunks is typical of many bark beetles in those parts of their range where the air is drier and the summer temperature higher. If this is general (similar phenomena have been observed in a number of groups), we conclude that the rule of habitat-exchange (Bei-Bienko 1930, 1962) may apply also to microhabitats. That rule states that an animal that is xerophilous in behaviour in colder and moister zones becomes meso- or hygrophilous in warm and dry zones.

To understand the action of climate on invertebrates we must also remember that we recognise two groups of invertebrates phylogenetically, leading open and cryptic modes of life respectively.

Invertebrate forest-dwellers leading a cryptic mode of life (for the whole or part of the developmental cycle) are very numerous. They include those living in the soil, in spiders' nests, in rolled leaves, in galls, in burrows, and in tunnels under bark and in wood. Evidently adaptation to such shelters is linked with procurement of a certain degree of isolation from external weather. Of course the atmosphere in each of these micro-habitats is connected with the external atmosphere of the biogeocoenose, and therefore the climate of the microhabitat is correspondingly connected with the climate of the biogeocoenose (its phytoclimate). The amplitude of climatic fluctuations is, however, much less in these more or less isolated shelters than outside. The more isolated and the smaller the shelter is, the greater is the influence of the dwellers in that microhabitat on the state of the atmosphere in it. We deduce that cryptic species of invertebrates have developed phylogenetically a mode of life whereby they are isolated, for the whole or part of their life cycle, from external (in particular, climatic) conditions.

Invertebrates leading an open mode of life are, in general, always exposed to the influence of the atmosphere and therefore to changes in it. Adaptation to specific climatic conditions has been phylogenetically developed by each species—more precisely, by particular populations belonging to each species. That determines the geographical (zonal) distribution of animals, within zones; their distribution by types of bio-geocoenose; and within biogeocoenoses, their distribution by synusiae and microhabitats. Forms with an open mode of life have phylogenetically developed a link with definite environmental conditions, and hence we conclude that forest-dwellers in particular are characterised by adap-tation to the smallest amplitude (as compared with biogeocoenoses in meadows, fields, and other open areas) of fluctuation in atmospheric phenomena.

In the phylogenetic development of each species, development of tropisms with regard to the principal meteorological factors is of primary importance, and thus positive or negative thermotropism has been developed.[1]

Tropisms existing in an animal guide it into zones of hygrothermal conditions nearest to the organism's requirements. Tropisms have been developed, not in a species as a whole but in separate populations, and vary in accordance with the rule of habitat (and microhabitat) exchange.

REACTIONS TO CLIMATIC FACTORS IN THE ONTOGENESIS OF INVERTEBRATES

As cryptic-living invertebrates are protected from most effects of climatic factors, discussion of reactions to the latter may be confined mainly to the species leading relatively open lives. It is usually believed that poikilothermal animals are distinguished by the absence of a mechanism for regulating their body temperature, and the latter fluctuates with external temperature (Naumov 1955). In spite of the lack of thermoregulation, the mechanism of which automatically maintains the required temperature within definite limits, poikilothermal animals still regulate the warmth of their bodies and evaporation by means of their environment. Moved by the reflex actions evolved in them, they make use of their ability to assume the temperature of their immediate environment.

These reactions are governed by the appropriate tropisms, which, as stated above, guide them to conditions close to the preferendum.[2]

Temperature therefore produces appropriate movements in invertebrates, sometimes limited to mere turning of the body to present a greater or smaller surface to its action (e.g., heating); Bodenheimer (1938) called the faculty for such movements 'pseudoheterothermy'. But an animal may be compelled to travel for various distances and at various speeds. In such cases it not only travels towards the preferendum, i.e., to a spot with more favourable (in particular, thermal) conditions, but actually raises its body temperature as a result of its muscular work (heterothermy, according to Bodenheimer).[3]

Weather not only produces reactions but also alters the rate at which they take place; the onset of adverse conditions produces torpor, enabling the animal to reduce loss of energy to almost zero.

[1] We may assume that positive phototropism is an adaptation enabling the animal to survive in conditions of insolation, and consequently heat is necessary to it, since in natural conditions light and heat are as a rule inseparable. Investigations of photoperiodism in insects by A. S. Danilevskii (1961) have shown that the lengthening of the day serves as a signal of change in seasonal living conditions (in particular, the heat regime), whereas light intensity produces no changes in physiological processes. That confirms the theory that the biological reason for the phototropic reaction is to ensure heat for the animal.

[2] In modern literature, 'preferendum' is defined as maximum attainment of the optimum possible in given conditions.

[3] N. P. Naumov (1955) gives the name of 'heterothermy' to lowering of the metabolic level during a period of reduced activity by vertebrates (hibernation, deep sleep). Apparently one should use that term to indicate fluctuation in heat-exchange as a result of change in the activity of an animal.

Invertebrates that live in the open are therefore constantly exposed to, but are not passive with regard to, the action of meteorological elements upon them, especially temperature and the relative humidity connected with it.

A deviation in certain meteorological factors, such as temperature and humidity, beyond the maximum or minimum for a specific population leads to the death of the latter. That again confirms the inadmissibility of applying the laws established for chemical reactions to the physiological processes of animals. B. P. Uvarov (1931), and in recent years Macfadyen (1963), have shown that in the relationship between temperature and rate of development of organisms the rule of van 't Hoff[1] is not applicable either as it stands, or with adjustments; the different curves proposed for determining the rate of development have also proved to be inapplicable. As B. P. Uvarov points out, we cannot compare the temperature in nature, which varies throughout the day, with the average laboratory temperature of experiments, since variations beyond optimal limits and below development thresholds disrupt the course of the latter and a return to the optimum cannot compensate for the disruption. The rate of metabolism depends on body temperature, the fluctuations of which, as stated above, do not always follow fluctuations in external temperature. Finally it must be stressed that in any conditions changes in the rate of development depend not directly on changes in temperature, but on a combination of separate reactions and the total behaviour of individuals.

Numerous investigations lead to the conclusion that in nature an organism reacts not to the action of a single factor but to combinations of external factors occurring at a given moment. In other words, its behaviour is determined not by a single reaction but by a combination of reactions, which in many cases leads to composite reactions not obviously dependent on any of the operative factors. Henson (1960), confirming that statement, cites observations according to which the action of temperature, accompanied by increased evaporation, produced movements, concentrations, and feeding in larvae of the Norway spruce leaf-roller (*Choristoneura fumiferana* Clem.); atmospheric evaporation affects their exudation of web-filament.

The combination of factors and of reactions also explains the fact that the thermal constant (Sanderson 1908, quoted by Frideriks 1932), or the rule of summation of temperature required by an organism for development, is often not confirmed. That happens because even in one locality similarity of conditions in different seasons is only relative. Different variations in the temperature fluctuations beyond the optimum (and even the sub-optimal levels) alter the significance of the temperature factor. Therefore the required sum of temperatures varies.

Similarly, we cannot regard properties such as frost-resistance as being

[1] The rate of reaction increases with temperature, and the coefficient Q_{10}, which denotes the increase quotient, has a value between 2 and 3 for each interval of $10°$.

independent. That property varies with wintering conditions (especially depth of snow cover), with the amplitude and frequency of fluctuations in winter temperatures, with 'autumn hardening', and with other factors.

In an article by N. N. Egorov, I. N. Rubtsova, and T. N. Solozhenikina (1961) it is stated that a population of green oak roller moths was depressed by the February frosts of 1956, which reached −35°C, whereas other investigators (perhaps erroneously) considered the critical temperature to be −30°C. Frost-resistance in invertebrates must not be regarded as a specific and absolute characteristic; it varies in different geographical localities, and therefore is characteristic not of species but of populations, but even in them it fluctuates in accordance with internal and external wintering conditions, with the nature of variations in temperature and humidity, with the state of the snow cover, with the state of the organism at the onset of diapause, etc. Therefore frost-resistance is a property ensuring that, although the winter retreat protects the population in some ways, the frost within it will not reach the extreme limits of tolerance. Thus while many factors and multiform reactions exist, their interactions are mutually compensatory.

Effects of weather on mass pest reproduction. It is well known that mass reproduction of many pests follows the arrival of dry weather, making it possible to foresee the danger of mass pest reproduction in certain natural conditions (Henson 1960). The occurrence of mass pest reproductions in a particular habitat, however, depends on the microclimatic conditions of the forest biogeocoenose (Rafes 1961, 1964).

In many areas the weather takes distinctive forms in different types of biogeocoenoses, where their own 'microweather' is created.

P. M. Raspopov (1961) observed successive outbreaks of mass reproduction of the black arches moth first in the forest-steppe, then in the foothill, mountain, and even sporadically in the mountain-taiga zones of Chelyabinsk province. With a succession of hot dry years even the cooler and moister habitats become dry, inducing outbreaks; the simultaneous fall of the water table weakens the trees, which also favours the reproduction of forest pests. Mass reproduction of the pine-shoot moth also depends on soil conditions and the water regime (Schimitschek 1962).

Analysis of mass reproduction of the larch tortrix moth (*Zeiraphera griseana* Hb.) in the Alps from 1910 to 1960 shows that it always breaks out first in unmixed mature larch stands on well-lighted valley slopes, and later (with lower maximum population density) in mixed larch and Siberian stone pine woods on shady slopes (Baltensweiler 1962). Intensive reproduction of the spruce-needle tortrix moth (*Epiblema tedella* Cl.) in northern Germany occurs only in habitats with poor soil and deep water table, where the water balance of the spruce is upset by excessive density or increased evaporation on slopes of unfavourable exposure (Führer 1963).

In many cases mass multiplication depends not only on the direct effects of weather on a phytophage, but also on changes in a number of environmental conditions. For instance, phenological coincidence of seasonal changes in the biochemical composition of plants with the requirements of insects at certain stages of development is one reason for outbreaks of mass multiplication, and divergence between these factors leads to the death of the insects and depresses their reproduction (Thalenhorst 1951, 1960, Schütte 1957, 1959, Edel'man 1963).

THE PLACE OF GROUPS OF PHYTOPHAGOUS INSECTS AND THEIR SIGNIFICANCE IN A FOREST BIOGEOCOENOSE

We have discussed the links between phytophagous insects and various other animals, i.e., within a component, and also the links with plants, micro-organisms, soil, and atmosphere, i.e., with other components of a biogeocoenose. We have surveyed the interactions (co-actions) in which phytophagous insects take part, and also the relationships arising both from these interactions and from other processes taking place in a biogeocoenose. In other words, we have examined the relationships that link invertebrate populations with the environment. We have, however, constantly pointed out that study of separate links cannot give a fully satisfactory explanation of either the permanence of the links of any species with a given biogeocoenose (or type of biogeocoenose) or of the laws governing its behaviour. Such an explanation can be provided only by study of the links between a species and a biogeocoenose as a whole, and above all by discovering its place in the exchange of matter and energy, i.e., in the biogeocoenotic processes taking place in a community.

The place of groups of phytophagous invertebrates in a forest biogeocoenose is decided primarily by the trophic relationships of each species. The absence of a food product inhibits the entrance of a phytophage into a given biogeocoenose, and as a rule a phytophage enters a biogeocoenose as a member of a consortium. Depending on the specific conditions of syngenesis of a given biogeocoenose, the established ecological links between a plant and a phytophage are subject to alteration. Thus if a plant in a new biogeocoenose is subject to new growing conditions, the conditions may also be changed for the phytophage linked with it. We have given examples showing how changes in the biochemical properties of a plant have changed the extent of a phytophage's preference for it, and how changes in their hygrothermal regime have changed the place of habitation of trunk-dwellers (exchange of microhabitat). Instances have been quoted where extreme fastidiousness of an organism with regard to certain conditions excludes it from consortia in a new environment. Therefore the specific environmental conditions of a biogeocoenose determine firstly the possibility of a species entering it, and secondly its mode of life in these conditions. Each condition is affected by its combination with others: shortage of precipitation is not so severely felt if there is plentiful soil moisture, low air humidity has less effect in low temperatures, and so on.

Of course types of biogeocoenoses (forest types) with similarity in a number of conditions, especially similarity in phytocoenose composition, have similar groupings of phytophagous invertebrates.

V. N. Stark prefaced his definitive work *Forest insect pests* (1931) with a quotation from G. F. Morozov's *Study of forest types*, allotted a special section to 'Entomofauna and forest types', and remarked that as early as 1922 he had thought it possible to study forest insect pests only

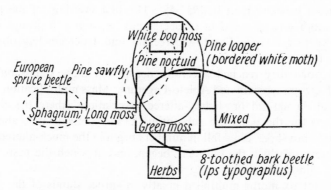

Fig. 44. Distribution of the chief types of insect pests by types of forest (Vorontsov 1960).

on the basis of forest types. G. K. Pyatnitskii (1929) presented material on spruce bark beetles by forest types, in the sense given to that term by V. N. Sukachev. Later a number of investigators have discovered links between entomofauna and forest types: a survey of such works for the last thirty years has been made by A. I. Vorontsov (1960). Besides published examples, Vorontsov presents data obtained by himself on numbers of insect pests in different types of pine forests; the author graphically shows how the distribution of certain species is located on Sukachev's generalised diagram of forest types (Fig. 43). Study of the links between groupings of phytophages and forest types is continuing (Malozemov 1962, Patoćka *et al.* 1962).

Thorough investigation of the role of bilberry in pine-bilberry forests reveals that it creates an intimate network of coenotic links that positively affect the well-being of the forest (Karczewski 1962). Out of 249 species of entomophagous insects 30% feed on pine pests and the remainder on bilberry pests.

There are frequent instances of forest biogeocoenoses or forest types differing not in species-composition but in density of population of a certain species of phytophage. New data supporting this statement have appeared in many publications (Ozols 1961, Padii 1962, Lugovoi 1963, Kangas 1963, Postner 1963, *et al.*

The dependence of mass multiplication on forest types and on character-istics of biogeocoenoses is determined, obviously, not only by the species-composition of phytophages but also by their population density in specific environmental conditions.

Stolina (1959) devoted a special article to the application of forest types in insect ecology. It contained an analysis of mass multiplication of tortrix moths in Slovakia, spruce (*Choristoneura murinana* Hb.) and larch (*Zeiraphera griseana* Hb.) in 1957-58. The first breeding of the spruce tortrix moth was observed in spruce stands belonging to groups of the derivative forest types Fageto-Quercetum and Querceto-Fagetum (ac-cording to Zlatnik).

Only when they reached a certain density on these stands did they attack Fagetum typicum and Abieto-Fagetum. Mass multiplication began to fall off in natural or slightly-altered associations (there the climatic conditions for the pests are sub-optimal). Only much later, and only when food ran short, did a fall in numbers begin in the much-altered plant associations in which the outbreak began, and in which the pests found optimal climatic conditions.

Larch tortrix moths multiplied greatly in spruce stands of the groups Abieto-Fagetum and Sorbeto-Picetum; there, where spruce grows natur-ally, 74% of the trees were infested. In other types infestation was much less: Fagetum abietino-picetosum 12%, Fageto-Abietum 9%, Abieto-Fagetum 5%. The behaviour of the species differed in that the spruce tortrix moth rapidly infested stands of adjacent groups of types, although they were climatically less favourable; the larch tortrix moth did not attack spruce stands of adjacent groups although they were climatically equally favourable, even if its food was insufficient. The larch tortrix moth is less dependent on climatic conditions than on other environmental factors.

N. G. Kolomiets, in a collective monograph on the Siberian pine moth (Zhokhov *et al.* 1961) has superimposed the optimal habitats of the pest on a diagram of forest types in Western Siberia. A number of other works have also described the links between the Siberian pine moth and the conditions in certain types of biogeocoenoses (Kolomiets 1957, 1960 a, b, Kostin 1958, Galkin 1960, 1963, Ivliev 1960, Petrenko 1961).

The division of foci of mass reproduction into primary, secondary, and tertiary is widespread in entomological literature. In the primary type the regulators of pest numbers are least powerful, and pest population density increases in a minimum of favourable conditions; in the other two increase in pest numbers is correspondingly less intensive. The connection between primary foci and the zone of maximum pest damage is very evident, as are the connections between secondary foci and the zone of mass distribution and between tertiary foci and the remainder of the range. The discovery of the connections between certain forest types and optimal and sub-optimal habitats of any pest provides the soundest biogeo-coenotic explanation of mass multiplication.

The effect of mass pest multiplication on a biogeocoenose. Outbreaks of mass pest multiplication arise as a result of a concurrence of circumstances specific for each biogeocoenose, and as they run their course they produce similarly specific changes in biogeocoenotic processes.

In normal conditions, consequently, population densities of mass pests remain 'tolerable'. As mass multiplication increases, they become 'critical' and later 'intolerable'. Regarding this process from the point of view of the biological productivity of a biogeocoenose, we note that the capacity of the food base becomes exhausted; 'primary production' (matter accumulated by green plants) for the year is entirely consumed by the pests, i.e., goes to form 'secondary production', and the soil is deprived of its accession of plant litter. As mass multiplication of the pests increases, not only do they consume the matter produced by the plants during the year but they destroy new matter as soon as it appears. In order to form green matter in such cases the plants use up not only their annual resources but also their physiological reserves, which exhausts their stocks of these and leads to weakening and death.

Therefore as mass multiplication develops there is at first a quantitative change in the circulation of matter and energy; the amount of secondary production increases, with the proportion of animal remains in the soil also increasing. Later the amount of plant remains begins to increase, not from shedding of the current year's production but from dying-off of the main stock of plant matter.

Numerical data for these phenomena are almost totally lacking. Studies of terrestrial biogeocoenoses in this direction are just beginning, and a number of calculations have been made on the basis of surmises and estimates. At the same time the biogeocoenotic importance of mass pest multiplication is evident from many examples.

The cycle of mass pest multiplication (from a suppressed or 'latent' state of the population to a maximum and again to another latent state) may be catastrophic for a biogeocoenose. Excessive consumption of its resources leads to the death of a tree stand, the chief source of primary production in a forest, or of its dominant species. In either case that leads to complete or partial death of the forest, or, in other words, to replacement by another biogeocoenose. Such cases have frequently been recorded in the history of forestry and there is no need to dwell upon them here.

In cases of less intensive outbreaks of mass reproduction, either they are suppressed by natural regulators or control measures may prevent the replacement of the biogeocoenose, but the biogeocoenotic processes undergo changes and the whole character of the biogeocoenose is altered. Only a few observations of this kind have been published; they are merely incidental, and often are restricted to statements that debilitation has taken place without going into detail or analysing the phenomenon.

It was remarked long ago that pests, by destroying green matter, injured tree stands not only by impairing photosynthesis but also by opening up the forest. In that way so-called secondary pests are enabled to infest the trees (resistance to them is lowered by the weakening of the trees) and also better conditions for their development and reproduction are created, since illumination is increased and the temperature is raised beneath the canopy.

Collins (1961) reports from Connecticut, u.s.a. that defoliation of trees by gypsy moth coincided with the peak period of growth of the trees, the trees being deprived of their leaves at the time of maximum day length. As a result of tree mortality, light penetrated beneath the canopy and favourable conditions were created underneath. In the forest under study the canopy was of oak (the first species to be defoliated) and the underbrush of red maple, which was subjected to attack later, and sometimes even retained its leaves when the food supply on the oaks lasted for the whole period of larval development. Detailed records were kept of the growth of maples in different parts of the forest, and in an open control area. Growth of the maples beneath the canopy slowed down and stopped when the oaks came into full leaf (the maples in the forest stopped growing six weeks earlier than those in the open); beneath an oak canopy defoliated by larvae the maples continued growing for two weeks longer than those beneath untouched canopy; when the maples themselves were stripped of leaves their growth ceased, but as they recovered they resumed growth until the oak leaves were regenerated. Collins rejects the idea that growth of the maples beneath defoliated canopy was due to elimination of competition by the oaks in transpiration, or to increase in nutrients in the soil as a result of accumulation of abundant larval excrement in the litter. Lessened transpiration by the oaks had no significance, since the maples increased their own transpiration. At the same time the nutritive value of larval excrement could not be utilised, since the maples ceased growing before arrival of the rains to leach it out. Therefore the improvement in the growth of the maples beneath defoliated canopy could result only from the removal of the shade cast by the oak canopy.

Destruction of leaves in Minnesota poplar forests during mass reproduction of the American lackey moth (*Malacosoma disstria* Hbn.) leads to increased illumination in the forests; as a result, growth increases in uninjured conifers in the underbrush, while it decreases in the poplars and hazels, and in consequence the forest type sometimes changes (Duncan and Hodson 1958).

In balsam fir, as a result of stripping of the needles by spruce tortrix moth larvae, formation of female flowers ceases and mass mortality of upper-storey trees takes place, with consequent improved development of firs in the underbrush (Ghent 1958).

In 1929-34, as a result of mass multiplication of fir loopers and later of black longhorn beetles in the Tubinsk forest area, patches of waste land were formed, on which dense herbaceous cover prevented germination of seeds and development of shoots from isolated fertile stone pines and birches and scattered groups of fir and spruce (Kutuzov *et al.* 1963).

A more thorough biogeocoenotic analysis was made by Rawlings (1961), who studied entomological and other factors in the life of plantations of *Pinus radiata* in New Zealand. A stand of that pine occupying 46,000 hectares was considered 'good' on 20,000 ha and 'satisfactory' on another 20,000 ha, but 6000 ha required replacement (the trees were stunted or too branchy, and many were polyconic). The pine is adapted to a mild climate with precipitation of 625-750 mm and a dry summer. Summer drought equivalent to 50 rainless days does no appreciable harm, but more intensive drought stimulates competition, to the advantage of taller trees. Moreover, more severe drought (such as that of 1946, equivalent to 60 rainless days) induces mass reproduction of the violet woodwasp (*Sirex noctilio* F.), leading to mortality among the shorter pines.

Wind is a very important climatic factor in these forests. The swaying of trees produces constant lashing of branches, as a result of which each tree-top becomes the centre of a circular space, which increases annually according to the local wind peculiarities. The lashing destroys buds and damages branches and terminal shoots, opening the way to fungal infection, mainly by *Diplodia pinea*. In trees that do not reach the upper canopy, shoots are suppressed by the lower branches of the taller pines. As a result, in dense stands the crowns are small, the branches point upwards, and the canopy is open. The wind breaks branches of tall trees. Trees with double crowns, which spread more widely than others, suppress their neighbours and break them down when they fall. On badly-drained soils trees with root systems near the surface are often blown down. On the whole the wind is a regulator of the form of the crowns, and favours the elimination of suppressed trees. Severe drought, as well as excessive moisture, favours dominants and kills suppressed trees. Drought stimulates mass multiplication of insect pests, and high humidity stimulates development of fungal diseases.

In one pine stand (planted in 1922), which was regarded as being in good condition, there was up to 1946 a deep layer of undecayed needles and twigs on the ground, with no ferns or other soil cover, although enough light and precipitation came through the canopy. Slender and relatively long crowns caught the rain, which flowed down the trunks and percolated into deep soil horizons. The structure of the soil beneath the litter was loose, and it contained few earthworms. A series of wet years had enabled fungi to develop and impaired the state of the crowns, so that by 1946 the stand was suppressed and artificial thinning was unnecessary. Dry weather favoured the wood-wasps, which became abundant and killed the weakened trees; by 1950 tree mortality had reached 30%, and in 1951 10% of the surviving trees perished. The natural thinning-out changed the forest remarkably: the branches grew out and bent so much that raindrops fell from their tips instead of running down the trunks; the litter decayed, the soil structure improved, earthworms multiplied, and white leached patches of soil began to disappear. Rich and varied ground cover appeared, including 30 fern species (there were 500-1000 tree-ferns per hectare). Clearly the state of the forest had changed greatly since 1946.

Many other examples might be cited. The above are enough, however, to demonstrate that qualitative changes take place in forests suffering from mass pest infestation. That results in quantitative changes in circulation, affecting one or several species of trees and one or several species of pests, and not only induces changes in the interactions between the species linked together but may also alter the composition of the biogeocoenose and the nature of reactions in its abiotic complex.

The inter-relationship of all biogeocoenose components is also displayed by the fact that each component not only is involved, to some extent, in any disruption of the normal (long-term average) interactions, but also is affected by it.

The role of phytophagous invertebrates in interactions between biogeocoenoses is determined by the degree of 'loyalty' of a species to a particular community. Populations whose life-activity does not go beyond the boundaries of a given biogeocoenose take practically no part in interactions with other biogeocoenoses. There may be exceptions in the case of forcible transportation of individuals, e.g., by wind. If a population is distributed over territory transcending the limits of a specific biogeocoenose (e.g., at forest margins) or including several biogeocoenoses, individuals constantly pass from one biogeocoenose to another. An example is provided by predators or parasites of phytophages that feed on both meadow and forest plants. Regular relationships between a species and two different biogeocoenoses may exist, e.g., sand wasps are linked by their breeding habits to the sands where they dig their burrows, but they may collect nectar on the outskirts of a forest or even within it. The movement of individuals from one biogeocoenose to another may take place irregularly or may be connected with daily or seasonal cycles.

Such movement of individuals is, in the first place, a movement of biomass, altering the balance of matter and energy in biogeocoenoses; and in the second place it is linked with participation in several metabolic cycles. At present, data on these subjects has only begun to be collected.

SOIL-DWELLING SAPROPHAGOUS INVERTEBRATES IN A FOREST BIO-GEOCOENOSE

Many terrestrial invertebrates are linked with the soil in their development, but only some are permanent soil residents. Soil-dwelling animals may be divided into 'active', i.e., taking part in the transformation of organic matter contained in the soil, and 'passive', using the soil only as a shelter from unfavourable conditions, or spending their dormant stages of development there (Gilyarov 1949). The activities of soil-dwelling phytophages, which by damaging plant roots may to some extent decrease the accumulation of organic matter in forest biogeocoenoses and thereby harm the forest economy, are not different in principle from those of invertebrates that damage the above-ground parts of plants, and therefore are discussed in the section on phytophagous invertebrates. But the majority (from 70 to 80%) of invertebrates that feed in the soil are saprophages (Gilyarov 1959). This group has been very little studied in comparison with phytophages, although if one looks on the activity of soil-dwelling invertebrates from the point of view of their participation in the circulation of matter and energy the role of saprophages is most significant. According to modern knowledge, the part taken by soil-dwelling invertebrates in the mineralisation of plant and animal remains consists mainly of indirect effects on these processes. Data given later will show what a great effect soil-dwelling invertebrates have on the rate of decomposition of organic matter. Soil-dwelling invertebrates also act upon synthesis of organic matter by plants; their action may be either positive (through increasing the fertility of the soil or through excretion of biostimulatory substances) or negative.

It is usual to divide soil-dwelling invertebrates according to their size into microfauna, which includes animals that are not easily observed or are invisible to the naked eye, and mesofauna,[1] i.e., larger invertebrates.

Microfauna consists of Protozoa, several worms (nematodes and Enchytraeidae), Acarina, and primitive wingless insects (Apterygota). It also includes small myriopod species, some small larvae of winged insects, etc. The numbers of these animals in forest soils frequently reach very high figures and usually surpass those of mesofauna. Their mass, however, is much less than the mass of the invertebrates classified as mesofauna (earthworms, molluscs, myriopods, insects and their larvae, etc.).

The usual method of censusing soil-dwelling invertebrates in the mesofauna group is to dig them up and sort them out manually. Microfauna are censused by other methods, which differ for different kinds of animals. Because the methods of censusing and further analysis are extremely laborious, only in rare cases are all groups of invertebrates counted; the figures obtained, apparently, are much too low, since the methods used to extract animals from the soil are still very imperfect. The data at our

---

[1] Macrofauna includes vertebrates living in the soil.

disposal show that the soil is inhabited by a multitude of animals belonging to all systematic groups of terrestrial invertebrates.

FORMATION OF COMPLEXES OF INVERTEBRATES IN SOILS OF FOREST BIOGEOCOENOSES

Many investigators both in our country and abroad are engaged in the study of complexes or of separate groups of invertebrates living in forest soils.

As a result of these studies, great differences have been observed in the composition of soil fauna in forests with coarse and slowly-decomposing litter (coarse humus of 'mor' type) and in forest types where litter quickly decomposes and mixes with the mineral part of the soil (soft humus of 'mull' type). The number of large soil-dwelling invertebrates is much less in forests with coarse-humus soils than in forests with mull soils, and therefore the total mass of animals in them, in spite of the higher numbers of small arthropods (Acarina, Collembola, etc.) is much less (Fig. 45).

In beech forests in Denmark, according to Bornebusch (1930), the mass of animals in coarse-humus soils is 16·6 g/m^2 (live weight), and in mull soils with soft humus 70·8 g/m^2. V. Ya. Shiperovich (1937) found that in waterlogged soils of sphagnum pine forests the weight of invertebrates

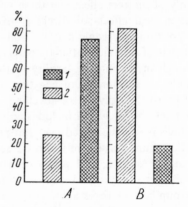

Fig. 45. Relationship between numbers (A) and biomass (B) of soil-dwelling invertebrates (in percentages of the total values for these items) (Murphy 1955). 1—in coarse humus; 2—in mull (soft humus).

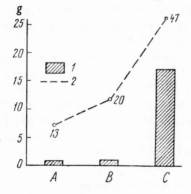

Fig. 46. Relationship of biomass of soil fauna to degree of decomposition of peat (Kozlovskaya 1959). 1—biomass (g/m^2); 2—degree of decomposition of peat (%); A—shrub-sphagnum pine forest; B—sphagnum spruce forest; C—boggy broad-leaved-herb spruce forest.

did not exceed 5-7 g/m². The mass of animals in the soils of dry pine forests was only slightly greater (16·4 in white moss and 24·6 in heather pine forests, etc.), whereas in the rich soils of composite pine forests the weight of fauna reached 55-62 g/m². He concluded that the slower the process of litter decomposition, the less is the weight of soil fauna. Later investigations (Zinov'eva 1955, Kozlovskaya 1959 *et al.*) have confirmed that conclusion (Fig. 46).

Differences in weight result mainly from differences in the numbers of large soil-dwelling invertebrates (especially earthworms). The numbers of several groups of small soil-dwellers (Apterygota, Acarina, Enchytraeidae), on the contrary, are higher in soils on whose surface there is always a layer of slowly-decomposing plant litter (Bornebusch 1930, Nosek 1954, Nielsen 1955).

Small arthropods are most numerous in the upper, less-decomposed parts of forest litter. With increase in weight (i.e., degree of decomposition) of the substrate their numbers, as is seen from Table 65, fall sharply (Nosek 1954).

Table 65. *Dependence of numbers of small arthropods on the degree of decomposition of litter (Nosek 1954).*

Character of litter	Average weight of 1 dm³ of substrate (kg)	No. of small arthropods per kg of substrate	
		Acarina	Apterygota
Beech-fir plantation with Oxalis *predominating in herbaceous cover*			
Slightly-decomposed litter	0·10	8114	3905
Semi-decomposed litter	0·27	2800	600
Much-decomposed litter, mixed with mineral part of soil	0·44	1300	50
Spruce plantation with dead soil cover			
Slightly-decomposed litter	0·12	24,950	10,500
Semi-decomposed litter	0·14	10,100	4300
Much-decomposed litter, mixed with mineral part of soil	0·51	1250	660

It has been demonstrated that the character of the soil population in a forest depends not only on the local habitat conditions but also on the composition and structure of the stand, and consequently each forest type has its own distinctive complex of soil-dwelling invertebrates. Nosek (1954) discovered great differences in the species-composition of the invertebrate fauna in the soil of a beech-fir plantation from that of a spruce plantation with dead ground cover (Table 66), growing in similar conditions (identical exposures and heights above sea level, same types of soil). He concluded that the succession of plant communities determines

Table 66. *Comparative characteristics of soil fauna of two forest types (Nosek 1954).*

Systematic composition of invertebrate fauna	No. per m²	
	Beech-fir plantation (Fageto-abietosum) with predominance of *Oxalis acetosella* in herbaceous cover. Brown forest soil	Mountain spruce plantation with dead ground cover. Loamy soil
Lumbricidae	40	10
Enchytraeidae	703	2672
Isopoda	2	2
Gastropoda	22	6
Pseudoscorpiones	21	32
Araneae	82	10
Acarina	17,104	40,436
Diplopoda	17	10
Chilopoda	251	60
Apterygota	7487	17,628
Psocoptera	5	—
Thysanoptera	11	—
Homoptera	8	—
Hymenoptera	27	14
Coleoptera	186	140
Lepidoptera	10	16
Diptera	930	1610

the succession of zoocoenoses (having in mind only soil-dwelling invertebrates).

These investigations were made in the Yablunsk mountains at a height of 700-890 metres above sea level, in the neighbourhood of Dol'naya and Gornaya Lumna (Czechoslovakia).

A similar conclusion was made by Thiele (1956), who believes that the soil population of any territory is divided into separate groupings (or complexes) corresponding to the distribution of plant communities.

The dependence of the distribution of soil-dwelling animals on the nature of the plant cover can be confirmed by many examples. It is shown by the changes in composition and numbers of soil fauna as a result of tree-felling and of change in stand composition (Table 67). Similarly, structural changes in the complexes of soil-dwelling invertebrates occur when forest plantations are established in belts. According to the data of V. M. Berezina (1937), in forest-steppe conditions substantial differences in the composition and quantitative ratios of insect species living in the soil beneath tree canopy (as compared with the entomofauna of adjacent treeless areas, fallow and fields) are observed as early as the eighth or ninth year after planting the forest belts. M. S. Gilyarov (1956), who made investigations in the north of the steppe zone, also reports the formation of 'secondary biogeocoenoses', differing from the soil fauna of the surrounding virgin and cultivated plakor steppe, in the soil beneath

Table 67. *Numbers of dominant groups of soil-dwelling mesofauna in habitats differing in type of plant cover (per 1 m²) in soddy weakly-podzolised soil (from soil samples taken in Rybinsk district of Yaroslav province).*

Systematic composition of invertebrates	*Oxalis*-fern spruce forest	Clearing in *Oxalis*-fern spruce forest	Goutwort-mixed-herbage birch forest succeeding *Oxalis*-fern spruce forest
DIPTERA			
Bibionidae	18·6	1·0	30·2
Tipulidae	4·1	0·1	—
Rhagionidae	1·8	5·0	4·3
Tabanidae	—	3·0	4·5
Dolichopodidae	—	5·0	1·5
Total (with other families)	25·6	18·6	43·2
COLEOPTERA			
Elateridae	7·4	1·5	10·5
Including:			
Athous subfuscus	6·5	1·2	1·6
Dolopius marginatus	0·3	—	5·0
Carabidae	2·3 (VIII)*	6·6 (XVI)	3·0 (X)
Staphylinidae	3·5	4·8	4·3
Cantharididae	1·3	0·6	2·7
Curculionidae	0·2	1·1	6·0
Total (with other families)	14·7	18·2	28·5
HETEROPTERA	0·3 (II)	9·3 (XI)	1·8 (II)
CHILOPODA	9·3	4·5	8·0
DIPLOPODA	3·5	5·0	6·1
LUMBRICIDAE	3·1	9·3	6·8
Total no. of invertebrates	73·0	85·6	110·5
No. of samples (0·25 m²) taken	24	24	24

* The number of species found is shown in parentheses (Roman numerals).

the canopy of strip-plantings. Many other authors arrive at the same conclusions.

The character of the effects of a tree stand on soil fauna depends to a great extent on the species-composition of the stand. This statement, previously expressed in general terms, has been proved by comparisons made by K. A. Gavrilov (1950) of the soils beneath plantations of uniform age and density, differing only in tree-species composition. Gavrilov's data are somewhat biased, being based on a comparatively small amount of factual material, but judging by later investigations are quite trustworthy. They show that the distribution of even such a group of invertebrates as earthworms, which live in the soil throughout their lives and are saprophagous in diet, clearly depends on stand composition (Table 68). The age and density of the stand are also important, apparently because they affect the degree of closure of the tree canopy. According to Berezina's (1937) observations, when a stand is thinned a number of insect species typical of open spaces appear in the soil beneath the canopy. Bornebusch (1932) records an increase in the numbers and variety of species of Lumbricidae as a result of thinning of a dense spruce plantation with dead ground cover. Investigations by several authors (Ronde 1951, Zrazhevskii 1957, Perel' 1958) have revealed differences in the numbers and quantitative ratios of species of Lumbricidae in the soil of tree plantations, uniform in tree species-composition but differing in age. A. G. Topchiev (1960) observed differences in the composition and density of Lamellicornia, Elateridae, and Melanosomata larvae in the soils beneath the canopy of pine plantations of different ages. The population densities and numbers of species of beetles of these families in the soils of plantations at the 'thicket' stage were much lower than those in pole-stage and medium-aged plantations.

Table 68. *Average numbers and species-composition of earthworms in tree plantations on leached chernozem* (per m²; 10 samples from each plantation).*

Species	10 spruce	10 pine	10 oak	10 birch
	Age 28, density 1·0	Age 31, density 0·9	Age 28, density 0·9	Age 28, density 0·8
Allolobophora caliginosa (Sav.) f. typ.	6·6	27·6	131·3	112·1
Lumbricus rubellus Hoffm.	few	9·7	36·6	69·8
Eisenia rosea (Sav.)	few	10·6	+†	63·5
Dendrobaena octaedra (Sav.)	few	21·0	few	+
Octolasium lacteum (Oerley.)	—	few	14·1	10·9
Lumbricus terrestris L.	—	—	+	+
Total	10·0	72·0	190·0	269·0

* Mokhovsk leskhoz, Orlov province (Perel' 1958).
† Not more than 5 per m².

The influence of the tree species composing the stand on soil-dwelling invertebrates is also exerted through their action on underbrush and ground vegetation (shrubs, herbage, and mosses). Data showing the effects of underbrush and herbaceous cover on the formation of complexes of soil-dwelling invertebrates may be found in the works of several authors; the degree of development and the species-composition of these layers are of importance not only for phytophages but also for such saprophages as earthworms (Bornebusch 1932, Ronde 1951).

The action of plant cover on soil-dwelling invertebrates takes many different forms. There are direct links between soil fauna and plants, e.g., the dependence of root-eating species on the abundance of roots in the soil and on the nutritive quality of the roots of different plant species. The distribution of a number of saprophage species belonging to various systematic groups (several species of earthworms, molluscs, Diplopoda, etc.) is strongly affected by the composition of forest litter, their main source of food, since the relationships of different animal species to the litter from any tree species are not identical. We may note that the direct effects of plants on soil-dwelling invertebrates are not limited to food relationships. Several published reports indicate that these effects are very complex. For instance, the emergence of beet eelworms (*Heterodera schachtii* Schmidt) from cysts is stimulated by root exudates from several plants (Korab and Butkovskii 1939), including some on whose roots they cannot feed (in which case the larvae perish). A. I. Zrazhevskii (1957) observed a toxic effect on earthworms caused by substances washed out of silver maple litter. Ramann (1911) found that earthworms avoid several plants, among which he names sage and mint. Later, water-soluble (and in sage, volatile) substances of phytoncide type, lethal to earthworms, were discovered in these plants (Drabkin and Balovnev 1952). Plant root exudates are also important in that they stimulate the development of bacteria, which are most numerous in the rhizosphere. The bacteria attract Protozoa, many of which feed on them, and as a result the number of Protozoa is much higher in the rhizosphere zone than at a distance from the roots, the root system (as has been clearly demonstrated in the cases of cotton and lucerne) having a definite selective property with regard to Protozoa. According to the observations of Andras (see Balogh 1959), specific communities of free-living nematodes also occur in the rhizosphere zone. Little study has yet been given, however, to the effects on soil-dwelling animals of substances secreted by plants in the course of their life metabolism and released by the decomposition of dead parts of the plants.

Plant cover also indirectly influences the form of complexes of soil-dwelling invertebrates, through changes in the hydrothermal regime and the physico-chemical properties of the soil. Observations in the steppe zone at the Derkul' research station of the U.S.S.R. Academy of Sciences Forestry Institute have shown that consumption of soil moisture there was due more to transpiration than to evaporation from the soil surface, since the soil under herbaceous cover dried up no less rapidly than that under cultivated tree stands. The difference is that trees take moisture mainly from chernozem layers deeper than 50 cm, whereas herbs mostly dry up the top 50-cm layer of soil (Zonn 1954b). The soil fauna beneath the tree canopy therefore has more hygrophilous species than the soil fauna of the open steppe, but consists mainly of invertebrate species linked in their development with forest litter and the topmost layer of the soil (Gilyarov 1956). Differences in the hydrothermal regime of the soil, due to the different nature of the plant cover, are displayed also in the seasonal activity of soil-dwelling invertebrates, as may be seen by comparing the vertical distribution of earthworms (*Allolobophora caliginosa*) during summer months in the 10-cm soil layer beneath the canopy of 28-year-old plantations differing in tree species-composition (percentages of that worm species in the total number of worms collected):

	1.VI	2.VI	3.VI	1.VII	3.VII
Oak plantation	75	64	58	67	96
Birch plantation	32	36	68	93	95

Table 69. *Some features of the soil fauna (mesofauna) of two soil varieties.**

Character of vegetation	Total no. in 1 m²	Nos. of dominant groups as % of all invertebrates collected		Frequency of strictly soil-dwelling invertebrates, % of occurrence in total no. of samples			
		Diptera	Coleoptera	Curculionidae larvae	Asilidae larvae	Octolasium (Lumbricidae)	Mermithidae
Soddy weakly-podzolized soil							
Oxalis-fern spruce stand	73·0	34·9	21·7	4·2	12·2	16·6	20·8
Goutwort-mixed-herb birch stand	110·5	45·5	25·9	62·5	33·3	25·0	12·5
Clearing in *Oxalis*-fern spruce stand	85·6	25·0	21·3	25·0	8·3	29·2	33·3
Slightly-soddy strongly-podzolized soil							
Bilberry-spruce stand	53·8	13·5	53·5	12·5	—	—	—
Bilberry-birch stand	51·3	10·7	32·8	8·3	8·3	4·2	—
Clearing in bilberry-spruce stand	72·1	14·6	25·4	—	—	4·2	—

* Rybinsk district, Yaroslav province.

The burrowing of worms to deeper layers (where they encapsulate and become dormant) when the topmost soil layer dries up, and their return to the active state when the rainy period begins in late June, occur earlier in the soil under birch plantations than under oak, where the dense canopy hinders both drying and wetting of the soil.

The structure and the mechanical composition of the soil also affect soil-dwelling invertebrates, since they govern the soil's humidity, its relative proportions of air and water, and its porosity (the latter is particularly important for animals that are incapable of active tunnelling). Moreover, the distribution of soil-dwelling invertebrates is also affected by the quality and quantity of humus, soil acidity, soil and subsoil contents of calcium, soil salinity, etc. (Gilyarov 1949). Evidently the properties acquired by a soil because of the character of its vegetation are bound to show their effects in the species-composition and numbers of its invertebrate fauna.

The influence of plant cover on the soil population must not, however, be considered to be paramount (as is done, for instance, by Grigor'eva 1950) in deciding the composition and numbers of the soil fauna in a certain location. Even in cultivated soils, in spite of the equalising effect of cultivation and planting with uniform farm crops, the composition and numbers of the soil fauna depend on the soil type (Gilyarov 1939, 1942).

That dependence is still more evident when one compares localities occupied by the natural plant formations found typically on the soils under study. As seen from Table 69, the population of each of two contrasted soil types is characterised by a number of features occurring throughout the soil under different kinds of plant cover. The soddy weakly-podzolised soils are distinguished from soddy strongly-podzolised soils by a higher number of soil-dwelling invertebrates, by different proportions of the dominant groups, by greater frequency of typical soil-dwelling forms, and by the fact that species more typical of more southerly districts occur in them more frequently and in greater numbers.

Thus the complexes of soil-dwelling invertebrates are formed according to the character of the plant cover, but only from species capable of living in conditions peculiar to the soil variety concerned. The soil-zoological method of soil diagnosis devised by M. S. Gilyarov (1953b, 1956a) is based on that fact.

Climate strongly affects invertebrate distribution. In the case of soil-dwelling invertebrates, the effects are altered by the composition and structure of the phytocoenose and the hydrothermal regime of the soil (which depends on the mechanical composition and structure of the soil, the level of underground water, surface flow, etc.).

Zonal variation in soil-climate is accompanied by regular variations in numbers, mass, vertical distribution, and group-composition of soil-dwelling invertebrates,

Table 70. *Numbers of soil-dwelling invertebrates, censused by the soil-sample method, in different soil-climatic zones (per m²).*

Invertebrates	Taiga subzone					
	Vologda province		Yaroslav province			
	Spruce-bilberry on podzol	Spruce-herb-green-moss on peat-humus soil	Spruce-bilberry	Birch-bilberry	*Oxalis-*fern spruce stand	Goutwort-mixed-herb birch stand
			Slightly-soddy strongly-podzolised soil			
					Soddy weakly-podzolised soil	
Earthworms	5	8	2	2	3	6
All soil-dwelling invertebrates	38	41	53	51	73	110

In the tundra zone soil-dwelling invertebrates concentrate in the thin surface layer of the soil; sometimes only the moss carpet is inhabited. The surface-layer distribution of animals in the tundra soil profile is due to the permafrost and the heavy waterlogging, which creates an oxygen deficit in the soil. As I. V. Stebaev (1962) has noted from soil samples in Salekhard district, at a depth of 4-5 cm in tundra soil invertebrates are often found in a cold-numbed state, especially in early summer. Small forms (Enchytraeidae, Collembola, and apparently also Oribatidae) are most widespread there. They often form the main mass of soil fauna, which in such cases does not exceed 1 g/m². The number of earthworms is usually small. Among higher insects Diptera larvae predominate (Kozlovskaya 1957 a, b, Stebaev 1959, Chernov 1961).

The mass of invertebrates is somewhat higher in soils of the taiga subzone (Shilova 1950, Kozlovskaya 1957a, 1959) and soil-dwelling animals do not penetrate deep into the soil. They are represented mainly by litter forms. Among mesofauna, larvae of Diptera and Elateridae are the most numerous.

In the coniferous-broad-leaved forest subzone the mass of soil-dwelling invertebrates reaches comparatively high figures; in the forest soils of Denmark (Bornebusch 1930) 70-100 g/m², and in the composite pine forests of the European part of the u.s.s.r., as noted by V. Ya. Shiperovich (1937), 55-56 g/m². According to our data, in the oak groves in the Tula reserves the weight of earthworms alone reaches 81·2 g/m². In the coniferous-broad-leaved forest subzone and the forest-steppe earthworms constitute the main biomass of the soil fauna. Typical soil-dwelling forms are widely represented among all groups of soil-dwelling animals in these districts. In favourable conditions invertebrates usually concentrate in the upper 10 cm of the soils, but when the upper layers dry out or

Table 70 (*continued*)

Coniferous-broad-leaved forest subzone					Forest-steppe			Steppe
Minsk province*		Moscow province			Tula province	Kursk province*		Lugansk province*
Pine stand with oak on weakly podzolised loam	Oak stand with ash on weakly podzolised loam	Sedge-*Oxalis* spruce stand with oak	Hairy-sedge spruce stand with oak	*Oxalis-Lamium* spruce stand with oak	Ash-linden-goutwort oak stand. Grey forest soil	Oak belt on chernozem	Unknown (reserve) steppe	Plakor virgin steppe
110	124	69	120	285	209	102	39	12
216	232	107	172	308	—	605	463	64

* From published data: for Minsk province from L. A. Zinov'eva (1955); for Kursk and Lugansk provinces from M. S. Gilyarov (1956, 1960).

freeze they go farther down, migrating to considerable depths (1 metre and more).

The numbers of mesofauna animals are much smaller, and consequently so is the mass of invertebrates in steppe-zone soils (Table 70). Insect larvae (mainly Coleoptera) predominate in the mesofauna group. In spring invertebrates approach the soil surface, but as the upper soil layers dry out they gradually move deeper (Gilyarov 1949).

Thus even when comparisons are made on a zonal scale the highest values for the mass of soil-dwelling invertebrates coincide with the regions of most intensive decomposition of plant remains.

PARTICIPATION BY SOIL-DWELLING IN VERTEBRATES IN DECOMPOSITION OF FOREST LITTER

In the forest the greatest quantities of plant remains reach the soil in the form of shed parts. The fertility of soils beneath the forest canopy depends on the rate and nature of the decomposition of forest litter. Forest litter not only is the source of organic and mineral matter that enters forest soil, but also affects the hydrothermal regime and aeration of the soil. Experiments made in the Central Chernozem reserve (Kursk province) by G. F. Kurcheva (1962) show how great the role of invertebrates is in the decomposition of forest litter. In test plots in an oak forest, the participation of invertebrates in the decomposition processes was prevented by adding naphthalene, which drives away animals but does not inhibit bacterial and fungal activity. As a result, litter decomposed there at only a fraction of the rate in control areas where naphthalene was not used. In the first year of observations (with weather conditions favourable for

FFB X

invertebrates) about 55% of the litter on the control plots decomposed during a warm period of 140 days, but only 9% where invertebrate activity was excluded. In the next two years of observations (with drier and hotter summers), during the same periods 37% of the litter decomposed on the control plots and only 10% on the test plots.

In the opinion of several authorities (Bornebusch 1930, Dunger 1958a), in forests where there is a combination of favourable conditions for soil-dwelling invertebrates, the entire mass of shed parts is worked over by these animals in the year of litter fall. According to Hungarian investigators (Dudich *et al.* 1952), in the forests of Central Europe soil-dwelling arthropods alone work over about 40% of the shed matter each year, i.e., a minimum of 240 g/m^2. In Dutch oak forests (van der Drift 1949), members of one group of soil-dwelling saprophages (Diplopoda) consume not less than 10% of the litter, and small arthropods (Acarina and Collembola) approximately 50%. These figures, however, are very rough approximations, as they were obtained from analysis of laboratory experimental data.

The rate at which soil-dwelling invertebrates consume litter depends on many factors. The composition of the litter is one of the principal of these. It is known that soil-dwelling saprophages have selective ability regarding offal of different tree and shrub species.

That is seen, for instance, from the results of experiments made by Dunger (1958b) with soil-dwelling invertebrates belonging to several systematic groups (Diplopoda, Isopoda, Pulmonata, Oligochaeta). Having arranged nine tree species selected for the experiment in an 'order of preference' drawn up on the basis of the amounts of litter eaten by the animals, Dunger observed that the order coincided for all the animals studied. The animals used the greatest amounts of litter from linden, ash, and black alder. They were least satisfied with litter from oak and beech (Fig. 47). Similar results have been obtained by other investigators (Lindquist 1941, Lyford 1943, Wittich 1953, *et al.*). It is not yet quite clear what decides the degree of suitability of litter for saprophages. It is believed that the chemical composition (nitrogen content, carbon-nitrogen ratio, calcium content, presence of flavoured and tannic substances), mechanical properties, and moisture content of the litter are of significance (Dunger 1958b).

The relationship of every species of soil-dwelling invertebrates to the litter of any tree species is not always the same, although it does coincide in many cases. For example, the litter[1] of several coniferous species (pine, spruce, and some others) is not generally used as food by many invertebrates, but it is successfully consumed by several bibionids (Schremmer 1958) and some species of testaceous Acarina. In experiments with earthworms (Lindquist 1941) it was observed that only one out of seven species tested (*Lumbricus terrestris*) was able to eat oak litter, worms of

---

[1] This refers to litter little decomposed, when the needles still exist separately.

that species (as our observations have shown) increasing in weight as much as worms fed on linden and hazel litter.

The rate at which soil-dwelling invertebrates consume litter depends also on temperature and the amount of moisture in the litter (van der Drift 1951, Gere 1956). According to the data of these authors, seasonal variations are observed in the activity of soil-dwelling invertebrates. In experiments with Diplopoda, with uniform moisture and temperature, the animals consumed litter much more rapidly in spring and early summer than in autumn and winter.

It has been demonstrated that in many species of soil-dwelling sapro-phages[1] small individuals consume food in relatively larger quantities (as

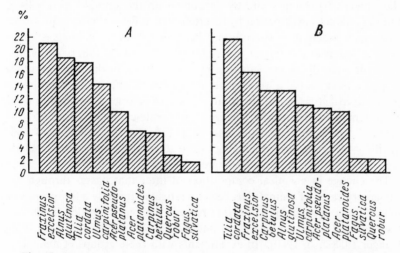

Fig. 47. Amount (%) of litter of nine deciduous tree species eaten by invertebrates of 14 different species (Dunger 1958). A—Fresh offal; B—overwintered litter.

percentage of body weight) than do large individuals of the same species. Food consumption varies in proportion to the active surface of the body and is therefore proportional to $\sqrt[3]{W^2}$, where W is the weight of the animal (van der Drift 1951, Gere 1956).

Apparently that is also true to some extent of invertebrates belonging to different species and even different groups, although we cannot entirely agree with Gere (1956), who believes that rate of food consumption does not depend on the species of the animal. Small free-living nematodes, for instance, consume ten times as much oxygen per unit of mass as do earth-worms in the same conditions (Overgard 1949), which is indicative of their much higher metabolic energy.

We thus have grounds for believing that small invertebrate species in the mesofauna group consume relatively more litter than do larger ones.

[1] This does not apply to molluscs and fully-metamorphosed insects (Balogh 1959, Dunger 1960).

Balogh (1959) believes that small arthropods (Collembola and Acarina), numbering 200,000 per m^2 and with a total weight[1] of 1·1 g, would consume forty times as much litter as would mesofauna animals with a total weight of 6 g in the same length of time. As the author points out, that is for the present purely hypothetical.

In evaluating the part taken by soil-dwelling invertebrates in the decomposition of forest litter, important factors are not only the amount of litter eaten by the animals, but also the qualitative changes in the chemical composition of the litter due to their activities.

When we survey the effects of soil-dwelling invertebrates on the process of chemical change of plant litter, we must distinguish two groups of phenomena: (*i*) changes directly dependent on the animals' metabolism, and (*ii*) indirect participation by invertebrates in decomposition processes through their action on microflora, where the role of the animals is 'more like the action of a catalyser in its nature' (Macfadyen 1961). According to recent data, direct participation by invertebrates in the decomposition of plant litter is on a very small scale. From approximate estimates (Macfadyen 1961), out of the total amount of calories contained in dead plant remains entering the soil in a temperate-climate pasture 85% is freed by microfloral activities and only 15% by the activities of soil-dwelling invertebrates, including Protozoa; these estimates include predators and invertebrates feeding on fungi and bacteria.

The change in chemical composition of litter that passes through the alimentary canal of animals has been investigated for only a few groups of soil-dwelling saprophages. In this respect the most detailed study has been devoted to Myriopoda and woodlice (van der Drift 1951, Gere 1956, Dunger 1958 a, b, 1960). In experiments with these groups of animals it has been found that their excrement immediately after excretion differs little in chemical composition from the litter eaten. Only with a diet of particularly-preferred, slightly-decomposed litter (e.g., litter of elder) did the humic acid content of the excrement increase considerably. That was not observed when litter of many other species was consumed. It was concluded that soil-dwelling saprophages rarely take part in the humification of plant litter (Dunger 1960). In Dunger's opinion, the direct role of invertebrates in humification consists in the fact that by eating plant litter they release part of the lignin (Dunger 1958a), which becomes available for chemical reactions. We question the correctness of such a broad generalisation, based on the results of experiments with only two groups of soil-dwelling invertebrates. Microscopic analyses made by Dunger (1958 a, b) of the excrement and the gut contents of Diplopoda and woodlice showed that in most cases these saprophages broke down plant tissue only to a small extent. At the same time there were variations in the degree of trituration of ingested litter by different species, the changes in

[1] These calculations were based on the average weight and the average density of Collembola and Acarina, obtained as the result of a census made in Hungarian forests.

humus content with passage of litter through the animals' alimentary canals being correlated with the degree of breakdown of plant tissue. According to Kubiena's data, much more vigorous trituration and breakdown of litter takes place in the alimentary canal of earthworms and of several other invertebrates than in that of Myriopoda. It is quite probable that members of these groups of animals subject litter to more thorough chemical alteration than do Diplopoda or woodlice. It has been observed in experiments that the worms *Allolobophora caliginosa*, when feeding on soil containing finely-divided plant litter, excreted about 6% of nitrogen (which had been in a form unavailable to plants) in the form of easily-assimilable soluble compounds. Therefore the opinion, nowadays fairly widespread, that direct participation by invertebrates in the decomposition of organic matter is comparatively small, needs revision.

The indirect action of soil-dwelling invertebrates on the process of decomposition is well illustrated by Anstett's (1951) experiments. Being convinced that grape-vines decomposed much more quickly in the presence of earthworms (*Eisenia foetida*), Anstett removed worms from several test boxes. He found that the rate of decomposition after the worms were removed, after having been in the test boxes for a long time, was still much higher than in the control boxes where the worms had never been (Table 71). Microbiological analysis showed that in the test boxes (with worms) the population density of microflora was four or five times as high as in the controls (without worms).

Table 71. *Effect of earthworms* (Eisenia foetida) *on rate of grape-vine decomposition* (*Anstett 1951*).

Date of examination	% of ash content in dry matter			
	Test without worms	Test with worms	With worms until 1/IX, afterwards without	With worms until 15/IX
17.VII	12·8			
7.VIII	13·1	15·2		
1.IX	13·4	17·9	17·9	
15.IX	13·5	21·3	21·3	21·3
15.XII	13·9	25·1	24·7	24·9

Considerably higher intensity of decomposition of oak litter (as may be seen from the changes in carbon and ash content shown in Table 72) was observed by D. F. Sokolov (1957) in experiments with black millipedes (*Iulus terrestris*). Apparently in this case also the effect of invertebrates on the decomposition process must be attributed to stimulation of the activity of microflora.

Investigations by several authors have proved that the number of micro-organisms is much higher in invertebrate excrement (Table 73) than in the soil (Ruschmann 1953, Zrazhevskii 1957, Stebaev 1958a, van der Drift and Witkamp 1960). It is not clear where the increase in reproduction of the micro-organisms takes place, whether in the alimentary canal

Table 72. *Changes in content of biogenous and mineral elements during decomposition and mineralisation of organic matter in laboratory experiments with black millipedes (Sokolov 1957).*

| | Amount of dry matter (g) | Content, % of dry weight | | | | | |
| | | Carbon | | Nitrogen | | Salts and mineralised residue | |
		a*	b	a	b	a	b
In control containers							
Oak leaves	100	51·03	50·16	0·918	1·241	6·83	10·02
Soil	1000	4·43	4·46	0·476	0·429	87·47	83·26
In containers with black millipedes							
Oak leaves	100	51·03	40·62	1·918	1·807	6·83	18·59
Soil	1000	4·43	4·45	0·467	0·505	87·47	83·88

* *a*, in original material; *b*, at time of analysis (after 100 days).

of the animals or in their excrement (Day 1950), which affords a favourable environment for the development of microflora.

When feeding on forest litter and other plant litter invertebrates break it up finely and mix it with soil, and so create conditions for more intensive microbial activity. The work of small arthropods is particularly effective in that respect. According to several estimates (quoted by Nef 1957), beetle mites (Oribatidae), when feeding on pine needles, increase the surface of the eaten material to approximately 10,000 times the original figure.

Soil-dwelling invertebrates have still further effects on microflora. When invertebrates are introduced into a soil culture, widespread senile deterioration of fungi and bacteria (due, it is believed, to the presence of antibiotics in the soil) is arrested. For instance, very vigorous germination of spores was observed when *Trichoptera* or *Glomeris marginata* were introduced into a culture (Witkamp 1960, quoted by Macfadyen 1961). The soil is inhabited by a multitude of animals (small arthropods, nematodes, Protozoa, etc.) that feed on micro-organisms, but they are capable of beneficial action on the activity of the microflora. For example, when a combined culture is made of Protozoa (infusoria and amoebae) and *Azotobacter* much more nitrogen is accumulated and the bacteria are represented by young, actively-developing cells. The reason is that the

Table 73. *Total numbers of micro-organisms in soil and in earthworm excrement (10^3 per gramme of oven-dry weight) (Ruschmann 1953).*

Location	Carrot agar	Soil agar	Acid root agar	Mannite agar
Soil	3800	6300	44	420
Excrement	13,300	44,000	550	530

Protozoa excrete substances that stimulate bacterial development. It is well known that colonies of bacteria and fungi, after the first phase of growth and the peak of activity, decline. It is supposed that animals that eat microbes are able to intensify the activity of microflora, i.e., to accelerate energy-exchange (Macfadyen 1961), by breaking up the compactness of these colonies.

Soil-dwelling invertebrates also affect the composition of microflora. In the excrement of earthworms the numerical proportions of Actinomycetes, fungi, and bacteria were found to be different from those in the soil (Ruschmann 1953, Kozlovskaya and Zhdannikova 1961). In the excrement of tipulids, with a total content of micro-organisms higher than that in the soil, the number of fungi was several times less (Stebaev 1958).[1]

In addition, soil-dwelling invertebrates aid in the distribution of micro-organisms. Experiments with four species of fungi have shown (Hutchinson and Kamel 1956) that the rate of dispersal of micro-organisms in sterile soil was substantially higher when earthworms were introduced. Similar results were obtained in experiments with soil-dwelling Oribatidae, which, it was discovered, inoculate the soil with micro-organisms, carrying them on the body surface and in the alimentary canal. It has been proved that fungus spores remain viable after passing through the intestinal tract of Collembola (Poole 1959). Soil-dwelling invertebrates also depend on the activities of microflora.

Apparently a few invertebrate species are able to feed on slightly-decomposed forest litter. For instance, in experiments with earthworms (*Dendrobaena octaedra*) and with three species of Diplopoda (van der Drift 1951), the animals consumed litter taken from a lower layer (F_x) of litter in considerably larger amounts than litter from the less-decomposed upper layer (F_0). When they consumed litter from the upper layer the animals grew more slowly and mortality was higher than in tests with the more-decomposed part of the litter. Similar results were obtained (Gere 1956) in experiments with several species of Myriopoda and woodlice. Therefore micro-organisms prepare food for soil-dwelling saprophages, which in turn create conditions for more intensive activity by microflora.

The undigested organic matter remaining in the excrement of soil-dwelling invertebrates not only undergoes microbial decomposition but is used by invertebrates belonging to other species and groups. The process of decay of litter and transformation of it into amorphous 'humus' is accompanied, apparently, by successive changes in the composition of the organisms that feed on it (Birch and Clark 1953), similar to the changes observed during the decay of wood (Mamaev 1961). Our knowledge, however, is still insufficient to enable us to define the separate stages in the succession.

[1] It is possible that with other species and in other soils the composition of microflora would change in a different way.

ACTION OF INVERTEBRATES ON THE PHYSICO-CHEMICAL PROPERTIES OF
THE SOIL

All soil-dwelling invertebrate saprophages, while feeding on forest litter
and other plant litter, ingest at the same time a certain amount of soil
particles. Grains of sand and particles of soil may be observed in the
intestines of Collembola, black millipedes, earthworms, and other members
of soil fauna (Gilyarov 1939). Different species of saprophages ingest soil
and plant litter in different proportions. Approximately two-thirds of the
food of certain black millipedes (*Sarmatiulus kessleri* Lohm.) consists of
fallen leaves and other decomposing parts of plants (Sokolov 1957),
whereas many species of earthworms can exist merely on the humus contained in the soil. Soil-dwelling invertebrates not only break down plant
litter and mix it with the mineral part of the soil but also, as stated above,
change its chemical composition. Their excrement, as experiments with
earthworms have demonstrated, contains a greater total amount of
organic matter, of loosely-connected humates, of available forms of
nitrogen, and of mineral nutrients required by plants, than does the soil,
and it also contains a higher total of assimilated elements (A. A. Sokolov
1956, Zrazhevskii 1957, Ponomareva 1949, 1950). The composition of the
excrement depends largely on the nature of the litter or other organic
remains consumed. Thus, according to Tyurin, in experiments with
Lumbricus rubellus the additional humus in the excrement, as compared
with the soil, was 121·2% with a diet of elder litter, 101·9% with linden
litter, and 75·8% with English oak (Zrazhevskii 1957).

As they move about and leave their excrement at different depths in the
soil, invertebrates cause even distribution of organic matter in the soil
and deepen the humus horizon. Large invertebrates (earthworms, large
insect larvae, Diplopoda, etc.) take a considerable part in forming the soil
profile. Some of them are able to penetrate to a great depth. Where
invertebrates are represented mostly by small forms (Acarina, Collembola,
and other microfauna), which live mainly in litter and the surface layer
of humus, their role in transporting organic and mineral matter is negligible and is limited to a depth of a few centimetres in the topmost soil
layers. The depth to which the activity of soil-dwelling invertebrates
extends is correlated with the thickness of the humus layer (Gilyarov 1947,
1949, Fenton 1947, Shilova 1950). The animals may go down to the subsoil, later bringing to the surface horizons mineral particles containing
elements of value for plant nutrition. According to the observations of
A. A. Sokolov (1956), where carbonates lie not far from the surface
earthworms bring them to the topmost layers. In addition they enrich
the soil with secretions from their oesophageal (lime) glands (Ponomareva
1950), which help to neutralise acid soils.

The ability to change soil reaction is observed also in ants. I. A.
Krupenikov (1951) found that ants neutralise acid reaction in soils and

weaken alkaline reaction; he explains that by their excretions having a weak alkaline reaction.

By moving about in search of food, or in migrations due to unfavourable climatic conditions, etc., invertebrates increase the porosity of the soil. In this respect the activities of animals of different species vary. A number of species of wireworms use spaces already existing between soil grains, merely widening them slightly; others actively dig burrows (Gilyarov 1949). The depths to which invertebrates of different species penetrate also vary, there being some species in the mesofauna that remain mostly in forest litter, whereas others may penetrate the soil to a depth of several metres (the burrows of *Dendrobaena mariupoliensis*, according to Vysotskii's observations, reach a depth of 8 metres). The burrows of large earthworms enable tree roots to penetrate deeply, making moist subsoil horizons accessible to them (Goethe 1895, Vysotskii 1899). At the same time the animals use spaces left by decayed tree roots to travel through the soil. Small arthopods also may help to increase soil porosity by eating decayed plant rootlets (Gilyarov 1939). According to Vysotskii's estimates (1899), made in the Velikoanadol'skaya forest reserve, the space occupied by earthworm burrows alone was $1.57 \ 1/m^3$ of soil. Field experiments made by A. A. Sokolov (1956) in a poplar forest on light-grey forest soil have shown that the water-permeability of the soil is approximately doubled as a result of earthworm activity. Invertebrates improve soil aeration by digging tunnels, thus creating conditions for aerobic microbiological processes and leading to more rapid and complete decomposition of organic matter. A further effect of invertebrates in increasing the fertility of forest soils is connected with their share in changing soil structure. Mineral particles of soil, mixed in the animals' alimentary canals with finely-ground and digested plant matter, are excreted in the form of small lumps that constitute durable structural aggregates. The creation of coarse-grained soil structure is aided by various groups of soil-dwelling saprophages and phytophages: insects, Myriopoda, woodlice, earthworms, etc. (Romell 1935, Jacot 1936, Gilyarov 1939, 1957, Ponomareva 1953, Stebaev 1958b). The form and dimensions of structural elements of the soil resulting from the activities of soil fauna depend on the species of the invertebrates (Stebaev 1958b). The nature of the soil structure, as a rule, is determined by the resistance of the soil lumps to breakdown by water. Investigations show that insect excrement soaked in water retains its shape for several hours (Gilyarov 1939); earthworm casts also have great water-stability (Ponomareva 1953), as has the excrement of Diplopoda, woodlice, and other soil-dwelling invertebrates (Stebaev 1958b). The water-stability of excrement (judging by experiments made with earthworms) depends on its organic-matter content and on the mechanical composition of the ingested soil (Lindquist 1941, A. A. Sokolov 1956, Zrazhevskii 1957).

The amount of excrement discharged by black millipedes during a single growing season, in the forest belts planted in the north of the steppe zone, reaches 686 kg/ha (Sokolov 1957). According to the estimates of A. A. Sokolov (1956), made on the basis of S. I. Ponomareva's data, earthworms in an oak forest on soddy medium-podzolised soil (in Moscow province) are able to work over the entire mass of soil in the 0 to 20-cm layers in the course of six years. The part of their excrement discharged as casts on the soil surface alone amounted to 16 ton/ha there (Ponomareva 1950).

The activity of soil-dwelling invertebrates, therefore, does not consist solely in taking part in the decomposition of plant litter. By increasing the fertility of forest soils in different ways the animals ensure suitable conditions for more energetic synthesis of organic matter by plants.

INFLUENCE OF SOIL-DWELLING INVERTEBRATES ON THE NATURE AND CONDITION OF VEGETATION

The soil-forming activity of invertebrates is reflected in the condition and nature of the tree, shrub, and herbaceous cover. Relevant publications are still very few, and mostly concern the effects on plants caused by earthworms: a group to which the attention of scientists was called by Charles Darwin's observations.

The activities of soil-dwelling fauna not only lead to increased fertility of the soils in a forest (Zrazhevskii 1957), but also affect the evolution of forest stands (Sibiryakova 1949). In a survey of the distribution of spruce seedlings it was found that about half of the seedlings (and in most cases even more) were concentrated on soils that had been loosened by animals; while in broad-leaved plantations, where there were larger numbers of soil-dwelling invertebrates, the effects due to invertebrates were greater than those due to vertebrates (Table 74). On old burnt areas all the tree shoots observed were located on soil loosened by the activities of soil fauna.

Table 74. *Number of spruce seedlings concentrated on areas of soil loosened by animal activities (% of total number) (Sibiryakova 1949).*

Type of loosened soil	Composition and density of stand			
	10 birch; 40 years	7 poplar, 3 birch; 25 yr	Pine with mixture of birch; 0·4; 60 yr	8 spruce, 2 birch; 0·7
Soil loosened by invertebrates (earthworms, beetles and their larvae, ants)	53	39	44	10
Soil loosened by lemmings above their burrows and along lizard burrows	11	7	44	64

Some data exist showing the effect of soil-dwelling invertebrates on the species-composition of the herbaceous cover. According to observations by the Danish investigator P. E. Møller (1907) in beech forests, the distribution of many rhizomatous plants, which compose the greater part of the herbaceous cover, depends largely on the degree to which the soil is populated with earthworms. The casts of large earthworms (apparently *Lumbricus terrestris*) lying on the soil surface protect the young rhizomatous shoots, thus making possible the development of new young plants.

By loosening the surface layer of soil the worms enable the tender rhizomes of the plants to penetrate in all directions; and a reduction in the number of earthworms causes the rhizomatous plants to become fewer or disappear altogether.

Møller concluded that reduction in the number of earthworms would be enough to cause other plants to overrun the area and to change the principal features of the biological character of the vegetation.

Field observations and laboratory experiments show that the activities of soil-dwelling invertebrates also affect the condition and growth of plants. That refers mostly to cultivated annual plants, but some data exist concerning tree species. For instance, in 10-year-old spruce plantations on former arable land where in wet weather there were on an average 68 earthworms per m^2 the average height of the trees was 3·5 m; where the number of earthworms was 46 per m^2, it was only 2·5 m (in the same type of plantation). Respective survival rates were 85% and 65% (Sibiryakova 1949). The results of experiments with pine seedlings (Zrazhevskii 1957) show that the beneficial effects of earthworms' activity appeared as soon as the end of the first vegetative season (Fig. 48). Seedlings in containers with worms were distinguished from the controls not only by their growth but also by deeper colour and greater length and thickness of needles. Similar results were obtained by Zrazhevskii in experiments with other tree species. It therefore appeared that earthworms eliminated the unfavourable effect on seedlings due to fresh forest litter placed on the soil surface of the containers as fertilisers. In most tests with worms the weight

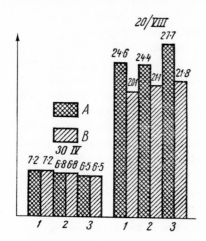

Fig. 48. Growth of pine seedlings (in cm) in relation to earthworm activity (from Zrazhevskii, 1957). 1—sand without nutrients (control); 2—sand with peat; (3) sand with peat and shale; A—with worms; B—without worms.

of seedlings in containers with litter was no lower than in the controls without litter (Table 75). As Zrazhevskii remarks, his results agree with those of M. P. Arkhangelskii (1929), in which the oat and barley yields in controls without fertiliser were higher than those in tests with organic fertiliser (straw, lupines, etc.). At the same time, in each test using fertiliser the yield in containers with worms was considerably higher than that in containers without worms.

Table 75. *Effect of earthworms on oak and ash seedlings in relation to composition of forest litter added to test containers (Zrazhevskii 1957).*

Composition of litter (leaves)	Total weight, g (average per plant)		Difference, %
	With worms	Without worms	
English oak			
Oak	18·1	19·8	− 8·5
Ash	24·7	20·4	21·1
Maple	25·1	19·9	26·1
Oak + maple	24·7	21·1	17·1
Oak + ash	25·2	19·4	29·9
Oak + maple + ash	21·6	23·1	− 6·5
Without litter	22·5	23·9	− 5·9
Green ash			
Ash	23·6	20·6	14·6
Oak	16·0	19·0	− 15·7
Maple	23·4	17·3	37·0
Ash + oak	21·6	21·6	0·0
Ash + maple	22·3	20·2	10·4
Ash + oak + maple	22·2	20·5	8·3
Without litter	21·8	22·3	− 2·2

According to Schönbeck and Brüsewitz (1957), earthworm activity leads to more rapid breakdown of substances entering the soil from grain-crop straw, which impede germination and development of seeds. It appears that something of that kind also took place in the above experiments with the application of organic fertiliser or fresh forest litter.

Recent investigations have shown that soil-dwelling invertebrates are also able to affect plants directly by secreting and discharging into the soil biologically-active substances. It has been discovered that earthworms produce biostimulators of group 'B', which enter the soil with the secretions of the glands in the musculo-cutaneous sac and the metanephridia and with their casts (Gavrilov 1950, Brüsewitz 1959). The increase in seed germination[1] observed in experiments with the application of a certain amount of earthworm casts to the soil is attributed to the ability of earthworms to excrete biologically-active substances (Zrazhevskii 1957). The accelerated germination of cotton seeds, which have previously been acted upon by some groups of Protozoa, is also attributed to the effect of stimulating substances. That apparently also explains why, when culti-

[1] The seeds were immersed for 1 hour in a liquid culture of Protozoa.

vated plants are grown on earthworm casts, the vegetative period in the test containers ends about two weeks earlier than in the controls, where the soil used was taken from the same spot where the casts were collected (Ponomareva 1958).

Thus soil-dwelling invertebrates affect the growth and development of plants not merely by improving the properties of the soil. The mechanism of their action is more complex, but has not yet been well studied.

The above data show how varied and extensive is the influence of soil-dwelling invertebrates on other components of forest biogeocoenoses. The extent and the nature of that influence differ in different forest types, since the complexes of soil fauna dwelling in them are different.

We are still far from being able to evaluate fully and to express in comparable terms the work done by soil-dwelling invertebrates in forest biogeocoenoses. At least it is clear that without taking proper account of their activities we cannot correctly interpret the nature of the energy-processes in the forest. Study of the problem is not merely of theoretical interest; unquestionably it is of value for practical forest management.

It is known that soil liming and improvement of swampy forest areas create conditions for more intensive activity of soil fauna. The process of building up the soil fauna in such cases does not always advance quickly enough, and therefore it is recommended that the soil should be artificially colonised with useful invertebrates (Wittich 1952). The same apparently should be done when tree stands are created artificially, especially in unforested districts; in selecting the tree species for planting due consideration should be given to their effects on soil fauna (Zrazh-evskii 1957). Such measures should be based on thorough study of the distribution and activity of invertebrates in forest biogeocoenoses. Knowledge of the complexes of invertebrates that live in forest soils of different types can be used in forest typology (Thiele 1956, Vaněk 1959) as an additional criterion for solving controversial problems.

VERTEBRATES IN FOREST BIOGEOCOENOSES

The significance of living organisms in the life of biogeocoenoses is decided by the part they take in the conversion of matter and energy, and depends on the following five processes:

1. the part taken by the organisms in the accumulation of solar energy;
2. the effects of the organisms on the gaseous composition of the air;
3. the effects of the organisms on the movement of water;
4. the effects of the organisms on the transformation and movement of mineral and organic matter in the soil and litter;
5. the creation, transformation, and transportation of living matter by the organisms.

The role of vertebrates in the accumulation of solar energy and formation of the gaseous composition of the air is very small. Therefore the significance of this group in the life of biogeocoenoses depends on its participation in the other three processes. At the present level of our knowledge that can be described only in the most general terms.

VERTEBRATES AS A COMPONENT OF A FOREST BIOGEOCOENOSE

The effects of vertebrates on the transformation of organic matter and energy are closely linked with the inter-relationships existing between them and the other components of a biogeocoenose. At the foundation of these inter-relationships lie the food requirements of the animals. In satisfying these, vertebrates form so-called food chains (Elton 1934). Every food chain contains: a food plant; phytophagous animal species that eat it; predators and parasites that feed on the phytophages; scavengers that eat carcases and excrement.

As a rule food chains advance from smaller to larger forms. In most cases the vegetarian animals near the bottom of a food chain are relatively small in size. Predators are usually considerably larger than their prey. But since the sizes of animals cannot go on increasing indefinitely, the number of links in a chain is limited. Usually there are no more than six. At the same time the number of individuals in each link of a chain is much higher than the number in the following link. Only in that way can species survive mortality from predation and other causes. The consequence is that in each successive link the animals are larger but their numbers are much smaller. This law, which is manifested in every biogeocoenose, has been given the name of 'the pyramid of numbers' (Elton 1934).[1] It follows that the biomass of each successive link in a food chain is smaller than that of the preceding link. Consequently the total mass of plant organisms in a biogeocoenose is always greater than the mass of phytophagous forms, and the latter exceeds the mass of insectivores and carnivores.

Frequently different food chains have common links or links that may take each other's place. Such chains form food cycles. An example of a food cycle is the eating of stone pine seeds by taiga vertebrates (Fig. 49).

As a rule there is a considerable number of food chains in every biogeocoenose, as a result of which the trophic links built up in it present a very complex picture (Fig. 50). Not all food chains are of equal importance in the life of the biogeocoenose. Many of them are composed of species represented by small numbers of individuals, and play no significant role in the general process of matter and energy transformation. Only a comparatively small number of species of birds and mammals, distinguished by high population numbers, form the core of the animal population and determine the share taken by vertebrates in the entire balance of the biogeocoenose.

In forest biogeocoenoses the vertebrate population has distinctive

---

[1] The law called 'the pyramid of numbers' is not applicable to links that contain parasites, which with each successive link become smaller but more numerous (Elton 1934).

features. Whereas on the tundras and steppes it is all concentrated on the surface of the earth and in the upper soil horizons, in a forest the complex synusial plant cover enables vertebrates to live at several levels. For that reason the main body of the vertebrate population in a forest biogeo-coenose attains the greatest diversity, and its trophic relationships the greatest complexity. At the same time the good protective conditions cause the number of forest-dwelling forms that breed underground or lead an underground mode of life to be relatively few (Kashkarov 1945).

The distinctive peculiarities of forest animal life are displayed in their use of plant food (Voronov 1959, Dinesman 1961). The food value of forest herbs, especially in northern districts, is very low. Moreover, it falls sharply at the close of the growing season because of leaching-out of mineral elements by autumn rains. The deep snow cover typical of the forest zone makes the herbaceous layer unavailable to most vertebrates during a considerable part of the year. On that account herbaceous plants are much less heavily eaten by mammals and birds in forest biogeocoenoses than on the steppes. The synthate built up in summer by the leaves of trees and shrubs passes in autumn into branches, trunks, roots, and bark, so raising their food value. The vegetative parts of trees and shrubs become accessible to animals during periods of deep snow. They provide many mammals and birds with valuable food, consumption of which enables them to survive winter shortages and to live in the forest zone. For that reason the group of consumers of vegetative parts of trees and shrubs is prominent in forest zoocoenoses. Among vertebrates in our country its chief representatives are moose, reindeer, roe deer, beaver, hare, common vole, large-toothed red-backed vole, northern pika, and birds of the grouse family. The vegetative parts of trees and shrubs are eaten to a lesser degree by many other animals, bears, wild boars, squirrels, birch mice, chaffinches, buntings, etc.

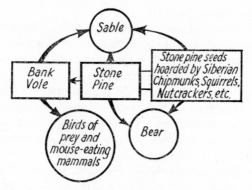

Fig. 49. Use of stone pine seeds by taiga vertebrates (from Folitarek 1947). Example of a food chain. The arrows denote consumers of the food.

The seeds and fruits of forest trees and shrubs form a valuable concentrated food. Because of the abundance of these, vertebrates that feed largely on seeds are widely distributed in forest biogeocoenoses. Seed-eaters form a conspicuous group in forest biogeocoenoses in the U.S.S.R., their chief representatives being squirrels, Siberian chipmunks, birch mice, dormice, bank voles, northern red-backed voles, nutcrackers, and cross-bills. Seeds play a large part in the diet of other vertebrates, including eaters of the vegetative parts of plants (reindeer, roe deer), predators (badger, bear, sable), and insectivores (shrews, jays, chaffinches, buntings, tits, and other small birds).

Finally, the abundance of insect food is an outstanding feature of forest zoocoenoses.

The great mobility of vertebrates enables them to cover a wide territory easily and profitably. Therefore the movement of animals from one forest type to another in the course of the year is a common occurrence, and the vertebrate populations of separate forest types or even groups of types are distinguished not by their species-composition but by their numbers of individuals. For instance, during the breeding season small insectivorous birds are most numerous in types of forest with abundant undergrowth, where they find the best conditions for nesting and feeding their young (Fig. 51).

The numbers of mouse-like rodents are highest in forest types with the best food conditions, seasonal fluctuations in the numbers of these animals in different forest types being marked by sharp upturns due to the ripening of certain foods (Simkin 1961, Shmal'gauzen 1961). There are definite seasonal migrations of ungulates from one forest type to another (Fig. 52).

Seasonal migrations of vertebrates from one forest type to another are due not only to the peculiarities of distribution of food and suitable nesting sites, but also to many other factors: differences in light, in depth of snow cover, in shelter conditions, in the abundance of predators and blood-sucking parasites, and in competition with species having similar habitat and other requirements.

Besides local movements from one biotope to another, some vertebrates undertake migrations over long distances. Well-known examples of such migrations are those of birds in spring and autumn and the migrations of squirrels, ungulates, and other mammals.

The part taken by vertebrates in the transformation of organic matter and energy is closely connected with seasonal changes and long-term fluctuations in their numbers. Vertebrate population numbers may multiply tenfold or many times that amount in long-term fluctuations, which are found in every biogeocoenose. The reasons for these fluctuations have not been completely clarified but, beyond doubt, for seed-eaters in forest biogeocoenoses they are primarily determined by the periodicity of the seed production of trees and shrubs. With good seed crops of oak, beech, stone pine, spruce, and other food plants, favourable

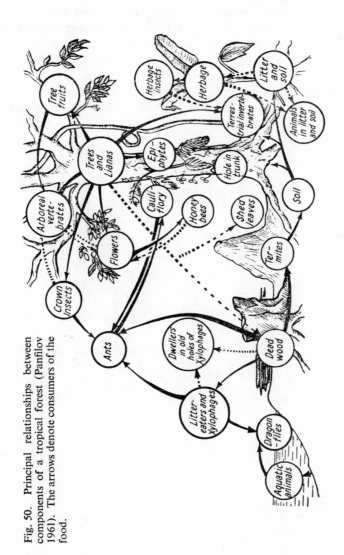

Fig. 50. Principal relationships between components of a tropical forest (Panfilov 1961). The arrows denote consumers of the food.

conditions are created for multiplication of the animals that feed on them, whose numbers rise quickly. In the next year the stocks of seeds are soon exhausted. The animals that have multiplied so greatly begin to suffer from food shortage and take to eating substitute foods of little value. Their numbers then fall, and some migrate to other districts. A coniferous crop failure, for example, induces squirrel migrations and flights of Siberian nutcrackers to Europe (Formozov 1933); and a large number of wax-wings appear in the central belt in years when mountain ash in the north fails to produce a berry crop (Siivonen 1941). The close connection between the numbers of seed-eaters and seed crops, it would appear, is seriously disrupted only in districts with very unfavourable climatic con-ditions and poor seed production by trees. Such disruptions have been

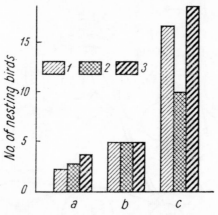

Fig. 51. Numbers of birds in forests of different types (Dubinin and Toropanova 1960) 1—in pine forests; 2—in spruce forests; 3—in birch forests; a—groups of forest types without undergrowth; b—groups of forests types with undergrowth; c—mixed forest types.

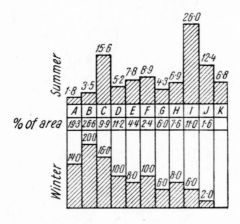

Fig. 52. Distribution of moose in Karelia (% of observations) (Nikiforov and Gibet 1959). A—lichen pine forests; B—berry-green-moss pine forests; C—sphagnum pine forests; D—pine-spruce forests; E—green-moss spruce forests; F—spruce planta-tions; G—broad-leaved forests; H—clearings and burned areas; I—sphagnum bogs; J—sedge bogs; K—lake margins.

observed among mouse-like rodents in the northern taiga of the European part of the u.s.s.r. (Koshkina 1957) and among squirrels in Yakutia (Egorov 1961).

In forest biogeocoenoses weather generally has a much stronger effect on long-term fluctuations in the numbers of animals that eat the vegetative parts of plants. The numbers of birds of the grouse family fluctuate in accordance with temperature conditions in spring. In cold springs their nestlings perish and the increase in their population is negligible (Semenov-Tyan-Shanskii 1960). The chief reason for long-term fluctuations in the number of blue hares in the European part of the u.s.s.r. is the death of large numbers of these animals from epizootics, which usually develop in cold and rainy years (Naumov 1947). In taiga districts the numbers of voles rise in warm dry years with late autumns, when because of favourable conditions the breeding season of these animals is prolonged (Bashenina 1951).

The effects of predation on the population dynamics of eaters of seeds and vegetative foods have not been proved, and different investigators have arrived at different conclusions. There is no doubt that they are very often governed by meteorological factors and depend on the state of vegetation, the depth of snow cover, thawing conditions, etc. Variations in the numbers of the predators in a forest, as in biogeocoenoses of other types, are caused mainly by variations in the numbers of their prey animals.

Vertebrate populations undergo massive fluctuations as a result of successional changes in forest vegetation. The most important of these changes are produced by tree-felling and fires. When a clearing replaces a mature forest the composition of the animal population changes abruptly. Many typical forest-dwellers, whose lives are linked with tall trees and tree seeds, go away and are replaced by open-country species. Later, as shrubs and tree seedlings grow on the cut or burned areas, favourable conditions are created for consumers of berries and twigs. The large amount of cut and burned areas favours an increase in the numbers and distribution of voles, hares, deer, roe deer, and several other mammals. As the seedlings pass into the pole stage the nutritive quality of the herbaceous cover deteriorates because of overshading, and a considerable number of the young tree shoots become inaccessible to the majority of twig-eaters, whose numbers then fall.

For the animal population the pole stage is the poorest stage in the development of a forest. Later, as the tree stand continues to grow, the vertebrate population of a virgin forest is gradually restored. It acquires the typical appearance of a forest fauna only when the cut area reaches the stage of a sub-mature stand (Danilov 1934, Larin 1955, Kerzina 1956, Leble 1959, Cringan 1958).

Geographical variations of the above process have been found in forests of the temperate zone (Raimers 1958), but they are not extensive.

Seasonal and long-term fluctuations in numbers, migrations and local movements of vertebrates make determination of the part taken by this group in the transformation of matter and energy a very complex problem. They certainly affect the dynamics and the distributional features of the whole process. We cannot, however, evaluate their significance in the energy balance of biogeocoenoses, even approximately. The available material reveals only in a general way the role of mammals and birds in the transformation of organic matter and energy in those conditions where the activities of vertebrates are most clearly displayed.

EFFECT OF VERTEBRATES ON WATER MOVEMENT AND ON THE
TRANSPORTATION AND TRANSFORMATION OF ORGANIC AND
MINERAL SUBSTANCES IN THE SOIL

The effect of vertebrates on water movement and on the transportation and transformation of organic and mineral substances in the soil and in forest litter has not yet been well studied. In fact we possess only isolated items of factual information, but these show the great importance of this aspect of the life-activity of animals.

An excellent example of the effect of vertebrates on water movement is provided by the work of the river beaver. That animal inhabits small

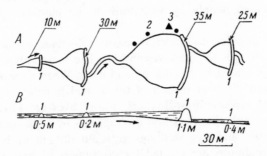

Fig. 53. Beaver dams (from Barabash-Nikiforov 1957).
A—plan; B—profile; 1—dams; 2—burrows; 3—lodge.
Upper row of figures—length of dams; lower row—height of dams.

forest streams, constructing a system of dams on them. The dams raise the water level and increase the area suitable for growth of the aquatic plants eaten by beavers (Fig. 53). At the same time the resultant flooding causes the death of some trees. Besides building dams, beavers dig a large number of canals, along which they float the branches they eat to their lodges. The canals drain the adjacent territory, which also affects the condition of the trees. The dams and other structures built by beavers control water flow, the process being on such a scale that, according to abundant data, rivers inhabited by beavers may be considered to be completely regulated (Horton 1948, Hinze 1950).

Because of the scarcity of beavers in our country, their constructive work is not of great significance now. Apparently in the past it was a substantial factor in the water balance of forest types adjoining streams.

Burrowing by vertebrates has a considerable effect on water movement in the soil. By constructing shelters and digging in search of food they break up forest litter and plant cover and dig out complex tunnel systems. Large mammals such as badgers and foxes make burrows going down as far as 2 to 2·5 metres. The area covered by their diggings, however, is not large. In broad-leaved forests there are only a few burrows of these

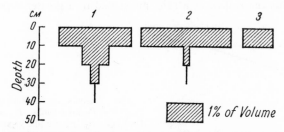

Fig. 54. Volume of burrows of small mammals, as % of soil volume in coniferous-broad-leaved forest (from Popova 1962). 1—in oak stand; 2—in pine stand; 3—in forest glade.

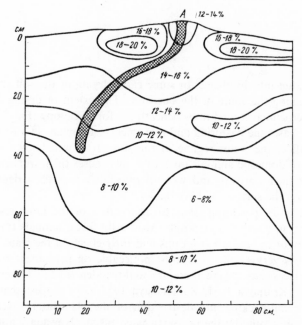

Fig. 55. Effect of burrows of small rodents on distribution of moisture in soil beneath forest canopy (from Popova 1962). Soil moisture as percentages. A—bank vole burrow.

animals in every 1000 hectares (Likhachev 1959), so they cannot have much effect on moisture distribution. The greatest role is played by the burrows of small insectivores and rodents, whose numbers may reach several hundreds per hectare.

According to observations in the coniferous-broad-leaved forests in Moscow province (Popova 1962), in a year of high numbers of bank voles the burrows of these animals covered an average of 10 to 15% of the area of pine and oak stands. The uppermost 10-cm layer was most thoroughly riddled with burrows. At a depth of 40-50 cm the burrowing activity of the animals practically ceased (Fig. 54). Underneath the burrows of small rodents the soil is, as a rule, moistened to a greater depth than in

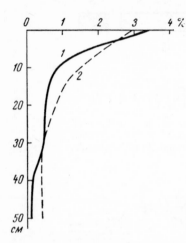

Fig. 56. Distribution of humus in forest soil (from Popova 1962). 1—in places not disturbed by mouse-like rodents; 2—in places burrowed into by bank voles.

areas not dug into by them. In some cases drying of the soil surface is observed along the route of the burrows. The extent of drying depends on the microclimatic conditions beneath the forest canopy (Fig. 55).

In some cases the pasturing and movements of animals in a forest may strongly affect the water regime of the soil. It is well known that the pasturing of domestic animals on heavy soils causes them to pack down, destroying their porosity and water-permeability to such a degree that water lies on the surface (Fal'kovskii 1929).

In loose soils, pasturing leads to destruction of the moss and plant litter cover, so causing drying-out. Most wild forest animals travel and graze singly or in very small groups and their effect on the water regime of the soil is very small. It is most noticeable on permanent routes to watering-places, etc., where there are well-trodden tracks. Observations in reserves (Krainova 1951, Zablotskaya 1957 a, b) suggest that in the past severe soil-packing and destruction of the plant-litter cover could have been done only by the now-extirpated bison. Nowadays noticeable trampling of the soil in U.S.S.R. forests results only from the pasturing of domestic animals.

The deeper moistening of the soil in places where animals dig burrows inevitably results in the migration of various water-soluble substances within the soil profile. Unfortunately there are practically no relevant data relating to forest biogeocoenoses, but it is known that in places where rodents burrow the humus extends into deeper soil horizons (Fig. 56).

Apparently the digging activities of shrews have considerable effect on all processes that take place in the soil. A large proportion of the tunnels of these animals are made in forest litter. In their search for food shrews mix forest litter with the upper soil layer, improving aeration and helping moisture to penetrate.

PARTICIPATION BY VERTEBRATES IN THE TRANSFORMATION AND TRANSPORTATION OF LIVING MATTER

Creation of living matter is performed by autotrophic organisms, and the activities of vertebrates are limited to transformation of it. The part taken by vertebrates in the transformation of living matter may be direct or indirect. Their direct action consists in eating food and converting it into more complex forms of living matter (into the biomass of the organisms) and energy, and in returning part of the organic matter to the soil in the form of carcasses and excrement; their indirect action consists in the effect of these processes on the environment.

Direct transformation of living matter by vertebrates takes place in several stages, each of which corresponds to definite links in food chains (Dinesman and Shmal'gauzen 1961). The first stage consists of conversion of plant matter by animals; the second, conversion of animal matter by predators, scavengers, and parasites; the third, conversion of excrement by coprophages; and the fourth, transformations of matter and energy taking place with the decomposition of carcasses and excrement.

In nature the above cycles are so intimately bound together that they often cannot be separated. Moreover, they are complicated by animal movements, as a result of which a cycle commenced in one biogeocoenose may end in a biogeocoenose of another class or even type.

Therefore quantitative evaluation of the role of separate cycles in the energy balance of biogeocoenoses is an exceptionally complex task. At present it can be done only approximately.

Starting with the 'pyramid of numbers', we find that in the processes of transformation of matter and energy the leading place among vertebrates is taken by herbivores. Consequently the role of animals in matter circulation is primarily determined by the use of vegetation by herbivores, and the stages of coprophagy and decomposition of excrement connected with it. These stages are the principal zoogenic stages in the transformation of matter and energy. The correctness of this view is confirmed by the data in Table 76, in which we present the biomass of different groups of vertebrates in a broad-leaved forest and of the food eaten by them. As the table shows, in a forest-steppe oak stand the dry weight of vertebrates is only 2·5 kg/ha, about 90% of that amount being attributed to herbivorous animals, ungulates and rodents. The biomass of small birds, which feed mainly on insects, is much less—some hundreds of grammes per hectare. The biomass of predatory birds and mammals is very small and is expressed in tens of grammes per hectare.

Table 76. *Biomass of the chief vertebrate groups in the Tellerman forest and of the food eaten by them (dry weight).**

Group of animals	Biomass (kg/ha)	
	Animals	Food eaten during the year
Ungulates	1·3	78·7
Mice and voles	0·9	247·0
Squirrels and hares	<0·1	1·7
Total	2·2	327·4
Small predatory mammals	<0·1	0·1
Other predatory mammals	0·1	0·3
Chief groups of mammals	2·3	327·8
Small birds (passerine, woodpeckers, etc.)	0·2	16·7
Predatory birds	<0·1	0·4
Total	2·5	344·9

* In compiling the table we have used data on vertebrate numbers in the Tellerman forest placed at our disposal by G. E. Korol'kova, and also data from various published sources on the weight of individuals of different species, the amount of food eaten by them, and the water content of the live organisms. The figures thus obtained, naturally, cannot claim great accuracy and only represent orders of magnitude.

The weight of food eaten during the year by all predators does not exceed 1 kg/ha, which represents about 40% of the biomass of herbivorous mammals and small birds. Actually the animal biomass consumed by them scarcely reaches that figure, since many predatory mammals include vegetable food in their diet.

It is noteworthy that the weight of food consumed during the year by herbivorous mammals is many times greater than the biomass of the animals themselves. It amounts to several kg/ha. That figure, however, is insignificant in comparison with the total biomass of vegetation and its annual increment and litter fall, which in forests are calculated in tons per hectare. Therefore vertebrates consume a very small proportion of the matter brought into circulation by forest vegetation.

This conclusion is supported by direct observations, admittedly made in other types of forest. According to O. I. Semenov-Tyan-Shanskii's (1960) observations, in the Lapland reserve gallinaceous birds (a prominent component of the zoocoenoses of taiga forests), with a population density

Table 77. *Consumption of biomass of food plants by gallinaceous birds in the forests of Kola peninsula (from data of Semenov-Tyan-Shanskii 1960).*

Item	Green parts of herbs and shrubs	Berries	Pine needles
Total stocks, kg/ha	2570	200	3000
Annual consumption by the whole population of gallinaceous birds, kg/ha	3	2·5	4·2
% of total stocks	0·1	1·2	0·1

of from 8 to 26 per km², consume annually not more than 1% of the biomass of the herbage-shrub layer, needles, and berries (Table 77).

In the coniferous-broad-leaved forests of the European part of the U.S.S.R. one of the chief consumers of vegetable food is the moose (European elk). Investigations made in a pine forest near Moscow have shown (Dinesman and Shmal'gauzen 1961) that these animals, which feed mostly on pine and mountain ash, consume during the winter only 8% of the biomass of the undergrowth of their food species in their wintering grounds. Of the 8% consumed by them, 5% reaches the soil surface in the form of excrement and only 3% is used in the animals' expenditure of energy. With a population density of three moose per km², the weight of shoots eaten by them amounts to some tens of kg/ha, whereas the annual increment and litter fall of these tree species amounts to several tons per hectare.

As the above example shows, a considerable proportion of the matter contained in food eaten by herbivores is not assimilated by the organism and returns to the soil surface in an altered form. A characteristic property of such matter after its alteration by vertebrates is that it is rich in biogenic organic and mineral compounds. The formation of these compounds and their concentration constitute one of the features of the final zoogenic stages of transformation of matter and energy. They probably have a substantial effect on the magnitude and rate of entrance of matter into the soil.

The significance of the concentration of biogenic compounds by vertebrates calls for investigation. But it is known that substances contained in fresh dung (Nikitin 1961) cause, at first, scalding and death of plants. In places where excrement is deposited bare spots are formed, which grow over again only after two or three years. Grasses, brambles, elders, and meadow and weed herbs grow there, distinguished from the surrounding vegetation by their more flourishing condition. Only in a few cases, where dung is regularly deposited on a certain spot, does the change in plant cover assume a permanent character. Such changes are produced by bird colonies, such as those of the common heron in forest districts. That bird settles in small colonies, building up to eight nests in a single tree. Heronries are distinguished by high humus content of the soil and vigorous growth of nitrophilous plants: Solomon's seal, nettles, elders, etc. (Tkachenko 1955, Brinkmann 1956, Novikov 1959). Such permanent effects of droppings, however, are unimportant in the life of the forest, as they appear only on very restricted areas.

Therefore only an insignificant part of the matter brought into circulation by vegetation is involved in the chief zoogenic cycles of matter and energy transformation. A still smaller amount of matter enters the other zoogenic cycles. Consequently direct treatment of living matter by vertebrates plays a very small part in the exchange and transformation of matter in forest biogeocoenoses.

The indirect effects of vertebrates on the transformation of living matter are mainly determined by the changes in the plant cover, and in the transformation of matter and energy by separate synusiae, caused by animals. Such changes often have a very strong effect on the life of forest biogeocoenoses.

One of the widespread forms of indirect influence by vertebrates on the transformation of living matter is seen in the results of their feeding on the vegetative parts of trees and shrubs.

As trees and shrubs grow older the food value of their vegetative parts diminishes. In old trees and shrubs they are inaccessible to the majority of animals and are eaten mainly by birds. As a rule, mammals procure twigs for food from young plants.

The eating of vegetative parts of trees and shrubs is injurious to the plants. The most common type of injury is the eating of bark and shoots.

Various species of mouse-like rodents eat the bark from the root collar or near it, up as far as the snow surface. Hares and ungulates gnaw round the parts of trunks and branches not covered with snow. Bark is accessible to ungulates only at the beginning and end of winter since they do not touch frozen bark. Continuous eating of bark round the whole trunk leads to the death of plants and withering of their tops. Damage to bark involving only a part of the circumference is less dangerous, but still noticeably affects the growth of the plant (Fig. 57). The eating of shoots, especially terminal shoots, is a more serious injury, and as a rule it lessens the height of a sapling and retards its growth for several years (Fig. 58). Moreover, the eating of shoots makes the stem of a sapling crooked, causes a second top to grow from one of the side shoots or dormant buds, and leads to the branching of shoots and distortion of the normal crown shape. According to the observations of Yu. A. Isakov, geese eat off all the terminal shoots of the young willows on the sandbanks of the Volga delta. Therefore the adult trees of that species there are generally forked at the lower part of the trunk. Regular stripping of pine needles and buds by birds of the grouse family disrupts normal tree growth and crown development. Sometimes, in spite of the small numbers of these birds, such damage is observed over a considerable area (Nef 1959 *et al.*). Similar results are produced by heavy feeding of ptarmigan on willow twigs; and constant feeding by reindeer on young birch shoots causes formation of peculiarly-shaped specimens of that species, called 'khodyli' (Tikhomirov 1959).

Damage to seedlings of trees and shrubs by mice, voles, and birds must be looked upon as a special case of damage to terminal shoots. The

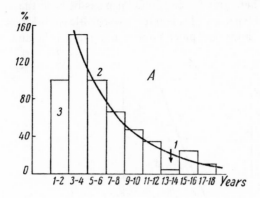

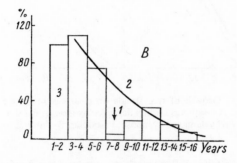

Fig. 57. Effect of moose damage to bark on the rate of growth of trees. A—pine; B—poplar; 1—time of injury to bark; 2—average rate of growth; 3—rate of growth for 2-year period.

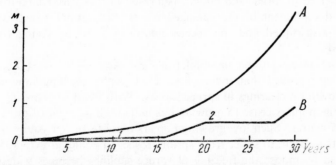

Fig. 58. Effect of eating of terminal shoots by blue hares on the growth of Daurian larch. A—curve of growth of uninjured saplings. B—curve of growth of saplings when 6-year-old (1) and 8-year-old (2) shoots were eaten.

animals eat them down to the ground (Petrov 1954), and birds pull off the tips or pull the seedlings right up. Nutcrackers can damage up to 84% of stone pine seedlings (Bibikov 1948).

The species-composition of the trees and shrubs whose vegetative parts are damaged by vertebrates is most varied in broad-leaved forests (Dinesman 1961). There animals eat different parts of both the main trees and the associated species. In coniferous-broad-leaved forests the broad-leaved species and also pine, poplar, mountain ash, and willow suffer

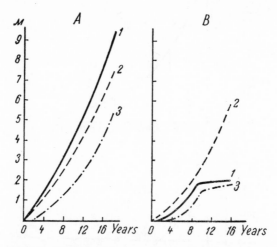

Fig. 59. Growth of tree stands on clearings in spruce forests. A—saplings slightly damaged by moose; B—saplings heavily damaged by moose; 1—poplar; 2—birch; 3—spruce.

from damage caused by vertebrates. In other taiga forests vertebrates show preference for the bark and shoots of willow, mountain ash, and poplar. They eat pine, fir, and larch less readily, although these species are preferred to spruce and birch. As a result of that selective retardation of the development of the damaged plants, the normal growth of the stands is disrupted and the species-composition of the young trees is changed.

Special investigations in coniferous-broad-leaved forests (Dinesman 1961) have proved that moose first select poplar seedlings among the new growth on clearings in spruce forests. With frequent browsing upon them, the poplars cease to grow and cannot reach a height of more than 2 to 2·5 metres. Such low-growing poplars cease to shelter young spruce, which then begin to suffer from spring frosts. In young growth seriously damaged by moose the freezing of spruce saplings becomes widespread and the growth rate falls sharply (Fig. 59), as a result of which their development falls markedly behind that of their congeners in less-damaged stands.

Unlike spruce, birch seedlings are rarely damaged by moose and are not harmed by frosts. In young stands suffering from moose browsing the growth rate of birch is not affected and it remains for a long time in the upper storey, from which poplar disappears. The practical importance of that situation depends on the proportion of birch in the young stands. Similar results follow browsing by moose on poplars growing in clearings in oak forests. Where birch and pine are sprouting in clearings in pine forests, moose do most damage to the pine. That causes severe suppression of the growth of the pine, and its entry into the upper storey of the forest is retarded, whereas birch holds the dominant position (by reason of greater height) for a long time.

In the opinion of many investigators (Vrublevskii 1912, Kaplanov 1948, Bannikov and Lebedeva 1956, Sablina 1959) selective browsing on tree species by red deer in the Belovezh bush territory causes replacement of broad-leaved species by spruce, and in the Far East it causes the decline of Amur oak and aralia. In some cases the activities of deer may completely stop forest regeneration.

In some broad-leaved stands in the Bashkir forest-steppe, where damage by brown hares is heavy, oak seedlings are frequently gnawed by these animals and become mere stumps, and Norway maple cannot reach the upper storey and remains in the undergrowth (Fedorako 1940). In larch-forest clearings in central Yakutia the activities of blue hares seriously retard the growth of larch and enable birch to remain in the upper storey of young stands (Dinesman 1959b).

Damage to vegetative parts by mouse-like rodents can kill up to 70% of seedlings and young growth of trees and shrubs. In summer voles, which are not very fastidious about the species-composition of their green fodder, damage seedlings of trees and shrub species in proportion to their relative abundance. Therefore the summer activities of these animals have little effect on the species-composition of tree stands. In their choice of ligneous food voles are more particular. Therefore in winter some species suffer from voles more than others, and vole damage may markedly alter the species-composition of regenerating forest. For instance, in the broad-leaved forests of the Tula forest reserve (Sviridenko 1940 a, b) damage to seedlings by common voles caused a heavy decrease in the proportion of Norway maple and linden, while there was an increase in the proportion of ash, and complete elimination of elm.

Damage to fruits and seeds of trees and shrubs by birds and mammals does not need detailed description. In most cases the fruits and seeds are eaten as a whole or so seriously damaged that they lose viability. Only the fruits and seeds of a few species, in which the embryo and endosperm are protected by a durable coating, pass through the alimentary tract of vertebrates without losing the power of germination.

Most important in the life of forest vertebrates are stone pine seeds, oak acorns, beechmast, filberts, and the fruits of pear, apple, and mountain ash. They are consumed by carnivores and ungulates, and to some extent by insectivorous mammals and many birds. The seeds of spruce, pine, fir, larch, birch, alder, linden, elms, and maples are valuable foods for birds, rodents, and shrews. The seeds of other trees play a small part in the life of forest zoocoenoses. Forest vertebrates (especially birds) eat large amounts of berries and fruits of shrubs, honeysuckle, spindle, buckthorn, etc.

Destruction by vertebrates of the seed crops of the chief forest-forming species has been studied in stone pine, pine, spruce, larch, oak, and beech forests, and fairly similar results have been obtained everywhere (Table 78). In coniferous forests from 30 to 90% of the cones were thrown to the ground by birds (nutcrackers, crossbills, woodpeckers). The fate of the seeds in fallen spruce cones differs in different districts. In the Kola peninsula they are scattered on the ground during the next spring, whereas in the central zone of the European part of the U.S.S.R. the seeds remain in the cones, and are used as food by animals over a long period (Danilov 1937, Novikov 1956, Vorontsov 1956).

Birds consume up to 75% of the seed crop, and squirrels not more than 38% (Table 78). In both coniferous and broad-leaved forests the majority of the seeds that fall to the ground are eaten by rodents, and only a very small part of the crop germinates and produces seedlings. In years of

Table 78. *Destruction by birds and rodents of seeds of the chief forest-forming trees.*

Forest formation	Knocked down by birds and squirrels %	Destroyed			Author
		% of total crop		By small rodents, % of amount of seeds falling to ground	
		by birds	by squirrels		
Stone pine forests	Up to 90	22–75	Up to 17	85	Bibikov 1948 Reimers 1956, 1958
Pine forests	38		24	2*	Danilov 1944
Larch forests		10–40†		18–40	Leshchinskii and Reimers 1959
Spruce forests	Up to 98	Up to 42	Up to 38	Up to 83	Danilov 1937, 1944; Molchanov 1938; Kruglikov 1939
Oak forests				Up to 100	Yunash 1940; Obraztsov and Shtil'mark 1957; Novikov 1959; Ashby 1959
Beech forests				Up to 100	Zharkov 1938

* % of total crop.

† As the authors state, probably part of the seed destruction attributed to birds was the work of Siberian chipmunks.

heavy seed production that fraction is quite sufficient for normal forest regeneration. In poor crop years small rodents consume all the seeds that fall to the ground and there is no forest regeneration (Yaroshenko 1929, Taylor and Gorsuch 1932, Pivovarova 1956, Nesterov and Nikso-Nikkochio 1951, Obraztsov and Shtil'mark 1957).

When the chief rodent-food species form only a small part of the stand, the fallen seeds are entirely eaten up by the animals, regardless of the size of the crop (Snigirevskaya 1954, Pivovarova 1956).

The preference given by small rodents to the seeds of certain species causes replacement of oak by ash, of beech by hornbeam, and of seed-grown by sucker-grown linden stands (Ivanova 1950, Sviridenko 1951, Obraztsov and Shtil'mark 1957).

As the observations of L. V. Zablotskaya (1957a) have shown, shrews may have a marked effect on the fate of the seed crops of several tree species.

Judging by the results of investigations made in the Belovezh bush (Lebedeva 1956), a considerable proportion of fallen acorns may be eaten by wild boars. In years of high acorn production, however, the percentage consumption of acorns falls.

It is very difficult to estimate the amount of fruits and seeds eaten by deer, roe deer, and carnivorous mammals, but the effect on the seed crop of the chief tree species is small, since a large proportion of the seeds fall to the ground and are eaten by small rodents.

Closely connected with the eating of fruits and seeds by vertebrates is *zoochoria*—a phenomenon that plays a prominent role in the life of forest biogeocoenoses. Among all kinds of zoochoria, that most characteristic of forests is the active gathering of fruits and seeds by animals, *synzoochoria*, and the dispersal of fruits and seeds after passing through the animals' alimentary canals without losing viability, *endozoochoria* (Levina 1957).

Seeds whose embryo and endosperm are mechanically protected by a tough coating are dispersed by endozoochoria. From 75 to 92% of their viability is retained after traversing the stomach and intestines of animals[1] (Obraztsov 1961). That feature is characteristic of all succulent fruits and seeds. Endozoochoria is connected pre-eminently with the activities of small birds (such as thrushes and warblers) having weak-muscled crops, of carnivorous mammals (bear, fox) and ungulates (wild boar). As a rule, seeds are disintegrated when eaten by rodents.

At present a considerable number of trees and shrubs are known to be disseminated by endozoochoria, most of them being species subordinate to the chief forest-forming species (Mazing 1957). There are a few exceptions to this rule, including, for instance, the dispersal of stone pine seeds by brown bears, described by F. D. Shaposhnikov (1949).

---

[1] Data exist showing that the germinative ability of some seeds is increased after they pass through the alimentary tracts of animals.

Largely on account of endozoochoria, many species with succulent fruits and seeds have penetrated into the steppe zone (Formozov 1950, Obraztsov 1961).

In forest conditions synzoochoria is associated with the gathering and storing of fruits and seeds by rodents and birds, which store away seeds of both subordinate and dominant forest-forming species: stone pine, oak, beech, pine, spruce, hazel, birch, etc. It is characteristic that most of the seeds stored by small rodents (mice, voles, Siberian chipmunks) are eaten by the animals or placed in situations where they cannot germinate. Only a very small percentage of stored seeds, including seeds stored by squirrels, can produce seedlings. Squirrels hide their hoards under fallen leaves and in holes and crevices in trees; they eat 99% of them later (Richards 1958).

Mammals do not carry seeds and fruits very far, but store them near their burrows. Many birds (various species of tits, nuthatches, rooks, jays, nutcrackers) store tree and shrub seeds, hiding them in cracks in bark or under lichens, moss, or leaves, or even burying them in the soil. These caches are of great value for winter survival of birds. Those stored by nutcrackers and jays play an especially important role in the life of our forests. Siberian nutcrackers store pine-nuts (*Pinus cembra*), which they often hide several kilometres from the parent trees (Gorodkov 1916). The nutcrackers hide their caches under mosses or lichens. Many of them are located on burned or cut-over areas or even on mountain tundra. Each cache usually contains from 6 to 12 pine-nuts. With a total number of from 833 to 3334 caches per hectare, the birds store from 4000 to 34,000 pine-nuts per hectare (Bibikov 1948). Siberian nutcrackers use only half of their winter stores (Raimers 1956). Some of them are eaten by rodents. Nevertheless the pine-nuts hidden by birds help in forest regeneration.

Jays perform similar work in oak forests, hiding acorns. In young pine stands the number of oak seedlings 'planted' by jays reaches several hundreds or even thousands per hectare (Formozov 1950, Kholodnyi 1957b, Novikov 1959).

The activities of jays and Siberian nutcrackers aid considerably in the regeneration of stone pine and oak on burned and cut-over areas and in the distribution of these tree species.

The digging activities of animals strongly affect forest vegetation. We have already mentioned their scale and their effects on the distribution of soil moisture; but the role of digging in the life of biogeocoenoses is not limited to that feature. On molehills, which are thrown up on the surface and usually occupy 2 to 3% of the area, germination of oak and maple seeds is twice as high as in soil not disturbed by moles. Poplar, birch, oak, elm, and maple grow on molehills. In the course of time the molehills are turfed over and their invasion by tree seedlings ceases (Voronov 1953,

1958). The effect of burrowing by rodents on forest regeneration is very marked. On clearings in the Tellerman forest soil-digging by common voles prevented the establishment of turf. Therefore oak seedlings developed better on the dug-up places than on the rest of the clearings (Obraztsov and Shtil'mark 1957).

In digging their burrows small mammals damage the roots of herbs. In places riddled with their burrows the green parts of herbs are reduced by from 35 to 52%, and favourable conditions are created for seed-germination and suckering of tree species. Forest regeneration is more intensive there than in adjacent areas (Sibiryakova 1949, Voronov 1958).

In the Forestry Laboratory of the Serebryanoborsk experimental forestry station (northwestern outskirts of Moscow), according to N. N. Popova's estimates, the area occupied by small mammal (mostly bank vole) burrows covered 10% of an oak stand. Only 34% of oak seedlings grew in the area not occupied by the rodent burrows. In 20% of the seedlings the root collar was covered with the soil thrown up, and in 12% the roots were damaged by burrow-digging. Obviously both of these groups of seedlings had started to grow on undisturbed soil, and the effects of the burrows and the scattered soil developed later. About 10% of the seedlings were growing above rodent and mole tunnels; their root systems curved round the tunnels, with some rootlets coming out through the walls. About 15% of the seedlings were growing above tunnels that had caved in or collapsed, and 9% on freshly-dug soil. Thus 34% of the seedlings were growing on the area dug up by the animals, which occupied only 10% of the total area, and 32% of them were suffering in one way or another from the effects of burrowing.

A census made by N. N. Popova in a pine stand showed that 35% of the pine seedlings there were growing above small-rodent burrows. The area occupied by the burrows and the dug-up soil did not exceed 15% of the whole. As Fig. 60 shows, the abundance of young pines, with a stand crown density of 0·5 to 0·6 or less, is closely connected with the size of the area burrowed into by rodents. With higher stand density, the heavy overshading by the canopy apparently checks the development of seedlings, and the connection is disrupted.

Rooting by wild boars plays an important part in several forest biogeocoenoses. By digging in litter and the uppermost soil layer they bury and scatter some of the fallen acorns and nuts, thus aiding in natural forest regeneration (Sludskii 1935, Abramov 1954, Lebedeva 1956). At the same time the wild boars damage the roots of seedlings and root out and kill young trees, sometimes thereby causing heavy losses (Dinesman 1959a). By digging up the soil they lessen the competition offered by low-growing plants to trees and shrubs. In young spruce stands in Tian-Shan, regeneration takes place only where animals have destroyed the thick moss cover that prevented seeds from germinating. In the Belovezh bush, seedlings of spruce, maple, hornbeam, and ash, and in well-lighted places

FFB Z

birch, appear where wild boars have been rooting. Repeated rooting is seldom observed among spruce seedlings in the Belovezh bush, but is common among broad-leaved seedlings and causes their death. That leads to replacement of broad-leaved species by spruce (Lebedeva 1956). Cases are known where rooting by wild boars has caused birch to replace oak-beech stands (Vietinghoff-Riesch 1952). By constantly devouring

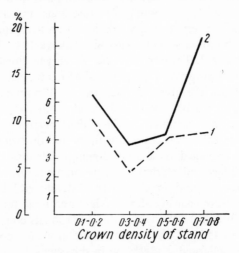

Fig. 60. Distribution of digging by mouse-like rodents and of pine seedlings in a hazel pine stand (from Popova 1962). 1—average number of pine seedlings per m²; 2—area occupied by rodent burrows (%).

beechmast and oak acorns the animals prevent regeneration by these species and at the same time their rooting creates conditions favourable to birch.

There have been frequent cases on the steppes of pioneer growth of forest vegetation in places dug up by animals (Vinogradov 1937, Obraztsov 1956). It is true, of course, that that occurs only where their establishment on the dug-up soil is not prevented by climatic conditions.

The consequences of the eating of forest herbs by animals have not yet been thoroughly studied. One may often find, under forest canopy and in forest glades, patches grazed by wild mammals. The scale of such occurrences and their effects on tree-herb inter-relationships are not clear. Only a few facts regarding this question have been recorded. In the Caucasus, for instance (Krainova 1951), the grazing of bison led within a short time to the elimination of cowparsnip and the growth of spiraea, comfrey, goat's rue, and other herbaceous plants. The bison eat mixed grasses first. Therefore in places grazed by them the amount of herbs and sedge increases (Zablotskaya 1957b). A similar phenomenon is observed with regard to axis deer kept in enclosures (Ryabova 1938).

Heavy grazing by moose on several species of mixed grasses has been recorded (Aleksandrova and Krasovskii 1957), and also destruction of lichens, grasses, and shrubs by voles (Koshkina 1957, Novikov 1956, 1959). In forests, however, that aspect of animal activity probably never reaches such a scale and such significance as on tundras, steppes, and semi-deserts.

The activities of carnivorous and insectivorous vertebrates have a notable effect on the state of forest vegetation. Predators consume up to 40% of the biomass of herbivorous forms and are bound to affect the participation of the latter in the transformation of matter and energy. Investigations in recent years have revealed that insectivorous forest birds reduce the numbers of insect pests (gypsy moth, lackey moth, brown-tail moth, winter moth, and various xylophages) by 40-70% (Knight 1958, Korol'kov 1963). That markedly affects the growth of trees and shrubs. In the Tellerman experimental forest, on areas of young growth protected from birds by nets throughout the summer, growth in height of oaks was reduced by half (Korol'kova 1963). Unfortunately there is no similar information regarding other insectivorous or carnivorous vertebrates. There are better records of the effect of separate species on the numbers of the animals that they prey upon.

This brief review of the various forms of action by vertebrates on forest vegetation shows that birds, and mammals even more so, have a considerable effect on the accumulation of vegetable mass in a biogeocoenose, retarding or accelerating the process, and in many cases even determining the type of accumulation. This constitutes the predominant role of vertebrates in the transformation of living matter. Therefore control of their activities can strongly affect the productivity of biogeocoenoses.

CHARACTERISTICS OF ANIMAL ACTIVITY IN DIFFERENT TYPES OF BIOGEOCOENOSES

We must now try to discover how far the above forms of participation by vertebrates in the transformation of organic matter are specific for forest biogeocoenoses.

Without doubt such specificity is possessed by *animals' digging activities and their consequences*. Much less digging by vertebrates is evident in forests than on steppes and tundras. That is due to the profusion of natural shelters in forests (Table 79). The principal burrowing animals in forests (mice, voles, moles, and shrews) construct their tunnels in the upper soil horizons. Deeper soaking of the soil and washing-down of humus takes place by way of the tunnels, and forest vegetation grows and turf forms on the soil thrown up by the animals.

Table 79. *Digging activities of mammals (Kucheruk 1960).*

Zone	Number on one hectare			
	Entrances to burrows of mice, voles, gerbils	Entrances to burrows of Siberian chipmunks, susliks, pikas, hamsters	Entrances to burrows of Arctic foxes, red foxes, badgers, marmots	Mounds made by zokors, moles, mole-voles
Tundra*	1000	—	0·001	—
Taiga	500	10	0·001	100
Broad-leaved forest	1000	100	0·01	1000
Steppe	5000	1000	50	?
Desert	1500	100	0·001	100

* In the tundras of the north-eastern part of the U.S.S.R. there are numerous burrows of the long-tailed Siberian suslik and some other rodents, whose digging activities are not shown in the table.

On the steppes much digging is done, not only by voles, but also by animals with deeper and more complex burrows, susliks (ground squirrels) and marmots. The mounds and burrows of these animals are character-istic features of steppe landscapes.

Susliks and marmots bring to the ground surface soil layers of very different chemical composition, which markedly affect soil-formation processes and lead to the development of their own distinctive vegetational complexes (Voronov 1959).

Lemmings and voles are the chief burrowing animals of the tundra. Their burrows are found there only in well-drained areas, and the great majority are located in the uppermost soil horizons. The principal result of the digging of these animals is the creation of well-drained mounds, free from vegetation, causing development of mosaic-patterned plant cover, and in some cases also development of meadow associations (Voronov 1959, Tikhomirov 1959).

The action of animals on vegetation has many distinctive features. In forests vertebrates consume a negligible part of the plant biomass, and the role of zoogenic cycles of matter and energy transformation is very small. Cases occur on the steppes where wild graminivores eat up practi-cally the whole plant mass, and consumption of 25-30% of its total amount is a common occurrence (Khodashova 1960). Animals destroy a considerable part of the plant cover on the tundra also. With a total stock of above-ground plant mass amounting to 73 ton/ha (Lavrenko *et al.* 1955), the remains of herbage eaten by a few lemmings during the winter weigh 20 ton/ha (Tikhomirov 1959).

Obviously zoogenic cycles of matter transformation play an incomparably greater role in the energy balance on the steppes and the tundras than in the forest. Most steppe and tundra vertebrates eat only a very small part, but the most nutritious part, of the vegetation that they destroy. By far the greater part of the plant mass 'mown' by animals falls directly to the soil surface.

Forest biogeocoenoses are distinguished by the forms of use of their plant mass by vertebrates. In forests vertebrates typically eat fruits and seeds of trees and shrubs rather than herbage (Voronov 1959). In winter, because of the low food value and reduced accessibility of grasses and other low-growing foods in forest biogeocoenoses, a diet of the vegetative parts of trees and shrubs is very common among vertebrates. On the steppes grasses retain a high food value even in winter, and in many districts are accessible to many animals throughout the whole year. There vertebrates feed on twigs only when forced to do so by the grass being in poor condition, by it being covered with sheet ice, or by an unusual winter of heavy snowfall. That is often accompanied by the death of the animals from starvation and a fall in their numbers (Dinesman 1961).

Many tundra animals eat the vegetative parts of shrubs and bushes in summer as well as in winter, but at the same time eat a considerable amount of the herbaceous layer. The reasons therefor are not clear, but it is possible that the buds and leaves of Arctic plants are rich in vitamins.

Forests differ from other types of biogeocoenoses in the comparatively small amount of digging by animals and in their effects on soil and vegetation; in the insignificant role taken by zoogenic cycles of transformation of living matter; in the widespread eating of seeds, fruits and vegetative parts of trees and shrubs by animals, with less feeding on the herbaceous layer.

The role of vertebrates in the life of different forest formations varies. In this respect virgin temperate-zone forests may be divided into two large groups. One of these includes nut-bearing, beech, oak, and stone pine forests, in whose regeneration zoochoria is important. On the other hand, zoochoria plays a limited role in spruce, fir, larch, and pine forests; it helps mainly to distribute subordinate species in these forests. In the second group animals feed on the vegetative parts of the trees, but while in pine and larch forests the shoots of the dominant species are heavily eaten, in spruce and fir forests the chief source of twigs as fodder is the subordinate species.

Data are as yet inadequate for more detailed classification of forest biogeocoenoses according to the role of vertebrates in their life. That role varies even in different forest types. Examples are the difference in the digging activities of small mammals in the pine and oak stands of the Serebryanoborsk forest (see Fig. 54), and the close dependence of the intensity of damage to stands by mammals on the nature of the herbaceous cover (Dinesman 1961) and the forest quality class (Kaletskaya 1959).

FOREST USE AND THE ROLE OF VERTEBRATES IN THE LIFE OF BIOGEO-COENOSES

The part taken by vertebrates in the life of forest biogeocoenoses has developed in the course of a prolonged evolutionary process. Its different forms, built up over millennia, have during recent centuries been much disrupted by man's economic activities. Such disruptions of natural biogeocoenotic relationships are now widespread over the globe (Elton 1960).

Alteration of the role of vertebrates was first caused by the most ancient types of forest use, hunting and pasturage. For example, as a result of excessive hunting the range of wild boars has been greatly reduced in the European part of the u.s.s.r., and in many districts their rooting activities have ceased to affect forest regeneration. The almost universal extirpation of beavers has completely eliminated their complex dam systems, and that, combined with forest clearing, has apparently seriously affected the hydrological regime of rivers. Red deer and roe deer have, like beavers, long ago become scarce in the European part of the u.s.s.r., and browsing by them on forest vegetation is now uncommon except in parks and game preserves. Many similar examples might be adduced, but unfortunately their significance in the life of the forest can only be guessed at in most cases.

In spite of the low food value of forest pasture, the pasturing of livestock in forests is still a prominent feature of animal husbandry. Only a few years ago up to 20% of the forest area in several provinces in the European part of the u.s.s.r. was used for pasturage (Obozov 1954, 1957). Domestic livestock eat up to 45% of the herbage in forests, and up to 85% in glades and clearings (Dekatov 1957). Pasturing animals trample down tree seedlings, eat and break off tips and branches of trees, damage bark and roots, destroy underbrush, and destroy the normal water regime of the soil; finally the herbaceous cover and forest types are altered (Lashchinskii 1958, Dekatov 1959, Nikitin 1961).

According to P. K. Fal'kovskii (1929), in the Ukrainian oak forests excessive pasturing first destroys underbrush and later causes elimination of ash. Maple and elm, which are in the second storey, are destroyed by continued pasturing and a light-loving herb and sod layer develops. Oaks begin to wither at the tips. The destruction of the second storey in oak forests by cattle often removes lateral shade from the trunks. The oaks, deprived of undergrowth, grow crooked (Dekatov 1959). In some cases the pasturing of livestock in forests may produce permanent breeding-places for insect pests (Lashchinskii 1958).

The greatest damage is caused when goats are pastured in a forest. Even when grass is plentiful they readily eat tree leaves and twigs. The least harm is done by cattle. When fodder is short, the latter browse first on linden, elm, maple, hornbeam, poplar, and ash; they eat oak and birch less readily.

Thus the pasturing of domestic animals in a forest has much in common with the activities of the wild mammals that eat the vegetative parts of plants. The pasturing of domestic animals, however, may cause more profound changes in a forest biogeocoenose because of the more concentrated feeding that is characteristic of pasturing by domestic animals. Pasturing by cattle has no noticeable effect on a forest if there are at least five hectares of pasture per head. In mixed forests of density 0·6 to 0·7, each head of cattle should have 3 hectares of pasture; in deciduous forests, 2 ha; and in glades and clearings, 1 ha, but it is usual in unsystematic pasturing to have from 0·8 to 5 ha per head (Obozov 1957). It is interesting to compare these figures with the feeding areas used by one of the principal browsers on twigs in the forest zone, moose. To judge by the data of the 1954 census (Isaev 1959), even in the provinces with the highest population density of these animals (3 to 4 per 1000 ha) each has not less than 200 ha of forest. If one takes the results of the moose census in the Darwin reserve (Kaletskaya 1959), where the population density in 1957 was very high (6·9 per 1000 ha), it appears that at that time each animal there had 145 hectares of forest, including 7·2 ha of young pines of growth class I. Thus even at their highest population density the pastures used by moose far exceed the norms accepted for the pasturage of domestic animals.

Besides its detrimental effect on the forest, cases are known where pasturing by domestic livestock, by destroying ground cover, has aided in forest regeneration from seed (Dylis 1947).

The role of pasturing in the life of forest biogeocoenoses is not limited to its effects on forest regeneration. Pastured areas often become unsuitable for habitation by many forest mammals and birds, which changes the effects of wild vertebrates on forest vegetation. For instance, the pasturing of livestock and hay-mowing reduce the food supplies of wild ungulates and increase their browsing on trees and shrubs (Julanger 1955, Hoskins and Dalke 1955, Chilson 1955, Kaznevskii 1959).

It is typical that in the Belovezh bush damage to the forest by deer increased markedly after the meadows were mown for hay (Kartsev 1903). In Austria the intensity of damage by these animals is closely related to the abundance of meadows and pastures (Schönwiese 1958).

Clear-cutting, by changing living conditions for forest animals, very strongly influences vertebrate activities, radically changing their effects on vegetation. In this connection close study has been given to the serious injury caused nowadays by moose to forests in the European part of the u.s.s.r (Dinesman 1959a). The intensity of moose damage to pines is determined by the number of pine seedlings per animal, and the density of the moose population is determined by the abundance of winter food (young pine and poplar growth). Thus the greater the total area of young growth and the smaller the proportion of pines in it, the more heavily the pines are damaged by moose.

New growth in clearings in dark-conifer taiga forests is mostly composed of birch and poplar, and in pine forests of pine and birch. For that reason forest-cutting leads to wide distribution of poplar in a narrow belt stretching from the Gulf of Finland and Lake Ladoga to the Southern Urals. Almost all cases of serious moose damage to pines are found in that belt. Therefore the serious injury done to forests by moose at the present time has a clearly-shown origin in human activity.

The extent and methods of forest-cutting, and the character and area of the forest environment of the animals, have a close connection with the effects on forest vegetation caused by small rodents, hares, red deer, and various birds (Dinesman 1961).

Evidently in modern conditions the role of wild vertebrates in forest biogeocoenoses is largely decided by the methods of forest exploitation. Often it results from new biogeocoenotic relationships formed by the vast changes made in the forests by man.

The study of these new relationships, and the devising of means of regulating them, constitute one of the chief tasks of forest zoology.

REGULATION OF VERTEBRATE ACTIVITIES IN THE FOREST

Regulation of vertebrate activities in the forest is still in the earliest stages of development. Up to the present investigators and economic forestry organisations both in the U.S.S.R. and abroad have paid attention only to methods of protecting tree stands from damage. Apparently there have been no serious attempts to regulate other effects on forests caused by vertebrates.

The various methods of protecting tree stands from vertebrates may be divided into three groups. The first group includes methods of direct control of vertebrate numbers. The second includes methods of altering biotopes in order to reduce the amount of damage by animals. The third includes methods of restricting access by animals to the vulnerable plants.

Many methods have been used for directly controlling vertebrate numbers. The numbers of valuable game animals can be most appropriately regulated by economic use of them. In that case one should try to keep the population density at a level where the harmful activities can be prevented by simple biotechnical measures; that population density has been called 'economically permissible'. It varies widely according to the species-composition, age, productivity, and nature of economic use of tree stands. Methods of determining the economically-permissible population density of moose have been proposed for the forests of our country (Kozlovskii 1960).

In order to restrict the numbers of animals of no economic value, mechanical, biological, and chemical methods of elimination have been suggested. The mechanical method is very laborious and therefore is little applied in forests. Chemical methods of eliminating animals are highly effective, but when they are used not only pests but also useful animals often suffer from them. Therefore chemical methods are permissible only in limited areas where the animals present a definite menace to seeds, seedlings, and young growth of tree species. The techniques of application of chemical methods are varied, and are decided in any particular case mainly by ecological features, among which the most important is the movements of the animals in the territory.

Regulation of the numbers of forest vertebrates by biological methods is apparently possible, but the theoretical and practical problems involved are still far from being solved. The best-studied possibility is the use of birds to control insect pests. Attracting birds by the construction of bird-houses and nesting-places has succeeded in reducing the numbers of insect pests by from 33 to 70%. To obtain such results one must attract from 15 to 20 pairs of insectivorous birds to one hectare of forest (Pfeifer and Ruppert 1953, Bruns 1955, Pyl'tsina 1956, Korol'kova 1961). Other means of using these vertebrate groups to keep down insect pest numbers are possible; stirring up forest litter, for instance, considerably increases the numbers of cocoons and pupae in it that are destroyed by birds (Il'inskii 1949, Formozov 1950).

The activities of birds are in most cases notably less effective than chemical methods of destroying pests. Nevertheless the consumption of pests by birds has a positive effect on the growth of tree species. Therefore the provision of bird-houses and nesting-sites in insect pest breeding-places must be considered a valuable forest-economy measure.

The close dependence of bird and mammal numbers and of the nature of their activities on the composition and structure of tree stands, on the characteristics of the herbaceous cover, etc., has led to the idea of regulating the effects of animals on vegetation by altering the biotope. It has been proposed to reduce the numbers of small rodents by mowing the herbaceous cover (Sviridenko 1951), by gathering up and destroying slash after tree-felling (Pershakov 1934, 1939, 1940, Obraztsov and Shtil'mark 1957), and by cutting underbrush (Obraztsov and Shtil'mark 1957, Paaver 1953). More than once it has been recommended that the activities of ungulates and hares should be controlled and that insectivorous birds should be attracted by changing the composition of stands in various ways (Pershakov 1939, 1940, Volchanetskii 1950). In many cases such measures have produced noticeable results but in most cases they have caused substantial changes in the structure of tree stands that have been far from meeting the requirements of forest economy, a fact seriously restricting the possibilities of regulating animal activities by changing biotopes.

Finally, the various methods of restricting the access of animals to plants consist in the protection of trees and shrubs by fencing them or covering them with netting, paper, and other materials. Most of these methods give entirely satisfactory results, but because of the labour involved they are applicable only to very limited areas.

In this group we should include a method enjoying unmerited popularity, that of repelling birds and mammals from trees by smearing the trees with protective pastes (repellents). The efficacy of these pastes varies widely, depending on the habits and the numbers of the animals, on their food supplies, and on other ecological conditions (Siegel 1956). They cannot give reliable results.

CHAPTER V

Micro-organisms as a component of a forest biogeocoenose

The community of micro-organisms is a basic component of a forest. Therefore in studying the natural laws of the development of a forest knowledge of the processes carried out by micro-organisms and of the factors creating the microflora of forest soils is important. The entire forest biogeocoenose, in which the processes of matter and energy exchange are constantly taking place, is essentially the chief factor forming the microflora of forest communities. The exchange of matter and energy between micro-organisms and other components of a forest has a specific character in each case and at each moment of time.

To discover the role of micro-organisms in a forest biogeocoenose we must trace all the lines of inter-relationship between microflora and the other forest components.

The soils are inhabited by a multitude of micro-organisms of extremely diverse composition. They include bacteria, Actinomycetes, fungi, algae, Protozoa, bacteriophages, etc.

The total number of micro-organisms may be determined by direct cell counts or by the method of growing cultures in artificial nutrient media.

By direct counts the number of micro-organisms in a definite soil sample is determined. That gives figures many times higher than those obtained by the culture method. The discrepancy between the figures obtained by direct count and by culture in nutrient media has a number of causes. It is not possible to create in the laboratory a universal nutrient medium in which all groups of micro-organisms can grow. Media suitable for the culture of heterotrophic organisms do not suit autotrophic organisms, and vice versa. On the other hand direct counts may include dead as well as living cells of micro-organisms, and also stainable particles of the soil itself. The method does not enable one to determine the systematic position of micro-organisms and consequently the proportions of separate groups.

In practice, methods of culture in special media are more often used. These methods differ both in the kind of culture and in the nature of the medium. Culture in an agar-based substrate makes it possible to determine the systematic position of the micro-organisms. Existing methods of counting micro-organisms in soils give only comparative indexes of microbial population density.

Published data on the numbers of micro-organisms in soils are in most

355

cases obtained from cultures in agar-based media of definite composition. Bacteria are the most numerous. One gramme of soil from the uppermost horizon contains from some tens of thousands to several millions of bacteria, depending on the soil, the vegetation, and other factors.

The biomass of micro-organisms in the soil attains substantial figures, although varying estimates of it are made by different investigators. According to N. A. Krasil'nikov (1944a, 1958), in grey soils under lucerne the total live bacterial mass is over 8000 kg/ha. In podzolic soils under clover it is 1500-4000 kg/ha, and under wheat 1100 kg/ha. In slightly-cultivated soil under wheat it is only 100-150 kg/ha.

I. V. Tyurin (1946) believes that the estimates of the bacterial mass for grey soils are exaggerated, since the increment of organic matter in these soils does not enable it to accumulate to that extent. According to his estimates, the live bacterial mass in grey soils is 1220 kg/ha, in chernozems 1800-2400 kg/ha, and in podzolic soils 740-1400 kg/ha. E. N. Mishustin and M. I. Pertsovskaya (1954), on the other hand, believe that the bacterial mass in soil is approximately twice as high as Tyurin's estimates, since the products of humus breakdown serve as a supplementary food source for micro-organisms that decompose plant and animal remains. More-over, in Mishustin's opinion we should include the considerable number of autotrophic bacteria and algae in the soil.

J. Pochon and H. de Barjac (1960) suggest that on an average the total biomass of bacteria in good fertile soil exceeds 500 kg/ha.

Strugger (1948) estimates that the total bacterial mass amounts to 0·03-0·28% of the weight of the soil.

Other micro-organisms, also of considerable mass, inhabit the soil in large numbers. For instance, according to Pochon and de Barjac, the live biomass of fungi amounts to 1000-1500 kg/ha; of Actinomycetes, up to 700 kg/ha; and of Protozoa, 100-300 kg/ha.

N. A. Krasil'nikov estimates that the biomass of fungi, Actinomycetes, and Protozoa amounts to 5-10% by weight of the biomass of bacteria.

The biomass of micro-organisms is active living matter, with a tre-mendous potential for transformation of matter and energy. It exists in a state of constant development, frequently undergoing change and renewal.

As a result of the activity of micro-organisms, diametrically-opposed processes of organic matter transformation, mineralisation, and synthesis take place in the soil, due to differences in their physiological functions. These processes are carried out by micro-organisms by means of successive and intimately-interlinked reactions.

By excreting various enzymes into the environment micro-organisms break down complex organic compounds into simpler ones that can be used both by the micro-organisms themselves and by plants.

The energy necessary for body-building, growth, and reproduction is derived by micro-organisms from oxidation (by aerobes) and fermentation (by anaerobes) of various substances.

The part played by micro-organisms in a forest biogeocoenose has many aspects.

One major task fulfilled by micro-organisms is the supply of food elements to plants. In natural conditions most nutritive matter exists in complex organic and mineral compounds unavailable to plants. Micro-organisms mineralise the organic matter and gradually transform mineral compounds of low solubility, through formation of acids and CO_2, into forms available to plants.

As organic matter decomposes in the soil a gradual change in the microflora is observed. At first rapidly-multiplying non-spore-forming bacteria and fungi develop the greatest activity. The dominant role of spore-forming bacteria and Actinomycetes appears in later stages of decomposition; they possess more powerful fermentative apparatus and are able to digest more stable forms of organic compounds.

The group of sporiferous bacteria linked with the transformation of organic matter in the soil includes members differing in their physiological, biochemical, and other characteristics, which to a great extent determine their ecological features.

Actinomycetes colonise semi-decomposed remains after bacteria and fungi have consumed all easily-digested matter.

In forest biogeocoenoses the preparation of nutritive matter for plants by micro-organisms takes place very actively in litter, and also in the soil, where the main mass of organic matter is concentrated. A certain proportion of the nutrients transformed by micro-organisms is consumed by them in body-building, and so is retained in the litter and the soil.

But that is a transitory phenomenon. After the microbes die the matter enclosed in their cells decomposes and is again converted into compounds available to plants. Many micro-organisms form 'supplementary' plant food in the soil: various vitamins, growth substances, amino acids, etc., which are absorbed by plants and stimulate their growth. The symbiosis of nodule bacteria with leguminous plants, e.g., with acacia, is very typical. The bacteria penetrate the root tissues and multiply there, and (because of their ability to fix atmospheric nitrogen) improve the nitrogenous nutrition of plants, while obtaining from them mineral matter and carbohydrates.

Mycorrhizae form on the roots of many forest plants—a symbiosis of roots and certain fungi. With the aid of mycorrhiza-forming fungi trees absorb nutrients from mineral and organic compounds of low solubility in the soil. In addition, mycorrhiza-forming fungi help plant growth by producing 'supplementary' nutrients.

Micro-organisms take a very active part in the formation of humus in the soil and in maintaining soil structure. They also assist actively in forming CO_2, and in the retention of nutritive elements in the soil by means of biological bonds.

Micro-organisms are important in breaking down injurious products of the metabolism of microbes and plants, dead plant parts, animal carcases, etc., and also toxic substances produced by certain microbes.

Micro-organisms can be used in the control of pests and diseases of trees. For instance, a method of using bacteria to control fungal diseases of pine cones has been suggested. In recent years a bacteriological method of control of gypsy moth larvae, etc., in Western Siberia has been devised and used with success.

CONDITIONS OF ACTIVITY OF THE MICROBIC POPULATION OF THE SOIL

Soil micro-organisms may live in soil solutions, on the surface of soil particles washed by these solutions, or within soil particles. Most of them, however, are connected with the solid part of the soil. According to the data of D. M. Novogrudskii (1936a), a soil solution from the topmost horizon of podzolic soil was relatively sterile containing only 0.1% of the bacteria, the remaining 99.9% being in the solid part of the soil. Fungi and Actinomycetes were not found in the solution. At greater depths the number of bacteria in the soil solution increased, and fungi and Actinomycetes sometimes appeared.

The absence of micro-organisms in a soil solution is due to the fact that soil particles have the property of retaining micro-organism cells (adsorption), and not to any unsuitability of the solution for the life-activity of the micropopulation.

Krasil'nikov (1958) found that bacteria of the same species develop differently in soil solutions obtained from different soils. Their intensity of development is greater in solutions obtained from fertile soils than in those from less fertile soils. Bacteria introduced into sterile soil solutions develop better in solutions of chernozem or of garden (fertilised) soddy-podzolic soil than in a solution of field soddy-podzolic soil, low in humus. Bacteria react in different ways to the nutrient content of the solutions.

ADSORPTION OF MICRO-ORGANISMS

The adsorption of micro-organisms by soil particles was discovered by N. N. Khudyakov (1926) and his pupils N. V. Dianova and A. A. Voroshilova (1925), N. S. Karpinskaya (1925), and others. According to their data, one gramme of soil can adsorb up to 4350 million bacterial cells, depending on the soil type and the specific characteristics of the bacteria. Novogrudskii (1936b) discovered that soil adsorbs not only bacteria but also the spores of microscopic fungi and Actinomycetes.

Krasil'nikov (1958) found that medium-loamy carbonate chernozem adsorbs two or three times as many *Azotobacter* cells as does a similar layer of soddy-podzolic forest soil. The upper soil layers, which contain more colloidal matter, have higher adsorbing capacity than the lower layers.

It has been found that the adsorption of microbic cells by soil is directly proportionate to the abundance of clay and dust particles in it. Small soil particles (0·0015-0·01 mm) adsorb microbic cells most actively, while larger particles (0·05-1·0 mm) do not adsorb them well.

The degree of adsorption of microbic cells varies, depending on the specific characteristics, the state of life-activity, the age, and other features of the micro-organisms themselves.

The pH of the soil strongly affects the amount of adsorption of micro-organisms. Spores of *Bac. mycoides* are adsorbed more by acid soils (pH = 4·5). As the pH of the environment increases to neutral the percentage of adsorption falls, and when it moves into the alkali zone the amount of adsorption remains at approximately the same level or rises slightly.

The adsorptive capacity of the soil undergoes seasonal fluctuations. According to Novogrudskii (1937), it is lower in early spring and late autumn than in summer. With low humidity and high temperature it rises, and with high humidity and low temperature it falls. The adsorptive capacity of different soils varies, depending on their moisture content. In optimal conditions for the development of microflora the soil does not adsorb them, but on the contrary releases them into the surrounding solution.

According to Khudyakov and his colleagues Dianova, Voroshilova, and Karpinskaya, and also Peele (1936) and others, bacterial cells adsorbed by soils maintain their life-activity, but their biochemical activity declines or ceases altogether. According to L. I. Rubenchik and his colleagues (1934), reduced activity was shown by nitrifying bacteria and increased activity by sulphate-reducing bacteria when adsorbed by estuary mud. In D. G. Zvyagintsev's experiments (1959), activity by adsorbed bacterial cells was observed on soil particles, whose surface was covered with biologically-active matter.

Krasil'nikov (1958) and Zvyagintsev (1959) found that *Azotobacter* cells in the adsorbed state actively reproduced and consumed oxygen. Krasil'nikov believes that the process of adsorption of microbic cells by soil particles should not be regarded merely from the point of view of its physical and chemical action; he suggests that it also has a biological character.

Krishnamurti and Somon (1951), analysing published data and those of their own investigations, concluded that the phenomenon has a specific character. The percentage of adsorption of cells depends on the properties of the adsorbent and on the species of microbe. In defined conditions the coefficient of adsorption is strictly constant.

An assemblage of many other factors affects the activity of soil micro-organisms. These include humidity, temperature, acidity, aeration, vegetation, and others.

ACIDITY OF SOIL SOLUTIONS

The acidity of soil solutions has great influence on the life-activity of the soil's microbic population. Different groups of the micropopulation have different reactions to environmental acidity (Table 80).

Table 80. *Reaction of micro-organisms to environmental acidity (E. N. Mishustin 1950).*

Micro-organisms	pH values within which the micro-organisms can develop		
	Minimum	Optimum	Maximum
Saprophytic bacteria	About 4·5	About 7·0	About 9·0
Nodule bacteria	About 4·3	About 7·0	About 10·0
Azotobacter	About 5·0	About 7·0	About 9·0
Nitrifying bacteria	About 4·0	7·8-8·0	About 10·0
Sulphur-oxidising bacteria	1·0-5·0	—	About 10·0
Actinomycetes	About 4·5	About 7·0	About 9·0
Moulds	About 1·5	About 7·0	About 9·0
Protozoa	About 3·5	About 7·0	About 9·0

The widest range of development is possessed by microscopic (mould) fungi, followed by Protozoa, bacteria, and Actinomycetes, although optimal conditions for all groups are found at approximately the same pH level.

Micro-organisms of the same systematic group do not always react in the same way to environmental acidity. The majority of soil-dwelling bacteria do not develop with pH below 4·5-5·0, but sulphur-oxidising bacteria are more acid-resistant. Among other bacteria adapted to low pH conditions are *Rhizobium japonicum*, *Azotobacter indicans*, *Aerobacillus macerans*, and several others.

There are also some microscopic fungi and Protozoa that fail to develop in an acid environment. Certain moulds (*Humicola griseae*, *Fus. sambucium*, *Cephalosporium*, etc.) are found only in soil with a pH of 6·4-8·0. Several Protozoa (*Acanthocystis acullata*) are able to develop only in alkaline media; on the other hand, *Carteria abtusa* and *Calpidium campylum* do not develop in alkaline conditions. Even different races of bacteria and different mould species of the same genus react differently to environmental acidity.

As a result of the varying acidity of the soil, microflora found in its microzones exist in different pH conditions. In acid or alkaline soils there are microzones favourable to the development of micropopulations with different reactions to acidity. The liming of acid soils enables more desirable microbiological processes to take place in these soils, with bacterial activity predominating in them.

SOIL MOISTURE

Soil moisture unquestionably exerts outstanding influence on the numbers and activities of the microbic population.

Individual groups in the soil microflora have different critical moisture thresholds. The lowest water content of soil in which several fungi and Actinomycetes can develop is approximately 80-85% of the maximum water-holding capacity. With soil moisture approaching maximum water-holding capacity they are able to maintain to a slight extent the processes of ammonisation and biological fixation of mineral nitrogen. With soil moisture at the maximum of water-holding capacity several species of bacteria multiply slowly (Novogrudskii 1946 a, b). At low soil moisture levels microflora show very weak biochemical activity. Intensive development of the microflora, and of the processes induced by them, proceeds in fairly moist soil, reaching the optimum at a level near that for higher plants: 40-70% of full moisture-capacity, depending on the microbes (Lipman and Brown 1908, Dushechkin 1911, Munter and Robson 1913, Greaves and Carter 1916, 1920, Kudryavtseva 1925, Genkel' and Butylin 1935, Feher and Frank 1937, Novogrudskii 1947, Enikeeva 1947, 1952).

Further increase in soil moisture has a detrimental effect on the development of most aerobic fungi, bacteria, and Actinomycetes, on account of the expulsion of air from the soil.

Movement of water from the soil into microbial cells can take place when the intracellular pressure is higher than the osmotic pressure of the soil solution. Osmotic pressure is very high within the cells of certain microbes (especially in fungi and Actinomycetes), sometimes reaching 200-300 atmospheres (Czapek 1924). According to Czapek's observations (1924) intracellular pressure is usually from 3-4 atmospheres and seldom exceeds 20 atmospheres.

S. P. Kostychev and I. Kholkin (1929) discovered that ammonising bacteria were active in soils of Central Asia with osmotic pressure of 80 atmospheres and upwards. The nitrifying process was not observed in these conditions.

E. N. Mishustin (1947) discovered that the osmotic pressure in the cells of soil bacteria rises with transfer from moister to drier zones. For instance, the osmotic pressure of *Azotobacter* and *Bac. mycoides* in podzolic soil does not exceed 4 atmospheres, but with transfer to drier districts it gradually rises to 15-18 atmospheres (in chernozem soil).

The regular increase in osmotic pressure in microbial cells is regarded as a unique adaptive reaction to environmental conditions.

The ability of microbes to survive changes in soil moisture varies within wide limits. Short-term changes in soil moisture, within definite limits, may not affect the numbers of micro-organisms very much, but their activity varies greatly. During periods of soil moisture deficit various enzymes remain fairly active. When soil moisture rises the activity of these enzymes rises still more rapidly (Enikeeva 1947).

OXIDATION-REDUCTION CONDITIONS

The possibility of aerobic and anaerobic processes taking place in a particular area of soil depends not only on its oxygen regime but also on the state of other elements contained in it.

The oxidation-reduction potential decides the direction and character of biochemical reactions and the solubility of products of microbial metabolism.

Oxidising processes take place vigorously in the surface layer of the soil, where oxygen penetrates and where aerobic microbes—most bacteria, fungi, Actinomycetes, algae, Protozoa, etc.—develop actively.

With increasing soil depth the oxygen content of soil solutions falls, and they lose their oxidising property. Below the oxidation-reduction limit (which varies in different soils, depending on moisture, temperature, soil cultivation, and other factors) anaerobic processes take place. Anaerobic micro-organisms (mainly bacteria) are most numerous in the upper soil horizons, mostly on account of the large amount of organic matter there and also the number of microzones in which the development of anaerobic bacteria is possible.

At the same time aerobic microbes (bacteria, Actinomycetes) can exist in conditions of relatively low oxygen (facultative anaerobes); and therefore in the deeper soil horizons aerobic processes take place together with anaerobic, although much less vigorously. A high oxygen content stimulates ammonising and nitrifying processes and benefits the development of fungi, Actinomycetes, bacteria, and other organisms that break down organic matter with formation of CO_2.

SOIL TEMPERATURE

The activity of micro-organisms in the soil depends on its temperature as well as on its moisture and aeration. That activity is stimulated by rise in temperature up to a definite maximum, the temperature limit for biological reactions being from 70-80°C.

For most soil micro-organisms the optimal temperature is from 25-35°C. Most soil micro-organisms have the ability to adapt themselves to gradual changes in temperature. Mishustin (1950) discovered that in warmer districts several soil bacteria have a higher temperature optimum than individuals of the same species taken from colder soils.

The temperature adaptations of soil bacteria to climatic conditions have been described by L. A. Garder (1927), B. L. Isachenko and T. N. Simakova (1934), A. I. Rogacheva (1947), and others.

Micro-organisms easily survive low temperatures, but when soils freeze many microbiological processes weaken or cease altogether, except those induced by the microbial groups adapted to life in low temperatures. Such soil organisms include, for example, *Fusarium nivale* and *Lanosa nivalis*. Several Protozoa are found to be active in temperatures below 0°C. They have been found encysted under ice, reproducing in the snow on mountain heights, etc.

According to the observations of F. M. Chistyakov and G. L. Noskova

(1938), some bacteria develop at temperatures down to $-5°C$, but the minimum temperature for most bacteria is $+2°C$. For some soil fungi the minimum temperature at which development is possible is below $0°C$, but their growth rate at that temperature is extremely low (Chistyakov and Bocharova 1938, Panasenko 1944).

Data exist showing increased activity of micro-organisms under the influence of winter frosts. Nodule bacteria become more active and virulent, *Azotobacter* develops and reproduces more rapidly, and so on.

Thermophilic micro-organisms have a temperature optimum between 50 and 60°C. They are distributed among bacteria (mostly spore-forming bacteria) and Actinomycetes, and occur rarely among fungi. According to Mishustin's data (1950), ecologo-geographical conditions do not affect the distribution of thermophilic micro-organisms. Their numbers are up to 10% higher in manured soils; virgin forest soils contain very few of them (from 100 to 1000 cells per gramme of soil). Thermophilic microflora are not characteristic of soils and are mainly brought there with organic manure in the course of cultivation.

A number of measures could be used to stimulate microbial activity in forest soils: improvement by drainage of damp and boggy soils, thinning or improvement cutting, neutralisation of acid soils, etc.

PROCESSES INDUCED BY MICRO-ORGANISMS IN THE SOIL

DECOMPOSITION OF NITROGENOUS ORGANIC MATTER

Protein decomposition—ammonisation

Considerable amounts of nitrogenous organic matter enter the soil with dead plant and animal remains. They constantly undergo breakdown into simpler compounds, with the formation of ammonia. The decomposition of proteins is accomplished by ammonising micro-organisms (bacteria, Actinomycetes, and fungi). The ammonising process may take place in either aerobic or anaerobic conditions. The group of micro-organisms that decompose proteins in aerobic conditions is very numerous and its members have varied activities. It includes many forms of bacteria, Actinomycetes, microscopic fungi, etc. In anaerobic conditions decomposition is performed mainly by bacteria.

The action of microbes on a complex protein molecule begins with hydrolysis, achieved by means of proteolytic enzymes by the micro-organisms. Micro-organisms that do not produce such enzymes can only use the products of protein hydrolysis. Various amino acids are obtained as the end-products of hydrolysis.

Most of the decomposed protein serves as a source of energy for microbes. Some products of protein breakdown are partly used for body-building of microbial cells.

The accumulation of free ammonia in the environment when protein is decomposed depends on the proportions of nitrogenous and carbonaceous

food for micro-organisms in the protein. If the substrate contains substances relatively rich in nitrogen, free ammonia collects. In an environment rich in carbohydrates the micro-organisms consume the freed
nitrogen and it undergoes 'biological fixation'.

Ammonisation of urea and of uric and hippuric acids

Micro-organisms cause hydrolytic breakdown of urea and of uric and
hippuric acids. Urea is decomposed by numerous and widely-distributed
bacteria, forming a separate group of urobacteria.

A characteristic feature of urobacteria is their ability to develop in a
strongly alkaline substrate. Most urobacteria can induce intensive decomposition of urea in both aerobic and anaerobic conditions, but the
process takes place mostly in aerobic conditions. Urea serves urobacteria
solely as a source of nitrogen. The de-amination reaction is induced by
the urease enzyme.

Uric acid is decomposed by a number of bacterial organisms (micro-
cocci, non-spore-forming and spore-forming bacteria), which use it as a
source of carbon and nitrogen or merely as a source of nitrogen. The
breakdown of uric and hippuric acids may also have energy significance.

Hippuric acid is decomposed by bacteria and fungi. The benzoic acid
formed as a result of hydrolysis, and glycocoll, serve as a good source of
carbon and nitrogen and are therefore further broken down by various
groups of micro-organisms.

Breakdown of chitin

The breakdown of chitin is performed by several non-spore-forming
bacteria, Actinomycetes, mycobacteria, and fungi. It consists of two
successive processes. Hydrolysis takes place under the influence of the
exoferment chitinase or chitase. The glucose and acetic acid thus formed
are used by chitin-decomposing bacteria and other micro-organisms as
sources of carbon and nitrogen.

DECOMPOSITION OF NON-NITROGENOUS ORGANIC MATTER

Breakdown of cellulose

The breakdown of cellulose is very widespread in nature. Various specific
micro-organisms break it down fairly easily into simpler carbon compounds; the end-products are CO_2 and water.

Among anaerobic bacteria *Bac. omelianskii* breaks cellulose down with
considerable vigour.

A major role in the breakdown of cellulose is played by aerobic cellulose-
decomposing micro-organisms, among which are widely-differing groups:
(*i*) bacteria: (*a*) myxobacteria of the genera *Cytophaga*, *Polyangium*, and
Myxococcus, (*b*) vibrios of the genera *Cellvibrio* and *Cellfalcicula*, which
give yellow-ochre and green tints to cellulose; (*ii*) fungi of the genera
Trichoderma, Aspergillus, Penicillium, Fusarium, Demathium, Clado-

sporium, Polyporus, Chaetomium, etc., and also Hymenomycetes; (*iii*) Actinomycetes: *Act. cellulosae, Act. violaceus, Proactinomyces cytophaga*, and many others.

The breakdown is achieved most actively and rapidly by finely-specialised bacteria, whose development typically takes place in environments without cellulose.

The breakdown of cellulose begins with its hydrolysis under the influence of the enzyme cellulase, secreted by the cell. The products of hydrolysis are oxidised to form high-molecular organic acids, available to microbes as a source of carbon and energy-producing material.

The breakdown of cellulose by aerobic bacteria proceeds most actively in the presence of nitrogen in the form of nitrates with a neutral or weakly alkaline reaction, and with optimal moisture, temperature, aeration of substrate, etc.

The species-composition of cellulose-decomposing micro-organisms in the soil depends on a number of factors and may vary considerably. In more acid and infertile soils cellulose-decomposing fungi predominate. Bacteria take first place in soils that are neutral or relatively rich in nitrogen. In chernozem soils, for example, the chief decomposers of cellulose are bacteria of the genera *Cytophaga* and *Cellvibrio*.

Cellulose-decomposing bacteria are described in detail in the monographs of A. A. Imshenetskii (1953) and the works of Z. F. Teplyakova (1952, 1955).

Breakdown of lignin and pentosans

Lignin is one of the substances most resistant to microbial activity. The most active destroyers of lignin are fungi of the genus *Merulius*, some species of the genus *Cerastomella*, and (among mould fungi) *Mucor chlamydosporus racemosus*.

When infested with fungi of the genus *Merulius* wood turns brown, is covered with numerous cracks, and becomes very brittle; fungi of the genus *Cerastomella* produce blue rot. *Mucor chlamydosporus racemosus* breaks lignin down vigorously, but not cellulose, so that the wood acquires a jelly-like consistency of yellowish colour.

The chemistry of lignin decomposition has not been studied well. It is believed that the first stage of the fungal action is a process of hydrolysis, with subsequent oxidation.

Pentosans are vigorously broken down by aerobic bacteria—*Bac. volatus, Bac. subtilis*, and *Bac. flavigena*—and by fungi—*Aspergillus flavus, Asp. niger, Mucor stolonifer*, etc. The chemical nature of the breakdown of pentosans remains undiscovered.

Breakdown of pectins

The complex compounds of intercellular pectin matter also are broken down by specific micro-organisms.

As a result of hydrolysis (under the influence of the enzyme pectinase) galacturonic acid, galactose, arabinose, xylose, acetic acid, and methyl alcohol are formed, and are further oxidised to CO_2 and water.

The compounds obtained by hydrolysis of pectic acid may be fermented by anaerobic bacteria, predominantly spore-forming: *Clostridium pectinovorum, Bac. amylobacter, Cl. felsineum*, etc.

Among aerobic micro-organisms that oxidise the hydrolytic products the most important are bacteria and fungi. The best-studied of the bacteria are *Bac. macerans, Bac. mesentericus, Bac. subtilis, Bac. asterosporus*, etc. Pectins are vigorously attacked by several fungi: *Mucor stolonifer, Aspergillus niger, Cladosporium*, etc. Fungi are particularly active in forest soils, where the mycelia penetrate forest litter and rapidly decompose it. Moreover, fungi are able to develop with relatively low temperature and moisture and relatively high acidity of the substrate.

Breakdown of sugars, starch, and organic acids

Readily-soluble monosaccharides and disaccharides are easily and rapidly broken down by widely-differing micro-organisms. Many groups of micro-organisms possessing the enzyme amylase can convert starch into dextrin.

Another, more specialised group of microflora hydrolyses starch to form acids, alcohols, and gases. Aerobic, facultative anaerobic, and anaerobic micro-organisms take part in that process. *Clostridium* spp., *Bac. amylobacter, Endosporus filamentosus, Bac. cereus*, and others have high amylolytic capacity.

Many bacteria that break down cellulose and pectins are also able to break down starch. That faculty is possessed by *Azotobacter*, which fixes atmospheric nitrogen.

The acids, alcohols, and dextrin that are formed readily undergo further decomposition by other micro-organisms into carbon dioxide and water.

Breakdown of fats

The first stage in fat breakdown is its hydrolysis under the influence of the enzyme lipase. The glycerin and fatty acids so obtained are further oxidised to form CO_2 and water. The hydrolytic breakdown of fat may take place in both aerobic and anaerobic conditions, with a wide range of acidity. The process is most intensive when oxygen is present.

Fats are actively broken down by non-spore-forming bacteria: *Pseudomonas fluorescens, Bac. pyocyaneum, Bac. prodigiosum, Bac. lipolyticum*. Among fungi, *Oidium lactis* and various species of *Penicillium, Aspergillus*, etc., take part in the process. Many species of Actinomycetes also take an active part in the decomposition of fats.

Oxidation of carbohydrates

Carbohydrates of the aliphatic and the aromatic series and their compounds can be oxidised to CO_2 and water by various micro-organisms.

The mineralisation of carbohydrates and their immediate derivatives is very important both in the general carbon cycle in nature and in the disposal of compounds that are poisonous to organisms.

Oxidation of carbohydrates takes place most intensively when oxygen is freely available, and is possible within a wide pH range.

Micro-organisms that oxidise carbohydrates of the fatty series include widely-distributed bacteria (*Ps. fluorescens, Bac. aliphaticum*), myco-bacteria (*Mycob. album, M. rubrum, M. luteum*), and fungi (certain species of the genera *Penicillium* and *Aspergillus*). Carbohydrates of the aromatic series are oxidised by a number of bacteria (spore-forming and non-spore-forming) and fungi.

Micro-organisms can also oxidise a small number of typical 'anti-septics'. Apparently all organic compounds found in nature as products of the life-activity of organisms can be oxidised by specific groups of micro-organisms in appropriate circumstances.

OXIDATION OF MINERAL COMPOUNDS

Nitrification

Ammonia, which is formed by the breakdown of organic nitrogenous compounds, undergoes further oxidation in the soil. It is first oxidised to nitrous acid, and then the nitrous acid is oxidised to nitric. The process is the result of the successive action of two groups of micro-organisms.

The oxidation of ammonia into nitrous acid is induced by different species of nitrose bacteria (*Nitrosomonas* spp.), and the oxidation of nitrous to nitric acid by bacteria of the genus *Nitrobacter*.

Nitrifying micro-organisms and the phenomenon of nitrification were first described by S. N. Vinogradskii at the end of last century. Nitrifying bacteria are autotrophic and are very specific with regard to the substrate oxidised by them. For them the oxidising processes have energy-releasing significance.

Nitrifying bacteria are widely distributed in nature. Their activity varies in different soils, depending on various factors.

Nitrifying bacteria are aerobes. The acidity of their environment strongly influences their development. Their optimal pH is from 8 to 9, but some of them are more resistant and can develop with a pH as low as 4·1-4·6. In conditions favourable for the development of nitrifying bacteria (humidity, temperature, pH, aeration, etc.) most of the mineral nitrogen in the soil exists in the form of nitrates. All measures intended to benefit the metabolism of these organisms will increase the amount of nitrates in the soil. Nitric acid improves phosphorus-nutrition conditions by increasing the solubility of phosphates.

Sulphuration

The oxidation of hydrogen sulphide (which is formed by the breakdown of proteins and by other chemical processes in the soil), of sulphur, and of its thio- and tetra-compounds to sulphuric acid is induced by a unique group of micro-organisms, sulphobacteria. Sulphuric acid makes possible the transformation of difficultly-soluble phosphates into soluble compounds, as a result of which the amount of mineral compounds available to plants is increased and the nutrition conditions for the plants are improved.

Sulphobacteria are widely distributed in nature: they occur in the soil, in sulphur springs, in stagnant water, in lakes, and elsewhere. These micro-organisms develop in a high concentration of hydrogen sulphide, which serves them as a source of energy and by the aid of which they assimilate CO_2.

Sulphobacteria carry out intracellular oxidation of H_2S into sulphuric acid in two stages. When H_2S is in excess they oxidise it to sulphur. The sulphur accumulates in their protoplasm as food reserves. As the H_2S is used up, these reserve deposits of sulphur begin to be oxidised and sulphuric acid is formed.

Sulphuric acid formed in the course of sulphuration is neutralised within the cells of sulphobacteria by calcium bicarbonate. It is converted into gypsum, in which form it is diffused from the cells into the environment. That process is performed by numerous members of the sulphobacteria: colourless (*Beggiatoa, Thioploca, Thiotrix*) and purple (*Chromatium okenii*, etc.). The purple bacteria can obtain energy both by oxidising sulphur and its compounds and by photosynthesis.

Besides sulphobacteria, which oxidise H_2S through the intermediate stage of sulphur, there is a unique group of thiobacteria, which oxidise H_2S, sulphur, and thiosulphates into sulphuric acid. The thiobacterial group is an extensive one, including some strictly autotrophic (*Thiobact. thiooxidans*), some facultatively autotrophic (*Thiobact. novellus*), some strictly anaerobic (*Thiobact. denitrificans*), and others.

Many widely-distributed heterotrophic micro-organisms also can oxidise thiosulphates to sulphates: *Ps. fluorescens, Achromobacter stutzeri*, etc.

REDUCTION OF MINERAL COMPOUNDS

Denitrification

Denitrification is performed by anaerobes and facultative anaerobes, mostly non-spore-forming bacteria. Denitrifying bacteria use the oxygen of nitrates to oxidise organic matter, reducing nitric acid to nitrous acid or free nitrogen. The most important are *Bac. stutzeri, Bac. denitrificans*, and *Ps. flourescens*, which are widely distributed in nature and are constantly found in the microflora of soils and plant rhizospheres. Nevertheless, no catastrophic loss of nitrogen is observed in the soil. Denitrification requires anaerobic conditions and also an adequate amount of nitrates and available organic matter; without these conditions, denitrifying bacteria merely assimilate nitrates and do not break them down. The importance of denitrification in soils increases considerably when moisture is excessive.

Besides 'real denitrification', the faculty of reducing nitrates to nitrites is widespread among saprophytic organisms (bacteria, Actinomycetes, fungi). They absorb nitrogen from nitrates and transfer it into nitrogenous organic compounds. Some microbes use nitrates as acceptors of hydrogen during oxidation of organic matter, thus providing energy. For other microbes the reduction of nitrates may be merely a preparatory stage in their absorption of nitrogen.

Reduction of sulphates to hydrogen sulphide

In anaerobic conditions salts of sulphuric acid and less-oxidised sulphur compounds can be reduced to H_2S, which is poisonous to living organisms.

Reduction of sulphates is carried out by various bacteria, of which *Vibrio desulfuricans*, *V. hydrosulfurens*, and *V. termodesulfuricans* are among the most active.

Desulphuration takes place with simultaneous breakdown of organic compounds from which microbes derive energy. When H_2S is formed in alkaline conditions, reduction of sulphates may take place even in the absence of organic matter if molecular hydrogen exists in the environment. In that case sulphides are formed as end-products, the hydrolysis of which leads to increasing alkalinity. Reduction of sulphates may take place with pH values from 4·5 to 8·7.

Even in anaerobic conditions sulphates may be reduced in very small amounts by certain widely-distributed ammonising bacteria. H_2S may accumulate not only in bodies of water but also in soil solutions if the soil is flooded for a long time.

Absorption of atmospheric nitrogen

Biological fixation of nitrogen is one of the principal biological processes taking place in the soil. It has theoretical interest and great practical importance.

Fixation of atmospheric nitrogen is of great importance to sylviculture. Whereas in the cultivation of farm plants mineral and organic fertilisers containing nitrogen are added to the soil, over the vast territories occupied by forests fertilisers are rarely applied and the forest has to use nitrogen fixed by micro-organisms.

The role of nitrogen-fixing micro-organisms in the soil is outstanding. According to American investigators, the amount of nitrogen added to all the soils of the U.S.A. by free-living nitrogen-fixers alone is greater than the amount added in fertilisers. They estimate that the total amount of fixed nitrogen added annually is divided up as follows: in organic manures, 2·57 million tons; in mineral fertilisers, 0·48 million tons; in rain, 3·57 million tons; resulting from the activity of free-living nitrogen-fixing bacteria, 4·37 million tons; and from the activity of symbiotic bacteria, 5·46 million tons.

The nitrogen-fixing bacteria ensure the uninterrupted course of the nitrogen cycle in nature.

Micro-organisms capable of absorbing atmospheric nitrogen are divided into two large groups: the so-called free-living soil-dwelling nitrogen-fixers, and nitrogen-fixers living symbiotically with plants—nodule bacteria.

Nodule bacteria

Nodule bacteria are able to absorb molecular nitrogen only in symbiosis with a root system. They form nodules on the roots of false acacia and caragana, and also on the roots of herbaceous leguminous plants.

The bacteria penetrate the root tissues and cause swellings to form there, and the root cells gradually fill up with reproducing bacteria. The bacteria obtain minerals and carbonaceous matter from the plants and provide the plants with nitrogenous compounds assimilated by them so that the plants contain more nitrogen in their roots and leaves. Their litter provides mineral nutrients for the soil microflora and enriches the soil with nitrogen. During a single growing season the accumulation of nitrogen by leguminous plants may exceed the normal range of 40-300 kg/ha.

Besides leguminous plants, other plants belonging to distinct systematic groups may enter into symbiotic partnership with nitrogen-fixing micro-organisms. Nodules are known to form on the roots of alder, *Elaeagnus*, wax myrtle, sea-buckthorn, sumach, and other plants. The best-studied of these symbiotic relationships is that between nitrogen-fixing micro-organisms and alder. Nodules are formed on alder roots by microbes resembling Actinomycetes (*Proactinomyces*). In partnership with these alder develops normally on nitrogen-poor soils. In one year speckled alder may fix from 50 to 100 kg/ha of atmospheric nitrogen. In some countries speckled alder is used as a nitrogen-fertilising crop.

Micro-organisms that fix atmospheric nitrogen in symbiosis with plants may form nodules on leaves as well as on roots. On the leaves of the tropical plant *Pavetta indica* such nodules contain bacteria (*Bac. rubeaceum*) capable of fixing atmospheric nitrogen. Plants whose leaves bear nodules develop much better than those without them. Similar nodules occur on the leaves of certain species of the family Dioscoreaceae.

Mycorrhiza fungi of the genus *Phoma* (*P. betae, P. cansaria*) living in symbiosis with the roots of heather (Ericaceae) have the faculty of fixing nitrogen. Like nodule bacteria in symbiosis with leguminous plants, mycorrhiza fungi obtain carbon-containing nutrients prepared by the plant, and some of the nitrogen assimilated by them apparently passes into the root system. Several other mycorrhiza-forming fungi, belonging to various species and capable of fixing atmospheric nitrogen in symbiosis mainly with herbaceous (farm-crop) plants, have been described.

Endotrophic mycorrhiza accompanying trees are widely distributed in nature, but their capacity for fixing atmospheric nitrogen is practically unstudied and cannot be considered to be definitely established.

Free-living nitrogen-fixers

Nitrogen-fixing bacteria living freely in the soil (*Clostridium pasteurianum*) were first isolated and studied by S. V. Vinogradskii in 1893. Later, in 1910, Beiernik isolated an aerobic fixer of nitrogen, *Azotobacter chroococcum*. Since then the number of micro-organisms known to be able to fix atmospheric nitrogen has greatly increased.

Azotobacter is an aerobe with a complex life-cycle. It is widely distributed in upper soil horizons (cultivated and fertile) with a pH of 5·8-7·0, and in bodies of water, but has also been found in more acid soils (pH 5·0-5·5). *Azotobacter* is more fastidious about the acidity, humidity, and substrate of soil than many other micro-organisms. It requires phosphorus, calcium, molybdenum, sulphur, iron, and a number of other elements. In order to ensure maximum development of *Azotobacter* in an environment lacking in fixed nitrogen, the following must be available: phosphorus 0·1 mole, calcium 0·001 mole, molybdenum and iron 0·0001 mole, etc. *Azotobacter* is used as a biological test-organism to determine the richness of a soil in phosphorus and calcium.

Azotobacter fixes molecular nitrogen in a pure culture only in the absence of fixed nitrogen. As the concentration of fixed nitrogen in the environment increases its nitrogen-fixing capacity falls or ceases altogether, but *Azotobacter* may continue to develop in such circumstances.

A low concentration (0·003%) of fixed nitrogen and a small amount of humic matter have a favourable effect on the assimilation of atmospheric nitrogen by *Azotobacter*.

Azotobacter obtains the energy for fixation of atmospheric nitrogen by the oxidation of various non-nitrogenous organic substances. In the soil in natural conditions *Azotobacter* uses the products of decay of organic matter and compounds formed as the result of the mineralisation of humus. This takes place when *Azotobacter* develops in association with other micro-organisms, and the breakdown of organic matter by different microbes produces intermediate substances used by *Azotobacter*.

The most widespread species is *Azotobacter chroococcum*. Some cultures have been found to assimilate 20 mg or more of atmospheric nitrogen per gramme of sugar decomposed. According to most investigators, the amount of nitrogen fixed by this species fluctuates from 10 to 50 kg/ha.

There are some other species of *Azotobacter* capable of vigorous fixation of atmospheric nitrogen, e.g., *A. agile*, *A. vinilandi*.

A strict anaerobe, *Clostridium pasteurianum*, is a typical agent of butyric fermentation. In the absence of fixed nitrogen in the environment and with a low carbon concentration it fixes molecular nitrogen. In optimal conditions its production of fixed nitrogen is equal to that of

Azotobacter. In natural conditions it apparently develops in association with various saprophytic bacteria. *Clostridium pasteurianum* can develop with lower pH values (to 4·7) than can *Azotobacter.* It is distributed everywhere in nature, including forest and water-saturated soils. Its optimal temperature is 25°C.

The ability to fix nitrogen in quantities smaller than those of *C. pasteurianum* is found in many other species of the genus *Clostridium*, which live in soils and have an important role there. Such anaerobic bacteria include *C. pectinovorum, C. felsineum* (both active in the breakdown of pectins), *C. acetobutilicum* (which takes part in the breakdown of pectins and cellulose), *C. buturicum* (an active disintegrator of starch), and others.

Other free-living nitrogen-fixers

Among other micro-organisms for which ability to fix nitrogen has been recorded are *Vibrio desulfuricans* (a reducer of sulphates), *Azotomonas insolita, Azotomonas fluorescens, Bac. asterosporus, Bac. hydrogens,* various species of *Pseudomonas* and *Aerobacter*, mycobacteria, and Actinomycetes. Some of them are able to fix up to 11-15 mg of nitrogen per 100 ml of culture (*Azotomonas insolita, Actinomyces* spp.). Most fix nitrogen at the rate of 1-2 up to 5-7 mg/g of carbon used. The amount of nitrogen fixed may be increased two or three times by growing them in mixed cultures.

Similar properties are possessed by several photosynthesising bacteria and the group of oligonitrophils, which have been little studied, and which fix nitrogen in small quantities.

Oligonitrophils are found in large numbers in soils everywhere and also in rhizospheres. These micro-organisms, united in a single physiological group, belong to various systematic units, and have the ability to develop in an environment with an insignificant amount (0·0014-0·03 mg/l) of nitrogen.

Algae

At present about 14 species of algae able to assimilate atmospheric nitrogen are known. They belong to the Nostocaceae (*Nostoc muscorum, N. punetiforme*, etc.). It is believed that that faculty is also possessed by some species of Oscillariaceae and Chroococcaceae.

Some scientists believe that the assimilation of atmospheric nitrogen by free-living organisms cannot be of great importance. In natural conditions, however, where micro-organisms exist in complex inter-relationships within microbic coenoses, the physiological and chemical properties of various microbes may be different from those observed in the laboratory; some species may suppress and others stimulate the process. It is known that the fixation of nitrogen by *Clostridium pasteurianum* and *Azotobacter* is increased in mixed cultures including micro-organisms that break down cellulose. At the same time the breakdown of cellulose is stimulated. An increase in nitrogen fixation is also observed in mixed cultures of nitrogen-fixers with algae, Protozoa, oligonitrophils, and other micro-organisms.

Most investigators believe that *Azotobacter* is (after nodule bacteria) the principal nitrogen-fixer in the soil, a view that may be due to its having been studied more than other nitrogen-fixers.

In optimal laboratory conditions *Azotobacter* fixes more nitrogen than other micro-organisms. It is quite widely distributed in the upper horizons of cultivated and fertile soils.

Azotobacter is almost entirely absent from forest and virgin soils. The distribution of this micro-organism is limited by its environmental requirements. In particular, it is not observed in soils with less than 14 kg/ha of available phosphoric acid. In the soils of South Australia it is scarce because of the small supply of calcium and the highly acid reaction; it is extremely rare in the acid peat soils of Sweden, etc. Nevertheless it appears that nitrogen accumulates in forest and virgin soils, and that cannot be due only to symbiotic micro-organisms (nodule bacteria). Leguminous plants and alders are far from being universal in occurrence.

Apparently the accumulation of nitrogen results mainly from the activities of ubiquitous but very-little-studied micro-organisms (fungi, algae, bacteria, and Actinomycetes) that are able to fix small quantities of nitrogen. If their abundance in the soil (especially in the rhizospheres of plants) is taken into account, the total amount of nitrogen accumulated by them may reach considerable figures.

The fixation of atmospheric nitrogen by micro-organisms in the soil has recently been confirmed directly in field conditions by the use of tagged nitrogen. The nitrogen-fixing ability of several micro-organisms (*Clostridium pasteurianum, Azotobacter, Phoma*, algae, etc.) has also been proved by the use of the isotope N^{15} in pure cultures.

Detailed data on the mechanism of fixation of atmospheric nitrogen have been presented by M. F. Fedorov (1952).

TRANSFORMATION OF PHOSPHORUS AND IRON COMPOUNDS

Vast quantities of organic compounds and of mineral compounds of phosphorus almost unavailable to plants are transformed by micro-organisms in the soil. These processes consist mainly in mineralisation of organic phosphorus, conversion of phosphates from less- to more-soluble forms, and reduction of phosphates.

Organic phosphorus compounds

These compounds undergo breakdown (mineralisation) and are converted into salts of phosphoric acid—a form available to plants. The process takes place under the influence of different groups of micro-organisms (bacteria, yeasts, fungi).

In recent years a number of specific bacteria active in phosphorus transformation have been isolated from the soil. In *Bac. megatherium* var. *phosphaticum*, for instance, the vigour of breakdown of organic phosphorus compounds much exceeds the requirements of the bacteria themselves, so that phosphoric acid accumulates in the environment. Since the mineralisation of organic phosphorus is very important for the development of higher plants, these specific bacteria are used as fertilisers —'phosphorobacterin' (*Bacterial fertilisation*, 1961).

Simultaneously with the mineralisation of organic phosphorus compounds in the soil, the reverse biological process takes place: re-inclusion of assimilated phosphorus (phosphoric acid) in organic compounds and its incorporation in the bodies of micro-organisms. Therefore if an excessive amount of substances that are too poor in phosphorus compounds enters the soil, a temporary phosphorus famine may ensue.

Phosphoric acid ions formed in the process of mineralisation of organic phosphorus compounds are absorbed by plant roots and micro-organisms. The phosphoric acid is partly retained in the soil, mostly passing into salts that are only slightly soluble and almost unavailable to plants.

The solution of phosphates in the soil is also aided by micro-organisms that create various acids (carbonic, sulphuric, nitric, etc.). Acids formed as a result of the life-activity of micro-organisms (e.g., nitrifying and thionising micro-organisms) transform tricalcium phosphate into soluble forms available to plants.

Transformation of iron compounds

Iron compounds in the soil also undergo a number of changes under the direct and indirect influence of micro-organisms.

Micro-organisms that excrete acids in the course of their metabolism change iron salts into soluble forms. Those that release hydrogen also induce the appearance of reduced forms of iron. The reduction of iron oxides into lower oxides, a process most intensive in the upper layers of gleyed soil horizons, is observed in anaerobic conditions.

Only specific 'ferrobacteria' are closely linked with the transformation of iron compounds, and oxidation of lower oxides of iron is for them a form of respiration. When an iron salt is oxidised in the cells of ferrobacteria, a soluble iron oxide is formed. The latter, when excreted from the cell into the environment, is converted into a less soluble state and becomes a hydroxide of iron.

Ferrobacteria are aerobes that occur in the soil, more often in water, mud, or waterlogged soils and have been described in more detail by N. G. Kholodnyi (1957a).

MOBILISATION OF ASSIMILABLE POTASSIUM

Bacteria are able to break down aluminium silicates and to free the potassium contained therein. The most active disintegrator of silicates (*Bac. mucilaginosus*) is recommended as a bacterial fertiliser. Its numbers depend on its living conditions, acid soils being unfavourable for its development, but liming of acid soils favours its multiplication.

According to published data, silicate bacteria are able to bring silicon as well as potassium into solution from silicates. They are able to separate phosphorus from tricalcium phosphate, to synthesise vitamins of the B group, and to increase the resistance of plants to fungal diseases (*Bacterial fertilisation*, 1961).

Under the influence of micro-organisms compounds of manganese, calcium, silicon, etc., undergo various transformations, but these have been studied very little.

FORMATION OF BIOTICALLY-ACTIVE SUBSTANCES

Biotic substances ('growth factors' or 'growth substances') are formed by various micro-organisms—bacteria, yeasts, Actinomycetes, fungi, algae, etc. They are vitamins and vitamin-like substances that stimulate the growth of plants and microbes, and affect cellular metabolic processes, structure, etc.

Widely-distributed autotrophic organisms (e.g., nitrifiers, hydrogen-oxidising bacteria, members of the group of thionising bacteria, etc.) and heterotrophic micro-organisms are active formers of biotic substances. They include most micro-organisms that inhabit the rhizospheres of plants and soils: *Azotobacter*, nodule bacteria, oligonitrophils, *Bac. denitrificans*, *Bac. mycoides*, *Bac. mesentericus*, species of the genera *Pseudomonas*, *Vibrio*, and *Mycobacterium*, and also Actinomycetes, yeasts, fungi (including mycorrhiza-formers), and many others.

According to the estimates of M. N. Meisel', in grey desert soils (fertile and irrigated) the microbes in a single hectare can synthesise annually about 400 g of vitamin B_1, 300 g of vitamin B_6, and 1 kg of nicotinic acid.

The rate of formation of biotic substances by soil micro-organisms depends on the associated micro-organisms, some stimulating and others depressing the synthesis of these substances.

Several substances synthesised by micro-organisms are used by the microbes themselves. The surplus, often in considerable amounts, is excreted into the external environment while the microbes are still alive. Biotic substances entering the soil are broken down and again synthesised by micro-organisms. While the micro-organisms live in the soil a constant exchange of these substances takes place, the amount depending on the rate of synthesis, the amount entering the soil, and the rate of decomposition.

Without the formation of vitamins and other biotic substances the development of many microbiological processes in the soil would be retarded. Many vitamins and similar substances are created by green plants themselves and the additional amounts provided by soil micro-organisms accelerate plant growth and development.

Micro-organisms also excrete antibiotic substances, whereby they suppress the multiplication or metabolism of other microbes (bacterio-static action) or even kill them (bactericidal action). The antagonistic action of micro-organisms upon each other is widespread in the soil, especially in plant rhizospheres. Antibiotic substances are absorbed by the root systems of plants and protect them against diseases. These sub-stances vary widely in their action and characteristics, and have selective effects.

FORMATION AND ASSIMILATION OF CARBON DIOXIDE

Soil (including rhizosphere) micro-organisms play a major role in the production of CO_2 gas. Micro-organisms that break down organic matter are constantly releasing it into the atmosphere. Lundegårdh (1924) believes that two-thirds of all the CO_2 in the air contained in the soil is formed by bacterial action, and the remaining one-third is exhaled by roots.

According to Bond (1941), the respiration of nodules on soybean roots is three times as high as that of the roots per unit of dry weight. The total mass of nodules exhales considerably more CO_2 gas than the whole mass of roots of the plant. P. Barakov (1910) estimates that the amount of CO_2 exhaled by micro-organisms is less than that exhaled by respiration of the plants. According to N. A. Krasil'nikov, A. V. Rybalkina, A. E. Kriss, M. A. Litvinov, and others (1934, 1936), the greatest exhalation of CO_2 coincides with maximum development of root microflora. B. N. Makarov (1953) has discovered that exhalation of CO_2 depends on the humidity and temperature of the soil and on the development of micro-organisms. Frereks (1954), Katznelson and Stevenson (1956), V. N. Smirnov (1953, 1954, 1955), Weissenberg (1954), V. N. Mina (1957), and others regard soil respiration as an index of microbial activity.

Part of the CO_2 formed escapes from the soil into the air and part is absorbed by plant roots and microbes. Micro-organisms not only exhale CO_2 but also absorb a certain amount of it to build their plasma (mostly autotrophic bacteria and algae).

It has been discovered that several saprophytic micro-organisms absorb CO_2 together with organic matter (Lebedev 1921).

Besides CO_2, soil air contains many other gases formed as a result of the metabolism of organisms and plants. It is believed that the peculiar smell of earth (or soil) is due to volatile products of the metabolism of microbes, mostly Actinomycetes.

Volatile organic substances contained in soil are to some extent a source of nutriment for, and also a factor inhibiting the development of, various species of microbes and plants. According to N. G. Kholodnyi (1944 a, b, c, 1951 a, b, c, 1957b) and his colleagues (1945), several bac-teria, fungi, and even isolated plant roots grew in a chamber where the only source of nutriment was soil exhalations. Vitamins released into the air were absorbed by the micro-organisms and plants. The use by micro-

organisms of vitamins and their debris from the air has been described by
M. N. Meisel' and N. Trofimova (1946, 1950).

The presence of both nutritive and toxic substances in soil air was discovered by Krasil'nikov (1958). In his experiments, the exhalations of
soddy-podzolic forest soil were more toxic to *Staphylococcus aureus* than
those of garden soil or chernozem. The depressing action of exhalations
from the soil varies according to the plant cover.

FORMATION AND BREAKDOWN OF HUMUS

Organic compounds entering the soil are partly mineralised by the activity
of micro-organisms and partly transformed into complex stable compounds known as humus.

The chemistry of the formation of humus compounds has not been fully
studied, but some facts have been fairly well clarified. Humus compounds
may be formed as a result of the interaction of compounds synthesised by
micro-organisms and products of the decay of plant remains in conditions
of partial prevention of the access of oxygen. In the opposite case, the
organic matter and humus compounds formed undergo full oxidation by
the microflora.

Some micro-organisms form a number of dark-coloured substances,
very close in their nature to humus, within their cells.

Humus is broken down also by the soil microflora and different components of humus decompose in the soil at varying rates.

Humic acid, which has a very complex cyclic structure, undergoes decomposition with the greatest difficulty, but it is not unavailable to many
common soil microbes, as was earlier supposed. In small doses humic
acid stimulates the breakdown of organic matter.

In recent decades various groups of micro-organisms have been described
as being able to use humic matter as a source of carbon and nitrogen.

Isolated microbes break down humic acid much more slowly than does
a mixture of different soil micro-organisms. Apparently in natural conditions an aggregate of micro-organisms, capable of disintegrating complex cyclic compounds, takes part in the decomposition of humic matter.

MICROFLORA OF FOREST SOILS

The transformation of matter and energy by micro-organisms in the soils
of forest biogeocoenoses has its own distinctive features. The nature and
direction of microbiological processes there are different from those in
meadow or cultivated soils, since a forest plant community influences the
microflora of forest soils.

Experimental data compiled up to the present and obtained by a single
method enable us (admittedly, still far from completely) to describe the
microflora of forest biogeocoenoses and the interdependence of microorganisms, soil, and arboreal vegetation.

Many investigators have noted differences between the microflora of forest soils and those of steppe, meadow, and cultivated soils. According to E. N. Mishustin and his colleagues (1951), soddy-podzolic forest soil under oak stands is much richer in fungi and poorer in bacteria than are meadow soils (Table 81). The bacterial population of forest soils is distinguished by the complete absence of *Bac. megatherium* and suppressed development of the non-spore-forming pigmented forms of bacteria, *Bac. mycoides*, and nitrifying bacteria. The number of *Bac. virgulus* is higher

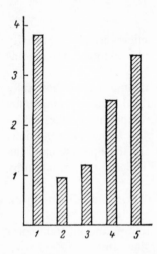

Fig. 61. Total number of micro-organisms—average for season (millions per g of oven-dry soil) (from Terekhov's data). 1—fallow land; 2—pine plantation; 3—mountain-ash-hazel pine stand; 4—hairy-sedge oak stand; 5—hairy-sedge linden stand.

in forest soils than in meadow soils, whereas *Bac. mesentericus* is absent from forest soils. No substantial difference is observed in the composition of Actinomycetes. Among microscopic fungi in forest soils one often finds distinctive forms such as *Micromucor*.

O. S. Terekhov and M. G. Enikeeva (1964) have observed differences in

Table 81. *Microflora of soils occupied by forest and meadow vegetation (10³ per gramme of soil)* (*Mishustin 1951*).

| Horizon | Depth, cm | Total number of bacteria | Non-spore-forming pigmented forms | Bacillar forms of bacteria | | | | | | | Total number of Actinomycetes | Total number of microscopic fungi | Energy of nitrification |
				Total number of spores	Bac. mycoides	Bac. mesentericus	Bac. megatherium	Bac. cereus	Bac. virgulus	Bac. idosus			
						Forest soil							
A₁	6-12	290	0	64	9	0	0	7	23	7	75	30·0	±
A₂	20-25	142	17	51	10	0	0	15	8	3	37	1·6	±
						Meadow soil							
A₁	2-10	780	52	175	22	0	60	28	few	10	61	2·1	+
A₂	20-30	168	8	59	7	0	10	20	0	0	33	0·9	+

the quantitative and qualitative composition of microflora in soddy-podzolic soils in fallow land and in forests. Average figures for the total number of micro-organisms during the vegetative period are given in Fig. 61. In the soil under a pine plantation and also in that under a 120-to-140-year-old mountain-ash-hazel pine stand, the total numbers of spore-forming and non-spore-forming bacteria and of Actinomycetes were less than in fallow soil, but the numbers of microscopic fungi were somewhat higher.

When the total numbers of micro-organisms are compared, the smallest numbers are found in the soil under pine stands. The fallow land and the land under the pine plantation were formerly ploughlands.

As a result of 20 years of pine growth the total number of micro-organisms in the soil had fallen sharply and was comparable to the number of micro-organisms in the soil under the 120-to-140-year-old pine stand.

In their qualitative, and partly in their quantitative, composition the micro-organisms in the pine-plantation soil were intermediate between those of the fallow land and the mountain-ash-hazel pine stand.

The total numbers of micro-organisms in the soils of a hazel-hairy-sedge oak stand (70-80 years old) and in a hairy-sedge linden stand (60-70 years old) were greater than those in the mountain-ash-hazel pine stand but less than those in fallow soil. The number was greater in the soil of the linden stand than in the soil of the oak stand.

Forest soils also differ from fallow land in the composition of their spore-forming bacteria (Table 82).

The dominant species in fallow soil were *Bac. idosus*, followed by *Bac. megatherium* and *Bac. cereus*. In forest soils *Bac. idosus*, and to a smaller degree *Bac. cereus* predominated. *Bac. megatherium* was found in small numbers in the soil of the mountain-ash-hazel pine stand, and absent in the soils of the oak and linden stands. No great differences were observed in numbers of *Bac. mycoides* in fallow and forest soils; they were somewhat greater under pine and linden. The numbers of *Bac. agglomeratus* in the soils studied were small, increasing only in the mountain-ash-hazel pine soil. *Bac. mesentericus* was practically absent from all of the soils. The differences in relative numbers of spore-forming bacteria in forest

Table 82. *Proportionate numbers of certain species of spore-forming bacteria in the upper layer of soddy-podzolic soil (% of total number of spore-forming bacteria) (average data from four analyses).*

Location	Bac. mycoides	Bac. mesentericus	Bac. megatherium	Bac. cereus	Bac. idosus	Bac. agglomeratus
Fallow land	5	2	33	13	44	2
Mountain-ash-hazel pine stand	9	0	12	20	26	12
Hazel-hairy-sedge oak stand	5	0	0	25	55	2
Hairy-sedge linden stand	8	2	0	31	38	2

soils under different tree stands apparently point to differences in the kind of organic matter in these soils.

The smallest number of nitrifying bacteria was observed in the mountain-ash-hazel pine soil, and the largest in the hairy-sedge linden soil. The nitrifying capacity of the fallow soil was lower than that of the hairy-sedge linden soil.

The specific figures for microflora in the soils under forest vegetation and fallow land apply also to the proportionate numbers of Actinomycetes growing in meat-peptone and starch-ammonia agar media (Table 83).

In the soil under the mountain-ash-hazel pine stand, as compared with fallow soil, the ratio between the number of Actinomycetes growing in SAA and MPA respectively is very small, indicating differences in the physiological characteristics of the Actinomycetes inhabiting these soils.

There are differences in the absolute and relative content of microscopic fungi in forest soil and in fallow soil.

The content of several species of microscopic fungi in fallow and pine-stand soils is shown in Table 84. The soil under the mountain-ash-hazel pine stand contains many more fungi of the genus *Mucor* and fewer of the genus *Penicillium* than does fallow soil. Among the Mucoraceae, the number of individuals of *M. ramannianus* is higher and that of *Zygorhynchus* is lower in the pine-stand soil. *Fusarium* occurs rarely and in small numbers in the soil of fallow land and in that of the hairy-sedge linden stand.

S. Ya. Mekhtiev (1953, 1957) studied the microflora of soddy-podzolic soils at different times during each of two years and discovered remarkable differences in microfloral composition in the soils of a deciduous forest (birch and poplar), of its outskirts, and of a cultivated field.

Variation in species-composition of spore-forming bacteria in the soils was clearly displayed (Table 85). The forest soil contained large numbers of *Bac. cereus*, but *Bac. megatherium* was poorly represented there. In the cultivated soil the content of *Bac. megatherium* was several times as high and the number of *Bac. cereus* fell. Less-worked soil (as compared with soil more thoroughly cultivated) contained fewer cells of *Bac.*

Table 83. *Proportionate numbers of Actinomycetes growing in starch-ammonia-agar (SAA) and meat-peptone agar (MPA) (Terekhov and Enikeeva 1964).*

Horizon	Number of Actinomycetes (10^3 per gramme of dry soil)						Ratio SAA:MPA (MPA taken as unity)		
	Fallow land		Pine plantation		Mountain-ash-hazel pine stand		Fallow land	Pine plantation	Mountain-ash-hazel pine stand
	SAA	MPA	SAA	MPA	SAA	MPA			
A_1	1516	33	161	5	240	36	46:1	32:1	8:1
A_2	1397	26	269	20	197	202	54:1	13:1	1:1

Table 84. *Content of certain species of microscopic fungi in soddy-podzolic soil (% of total number of fungi).*

Location	Horizon	Mucoraceae			Penicil-lium	Tricho-derma	Yeast-like fungi
		Total	Mucor ra-mannianus	Zygo-rhynchus			
Fallow land	A_1	8	4	2	64	7	1
	A_2	11	5	3	63	8	0
Pine plantation	A_1	18	14	0·5	37	1	1
	A_2	33	23	0	51	2	0
Mountain-ash-hazel pine stand	A_1	56	51	0	2	4	2
	A_2	64	63	0	19	0	3

megatherium and more of *Bac. cereus*. The soil of the forest outskirts was intermediate in bacterial content between forest and cultivated soils.

The soils of the forest and its outskirts contained more fungi of the genus *Mucor* than did cultivated soil. The forest soil contained a larger number of *Micromucor* fungi, and the forest outskirts more of other species. *Micromucor* fungi were absent from the cultivated soil. The forest-outskirts and cultivated soils differed from the forest soil in containing *Fusarium* spp., which were not found in the forest soil.

The nitrifying capacity of the cultivated soil was considerably higher than that of the forest and forest-outskirts soils.

More southerly soils (both virgin and cultivated) are richer in micro-organisms, but there also the distinctive features of forest influence are observed. O. I. Pushkinskaya (1951, 1953, 1954a) observed differences in the microflora of forests and of clearings. She studied the microflora of the dark-grey forest soil of an ash-oak stand (220 years old) with goutwort, of the soil of an ash-oak forest clearing (area lying fallow), and of leached chernozem (ploughland). Average figures (for horizon A) of 4 to 7 analyses, made at different times during the vegetative season, are given in Table 86.

There are differences among the soils of Table 86 both in the total number of bacteria and in the composition of separate physiological

Table 85. *Species-composition of spore-forming bacteria in soddy-podzolic soil in a forest, on the forest outskirts, and in a cultivated field (from Mekhtiev).*

Location	Spores (10³ per gramme of abs. dry weight)	Content (% of total number of spores)		
		Bac. megatherium	Bac. cereus	Other spp.
Forest	757	3·0	37·7	59·3
Forest outskirts	942	8·2	38·2	53·6
Cultivated field	1040	34·7	7·7	57·6

groups. In the clearing there was a somewhat smaller total number of bacteria than in the forest; fluorescent bacteria, typical of litter microflora, were much fewer there, as also were anaerobes and *Azotobacter*. Actinomycetes, promyxobacteria, and bacterial spores were more abundant.

Among spore-forming bacteria, *Bac. mesentericus*, *Bac. mycoides*, *Bac. agglomeratus*, and *Bac. megatherium* were more numerous in the fallow soil in the clearing than in the forest, whereas the numbers of spores of *Bac. idosus* and *Bac. virgulus* were smaller. The numbers of *Bac. cereus* were the same. The development of pigmented forms of non-spore-forming bacteria was less in dark-grey forest soils than in leached chernozem ploughland. The greatest numbers of *Bac. megatherium* were observed in the leached chernozem, but their development was depressed in the soil of the ash-oak stand. *Bac. virgulus* was not observed in the leached chernozem but was abundant in the forest soil. The numbers of *Bac. agglomeratus* were about the same in forest soil and in leached chernozem.

Differences in the qualitative composition of micro-organisms in the soils of tree plantations and of fallow clearings were observed by E. V. Runov and I. E. Mishustina (1960) on leached chernozem. Among Mucoraceae, for instance, *Mucor racemosus*, *Zygorhynchus* spp., and *Micromucor* spp. predominated in forest soils and *Rhizopus* in cultivated soils.

According to Runov (1954), in the chernozems of the southern part of the steppe zone cultivated soils, virgin steppe soils, and soils under forest strips differ quantitatively and qualitatively in their microflora composition (Fig. 62, Table 87).

Table 86. *Microflora of dark-grey forest soil under an ash-oak stand and in a clearing (10^3 per gramme of abs. dry weight).*

| Depth, cm | Total number of bacteria | Non-spore-forming bacteria | | | Spore-forming bacteria | | | | | | | | Total number of Actinomycetes | Total number of promyxobacteria | Number of anaerobes (10^3 per g of soil) | Number of *Azotobacter* colonies in soil layers |
		Pigmented	Fluorescent	Mycobacteria	Total number of spores	*Bac. mycoides*	*Bac. mesentericus*	*Bac. megatherium*	*Bac. cereus*	*Bac. idosus*	*Bac. agglomeratus*	*Bac. virgulus*				
Dark-grey forest soil under forest																
3-10	8237	87	1230	990	766	4	4	14	86	615	15	28	1340	30	1420	2
10-20	12,200	230	1140	1840	867	6	4	8	63	770	12	4	975	12	1120	60·5
Dark-grey forest soil in clearing (fallow)																
0-10	8427	242	0	435	1676	38	810	109	89	342	287	1	2170	325	830	1·3
10-20	8884	49	1440	105	29	17	82	33	68	435	85	2	1180	320	530	1·3
Leached chernozem (ploughland)																
0-15	5915	695	710	670	351	2	11	158	31	118	31	0	870	0	1420	7·5

Cultivation stimulates the activity of microbiological processes. In the presence of moisture even virgin steppe soil possesses high biological activity. Where forest vegetation is established the soil microflora gradually acquires distinctive features; for instance, the numbers of fluorescent bacteria fall sharply in forest soil, *Azotobacter* is not observed, and the

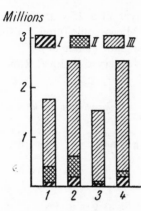

Fig. 62. Variations in numbers of non-spore-forming bacteria in chernozem soils (from E. V. Runov's data). 1—steppe soil; 2—between forest belts (fallow); 3—forest belt 15-16 years old; 4—forest belt 48-50 years old; I—pigmented bacteria; II—fluorescent bacteria; III—other non-spore-forming bacteria.

numbers of pigmented non-spore-forming bacteria increase. New forms of fungi appear, e.g., *Stysanus*, which is absent from cultivated soils; but the numbers of *Aspergillus*, etc., are much smaller.

E. I. Mishustin and his colleagues (1951) observed that cultivated chernozem soils that had been occupied by oak stands for 60 years had changed their microflora substantially, and in the appearance of their microbic coenose they approached virgin chernozem soil.

The activity of higher fungi increases in soils under forest vegetation. According to Runov and Mishustina (1960), 47 times as many fungal

Table 87. *Microflora of virgin steppe and forest strips on chernozems (average data for vegetative period, in 10^3 per g of dry soil) (Runov 1954).*

Location	Total number of bacteria	Spore-forming bacteria								Total number of				
		Total number of spores	*Bac. mycoides*	*Bac. mesentericus*	*Bac. megatherium*	*Bac. cereus*	*Bac. idosus*	*Bac. agglomeratus*	*Bac. virgulus*	Actinomycetes	Mycobacteria	*Azotobacter*	Butyric-acid bacteria	Microscopic fungi
Virgin steppe	2620	744	0	1	78	87	490	51	4	3316	528	33	67	61·5
Interstrip fallow	3077	959	12	34	126	118	558	27	5	2831	676	15	85	30·4
Forest strip 15-16 years	1846	916	24	5	133	232	420	22	24	1041	679	0	97	20·5
Forest strip 48-50 years	3226	673	4	13	71	53	463	42	0·6	2511	884	0	95	39·5

Table 88. *Number of pieces of hyphae and microscopic fungi in forest and cultivated soils (10^3 per g of dry soil).*

Type of soil	Pieces of hyphae	Microscopic fungi
Under forest	565	27
Cultivated glade	12	42

mycelia[1] were observed in the 0-70-cm soil layer in a 25-year-old tree plantation as in the soil of a cultivated glade. Microscopic fungi, on the contrary, were more numerous in the soil of the glade (Table 88).

An increased number of pieces of mycelia was observed by E. V. Runov and S. A. Valeva (1962) in forest podzolic soils.

Table 89. *Numbers and group-composition of micro-organisms in the uppermost soil layer under oak stands in different types of soil (average data).*

Soil	Forest type	Investigator	Total no. of micro-organisms	10^3 per g of soil				Proportions %			
				Non-spore-forming bacteria	Spore-forming bacteria	Actinomycetes	Microscopic fungi	Non-spore-forming bacteria	Spore-forming bacteria	Actinomycetes	Microscopic fungi
Soddy-podzolic (Moscow province)	Hairy-sedge oak, 70-80 years old	O. S.Terekhov, M. G. Enikeeva	2549	1292	654	537	66	50·6	25·6	21·2	2·6
Brown forest soil (Moscow province)	Hazel-hairy-sedge oak, 70-80 years old	O. S. Terekhov, M. G. Enikeeva	2506	1302	560	606	38	51·9	22·3	24·2	1·5
Leached chernozem (Orlov prov.)	Oak plantation, 60 years old	E. V. Runov, I. E. Mishustina	3996	3150	180	630	36	78·8	4·9	15·9	0·3
Dark-grey forest soil (Voronezh province)	Ash-oak with gout-wort, 220 years old	O. I. Pushkinskaya	8826	6657	766	1340	63	75·1	8·5	15·2	0·7
Podzolised-solodised solonetz (Voronezh province)	Oak stand, 50-70 years old	O. I. Pushkinskaya	5445	4389	189	750	117	80·6	3·5	13·8	2·1
Low-humus chernozem (Lugansk province)	Forest belt, 50-80 years old	E. V. Runov	5788	2565	673	2510	40	44·3	11·6	43·6	0·7
Cinnamonic forest soil (Bulgaria)	Oak stand	E. Kolcheva	6391	2320	380	3600	91	36·6	5·9	56·3	1·4

[1] A count of pieces of fungal mycelia gives some idea of the number of Basidiomycetes in the soil.

The specific effect of forest vegetation on the microbial population of forest soils has also been noted by S. A. Samtsevich and his colleagues (1949, Samtsevich 1955b), A. P. Vizir (1955), Timonin (1935), Bernat (1955), and others. According to these investigators, forest vegetation affects not only the numbers but also the qualitative composition of microflora.

From the examples quoted it is evident that the microflora of forest soils has its own characteristic features. A change in the qualitative and quantitative composition of the microflora in forest soils takes place under the constant and specific action of forest vegetation.

Meadow and steppe soils, and also cultivated soils, change their microflora substantially when they are afforested. When the forest is felled and the forest soil is again cultivated the soil apparently retains its peculiar features for a long time, while at the same time acquiring new features restoring similarity to the original ones.

The character of microbiological processes in the soils of forest biogeocoenoses is primarily determined by soil-climatic conditions. The investigations of E. N. Mishustin and his colleagues (1941, 1946 a, b, 1949, 1953, 1954, 1960, 1961) into the ecological characteristics of the distribution of micro-organisms have proved decisively that each type of soil has its own microbial associations. Against that background, the forest community determines the microflora of forest soils according to its typical tree stands. Tables 89 and 90 show the relationships between

Table 90. *Numbers and group-composition of micro-organisms in the uppermost soil layer under spruce stands in different kinds of soil (average data).*

Soil	Forest type	Investigator	Total no. of micro-organisms	10^3 per g of soil				Proportions %			
				Non-spore-forming bacteria	Spore-forming bacteria	Actinomycetes	Microscopic fungi	Non-spore-forming bacteria	Spore-forming bacteria	Actinomycetes	Microscopic fungi
Podzolic (Vologda prov.)	Spruce-bilberry with birch and ash, 120-140 years	E. V. Runov, S. A. Valeva	719	632	2	80	5	87·8	0·3	11·0	0·7
Soddy-podzolic (Moscow province)	Spruce-hazel-*Oxalis*, 70-80 years	O. S. Terekhov, M. G. Enikeeva	1650	1092	279	210	69	66·2	16·9	12·7	4·2
Leached chernozem (Orlov province)	Spruce plantation, 60 years	E. V. Runov, I. E. Mishustina	4403	2818	322	1130	73	64·0	7·3	25·7	1·6

soil types and the character of the microflora in oak and spruce stands.

The character of microbial coenoses in the same soils and in identical climatic conditions is affected not only by the tree stand but by the forest biogeocoenose as a whole.

V. G. Razumovskaya and N. N. Mustafova (1959) and Mustafova (1959) observed more spore-forming bacteria in the podzolic soil of a spruce-bilberry stand than in the same soil in a spruce-*Oxalis* stand. The dominant species were *Bac. mesentericus* in the spruce-bilberry stand and *Bac. subtilis* and *Bac. mycoides* in the spruce-*Oxalis* stand, indicating a difference in the organic-matter composition in the soils of these spruce stand types. The intensity of cellulose decomposition was higher in the spruce-*Oxalis* stand with herbaceous cover. Myxobacteria took part in it, being more active in breaking down cellulose than fungi and Actinomycetes, which predominated in the spruce-bilberry stand with moss cover. Such a difference in the cellulose-decomposing microflora indicates a difference in the composition of plant remains. Biological activity and nitrification were more vigorous in the spruce-*Oxalis* than in the spruce-bilberry stand.

A. I. Bursova (1955), studying the microflora in different types of spruce stands in Leningrad province, found the soils of composite spruce stands richer in total micro-organism content and in the number of physiological groups than the soils of spruce-bilberry and spruce-*Oxalis* stands.

O. I. Pushkinskaya (1954b), in studies of the microflora of soils in different types of forest (goutwort-sedge oak, linden-goutwort oak, hedge-maple oak, and hazel-elm) adapted to different relief conditions, also found considerable differences in the species- and group-composition of the micro-organisms.

Microbiological processes are somewhat depressed in dark-grey forest soils under oak stands with sedge (Runov and Egorova 1962). In the soils of oak stands, as compared with those of sedge-ash and sedge-poplar stands, the development of fluorescent bacteria and of several other micro-organisms is reduced. On the other hand, ash and poplar help to enrich the soil with fluorescent bacteria. Higher predominance of spore-forming bacteria is observed under oaks than under ash and poplar.

According to E. V. Runov and E. S. Kudrina (1954) and E. V. Runov (1960), micro-organisms are much fewer beneath pure oak plantations than in soils of mixed plantations (Table 91). A mixture of oak and Norway maple stimulated micro-organism development in the soil more than a mixture of oak and red-leaved ash.

Selection of species with different ecologo-biological characteristics may improve microbiological processes in forest soils and affect the development of the principal species.

Greater numbers of certain groups of micro-organisms in the soil under deciduous trees than in the soil under conifers have been observed in various soil types in natural conditions by Chase and Barker (1954),

Table 91. *Micro-organism content of chernozem soil under oak and mixed plantations* (*10^3 per g of abs. dry soil*) (*Runov 1960*).

Forest species	Bacteria	Actinomycetes	Microscopic fungi
Oak	1365	1460	32
Oak, Norway maple, caragana	4049	2720	99
Oak, red-leaved ash, caragana	2910	2265	59

Zimny (1960), E. V. Runov and S. A. Valeva (1962), O. S. Terekhov and M. G. Enikeeva (1964), and others.

The character of the micropopulation also is affected by the species of the tree stand. There are different proportions of physiological groups of micro-organisms under different tree species, each species making its own contribution to the microbial community. Differences in the soil microflora occur under different tree species.

M. A. Glazovskaya (1953) discovered differences in the numbers of Actinomycetes and yeast fungi and in the species-composition of bacteria and cellulose-decomposing micro-organisms in soils under Schrenk spruce and under Norway spruce. That author attributed differences in soil-forming processes to differences in the ash content of these species.

Higher fungi are more abundant under conifers than under deciduous trees (Table 92).

Table 92. *Number of pieces of hyphae* (*10^3 per g of dry soil*) *in the soil profile under pure stands on leached chernozem* (*Runov and Mishustina 1960*).

Depth, cm	Stand			
	Oak	Larch	Spruce	Pine
0-10	460	520	2540	970
10-20	290	340	1200	450
20-30	180	200	460	200

In leached chernozem soil the number of hyphal parts was two to three times as high under Norway spruce as under Engelmann spruce. No difference was observed in soils under Scots pine and under eastern white pine.

The differences were most evident under pure stands. According to Runov and Mishustina (1960), in equal-aged pure stands on leached chernozem the bacterial flora developed more vigorously under conifers than under deciduous trees; Actinomycetes, on the contrary, were more plentiful under conifers.

Soils under larch approach those under deciduous broad-leaved trees in their bacterial content (Table 93).

Soils under oak and birch are considerably richer in fluorescent and yellow-pigmented bacteria than soils in coniferous forests. The differences in the biogenic features of soils under deciduous and coniferous species have also been traced in the fermentative action of the soils. Catalase, saccharase, and peroxidase activities are higher in soils under deciduous species than in those under conifers (Runov and Mishustina 1960, Runov and Terekhov 1960).

Microbiological processes take place most actively in the upper humified layers of the soil, the number of micro-organisms falling sharply with increasing soil depth. The distribution of micro-organisms in the soil profile depends on the amount of organic and mineral matter available to the micro-organisms, the soil-solution reaction, the water-air regime of the soil, and other factors. The relationships between humus supply, pH, and number of micro-organisms through the profile of dark-grey forest soil in an ash-goutwort oak stand are shown in Table 94.

The smaller microbiological activity in the deeper soil horizons is due mainly to the reduced supplies of organic compounds available to micro-organisms and to unfavourable conditions of soil aeration. When water-soluble organic compounds are washed from the litter and upper soil layers into deeper soil horizons, some increase in micro-organism numbers occurs (Timonin 1935, Rybalkina 1957).

Table 93. *Composition of soil microflora under coniferous and deciduous stands (10^3 per g of dry soil).*

Stand	Total no. of bacteria		Total number of spore-forming bacteria	Fluorescent bacteria	Yellow-pigmented bacteria of type *Ps. herbicola*		Myco-bacteria		Actino-mycetes		Microscopic fungi
	MPA	SAA			MPA	SAA	MPA	SAA	MPA	SAA	
Oak	2880	7400	314	980	170	630	235	431	111	1034	56
Birch	3680	5700	356	890	390	430	263	245	121	1307	57
Spruce	1550	4300	252	340	80	70	108	390	207	1576	41
Pine	1700	3100	287	90	70	15	115	130	272	1915	42
Larch	2950	6400	276	610	130	190	198	119	139	2252	37
Open glade	3730	6880	463	650	140	825	35	410	60	1815	70

Table 94. *Relationships among humus supply, acidity of soil solution, and number of micro-organisms (O. I. Pushkinskaya).*

Horizon	Depth, cm	Humus, according to Tyurin, %	pH of aqueous infusion	Micro-organisms, 10^3 per g of dry soil		
				Bacteria	Fungi	Actinomycetes
A_0	0-3	62	7·4	109,850	934	17,700
A_1	3-10	12·2	7·3	8237	63	1340
A_1	10-20	5·5	6·8	12,220	51	975
A_2	20-30	4·75	6·7	5025	40	640
B_1	30-40	4·0	6·8	1926	23	355

Plant roots strongly affect the distribution of the microflora through the soil profile. A plant excretes a certain amount of organic compounds through its roots, and also sheds root filaments. That aids the multiplication of micro-organisms and their penetration into deeper soil horizons. According to Runov and Mishustina (1960), 34,000 micro-organisms were observed per gramme of soil at a depth of 2·5 metres (horizon C in leached chernozem), but 2,106,000 per gramme of soil alongside spruce roots. The distribution of micro-organisms in the soil is also strongly affected by worms, insects and their larvae, burrowing rodents, etc. S. A. Samtsevich and his colleagues (1949) observed that molehills, worm-holes, and dead roots in forest soils contained considerably more micro-organisms than the surrounding undisturbed soil.

E. V. Runov and I. E. Mishustina (1960) observed from 1·5 to 6 times as many bacteria at a depth of 30-70 cm as in the uppermost soil layer. These authors attribute such an increase in the numbers of bacteria in the deeper soil layers under 25-year-old birch and spruce plantations and in forest glades to the activity of earthworms.

The intensity of microbiological processes in forest soils undergoes seasonal fluctuations, which vary in different soil-climatic conditions and under different tree species.

In podzolic soil under a spruce-bilberry stand and under birch mixed with spruce, temperature and moisture conditions being favourable, the total number of micro-organisms increases from spring to autumn in all horizons (Runov and Valeva 1962). Along with the autumnal increase in separate groups of micro-organisms (bacteria, Actinomycetes, etc.) there is an increase in the proportion of cellulose-decomposing bacterial flora, and saturation of the soil with mycelia of higher fungi. When weather is unfavourable in autumn the number of micro-organisms falls.

The numbers of micro-organisms reach maxima in June and November in soddy-podzolic soils under hazel-*Oxalis* spruce stands, hairy-sedge oak stands, and hairy-sedge linden-oak stands (Terekhov and Enikeeva 1964). The fluctuations in micro-organism numbers there are correlated with fluctuations in soil moisture. Under a mountain-ash-hazel pine stand in the same conditions maximum development of micro-organisms is observed in July. In this case fluctuations in numbers of micro-organisms do not coincide with fluctuations in soil moisture (Fig. 63). Separate groups of micro-organisms do not develop equally during the course of the season. Non-spore-forming bacteria reach maximum development in the soil under linden stands in June, under spruce and oak in November, and under pine in July. Spore-forming bacteria and Actinomycetes reach their maximum numbers in the soil under conifers in autumn and under deciduous species in early summer.

A. V. Rybalkina (1957) observed that in soddy-podzolic soils, under young oak forest mixed with poplar and having a spruce understorey and hazel underbrush, bacteria occurred in maximum numbers in August. In oak forest and spruce-*Oxalis* stands, with the same soil conditions, the greatest numbers of bacteria were observed in June; and in a spruce stand on strongly-podzolised soil with dead ground cover they were observed in September.

T. V. Aristovskaya and O. M. Parinkina, studying the microbic landscape of ferro-humic podzols under mixed forest, observed definite fluctuations in the microflora during the growing season. In spring and

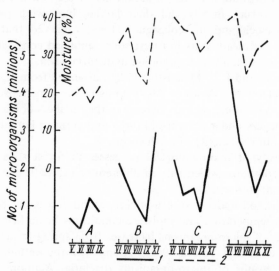

Fig. 63. Seasonal fluctuations in microflora and moisture in the upper soil layer, by months (from Terekhov and Enikeeva). 1—total number of micro-organisms; 2—soil moisture; A—pine stand; B—spruce stand; C—oak stand; D—linden stand.

autumn, with adequate soil moisture, various forms of bacteria, Protozoa, and micro-organisms were observed in abundance, accumulating iron and manganese. Profuse development of algae within the microbic landscape was observed in autumn. During the summer dry period fungal mycelia predominated and Actinomycetes appeared.

O. I. Pushkinskaya (1953), tracing the monthly development of microflora in dark-grey forest soil under a stand of ash and oak with goutwort, observed two peaks of development of micro-organisms during the growing period (Table 95). The high micro-organism content of Horizon A_1 in April was sharply reduced in May, with another rise in August, the same picture being observed in bacterial development. Almost the reverse was observed with regard to Actinomycetes and microscopic fungi in these soils. The numbers of Actinomycetes and fungi are less in April than in May, and their maxima occurred in July for Actinomycetes and in June for fungi. In Horizon A_2 no increase in the numbers of bacteria was observed after the spring fall; the greatest numbers of Actinomycetes were observed in April and July, but the numbers of microscopic fungi did not change throughout the season. In Horizon B_2 maximum numbers of bacteria were found in June, of Actinomycetes in May, and of microscopic fungi in August.

According to Feher (1933), in loamy soils in coniferous and broad-leaved forests in Finland bacteria reach maximum numbers in summer, but fluctuations in the numbers of separate physiological groups did not coincide with fluctuations in the total numbers of bacteria. The numbers of nitrogen-fixing bacteria reached a peak in late summer and autumn; nitrifying bacteria had two peaks, in spring and autumn; and the fluctuations in the numbers of denitrifying bacteria were opposite to those of nitrifying bacteria. Several differences were observed in the numbers of bacteria in the soils of deciduous and coniferous forests in the same conditions of temperature and precipitation. Cellulose-decomposing bacteria had maximum development in coniferous forests in early spring and in deciduous forests in summer.

From the data quoted it is seen that seasonal fluctuations in micro-organism numbers are different in different soil-climatic zones, and in coniferous and deciduous forests in the same zone. Seasonal variations in micro-organism numbers are primarily due to changes in the ecological and physico-chemical properties of the soils (mainly in their moisture and temperature). Also important are the rate of root exudation and the periodicity of dying-off of roots and shedding of leaves and needles, which provide saprophytic micro-organisms with a source of food.

In humid tropical conditions, according to Korbe, no seasonal fluctuations are observed in the numbers of bacteria and fungi in forest soils.

Higher plants may affect micro-organism numbers by causing variations in the bactericidal properties of roots and soil.

TOXICOSIS OF FOREST SOILS

Toxicosis of forest soils plays a definite role in the life of plants and micro-organisms in forest biogeocoenoses, but has not been well studied. Exhaustion of soils under farm crops and fruit trees has been studied in much more detail, but still inadequately. The subject of toxicosis of forest soils, nevertheless, is linked with urgent theoretical and practical

Table 95. *Dynamics of development of micro-organisms in horizon* A_1 *in dark-grey forest soils in stands of ash and oak with goutwort* (10^3 *per g of dry soil*) (*O. I. Pushkinskaya*).

Month	Moisture %	Bacteria	Actino-mycetes	Microscopic fungi	Total number of micro-organisms
April	No data	65,232	11,382	67	76,681
May	25	5440	12,814	132	18,386
June	20	12,937	7140	177	20,254
July	20	7405	14,624	83	22,112
August	22	25,210	9962	107	35,279
September	18	7280	3604	111	10,994

problems of sylviculture. The discovery and investigation of toxicosis of soils under tree stands will facilitate analysis of the complex interactions between separate components of a forest, especially between trees, soil, and micro-organisms. Study of the toxic properties of soils will help to explain the processes of forest regeneration and succession of tree species, toxicosis of soils probably being one aspect of the effect of trees on soil.

Toxicosis of soils does not occur in all types of soil and forests, but it may appear in particular cases with specific conditions and specific interactions between forest components.

In a forest biogeocoenose soil toxicity shows up much more than in soils under herbaceous cover or in cultivated soils. It is most clearly evident in the podzol zone, where forest soils are more toxic than unforested soils, the tree species strongly affecting the degree of soil toxicity. Podzolic soil under spruce is most toxic (and that under pine and poplar least so) for microbes (*Azotobacter*) and herbaceous plants (wheat, beets). Soils under birch and oak are only slightly, or not at all, toxic. Soils become less toxic after forest-cutting and ploughing (Krasil'nikov *et al.* 1955). Higher toxicity of forest soils, compared with those under herbs or cultivated, has been noted by N. Rayner and V. Neilson-Jones (1949), E. V. Runov and M. G. Enikeeva (1959), and others. R. A. Zhukova (1959), while studying toxicity in the soils of Kola peninsula in connection with cellulose-fermenting aerobic bacteria, observed differences in the degree of soil toxicity depending on the tree-stand species. Ferrous podzol developing under suppressed pines with lichen cover is slightly toxic. The same podzol under scrub birch with a mixture of spruce and juniper, and also with heather, cowberry, and bilberry, contains many toxic substances that suppress the development of cellulose-fermenting aerobic bacteria. The soddy meadow soils of Kola peninsula, according to Zhukova, are almost non-toxic. Most fenlands are slightly or not at all toxic, but in contrast raised bogs and transitional bogs similar to them are highly toxic.

Toxicity (to *Azotobacter* and wheat) of chernozem soils under tree plantations in the arid-steppe zone was described by Runov and Enikeeva (1959). Chernozems are more toxic under unmixed oak stands than in mixed plantations (oak-ash-acacia). In the arid-steppe zone soil toxicity is found to increase during the dry period and to fall when the soils are moistened. Toxicity in chernozems is displayed towards both *Azotobacter* and shoots of higher plants.

No definite regularity is observed in the distribution of toxicity through the soil profile. Rayner and Neilson-Jones state that toxicity towards fungi appears in the 0-30 cm layer but disappears lower down.

E. P. Gromyko (1960) recorded greatest toxicity in podzolic soils in the 30-60-cm layer. R. A. Zhukova (1960) mentions the irregularity of distribution of toxicity through the soil profile. According to her, in the podzol zone the upper soil horizons are in most cases toxic.

Toxicity may fluctuate during the vegetative period. Gromyko records high toxicity for *Azotobacter* in spring and early May, with a fall by autumn. Runov and Enikeeva observed greater toxicity of arid-steppe chernozems during the dry period than during wet periods. According to N. A. Krasil'nikov and his colleagues, toxicity is found in podzolic soils in summer and autumn; in late autumn and winter it decreases, and it is at a minimum by spring. Rayner and Neilson-Jones report maximum toxicity in autumn and winter, when soil moisture is highest.

These facts indicate that the nature of toxic substances in the soil and their properties and origin may be extremely variable. They depend on the characteristics of the soil, its micropopulation, climatic conditions, etc. Toxicity may be due to substances of plant or microbial origin, or to certain mineral substances. In soils with low pH toxicity for micro-organisms is due to the presence of iron and aluminium salts (Katznelson 1940, Gromyko 1960).

N. Rayner and V. Neilson-Jones have pointed out that the toxicity of the soils of Wareham forest (England) is due to the formation and ac-cumulation of hydrogen sulphide as a result of the activity of sulphur-reducing bacteria. The H_2S so formed suppresses the development of mycorrhiza-forming fungi and the growth of pine seedlings, and conse-quently afforestation of these soils is unsuccessful, tree plantations de-veloping poorly. Soil toxicity due to accumulation of quantities of H_2SO_4 was found by Subramony (1960) in acid peat soils in Kerala (India). It appeared that these soils, the subsoil, and the bedrock had a high sulphur content. After comparing the results of chemical analysis of different forms of sulphur with those of bacterial investigations, that author concluded that there was intensive bacterial oxidation of sulphur and its products in these soils.

The investigations of Duchaufour and Rousseau (1959) in fir and spruce plantations in the mountain districts of the Vosges and Jura (France) showed that poisoning of coniferous seedlings was due to a high manganese content in the soils. Toxicosis has been discovered both in acid soils (pH 3·4-4·7) and in those with pH 5·0-6·4. In the latter case, with parent rock rich in calcium, toxicosis is more rare. Dense herbaceous cover, especially of grasses, increases toxicity to seedlings, whereas mossy cover decreases it. The authors recommend that reafforestation should be accompanied by the elimination of grass cover and the removal of strips of the toxic humic layer of the soil.

Other investigators attribute toxicity of forest soils to the organic part of the soil, with toxic properties in its humus.

Micola (1952) observed that forest humus was toxic to the fungus *Pythium ultimum*, toxicity being most evident in extracts from humus under pines; beech humus came next, and the extracts from birch humus were least toxic.

Bublitz (1954) believes that substances are formed in coniferous forest soils that suppress microbiological processes and humification of the soil. By the chromatographic method he proved that toxic substances in the humus part of the soil are more active than those in needles. The nature of the toxic substances in the humus has not been precisely determined; Bublitz classifies them as terpenes. The antimicrobial substances in humus are partly obtained from needles and partly formed during the humification process.

The breakdown of toxic substances in needles and the formation of toxic substances in humus are due to the activities of fungi.

R. Moreau (1959) investigated the toxicity of aqueous extracts from the humus layer of forest soils. He studied the effect of an aqueous extract from the soil of an old (200 years) fir forest on seed germination and on the microflora of young fir-forest soils and found that the soil extract from the old forest had a toxic effect. Germination of garden cress seeds was 15% lower in the extract than in the control. Extracts from the soil of the old fir forest, and also from that of a beech forest, markedly depressed the development of the microflora of soil from a young forest. The author attributes the difference to the humus composition.

The toxic properties of certain soils are largely due, it appears, to biological factors. A number of investigators (Vorob'eva and Shchepetil'nikova 1936, Gorlenko 1946, Krasil'nikov 1946, Krasil'nikov et al. 1955, Dobbs and Hinson 1953) regard toxicosis as the result of accumulation of toxic micro-organisms and the products of their life-activity in the soil.

According to N. A. Krasil'nikov, toxicity of soils is largely due to the activity of inhibitor-microbes, which produce toxic substances that can accumulate in the soil. Such substances can be produced by bacteria, fungi, and Actinomycetes. Toxicity in soddy-podzolic soils is due to the development of toxic species of fungi, belonging mostly to the genus *Penicillium*. Some species of the genera *Fusarium* and *Trichoderma* are also toxic.

The old theory of soil toxicosis is worthy of attention: it was based on the possibility of saturation of the soil with toxic substances produced by the plants themselves during their growth (De Candoll 1913, Ishcherekov 1910, Rippel 1936, Mishustin and Naumova 1955, Winter 1952, Martin 1948, 1958). In that case the sources of toxic substances in the soil would be dead roots, root exudates, and shed leaves, bark, and other parts of plants, saturation of the soil with toxins from the plants being due to the one-crop system and therefore closely linked with the problem of soil exhaustion. Soil exhaustion has been studied mostly in connection with the prolonged growing of farm crops without rotation.

In forestry the problem of soil exhaustion has been practically untouched, although nowadays facts are available pointing to possible soil exhaustion under tree crops also. Soil exhaustion is often observed under fruit trees.

Kemmer (1935) observed soil exhaustion occurring frequently under apple trees. Terminal growth ceased in young apple trees in such soils within two years. After transplantation of the ailing trees into healthy soil the condition disappeared. Soil exhaustion is also known in connection with other species: cherries (Bronsart 1949), citrus fruits, figs, and peaches. In California and Canada growers often have great difficulty in regenerating old peach orchards. Young peach trees in old soils are stunted and sometimes die off completely. The external pathological symptoms are retarded growth, stunting, and chlorosis.

Patrick (1955) demonstrated the connection between the chemical composition of peach trees and the toxic effect of the soil in peach orchards.

Use of mineral fertilisers, and of micro-elements in varying quantities and combinations, does not eliminate soil exhaustion. Placing peach roots in virgin soil leads to suppression of their growth.

There are many theories giving different reasons for soil exhaustion (impoverishment of the soil in elements of plant nutrition, or accumulation of phytopathogenic organisms (nematodes, etc.) in the soil). The causes may apparently vary, depending on the circumstances.

Soil toxicity in oak plantations and stands was studied by E. V. Runov and his colleagues (1956, 1963) in connection with regeneration, succession of species, and planting of oaks in the forest-steppe zones of the Soviet Union. The investigations were made in the arid-steppe zone, on ordinary low-humus chernozems (Derkul', Lugansk province) and in oak stands of different ages on dark-grey forest soils (Tellerman forest station, Voronezh province).

Oak roots, as well as above-ground parts, contain phytoncidal substances that suppress the development of many micro-organisms and of higher plant seedlings (Runov and Egorova 1958, Egorova 1962, Runov and Enikeeva 1963). The toxic action of these substances is due, apparently, to the presence of phenol compounds in tanning substances. Entering the soil, these substances, either as they stand or after breakdown by microbes, may accumulate and make the soil toxic. In the arid-steppe and forest-steppe zones periods of drought enable these substances to remain in the soil, whereas during wet periods they are broken down by soil micro-organisms. In fact, as investigations have shown, the greatest soil toxicity is observed during drought periods, and it declines when the soil is moistened.

The activity and abundance of microflora play a major role in soil toxicosis in oak plantations and stands. On analysis of data on the microflora of the litter, soil, and rhizosphere of forest trees, it is easily seen that micro-organisms are as a rule fewer under oaks than under other species, birch, ash, poplar, etc. Microflora is also less developed in pure oak stands than in mixed stands (oak and ash and acacia). Such depression of microflora indicates that oak may in certain circumstances have a toxic effect on soil.

The fact that as oak stands grow older the toxicity of the soil increases

(Runov and Egorova 1963) is worthy of attention. That affects both micro-organisms and, what is more important, oak seedlings. Observations on seedling development in soil under oaks of various ages (25, 52, 70, 220 years) have shown that, as the age of the stand increases, so does the suppression of seedling growth. The soil toxicity apparent in old oak stands may be one of the factors restricting natural oak regeneration.

The above facts indicate the complex nature of soil toxicosis, which may appear in different forms, depending on conditions, or may be completely absent. It may be transitory, and its depressing effect on the growth of plants and microbes may appear for only a brief period. It has a different form in every biogeocoenose. When the soil is soaked with snow melt-water, or in time of frost or drought, or, on the contrary, during wet periods, or when organic matter or lime is added, soil toxicity may decline or even disappear. Further analysis of the problem of toxicosis is required in order to understand the interactions of the forest with soil and microflora.

INTER-RELATIONSHIPS OF MICRO-ORGANISMS AND TREES

Climatic conditions and soil types are of great importance in determining the species-composition of the microflora and the direction of micro-biological processes in forest soils. Another powerful factor affecting the microflora is the forest biogeocoenose, in which arboreal vegetation is dominant. Under the influence of different trees distinctive microbial populations develop in the soil, depending on stand composition, forest type, age of trees, etc.

In recent years a mass of data has been accumulated demonstrating the specific effects of plants on soil microflora. Besides the physico-chemical features of the soil, its content of organic and mineral nutrients, and climatic conditions, plants also have a strong selective influence on micro-organisms.

During studies of the microflora of forest soils definite regularities in the distribution of micro-organisms have been observed, depending on the species-composition of tree stands in identical soil-climatic conditions, indicating that in a forest biogeocoenose the exchange of matter and energy between vegetation, soil, and micro-organisms depends directly on the species of the trees. The varying qualitative and quantitative composition of the microflora under different tree species is also evidence of distinctive forms of the processes of humification and transformation of mineral matter in the soils of different tree stands, which no doubt also show their effects in the formation of forest soils and in plant nutrition.

The effect of forest vegetation on the microflora is most clearly displayed in the features of the microbial population of forest litter and of the rhizospheres of forest plants. Forest litter and tree root systems are leading factors in the creation of specific microbial coenoses in forest soils, since that is where the principal microbiological processes take place because of the accumulation of organic matter. We must not overlook another aspect of that situation, namely, the role of micro-organisms in the decomposition of plant remains and the influence of microflora on plant growth. The principal transformations of organic and mineral matter are carried out in litter and rhizospheres by micro-organisms.

Data on the mutual influence of trees and micro-organisms are comparatively few. Much more is known of the inter-relationships of micro-organisms and herbaceous plants (mainly farm crops). Descriptions of the inter-relationships between trees and microbes are therefore mostly based on analogies with herbs, but the laws relating to trees may be entirely different.

MICROFLORA OF FOREST LITTER

The amount and the characteristics of forest litter are constantly varying, both during the growing period and with the age of the tree stand; it is affected by a number of factors, among which microbiological processes are outstanding.

A considerable proportion of the organic matter formed by trees returns to the soil during their lifetime in the form of shed leaves, needles, branches, and fruits, dead roots, and various exudates. It all undergoes mineralisation and transformation by various representatives of the microflora, whose species-composition is determined by the biochemical features of the tree species typical of a given biogeocoenose. As a result of the activity of micro-organisms great masses of organic matter in the litter are decomposed, so that the soil is enriched with mineral elements. The rates and periods of that enrichment depend on the conditions and extent of decomposition of the litter under microfloral action. As the litter breaks down various organic products are formed, changing the chemical composition of humic matter in the soil.

The microbial population of litter is extremely diverse and has quantitative and qualitative features distinguishing it from the soil microflora. In every gramme of dry weight of litter there are tens and hundreds of times more micro-organisms than in a corresponding weight of soil. As well as bacteria, which are the chief groups of micro-organisms, microscopic and higher fungi, Actinomycetes, algae, Protozoa, and also a number of invertebrates develop there. For all these organisms litter is a source of nutrients necessary for their constructive and energetic metabolism. The litter is colonised by distinctive species of micro-organisms, depending on its biochemical composition and the soil-climatic conditions, and consequently its decomposition proceeds in a definite way and at a definite rate.

There are very great differences in the microflora found in the litter of coniferous and of deciduous forests, which have types of litter differing in their content of organic matter and mineral elements. In stands of equal age (25 years) of leached chernozem (Orlov province) detailed comparisons between the microfloras of spruce, pine, larch, oak, and birch litter have been made by Runov and Mishustina (1960).

Birch litter has most bacteria present; oak litter is much poorer in micro-organisms. Among conifers, spruce and pine litter have the smallest numbers of bacteria, with larch approaching the deciduous broad-leaved species (Table 96).

The most numerous group taking part in the decomposition of forest litter is non-spore forming bacteria, which constitute 80-90% of the total number of micro-organisms. The greatest numbers of non-spore-forming bacteria inhabit deciduous forest litter, the bacterial flora of which is from two to seven times as abundant as that under spruce and pine. The numbers of spore-forming bacteria and Actinomycetes are higher under conifers. The greatest proportion of Actinomycetes is found in spruce and pine litter and the least in birch litter. Microscopic fungi are fewer under spruce and pine than under the other tree species.

The microfloras of the litter of different tree species differ both in their total numbers and in the proportions of separate species and groups. Birch and larch litter are comparatively rich in yellow-pigmented bacteria of the *Pseudomonas herbicola* type, fluorescent bacteria, and mycobacteria. Yellow-pigmented bacteria and mycobacteria are comparatively few under oak. Under spruce and pine representatives of these three groups rarely occur, and there is also selective representation of forms of microscopic fungi. One of the most distinctive groups is the *Penicillium* group; along with it, *Cladosporium* occurs in large numbers under birch, pycnidial fungi under larch, and *Trichoderma* under pine.

In the broad-leaved forest zone the litter microflora also shows substantial differences according to the tree species. As a rule, ash litter is the richest in micro-organisms and oak the poorest (Vizir 1956, Runov and Egorova 1962). Both qualitative and quantitative differences are observed.

The degree of micro-organism colonisation of the litter of different deciduous species is correlated with the availability of mineral elements.

Table 96. *Micro-organism content of litter of different forest stands (10^3 per g of oven-dry matter)* (*Runov and Mishustina 1960*).

Stand	Total no. of micro-organisms	Non-spore-forming bacteria	Spores of bacteria	Actinomycetes	Microscopic fungi	Percentages			
						Non-spore-forming bacteria	Spores of bacteria	Actino-mycetes	Fungi
Oak	84,388	77,200	258	6150	780	91·5	0·30	7·3	0·9
Birch	240,764	236,200	104	2200	2260	98·1	0·04	0·9	0·9
Spruce	44,414	35,680	736	7500	498	80·3	1·60	16·9	1·1
Pine	35,351	30,170	529	3900	752	85·3	1·50	11·0	2·1
Larch	185,944	175,600	1352	7800	1192	94·4	0·70	4·2	0·7

Mineralisation takes place most rapidly in ash litter (Utenkova 1956, Nykvist 1959a, Gilbert and Bocock 1960) and birch litter (Micola 1960), which are rich in the chief nutritive elements (N, P, Ca, K). In oak litter the majority of mineral elements remain insoluble until summer, and only in summer does the rate of their mineralisation increase.

The breakdown of composite litter and of mineral elements leached from it proceeds more rapidly than that of litter from oak leaves alone (Utenkova 1956, Pokhiton 1958). Litter from the conifers (pine and spruce) mineralises slowly (Micola 1960, Nykvist 1959b).

The numbers of micro-organisms in litter vary widely according to the soil-climatic conditions of the locality. Consequently the rate of decomposition of litter of the same tree species is different in different biogeocoenoses (Table 97).

Table 97. *Number of micro-organisms in forest litter in different soils (number of micro-organisms growing in MPA) (10^3 per g of dry matter, average data).*

Soil type	Oak		Spruce		Author
	Litter	Soil	Litter	Soil	
Podzolic (Vologda province)	—	—	4886	716	E. V. Runov, 1962
Soddy-podzolic (Moscow province)	127,342	1292*	54,638	1092*	E. V. Runov, 1962
Leached chernozem (Orlov province)	84,388	2480	44,414	1681	E. V. Runov, 1962
Dark-grey forest soil (Voronezh province)	67,400	2505	—	—	O. I. Pushkinskaya, 1954
Ordinary chernozem (Lugansk province)	21,345	1104	—	—	E. V. Runov, 1954
Southern chernozem (Kirovograd province)	11,434	1256	—	—	A. P. Vizir, 1958

* Non-spore-forming bacteria in MPA.

In moderately moist conditions with adequate warmth, abundant development of microflora is observed in oak litter; with shortage of moisture (in more southerly districts) oak litter contains fewer micro-organisms. In such conditions the effects of the products of litter decay on soil and on plant nutrition are apparently quite different, and their passage into the soil, due to microfloral development, proceeds at different rates. In spite of the adequate amount of moisture in northern districts, colonisation of conifer (spruce) litter there is at a low level. Evidently in that case other factors (lack of heat, short growing period, etc.) restrict the development of microflora in litter.

The numbers and species-composition of micro-organisms in forest litter fluctuate considerably during the growing period, causing an uneven rhythm of litter decomposition and of passage of mineralisation products into the soil. That is due to the periodicity of arrival of fresh organic matter (litter fall), to the successive stages of its breakdown, and to the temperature and moisture regimes characteristic of the geographical zone.

Depression of microbiological processes in litter at the beginning of the growing period (in spring) is characteristic of northern districts (Table 98). Litter decomposition results from autumnal development of microflora, a result of the late warming of the soil.

Table 98. *Seasonal dynamics of litter microflora on podzolic soils in Vologda province* $(10^3$ *per g of dry matter*) (*Runov and Valeva 1962*).

Soil type	Time of sampling	Horizon	Non-spore-forming bacteria	Spore-forming bacteria	Actino-mycetes	Micro-scopic fungi
Humus-peaty	June	A'_0	44,689	79	2625	132
		A''_0	10,228	38	552	27
Humus-peaty	August	A'_0	106,835	23	6309	9·7
		A''_0	15,687	31	1825	1·8
Humus-peaty	September	A'_0	141,348	144	14,495	165
		A''_0	27,585	111	4550	28
Podzolic	June	A_0	2028	10	52	36
Podzolic	August	A_0	16,407	6	1398	83
Podzolic	September	A_0	17,172	110	1944	181

At the southern limits of spruce range, litter decomposition, accompanied by maximum microfloral development, takes place in spring and early summer.

Bacteria take part in the decomposition of oak litter on the dark-grey forest soil of Voronezh province throughout the growing period. By autumn, however, the number of microscopic fungi increases considerably (from 250,000 per gramme of soil in April to 1,504,000 in September), and also the number of Actinomycetes (from 70,780,000 in April to 324,460,000 in September) (Pushkinskaya 1953).

In more southerly districts (Lugansk province), on ordinary low-humus chernozems, bacterial breakdown of litter in oak stands takes place mostly during the spring months (Table 99). By autumn the numbers of bacteria fall, but the numbers of Actinomycetes rise (Runov 1954).

In the broad-leaved forest zone the quantity of bacteria, particularly non-spore-forming bacteria, increases in forest litter in autumn after the leaves fall (Runov 1954, Vizir 1956). That development is due to the arrival of fresh organic matter (with fallen leaves) and to increase in moisture. An autumnal increase in the numbers of Actinomycetes in the litter of various tree stands has also been recorded from the leached chernozem of Orlov province and the podzolic soils of Vologda province.

In summer there are fluctuations not only in the total numbers of micro-organisms and of the separate physiological groups, but also in the species-composition of micro-organisms within the same group. For instance, on leached chernozem different stages of the fungal process of litter decomposition are carried out by different species of microscopic fungi. The numbers of microscopic fungi reach a peak in summer and are low in spring and autumn. In spring, in the first phases of microfloral development, yeast-like forms and pycnidial fungi of the *Phoma* type occur in coniferous and deciduous litter. In summer, during the period of maximum development of fungal microflora, *Penicillium* spp. and *Cladosporium* spp. predominate. Mucoraceae develop mainly in late autumn. The tendency of microscopic fungi to develop at distinctive stages of the decomposition of plant remains is due to their physiological characteristics. Members of the genus *Penicillium* break down cellulose vigorously (Khalabuda 1948), whereas Mucoraceae are able to break down relatively unstable organic compounds and proteins, and evidently (as V. Ya. Chastukhin suggests) develop by use of products of microbial synthesis. A similar pattern of succession of microscopic fungal species was observed by V. Ya. Chastukhin (1948) in studying the breakdown of logs in spruce and pine forests.

We may assume that definite stages of litter decomposition correspond to the phases of maximum development of specific groups of micro-organisms. This assumption is based on the known fact of succession of groups of micro-organisms during the decomposition of plant remains: at first yeast fungi and non-spore-forming bacteria develop, followed by spore-forming bacteria, then by cellulose myxobacteria, and finally by Actinomycetes (Kononova 1951).

As a result of fluctuations in the numbers and species-composition of micro-organisms in the course of litter decomposition during the vegetative period, the liberation of mineral elements from the litter of different tree species takes place irregularly (Stepanov 1929, 1940, Shumakov 1941, Boswell 1956, Nykvist 1959 a, b, Slovikovskii 1960).

The breakdown of litter by micro-organisms does not take place uniformly throughout its depth. The higher and lower layers of coniferous

Table 99. *Seasonal dynamics of microflora in forest litter on ordinary chernozem (10^3 per g of litter)* (*Runov 1954*).

Nature of stand	Time of sampling	Total no. of bacteria per g of soil	Non-spore-forming bacteria	Total number of spores	Actinomycetes	Microscopic fungi	Cellulose-decomposing micro-organisms (% of fragments infested)
Forest area, 15-16 years old, ash, oak, caragana	Spring	52,731	51,273	337	35,356	537	100
	Drought period	43,222	43,181	127	22,663	307	50
	After rain	23,699	22,121	487	44,351	138	20
Forest belt, 48-50 years old, oak, ash	Spring	53,376	52,816	269	19,918	537	100
	Drought period	37,807	37,722	68	12,602	192	10
	After rain	32,071	31,018	265	25,645	409	15

and deciduous litter differ greatly from each other in water-air conditions, in ash content, and in the content and form of organic matter. The process of mineralisation of organic matter varies according to the succession of microflora from top to bottom of the litter.

In coniferous stands on leached chernozem the upper, slightly-decomposed layer of litter is occupied mainly by bacteria—fluorescent and yellow-pigmented bacteria of the *Pseudomonas herbicola* type—and mycobacteria. Apparently these forms begin the breakdown of the needles, which are rich in resin, terpenes, etc. (unavailable to most micro-organisms) as well as in the usual organic and mineral compounds. In the lower humified layers of the litter there is an increase in the numbers of Actinomycetes, microscopic fungi, and spore-forming bacteria (Table 100), and nitrifying bacteria appear. Variation in the species-composition of the micro-organisms, according to the type of litter layer, is also observed. Breakdown of cellulose takes place slowly in the upper layers. Most of the cellulose-decomposing micro-organisms (bacteria, fungi, Actinomycetes) are found in the lower layers of coniferous litter. The intensity of cellulose breakdown in the lower layers of litter is also noted by Golley (1960). The presence of easily-available forms of nitrogen in the substrate is necessary for the development of cellulose-decomposing bacteria.

Table 100. *Microflora of different layers of litter in coniferous stands (10^6 per g of oven-dry matter) (Runov and Mishustina 1960).*

Micro-organisms	Norway spruce		Scots pine	
	A'_0	A''_0	A'_0	A''_0
Number of bacteria (MPA)	28·35	8·85	9·12	3·72
Bacterial spores (MPA and grape-must agar)	0	0·13	0	0·11
Yellow-pigmented bacteria of *Ps. herbicola* type	0·26	1·32	0·60	0·03
Fluorescent bacteria	13·91	5·52	2·04	1·32
Mycobacteria	1·30	0·24	1·32	0·36
Actinomycetes	1·60	8·28	0·24	0·72
Microscopic fungi	0·12	0·24	0·34	0·63
Nitrifying bacteria (titre)	0	0·1	0	0·01
Number of fragments of fungal hyphae	1·35	6·30	1·45	4·32

Mycelia of higher fungi usually develop in the lower layers of litter, being much fewer in the upper layers. Distinctive species of microscopic fungi are characteristic of the separate layers of litter (Brezhnev 1950b, Shilova 1951, Kendrick 1959). Brezhnev, studying the microflora of the 'Les na Borskle' reserve, observed *Mucor* and *Rizopus* in the upper layers and *Tieghemella* and *Zygorhynchus* in the lower. Kendrick records predominance of *Lophodermium pinestris* in the undecomposed layer of pine litter, the same in the slightly-decomposed layer, and *Trichoderma* and *Penicillium* in the much-decomposed layer.

As a result of the life-activity of micro-organisms and the successive changes in microflora displayed in their layer-distribution, organic and mineral matter in litter undergo substantial transformations. With transition from fresh litter to litter of varying degrees of decomposition and finally to mineral horizons of the soil, organic matter becomes poorer in available substances, becomes more stable, and contains smaller proportions of carbon and nitrogen (Runov and Sokolov 1958 (Table 101).

In the lower layers of litter in coniferous stands the amount of water-soluble organic compounds decreases considerably. Mineralisation of humified, poorly-soluble organic matter is performed not by non-spore-forming bacteria but by Actinomycetes, fungi, and spore-forming bacteria, which predominate in the lower layers.

The micro-organisms that break down the litter of deciduous trees also vary in different layers (Table 102).

In the litter of deciduous stands the upper layers are most biogenic

Table 101. *Biochemical composition of organic matter in litter of various tree stands on leached chernozem* (% *of dry non-ash matter*) (*Runov and Sokolov 1958*).

Sub-horizon or horizon	Water-soluble compounds				Wax-resins	Hemi-cellu-lose	Cellu-lose	Unhydro-lysable residue	Proteins	Total
	Dry residue (non-ash)	including								
		sugars	C	N						
Larch										
A'_0	7·84	1·46	4·24	0·155	7·33	15·04	18·29	36·79	8·28	93·57
A''_0	5·98	0·26	2·92	0·158	5·07	14·98	15·16	43·43	9·85	94·47
Spruce										
A'_0	9·86	0·46	5·21	0·181	6·00	15·38	17·64	32·30	6·57	87·75
A''_0	9·70	0·57	4·74	0·293	4·15	13·86	11·39	37·95	11·59	88·04
Pine										
A'_0	8·87	1·31	4·81	0·150	9·83	15·09	13·13	34·47	6·35	89·74
A''_0	8·14	0·92	4·21	0·215	7·60	14·79	21·67	33·55	8·85	94·60
Oak										
A_0	8·94	1·55	4·24	0·238	5·12	16·32	17·08	32·75	10·65	90·41

Table 102. *Microflora in forest litter and soil on southern chernozems* (10^3 *per g of litter or soil*) (*Vizir 1955*).

	Oak			Ash		
	Total number	Spore-forming	Fungi	Total number	Spore-forming	Fungi
Litter						
Upper layer	4320	1470	60·7	7230	990	278·1
Lower layer	3630	1320	188·8	4104	1210	192·5
Soil 0-3 cm	1275	369	45·0	1731	627	25·0

(Pushkinskaya 1953, Vizir 1958). Micro-organisms are fewer in the lower layers, indicating relatively rapid and complete mineralisation of deciduous litter.

A very similar type of litter decomposition by layers in beech and spruce forests in Germany, depending on tree species and soil type, has been studied by Meyer (1960). In comparing brown and podzolic soils he observed that there were more bacteria than fungi in the litter of both species on brown soils. Different layers of litter had distinctive species of bacteria and fungi. The numbers of micro-organisms in different layers, as well as the rate of respiration, depend on the organic-matter content (Tables 103, 104). In stands of both species the greatest numbers of micro-organisms were found in the decomposing layer. In the topmost layer there were few microbes, and in the lower layers also the numbers were small. There were more fungi under spruce, on account of a difference in pH: spruce humus has a lower pH than beech humus.

Table 103. *Microflora and respiration in different layers of litter on different types of soil.*

Layer of litter	Bacteria (10^6)		Fungi (10^6)		Intensity of respiration, microlitres of oxygen per g of organic matter per hour	
	Beech	Spruce	Beech	Spruce	Beech	Spruce
Brown soil						
L	6·9	5·7	0·9	4·0	298	226
F_1	97·5	93·7	2·5	3·8	370	152
F_2	81·9	66·1	4·6	4·4	202	64
H_1	80·0	16·6	6·3	3·6	80	45
H_2	4·3	49·9	6·3	11·5	38	24
Podzolised brown soil						
L	20·8	5·3	2·5	4·9	231	192
F_1	107·7	40·8	4·3	3·5	285	117
F_2	62·3	15·7	12·0	3·5	156	36
F_3	—	12·2	—	7·3	—	32
H_1	89·5	27·3	5·2	4·7	73	19
Pseudogley podzol						
L	10·5	7·8	3·4	4·4	248	222
F_1	112·2	72·0	4·3	5·6	266	98
F_2	14·2	13·5	5·6	3·7	126	40
F_3	12·2	2·6	4·1	3·7	47	23
H_1	15·8	5·2	3·2	1·0	37	6
H_2	6·8	2·0	0·3	1·0	17	3
Podzol						
L	2·6	13·9	2·3	8·1	220	188
F_1	2·1	31·2	6·8	6·7	290	76
F_2	7·4	1·1	4·0	5·3	150	33
F_3	3·7	0·4	4·0	3·2	49	16
H_1	2·3	0·3	2·3	1·6	19	6
H_2	1·9	—	0·8	—	9·9	—

In the decomposing layers the intensity of microbiological activity depends more on tree species than on soil type. Under spruce the activity is 22-40% lower than under beech. In the humified layers of litter, on the other hand, activity depends more on soil type than on tree species. Greater intensity of microbiological processes is observed in brown soil than in podzolic soil.

From study of the amount of respiration Meyer concludes that the time required for litter decomposition to be completed depends on the soil type. Litter becomes fully mineralised most rapidly in brown soil, more slowly in podzolic brown soil. Litter decomposition is still slower in pseudogley podzol and slowest of all in ferruginous podzol.

Kendrick (1959) believes that litter decomposition in English pine forests is completed in approximately ten years. Shed pine needles are

Table 104. *Thickness of layer, pH, organic matter, nitrogen, and microflora in different types of soil in a spruce forest.*

Layer of litter	Thickness of layer of litter (cm)	pH	Organic matter (% of dry matter)	Total nitrogen (% of organic matter)	C/N	Microflora (10^6)	
						Bacteria	Fungi
Brown soil							
L	1	4·0	90·5	1·17	49·8	5·7	4·0
F_1	2	4·6	82·3	1·93	30·1	93·7	3·8
F_2	2	4·0	65·8	1·89	30·9	66·1	4·4
F_3	—	—	—	—	—	—	—
H_1	1	3·8	30·1	3·20	18·2	16·6	3·6
H_2	4	4·0	5·2	6·42	9·1	49·9	11·5
Podzolised brown soil							
L	1	3·8	85·8	1·21	48·0	5·3	4·9
F_1	2	4·0	79·6	2·05	28·4	40·8	3·5
F_2	2	4·0	71·7	1·98	29·4	15·7	3·5
F_3	3	3·8	71·3	2·01	29·0	12·2	7·3
H_1	1	3·8	27·5	2·68	21·7	27·3	4·7
H_2	—	—	—	—	—	—	—
Pseudogley podzol							
L	1	3·0	88·5	1·56	37·3	7·8	4·4
F_1	2	4·1	87·5	2·13	27·4	72·0	5·6
F_2	3	3·4	81·1	2·00	29·2	13·5	3·7
F_3	6	3·3	82·2	1·99	29·2	2·6	3·7
H_1	6	3·2	69·3	2·26	25·7	5·2	1·0
H_2	5	3·2	75·3	1·83	31·7	2·0	1·1
Podzol							
L	1	3·6	88·7	1·35	43·1	13·9	8·1
F_1	2	4·0	83·8	1·94	29·9	31·2	6·7
F_2	3	3·4	80·4	2·09	27·9	1·1	5·3
F_3	7	3·1	56·4	1·71	34·0	0·4	3·2
H_1	4	3·1	31·5	1·87	32·7	0·3	1·6
H_2	—	—	—	—	—	—	—

slightly decomposed in six months, reach a medium state of decomposition in two years, and are much decomposed after seven years.

The thickness of the litter layers and the intensity of accumulation of humus and of mineralisation of matter in the soil depend both on soil type and on tree species.

Almost all microbiological processes take place more intensively in litter than in soil. In this connection the microflora concerned with transformation of nitrogen and carbon has been most thoroughly studied. In the coniferous forests in the European part of the U.S.S.R, there is greater concentration of the micro-organisms taking part in nitrogen circulation (ammonisers, denitrifiers, anaerobic nitrogen-fixers) in litter than in soil.

According to N. N. Mustafova (1959), in spruce forest litter (Leningrad province, Severskii leskhoz) ammonisers, predominantly representatives of *Pseudomonas fluorescens* and *Achromobacter*, number up to 6 million. A greater rate of ammonisation and denitrification in litter than in soil has been recorded by I. I. Dobrogaev (1939) in coniferous forests in Moscow province; the rate increased from spring to summer and was especially high by autumn.

The nitrification processes are generally less active in northern soils (Mishustin 1954, Mazilkin 1956). A very small number of nitrifying bacteria has been recorded in the litter of coniferous (especially spruce) and deciduous species on leached chernozem (Runov and Mishustina 1960) and on podzolic soils in Vologda province (Runov and Valeva 1962). Some investigators, however, have observed more vigorous nitrification in litter than in soil: V. S. Shumakov (1948) in coniferous forests in Moscow province, O. I. Pushkinskaya (1953) in oak stands in Voronezh province, and Nemec and Kvapil (1927) in forests in Germany.

As for free-living nitrogen-fixers, *Azotobacter* is practically never seen in either coniferous or deciduous litter, but *Clostridium pasteurianum* is generally found there, especially in northern districts with abundant moisture.

Decomposition of cellulose is the most important form of carbon transformation. Litter is much more heavily populated with cellulose-decomposing micro-organisms than is soil. Microscopic fungi take part in cellulose decomposition in northern coniferous forests (Dobrogaev 1939, Runov and Valeva 1962). By autumn a greater share in cellulose decomposition is taken by bacteria, which are very demanding with regard to available nitrogen and phosphorus. The breakdown of cellulose in litter is correlated with the presence of easily-available forms of nitrogen.

In chernozems in more southerly districts the bacterial type of cellulose decomposition is predominant in the litter of oak and other species. Cellulose and other high-molecular carbon compounds are broken down by higher fungi as well as by bacteria, microscopic fungi, and Actinomycetes.

THE ROLE OF HIGHER FUNGI IN DECOMPOSITION OF PLANT REMAINS IN
THE FOREST

The litter fall and surface litter of forest vegetation, as well as wood,
contain large amounts of substances difficult to dissolve—lignin, pectin,
cellulose. The decomposition of these substances is performed mainly
by higher fungi. Breakdown of lignin by bacteria apparently takes place
to a negligible degree. Basidiomycetes are mainly responsible for breaking
down lignin. The decomposition of lignin in litter is achieved mostly by
Hymenomycetes (*Marasmius peronatus, Clavaria gracilis*), for which the
lignin in litter is the sole source of energy. The lignin in wood is broken
down, as a rule, by other species of fungi (*Polyporus versicolor, Armillaria
mellea*, etc.) (Goetlieb and Pelczar 1951). The fungi that form mycorrhiza
on trees do not usually consume lignin and cellulose. That faculty is
possessed by the fungi that form mycorrhiza on Orchidaceae. Melin
divides forest fungi into two groups: (*i*) saprophytic, breaking down lignin
and cellulose and aiding decomposition of forest litter, and (*ii*) mycorrhiza-forming, most of which are unable to break down these compounds.

In natural conditions the fruit-bodies of both saprophytic and mycorrhizic fungi occur in forest litter. The microflora, like the microbial
population, has its own specific features, depending on the composition of
the tree stand.

In deciduous forests (oak forests in particular) litter saprophytes
consist of a wide variety of species, specific for each biogeocoenose and not
occurring in mixed, still less in coniferous, forests (Chastukhin and
Nikolaevskaya 1953). In a sedge oak stand the litter contains in greatest
numbers *Collybia confluens, Cortinaria hinnuleus*, a number of species of
Inocybe, Mycena pura, and *Clitocybe* spp. Characteristically, the litter of
oak forests contains a large number of higher Ascomycetes and small
gill mushrooms (Agaricaceae family, *Mycena debilis, Mycena bescens*,
etc.) and some *Cortinarius hinnuleus*. 'Honey fungus' (*Armillaria mellea*)
is widely distributed. In oak stands one also finds mycorrhizic fungi
linked to the trees—*Amanita* spp., *Boletus* spp., *Tricholoma* spp., etc.—as
well as the specialised forms of species of fungi that grow in both deciduous and composite forests (*Mycena pura, Clitocybe infundibuliformis,
Lactaria laccata*, etc.).

In oak forest litter several layers differing in microflora composition
may be distinguished (Chastukhin 1952). The microscopic fungi *Alternaria*
and *Cladosporium* are characteristic of the top layer of fresh-fallen leaves.
The next layer usually contains mycelia of higher basidial fungi and
various associated microscopic fungi, *Trichoderma, Penicillium, Chaeto-
mella*, etc., which are able to break down cellulose. In the third, fully-
decomposed layer one finds mycelia of basidial fungi and microscopic
fungi of the Mucoraceae, *Mycogone*, etc. Consequently the breakdown
of different layers is performed by different fungi (both higher and microscopic). According to laboratory experiments, the decomposition of oak

litter by pure cultures of basidial fungi (*Collybia dryophila*, the most widespread of oak-forest fungi) proceeds much more intensively than that by microscopic fungi.

In coniferous forests the organism that breaks down spruce litter is *Marasmius* (Chastukhin 1945), but other fungi may be plentiful in the litter, depending on the circumstances. In mixed spruce stands the litter is broken down mainly by *Mycena vulgaris* and other *Mycena* spp. as well as by *Marasmius*. One also finds there *Collybia*, *Psathyra*, and *Pluteus*; they feed mainly on leaves of deciduous trees falling into mixed spruce forest litter.

The species of higher fungi that break down plant remains and form mycorrhiza vary with the age of the stand (Chastukhin 1948). In young pine plantations (four years old) butter mushrooms (*Boletus luteus*), the pioneer in pine forests, are most abundant. In 15-to-20-year-old pine stands, besides *Boletus luteus*, there occur *Lactarius rufus*, *Telephora terrestris*, *Clitocybe*, *Lycoperdon*, and *Mycena*. In a 50-year-old stand the fungus species typical of young stands have disappeared and are replaced by *Amanita muscaria*, *Boletus variegatus*, *Boletus bovinus*, and *Boletus edulis*.

The decomposition of plant remains, other than litter, in a forest—large branches, stumps, logs, etc., i.e., woody material—is of importance. The breakdown of wood, which consists largely of cellulose and lignin, is achieved mainly by wood-attacking fungi. These include members of the genera *Merulina*, *Polyporus*, *Fomes*, *Stereum*, etc. The breakdown of wood by the enzymes of these fungi occurs in two ways: (*i*) The fungi first attack lignin and pentosans, destroying their bonds with cellulose. Simpler compounds are formed from lignin and further consumed by microscopic fungi and bacteria. (*ii*) In the so-called brown rot type of breakdown, the fungi act mostly on cellulose. There is one group of fungi that acts simultaneously on cellulose and lignin. Different types of fungi attack the litter of coniferous and of deciduous trees. Spruce logs in composite spruce forests are broken down by *Fomes pinicola* and *Armillaria mellea*. These two species reduce spruce logs mainly to brown rot. Members of other species of wood-attacking fungi rarely occur in mixed spruce forests. The second stage of decomposition of spruce wood is achieved by *Omphalia campanella* (Chastukhin 1945, 1948, Nikolaev-skaya and Chastukhin 1945).

Fomes fomentarius is the chief species found on birch logs in mixed spruce forests.

The wood of ash, mountain ash, poplar, and other deciduous species in mixed spruce forests is broken down not by members of the Polyporaceae family but by various Agaricineae with fleshy, rapidly-disappearing fruit-bodies.

The decomposition of wood in oak forests (Voronezh province) follows a different course. The most widespread attackers of oak wood are

Daedalea querina, Armillaria mellea, Polyporus sulphureus, and *Stereum hirsutum,* which perform the first stage of breakdown. The final stages are performed by *Hymenochaetea rubiginosa, Mycena polygramma,* and *Hypholoma sublateritium.*

Saprophytic fungi attacking litter and wood are able also to create soil humus, because of the comparatively large mass of mycelia and the formation of humin-like compounds. In addition, wood-attacking fungi secrete antibiotic substances that suppress the growth of several micro-organisms (Shivrina 1961) and microscopic fungi (Nikolaevskaya 1957).

PHYTONCIDAL PROPERTIES OF FOREST LITTER

In the inter-relationships between forest tree species and micro-organisms a substantial role is played by the content of phytoncidal substances in plants, as one of the factors determining the composition of the microflora of forest soils by its ability to restrict the development of some of its members.

Fresh litter of most trees (both coniferous and deciduous) contains phytoncidal substances that suppress the development of certain micro-organisms (Runov and Enikeeva 1955). The selective effect of the litter of different tree species and the degree to which it affects soil microflora vary. The most toxic properties are possessed by oak and maple litter, and the least by ash and honeysuckle litter (Table 105).

Table 105. *Effect of litter extracts on the microflora of chernozem soils (number of microbes, 10^3 per g of abs. dry soil*) (Runov and Enikeeva 1955).*

Extract from litter of	Non-spore-forming bacteria	Spore-forming bacteria	Actino-mycetes	Micro-scopic fungi
Control	5041	2197	4264	34·7
Oak	0	0	0	35·4
Maple	0	0	0	38·3
Ash	2522	1157	936	35·7
Wych elm	202	988	0	46·1
Acacia	611	1092	195	42·2
Honeysuckle	3172	1430	299	46·7

* Culture from the soil in nutritive medium with added extract (2 ml to 15 ml of agar medium).

The microscopic fungi in the experiments were absolutely immune to the phytoncidal substances in the litter, their development actually being stimulated in some cases. The Actinomycetes group was heavily suppressed. Bacteria reacted unequally to litter phytoncides. Bacteria of the *Pseudomonas herbicola* and *Bac. agglomeratus* types proved to be the most sensitive.

The phytoncidal effect of fresh litter (litter fall) is much stronger than that of old litter (ground litter) (Table 106).

Litter from trees and shrubs suppresses the development of *Azotobacter*

Table 106. *Effect of extracts from oak litter fall and oak ground litter on soil microflora* (*number of microbes, 10^3 per g of abs. dry soil**) (*Runov and Enikeeva 1955*).

Extract	Non-spore-forming bacteria	Fluor-escent bacteria	Yellow-pigmented bacteria	Myco-bacteria	Spore-forming bacteria	Actino-mycetes	Micro-scopic fungi
Control	2562	526	15	170	1568	811	29·1
Oak litter fall	19	0	0	0	215	0	36·2
Litter A'$_0$	1121	259	0	188	1381	124	31·0
Litter A$_0$	1651	379	0	96	1444	694	42·0

* Culture from the soil in nutritive medium with added extract (2 ml to 15 ml of agar medium).

(Vizir 1956). Aqueous extracts from pine, oak, poplar, and alder litter, when applied to podzolic soil, suppressed the development of *Azotobacter* introduced there artificially. Not all extracts act in the same way. Those from caragana, European elder, birch, and ash litter favour the development of *Azotobacter*. There are also species of bacteria immune to the action of phytoncides in litter. Winter and Willeke (1952) used fallen leaves, macerated in water, from birch, beech, Persian walnut, oak, willow, chestnut, maple, linden, hazel, and alder, and obtained no suppressing effect on *Bac. subtilis*, *Esch. coli*, and *Staphylococcus aureus*.

Litter does not affect microscopic fungi. Higher fungi (mycorrhiza-formers and litter saprophytes) react differently to phytoncidal litter (Melin 1959). Mycorrhiza-forming fungi show high sensitivity to the action of aqueous extracts from birch, beech, elm, oak, poplar, and pine litter. Large doses retard and suppress their development, but small doses stimulate it. Litter saprophytes are not suppressed by the extracts; on the contrary, their growth is stimulated.

Available data do not yet enable us to describe the selective action of litter, but unquestionably it exists. We may assume that the abundant development of fungi in forest soils is due to their low sensitivity to phytoncides contained in the leaves and litter of various tree species. Fungi that are insensitive to phytoncides should be the pioneers in breakdown of plant remains, as is often observed in nature; but this calls for further detailed investigation.

In natural conditions, apparently, phytoncides in litter do not last long. They are partly leached away and partly decomposed.

Litter formed from the litter fall also contains substances that suppress the growth of micro-organisms. Aqueous extracts from coniferous and deciduous litter act differently on microflora (Table 107). Extracts from coniferous (especially spruce) litter act most strongly. Deciduous litter is less toxic. Larch litter is similar to that of broad-leaved deciduous trees in its antibacterial action.

Phytoncides are not evenly distributed throughout the depth of litter. The upper, undecayed, layers of coniferous litter are more bactericidal

Table 107. *Effect of litter extracts on meadow soil microflora (number of micro-organisms, 10^3 per g of oven-dry soil) (Runov and Egorova 1958).*

Extract from litter of	Total no. of bacteria	Spore-forming bacteria	Actino-mycetes	Microscopic fungi
Spruce	1420	392	1135	44·2
Pine	1985	274	812	54·7
Larch	2770	369	668	23·8
Oak	2270	487	935	20·2
Birch	2610	440	977	26·1
Control	3785	628	1160	68·9

than the lower ones, which are in a state of intensive decomposition (Table 108).

Micro-organisms in forest soils are more resistant to litter extracts than is the microflora of cultivated soils, and the soil microflora from any stand is less sensitive to an extract from the litter of that stand than to that from a stand of another tree species. We receive the impression of a definite adaptation of the micro-organisms to their living conditions, which throws light on the formation of forest microbic coenoses.

The phytoncidal effect of litter depends on its composition, and consequently to a considerable extent on the forest type. The effect of litter on the microbiological activity of the soil is not restricted to the action of products of its decomposition or to alteration of the temperature, moisture, and air regimes of the soil. The presence of toxic products radically alters the course of microbiological processes in the soil. In some cases the presence of toxic substances in litter and tree litter fall appears to explain impairment of the nitrification process, which is so important in the circulation of nitrogen. Nitrifying capacity and the number of nitrifying organisms are at a lower level in forest soils than in meadow or cultivated soils, and depend largely on the composition of forest vegetation (Hesselman 1916-17, 1926, Nemec 1930, Remezov 1941).

Nitrification is retarded by resin, wax, tar, and tannin, i.e., by typical forest (especially coniferous forest) products (Koch 1914, Nemec 1930,

Table 108. *Effect of extracts from different layers of spruce and pine litter on meadow soil microflora (number of micro-organisms, 10^3 per g of oven-dry soil) (Runov and Egorova 1958).*

Extract from litter of	Non-spore-forming bacteria	Spore-forming bacteria	Actino-mycetes	Microscopic fungi
Spruce A'_0	706	405	615	25·1
Spruce A''_0	1234	505	900	23·9
Pine A'_0	912	131	310	32·6
Pine A''_0	1610	322	823	22·6
Control	3085	632	1120	73·5

Remezov 1941, Shumakov 1948). A tar content of over 5% in forest soil stops nitrification (Nemec 1930).

According to Shumakov's data, water-soluble products of forest litter and especially of litter fall have a strong toxic effect on nitrifying organisms in forest soils; these organisms are more severely affected by aqueous extracts from litter than by tar. The extent of the toxic action of the extracts depends on the composition of the litter: an aqueous extract from conifer-moss-cowberry litter is more toxic for nitrifiers than an extract from conifer-deciduous litter. Aqueous extracts from pine-cowberry litter have a particularly strong toxic effect on nitrification. The differences in toxic effect of aqueous extracts from litter of varying composition are largely due to differences in the nitrification activity in different forest types: a low level of that activity occurs in pine-cowberry stands and a high level in pine-linden stands (Moscow province).

The presence of phytoncidal substances in the litter fall and surface litter not only limits the distribution and life-activity of several groups of micro-organisms, but also apparently has great biological significance in the life of the forest, since toxic substances in litter retard germination of seeds and development of shoots of higher plants (Winter and Bublitz 1953b, Bublitz 1953).

The germination of pine and spruce seeds is retarded in coniferous forest litter. According to Bublitz (1953), germination of pine and spruce seeds is seriously retarded in spruce litter but not in an aqueous extract from it. According to Runov and Egorova, the layers of litter have differing degrees of toxicity for germination of pine and spruce seeds (Fig. 64). The topmost layer is most toxic and the lower, less-decomposed layers less toxic. Pine seeds are more sensitive than spruce seeds to extracts from spruce litter. The phytoncidal substances in litter may probably impede natural restoration of coniferous species by retarding the germination of seeds and the growth of seedlings. This problem certainly is of both theoretical and practical interest, but has been little studied. Thorough study should also be given to the question of the effect of litter on various species of herbaceous plants as a barrier to the penetration of such plants into a forest. According to E. V. Runov's (1957) investigations, the lower layers of deciduous (especially oak) litter, which are in an advanced state of decomposition, retard the germination of seeds and growth of shoots of herbaceous plants.

The phytoncidal properties of forest litter may be due, apparently, to various factors.

The first of these is the content in litter fall and surface litter of substances of plant origin that suppress the development of various micro-organisms and also inhibit the germination of seeds and the development of shoots of higher plants. As Bublitz (1954) has shown, substances contained in coniferous litter suppress the growth of bacteria and seeds not by raising the pH but by their own antibiotic action. Such active sub-

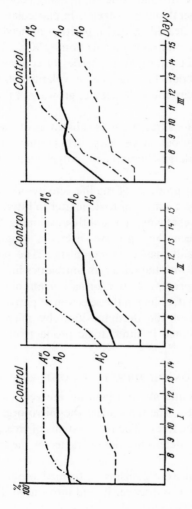

Fig. 64. Effect of coniferous litter on germination of spruce and pine seeds (% of controls) (from Egorova's data). I—germination of spruce seeds in pine litter; II—germination of pine seeds in pine litter; III—germination of pine seeds in spruce litter.

stances may include waxes, tars, terpenes, and tanning substances. Nemec (1930) states that nitrification proceeds in the presence of 3% of tar, extracted with alcohol from coniferous litter, but with 5% nitrification ceases. There are comparatively large amounts of these tars in coniferous litter—8-18% in pine litter and 2·7-10% in spruce litter. Koch (1914, Koch and Oelsner 1916) attributed the suppressing action of coniferous litter on microbes in general and on nitrifiers in particular to the presence of spruce and pine oil, released by decomposition of needles. According to his data, 2% of spruce oil lowers nitrification by 90%. Later investigations have shown that the antiseptic properties of conifers (in the above case, of needles of Siberian fir) are due to volatile oils. The greatest amounts of volatile oils are found in needles growing in the shade, which also are the most toxic.

In the leaves of deciduous trees various substances of the phenol group have toxic properties. In oak leaves they include quercetin and tanning substances. There are large amounts of polyphenols in beech litter, especially in the leaves.

A second factor is that products of the life-activity of micro-organisms and higher plants may be a source of toxic matter in litter. Micro-organisms in litter form a changing, constantly-developing coenose. The inter-relationships between their groups—bacteria, fungi, and Actinomycetes—are complex and may be antagonistic. The phenomenon of antagonism is connected with the release of antibiotics into the surrounding environment. In the process of litter formation, antibiotics of microbic origin may accumulate in it, aiding in the selection of particular organisms. In addition, as a result of the decomposition by micro-organisms of various substances contained in litter, toxic products of decomposition may be formed.

EFFECTS OF TREE ROOTS ON THE MICROFLORA OF FOREST SOILS

The interactions of trees and micro-organisms are very clearly displayed in the region of the roots. The trees affect the micro-organisms and the micro-organisms affect the trees. The root systems of trees, together with the litter, help largely to determine the character of the microbial communities in forest soils.

A zone with many micro-organisms is formed around plant roots. Hiltner (1904) was the first to discover it, and introduced the term 'rhizosphere', meaning the zone in which plant roots affect the biological system of the surrounding environment. His discovery was later confirmed by many investigators (Krasil'nikov 1958).

The number of micro-organisms in the rhizosphere is variable. It usually falls off as the distance from the root surface increases. E. F. Berezova (1950) divides the rhizosphere into three zones: (*i*) root zone, micro-organisms on the root surface and within the root tissues; (*ii*) root-adjoining zone, micro-organisms in the soil directly adjacent to the roots;

and (*iii*) rhizosphere zone at a distance of 0·5 mm or more from the roots.

Among the many works devoted to the microflora of plant rhizospheres, very few relate to trees. These are mostly works by Soviet micro-biologists. Among foreign investigations, we know of a work on micro-flora in the rhizosphere of yellow birch seedlings (Ivarson and Katznelson 1960).

Soviet investigators have studied the microflora of tree root zones in various districts in the European part of the u.s.s.r. A group of micro-biologists of the Forestry Laboratory under the leadership of E. V. Runov have studied the subject in various soil-climatic conditions: in the ordinary low-humus chernozem of south-eastern Ukraine (Runov and Enikeeva 1961), in the leached chernozem of Orlov province (Runov and Mishustina 1960), in the soddy-podzolic soil of Moscow province (Bol'shakova 1964), and in different soils of the central taiga of Vologda province (Runov and Zhdannikova 1960). The same method was used throughout, the objects of study being the roots and the rhizosphere soil zones of various deciduous and coniferous species (English oak, Norway maple, ash, caragana, birch, larch, pine, and spruce).

Similar investigations were carried out by O. I. Pushkinskaya (1951) with two- and three-year-old plantations of oak, maple, and ash in the dark-grey forest soils of the Tellerman experimental forestry station in Voronezh province and on the ordinary chernozem of the Derkul' re-search station. S. A. Samtsevich and his colleagues (1952, Samtsevich 1956) studied the rhizosphere microflora in oak, ash, larch, and spruce plantations in the northern and southern forest-steppes of the Ukraine. A. Ya. Tribunskaya (1955) studied the rhizosphere microflora of 5-month-old pine seedlings (with and without mycorrhiza) in grey forest soil. A. Ya. Manteifel' and his colleagues (1950) and E. I. Kozlova (1955, 1958) studied the micro-organism composition in oak rhizospheres in various soils (light-brown, dark-brown, and soddy-podzolic) in Volgograd and Moscow provinces.

Several investigations have been devoted to separate groups of rhizo-sphere microflora. N. A. Krasil'nikov (1944b) studied the possibility of *Azotobacter* survival in nursery soils under different tree species—ash, maple, willow, and acacia; I. T. Nette (1955) studied the denitrifying bacteria in oak rhizospheres, and T. P. Sizova and E. A. Itakaeva (1956) the microscopic fungi in birch rhizospheres.

All these studies confirm that tree rhizospheres have a higher content of micro-organisms than the surrounding soil. According to Samtsevich's (1956) data, in the rhizospheres of oak, ash, larch, and spruce on ordinary podzolised chernozem the number of micro-organisms is 1·5 to 2 times as high as in the soil beyond the rhizosphere. A. P. Vizir (1955) found that ammonising micro-organisms were 2 to 2·5 times as numerous in oak rhizosphere as in control soil. O. I. Pushkinskaya (1951) presents similar data for oak.

Investigations by the microbiologists of the Forestry Laboratory also reveal higher micro-organism content in root zones. The number of micro-organisms is usually higher in the rhizospheres of deciduous trees than in those of conifers (Table 109).

An increased number of micro-organisms in oak rhizospheres has been observed both with young seedlings (Manteifel' *et al.* 1950) and with mature trees. The number of micro-organisms in the rhizospheres varies according to the age of the stand. Sizova and Itakaeva (1956), studying microscopic fungi in the rhizospheres of birch of various ages, observed that their numbers increased until the birch was 50 years old and then decreased. E. V. Runov and S. V. Egorova (1962) observed an increase in micro-organism numbers in the rhizospheres of oak trees between the ages of 5 and 100 years and lower numbers at the age of 220.

The so-called rhizosphere effect (the ratio of micro-organism numbers in the rhizosphere to those in the surrounding environment) varies in different soil horizons: with birch it is greater in horizon B than in horizon A (Ivarson and Katznelson 1960).

The rate of micro-organism reproduction in any environment is governed mainly by the abundance of nutritive and energy-providing matter there. Micro-organisms find such matter in greater quantity near the roots than in the surrounding soil, one source being root decay. In the metabolic process of trees certain parts of their roots (cortex, root hairs in non-mycotrophic species, suction roots) are periodically renewed. A. Ya. Orlov (1955a, 1960a) believes that dead suction roots form the chief source of organic matter in forest soil. According to his data, the duration of life of suction roots is 3 to 4 years; he estimates that the soil of a 50-year-old spruce stand receives 0·5 ton/ha of dead spruce roots every year, and that of a 25-year-old spruce stand 2·0 ton/ha.

The chief reason, however, for concentration of micro-organisms in the rhizosphere is the exudates of living roots. Exudation of water by roots

Table 109. *Development of micro-organisms in tree rhizospheres in different soils.*

Soil type	Number of micro-organisms (10^3 per g of soil)				Rhizosphere effect			
	Birch	Oak	Spruce	Pine	Birch	Oak	Spruce	Pine
Podzolic (Vologda prov.)	9300	—*	4800	—	7	—	3	—
Peaty-humus (ibid.)	12,400	—	7700	—	4	—	2	—
Soddy-podzolic (Moscow province)	1400	1800	1200	800	2	2	1·3	1·2
Leached chernozem (Orlov province)	466,600	279,000	160,000	270,000	233	139	107	118
Ordinary chernozem (Lugansk province)	—	7100	—	—	—	2·5	—	—

* A dash means 'No data'.

and moisture of the surrounding soil were first noticed in grasses (Breazeale 1930). According to these data, roots may discharge water into dry soil if they are over-supplied with water from damp soil layers. B. A. Chizhov observed increases in moisture in dark-brown soil around the roots of a spring wheat crop, in one experiment from 9 to 13-17%, and in another from 4 to 14-16%. Similar moistening of dry soil around tree roots is possible. That is shown by Samtsevich (1956), who observed an increase of moisture in the rhizosphere under oak, ash, and larch amounting to 2% in September and to 3-10% in December.

There is usually a difference in the acidity of the soil near roots and at a distance from them. In the root zone of grasses partial neutralisation of both acid and alkaline soils takes place. E. M. Baraskina (1959), while studying the root exudates of oak and ash trees from 50 to 60 years old, noted considerably higher soil acidity near the roots, a period of acidity of root exudates alternating with a period of alkalinity. Samtsevich (1956) recorded acidity of the rhizosphere soils of ash (pH 6·6-6·4) and of larch (pH 5·9-5·6), and alkalinity in the rhizosphere soil of spruce (pH 4·4-5·4).

Mineral compounds are found in plant root exudates. M. V. Zhuravleva (1953) reports that tree roots (oak, ash, pine, larch, and maple) excrete phosphorus, potassium, and ammoniacal nitrogen. A. I. Akhromeiko and V. A. Shestakova (1958) have demonstrated by the tagged-isotope method the ability of oak and ash seedling roots to excrete compounds containing phosphorus. E. M. Baraskina (1959) found compounds of calcium, magnesium, and aluminium in oak and pine root exudates.

The organic part of root exudates plays a major role in the concentration of micro-organisms near roots. According to Samtsevich (1956), the amount of soluble organic matter in the rhizospheres of oak, ash, larch, and spruce is approximately twice as great as in the surrounding soil.

Very little study has been given to the composition of the organic matter in tree root exudates. It must certainly be close to the composition of that in herbaceous plant root exudates, which have been described in some detail. The latter have been found to contain large amounts of organic acids, including acetic, formic, and oxalic (Stoklase and Stoklase 1902), and tartaric and malic (Bkhuvanasvari and Subbo-Rao 1959). Sugars (glucose, maltose, etc.) and amino acids have also been found in root exudates.

In the study of root exudates in sterile cultures by the chromatographic method (Frenzel 1957), it has been discovered that maximum root secretion of amides and amino acids takes place in optimal conditions for plant growth. The author points out that these secretions are produced by living plant cells.

Root exudates contain various enzymes. V. F. Kuprevich (1949) has observed a large collection of enzymes in the exudates of tree roots (oak, birch, aspen, willow, pine, and spruce), including catalase, phenolase, tyrosinase, asparaginase, urease, amylase, protease, and lipase. The degree of activity of enzymes varies in the exudates of different plants. Roots without mycorrhiza have a smaller variety of enzymes than mycorrhizal roots (Kuprevich 1952).

There are few data on the exudation of growth substances and vitamins by plant roots. Their presence in the rhizosphere or the soil is usually due to microbial synthesis (Cook and Lochhead 1959).

All of the above-mentioned compounds serve as good energy-providing material for the micro-organisms that develop in the rhizosphere. The

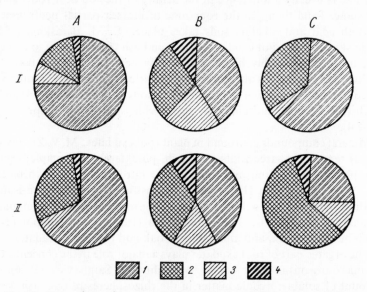

Fig. 65. Proportions of groups of micro-organisms in rhizosphere of birch (I) and spruce (II) (from Bol'shakova's data). 1—non-spore-forming bacteria; 2—Actinomycetes; 3—spore-forming bacteria; 4—microscopic fungi; A—podzolic soil; B—soddy-podzolic soil; C—leached chernozem.

concentration of these substances is due to the more intensive activity of microbes around roots.

Root exudates also contain substances that inhibit the development of micro-organisms (phytoncides), which will be discussed later.

The various substances occur in different amounts in the root exudates of different plants. Sugars and organic acids predominate in grasses, and amino acids and other nitrogenous compounds in leguminous plants (Virtanen and Torniainen 1940).

The root exudates of mycotrophic trees must be distinguished from those of other plants. Before entering the soil they pass through a dense sheath of mycorrhiza, which intercepts some of the compounds required for the development of mycorrhizal fungi and which discharges into the surrounding soil products of its own metabolism.

Around the roots of grasses micro-organisms also develop partly at the expense of replaced root hairs. That source of energy-providing material is lacking in the rhizospheres of trees with well-developed mycorrhizae.

The above features affect the composition of micro-organisms in the zone surrounding tree roots.

In the rhizospheres of farm crops, regardless of climatic and soil conditions, non-spore-forming bacteria are the dominant group of micro-organisms. Actinomycetes form only 1-4%, and fungi only a fraction of 1%, of the total. In tree rhizospheres the part taken by spore-forming bacteria and Actinomycetes is much greater (Fig. 65), indicating the presence in tree root exudates of organic compounds difficult of solution.

Under the same tree species the proportions of groups of micro-organisms differ in different soil-climatic conditions. The number of Actinomycetes in the rhizospheres of tree species increases from north to south. In the soddy-podzolic soils of Moscow province the percentages of spore-forming bacteria and microscopic fungi in tree rhizospheres is higher than in the rhizospheres of the same species in different soil-climatic conditions.

The non-spore-forming group of bacteria in tree rhizospheres is quite varied in species-composition. It contains species of *Pseudomonas* and *Achromobacter*, as well as nitrifying and other bacteria. The first two groups are the most widely distributed. Among them are species that develop in meat-peptone agar; others prefer synthetic media with mineral nitrogen and carbohydrates; a third group develops equally in both media. We have no reliable method of distinguishing these bacteria, and therefore it is difficult to say anything about their characteristics. We can only point out that the role of the bacteria that develop in a medium containing mineral nitrogen and carbohydrates is higher near roots than at a distance from them. That is seen from Table 110, which is taken from Runov and Mishustina (1960). Similar results have been obtained for the soddy-podzolic soils of Moscow province (Bol'shakova 1964).

Table 110. *Ratio of numbers of bacteria in starch-ammonia (SAA) and meat-peptone (MPA) agars.*

Tree species	Ratio of groups (SAA:MPA)		
	Root	Soil adjacent to root	Control soil
Oak	7·7	2·7	2·7
Birch	22·9	1·6	2·0
Larch	8·7	2·3	1·8
Pine	3·1	3·4	2·0
Spruce	3·4	6·6	0·9

Data exist showing the presence in tree rhizospheres of large numbers of oligonitrophilous micro-organisms, which are able to satisfy their needs with traces of nitrogen (Chastukhin and Nikolaevskaya 1962, Samtsevich 1956, Mishustina 1960, Bol'shakova 1964). A large proportion of these bacteria are non-spore-forming.

Spore-forming bacteria form a fraction of 1% of the total number of micro-organisms in the rhizospheres of farm crops. They compose 10% of the total in tree rhizospheres, according to the data of the Forestry Laboratory microbiologists. A similar relation may be obtained from analysis of the material of A. Ya. Manteifel' and his colleagues (1950) and S. A. Samtsevich (1956).

The group of spore-forming bacteria is easily differentiated, and therefore the specificity of microflora can easily be traced in it. The composition of the group of micro-organisms in rhizospheres is decided mainly by the soil type.

Table 111 shows the development of different species of these bacteria in the rhizospheres of birch and spruce in the soils of different climatic zones. *Bac. agglomeratus* predominates in the rhizospheres of trees growing in the podzolic soils of Vologda province, but in the rhizospheres of the same species growing in soddy-podzolic soil and leached chernozem that species of spore-forming bacteria amounts to scarcely 4% of the total. *Bac. idosus*, on the contrary, develops better in soddy-podzolic and chernozem soils. *Bac. mycoides* does not occur in the rhizospheres of trees growing on podzolic soils, but is abundant in soddy-podzolic soils. That distribution of spore-forming bacteria in tree rhizospheres corresponds to their development in soils of these types (Mishustin 1954).

Table 111. *Composition of spore-forming bacteria in the rhizospheres of birch and spruce in soils of different climatic zones.*

Soil type	Tree species	Number of spores (10^3 per g of soil)	Percentages of groups					
			Bac. agglo- meratus	*Bac. mycoides*	*Bac. cereus*	*Bac. mesen- tericus*	*Bac. mega- therium*	*Bac. idosus*
Podzolic (Vologda	Birch	908	65	0	9	0·3	4	12
province)	Spruce	308	41	0	12	0·5	7	17
Soddy-podzolic	Birch	251	1	14	59	0	1	20
(Moscow province)	Spruce	205	9	7	13	0	0	68
Leached chernozem	Birch	4400	4	5	22	5	2	62
(Orlov province)	Spruce	11,300	1	11	14	2	5	61

There are few published data on the effect of tree-stand composition on the development of spore-forming bacteria. In soddy-podzolic soils an increase in the content of *Bac. cereus* has been observed in the rhizospheres of deciduous species (birch and oak) as compared with conifers.

Actinomycetes develop in considerable numbers in tree rhizospheres. The composition of this group of micro-organisms has been very little studied. E. V. Runov and I. E. Mishustina found considerably more pigmented forms of Actinomycetes in the rhizospheres of conifers than in those of deciduous species. V. S. Bol'shakova observed Actinomycetes of various systematic groups in the root zone of different tree species. It appeared that Actinomycetes of the Griseus series (system of Gauze *et al.* 1957) develop better in the rhizospheres of oak than in those of other species, and Actinomycetes of the Chrysomallus series are well represented only in spruce rhizospheres.

Fungi participate actively in the transformation of matter in forest soils. Microscopic fungi usually form up to 10% of the total number of micro-organisms in tree rhizospheres. Considering their large mass (as compared with the mass of bacteria and Actinomycetes) we must conclude that they have a significant role in the transformation of matter in the soil around the roots. The numbers of these fungi in the rhizospheres of farm crops are not large. According to the data of O. I. Pushkinskaya (1951), there are only one-tenth as many fungi around the roots of sunflowers, sugar beets, and potatoes as in tree rhizospheres.

It is known from published data that the number of microscopic fungi is higher in tree rhizospheres than in the surrounding soil. Only in oak rhizospheres is their number the same as, or even less than, that in the control (Manteifel' *et al.* 1950, Bol'shakova 1964).

The rhizosphere effect for trees in the south-western U.S.S.R. (in ordinary and leached chernozems and grey forest soil) results in a concentration of fungi equal to 10 or more times that in the outer soil; in soddy-podzolic and podzolic soils it is considerably less.

The dependence of fungal development in the rhizosphere on soil type has been noted in the work of Runov and Zhdannikova (1960). In the central taiga conditions of Vologda province they found from four to eight times as many fungi in spruce rhizospheres in boggy soils as in peaty-humus and podzolic soils.

The numbers of microscopic fungi vary in the rhizospheres of different tree species. E. V. Runov and M. G. Enikeeva (1961) observed suppression of fungal processes in the root zone of oak as compared with those in the root zones of maple, ash, and acacia. In the soils of Moscow and Vologda provinces it was observed that these fungi develop better in the rhizospheres of deciduous species than in those of conifers (Runov and Zhdannikova 1960, Bol'shakova 1964).

Apparently there are antagonistic relationships between microscopic fungi and higher basidial fungi in the rhizosphere. More hyphae of basidial fungi are found in coniferous than in deciduous rhizospheres. In ordinary chernozems they are more numerous in oak rhizospheres than in the root zones of ash and other species.

An increased content of fungi in the rhizospheres is observed in trees of all ages. Sizova and her colleagues (1956, 1961) compared the numbers of microscopic fungi around the roots of birch and oak at the ages of 1,

4, 6, and 50 years and greater ages, and concluded that the numbers of fungal embryos around the roots increase with the age of the tree, reaching a maximum at 40-50 years and falling later. Similar data were obtained by Pushkinskaya (1951) for oak on ordinary chernozems.

The majority of microscopic fungi (70-90%) in root-zone soils (ordinary chernozem, leached chernozem, grey forest soil) consist of *Penicillium* spp. Representatives of *Aspergillus*, *Fusarium*, and *Altermaria* also occur in small numbers. Fungi of the Mucoraceae family and *Trichoderma* spp. form another small percentage. The authors observed no considerable differences in fungus composition among tree species on these soils. In tree rhizospheres in soddy-podzolic and peaty-humus soils the numbers of fungi of the Mucoraceae family, especially *Mucor ramannianus*, were much higher. In the root zones of pine and spruce *Mucor ramannianus* constituted 40% of the total number of fungi, but in those of deciduous trees (birch and oak) only 2-4%.

The aerobic nitrogen-fixer *Azotobacter* is not, as a rule, observed in tree rhizospheres. This is stated in the works of the Forestry Laboratory microbiologists (Runov, Samtsevich, and Tribunskaya). *Azotobacter* is observed only in the rhizospheres of a few young tree seedlings in nurseries. N. A. Krasil'nikov (1944b) records stimulation of *Azotobacter* development in acacia rhizospheres, but at the same time points out that he found it completely absent from the rhizospheres of seedlings of other species (ash, maple, and willow) in their fourth year of life. Manteifel' and his colleagues (1950) found *Azotobacter* in the rhizospheres of one- and two-year-old oak seedlings. There are very few data about the development of anaerobic nitrogen-fixers (*Clostridium pasteurianum*) in tree rhizospheres; Manteifel' and his colleagues (1950) refer to the presence of that group in oak rhizospheres, but their numbers there were no higher than in the surrounding soil.

In Table 112, compiled from the records of a group of Forestry Laboratory microbiologists, we present data on the development of nitrifying bacteria in the rhizospheres of several species. It is seen from the table that the number of nitrifiers decreases with increasing proximity to the root, reaching a minimum on its surface. The number of nitrifiers on the rhizospheres of trees growing in more southerly districts (on chernozem and grey forest soils) is higher than that in soddy-podzolic soils; and in the central taiga nitrifiers were not found at all in birch and spruce rhizospheres. Nitrifying bacteria develop less vigorously under conifers than under deciduous species. Somewhat different results were obtained by Manteifel' and his colleagues (1950), who found better development of nitrifiers in the rhizospheres of young oaks than in the surrounding soil; nitrifying activity was higher in ten-year-old oaks than in one-year-old ones.

The question of distribution of denitrifying bacteria in tree rhizospheres is almost completely unstudied. Only the works of Manteifel' and his

Table 112. *Number of nitrifying bacteria in tree rhizospheres.*

Soil	Tree species	Root zone			Zone surrounding roots		
		May	July	October	May	July	October
		Vologda province					
Podzolic	Birch	Not			Not		
	Spruce	observed			observed		
Peaty-humus	Birch	Not			Not		
	Spruce	observed			observed		
		Moscow province					
Soddy-	Birch	0	0	0	10	10	100
podzolic	Oak	0	0	10	10	10	1000
	Pine	100	0	0	10	10	0
	Spruce	0	0	0	0	10	1000
		Orlov province					
Leached	Birch	0	100	1000	0	0	1,000,000
chernozem	Oak	100	0	1000	0	0	100,000
	Larch	0	0	1000	0	0	10,000
	Pine	0	0	100	0	0	10,000
	Spruce	0	0	100	0	0	10,000
		Lugansk province					
Ordinary	Oak	0	0	—*	100	100	—
chernozem	Maple	100	0	—	100	1000	—
	Ash	0	0	—	100	100	—
	Acacia	0	0	—	1000	1000	—

* A dash means 'No data'.

colleagues (1950) and Nette (1955) on their development in oak rhizospheres are known. The former found no substantial difference in the content of them in the rhizosphere and in the soil, and the latter reported higher numbers (10-100 times as many) of denitrifying bacteria near the roots. The denitrifying bacteria isolated by Nette from oak rhizospheres were small non-spore-forming bacilli of the genera *Pseudomonas* and *Achromobacter*, most of them being also ammonisers.

Cellulose is broken down in the rhizosphere, as in the soil, by bacteria (aerobic and anaerobic), fungi, and Actinomycetes. Data on this subject can be found in the works of Runov and his colleagues, Samtsevich (1956), and Manteifel' and his colleagues (1950).

Most of these studies found greater numbers of aerobic cellulose-micro-organisms in the rhizosphere than in the surrounding soil. But in the dark-brown soils in Kamyshin (Manteifel' *et al.* 1950), in the soddy-podzolic soils of Moscow province (Bol'shakova 1964), and in the boggy and podzolic soils of the central taiga of Vologda province (Runov and Zhdannikova 1960), no increase was observed, the numbers of cellulose-micro-organisms being the same in the rhizospheres and in the surrounding soil.

Cytophaga is most often observed among cellulose-decomposing bacteria in rhizospheres, and *Demathium* among fungi. The development of these micro-organisms proceeds differently under different tree species. E. V. Runov and M. G. Enikeeva (1961) observed bacterial decomposition of cellulose predominating in maple rhizospheres, and fungal in oak and ash rhizospheres. E. V. Runov and I. E. Mishustina (1960) recorded predominance of cellulose-decomposing bacteria in the rhizospheres of all species under study on leached chernozem. Fungi and Actinomycetes of the same group were much better developed in oak and coniferous than in birch rhizospheres.

Comparing the development of cellulose-decomposing micro-organisms under birch and spruce in podzolic soils, E. V. Runov and E. N. Zhdannikova (1960) observed better development in spruce rhizospheres. Cellulose-micro-organisms were represented there mostly by bacterial forms. According to Samtsevich's data (1956) cellulose-decomposers are much better developed under spruce than under oak, ash, or larch.

Together with compounds easily used by micro-organisms in constructive and energetic metabolism, plant roots contain antibiotic substances that restrict the development of certain micro-organisms. This is a powerful factor both in the selection of specific microflora in tree rhizospheres and in the process of breakdown of root remains. Phytoncides may be included in root exudates of grasses, trees, and shrubs.

The role of the phytoncides exuded by grass roots was studied by Metz (1955). He obtained root juices from several plants (*Ranunculus acer, Mercurialis annua*, etc.) and observed their inhibiting effect on microflora. Micro-organisms obtained from the surface of roots of plants which possessed strong phytoncidal action were not, as a rule, inhibited by the root juices of these plants.

The very toxic plants had an inhibiting effect both on micro-organisms taken from the soil and on those taken from plant roots not containing inhibiting substances. Sterile cultures of plants with susceptible and resistant micro-organisms gave the same results as root juices.

Metz concludes that toxic substances are exuded from roots into the environment, and the specific composition of rhizosphere microflora is determined not only by varying utilisation of stimulating and absorbable substances in root exudates but also by the selective action of inhibiting substances that restrict the development of certain groups of micro-organisms.

There is comparatively little information on the nature of the phytoncidal substances in tree and shrub root exudates. The roots of guayule secrete trans-cinnamonic acid, which is toxic to the guayule itself (Bonner 1950). There have been more detailed investigations of inhibiting substances in fruit trees, because of poor growth in citrus, peach, and other orchards. Amygdalin is contained in peach roots (Patrick 1955), there being more in small than in large roots (Ward and Durkee 1956). Floricin (polymerised castor oil) is contained in apple roots, and suppresses micro-organism development. Intermediate products of its decomposition inhibit apple seedling growth (Börner 1958). Juglone, which is contained in walnuts, retards bacterial development and is toxic to walnut roots (Drobot'ko and Aizenman 1958).

The study of phytoncides in various tree and shrub roots has been carried on in the former Forestry Institute of the U.S.S.R Academy of Sciences by Runov and his colleagues.

Investigations of the reaction of micro-organisms to root juices or aqueous extracts from roots give some idea of the mechanism of the action of the root systems of various species on microflora.

It has been shown that the roots of equal-aged (25-year-old) trees of coniferous and deciduous species on leached chernozem contain substances which are toxic to micro-organisms in varying degree. The roots of larch and deciduous species (oak and birch) have a strong phytoncidal action. The roots of conifers (spruce and pine) are less phytoncidal, especially pine. Among tree species growing in the steppe zone (Lugansk province), ash and oak root extracts have the strongest antimicrobic action, followed by elm and maple. Shrub (acacia and honeysuckle) roots are very weakly phytoncidal.

Different representatives of the soil microflora react differently to the phytoncidal properties of roots (Table 113). Actinomycetes are the most sensitive group. Microscopic fungi, as a rule, are not suppressed by the extracts, and sometimes their growth is even stimulated. Non-spore-forming bacteria are more resistant to phytoncides than are spore-forming bacteria.

The phytoncidal properties of plant roots depend on a number of factors: soil-climatic conditions, the habitat of the plants, their phase of development, etc. Metz remarks that the toxic substance content of *Ranunculus acer* is much lower in the vegetative period than in the fruit-

Table 113. *Effect of tree root extracts on soil microflora* (% *of culture suppressed by the extracts*).

Extract from roots of	Non-spore-forming bacteria			Spore-forming bacteria			Actinomycetes	Microscopic fungi			
	Yellow-pigmented type of *Ps. herbicola*	Non-spore-forming bacteria in MPA	Non-spore-forming bacteria in SAA	*Bac. mycoides*	*Bac. mega-therium*	*Bac. idosus*		*Mucor*	*Penicillium*	*Cladosporium*	*Trichoderma*
Spruce	71	45	82	100	100	100	87	0	0	0	0
Pine	48	31	65	100	50	91	55	0	0	0	0
Larch	84	79	100	100	100	100	100	30	0	0	0
Birch	55	43	88	100	60	58	92	10	0	0	0
Oak	77	67	94	100	100	100	95	50	0	0	0
Number of test cultures	31	58	17	14	10	12	60	10	32	7	7

bearing period. E. V. Runov and his colleagues observed fluctuation in the phytoncidity of root extracts, depending on the phase of development and the age of stands.

The qualitative and quantitative composition of the rhizosphere microflora varies with the phase of development of the plants, a relation being sometimes observed between it and the phytoncidity of the roots. No special investigations of the subject have been made, but we can assume that trees and shrubs regulate microbiological processes in their root zones by means of phytoncidal substances. Biological equilibrium between plants and micro-organisms is attained by means of the usable and phytoncidic substances contained in root exudates. Here we have another aspect of the interaction between plants and micro-organisms: on the one hand, many bacteria can break down certain toxic root exudates, and on the other hand microbes are important competitors for nutrients.

Not only is the phytoncidity of roots one of the factors determining to some extent the inter-relationships between micro-organisms and trees, but by studying it we are able to assess complex phenomena of the interaction of higher plants. The inhibiting action of tree and shrub root extracts affects a number of trees and herbaceous plants, as is shown by the failure of seeds of trees and herbaceous plants to germinate in extracts from roots of different tree species (Table 114).

Table 114. *Effect of tree root extracts on the growth of seedlings of wheat, spruce, and pine (% of control) (Egorova 1962).*

Plant	Extract from roots of				
	Spruce	Pine	Larch	Birch	Oak
Wheat:					
Shoots	61	58	37	82	74
Roots	45	44	15	48	59
Spruce:					
Shoots	83	108	63	110	93
Roots	65	70	20	60	55
Pine:					
Shoots	93	70	69	70	82
Roots	82	60	53	53	53

The inhibiting effect of the extracts appears most strongly on the growth of rootlets of wheat, spruce, and pine, and shoots are less affected. As in the case of micro-organisms, larch has the strongest suppressing effect on shoot development, and larch root extract heavily suppresses spruce seedling roots. Extracts from deciduous tree roots affect the development of spruce and pine seedlings rather more strongly than do coniferous root extracts. The development of wheat is more affected by extracts from coniferous than by those from deciduous roots.

The effects of root extracts on the development of shoots of various meadow and steppe plants are very interesting (Table 115). Tree and shrub root extracts suppress the growth of steppe herbs to varying degrees.

Table 115. *Effect of root extracts on growth of shoots of meadow-steppe grasses (% of control)* (*Runov and Enikeeva 1963*).

| Plant | Extracts from roots of | | | | | |
	Oak	Maple	Ash	Birch	Honey-suckle	Caragana
Couch grass						
shoots	75	38	80	108	25	42
roots	30	20	15	65	10	0
Divaricate brome grass						
shoots	54	46	0	76	0	0
roots	18	0	0	20	0	0
Meadow fescue						
shoots	73	64	64	64	36	36
roots	47	67	20	80	27	20
Sheep's fescue						
shoots	25	62	25	62	0	25
roots	28	82	0	46	18	9
European stickseed						
shoots	43	57	62	43	15	28
roots	17	56	6	11	6	28
Violet sage						
shoots	76	76	106	94	76	41
roots	22	55	13	42	42	3

Brome grass proved to be most sensitive, followed in descending order by European stickseed (*Lappula echinata*), sheep's fescue, couch grass, meadow fescue, and violet sage. The same degree of inhibiting effect on the sprouting and development of plants is not always observed; in some cases the extracts reduce the rate of seed germination but do not affect the length of shoots and roots, and vice versa.

The nature of the substances giving phytoncidal properties to root extracts has been little studied. The phytoncidal substances are thermostable: their inhibiting effect is not reduced by heating or autoclaving. Neutralisation of root extracts to a neutral or slightly alkaline reaction does not lower their toxic effect.

Preliminary investigations (using paper chromatographs to separate antibiotics) enables us to say that the toxic effect of root extracts is to a definite extent due to the presence of phenol compounds included in tanning. It is known that aqueous extracts from remains of herbaceous plants (mostly grasses) containing various phenol compounds, especially parahydroxybenzoic acid, have a depressing effect on the growth and development of higher plants (Winter and Schönbeck 1953, Börner 1955). While studying the action of that acid on micro-organisms Knösel (1958) observed that the development of most of the microscopic fungi that occur frequently in the soil is stimulated by use of p-hydroxybenzoic acid as a source of carbon. The development of Actinomycetes in a pure culture was seriously reduced when p-hydroxybenzoic acid was added. Spore-forming bacteria developed in a comparatively high concentration of that acid.

The effect produced on micro-organisms by tree root extracts containing a comparatively large amount of tannin (larch and oak) is essentially similar to that produced by p-hydroxybenzoic acid. That is another indirect indication of the connection between toxicity and the phenol group, but the toxicity of tree root extracts may be due to other substances, depending on the biological peculiarities of the species.

EFFECTS OF MICRO-ORGANISMS ON TREES

Micro-organisms in the soil and especially in the rhizospheres of trees carry out the most diverse processes of transformation of organic and mineral matter, which cannot but affect plant growth and development. Plant nutrition is intimately connected with the microflora of the root zone. The question of the role of micro-organisms in plant nutrition, in the creation of effective soil fertility, and in the raising of tree productivity (as in the raising of farm crop yields) is a topic of lively discussion. Experimental material is constantly being accumulated, convincingly proving the great importance of the microbiological factor in plant growth and development.

The chief role of micro-organisms in plant nutrition consists in the formation of mineral compounds available to plants and released in the process of breakdown of plant remains and root exudates. But that is not all.

Rhizosphere micro-organisms exert direct influence on plant growth through their metabolic products. Some microbes stimulate, others depress plant growth and development.

In their metabolism bacteria excrete the most diverse biotically-active substances: biotin, thiamine, riboflavine, auxins, vitamin B_{12}, nicotinic and pantothenic acids, and many similar compounds.

The ability of plants to absorb products of microbial metabolism has been shown experimentally by N. A. Krasil'nikov with regard to antibiotics. Plants can also absorb amino acids and other organic compounds.

The use of metabolic products of micro-organisms by trees has been described by A. I. Akhromeiko and V. A. Shestakova (1958). They added P^{32} to cultures of *Azotobacter chroococcum*, *Pseudomonas fluorescens*, and yeasts and then applied the cultures to a sandy substrate in which oak and ash seedlings had sprouted. After a short time P^{32} was found in the plant tissues, more phosphorus being found in the oak than in the ash.

Plants growing in sterile conditions and in the presence of micro-organisms feed differently and have different composition of matter in their tissues. Substantial differences have been observed in the content of amino acids and organic compounds of phosphorus (Ratner and Kolosov 1954, Kotelev 1955, Krasil'nikov and Kotelev 1956).

Plants absorb nutrients differently from substrates in sterile and non-sterile conditions. Micro-organisms not only supply plants with supplementary nutriment in the form of their metabolic products; in their presence absorption of nutrients by the root system is stimulated.

When tree seeds are treated with certain micro-organism cultures before sowing, seedling development is favourably affected.

According to P. E. Malyshkin (1951), after treatment of acorns with *Azotobacter* and *Trichoderma* the root system and underground parts of oak seedlings showed 20-40% better development than the controls. He points out that *Azotobacter* and *Trichoderma* not only survive in the rhizospheres of one-year-old oaks but develop excellently. Other micro-organisms (*Bac. mycoides*, bacteria of *Pseudomonas* type, and the fungus *Tieghemella glauca*) showed no noticeable effect on oak growth. *Pseudomonas* and the fungus soon disappeared from the rhizospheres of the young oaks.

Pre-sowing treatment with *Azotobacter* of seeds of spruce, pine, Siberian larch, black locust, and caragana increased their germination rate. Application of *Azotobacter* leads to greater accumulation of chlorophyll in tree leaves (A. V. Ponomareva 1951, 1953). Tribunskaya (1950) records stimulation of the growth of young seedlings of maple, ash, and acacia in grey forest soils, poor in organic matter, after the seeds were treated with *Azotobacter*.

Soil micro-organisms are able to use compounds that harm plant growth; these may be secreted by the plants themselves or by other micro-organisms.

Further study of the effect of micro-organisms on plant growth and nutrition should be very helpful in attempts to increase soil fertility and raise tree productivity.

INTER-RELATIONSHIPS BETWEEN HIGHER HYMENOMYCETES AND TREES

A substantial part in the exchange of matter and energy in a forest biogeocoenose is played by the higher Hymenomycetes, which form mycorrhiza on the roots of higher plants. Tree growth depends not only on soil type, water supply, nutrients, light, etc., but also on the development of mycorrhiza-forming fungi, which permit plants to use mycotrophic nutrition. By the formation of mycorrhiza, fungi are metabolic intermediaries between plants and soil. In most tree species mycorrhiza-forming fungi fulfil the function of root filaments.

Three types of mycorrhiza are distinguished, according to their anatomical structure: ectotrophic, endotrophic, and ecto-endotrophic. The great majority of trees (conifers: spruce, pine, larch, etc.; deciduous: oak, birch, poplar, etc.) have, as a rule, ectotrophic mycorrhiza, characterised by the formation of a fungal sheath, reduction of root filaments, and the presence of fungal hyphae among the cells of the cortical parenchyma. Endotrophic mycorrhiza (with which root filaments are not reduced and there is no fungal sheath, but the fungal hyphae penetrate the cells of the root cortex) occurs as a rule in grasses, more rarely in shrubs (acacia, heather).

The mycorrhizae of coniferous and deciduous trees are in most cases formed of Basidiomycetes, mostly Hymenomycetes and rarely Gastromycetales. Among Hymenomycetes the commonest mycorrhiza-formers are species of *Boletus*: *B. edulis* (Polish mushroom), *B. luteus* (butter mushroom), *B. bovinus* (bovinus mushroom), *B. scaber* (birch mushroom), *Lactarius* spp. (saffron milk-cap, *L. rufus*, etc.), *Amanita* (death-cap), etc. Among Gastromycetales, *Scleroderma* and *Rhizopogon* form mycorrhiza.

Most mycorrhiza-forming fungi are not specialised and can form mycorrhiza on different tree species. That applies, for instance, to *Amanita muscaria* and several species of *Boletus*, *Lactarius*, etc. Therefore a single tree may have mycorrhiza formed by different fungi. There are some strictly-specialised mycorrhiza-formers such as *Boletus elegans*, which forms mycorrhiza only on larch.

Summaries of morphological, ecological, and anatomical studies of mycorrhiza are ably presented in the works of Melin (1925), Harley (1959), Lobanov (1953), and Shemakhanova (1962).

The inter-relationships of mycorrhiza-forming fungi with trees are very complex and have been investigated less than anatomical and morphological features. Some investigators regard mycorrhiza-forming fungi as parasites on a plant-host (Kamenskii 1886, Weyland 1912, Romell 1939). Most authors consider that mycorrhiza-forming fungi benefit the plant (Baranei 1940, Vysotskii 1929, Melin 1925, Mishustin *et al.* 1951, Lobanov 1953, Shemakhanova 1962). Without mycorrhiza tree seedlings either develop poorly or perish. There is a third point of view, according to which mycorrhiza-forming fungi may play both positive and negative roles in plant life, depending on the physiological state of the plant-host (Kostychev 1933, Yachevskii 1933, Kuprevich 1947).

At present most investigators regard mycorrhiza on trees as a symbiotic association, from which both the plant and the fungi derive benefit.

Mycorrhiza-forming fungi have a beneficial effect on the mineral nutrition of trees, and stimulate the general metabolic activity of plant roots (McComb and Griffith 1946, Simkover and Shenefelt 1951, Harley and Brierley 1955, Melin and Nilson 1955, Morrison 1961, 1962 a, b). They aid in the provision of nitrogen, phosphorus, potassium, calcium, magnesium, and iron. Plant mycorrhizae contain, as a rule, more nitrogen, phosphorus, and potassium (Hatch 1936, Harley 1937, 1952, McComb 1943, Kramer and Wilbur 1949, Melin and Nilson 1950). The fungus *Boletus luteus*, which forms mycorrhiza on pine, has a very beneficial effect on the phosphorus nutrition of trees (Stone 1949). Young oaks with mycorrhiza absorb P^{32} more intensively than those without it (Shema-khanova 1955a, Tarabin 1961).

In addition, mycorrhiza fungi enable plants to obtain mineral nutrients from compounds usually unavailable to plants, e.g., phosphorus from apatite, felspar, and tricalcium phosphate, and potassium from mica and grits (Stone 1949, Wilde 1954, Eglite 1958, Shemakhanova 1962). According to Kuprevich (1952), plant roots without mycorrhiza have a smaller range of enzymes than those with it, and their activity is lower. Some investigators believe that assimilation of organic as well as mineral compounds is stimulated by mycorrhiza-forming fungi (Melin and Nilson 1953). The data of Melin and Nilson, obtained by use of P^{32}, show that unde-aminated glutamic acid in solution passes from the fungus to the plant host. Organic matter that is unavailable to the plant, according to many investigators, may be absorbed by mycorrhiza-forming fungi, but that has not been shown definitely.

The water supply of plants is improved by mycorrhiza (Khudyakov 1951, Runov 1955). According to Shemakhanova (1962), mycorrhiza-forming fungi supply plants with vitamins, mainly of the B group (biotin, pantothenic and nicotinic acids) and other supplementary growth factors.

The physiological functions of trees with mycorrhiza differ substantially from those of trees without it. Oak seedlings with mycorrhiza show a tendency to increased rate of transpiration (Shemakhanova 1955b, Samtsevich 1958). In pine seedlings (and to a lesser extent in oak seedlings) with mycorrhiza an increased chlorophyll content is found, and the activity of the root system is higher.

The fungi in turn receive simple carbohydrates from the plant, as shown by Melin and Nilson (1957) by use of the isotope C^{14}. The fungi also receive some other substances (amino acids, phosphatides, mineral matter, etc.). The close dependence of the fungi on the plant-symbiont is shown by the fact that the fruit-bodies of the fungi appear only when the plant component is present. In a pure culture, or when the connection with the plant is broken, the fungi do not produce fruit-bodies.

Most investigators remark upon the generally beneficial effect of mycorrhiza on plant growth. Wood increment, height, trunk diameter, and other tree dimensions are in direct ratio to the amount of their mycorrhiza. The greater the mycorrhiza development, the better is the tree growth. Mycorrhiza formation depends on a number of factors. Bilberry, for instance, provides more favourable conditions for development of mycorrhiza on pine and spruce roots than does mountain sorrel (Lobanov 1962). Soil type has a great influence on mycorrhiza formation. Mycorrhiza develop better in chernozem soils than in dark-brown and light-brown soils. On conifers mycorrhiza formation is better in less fertile soils. Fertilisation with mineral salts (nitrogen, phosphorus, potassium) depresses mycorrhiza formation on conifers. Excessive fertilisation leads to insufficient formation of mycorrhiza and consequent poor seedling survival on poor soils in northern districts (Björkman 1956, Lobanov 1960).

The N:P ratio is of importance in mycorrhiza formation. In experimental conditions, mycorrhiza forms on plants with moderate supply of nitrogen and phosphorus (Trubetskova *et al.* 1955, Shemanakhova 1962).

Mycorrhiza formation depends on light intensity, strong illumination stimulating it. A. A. Vlasov (1955), E. N. Mishustin (1955), and S. A. Samtsevich (1955a) remark on the need for adequate amounts of moisture for successful mycorrhiza-formation in steppe and forest-steppe districts.

The presence of mycorrhizae that improve growth in tree plantations is one of the conditions of successful afforestation. Many authorities believe it is necessary to enrich the soil with mycorrhiza-forming fungi when planting forests in steppe conditions (Lisin 1949, Krasovskaya and Smirnova 1950, Lobanov 1952, Shemakhanova 1962). Application of pure cultures of mycorrhiza-forming fungi benefits the growth of oak and pine seedlings (Runov 1952, 1955). Most investigators have recorded favourable results from adding mycorrhiza in the first years of life of seedlings. Shemakhanova (1962) presents data showing that that effect continues for up to nine years.

MICRO-ORGANISMS AND THE ATMOSPHERE

Atmospheric conditions have considerable influence on microbiological processes in forest biogeocoenoses. In northern districts these processes are restricted by inadequate heat, with adequate moisture. In zones of inadequate moisture, with optimal temperature conditions, the development of micro-organisms is limited by moisture shortage. It is not only the type of soil, i.e., its content of organic and mineral matter, that forms the soil microflora. The direction and rate of microbiological processes in soils with the same content of organic and other matter are closely dependent on hydroclimatic conditions.

The seasonal dynamics of microflora (i.e., the periodicity of intensity of microbiological processes in the soil) are linked directly with the climatic factor. In the central taiga zone (Vologda province) maximum microfloral development in forest soils is seen at the end of summer, after the soil has been warmed sufficiently. In forest soils of the southern taiga (Moscow province) and in the forest-steppe and steppe zones (Orlov and Voronezh provinces) microbiological processes decline mainly in spring, and to some extent in autumn; in summer the activity of micro-organisms diminishes somewhat when there is a decrease in soil moisture. In the southern steppes (Kirovograd province, Moldavia), with a comparatively warm and moist winter, large numbers of micro-organisms are found during the winter months.

The course of microbiological processes is, no doubt, affected by the microclimate under forest canopy and by its changes when trees are felled.

As a result of the activity of micro-organisms the lower layers of the air under forest canopy are enriched with CO_2, which certainly affects the carbon nutrition of plants.

Micro-organisms are carried by air currents into higher atmospheric layers. When the air is much polluted, bacterial survival is considerable at a great height. Fungi imperfecti have been observed at a height of 22,108 metres. There are far more bacteria in the air above cities than in the air above forest areas or fields. According to Ya. G. Kishko (1961), at a height of 250 metres above Lvov there are 1176 micro-organisms per m^3 of air, above a field 628, and above a forest 662. There are qualitative differences in the microflora above a forest and above a city. Above a city pigmented forms predominate; there are fewer spore-forming bacteria and still fewer fungi. The air above a forest is heavily laden with fungi (50% of the total number of micro-organisms), and has very few pigmented bacteria.

Air currents can carry micro-organisms for considerable distances—tens or even hundreds of kilometres.

The study of the epiphytic microflora occurring on the leaves, bark, and seeds of trees is of great interest. It is only recently that this question has begun to be studied in relation to tree species.

MICRO-ORGANISMS AND ANIMALS

Apart from pathogenic micro-organisms and microflora living in animal intestines, soil-dwelling micro-organisms have practically no direct effect on the life of animals. But after the death of mammals, birds, insects, etc., their carcases become the prey of micro-organisms that mineralise animal remains. Obviously the amount of mineral matter released into the soil by decomposition of carcases is proportional to the animal population of the forest.

An indirect inter-relationship between microflora and the metabolism of animals, especially herbivores and birds, exists through the effect of microbe numbers on the growth of plants that serve as animal food. There are, however, no precise data on that subject in the literature.

Animals affect microbiological activity in forest soils through their action upon litter and soil. Various animals, by mixing litter with the soil, digging up the soil, or burrowing through it, help the microflora to develop and thereby intensify the processes of decomposition of plant remains in the forest. As a result of the penetration of microflora along animal burrows deeper soil layers become involved in the circulation of organic matter and humification is accelerated.

Various insects play a major role in the decomposition of litter and wood. By consuming litter and gnawing wood they create favourable conditions for micro-organism development. When insects and earthworms are present, the decomposition of forest litter advances more rapidly.

EFFECT OF FORESTRY MEASURES ON FOREST SOIL MICROFLORA

The course of microbiological processes in forest soils is markedly altered by the economic activity of man. For successful afforestation in steppe and forest-steppe districts, thorough study of forest-growth conditions is essential. Study of microbiological processes in the soil before tree-planting enables us to work out theoretical principles of afforestation and to solve a number of practical problems connected with correct selection of species and with the effect of afforestation on soil fertility.

Microbiological investigations have shown that in soils under forest strips the processes induced by microbes take place with considerably more intensity than in virgin soils. The breakdown of organic matter in the soil and of dead plant parts, and the conversion of nutrients into forms available to plants, proceed there more rapidly, to the benefit of plant growth.

In places where trees are planted, deeper soil horizons are involved in the process of transformation of organic and mineral matter by micro-organisms than in uncultivated land. Under forest belts micro-organisms occur at a much greater depth than in virgin soil or ploughed land, because of the deep penetration of tree roots. For correct selection of tree species for planting the microbiological factor has to be taken into account. It has been found, for example, that microbiological processes are some-what depressed in pure oak plantations as compared with mixed stands. The decomposition of plant remains evidently proceeds more actively in mixed than in unmixed oak plantations.

The soil microflora in forest belts has been studied by I. E. Brezhnev (1950 a, b), A. Ya. Manteifel' et al. (1950), Z. M. Ryashchenko et al. (1952), A. Ya. Slesarenko (1952), S. F. Morochkovskii (1953), E. I. Kozlova (1955), M. A. Vinokurov et al. (1959).

Forestry measures connected with tree-felling lead to changes in the numbers and composition of microflora. Study of the course of micro-biological processes in the soil of clearings is intimately linked with the solution of reafforestation problems.

Tree-felling leads in most cases to intensification of microbiological processes, because of the enrichment of the soil with plant remains (Bursova 1955, Truhlář 1958, Tvorogova 1959, Shubin and Popov 1959, Pushkinskaya 1962).

The effect on microbiological processes due to thinning of coniferous plantations has been studied (Bernat and Novotna 1955). In thinned stands the number of fungi decreased from 1,400,000 to 800,000 per gramme of soil, but the number of ammonising bacteria rose from 13 to 36 million; the greatest increase was seen in the spore-forming bacteria.

In a study of the trenching method of reafforestation (Czechoslovakia) a decrease in the numbers of fungi was observed, with an increase in ammonising bacteria able to grow in meat-peptone agar (Bartlova et al, 1955).

In the arid-steppe zone, improvement cutting in oak-ash plantations with caragana underbrush leads to increased microbiological activity in forest soils. Removal of caragana from a plantation leads to a decrease in the numbers of cellulose-decomposing bacteria.

In burnt clearings occupied by hairgrass and willowherb, typical of the northern and central taiga subzones after concentrated clear-cutting, an increase in the numbers of microflora is observed (Tvorogova 1959).

In hairgrass clearings the content of ammonising bacteria in the soil increases in the first year and falls in subsequent years. In the soil of willowherb-burnt clearings there is a greater increase in microflora (ammonisers, nitrifiers) than in hairgrass clearings, persisting for eight years (Table 116).

In the litter of willowherb-burnt clearings the content of nitrifying and cellulose-decomposing micro-organisms increases considerably during the fourth or fifth year. Fire-clearing of cut areas strongly affects the soil microflora; in most cases the number of micro-organisms in forest soils increases after fires.

Table 116. *Micro-organism content in the soil of hairgrass and willowherb-burnt clearings of different ages (Tvorogova 1959).*

Horizon	Item (years after cutting of trees)	No. of microbe cells (10^3 per g of soil)						
		Bacteria in MPA	Fungi	Nitrifying bacteria	Denitrifying bacteria		Aerobic cellulose-decomposing micro-organisms	Butyric fermentation micro-organisms
					Phase I	Phase II		
	Hairgrass clearings							
A_0	New bilberry-spruce growth	230,000	40	0·14	3800	20	3	1300
	1 year	240,000	22	2·00	2800	20	28	800
	5 years	124,000	27	0·17	2200	5	0·37	380
	8 years	36,000	17	0·30	3400	4	3·3	2800
A_2	New bilberry-spruce growth	15,000	4·3	—	470	—	—	24
	1 year	96,000	0·22	—	1200	2·4	—	143
	5 years	30,000	—	—	1300	—	—	104
	8 years	33,000	—	—	900	—	—	27
	Willowherb-burnt clearings							
A_0	New bilberry-spruce growth	230,000	40	0·14	3800	20	8	1300
	1 year	360,000	4	20·0	3900	700	30	2500
	5 years	210,000	—	21·0	4000	3100	30	830
	8 years	96,000	15	—	2100	—	0·77	1300
A_2	New bilberry-spruce growth	15,000	4·3	—	470	—	—	24
	1 year	80,000	—	2600	850	116	—	140
	5 years	98,000	—	400	820	111	—	230
	8 years	42,000	1·9	—	1350	—	—	34

N. N. Sushkina (1931) studied the effect of fires in coniferous forests with a mixture of deciduous species on loamy and sandy-loamy soils in Leningrad province. After fires the processes of denitrification and butyric fermentation proceeded actively. Intensification of the nitrification process also occurred (Hesselman 1916-17, Sushkina 1933). According to E. V. Runov and E. N. Zhdannikova (1962), within a month of a fire microbiological activity was already increasing, the greatest increase being in the uppermost 0-1 cm layer. In deeper horizons little change in the microbial population was seen. Maximum increase in microbe numbers was recorded three months and a year after the fire, and then the numbers of microflora decreased. Among the micro-organisms the numbers of bacteria increased most rapidly initially. The numbers of microscopic fungi fell at first, and then began to rise after three months. During the first year after the fire cellulose-decomposing and nitrifying microorganisms developed vigorously.

According to investigations in the Belgian Congo, wood ash induces a rapid increase in the numbers of microbes, especially bacteria (Meiklejohn 1955). Immediately after a fire microbiological activity falls sharply— sometimes fungi disappear entirely—but within one or two months the numbers of micro-organisms exceed those in the controls. During the period immediately after a fire there is an increase in butyric fermentation and a decrease in nitrification. Corbet (1934) recorded a considerable increase in the numbers of bacteria, fungi, and Actinomycetes in Malaya after a forest fire. Such increases in microflora after a fire are due to the entry of nutrient elements into the soil, change in pH, and other factors.

Application of mineral fertilisers and lime produces beneficial effects on the microflora in forest soils similar to those resulting from fires (Mustafova 1958, Loub 1959, Hartmann 1960, Eglite and Yakobson 1961).

A. K. Eglite and Z. A. Yakobson (1961) state that the productivity of tree plantations is directly due to microbiological processes taking place in the soil. By use of various agrotechnical measures microbiological processes can be regulated and the productivity of infertile soils increased.

Such a technique as applying herbicides to the soil while forest belts are being cultivated after planting does not reduce the number of bacteria in oak seedling rhizospheres. The numbers of fungi increase, but those of Actinomycetes decrease (Klyuchnik and Petrova 1960).

Application of antibiotics (streptomycin) to sandy forest soils reduces the number of Actinomycetes and bacteria. The effect of streptomycin lasts for about six months, fungal development not being stimulated (Koszubiak 1961).

The use of bacterial fertilisers, especially *Azotobacter*, has found practical application in forestry in recent years, good results being obtained when pine seeds and acorns are treated with bacteria in nurseries (Shestakova 1962).

Drainage of waterlogged forest soils in the central taiga has had a beneficial effect on microbiological processes. By the end of the first year of drainage the activity of various groups of micro-organisms increases notably, causing more thorough decomposition of organic matter (Zhdannikova and Popova 1961).

CHAPTER VI

Soil as a component of a forest biogeocoenose

Soil is a complex natural formation, directly and diversely linked with other aspects of nature: atmosphere, mineral strata, vegetation, animals, water. Therefore soil is one of the principal components of the biogeosphere, determining—and displaying in its own features and properties —the characteristics of formation and development of separate biogeocoenoses. This view of the soil has been widely developed both in the U.S.S.R. and abroad. In our country it is indissolubly linked with the names of several eminent scientists: Vernadskii, Berg, Neustrev, Morozov, Vysotskii, Abolin, Polynov, and others, and it has been developed in Sukachev's works.

The foundation of study of soil as a biogeocoenose component is knowledge of all the multiform interactions between it and the other animate and inanimate components of biogeocoenoses, and primarily of the role of soil in the biogeocoenotic exchange of matter and energy.

From the biogeocoenotic point of view, 'soil' means *the upper layer of the lithosphere, which is involved in biological circulation and has acquired the features of a natural component, with its own characteristic direct exchange of matter and energy, which determines its structure, productivity, and laws of formation and evolution.*

From that definition it follows that the formation of soil is due not only to processes within itself, but also to processes throughout all the surface layers of the earth in which the life-activity of organisms takes place. Under the combined influence of these organisms and atmospheric agents, changes occur in the depths of soil-forming strata.

The vertical limits of the soil are: above, the external surface of forest litter in contact with the lowest layer of air; and below, the deepest penetration of tree root systems.

That thickness of soil provides organisms with moisture and nutrient elements, and at the same time is itself affected by them to an extreme degree. It possesses physical properties that very well fulfil the requirements of plants and animals (temperature, aeration, etc.).

The horizons lying below the soil occupied by roots are also involved in soil formation, but changes in them are mostly due to physical and physico-chemical processes resulting from the movement of water and substances not used by organisms. In these horizons there is a certain amount of movement of soil-forming products withdrawn from the biogeocoenotic exchange of matter and energy.

438

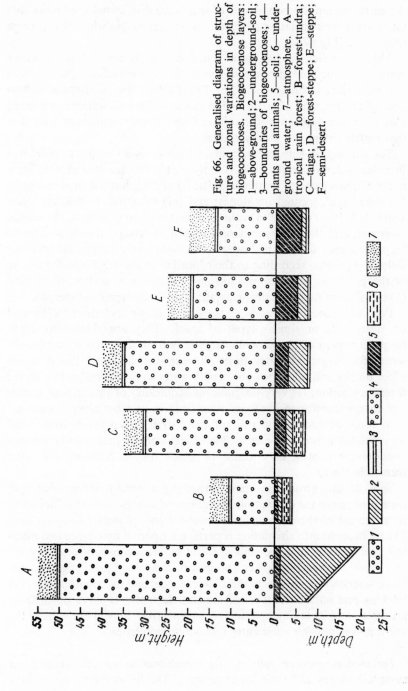

Fig. 66. Generalised diagram of structure and zonal variations in depth of biogeocoenoses. Biogeocoenose layers: 1—above-ground; 2—underground-soil; 3—boundaries of biogeocoenoses; 4—plants and animals; 5—soil; 6—underground water; 7—atmosphere. A—tropical rain forest; B—forest-tundra; C—taiga; D—forest-steppe; E—steppe; F—semi-desert.

The depth of the surface layer of the earth that takes part in soil formation corresponds in general to the depth of the biogeocoenotic cover of the earth, within the boundaries of which are distributed the bodies that take part in the circulation of matter and energy and in which that process takes place (Fig. 66).

Although relief, as has been stated earlier, is not a biogeocoenose component, the configuration of the earth's surface affects the processes that take place in biogeocoenoses. Relief affects the distribution of heat and light energy and of precipitation and underground water, and thereby affects the composition of the soils of forest biogeocoenoses and the way that matter enters and moves in them.

The connection between elementary landscapes and types of relief was first clearly demonstrated by B. B. Polynov (1956). He defined three main types of movement of matter in soils: (*i*) eluvial, located in watersheds; (*ii*) superaqual, located on slopes; and (*iii*) subaqual, in flooded areas. Later A. I. Perel'man (1961) called the first type autonomous, the second emergent, and the third submerged. Correspondingly the soils of forest biogeocoenoses formed in lowland conditions may be divided into the following groups, according to their location in the relief and the nature of the processes of transportation and migration of matter: (*i*) eluvial; (*ii*) transitional (accumulative-eluvial); and (*iii*) emergent-submerged.

Forest-biogeocoenose soils of the *eluvial* group include those formed on watersheds or similar types of relief. They are characterised by: (*a*) procurement of matter and energy only from the atmosphere; (*b*) absence of underground water, or its presence at such a depth that it has no effect on the soil and does not supply water and mineral matter to the forest vegetation; (*c*) very insignificant acquisition of products of matter and energy transformation from other surfaces or other biogeocoenoses.

In forest-biogeocoenose soils of the eluvial group the movement of accumulated matter and energy due to atmospheric precipitation reaches considerable depths, as a result of which illuvial horizons are often formed in them.

Soils of this group are characterised by a certain independence of formation, since the circulation of matter and energy in them is limited to procurement of these from the atmosphere and of matter from the soil. Matter brought into circulation is partly transported by surface and intra-soil flow of water into the soils of forest biogeocoenoses of other groups and partly washed into lower horizons of the soil. Forest vegetation in biogeocoenoses is in the highest degree adapted to counteracting the leaching and natural removal of mineral elements. It slows down these processes, and by taking up mineral elements from illuvial horizons of the soil it makes possible their entry into active circulation.

Forest-biogeocoenose soils of the transitional group are situated on mountain slopes and their lower regions. The following processes take place in them: (a) dispersal (in contrast to acquirement by the eluvial group) of atmospheric precipitation and of heat and light energy; (b) entrance of moisture and of various products of matter and energy exchange from adjacent biogeocoenoses, and dispersal of them by surface and intra-soil water flow; (c) additional accumulation of matter from underground water.

The above phenomena are sharply differentiated on slopes of different exposure (especially southern and northern), which produces individualisation of the soils in forest biogeocoenoses. The dispersal and transportation of matter and energy produces a form of circulation of these differing from that in the eluvial group. That is well displayed in the composition and external appearance of forest biogeocoenoses, in the physiological processes taking place in them, and consequently in the total mass of organic matter created. The intensity and course of biochemical reactions in the soils are so different that they lead to formation of special soil types.

Forest-biogeocoenose soils of the emergent-submerged group are situated in depressions that are regularly or periodically inundated by river flooding, surface flow, heavy precipitation, sea irruptions, etc.

The distinctive features of forest-biogeocoenose soils in such conditions include: (a) periodic enrichment with soluble or solid matter from the surface; (b) leaching of mobile nutrients (in cases of heavy atmospheric precipitation); (c) acquirement of excessive amounts of soluble mineral and organic matter, often harmful for forest vegetation, with which the latter must contend or to which it must adapt itself. Soils of such biogeocoenoses are distinguished by many peculiarities. Constant or temporary excessive moisture is predominant in them; organic matter of a special type (peat) accumulates; silt, and in some places even salts, are deposited.

Peatiness, silting, and salinity (together or separate) are characteristic of the soils of such biogeocoenoses. They are situated in river valleys, on the shores of land-locked or open lakes, on watersheds (swampy flat or depressed surfaces, mangrove swamps, etc.).

Whereas forest-biogeocoenose soils of the first group are characterised by a certain independence of development, not linked with addition of matter by solid or liquid inflow from outside, soils of the two latter groups may be considered more dependent, since their development is to a great extent connected with inflow of matter and energy from watersheds or from higher parts of the land, or from the waters of rivers, lakes, or seas.

Accumulative processes are clearly manifested in the soils of all three groups, but the character and extent of their manifestation are different in each group.

In soils of the eluvial group accumulative processes are limited, and are linked with soil-forming strata and the atmosphere. Accumulation of organic remains is greatest on the soil surface.

Accumulation is more intensive in soils of the transitional group, on account of the inflow of matter with surface and underground water flow and introduction of solids as a result of erosion. Accumulation of organic remains is on a smaller scale.

In soils of the emergent-submerged group accumulation is most evident, because of periodic floods and the upward growth of soils through alluvial deposits. Accumulation of organic remains is irregular; it may be at a maximum or a minimum as compared with the preceding groups.

The processes of outflow take place to an equal extent in soils of all three groups, but they are more intensive in the first; changes in character

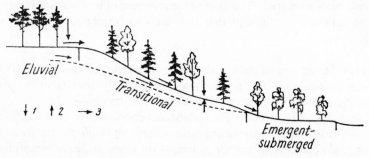

Fig. 67. Diagram of soil distribution as related to different groups of forest biogeocoenoses and relief. ↓ —entry of matter into biogeocoenoses from the atmosphere; ↑ —entry of matter into biogeocoenoses from underground water; → —entry of matter into biogeocoenoses from surface and underground water flow.

through leaching and erosion are characteristic of soils of the second and third groups.

We should point out that each group contains a great variety of types of forest biogeocoenoses. That is due not only to geomorphological differences in land areas but also to climatic differences, both zonal and local, including those on slopes of different exposures, etc.

One of the features of soil distribution in types of forest biogeocoenoses is the transitional phases: soils of the eluvial group combine with or merge into transitional soils, and the latter combine with or merge into the emergent-submerged group. That distribution is related to relief and is diagrammatically shown in Fig. 67.

The connection between types of biogeocoenoses and relief corresponds to the concept of soil-topographical profiles, or catenas. In mountain conditions the formation and distribution of types of forest biogeocoenoses is subject to the same laws as in lowland conditions. But abrupt changes in elevation within short distances, greater angles of slope, proximity of mineral strata to the surface, well-marked vertical zoning, great variations in the heat and moisture regimes of slopes of different exposures, and the well-developed migration of water and solid matter down slopes will lead to different intensities and manifestations of the processes that influence soil formation in mountain forest biogeocoenoses. Moreover, in mountain conditions soils are often not distinct from the lithosphere, since tree root systems not only pierce mineral strata but also bring matter and energy from them into the biological cycle.

In mountain districts, therefore, forest biogeocoenoses as a whole and their soils in particular call for special examination and for further study of the laws of their formation.

There is still another type of accumulation of matter and energy in nature, caused by the earth's internal forces breaking out into the atmosphere and subsequently exerting such substantial influence on the development of soils of forest biogeocoenoses that it requires to be separated into a special class, volcanic.

Soils of forest biogeocoenoses connected with volcanic action are classified as *aerial-volcanic*, and are characterised by accumulation of airborne ash. Such formations occupy extensive areas in volcanic districts. There the soils of forest biogeocoenoses (regardless of their development, whether on lowlands, watersheds, slopes, or depressions) are distinguished by such massive accumulation of airborne volcanic ash and scoria that it stops the natural processes whereby the eluvial, transitional, and emergent-submerged groups of soils are formed.

Their structure and characteristics show that soil formation proceeds irregularly, being restricted to the intervals between volcanic eruptions and successive showers of ash. Therefore it is proposed to call these processes *interrupted-eluvial*, *aerial-transitional*, and *water-transitional*.

The above are the basic processes and phenomena governing the development of soils in forest biogeocoenoses and their distribution by groups.

In studying soil as a component of forest biogeocoenoses, we must clarify the following problems: (*i*) the inter-relationships and interactions between soils and other components of forest biogeocoenoses, and the inter-relationships among them in different climatic zones; (*ii*) the effect of biogeocoenotic processes on soil formation; (*iii*) the effect of migration processes on the soils in adjacent forest biogeocoenoses.

ACTION OF FOREST VEGETATION ON SOILS

Clarification of the relationships between trees and soils involves discovery of the genetic properties of soils that make possible the formation of forest biogeocoenoses.

At the foundation of the study of soils is modern knowledge of soil origins, supplemented by consideration of such distinctive properties of forest soils as the composition and characteristics of forest litter and its role in humus formation.

It would, however, be incorrect to look for correspondence between forest types and soil types, since these taxonomic units have different scope and content. A forest type, or a type of forest biogeocoenose, is a lower unit of biogeocoenotic classification; a soil type, on the other hand, is one of the higher units of genetic soil classification.

A soil type corresponds to a wide group of forest types, combining an assemblage of tree stands of varying composition and origin. In this connection we must consider new data on the genesis of forest soils from Siberia, the Far East, Central Asia, and other regions, which make substantial changes in former ideas on the linkage between soil types and definite forest formations. That relates especially to the type of podzolic soils, whose extent nowadays is constantly shrinking because of the definition of such new soil types as cryogenic-taiga, soddy-forest, etc. In the study of forest soils as components of forest biogeocoenoses we must discover the connections between not only tree stands, but also all of their associated vegetation, and the sum of soil features and properties that govern their origin and development and correspond most fully to the level of their natural productivity. We must consider, from all points of view, those distinctive features of soil formation under forest vegetation that alter to some extent (as compared with conditions under steppe, meadow, and other types of vegetation) the natural circulation of matter and energy. Of special significance are the nature and properties of forest litter as being an accumulator of mineral matter required by forest vegetation and of nitrogen, and therefore being highly important in the life of the forest.

Moreover, the qualitative differences in the composition of vegetation and forest litter are linked with the intensity of the life-activity of animals and micro-organisms, since the latter are intimately dependent on vegetation as a source of food. Thus the soil represents both the cause and the effect of formation of distinctive types of forest biogeocoenoses.

The biogeocoenotic approach to study of the connection between forest types and soils is based on the achievements of painstaking genetic pedology, which takes into account all the diverse properties of soils.

The above observations are relevant to discussion of the inter-relationships between forest vegetation and forest soils.

RELATIONSHIPS BETWEEN PRINCIPAL FOREST-FORMING SPECIES AND SOILS

Although these questions properly relate to tree ecology, we touch on them here to the extent to which they concern biogeocoenology. Many data on them have been published. These data are scattered, however, through various publications, sometimes little known or not easily available. In addition, investigators do not always precisely define the tree species on which they make observations. Not only are species of a single genus distinguished by different ecological characteristics, but such differences occur in even finer taxonomic subdivisions.

The properties of soils are very diverse and vary not only horizontally but also vertically, even within the limits of the layer in which tree root systems are situated.

In nature it is extremely difficult to segregate the effect of any single soil characteristic upon trees. Therefore the study of such questions is of merely ancillary significance, since it is required for understanding the inter-relationships between soils and forest vegetation, in particular physico-geographical conditions.

We shall try to survey briefly the relationships between tree species and the following properties of soils: mechanical and mineral composition, physical properties, moisture, acidity, salinity, alkalinity, peatiness, frost, and gas regime.

It is customary to believe that the fertility of soils is closely connected with their mechanical consistency. It is least in sandy and highest in clayey soils. Sandy-loam and loamy soils hold an intermediate position. The fertility of soils decreases in direct proportion to their stoniness (accumulation of small-grained fragments of varying mechanical consistency).

The above picture is true only in the most general terms and needs much correction in detail. Sandy soils differ in their forest-vegetational characteristics, since they differ in mineralogical composition, being enriched in varying degrees with felspathic minerals containing calcium, magnesium, potassium, phosphorus, aluminium, iron, and other elements. The greater the content of the latter in sands and the less the SiO_2 content, the more suitable they are for forest growth, and vice versa. Soils may be divided into the following three groups according to their approximate SiO_2 content:

1. $SiO_2 < 80\text{-}85\%$;
2. SiO_2 from $85\text{-}95\%$;
3. $SiO_2 > 95\%$.

The first group are considered to be relatively rich, the second intermediate, and the third poor.

Sandy soils of the first group may evolve into sandy loam, but those of the second and third groups retain their original mechanical consistency, since they contain too little in the way of mineral reserves to be enriched by finer (especially clay) particles.

Sandy soils contain little nitrogen, to which circumstance (other things being equal) are due the differences in productivity of forests growing on them. In spite of the poor mineral content of sandy soils, argillaceous and ferruginous bands of ortsand or ortstein may develop within them at varying depths during the process of soil formation. If the former are present and the latter slightly developed, the forest-growing properties of the soil are improved in comparison with soils not possessing these. That improvement consists in the accumulation within the soils of mineral nutrients available to trees and in the improvement of water supply to plants because of a decrease in outflow of water and in leaching of nutrients. Even more important is the dichotomy of the soils in mechanical consistency. If the upper sandy layer, from 0·4-0·5 to 1·5-2·0 metres in depth, contains heavier (loamy or clayey) deposits the unfavourable properties of the sands are eliminated, which is outstandingly displayed in the growth of the more nutrient-demanding coniferous and deciduous trees on such soils. In evaluating other combinations of layers of different mechanical consistency through the soil profile, we must take into account the composition of the surface layer, the nature of the succession of layers and their thicknesses, and the hydrological characteristics of the soil.

There cannot be a single type of approach to evaluation of the forest-growing capabilities of such soils.

Only in uniform sandy soils more than 2 metres in depth is a homogeneous oligotrophic regime developed, and there pine predominates. That is due in the first place to the comparatively low nutrient and moisture requirements of pine, and to its ability to develop a dense, deep root system (often with a secondary concentration of absorbing roots) depending on variations in the soil's mechanical consistency. According to Vasil'eva's unpublished data, in the Serebryanoborsk forest near Moscow, where there is an ortsand-loamy layer at a depth of 150-200 cm, the total quantity of roots (and in particular those less than 1 mm in diameter) is much increased, as is seen from the following figures on the root-system distribution of 140-year-old pines in deep sandy soil (in kg/ha):

Depth, cm	0-50	50-100	110-150	160-200
Total roots	28,737	1026	432	1026
Roots < 1 mm	4425	132	25	237

With sandy soil up to 60-100 cm in depth and rich in minerals containing calcium, magnesium, phosphorus, potassium, etc., and also with moraine deposits (as is well seen in the North German lowland, Eberswald district), composite beech-pine forests thrive. In the same area, poor (leached) quartz sands are occupied only by pine of quality class not higher than III-IV (Ehwald et al. 1961).

Soils with textures ranging from sandy-loam to clay fulfil, as a rule, the mineral requirements of tree species in all conditions. Differences in the growth and productivity of trees growing on them are due not to their texture but to the physical and chemical properties of the soils. Therefore the texture of soils within this range is not a contra-indication for the majority of forest-forming tree species.

A stony-gravel soil in mountainous or plains regions does not prevent the growth of various tree species; it affects vegetation mostly through water supply and the chemical composition of mineral strata, the thickness of the upper fine-grained and less stony layer being important.

Stoniness has comparatively little effect on pine. All other species (spruce, oak, birch, European larch, etc.) are more sensitive, and if the fine-grained layer is less than 10 cm in depth their growth is usually much weakened and their productivity falls to quality classes IV and V. That is most clearly shown on stony soils developed on crystalline strata; on sedimentary strata, which are less compact, the effect is less evident.

Thus the direct dependence of the growth and productivity of various tree species on the texture of soils is manifested only in the most general form, and appears comparatively clearly only on sands and stony sub-strates. In all other cases no direct connection between the growth and productivity of trees and the texture of soils is observed. The influence of the latter is exerted through other soil properties.

The relationships between trees and the mineral composition of soils take many forms. The most clearly evident manifestations of these are the differences resulting from types of weathering of soil-forming strata in the boreal and tropical bioclimatic zones.

Soil-forming strata and soils formed from them are divided into two main groups according to their mineral composition: siallitic and allitic. The former are characteristic of cold and temperate, the latter of hot-arid and humid-tropical zones.

Soils and soil-forming strata of siallitic type are distinguished by rich-ness in SiO_2, CaO, and MgO, and relatively small content of Al_2O_3, Fe_2O_3, etc. In soils and soil-forming strata of the allitic type Al_2O_3 and Fe_2O_3 predominate, and they are poor in SiO_2, CaO, MgO, and other higher oxides (Table 117).

Table 117. *Gross chemical composition of soils and soil-forming strata of siallitic and allitic types of weathering.*

Type of soil	Depth (cm)	SiO_2	Al_2O_3	Fe_2O_3	P_2O_5	MnO	CaO	MgO	SO_3
Siallitic	2-10	78·72	10·26	2·86	0·10	0·48	1·34	1·09	*
	170-180	72·28	14·10	4·50	0·01	0·99	1·50	1·99	*
Allitic	0-10	25·48	41·74	16·98	0·18	None	0·09	0·07	0·10
	300-350	28·12	39·25	18·78	0·18	Traces	0·22	0·05	0·19

* Not determined.

Corresponding to that division of soils and subsoils, we may divide the tree species growing on them into two groups. Species linked with allitic soils are characterised by relatively higher requirements of Al_2O_3, whereas species that prefer siallitic soils consume more CaO and less Al_2O_3.

That difference is due to adaptation of the tree species to the chemical composition of the mineral part of the soils and strata, and is very well shown in the ash content of their needles and leaves. An example is provided by the ash content of the needles of certain pine species growing on soils of siallitic and allitic types of weathering (Table 118).

On strata and soils of siallitic type considerably more CaO and much less Al_2O_3 accumulates in the needles. On soils of allitic type the Al_2O_3 requirement is much higher.

The physiological role of that oxide, however, is not yet clear.

In certain climatic zones the effect of the mineral composition of soil-forming strata on vegetation is quite evident. In the taiga and forest-steppe zones it can be traced in variations in the ash content of leaves and needles (Table 119).

Not all tree species react in the same way to the composition and characteristics of soil-forming strata and soils formed from them. Pine and birch show little reaction to excess of $CaCO_3$. The CaO content of their needles and leaves is fairly stable even in soils with different carbonate

Table 118. *Ash content of needles of* Pinus *species growing on soils of siallitic and allitic types of weathering of granites (% of ash content) (from data of S. V. Zonn for* P. massoniana *(China) and of* Yu. Abaturov *for* P. silvestris *(Southern Urals)).*

Soil, type of weathering, species of pine	Ash content	SiO₂	Fe₂O₃	Al₂O₃	CaO	MgO	SO₃	P₂O₅	K₂O	MnO
Red, allitic, P. massoniana	1·43	9·0	2·7	16·1	18·8	12·6	22·3	17·4	27·2	2·0
Soddy, siallitic, P. silvestris	2·61	8·4	2·6	2·3	31·4	10·0	—	6·1	19·5	—

Table 119. *Variations in ash content of leaves and needles of some tree species according to the composition of the soil-forming strata (% of ash content) (S. V. Zonn).*

Nature of soil-forming strata	Ash content, %	SiO₂	Fe₂O₃	Al₂O₃	CaO	MgO	SO₃	P₂O₅	K₂O
Scots pine									
Crystalline	26·1	8·42	2·69	2·30	31·40	9·95	*	6·13	19·5
Limestones	1·33	2·76	0·89	—	24·34	9·35	*	16·12	*
Loam without carbonates	2·63	8·75	2·66	—	31·20	11·00	*	8·43	19·8
Loam with carbonates	2·19	20·20	0·91	9·13	29·63	12·30	9·60	6·85	11·4
Norway spruce									
Crystalline	4·74	19·40	5·48	26·4	25·7	11·6	4·46	6·96	—
Limestones	3·53	13·7	0·9	—	52·3	4·7	—	5·08	—
Loam without carbonates	4·72	31·2	7·0	5·92	40·0	3·38	3·60	2·54	3·39
Loam with carbonates	4·77	26·2	2·10	2·73	43·9	5·03	5·87	2·73	7·97
European white birch									
Crystalline	5·21	8·62	3·64	4·02	38·4	14·6	—	5·94	24·8
Loam without carbonates	3·99	11·52	4·77	4·52	42·12	13·52	5·76	5·52	12·27
Loam with abundant carbonates	6·11	1·31	0·16	5·07	31·9	18·2	3·76	3·76	20·45

* Not determined.

content developing in different chemical conditions (from taiga to arid-steppe), fluctuating from 24 to 31% in pine needles and from 32 to 42% in birch leaves. We may assume that pine and birch possess a clearly-expressed ability to regulate their absorption of calcium, enabling them to grow on strata and soils both without carbonate and heavily carbonated, without reduction in their productivity.

These tree species show a tendency to greater calcium consumption on soils and strata without carbonates than on those with carbonates.

Spruce behaves differently. When it grows on limestones, CaO predominates in its leaves (up to 52%), whereas on strata without carbonates or on carbonate-silicate (loam) strata the CaO content falls to 25-44%.

Such wide fluctuation in CaO accumulation in its needles when spruce grows on different strata points to a lack of ability to regulate its consumption of calcium. As a result spruce shows stunting of growth and decrease in productivity on carbonate soils and strata. A physiological explanation has been given by L. A. Ivanov (1936). He showed that as the ash content increases in leaves and needles there is a fall in photosynthesis and current increment of organic matter.

On limestones all tree species show small consumption of SiO_2, because of the small amount contained in the substrate. On carbonate-silicate strata (carbonate loams) consumption of SiO_2 by pine increases, most probably because it is dissolved by pine root exudates. Birch possesses that faculty to a smaller degree. The higher consumption of P_2O_5 by pine on limestones is worthy of note.

Spruce growth is retarded both by shortage of calcium (on crystalline strata) and by excess of it (on limestones); in both cases spruce stands are not highly productive.

The above examples show that studies in this field may provide much material for more thorough understanding of biogeocoenotic inter-relationships, valuable for assessment of the forest-growing properties of soils. It must be remembered that such inter-relationships are subject to substantial changes in different soil-climatic conditions.

It must also be pointed out that the growth and productivity of separate tree species may also vary with fluctuations in the chemical composition of soils and sedimentary strata. On dolomitised limestones (with higher magnesium content) spruce growth is somewhat better than on limestones with calcium predominant. On miaskites (with increased amount of magnesium) pine growth may be impaired, and so on. These problems, however, are almost unstudied and do not come within the field of view of sylviculturists.

Volumetric weight, as an index of soil density, is ecologically one of the most important physical features of soils. Volumetric weight is also an index of such soil properties as porosity, water-permeability, moisture capacity, etc. Almost no direct comparisons of the growth and productivity of tree species with these properties have been made. Their relationship to separate species is closely dependent on the nature of the distribution of the roots of these in the soil.

Looseness, correct texture, and high permeability to water and air are essential conditions for normal growth and high productivity of trees. Trees react fairly quickly by lower productivity to unsatisfactory combinations of these properties or to their deterioration as a result of human interference, and in extreme cases of deterioration tree stands perish and are replaced by herbage.

We may surmise that the depth of root penetration into the soil is largely a result of the adaptation of trees to physical properties of the soil. That is most evident in the case of Norway spruce. Its development is linked with podzolic soils with well-marked illuvial (more clayey and compact) horizons located at a depth of 30-60 cm, or with compact argillaceous deposits overlaid to some depth by soils lighter in mechanical structure, or with soils in close proximity to mineral strata. On such soils spruce develops a root system mostly near the surface, its depth of penetration being limited by the above-mentioned layers. Excess of moisture in the upper horizons, due to retention of water in the soil by the clay and compact layers, also frequently contributes to that result.

The shallow development of the root system of spruce in these circumstances has been taken for a regular characteristic of the species.

On leached chernozems, however, with friable and structural soil and slightly-developed illuvial horizons, the same species (planted) develops taproots and anchoring roots penetrating to a depth of 1 metre or more.

Still more striking is the variation in structure of oak root systems. Although oak possesses a root system that penetrates friable soils to a depth of 5-6 m (Veliko-Anadol'), on soils with dense horizons almost impermeable to water oak root systems become shallow and spread horizontally, their area exceeding by many times those on friable soils. This variation is very clearly displayed on the residually-alkaline soils of the Tellerman forestry station, as has been described by V. N. Mina and I. N. Elagin (1953). A similar phenomenon has been observed on soils with solid bedrock in close proximity.

There are also contradictory cases where, in spite of the friability and relatively high water-permeability of soils, root systems of many tree species have a well-marked shallow distribution (not deeper than 40-60 cm). That is most evident in many tropical species and is due to various causes, especially to features of the water and gas regimes of the soils.

A number of species (pine, birch, poplar, beech, hornbeam, etc.) have more adaptable root systems. In soils with extremely unfavourable physical conditions their distribution changes its nature. Beech, birch, and poplar, even in dense soils, not only develop a large surface root system but also penetrate to deeper horizons, and beech roots make their way into crevices in dense mineral strata, improving the soil and accelerating the weathering of these strata. Beech, moreover, grows well on moist but well-aerated soils.

A few species (e.g., black locust) have well-marked surface root systems and show little reaction to changes in the physical properties of soils.

The effect of the physical properties of soils on the productivity of trees is therefore very great. In some cases tree species adapt themselves to changes in these properties and their productivity does not fall, but in others (e.g., with development of alkalinity) it falls sharply.

In general we may consider that root systems develop less actively with deterioration in the physical properties of the soil, and the growth of the above-ground parts of trees declines.

In spite of the inadequacy of data concerning the effect of the physical properties of soils on the growth and productivity of separate species, the available data still show a close relationship, manifested differently according to the types of soils and the physico-geographical conditions of their formation.

Soil is the chief source of moisture supply for trees. It accumulates atmospheric precipitation; underground water passes through it before entering plants; and, finally, soils interact to some degree with the waters of rivers, lakes, temporary floods, and bogs. Soil moisture serves to some extent as an index of soil fertility, since it contains the nutrients that are most available to plants.

The study of water procurement by trees and the discovery of norms of optimal water supply to them are very important. Water uptake by trees is closely linked with the properties of the soil and the supply of moisture in it. Soil humidity and moisture supply depend on the geography of atmospheric humidity, on relief (which governs the distribution of water), and on soil properties (which determine the retention and release of moisture). Soils are divided into three groups according to their moisture supply: optimal, excessive, and inadequate moisture. Within each group there may be finer subdivision both by moisture supply and by the nature of the supply (atmospheric, underground, surface flow, stagnant, etc.).

The amount of soil moisture always serves as one of the chief indicators for classification of forests by types. Up to the present, however, visual methods have been mostly used for this purpose; they are subjective and imprecise, because of the great fluctuations in soil moisture during the vegetative period.

Adequate data have now been accumulated regarding the water regime of forest soils and expenditure of moisture by trees in transpiration. They make it possible to give quantitative figures for soil moisture supply and for water use by tree species, forest types, or stand types.

All tree species, indeed, grow most rapidly and show maximum productivity only in moisture conditions optimal for them.

With departure from optimal conditions moisture consumption falls, and as a result current increment and the productivity of trees decrease.

In the temperate zone the rates of such decreases are not the same for all species. Some species are more, some less, sensitive to a decrease in moisture supply. Excessive moisture presents a still more complex picture. Some tree species (e.g., spruce) are sensitive to shortage of oxygen in the soil and in the water in the soil. Besides, excessive moisture has different effects, depending on the climatic conditions.

As is well known, tree species are divided into two groups, drought-resistant and moisture-loving. In the temperate zone the most productive stands grow in the forest-steppe, which possesses the best conditions of heat and moisture regimes, which may be provisionally taken as being optimal for the most important tree species.

There, on leached chernozems at the Mokhovsk forestry station, tree stands of quality classes I and Ia at the age of 60 years grow with the following moisture supplies (average for 2-3 years) and total moisture consumption from a layer 3 m deep, according to the data of Zonn and Kuz'mina (1960) (Table 120).

Table 120. *Stocks and total consumption of moisture (in mm) by stands of different tree species on leached chernozem.*

Species	Stocks of moisture		Total moisture consumption
	Total	Available	
Norway spruce	710	338	255
Scots pine	736	338	284
Siberian larch	870	506	414
Oak	824	458	439
European white birch	855	489	498

According to these data larch, oak, and beech are more hygrophilous than spruce and pine. Actually, however, pine, oak, birch, and to some degree larch are more drought-resistant than spruce. The latter, as is well known, does not grow, even in plantations, south of the northern part of the forest-steppe zone. As for larch, Kurile Dahurian larch grows in Kamchatka with a moisture supply of about 400 mm and a total water use by it of about 290 mm, which indicates high drought-resistance in that species. Pine, oak, and European white birch are very resistant to soil aridity. The first can grow with moisture supply below 250 mm and with total water use by it of about 150-200 mm. Oak grows with consumption of moisture from the soil not exceeding 150-200 mm, and birch with 315 mm (Stepanets 1962). In such cases, however, growth, productivity, and longevity are reduced in all species. Even such a drought-resistant species as Chinese elm shows high moisture requirement in optimal moisture conditions. If in optimal moisture conditions the average duration of life of a tree species is 200-250 years, with lowering of water supplies below the optimum (according to Zonn 1959) the duration of life is reduced as follows:

with an annual water deficit of 80-100 mm to 40-60 years
with an annual water deficit of 200-280 mm to 25-40 years
with an annual water deficit of 260-300 mm to 20-30 years
with an annual water deficit of > 300 mm to 10-15 years

Unfortunately similar data for other species (beech, ash, hornbeam, maple, etc.) have not yet been collected. Neither are there any for the subtropic and tropic zones. We can only point out that the general relationship—decrease in growth and productivity with decrease in moisture supply—holds good in these conditions also; but the general rule needs substantial adjustments, depending on species-composition and other circumstances. For instance, European larch is very sensitive to both shortage and excess of moisture, whereas Dahurian larch is far from fastidious in this respect. It grows either with relatively little or with very abundant moisture.

Schrenk spruce differs from Norway spruce and many other species in its extreme drought-resistance, with minimal total annual consumption of moisture (judging by some indirect data, not more than 200 mm). These examples confirm that both the hygrophilous property in tree species and intensity of growth in relation to moisture, when assessed by visual methods, fail to express the trees' actual use of soil moisture.

Moisture supplies are determined by atmospheric precipitation, by the characteristics of genetic types of soil, and by specific differences in the tree species.

As stated above, the relationships between the growth of tree species and the degree of excess of moisture are more complex and less studied. The term 'excessive moisture' with regard to tree species requires a certain amount of clarification. We understand by it the accumulation of moisture in the soil above optimal amounts required for uninterrupted supply of moisture to the trees. Many tree species are injuriously affected not by excess of moisture as such but by the method of its accumulation, the duration of its retention, its degree of stagnation, and its composition.

The type of accumulation of excessive moisture in the soil may be (i) stagnant, (ii) flowing, or (iii) periodic. Variations in tree growth correspond to this classification. Stagnant water is the most dangerous for the majority of species, since it leads to shortage of oxygen in soil water and to formation of toxic compounds, organic and organo-mineral (ferrous iron, organic acids, etc.). Pine can survive such conditions longer than other species, but with severe loss of growth and productivity. Flowing water benefits some species, as is shown by their improved growth and productivity. Sylviculturists formerly attributed that to improved oxygen supply to the roots; but studies in recent years (Orlov 1960b) have shown that the chief cause is improved plant nutrition, taking place through the periodic influx of nutrients from the water into the upper root-occupied soil layer, which for part of the time is free from flooding. The beneficial effect of brief temporary falls in the level of soil and underground water, permitting atmospheric oxygen to enter the soil, has a secondary effect. The rate of outflow of oxygen-deficient water is important, as that outflow permits oxygen-rich atmospheric precipitation to penetrate the soil.

These processes are closely connected with the mechanical consistency of soils: when soils are more clayey or peaty the processes are slowed down, but in soils of lighter sandy-loam or loamy consistency they are accelerated. Abundant flowing water benefits spruce, pine, beech, and birch, raising their productivity. Oak, linden, poplar, and ash survive it, but their productivity falls to quality class I-II. European larch and fir quickly perish when moisture is excessive.

Periodic excessive moisture includes inundation of soils by floodwaters in river valleys, depressions in plains, etc. Tree species react differently to such flooding and to its duration. Some species of aspen survive the most prolonged flooding; oak, ash, maple, and other species are more sensitive and endure flooding for not more than 10-15 days. Conifers are injuriously affected.

Thus even the briefest survey of the relationships between the growth and productivity of various tree species and soil moisture shows their great complexity, especially since the latter depends not only on the mechanical consistency of the soil but also on the whole assemblage of characteristics of soils of a given genetic type.

The relationship between trees and soil acidity has not yet been adequately clarified. Some authorities believe that trees prefer an acid reaction in soils, that view being based mainly on observations relating to the taiga zone and its acid podzolic soils. More extensive observations have produced many new data, sometimes supporting that view and often contradicting it.

The reaction of soils is a property arising as a result of their interactions with trees growing in particular climatic conditions, i.e., it is a bioclimatic phenomenon. Available extensive data on the growth of various tree species in soils of different acidity enable us to compile the following list of pH ranges in which they thrive:

Spruce	3·5-7·0
Pine	3·0-7·5
Birch	4·0-7·2
Oak	4·5-8·0
Larch	4·0-5·5-6·0 (data scanty)
Beech	4·0-7·5

As a rule, optimal pH values for all species lie within the range 6·0-6·5. Within that pH range all species generally show the highest quality class in their stands. There are many exceptions to the rule, especially in the taiga zone, where maximum productivity of spruce occurs with a pH of 5·0-5·5. With a further increase in soil acidity, quality class usually falls.

Oak and beech are most productive with a pH of 5·5-6·5; ash, maple, and other broad-leaved species grow best with a pH of 6·0-7·0.

It is impossible, however, to make an absolutely definite statement regarding the dependence of tree growth on pH values.

In the tropic and subtropic zones trees grow on soils with a narrower pH range. In these conditions there is a certain differentiation: trees of the humid tropic and subtropic zones grow in soils with pH from 3·8 to 5·8; in monsoon regions, from 4·5 to 6·0; in savannah-deciduous regions, from 5·0 to 7·5; and in mangrove swamps, from 6·0 to 8·5. There are almost no published data on the relationship between the productivity of tree species in these zones and soil pH values.

Different species of the same genus are adapted to soils of different pH values. According to our data and those of Lyu Shou-Pue (1958), certain species of *Picea* thrive in soils with the following pH values: *P. excelsa*, 3·5-7·0; *P. likiangensis*, 4·6-5·8; *P. jezoensis*, 5·8-6·0; *P. schrenkiana*, 6·4-7·0. As may be seen, the ecological range of different spruce species (generally fairly wide) is narrower for certain species. It is significant that *P. schrenkiana* has a substantially different ecology and is adapted to neutral and even alkaline soils.

Quercus robur has the widest range of adaptation to pH among oak species, extending from 4·5 to 8·0, with optimum about 6·0 to 6·5; *Q. petraea* and *Q. cerris* have a narrower range, from 6·0 to 7·0.

Although pines do not react so much to soil pH, some differences may be observed among them: *Pinus silvestris* from 3·0 to 7·5, *P. banksiana* from 6·0 to 7·0, *P. strobus* from 5·8 to 6·5, *P. nigra* from 5·0 to 7·0, *P. leicodermis* from 6·0 to 7·5 (adapted to limestones).

There are differences between birch species. European white birch grows within pH 4·0-7·2, but Kirghiz birch within pH 6·5 to 7·5-8·0.

These data show that, besides species that do not react to soil pH, there are other species with fairly well-marked adaptations to more acid, neutral, or even alkaline conditions. Without knowledge of that fact, discovery of the interactions of these species with soils and creation of lasting tree plantations are impossible.

All tree species are included among the plants that prefer soils not containing toxic soluble salts. There is only an insignificant group of *salt-resistant species*, growing constantly in conditions of high salt concentration in the waters washing them and the soil. They include the mangroves *Rhizophora, Avicennia,* and others that grow in slimy and boggy soils regularly washed by sea-water during the daily flow of tides. In desert conditions also there are such species as black and white saxaul, tamarisk, and other shrubs, which have considerable amounts of soda and potash concentrated in them.

The majority of forest-forming tree species cannot endure even the most insignificant amounts of salinity. Only a few of them are adapted to a relatively high salt content, but their growth and longevity are decreased.

Among salt-resistant species we may include the pine and birch that grow in Naurzum (northern Kazakhstan). The birch there is distinguished as a separate species or variety under the name of Kirghiz birch. Scots pine grows in Naurzum to an age of 70-80 years in soils containing 0.106% Cl and 0.6% SO_4; and the birch normally grows in soils with 0.072% Cl and 0.36% SO_4.

More general statements have been made to the effect that a content of 0.5% of salts in soil is lethal for all tree species. Data exist showing that with a content of 0.178% of salts in the soil survival of one-year-old oaks is only 67%, and with 0.78% of salts survival is only 4%. Black locust is less sensitive to salts and survives with up to 0.840% of them in the soil. Laurel-poplar dies off when the soil contains 0.04-0.07% Cl and 0.16-0.44% SO_4; Chinese elm is seriously suppressed with a chloride content of 0.15-0.49% and a sulphate content of 0.07-0.31%.

From these (far from complete) data it is seen that Chinese elm is the most salt-resistant, followed by black locust, laurel-poplar, and oak.

The relationships between tree growth and soil salinity are, however, very complex and diverse; they are governed by the mechanical consistency of the soil, the way the salts are distributed, their composition, and many other factors. Often, even with high soil salinity at some depth below the surface, trees thrive because their root systems lie above the saline horizons. In addition, the effects of salinity on different species are not uniform. Soil salinity as a factor in plant growth requires special study in view of its great importance to steppe and desert afforestation.

Soil alkalinity is a feature linked with the introduction of sodium ions into the absorbent complex of soils. The presence of metabolic sodium gives soils an alkaline reaction, in some cases leading to formation of sodium carbonate, which is toxic to roots. In alkaline soils there is increased compactness in dry, and viscosity in moist, layers with heavy sodium concentration, making the soil less permeable and causing part of the water reaching it to be lost in surface and sub-surface outflow. Soils are divided by their degree of alkalinity into: slightly alkaline, with content of metabolic sodium 5-10% of capacity; medium alkaline, 10-15%; strongly alkaline, 15-20%; and solonetz, over 20%.

In slightly-alkaline soils the growth of such species as ash and maple begins to decline. Medium alkalinity affects all species except Tartar maple and wild pear. In strongly-alkaline soils decline in growth is at a maximum. On solonetz almost all species, except a few shrubs, perish in their earliest years. In cases of residual alkalinity oak, pear, and Tartar maple show the highest resistance.

The quality class of oak stands depends on the degree of alkalinity: on slightly-alkaline soils forests of quality class III grow; on solodised-alkaline soils, class IV; on solidised solonetz, class V; and on solonetz, class Va.

There are substantial differences in the weight of single oak trees (in kg of dry matter) at the age of 55 years on non-alkaline soils and on solodised solonetz, growing in the same climatic conditions (Tellerman forestry station). On non-alkaline soil the total weight of a tree was 844 kg, including 87 kg of roots. On solodised solonetz the total weight was 235 kg, and that of the roots 95 kg. The ratio of the weights of above-ground and underground parts was 9·3 in the former case and 2·5 in the latter.

The sharp fall in oak productivity on solonetz is due to lack of moisture and mineral nutrients, and to the toxicity of metabolic sodium.

Because of insufficient study of forests growing on soils with prolonged seasonal freezing or permafrost, little attention has been paid until recently to this factor in the growth and productivity of trees. In recent years, however, a certain amount of information has been accumulated in both pedology and sylviculture showing its very great and specific effect on tree species-composition and on the trees' supply of water and nutrients, etc. That is due primarily to the fact that freezing is a special factor in soil formation. It causes soil processes to develop differently from their course in unfrozen soils, as a result of which soils of a distinctive genetic type, taiga-frost soils, are formed. They are distinguished by absence of the podzolisation process and by other features.

Frozen soils are seasonally or permanently impermeable, restricting leaching and reducing the thickness of the soil layers that tree roots can occupy.

When atmospheric moisture is inadequate the presence of frozen layers improves the water supply of trees, since water that accumulates in a solid state in the frozen horizons liquefies when it thaws and is consumed by the plants.

The gas content of soils has until recently remained outside the sphere of attention of investigators. But in certain conditions it also can be important as a factor strongly influencing the growth and productivity of trees. Particularly important in this regard are the ratio between oxygen and CO_2 in soil-contained air, and accumulation of CO_2 in the soil. It has been discovered that if the content of CO_2 in soil air is more than 9-10% it may have a poisoning effect on tree roots. Accumulation of that amount of CO_2 is observed in tropical forests at depths below 50-60 cm (Zonn and Li Chen-Kvei 1960), with a corresponding fall in the amount of oxygen. It is possible that one of the reasons for the shallow penetration of tropical tree roots into the soil is the higher concentration of CO_2 there.

In temperate-zone conditions, with periodic excessive moisture and high water table, an increased CO_2 content up to 8-9% is sometimes observed (Remezov 1952), which may affect the growth of trees to some extent. The accumulation of CO_2 in soils may be due to many causes, among which the principal ones are the rates of release of CO_2 by roots and decomposing litter, and the nature and intensity of diffusion of CO_2

both in the soil and in the air layer adjacent to the soil. All of these questions require further study, since existing data (Matskevich 1953, Zonn 1954, Mina 1954, 1957) show that production of CO_2 depends on the tree species-composition, on the conditions of litter decomposition, on soil moisture, and on many other factors.

Study of the relationships between tree species and separate properties of soils is a prerequisite for discovery of the inter-relationships between forest vegetation and soils. It is practically impossible, however, to trace the effect of a single soil property on trees in natural conditions, since all the properties are interlinked and the extent of their manifestation depends on their varying combinations.

We shall dwell briefly on some features of tree species that depend primarily on the soil, namely, on the ash content of their above-ground parts (especially leaves and needles); on the amount and composition of their litter; on the structure, amount, and ash content of their roots; and on their moisture requirements, etc. Leaves and needles of trees are the chief accumulators of mineral matters and nitrogen. Their ash content reflects to a definite extent the content and mobility of assimilable compounds in the soil. Such a relationship can be fully revealed only by adhering to uniform methods of collecting leaves and needles for analysis, since their ash content fluctuates substantially during the vegetative period and varies with the age of the leaves or needles, with their location in the crown, etc. Although data on the analysis of leaves and needles collected by a uniform method are lacking, existing analyses of leaves and needles of trees growing on soils of contrasting qualities enable us to deduce some important laws.

With comparatively small fluctuations (within the range 3·53-4·77) in the total ash content of needles of Norway spruce growing on different soils, substantial variations are found in the accumulation of separate oxides (Table 121).

In order to reveal the differences due to soil characteristics, we have provisionally taken as the optima the figures for the content of the various salts in the needles of spruce growing on leached chernozem, since the quality class there is the highest (Ia). We may then assume that excess or shortage of any oxide, as compared with its content in leached chernozem, has definite significance.[1]

The extremely unfavourable conditions for spruce growth (with fall in quality class to V and Va) on humus-ferruginous podzol (Kola peninsula) are reflected in the shortage of iron, calcium, and especially potassium and nitrogen, accumulated in the needles; and those on humus-carbonate (Estonia), in the shortage of SiO_2, iron, and probably aluminium. On the former soil we find an excess of silicon and aluminium in the needles,

---

[1] The growth and productivity of spruce depend on the whole complex of soil properties and growth conditions, but this comparative approach (based on the biogeocoenotic concept of inter-relationships) is worthy of attention.

and on the latter an excess of calcium and phosphorus. That difference corresponds with the abundance or shortage of these oxides in the soils.

A similar picture appears in the ash content of needles on peaty-humus soils. When these soils are sandy or gleyed the relative weight of silicon increases and the content of aluminium, iron, and calcium falls; when they are loamy, with limestone fragments, the relative amounts of calcium, aluminium, and iron rise and that of silicon falls. We must point out that whereas the quality class IV of spruce on sandy peaty-humus soil may be due to shortage of mineral substances, the loamy soil contains adequate amounts of them. The cause of low productivity in the latter case is excessive moisture.

The only relationship missing is that between the ash content of needles and the nature of podzolic-type soils (medium-podzolised and soddy-podzolic) where the content of iron, calcium, phosphorus, potassium, and nitrogen shows wide variation. On soddy-podzolic soil the needles contain much more phosphorus and potassium and less iron. Therefore spruce growing on it is of quality class III, and on medium-podzolised soil of quality class IV. Such relationships need further checking.

The striking difference in the ash content of spruce on the dark-coloured soils of Bulgaria attracts attention. Although the bilberry-green-moss spruce stands of Bulgaria approach those of the taiga in their plant

Table 121. *Variations in ash content of Norway spruce needles in accordance with soil conditions (% of total ash content)*

Soil	Ash content, %	SiO_2	Al_2O_3	Fe_2O_3	CaO	MgO	P_2O_5	MnO	K_2O	SO_3	Total nitrogen, %	Quality class
Humus-ferruginous podzol (Kola peninsula)	3·64	39·9	5·0	0·4	33·8	4·8	3·2	2·9	0·6	3·2	0·60	Va
Medium-podzolised (Kadnikov forest, Vologda province)	4·72	31·1	6·0	7·0	40·0	3·3	2·5	2·9	3·3	3·6	1·22	IV
Soddy-podzolic (Moscow province)	3·65	23·0	8·5	0·4	29·0	4·6	8·5	3·2	18·0	7·6	1·15	III
Dark leached (Bulgaria)	4·74	19·4	26·3	5·4	25·7	11·6	6·9	—	—	4·4	0·72	III
Peaty-humus strongly-gleyed sandy (Vologda province)	4·92	40·2	1·2	0·2	28·4	4·8	8·1	3·2	18·3	5·2	1·06	IV
Peaty-humus loam (Vologda province)	4·37	17·6	5·7	5·9	54·9	3·8	3·4	1·3	2·7	4·3	1·52	IV
Leached chernozem (Mokhovoe, Orlov province)	4·77	26·4	3·5	2·10	43·8	5·0	2·5	—	6·9	5·8	1·13	Ia
Humus-carbonate (alvaric) (Estonia)	3·53	13·7	*	0·9	52·3	4·7	5·1	0·9	*	*		Va

* Not determined.

species-composition, there is a much higher aluminium content and a lower content of calcium and silicon. That difference is due to the fact that the dark-coloured soils in Bulgaria have developed in near-subtropic, very humid conditions, with very warm winters, and consequently there is accumulation of kaolin products of weathering and increased absorption of highly-mobile aluminium compounds. The process is aided by the shortage of calcium in the local soil-forming crystalline rocks. This explains one of the substantial biogeocoenotic differences between the spruce forests of Bulgaria and those of the taiga, despite their external similarity.

In general the above data permit us to say that the ash content of spruce needles reflects the characteristic features of the content and mobility of mineral nutrients in soils, and may be one of the biogeocoenotic indicators characterising the inter-relationships of spruce with soils of different origins.

Similar relationships are observed in study of the ash content of beech leaves (Table 122).

Table 122. *Variations in ash content of beech leaves according to soil conditions and altitude of growth (% of total ash content).*

Soil	Ash content, %	SiO_2	Al_2O_3	Fe_2O_3	CaO	MgO
Brown forest soil, on loam (town of Mashuk, N. Caucasus)	8·48	25·4	4·5	1·6	28·6	8·03
Brown forest soil, on eluvium of extrusive strata (town of Zheleznaya, N. Caucasus)	7·80	27·4	3·6	1·2	14·6	6·6
Brown forest soil, podzolised, on loam (N. Osetia)	9·15	28·1	5·1	3·4	20·1	6·3
Brown forest soil, on eluvium of crystalline strata (Bulgaria)	5·64	44·6	21·4	3·1	10·6	9·4
Mountain forest soil, brown, alt. 700 m, weakly podzolised (Bulgaria)	8·93	32·4	7·8	0·3	33·5	7·2
Mountain forest soil, brown, alt. 1550-1800 m (Bulgaria)	7·54	25·3	2·9	0·4	36·4	8·3
Mountain meadow-forest soil, alt. 1800-2000 m (Bulgaria)	6·05	27·2	10·0	0·3	40·3	7·7

On brown forest soils developed as products of weathering of sedimentary strata, beech leaves are comparatively rich in calcium and contain similar amounts of silicon, aluminium, iron, and magnesium. Beech productivity is higher on them (quality class II-III) than on brown forest soils developed on eluvium of extrusive strata (town of Zheleznaya). Beech leaves on such soils contain less calcium, aluminium, and iron than those on the soils just mentioned; the quality class falls to IV-V. Most distinctive are the leaves of beech growing on brown forest soils developed on loamy eluvium of crystalline strata. They have higher contents of silicon and especially of aluminium, with a much lower calcium content, and the quality class is III-IV.

The similarity in the ash content of beech leaves and spruce needles in Bulgaria (higher Al_2O_3 content) is noteworthy; it must be considered a typical feature of the region, resulting from the characteristics of the bioclimatic conditions.

Finally, the ash content of beech leaves falls sharply with growth at higher altitudes because of decrease in the content of all oxides. Differences in the ash content of the leaves, depending on the exposure of slopes, are observed. It is lower on eastern than on western slopes, probably on account of differences in water supply (Dzhafarov 1960).

On the whole the ash content of beech leaves, like that of spruce needles, reflects fairly clearly the differences in soils of the same origin but of different composition.

In birch leaves (see Table 119) on podzolic soils (loams, without carbonates) more SiO_2 and CaO are accumulated than on brown soils developed on crystalline strata, especially on dark-chestnut soils (strongly-carbonated loams). On the latter the ash contains the smallest amounts of SiO_2, Fe_2O_3, and K_2O, which also indicates the relationship of birch with soils and especially with soil moisture and the degree of availability of the required elements. (*Translator's Note:* Table 119 shows a much higher content of K_2O on these soils; the Russian text here makes the additional error 'K_2O_2'. There may be other printer's errors in the last sentence.)

Equally well-marked are the differences in ash content of oak leaves (Table 123) depending on soil conditions.

The analyses quoted clearly show a fall in the ash content of leaves from leached chernozems, through dark-grey forest soils, to podzolised grey forest soils. The optimal ash content, corresponding to the highest quality class, is found on the dark-grey soils of Tellerman forest. Varia-

Table 123. *Ash content of oak leaves in relation to soil conditions (% of weight of ash).*

Soil	Ash content, %	SiO_2	Fe_2O_3	Al_2O_3	MnO	CaO	MgO	P_2O_5
Leached chernozem (Mokhovoe, Orlov province)	5·28	11·3	0·2	3·6	*	35·6	8·1	4·7
Dark-grey (Tellerman forest, Voronezh province)	5·05	10·1	0·37	1·2	1·4	41·2	6·9	4·2
Grey forest, podzolised (Tula forest reserve, Tula province)	4·57	15·7	0·4	*	*	30·4	9·8	9·6
Dark-grey-brown sandy loams (Voronezh reserve, Voronezh province)	4·90	20·0	0·4	1·6	1·0	32·4	7·1	2·8
Residual solonetz (Tellerman forest, Voronezh province)	4·12	15·0	0·52	0·7	3·1	35·6	18·0	4·5

* Not determined.

tions from its total amount and composition on leached chernozem are insignificant, and the quality classes of oaks on the two soils are similar. The differences on grey podzolised soils are greater: on them the CaO content of leaf ash decreases and the SiO_2 content rises, corresponding to a certain poverty of these soils in lime and increased availability of SiO_2. On these soils the quality class of oaks is lower.

When oaks grow in the same climatic conditions but on soils differing in origin and properties, e.g., on dark-grey soils and on residual solonetz (Tellerman forest), there is considerable difference in the composition of the salts in their leaves. On solonetz the content of CaO decreases and that of MgO, MnO, and SiO_2 rises, corresponding to their availability in these soils. The quality class of oaks falls to Va. In dark-grey-brown loamy soils also the SiO_2 content rises and the CaO content decreases, corresponding to the content and availability of these salts in the soil.

The composition of the ash of tree roots may serve as one of the biogeocoenotic indicators of the inter-relationships between the trees and the soil. A few data supporting this statement are presented in Table 124.

The concentration of mineral elements in spruce roots fluctuates widely in accordance with the content of these elements in the soil. The richer the soil (leached chernozem), the more SiO_2 and Fe_2O_3 accumulates in the roots, and the less CaO, K_2O, and P_2O_5. In relatively poor

Table 124. *Variations in composition of ash of certain tree roots according to soil conditions (% of total ash content).*

Soil	Ash content, %	SiO_2	Al_2O_3	Fe_2O_3	CaO	MgO	K_2O	P_2O_5	SO_3	N
			Spruce roots							
Leached chernozem (Mokhovoe, Orlov province)	6·68	3·43	12·5	6·5	26·5	2·5	3·8	5·7	7·5	1·18
Soddy-podzolic (Nelidovo, Kalinin province)	2·63	4·1	11·4	1·1	32·3	4·9	24·7	9·5	8·7	0·75
Podzol (same locality)	2·90	4·5	14·4	4·1	29·0	3·1	17·2	10·0	5·9	0·35
Peaty-humus gleyed soil (Mordovsk reserve)	2·53	4·4	13·4	4·3	33·9	10·2	14·9	8·7	9·8	0·65
			Oak roots							
Leached chernozem (Mokhovoe, Orlov province)	6·40	21·5	13·4	3·2	21·0	10·3	8·0	9·2	5·1	0·95
Dark-grey-brown sandy loam (Voronezh reserve)	4·41	4·0	6·0	0·3	43·0	6·0	9·6	4·6	5·0	1·62
Soddy-podzolic (Nelidovo, Kalinin province)	2·20	4·1	23·6	0·9	36·8	5·0	20·0	10·9	11·6	0·54
Podzol (same locality)	2·42	5·3	19·8	11·9	28·5	3·3	16·1	9·9	9·0	0·53

soils (soddy-podzolic and podzol) there is much less SiO_2 and Fe_2O_3 and more CaO, K_2O, and P_2O_5. The composition of the roots of trees on peaty-humus gleyed soil corresponds to the latter case. Consequently in rich soils with abundant calcium (leached chernozem) roots accumulate the least amount of mineral elements, except for SiO_2. In poor soils, on the contrary, accumulation is more intensive, so that we may assume that in soddy-podzolic soils there is a sufficiency of mineral elements, and in podzols a shortage. The total nitrogen content in roots falls from chernozems to podzols, and rises only slighly (as compared with the latter) in peaty-humus soil. Oak roots have a similar mineral content. There is a certain difference in the fact that in the roots of oaks growing on leached chernozems not only the content of SiO_2 and Fe_2O_3, but also that of Al_2O_3 and P_2O_5, is higher than on the poorer grey-brown sandy-loam soil; on the latter, the roots contain much more CaO and K_2O. The nitrogen content of roots also is lower in the latter than in the former soil.

Another regular occurrence is observed in birch roots: increase in the content of SiO_2, Al_2O_3, Fe_2O_3, K_2O, and P_2O_5 and decrease in CaO in soils of podzolic type (as compared with dark-chestnut soils), which correlates well with the higher availability of these compounds in such soils and the shortage of calcium in them.

Thus the ash content of both leaves and roots of trees clearly displays the effect of genetic variations in soils. Determination of such relationships enables us, in the first place, to use ash content as an indicator of the quality of forest soils; and in the second place, they strongly suggest the existence of more profound relationships that should help us to understand the interactions of plants and soils that affect the nature of the circulation of matter and energy.

Diagnosis of biogeocoenotic tree-soil relationships through ash content has a promising future, especially for the discovery of nutrient trace elements. Such an approach, however, has not yet been recognised and applied in our country, although it is receiving increasing attention in Western Europe. It offers a possible method of revealing formerly-unknown changes in the characteristics of tree species as a result of soil conditions.

The amount of annual litter fall (consisting of varying quantities of shed leaves and needles, with which the current increment and productivity of trees are closely connected) reflects the life-activity of tree species. It is very difficult, however, to compare trees in this respect, since sufficient data are not available relating to trees of similar ages. Therefore we give some (more or less reliable) figures for litter fall as one indicator of the way in which growing conditions for trees vary in accordance with soil-climatic conditions. Such figures for spruce (*Picea excelsa*) offal are given in Table 125.

The data presented have only relative value, but they show that the richer the soil is in nutrients and the more favourable its physical properties

Table 125. *Variation in amount of spruce litter in accordance with soil conditions.*

Forest type, location	Age of spruce, years	Soil	Amount of litter (kg)	Author
Green-moss-*Oxalis* spruce stand (Central reserve, Kalinin province)	38	Podzol	2020	Remezov *et al.* 1959
Spruce plantation (Mokhovoe, Orlov province)	25	Leached chernozem	7553	Rozanova 1960
Fern-oak spruce stand (Mordovsk reserve, Mordovsk A.S.S.R.)	55	Soddy-podzolic	3211	Remezov *et al* 1959
Spruce plantation (Mokhovoe, Orlov province)	60	Leached chernozem	7900	Rozanova 1960
Herb-green-moss spruce stand (Kharovsk leskhoz, Vologda province)	45	Peaty-humus gleyed	3155	Parshevnikov 1958

are (leached chernozem), the more green matter is produced, and consequently the greater is the mass of litter. These quantities are least on the poorest soils; the amount of spruce litter on soddy-podzolic and peaty-humus soils is intermediate.

We may assume that the increase in amount of shed needles on leached chernozem is because the annual production of needles there is much higher than on podzol.

On all other soils needles remain on spruce for seven years or more, which may slow down physiological processes.

A clear connection between the amount of litter and soil conditions is also shown by oak. In identical climatic conditions (Tellerman) an oak forest, 50-60 years old, produced the following amounts of litter: on dark-grey soils 3288 kg/ha, and on residual podzolised-solodised solonetz 2246 kg/ha. The difference is due to the fact that on residual solonetz, because of conditions unfavourable for oaks, leaf mass is much reduced.

At ages up to 15 years oaks planted on these soils show no substantial differences in amounts of litter: on dark-grey soils 1109 kg/ha, and on residual podzolised-solodised solonetz 1174 kg/ha. That is because at these ages the moisture and nutrient requirements of oaks are met more easily than at the age of 50-60 years. The amounts of litter from 25-year-old oaks are as follows: on leached chernozem (Mokhovoe) 3784 kg/ha, and on ordinary chernozem (Derkul') 2706 kg/ha.

As may be seen, with a lower amount of moisture in ordinary chernozem (as compared with leached chernozem) the mass of litter, and correspondingly the productivity, decreases. At the age of 50-60 years the dependence of variations in the amount of litter on differences in soils is seen from the following figures: on leached chernozems (Mokhovoe) litter amounts to 3700 kg/ha, on dark-grey soils (Tellerman) 3288 kg/ha, and on grey-brown sandy-loam soils (Voronezh province) only 2072 kg/ha. In all these conditions oak experiences no lack of moisture. The decrease in amount of litter from leached chernozems to grey-brown soils is mainly related to the amounts of nutrients in the soils, which are in turn linked with their mechanical consistency and the degree of leaching, which depends on the water supply.

These examples exhibit a substantial direct connection between the abundance of leaves or needles on tree crowns, the amounts of litter, and the properties of soils.

In the past that connection either escaped notice or was considered impossible to demonstrate. But systematic application of the biogeo-coenotic approach, directed towards discovery of the relationships between soils and trees, has made it possible (for the present, admittedly, in only a few cases) to prove the existence of that connection.

Similar relationships may be traced between other tree species and soils.

An equally valuable indicator of variation in the features of tree species on different soils is provided by the root system, its structure and its mass. Developing directly in the soil, tree roots are most sensitive to variations in soil properties. Among the latter, as has been stated above, the most important are its richness in nutrients, its moisture supply during the growing period, and the presence of dense layers of differing origin. The relationship of the roots to these properties is evaluated by their total amount, by the nature of their distribution and extent, and by the amount of their most active parts (those less than 0·6 mm in diameter and absorbing tips).

In principle the combination of all these indicators should provide a complete picture of variations in the underground parts of trees. Complete data of this kind, however, have not been collected for separate tree species. It must be noted that the total root mass depends on the energy of growth of the tree, especially at younger ages, which should be taken into account in such studies and in analysis of data.

In the most favourable growth conditions, on leached chernozem (Mokhovoe), the above-ground and underground masses of single trees at the age of 25 years vary as follows (Table 126).

Table 126. *Variation in amounts of above-ground and underground parts of trees at age of 25 years on leached chernozem (Rozanova 1960).*

Tree species	Weight of a single tree, kg of oven-dry mass	
	Above-ground parts	Roots
Oak	2223	465
Spruce	5813	1859
Siberian larch	10,574	2582

Only after 30-40 years does the root mass reach the maximum amount typical of each species.

If the soil contains illuvial or other compact layers the roots do not penetrate so deeply and their horizontal extent increases. For instance, the amount of roots of a 55-year-old oak in 1 dm^2 in the 0-50-cm layer of dark-grey soil is 200·7 g, and the total length of roots less than 1 mm in diameter in that layer is 1695 metres. On podzolised-solodised solonetz the corresponding figures are 38·5 g and 4093 metres (Tellerman forestry station, from data of I. N. Elagin and V. N. Mina 1953).

In dark-grey soil the main mass of roots can be traced to a depth of 120-150 cm, their total weight in that layer is 2472 g, and their total length 3256 metres; on solonetz there are practically none deeper than 50 cm. The depth at which the illuvial horizon lies, its degree of compactness, and the amount of moisture in the horizons beneath it have a still stronger effect on the amount and nature of distribution of spruce roots less than 1 mm in diameter, as may be seen from Table 127.

On soddy-podzolic soil with a less well-developed illuvial horizon and without excessive moisture, roots of diameter less than 0·6 mm reached maximum abundance at a depth of 20 cm. They decreased rapidly in

Table 127. *Amount of roots under spruce stands on different soils (Kharovsk leskhoz, Vologda province) (Parshevnikov 1958).*

Soil	Depth, cm	Amount of roots kg/ha
Fern-Oxalis *spruce stand*		
Soddy-weakly-podzolised	3-10	1780
	10-20	2700
	20-30	720
	30-40	47
	40-50	14
	3-50	5261
Bilberry spruce stand		
Strongly-podzolised	2-8	4180
	8-10	1380
	10-20	300
	20-30	01
	30-40	0
	40-50	0
	2-50	5861
Horsetail-sphagnum spruce stand		
Peaty-podzolic-gleyed	0-4	140
	4-11	0
	11-16	0
	16-20	0
	0-20	140

amount lower down, and at a depth of 40-50 cm only a few scattered ones remained.

On strongly-podzolised soils with periodic soaking the maximum weight of the root-mass occurred at a depth of 2-8 cm; below that the amount decreased abruptly, and at a depth of 20 cm roots had almost disappeared. Finally, on peaty-podzolic-gleyed soil roots of the same diameter occurred in the litter (0-4 cm) in very small quantity. Only scattered specimens of spruce live on such soils.

Thus the depth of penetration of spruce roots less than 0·6 mm in diameter and their total amount are dependent both on the degree of podzolisation and on the moisture supply. As these increase, the depth of penetration diminishes and the concentration of roots in the soil comes nearer to the surface. The greater mass of roots in strongly-podzolised soil may be due to its smaller nutrient content as compared with soddy-weakly-podzolised soil. Consequently, even in its natural range, the development of the root system of spruce depends on soil conditions.

With the decrease in water supply from dark-grey to light-chestnut soils, the depth of penetration of the main root mass of oaks and other broad-leaved species changes as follows (Zonn 1959): dark-grey soils (Tellerman), 120-150 cm; chernozems (Veliko-Anadol'), 250-300 cm; southern chernozems (Belye Prudy), 170-180 cm; chernozems (Derkul'), 120-170 cm; light-chestnut soils (Zavetnoe), 100-120 cm. As may be seen, root depth is relatively small where water supply is greatest, i.e., on dark-grey soils. Where moisture is inadequate and there is no capillary fringe, the average depth of root penetration is least. It is determined by the average depth of moistening of the soils in spring (chernozems at Derkul' and light-chestnut soils at Zavetnoe).

Unfortunately, there are almost no comparable data on the distribution and total amount of oak roots in these soils. We present data only for the chernozems of Derkul' and the southern chernozems of Belye Prudy (Table 128).

Table 128. *Amount of roots in chernozems by horizons (in g/m²).*

Depth in cm and horizon	Chernozem (Derkul')	Southern chernozem (Belye Prudy)
0-50 (A)	1012	3068
50-80 (B)	352	362
80-120 (C)	247	545

Although these data refer to plantations of different ages, they are of interest because they show maximum accumulation of roots in the humus layer and smaller amounts in soil horizons B and C.

The depth of root penetration in chernozems is several times as great

as that in podzolic soils, because of shortage of moisture in the chernozems.

Equally important in the life of biogeocoenoses are the moisture requirements of trees and the degree of fulfilment of these when trees grow
on different types of soil. We have dealt with that problem earlier.
Therefore here we merely stress the following: the moisture requirements
of trees are not uniform on all soils. Tree species consume much more
moisture than they require for the normal course of their photosynthetic
activity.

The climatic conditions in which they grow strongly affect the moisture
consumption of trees. The drier and warmer these conditions are, the
more moisture trees use, other things being equal. That is convincingly
demonstrated by experiments made in arid conditions, where expenditure
of water in transpiration is many times greater with irrigation than
without it. Therefore most species are to a certain extent adapted to a
shortage of moisture. But that adaptation leads to a reduction in the
energy of growth in the above-ground parts of trees and an increase in
the underground parts, and as the soil-moisture deficit increases the
duration of life of the trees diminishes. Excess of moisture is reflected in
an equal degree: trees react to it by decreased growth in both their above-
ground and underground parts.

Excess of moisture, however, can be easily eliminated by appropriate
ameliorative measures, but shortage of it cannot be fully overcome even
by irrigation, since atmospheric dryness is not modified thereby. Supplementary watering improves the growth and productivity of trees only
partially, and not proportionately.

Moisture shortage begins to show itself in the southern chernozems,
and increases in the light-chestnut and brown semi-desert soils. Excessive
moisture has an unfavourable effect on podzolic and similar soils; on
grey forest soils and chernozems its effects are sporadic, since expenditure
of moisture in evaporation and transpiration increases in warmer conditions.

In general, the supply of moisture in soils occupied by tree stands serves
(as we have discovered (Zonn 1959)) as a valuable indicator of the soil
conditions to which tree species are adapted.

The above comments show the chief ways in which some features of
tree species change in accordance with the composition and properties of
the principal types of soil.

The above-described general relationships between tree growth and
productivity and soil conditions cannot be regarded as stable and permanently valid. They fluctuate under the influence of the constant (or
briefly interrupted by human activity in the form of tree-felling) action of
forest vegetation upon soils.

From the biogeocoenotic point of view it is of the highest importance
to understand these complex and many-faceted interactions of forest

vegetation and soils. They have a decisive effect on the evolution of bio-geocoenoses.

The life-activity of other living components—vertebrate and invertebrate animals, and micro-organisms—is linked with soils and vegetation. Vegetation and soils are for them both a habitat and a source of food. The animal and micro-organism populations not only destroy or trans-form the products of plant life-activity, but also, as a link in the vegetation-soil system of circulation of matter, help to retain the most available compounds in the biogeocoenotic cycle. Moreover, they often take an active part in the accumulation of certain elements (nitrogen, sulphur, etc.) from the atmosphere. In biogeocoenoses where the activity of animals and various micro-organisms is to some degree weakened and where for some reason they cannot make full use of the products of plant life-activity, degradation (in a broad sense) of forest biogeocoenoses may take place. An example of such degradation is the flooding of forest biogeocoenoses, with accumulation of peaty products of plant decom-position, leading ultimately to the formation of a peat bog.

The progressive development of forest biogeocoenoses, reflected in the complexity of composition and structure of its living components (bio-coenoses), is accompanied by accumulation of organo-mineral compounds and also by leaching of them and improvement of the physical state of the soil. These qualitative changes in the interactions of vegetation and soil, accompanied by more intensive circulation of matter and energy, are linked with the whole course of development of the soil-forming process and production of definite types of soil.

Vegetation naturally takes the leading role in the development of bio-geocoenoses and consequently of soils. It cannot, however, be studied in isolation from all other components. The activities of animals and micro-organisms, the influence of soil-forming strata and phytoclimate, cannot be overshadowed by vegetation. On the contrary, the activity of vegeta-tion may be substantially altered by changes in climate, in the animal population, or in soil-forming strata or soils, as a result of which quali-tatively and quantitatively different processes of matter and energy exchange are built up in biogeocoenoses, and consequently genetically-different soil types are formed under forest vegetation of identical or very similar composition.

Moreover, the biogeocoenotic role of each component of forest vegeta-tion must be considered separately. Not always are trees the decisive factor in the trend and nature of changes in biogeocoenotic processes. Far-reaching alterations in these may be caused by the shrubs, herbage, and mosses that form part of forest biogeocoenoses, with relatively little change in tree-species composition.

The above statement has particular significance with regard to our understanding of the role of producer plants in the life of forest bio-geocoenoses, since they reflect changes in soil properties.

We must not forget the importance of soil-forming strata in biogeo-coenotic processes. Their chemical composition (primary and secondary) has the greatest influence, being affected by weathering and by the com-pactness and water-permeability of the soil. These properties cause qualitative and quantitative variations in the circulation of matter and energy.

Hence we obtain the basic biogeocoenotic axiom, that the interactions of forest vegetation and of separate tree species with the soil in both natural and artificial biogeocoenoses are not simple and straightforward. The reasons include not only variations in the composition of forest vegetation but also the diversity of climatic conditions in which its inter-actions with the soil take place. The interactions of some components of forest biogeocoenoses that affect the development of soils are represented in Fig. 68.

Selection of litter as the principal forest-biogeocoenose component is based on the fact that, in the first place, it is formed from dead organic

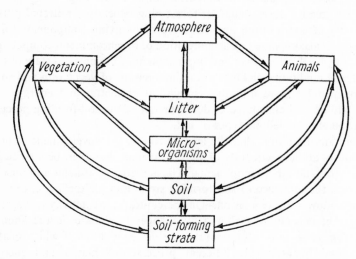

Fig. 68. Diagram of interactions of soil with other components of forest biogeocoenoses.

remains and is a result of the life-activity of organisms and not a product of the soil-forming process as a whole.

In the second place, litter, which contains larger reserves of heat energy than soil, and concentrates within itself mineral matter in the form of complex organo-mineral compounds, actively affects the soil and changes both the processes taking place in it and its characteristics.

Thirdly, litter regulates the eluvial outflow of available compounds by freeing heat energy and mineral matter through decomposition and mineralisation, and even aids accumulation of these compounds in the soil by forming more complex organo-mineral compounds and co-agulations.

Fourthly, litter is the horizon of most vigorous biogeocoenotic activity of vegetation and animal life, on account of the high concentration of necessary nutrients in it. As is well known, removal of litter from a forest biogeocoenose leads to its degradation. Litter is indissolubly linked with maintenance of the soil's physical state (aeration, water supply) at the level required by plants and animals.

Finally, forest litter is the basic source of humus-formation for the soils of forest biogeocoenoses, since the organic matter in tree roots is of secondary importance in that process.

In soils with herbaceous cover, on the other hand, formation of humus and the accumulation of nutrients result from the annual dying-off of root systems. Therefore in herbaceous biogeocoenoses the removal of above-ground matter does not lead to such serious loss of fertility and deterioration of the physical state of the soil as it does in forest bio-geocoenoses.

Litter, as a component of forest biogeocoenoses, plays a major role in their life as a whole, especially in interactions with the soil. Recognition of that property of litter is not new. V. S. Shumakov (1958) and others have made such a distinction between litter and soil. At present pedo-logists of the u.s.a. (Smith 1960) and the German Democratic Republic (Kundler 1961) and others propose to distinguish it as a horizon that is at the same time independent and linked with the soil. They denote it by the symbol 'O' to distinguish it from the accumulative soil layer, described as horizon 'A'.

Thus the biogeocoenotic concept of litter as the principal horizon accords with present trends in genetic pedology.

Naturally the interactions of separate components of forest biogeo-coenoses with the soil, as represented in Fig. 68, have only a generalised and fundamental significance. The roles of the separate components shown in the diagram may fluctuate in importance, and similarly one must differentiate the effects of different kinds of vegetation (especially forest vegetation), which are grouped together in the diagram.

It is possible for separate components to change places in their bio-geocoenotic roles as a whole and in their interactions with the soil. An example is provided by biogeocoenoses developing on soil-forming strata lying close to the surface, when the mineral part of the soil directly reflects the characteristics of the rock strata.

In such biogeocoenoses, which may develop in the same climatic conditions as biogeocoenoses with deep soil, the biogeocoenotic role of soil-forming strata (e.g., limestones lying less than 50-60 cm below the surface) is strengthened. In that event the role of climate is, as it were, absorbed or extinguished by the properties of the strata, and its effect is displayed in the species-composition and ash content of the vegetation and in its productivity, and consequently in the characteristics and typology of the soil.

These considerations enable us to assert that the picture of simple, monotypic action of trees on soil, without analysis of the complex biogeocoenotic interactions in specific (above all, climatic) conditions of biogeocoenose formation, does not correspond to the true situation. It is necessary to stress this point because some sylviculturists (Nesterov 1954, 1961) still believe that podzolic soils are the cause and the result of specific action by forest vegetation upon soil. That concept, originating from the incorrect views of V. R. Williams on forest vegetation, actually excludes the possibility of active and purposeful alteration of its interactions with soil by changing the tree-species composition and other measures.

In that way forest vegetation is quite unjustifiably looked upon as merely consuming and impoverishing soil activities, and the characteristics of biogeocoenotic circulation of matter and energy in forest biogeocoenoses are disregarded.

These characteristics consist primarily in the accumulation, and the retention in the bodies of living organisms, of a considerable amount of nitrogen and mineral elements, most of which return to the soil in the form of available organo-mineral compounds and phytolitharia. Only a small fraction is removed from forest biogeocoenoses once in 100 or 200 years, by tree-felling. These losses are replaced by use of mineral matter from deeper soil horizons and by mobilisation of such matter from primary undecomposed minerals and rock fragments contained in the soil.

At the same time the soil is enriched with large quantities of organic matter and nitrogen, which in the processes of decomposition are again used by developing vegetation.

Accession of products of biocoenose life-activity to the soil takes many forms, many aspects of which have not yet been well studied. We can only point out that it takes place continuously and by various means. It may occur, for example, through the washing-off and leaching of mineral elements, nitrogen, and soluble organic matter from the green parts and roots of plants when they are drenched with precipitation; through the addition of various substances from the atmosphere, carried by precipitation or wind; through fixation of atmospheric components by both above-ground and underground (in the case of nitrogen) parts of plants; through excretions of roots and animals; and by other means.

The principal method of accumulation of matter and energy in the soil, however, is the annual dying-off of animals and of above-ground and underground parts of plants. These cause formation of litter and originate all the features of soil processes that characterise forest biogeocoenoses.

The soils of forest biogeocoenoses annually acquire an average of 3-5 ton/ha of organic litter containing mineral matter and nitrogen. The mineral matter enters the soil completely; as for the organic matter, part of it (in the form of the decomposition end-product, CO_2) enters the atmosphere and part is converted into humus.

We may assume that accumulation in forest biogeocoenoses is accompanied not only by redistribution of mineral matter and nitrogen throughout the entire biogeocoenotic layer, but also by gradual vertical increase in the latter through the accession of humic matter and other decomposition products. The rates of such accumulation are variable. They depend on many factors, as is confirmed by existing data on accession and decomposition of products of life-activity in forest biogeocoenoses.

Opposed to accumulation are the processes of outflow from biogeocoenoses of mineral matter, nitrogen, and soluble organic compounds in the form of true or colloidal solutions. These processes are entirely due to the circulation of water in the biogeocoenoses. When more moisture is received than is consumed in transpiration and evaporation, it soaks through the soil layer (in certain seasons of the year, or possibly throughout the whole annual cycle) and carries a certain amount of various compounds away from the sphere of biogeocoenotic action.

That outflow of moisture into deeper soil horizons depends in the first place on the total amount of precipitation and its consumption in evaporation and transpiration, and in the second place on the physical and physico-chemical properties of the separate soil horizons. The general rule is, the higher the atmospheric humidity and the less the consumption of moisture in evaporation and transpiration by forest biogeocoenoses, the greater is its outflow beyond the boundaries of the biogeocoenotic layer, and vice versa. According to experimental data (I. S. Vasil'ev 1950, Rode 1952, Zonn 1954, 1959), the outflow of moisture from forest biogeocoenoses in different zonal conditions varies from 5 to 25% of the total amount received in atmospheric precipitation.

Increase in the amount of outflow is observed from south to north, from semi-desert forest biogeocoenoses to the forest-tundra. At the same time there is a variation in the leached matter, accompanied by regular variation in the depth of the biogeocoenotic horizons of the soils. In the driest (semi-desert) conditions, the biogeocoenotic layer of the soil increases mainly through leaching of the most soluble chlorides and sulphates. The less soluble carbonates undergo only negligible leaching.

In the soils of forest biogeocoenoses on the steppes chlorides and sulphates are washed still deeper, and carbonates descend farther, than in semi-desert conditions.

In the forest-steppe zone there are no chlorides or sulphates, and carbonates are almost completely leached out of the entire thickness of the biogeocoenotic layer (except in soils developed on limestones). Iron and aluminium become available to some extent, as do organic compounds.

In the taiga zone the soils of forest biogeocoenoses, as a rule, lack carbonates throughout their whole depth. All mineral and soluble organic compounds (especially those containing calcium, magnesium, iron, aluminium, etc.) are subject to migration in solution. They are partly washed into the lower part of the biogeocoenotic layers of the soil, where they are retained and aid in forming illuvial-accumulative horizons, and are partly carried out of the biogeocoenose into underground waters and thence into rivers and seas.

The soils of forest biogeocoenoses in the forest-tundra zone are somewhat less leached because of the smaller thickness of the biogeocoenotic layer of the soil and the less intensive circulation of matter. Qualitative change in mineral compounds as a result of excessive moisture (transformation from oxides into lower oxides), and accumulation of partly-decomposed organic matter, are more characteristic of them.

Corresponding to the above range in the chemical composition of soils, the nature and thickness of separate horizons vary, reflecting the trend of biogeocoenotic processes. Soil layers thus differentiated reflect the varying intensity of biogeocoenotic processes, which in turn is determined by the intensity of accumulation of the products of biocoenose life-activity and of their transformation in the soil, and by the extent to which the more soluble compounds are washed away.

In the soils of forest biogeocoenoses the following horizons are distinguished:

AA active accumulation of raw energy-material in the form of litter;
A accumulation of humic matter (corresponds to the thickness of the humus horizons of the soil);
IT intensive transportation of humic and the most soluble mineral compounds, with relatively greatest impoverishment in these (due to low activity of biogeocoenotic processes and scarcity of roots);
VA[1] secondary accumulation (illuvial deposit) of organo-mineral compounds, with alteration of their physical state;
GT deep transit or accumulation of organic and mineral compounds carried out of biogeocoenotic circulation.

The suggested classification is based on the genetic features of the separate soil horizons, as well as on the nature of the migration of products formed as a result of biogeocoenotic processes. The nature and the thickness of these horizons vary with the zonal location of biogeocoenoses. In Fig. 69 their zonal combinations are shown in conjunction with the degree of leaching of the soil.

The thickness of the horizons of active biogeocoenotic accumulation (horizons AA and A) is least in forest-tundra and taiga biogeocoenoses; it rises abruptly in the forest-steppe and falls again in the steppe and semi-desert zones. That is correlated with the degree to which the soil is leached of the most coagulated exchange-products of salts of calcium,

[1] *Translator's note:* The symbols 'VA' and 'GT' (used in Fig. 69) are transliterated Russian initials. The other symbols 'AA', 'A', and 'IT' are also transliterated Russian initials, but they happen to coincide with the initials of the English terms describing the horizons in question.

sodium, etc., linked with the anions CO_3, Cl, and SO_4. Only in forest-tundra conditions is reduction in the thickness of the leached horizon due to the depth of the water table or to permafrost, which restricts the outflow of soluble compounds for physical, not for biogeocoenotic, reasons.

The total thickness of biogeocoenoses is very small in the semi-desert and forest-tundra zones, mainly on account of the unfavourable physical

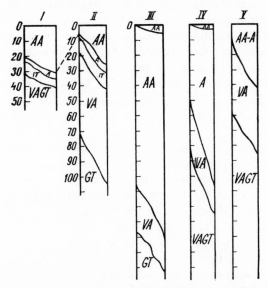

Fig. 69. Diagram of structure and thickness of soil horizons in forest biogeocoenoses in different zones (minimum thickness of horizons left, maximum on right). I—forest-tundra; II—taiga; III—forest-steppe; IV—steppe; V—semi-desert.

properties of the soils, which restrict the possibilities of deeper root-penetration (temperature, excessive moisture, compaction). The great increase in the biogeocoenotic thickness of soils in forest biogeocoenoses in the steppe and semi-desert zones is due mainly to insufficient moisture.

Both the general structure of soils and the biogeocoenotic interpretation of the thickness of their horizons vary correspondingly (Fig. 69).

The interactions of vegetation and soils in forest biogeocoenoses arise from the fact that soil is the source of plants' water and mineral nutrition and the accumulator of the products of all life processes that take place in biogeocoenoses. Therefore both the total amount of biomass and the separate parts or elements of it entering the soil depend on the nature and trend of interactions between forest vegetation and soils. This subject calls for more detailed examination.

AMOUNT OF BIOMASS OF FOREST BIOGEOCOENOSES AS A GENERAL INDICATOR OF THE INTERACTIONS OF VEGETATION AND SOILS

The total amount of biomass produced by phytocoenoses[1] is one of the chief indicators of variation in the processes of creation of organic matter and of the related exchange of matter in different forest biogeocoenoses.

Fluctuations in the total amount of biomass reflect, first, the general vigour of development of vegetation; secondly, variations in external (especially climatic) conditions of formation of forest biogeocoenoses; and thirdly, differences in the intensity of the interactions of their components. Knowing these separately and in total, we can obtain more profound knowledge of the essentials of the interactions of all components of biogeocoenoses and especially interactions between phytocoenoses and soils.

There are many difficulties, however, in that field, due to the diversity in type and age of forest biogeocoenoses and the variability of their separate components. The task is further complicated by its laborious nature, which can provide only an approximate qualitative figure for biomass.

For these reasons we possess very inadequate information on the total amount of biomass produced by forest biogeocoenoses.

Data of this kind have been collected mainly in the u.s.s.r.; they are much more scanty in the literature of Western Europe and the u.s.a. (consisting merely of estimates of the biomass in some pine and birch forests in England and in tropical forests).

The existing data make it possible to formulate some general laws concerning accumulation of biomass in forest biogeocoenoses, depending on their zonal location and species-composition.

In spruce forests, biomass accumulation increases from those growing on humus-ferruginous podzols, through podzols, to soddy-podzolic soils (Table 129). On humus-ferruginous podzols the amount of biomass in 200-year-old spruce stands is only 137·3 ton/ha, but on strongly-podzolised soils it reaches 315 ton/ha at the age of 93 years. We may assume that in the latter case the amount will at least double by the age of 200 years and reach 500-600 ton/ha. Consequently the intensity of biomass accumulation on strongly-podzolised soils is 3·6 to 4·3 times as high as on humus-ferruginous soils, and 1·5 to 2 times as high as on podzols. Still greater biomass production is observed in composite spruce stands on soddy-podzolic soils. There the amount reaches 383 ton/ha at the age of 115 years, i.e., is greater than on any other soil at ages up to 200 years. The greatest biomass is produced by riverside forests on peaty-humus soils (mixed-forest subzone); there at the age of 45 years the amount of biomass reaches 341 ton/ha, i.e., the same as at 200 years on the podzols of the northern taiga subzone.

The amounts of biomass accumulated in forest-tundra conditions are much lower. According to K. N. Manakov (1962), in Kola peninsula

[1] In comparison, the zoomass in forest biogeocoenoses is very small.

Table 129. *Total amount of biomass of different types of forest biogeocoenoses (ton/ha).*

Forest type, soils, and location	Age (years)	Needles or leaves	Herb-moss cover	Above-ground woody parts	All above-ground parts	Roots	Total mass	Authors
Spruce stands								
Mossy, humus-ferruginous podzol. Forest-tundra (Archangel province)	200	2·0	5·2	99·2	108·1	29·2	137·3	Marchenko and Karlov 1962
Green-moss-*Oxalis*. Strongly podzolised. Southern taiga (Nelidovo, Kalinin province)	93	10·0	1·0	239·5	250·5	65·4	315·9	Remezov *et al.* 1959
Fern-oak. Strongly podzolised (same location)	115	15·8	0·6	289·6	306·0	78·8	384·3	Remezov *et al.* 1959
Riverside. Peaty-humus, sandy. Broad-leaved forest subzone (Mordovsk reserve)	45	15·7	*	223·7	238·7	102·4	341·1	Remezov *et al.* 1959
Pine stands								
Cowberry. Soddy-podzolic, sandy soil. (Mordovsk reserve)	95	3·9	*	181·8	185·7	62·8	248·5	Remezov *et al.* 1959
Linden. Brown forest, weakly-podzolised soil (same location)	57	10·4	*	196·1	206·5	248·7	455·2	Remezov *et al.* 1959
Oak stands								
Sedge-goutwort. Dark-grey brown, sandy loam soil (Voronezh reserve)	130	3·3	0·6	227·9	231·8	110·3	342·1	Remezov *et al.* 1954
Sedge-goutwort. Dark-grey brown, sandy loam soil (Voronezh reserve)	48	3·0	*	184·6	187·6	62·2	249·8	Remezov *et al.* 1954
Sedge-goutwort. Dark-grey, clay loam soil (Tellerman, Voronezh province)	55	3·5	0·4	149·1	153·0	85·1	238·1	Mina 1955
Tropical forest								
Original forest (Ghana)	50	—	—	262·4	No data	25·2	287·6	Greenland and Kowal 1960
Secondary forest (Ghana)	18	5·6	—	No data	135·6	31·8	174·0	

* Not determined.

they may fall to 51 ton/ha, on account of the soil and climatic conditions of biogeocoenose formation (shortage of heat and shallow seasonal thawing, sometimes permanently-frozen soil).

The nature of the interactions of vegetation and soils is expressed not only in the amount of biomass produced but also in the share taken by different chemical elements in the circulation of matter and energy.

Data for two types of pine biogeocoenoses show that linden-pine stands produce almost twice as much biomass as cowberry-pine stands on soils of the same mechanical consistency but with a different level of fertility. The participation of linden in this case determines quantitative differences in matter and energy exchange, and consequently greater accumulation of organo-mineral compounds, which results in weakened podzolisation of the soils.

For a similar bilberry-pine stand in Archangel province (Obozerskaya), Molchanov (1961) presents similar data. In that pine stand the amount of growing mass at the age of 100 years was 262·5 ton/ha, of which bilberries and moss cover accounted for nearly 14·5 ton/ha. At the age of 60 years the corrresponding figures were 154·1 and 13·1 ton/ha. At the same age a linden-pine stand produced three times as much biomass.

The total productivity of the same biogeocoenoses, including all shed needles, the tree stand, and the herbage-shrub layer[1] was 284·0 ton/ha at the age of 60 years, 542·0 ton/ha at 100 years, and 1147·0 ton/ha at 240 years.

The production of biomass by oak stands at Tellerman and in the Voronezh reserve at the age of 48-55 years is very similar, in spite of the differences in mechanical consistency of the soils (Table 129). The weight of the biomass of typical oak trees differs, as is seen from the following figures:

Weight of a single oak tree in kg

	Age (years)	Trunk and branches	Leaves	All above-ground parts	Roots	Total weight
On sandy loam	45	300·5	4·3	304·8	80·5	385·2
On heavy clay-loamy soil	55	789·3	17·6	806·9	87·0	893·9

On clay-loamy soil the weight is more than twice as high as on sandy loam, which confirms not only the differences in soil fertility and in matter and energy circulation, but also the profound qualitative difference between these biogeocoenoses. The closeness of the figures for total biomass is due to differences in the number of trees per hectare. At Tellerman there are 1436 per ha, and in the Voronezh reserve 1775. Consequently the differences in soil productivity are, as it were, levelled out by the total mass of organic matter produced by the vegetation.

[1] Total of all growing matter

Data on biomass production by two types of tropical forest indicate the more favourable (as compared with the temperate zone) general combination of living conditions for trees at an early age, and the similarity to figures for oak forests at the age of 50 years.

Of not less biogeocoenotic significance are the comparative data on biomass production by different tree species in the same soil conditions. As an example we present data on the total amount of biomass in unmixed stands of oak, spruce, and larch at the age of 50 years on leached chernozems in Orlov province (Rozanova 1960).

The biomass in oak stands is 131·0, in spruce stands 169·3, and in larch stands 176·7 ton/ha. Such differences in biomass accumulation by these species in identical soil conditions show how much effect the natural characteristics of tree species have on their interactions with the soil.

Biomass accumulation by oak stands of different quality classes on sandy soils of varying depth proceeds at different rates (Table 130).

Table 130. *Weight of mass of oak stands of different quality classes (in ton/ha of dry weight)* (*Remezov 1961*).

Quality class	Age	Trunk	Branches	Leaves	All above-ground parts	Roots	Total weight
I	55	167·0	26·6	3·3	196·9	87·5	284·4
II	52	166·1	23·5	3·9	193·5	50·8	244·3
III	55	97·6	19·7	2·3	119·6	29·8	149·4
IV	47	53·9	15·0	2·6	71·5	39·1	110·6

The stands of quality classes I and II are located on sandy soil, underlaid by loam at a depth of two metres; those of quality classes III and IV on deep sandy soils. The difference is entirely due to shortage of nutrients for the oaks, although the pine stands that preceded them had higher productivity (quality class I-II). When the pines were replaced by oaks a disruption took place in the soil-formation process—soddy-podzolic soils evolved into grey forest soils.

In closing this survey of data on biomass production by forest biogeocoenoses, we would again point out that these data serve as a quantitative indicator not only of the life-activity in a biogeocoenose for a definite period of time but also of the intensity of that activity. At the same time they enable us to form an idea of the varying amounts of shed organic matter that enter the soil every year.

PLANT LITTER AND ITS SIGNIFICANCE IN THE LIFE OF FOREST BIO-
GEOCOENOSES

The annual shedding of organic debris by tree species on the soil surface is one of the outstanding characteristics of forest biogeocoenoses. The longevity and productivity of trees are directly dependent on the amount of dead organic matter that reaches the soil surface every year, containing elements of mineral and nitrogenous nutrition essential for trees.

Litter is the chief food source for most soil-dwelling animals and micro-organisms in forest biogeocoenoses. The latter synthesise various growth and toxic products, whereby the growth of some tree species is stimulated and that of others impaired, even to the point of killing them (Runov and Enikeeva 1955, Egorova 1962).

Humic matter formed as a result of animal and microbial activity takes an important share in alteration of the mineral part of the soil, in promoting the availability of nutrients, in forming new organo-mineral compounds, and in their accumulation and transportation out of the biogeosphere.

A considerable part of the mineral matter and nitrogen consumed by plants returns to the soil in litter fall, in compounds differing qualitatively from those existing in the soil. Litter is an energy-providing material, enabling soil-formation and other biogeocoenotic processes to develop.

Therefore it is important to know the amount, composition, and properties of litter, and also the timing of litter fall and decay, as it is a basic material containing supplies of radiant energy and affecting the development and evolution of biogeocoenoses.

The amount of litter entering the soil every year varies widely. It depends primarily on the composition and age of the tree layer, which provides most of it; also on the degree of development of other plant layers, on the climatic conditions, and on a number of other factors.

The average amounts of litter annually reaching the soil in forest biogeocoenoses in the European part of the U.S.S.R. are as follows: in larch forests 2·7, in oak forests 3·9, in pine forests 4·1, and in spruce forests 5·0 ton/ha (Zonn 1954a). Their dependence on the species of trees forming the biogeocoenoses shows up very clearly; it results from the character-istics of the species, and above all from the amount of leaves or needles that the trees bear.[1]

The same ratio is maintained when these species grow in identical conditions, at the same age, and with similar stand density, as is seen from data obtained by I. M. Rozanova (1960) at Mokhovoe in Orlov province on leached chernozem (plantations): for oak (25 years old, density 1·3) litter fall was 3·7 ton/ha; for larch (26 years old, density 1·0), 4·4 ton/ha; for pine (6 [sic; perhaps 26? Tr.] years old, density 0·9), 6·2 ton/ha; and for spruce (27 years old, density 1·0), 7·2 ton/ha.

With increase in age of tree stands the amount of litter fall fluctuates within narrow limits, since stand density and amount of leaves or needles offset each other. In Table 131 we show the fluctuations in the amount of litter with increase in age.

---

[1] The highest figures for annual litter fall are recorded for forest biogeocoenoses in tropical rain forests. According to the data of S. V. Zonn and Li Chen-Kvei (1962), in a Heronier forest biogeocoenose in the tropical rain forests of China the average amount of litter fall is 11·6 ton/ha. Litter is shed throughout the year, with the follow-ing seasonal rhythm: foggy season (November-February), 2·3 ton/ha; dry season (March-May), 4·7 ton/ha; rainy season (June-October), 4·6 ton/ha.

In spruce and pine stands a reduction in the amount of litter fall with increasing age is clearly seen. In oak stands a reduction in the amount of litter occurs only when the water regime of the soil is unsatisfactory for oaks (in the oak stand growing on solonetz). That relationship is also observed in other biogeocoenoses in which the interaction of components is weakened because of unfavourability of one or more factors.

It should be noted that the above data refer only to litter shed by trees. The total amount, including dead herbaceous cover, would be 200-500 kg/ha more. Still more striking are the fluctuations in the amount of litter fall from green-moss-*Oxalis* spruce stands growing in different zones (at the age of 100-200 years): in the forest-tundra zone (Archangel province, Marchenko and Karlov 1962) about 6·0, in the northern taiga about 5·0, in the southern taiga about 4·3, and in mixed forests about 3·6-5·7 ton/ha.

At first glance these figures may seem paradoxical, as they show the amount of litter fall increasing from south to north. That, however, is due to increase, in the same direction, of the proportion of moss and shrub litter, and of herbaceous cover in the mixed-forest subzone.

In Kola peninsula, according to V. I. Levina (1960), the total above-ground mass of moss and shrub cover is: in lichen pine forest 40·7, and in green-moss pine forest 43·5 ton/ha. The amount of annual litter fall in these circumstances, according to K. N. Monakov (1962), is: in bilberry spruce stands of growth class IV, quality class Va, 15 ton/ha; and in blue-berry pine stands of growth class V, quality class V, 23 ton/ha.

Such substantial differences in amounts of litter are due to the development of the biogeocoenose types as a whole, to the density of their tree stands, and to the zonal and provincial conditions of their formation.

The amount of annual litter fall varies greatly in forests growing in different climatic conditions. A bilberry pine stand in Bryansk province

Table 131. *Variations in amount of litter fall with increase in age of stands in different types of forest biogeocoenoses (ton/ha).*

Green-moss-*Oxalis* spruce stand (Remezov 1959)		Cowberry pine stand (Remezov 1959)		Sedge-goutwort oak stand (Remezov 1959)		Goutwort-sedge oak stand (Mina 1955)		Oak stand on solonetz (Mina 1955)	
Age years	Amount of litter fall	Age years	Amount of litter fall	Age years	Amount of litter fall	Age years	Amount of litter fall	Age years	Amount of litter fall
24	3·2	14	2·5	12-16	3·8	25	3·2	10-15	1·87
38	2·0	33	2·3	48-52	3·9	43	3·7	30-40	2·50
60	1·9	45	1·9	93-97	3·7	55	3·5	50-60	2·77
72	1·9	71	2·0	130-140	4·3	220	3·6	60-70	1·72
93	1·9*	95	1·3					160-200	1·40

* The figures are possibly too low, since in the same type and at the same age (according to Abramova, 1957) the amount of litter was 4·2-4·5 tons/ha.

produces up to 3·2 ton/ha; a pine stand with oak 3·0 ton/ha in Moscow province, and 1·3 ton/ha in Voronezh province; and maple stands in the same provinces 3·6 and 2·0 ton/ha respectively (Zonn 1954a).

In spruce stands (with bilberry and *Oxalis*) in Kalinin province the amount of annual litter fall is 3·7-4·2 ton/ha (Abramova 1947); in Moscow province 6·2-6·9 (Nesterov 1954); and in Bryansk province 4·9 (Zonn 1954a). As we see, the variation in spruce stands is greater than that in pine stands.

In spite of the scarcity of the data so far collected on the amount of annual litter falling in forests, it is fully evident that it depends to a very great extent on the composition of forest phytocoenoses and on their interactions with other components. Consequently the amount of annual litter reflects a definite intensity of the interactions between all the components of forest biogeocoenoses. At the same time the dependence of the amount of litter fall on the composition of forest biogeocoenoses enables us to regulate it in such a way as to regulate the interactions between phytocoenoses and soil, which is very important in conditions unfavourable for the life-activity of forest biogeocoenoses.

The quality of litter depends on the amount of heat-energy accumulated in it, on its content of mineral matter and nitrogen, and on the composition of the organic matter.

Up to the present we do not possess data on litter classified by its heat-energy content. Only Molchanov (1961b) gives the calorific value of bilberry-pine litter as 4700-5300 cal/g.

More study has been given to the accumulation of mineral matter and nitrogen in litter. Data are still far from adequate, in spite of their importance in the study of plant-soil interactions and in discovery of the conditions that provide for fullest use of soil fertility in regenerating the organic mass of forest phytocoenoses. A survey of the existing data shows that the accumulation of nitrogen and mineral matter in forest-biogeocoenose litter increases regularly from the northern taiga to the forest-steppe, and reaches a maximum in tropical rain forests (Table 132).

Still higher values for some elements brought into circulation in tropical forests are presented by I. A. Denisov (1962). For four phytocoenoses (*Brachystegia* sp., *Macrolobium* sp., *Mustanga smithii*, and mixed forest) they vary within the following limits: nitrogen 140-224, potassium 48-104, calcium 84-124, and phosphorus 4-9 kg/ha.

Therefore in tropical conditions also wide variations in matter and energy are observed, due to various causes.

These variations are due primarily to the amount of litter, and then to the content and availability of the elements in the soil.

Variations in the mass of litter within zonal phytocoenoses are due to variations in other components that help to form biogeocoenoses. Increase in the ash and nitrogen content of litter in northern taiga conditions from bilberry spruce stands to blueberry pine stands is attributable to the

Table 132. *Content of mineral matter and nitrogen in litter of forest biogeocoenoses (ton/ha).*

Nature of forest and location	Pure ash	Mineral elements									N
		SiO_2	Al_2O_3	Fe_2O_3	P_2O_5	CaO	MgO	MnO	K_2O	SO_3	
Northern taiga											
1. Bilberry spruce, Kola peninsula	39·39	9·14	1·88	0·49	1·90	13·33	2·69	1·46	3·29	2·60	11·78
2. Blueberry pine, same location	71·71	7·66	2·82	1·43	3·74	28·51	5·01	2·07	10·76	5·09	17·70
Southern taiga											
3. Green-moss spruce, Nelidovo, Kalinin province	66·7	19·6	5·00	0·90	4·9	21·4	3·8	3·00	4·9	3·2	18·6
4. Mixed spruce, same location	71·06	25·1	4·80	0·56	4·7	25·7	2·0	2·80	3·8	1·6	20·1
5. Cowberry pine, Mordovsk reserve	36·96	4·7	2·5	0·56	2·1	15·4	3·2	2·10	2·9	3·5	9·9
6. Linden pine, same location	88·17	12·4	6·1	1·36	6·7	36·6	6·5	1·71	10·5	6·3	26·3
Mixed-forest subzone											
7. Hazel-*Oxalis* spruce, Podushkino, Moscow province	168·4	52·0	7·7	4·6	4·9	52·3	6·2	9·2	*	†	
8. Hairy-sedge oak, Moscow province	267·8	42·9	8·9	5·6	13·5	92·8	10·7	12·6	*	†	
9. Hairy-sedge linden-oak, Moscow province	326·1	35·3	10·1	3·2	18·7	98·1	15·5	9·0	*	†	
Forest-steppe											
10. Spruce (planted), Mokhovoe, Orlov province	607·9	253·3	60·3	7·4	9·6	210·6	19·9	‡	21·10	28·7	92·8
11. Larch (planted), same location	224·0	22·1	30·7	7·6	5·4	52·9	24·3	‡	27·5	53·5	95·6
12. Oak (planted), same location	171·3	32·4	33·9	1·8	6·2	36·7	14·3	‡	*	24·9	35·1
13. Sedge-goutwort oak, Tellerman, Voronezh province	231·2	38·0	9·7	1·4	19·0	110·1	15·4	‡	21·5	16·1	40·0
14. Goutwort oak, same location	360·0	139·2	30·2	5·0	3·1	119·2	18·9	2·5	69·6	*	52·5
15. Oak on solonetz, same location	56·0	19·7	4·5	0·7	0·5	17·9	5·6	1·1		*	
Tropic zone											
16. Heronier Tropical rain forest, China	2030·3	1420·2	213·1	15·0	18·2	126·5	61·5	7·4	27·3	41·1	158·5
17. Tropical rain bamboo jungle; same location	2070·8	1316·3	407·2	32·4	13·9	78·1	46·4	2·2	36·9	43·4	110·4

* Not determined.
† Traces.
‡ None.

fact that these stands belong to different groups of forest biogeocoenoses —the former to eluvial and the latter to transitional. Therefore in the blueberry pine stand the soil and underground water accumulate a considerable amount of mineral matter and nitrogen, which are leached from the soil of the bilberry spruce stand.

Moreover, the data quoted show that humus-ferruginous podzol, a component of the bilberry spruce stand in the northern taiga, contains the smallest amount of available mineral matter and nitrogen, a situation that is reflected in the litter.

A similar situation (increase of ash and nitrogen content in litter) in the mixed-forest subzone is entirely due to the complexity of the floristic composition of the litter analysed.

The forest-steppe data are very significant. In the first place, they confirm the dependence of the amounts of mineral matter and nitrogen on the tree-species composition in artificial biogeocoenoses (analyses numbers 10, 11, and 12). With homogeneity of soils, the maximum amounts of mineral substances are found in spruce litter, with less being found in oak litter. Larch, which occupies an intermediate place, has the highest figure for return of nitrogen to the soil.

Secondly, in natural biogeocoenoses (analyses numbers 13 and 14) the use of soil reserves of mineral matter and nitrogen is far from complete, and is low in comparison with the artificial spruce biogeocoenose. Consequently there are unlimited possibilities for increasing circulation and production of organic matter by selection of species and increasing the density of stands.

Thirdly, differences in soil conditions (analyses numbers 13, 14, and 15) substantially affect the consumption and return of mineral elements and nitrogen. In this respect the physical texture and genetic properties of soils are most important. In the same type of soil (dark-grey), with sandy soils the return of mineral matter and nitrogen in litter (analysis number 13) is much lower than that with clay-loamy soils (analysis number 14). There is a still greater decrease in the amount of these items on solonetz as compared with dark-grey soils (analyses numbers 14 and 15). The marked fall in the amount of CaO brought into circulation on solonetz and humus-ferruginous podzol cannot ensure the accumulation of other elements in them, so that eluvial removal of them from these soils is prevalent. The differences in content of separate mineral elements in litter fall govern the intensity of their biological accumulation in the soil.

Fourthly, the data on mineral matter and nitrogen content in tropical biogeocoenoses show that, apart from the essentially different nature of return of elements to the soil (with predominance of SiO_2, Al_2O_3, Fe_2O_3, and nitrogen), that is where the soil resources are most fully used. That is due, to a certain extent, to the extremely complex composition of phytocoenoses and to the year-round consumption of nitrogen and mineral matter.

On the whole, then, very large quantities of mineral matter and nitrogen are brought into circulation in forest biogeocoenoses, especially when it is remembered that similar returns to the soil take place every year. If the average duration of life of biogeocoenoses is taken as 100 years, then all the figures are to be multiplied by 80 to represent the period from 20 to 100 years.

Unfortunately there are as yet no systematic data on the composition of organic matter in litter. It is not possible to discover the biogeocoenotic laws governing its fluctuations on the basis of the existing incomplete data.

FOREST LITTER AND ITS ROLE IN THE BIOGEOCOENOTIC PROCESS
IN THE FOREST

Litter is a distinctive organogenic forest formation, covering the soil surface with a layer of varying depth and subject to definite laws connected with both the living population of the litter (micro-organisms and animals) and atmospheric (phytoclimatic) conditions.

Litter on the ground is formed from shed organic material, being that part not decomposed during the vegetative period. Unlike fresh litter, which consists of dead parts of plants retaining their original form, ground litter consists of organic matter decomposed to varying degrees, and possesses definite structure and properties. Mineral matter returning to the soil from litter also accumulates in it.

Forest litter has a very high content of nitrogen and nutrients required by plants, especially phosphorus, calcium, potassium, magnesium, etc. Therefore the main mass of small roots is located in it, and it is one of the chief sources of nutriment for trees.

Representing a stage in the conversion of litter fall into humic matter, litter has a content of organic compounds similar to that of soil humus.

There are two types of conversion of the litter fall into humic matter: (i) litter fall-litter-humus; and (ii) litter fall-humus. The former is typical of forests in moderately cold and moderately warm regions, the second of warm dry and hot humid regions.

Accumulation of litter is a biogeocoenotic process depending not only on the amount of litter fall but also on interactions with microbiocoenoses and zoocoenoses. Therefore litter accumulation is more constant than shedding of litter. The amount of the latter, as is well known, fluctuates widely from year to year.

The intensity of accumulation of litter is linked with the intensity of a number of processes and phenomena occurring in biogeocoenoses. For instance, the accumulation of litter and its humification may serve as a definite indicator of changes in a whole assemblage of biogeocoenotic processes, caused by a decrease in the amount of solar energy received or in the activity of animals and micro-organisms when inflow of moisture is much in excess of its expenditure in evaporation.

Similarly, with excessive inflow of heat and insufficiency of moisture the mineralisation of organic matter is intensified. In that case there is no formation of litter useful for the life of biogeocoenoses. Whereas in the soils of herbaceous biogeocoenoses the density of their living population is linked with root distribution (Afanas'eva 1952), in temperate-zone forest biogeocoenoses that density is primarily linked with litter. Life in it is many times more intensive than in the soil, which ensures retention of organogenic elements in the surface layers of the soil.

In spite of the very great part taken by litter in the life-activity and the evolution of forest biogeocoenoses, there has been insufficient study of the amount of litter and of its content of mineral and humic matter.

Rather more data have been collected on the amounts of litter in different types of forest biogeocoenoses (Table 133). They show that the greatest amounts of litter accumulate in forest biogeocoenoses in the northern and central taiga, reaching 57-69 ton/ha. Particularly high figures are recorded for pine stands in the northern taiga zone: in bilberry stands up to 90 ton/ha, and in long moss up to 105 ton/ha. In similar pine stands in Moscow province (mixed forest subzone) the amounts of litter fall slightly and the corresponding figures are 72 and 103 ton/ha, and in bilberry stands in Bryansk province the amount is only 39 ton/ha.

Such high figures for litter accumulation point to weakening of biogeocoenotic interactions on account of an unfavourable phytoclimatic regime, which depresses the life-activity of the phytocoenose and the 'work' of micro-organisms and animals.

At the same time the accumulation of litter in the green-moss-bilberry and juniper spruce stands of southern Bulgaria (Zonn 1960), with phytocoenoses similar in composition to those in the central taiga, is only one-eighth or one-tenth as much, and is 6·6-8·5 ton/ha. These data confirm the influence of types of forest biogeocoenoses in determining the amounts of litter. Forest types also affect the amounts of litter in the mixed-forest subzone in similar climatic conditions. In the Belovezh bush (Utenkova 1961) the amounts are: in an oak stand with hornbeam 8·2-10·0 ton/ha, in an oak stand with spruce and bilberry 11·0-12·8 ton/ha, in a spruce stand with oak 17·9-19·7, and in a bilberry spruce stand 34·0-52·6 ton/ha. In the Moscow area (Polushkinsk forest) the differences in litter accumulation in different biogeocoenose types are still more marked: in a hazel-*Oxalis* spruce stand with oak 23·0 ton/ha, in a hairy-sedge oak stand 6·3 ton/ha, and in a hairy-sedge linden-oak stand 2·7 ton/ha.

Similar relationships are clearly shown also in data on amounts of litter accumulated in forest-steppe biogeocoenoses (see Table 133). The amounts of litter decrease sharply from spruce through larch to oak. In forest-steppe conditions less litter accumulates than in other zones, indicating high intensity of interaction between all biogeocoenose components. In artificial steppe biogeocoenoses the amount of litter may

Table 133. *Amounts of litter and its content of mineral elements, by types of forest biogeocoenoses.*

Type of biogeocoenose and location	Amount of litter (ton/ha)	Mineral element content (kg/ha)											
		Ash	SiO_2	Al_2O_3	Fe_2O_3	MnO	CaO	CaO	MgO	K_2O	P_2O_5	SO_3	N
Northern sparse taiga													
Lichen pine on ferruginous podzol (Kola peninsula)	34·5	694·8	206·5	118·9	58·0	6·7	142·6	26·9	*	45·7	28·2	*	*
Green-moss spruce; humus podzol (same location)	57·5	1627·3	587·7	251·4	189·5	36·1	207·8	61·2	—	112·8	60·7	*	*
Central taiga													
Bilberry spruce on medium-podzolised loam (Kharovsk leskhoz, Vologda province)	51·0	2986·0	1284·0	295·0	127·0	117·0	694·0	97·0	122·0	112·0	138·0	862·0	
Herb-green-moss spruce on peaty-humus gleyed soil (same location)	69·0	3623·0	1168·0	262·0	172·0	41·0	1160·0	234·0	166·0	186·0	234·0	1118·0	
Forest-steppe													
Goutwort oak on dark-grey soil (Tellerman, Voronezh province)	13·3-19·3	3002					No data						
Oak on solonetz (same location)	10·1-12·2	755					No data						
Spruce plantation on leached chernozem (Mokhovoe, Orlov province)	15·2	1016·6	439·0	103·5	22·3	*	425·3	27·3	21·3	28·8	48·6	206·0	
Larch (same location)	12·3	566·9	116·8	89·7	27·1	*	204·2	50·4	30·7	12·3	35·7	150·0	
Oak (same location)	5·3	221·3	84·8	33·2	8·9	*	122·8	11·6	32·1	5·8	22·1	77·5	
Mountain spruce stands													
Green-moss-bilberry spruce on dark-coloured soil (Bulgaria)	8·5	1412·4	827·9	258·1	33·1	*	136·8	65·4	*	17·8	73·1	79·0	
Juniper spruce (Bulgaria)	6·6	995·3	609·8	172·3	53·5	*	66·0	32·3	*	21·9	39·6	66·3	

* Not determined.

rise, as is shown by data for Derkul' station (Lugansk province), where the amount in a 60-year-old oak-ash strip varied from 36 to 40 ton/ha, and in a 15-to-16-year-old forest area from 15 to 18 ton/ha. One reason for the differences is retardation of litter decomposition as a result of drought and depression of animal and micro-organism activity through lack of moisture; another reason is the enrichment of litter with mineral particles as a result of the activity of soil-dwelling animals and wind transportation.

The above are a few data on differences in the nature of accumulation of litter, due to variations in interactions within the system of litter-vegetation-soil-micro-organisms and meso-organisms-atmosphere-litter. Variations in these interactions affect the amounts of mineral elements and nitrogen (see Table 133). The proportion of these in litter is much higher than in litter fall, because of the lower proportion of organic matter. The variations, however, differ in intensity, that circumstance being connected with zonal differences in the interactions of biogeo-coenose components.

The total ash content and the content of the chief mineral elements in litter increases from the northern taiga to the forest-steppe, on account of differences in the use of these substances by plants, and of reduction of their leaching from litter into the soil. In the northern and central taiga types of biogeocoenoses only iron, aluminium, and silicon are accumulated; all the other elements (manganese, magnesium, potassium, phosphorus, calcium) are in varying degrees washed out not only from the litter but also from the upper soil layers of biogeocoenoses. In forest-steppe biogeocoenoses magnesium, calcium, sulphur, aluminium, and iron are accumulated, and very small amounts of potassium, phosphorus, and silicon are carried into the upper soil horizon.

The spruce biogeocoenoses of Bulgaria differ markedly from those of the taiga in the nature of the accumulation of mineral elements: there SiO_2, Al_2O_3, and Fe_2O_3 predominate, with a great decrease in CaO and MgO. In mineral circulation, therefore, these biogeocoenoses are closer to the subtropical than to the taiga type. Similar accumulation of mineral elements takes place in the high-mountain spruce forests of Eastern Tibet (Zonn 1964).

Thus forest biogeocoenoses—possessing a highly-developed ability to accumulate on the soil surface, in the form of forest litter, organic matter containing mineral elements and nitrogen—create not only food reserves for plants and animals but also energy resources, under the influence of which both separate biogeocoenose components and biogeocoenoses as a whole undergo alteration. These alterations take place as a result of the soil's gradual acquirement of new characteristics and properties.

When there is lack of harmony among the interactions of separate components—leading, for example, to the predominance of accumulation over decomposition of litter, on account of insufficient heat and excessive moisture—the action of other components also is disrupted. Changes then take place more rapidly, and usually in the direction of regression of the forest biogeocoenose. That is very characteristic of the forest-tundra, where the soils of forest biogeocoenoses are unstable and often degrade into a boggy state. Such an alteration results from excessive conservation of energy-producing material and the leaching of mineral matter out of it.

In the southern taiga and the mixed-forest subzone similar phenomena appear to a lesser extent. There the process of biological accumulation of

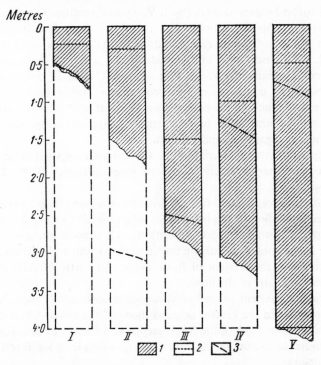

Fig. 70. Variation in some features of soils in forest biogeocoenoses in different zones. 1—total thickness of soil; 2—limit of active accumulation of humic matter; 3—limit of active leaching of mineral and organo-mineral compounds; I—forest-tundra; II—taiga; III—forest-steppe; IV—steppe; V—semi-desert.

matter and energy takes place not only in litter but also in a layer of soil of greater thickness (Fig. 70). It is accompanied by more intensive synthesis of organic and organo-mineral matter.

Under forest plantations in steppe conditions the soil-formation process shows periodical depressions, caused by the slackening of biogeocoenotic life-activity during drought years. But, on the whole, active biogenic accumulation takes place, accompanied by qualitative changes in the soils.

In spite of the fact that organic matter predominates in forest litter, its role in biogeocoenotic processes, and especially the significance of litter composition in the life of phytocoenoses and zoocoenoses and its role in changes in soil processes, have been very inadequately investigated.

The well-known classification of forest litter (according to the degree and nature of its decomposition) into peaty (raw-humus), coarse-humus (moder), and soft-humus (mull), is not based on objective qualitative features, which makes its use difficult.

Systematic study of the composition of humic matter in litter in different types of forest biogeocoenoses (by I. V. Tyurin's method) reveals several laws governing its formation. The nature of the interactions of the separate groups of humic compounds with the soil has not been fully clarified. The description of the podzolising action of fulvic acids, as developed especially by V. V. Ponomareva (1962), is not universally applicable. The composition and properties of fulvic acids vary, and depend not so much on the composition of plant litter as on variations in its interactions with other biogeocoenose components, especially atmosphere, micro-organisms, and animals.

The general laws of variation in the group composition of humic substances in litter in different types of biogeocoenoses (Table 134) are as follows.

The total carbon content falls off from the southern taiga to the steppe zone, corresponding to the increase in mineralisation of organic remains as compared with their decomposition. That basic process involves increase in the content of carbon linked with calcium and of humic acids, and decrease in the content of fulvic acids and partly of carbon in the undecomposed part of the litter.

The carbon content of easily-hydrolysable compounds rises from the biogeocoenoses in the central taiga to those of the mixed-forest subzone and rises again in those of the forest-steppe and steppe, on account of the very favourable conditions for litter decomposition found in forest-steppe biogeocoenoses.

Some variation from these rules—in particular, a fall in total carbon content in central and northern taiga conditions—results from the very intensive leaching of carbon compounds there. That shows up in a fall in the $C_H : C_F$ (humic acids to fulvic acids) ratio to 0·2. In the southern-taiga types of biogeocoenoses it rises to 0·5. Here we see the dependence of the composition of humic matter on the stand composition (spruce and pine), and within a single stand on the degree to which mosses and lichens participate in litter formation. The total nitrogen content is higher in pine biogeocoenoses than in spruce, on account of the greater solubility of nitrogenous compounds in spruce stands and their more intensive leaching.

In mixed-forest biogeocoenoses the $C_H : C_F$ ratio rises to 0·5-0·7, and in the forest-steppe to 0·8, indicating an increase in the accumulation of less-soluble humic matter in both litter and soil.

Table 134. *Composition of organic matter in litter.*

Type of biogeocoenose and location	Carbon content (%)	Carbon as % of total carbon							$C_H:C_F$	$C:N$
		Wax-resin	Decalcinate	Humic acids	Fulvic acids	Easily-hydrolysed compounds	Undecomposed residue			
Central taiga										
Bilberry spruce. Podzol (Vologda province)	38·18	9·7	1·6	6·7	22·5	3·1	47·6		0·19	*
Southern taiga										
Bilberry spruce. Shallow podzol (Yaroslav province)	49·43	11·0	2·0	6·7	22·1	3·9	54·3		0·30	27·2
Oxalis-fern spruce. Soddy-weakly-podzolised soil (same location)	52·69	9·9	2·2	5·2	22·5	3·0	57·2		0·23	25·7
Sphagnum spruce. Peaty podzol, gleyed (same location)	48·64	7·9	1·1	5·3	22·0	3·3	60·4		0·24	26·4
Lichen pine. Podzol of medium depth (same location)	46·12	15·4	1·1	5·1	22·1	3·8	52·5		0·23	42·0
Bilberry pine. Strongly-podzolised soil (same location	48·61	14·0	1·1	6·6	18·2	3·3	56·7		0·36	33·5
Long-moss pine. Deep podzol, gleyed (same location)	42·32	9·1	1·5	10·5	22·6	2·9	53·4		0·46	30·5
Mixed coniferous-broad-leaved forest										
Oxalis-green-moss spruce. Soddy strongly-podzolised soil (Moscow province)	35·82	10·5	1·7	9·2	17·0	21·9	39·7		0·54	*
Sedge oak. Soddy-medium-podzolised (same location)	42·08	11·6	2·5	6·0	11·6	17·5	50·8		0·52	*
Forest-steppe										
Sedge-goutwort oak. Grey forest soil (Ryazan province)	39·98	19·2	4·6	5·6	16·6	5·6	50·0		0·27	*
Ash-goutwort oak. Dark-grey soil (Voronezh province)	33·66	8·8	4·4	10·1	13·7	3·5	59·5		0·74	*
Oak on solonetz (same location	31·01	7·6	2·6	12·1	15·1	3·6	59·0		0·80	*
Steppe										
Ash-elm oak. Chernozem (Lugansk province)	30·65	8·2	5·7	13·2	15·0	4·2	53·7		0·87	*
Mountain forests										
Bilberry spruce. Dark-coloured soil (Bulgaria)	38·19	7·9	1·3	14·6	21·1	—	55·5		0·59	*
Cowberry pine (Bulgaria)	34·12	12·2	2·1	15·3	24·6	—	45·1		0·62	*

* Not determined.

The above data show that each type of forest biogeocoenose has its own characteristic composition of humic compounds in the litter, reflecting the zonal conditions of litter formation. In each subzone the types of forest biogeocoenoses are distinguished by their own characteristics of litter decomposition and composition of humic compounds.

The differences in organic-matter composition of litter between Bulgarian biogeocoenoses (bilberry-spruce and cowberry-pine) and biogeocoenoses in the central and southern taiga of the U.S.S.R. are most significant. In the nature and degree of litter decomposition, the forest biogeocoenoses of Bulgaria are close to forest-steppe biogeocoenoses. Therefore they cannot be considered analogous to taiga biogeocoenoses.

We may suggest a classification of litter types on the basis of differences in their content of separate groups of humic compounds. This classification is based on the ratio of humic to fulvic acids ($C_H : C_F$), since these acids are the most important in humus formation and in action upon the mineral part of the soil, reflect the character and depth of litter, and determine the types of matter and energy circulation in the soils of forest biogeocoenoses. The classification of forest litter types by their $C_H : C_F$ ratio is as follows:

Group	Range of $C_H : C_F$	Humus accumulation in Horizon A of soil	Action on mineral part of soil
1. Fulvate	< 0.2	Almost absent	Most aggressive
2. Humate-fulvate	0.2-0.5	Slight	Aggressive
3. Fulvate-humate	0.5-0.7	Medium	Slightly aggressive
4. Humate	> 0.7	Intensive	Accumulative

By aggressive action we mean destructive action by fulvic acids on the mineral part of the soil, accompanied by leaching of breakdown products and relative accumulation of SiO_2 in the soil; and by accumulative action, accumulation of humic matter, accompanied by transformation of the mineral part of the soil by relatively slight movement of the clay fraction through the profile.

Fulvate decomposition of litter is typical of the forest-tundra and northern-taiga types of forest biogeocoenoses; humate-fulvate of the southern-taiga type; fulvate-humate of the coniferous-broad-leaved type; and humate of the forest-steppe and steppe types. Naturally within the limits of each group there may be variations, due to variations in the interactions in forest biogeocoenoses.

The suggested classification should be made more precise and detailed as further data accumulate.

From the biogeocoenotic point of view the figures for total carbon and nitrogen content in litter, and for the amounts of separate groups of humic compounds, compiled by S. P. Koshel'kov (Table 135), are very important. They show litter reserves in relation to compounds of value to plants and the possible intensity of their action upon soils.

In addition, on the basis of these figures we can construct a new classification of litter, taking into account its biogeocoenotic and sylvicultural properties. The segregation (proposed by Koshel'kov 1962) of such litter types as weak-humus, coarse-humus, dry-peaty, and peaty is based on the qualitative and quantitative features of humic compounds produced by them. Although such a classification of litter has as yet been proposed only for spruce and pine biogeocoenoses, there are grounds for believing that no less substantial differences will be found in other types of biogeocoenoses.

Such a classification of litter types makes possible a deeper and more thorough analysis of their biogeocoenotic role. The importance of litter in the life of biogeocoenoses is more many-faceted than has been represented up to the present. It not only affects the soil and serves as a direct source of nutriment for plants, animals, and micro-organisms, but also acts upon the physiological processes taking place in plants, especially in their photo-synthetic activity, which depends on the exhalation of CO_2 by litter into the atmosphere.

The breakdown and mineralisation of organic matter in forest litter is accompanied by exhalation of gaseous products, among which the most important are CO_2, O_2, and NH_3. Exhalation of these depends on many factors, primarily on the activity within the litter of living organisms that hasten its decomposition and, by their respiration, enrich the phytocoenotic atmosphere and the soil-contained air with CO_2. Respiration by roots also assists in exhalation of CO_2. Chemical reactions accompanied by release of CO_2 and other gases are less important. According to V. N. Mina (1960), in forest-steppe conditions the relationship between the 24-hour production of CO_2 by litter and by roots less than 0·5 mm

Table 135. *Amounts of carbon (total and by groups) and nitrogen content in litter in southern taiga in Yaroslav province (kg/ha) (Koshel'kov 1961).*

Type of biogeocoenose and litter	Total carbon	Total nitrogen	Carbon					
			Wax-resin	Decalcinate	Humic acids	Fulvic acids	Hydrolysable	Remainder
Oxalis-fern spruce stands, slightly-coarse-humus	12,600	490	1240	280	650	2830	380	7200
Bilberry spruce stand, coarse-humus	24,400	860	2270	370	1570	5270	720	13,200
Sphagnum spruce stand, peaty	36,000	1360	2840	400	1910	7900	1180	21,770
Lichen pine stand, coarse-humus	5950	142	920	60	300	1320	230	3120
Bilberry pine stand, dry-peaty	18,500	550	2570	200	1240	3360	630	10,500
Long-moss pine stand, peaty	28,700	910	2530	420	2920	6280	810	14,840

in diameter in different tree stands was as follows. In a spruce stand litter produced 119 kg, and root respiration 40 kg; in an oak stand the corresponding figures were 169 and 61, and in a birch stand 137 and 46.

These data confirm that litter is the chief producer of CO_2. The amount of CO_2 exhaled by root respiration does not exceed 33-36% of the total amount exhaled by litter in the most favourable conditions of biogeocoenose life-activity. Where these conditions are worse (taiga, forest-tundra) we may assume that the proportion is still smaller, because of deterioration in the soil conditions for root respiration.

A large part of the CO_2 exhaled during litter decomposition enters the phytocoenotic layer of the atmosphere, and a smaller part, as shown by the investigations of I. A. Tyurlyun (1958), may be diffused in the soil. It has, however, been denied by several investigators that the CO_2 accumulated in the soil is derived from that source; they believe it to be a product of root respiration. The CO_2 diffused in the atmospheric layer nearest the ground is absorbed by the crowns of plants in the process of photosynthesis, as has been shown by the investigations of Zonn and Aleshina.

According to the data of V. N. Smirnov (1955), the production of wood in different biogeocoenoses is directly related to exhalation of CO_2 (Table 136).

Exhalation of CO_2 by litter also depends on the zonal conditions of formation of the litter.

CO_2 is exhaled from the surface of litter in much greater quantities than from the surface of soil without litter (Fig. 71). The activity of

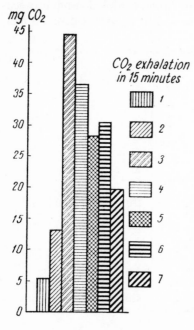

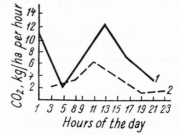

Fig. 71. CO_2 exhalation from surface of litter (1) and soil (2).

Fig. 72. Effect of black wireworms (*Iulus terrestris*) on CO_2 exhalation from soil, from soil with leaves, and from leaves alone (from Sokolov). 1—from soil; 2—from soil with wireworms; 3—from soil with oak leaves and wireworms; 4—from soil with oak and caragana leaves; 5—from oak leaves with wireworms; 6—from oak and caragana leaves; 7—from oak leaves.

Table 136. *Relation between forest type and CO_2 exhalation.*

Type of forest and growth class	Quality class	Amount of wood, m^3/ha	CO_2 (kg/ha per hour)
Lichen-moss pine, class II	III	51	0·86
Cowberry pine, class II	II	98	1·25
Long-moss pine, class II	III	81	1·10
Sphagnum pine, class II	V	20	0·65
Mixed spruce, 60-80 years	I	—	up to 3·68
Green-moss spruce, 40-60 years	II-III	—	up to 2·27

animals increases CO_2 exhalation by accelerating litter decomposition (Fig. 72). That activity in turn depends on the species-composition of the trees whose litter fall forms the litter; this is manifested very clearly in artificial biogeocoenoses. The following data (Mina 1957) show the amounts of CO_2 exhalation (in kg/ha) from litter surface at different periods during the vegetative season, in stands of different species:

	*Periods**		
	11-18.VI	12-18.VII	29-30.IX
Spruce	119	124	104
Larch	188	137	59
Oak	169	119	51
Birch	137	128	54

* Probably per day, but the original does not specify (*Tr.*).

The above data show not only the relation between CO_2 exhalation and litter composition, but also the dynamics of the process, resulting from weather and other conditions.

The effect of zonal conditions on CO_2 exhalation from litter is shown in Fig. 73. As may be seen, CO_2 exhalation is at a maximum in forest-steppe biogeocoenoses and falls away both to the north and to the south. North of the forest-steppe the intensity of CO_2 exhalation is reduced by

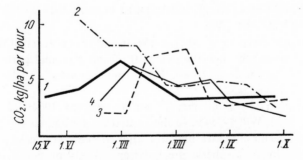

Fig. 73. Biological activity of different soils. 1—shallow chernozem (Derkul' station); 2—leached chernozem (forest-steppe, Mokhovskoe forest); 3—peaty-humus soil (central taiga); 4—shallow podzol (central taiga).

insufficiency of heat, excess of moisture, and lower numbers of animal organisms in the litter. To the south the process is impeded by excessive heat and inadequate moisture, inducing periodicity in the 'work' of the organisms.

In forest biogeocoenoses on the steppes, during periods favourable for decomposition, CO_2 exhalation from the soil surface reaches the following maxima (Zonn 1954a): in a 50-year-old oak-ash strip 17·3, in an oak-ash-acacia plantation 10·5, and in an unmixed oak stand 3·9 kg/ha per hour.

The altitude of biogeocoenoses strongly affects CO_2 exhalation. According to B. A. Dzhafarov (1960, Fig. 74), CO_2 exhalation increases from a mixed-herb beech stand (A, altitude 700 m) to a fescue beech stand (B, altitude 1550 m) and thence to a subalpine birch stand (C, altitude 2000 m), especially during the warm period. The increase is due to the reduced density of the stands with increased altitude, and also to differences in the quality of light and the amount of heat energy reaching the litter surface.

The distribution of CO_2 through soil-contained air is subject to the following laws: the amount increases with depth during the warm period and decreases somewhat during the cold period. Average CO_2 concentration (as % of the total) at depths below one metre varies as follows: in southern taiga biogeocoenoses about 1·5 to 1·7, in the forest-steppe 2·5, on the steppes 3·7-4·0, and in the tropics more than 10·0.

In the soil CO_2 has a double role. First, it aids the solubility and therefore the migration of mineral matter. Thus CO_2 is a direct and indirect participant in matter and energy circulation in the plant → animal → soil system. Secondly, it may be a severe impediment to the life-activity of organisms in biogeocoenoses. It is known that with CO_2 concentration higher than 10-12% roots die off. Such concentration may occur in peaty layers, which produce large amounts of CO_2. With excessive atmospheric humidity, high concentration of CO_2 may take place in the soil because its diffusion in the atmosphere is impeded.

LEACHING OF CHEMICAL ELEMENTS FROM PLANTS AND ITS ROLE IN THE BIOGEOCOENOTIC PROCESS

Until recently it was believed that matter and energy reached the soil by two main routes: through the dying-off of above-ground parts of plants (mainly in the forms of wood, bark, roots, needles, and leaves), and through root exudates.

Investigations by a number of authors (Arens 1934, Tamm 1951, Lausberg 1953, Will 1955) have shown that matter and energy also reach the soil as a result of the washing of matter from needles, leaves, branches, and trunks by atmospheric precipitation. Arens proposed giving the name of 'cuticular excretion' (*Kutikuläre Excretion*) to that method of removal of mineral matter from needles and leaves. In spite of the fact that quantitative data on that method of passage of matter and energy to

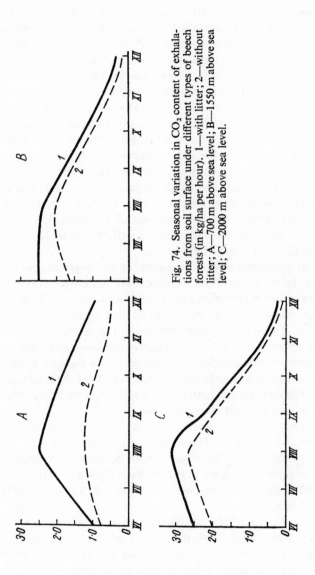

Fig. 74. Seasonal variation in CO₂ content of exhalations from soil surface under different types of beech forests (in kg/ha per hour). 1—with litter; 2—without litter; A—700 m above sea level; B—1550 m above sea level; C—2000 m above sea level.

the soil are still very few, its significance in the life of biogeocoenoses cannot be ignored.

It is known that in forest biogeocoenoses the amount of precipitation reaching the soil surface in liquid form is actively governed by the entire assemblage of plants in the phytocoenose. A part of the precipitation intercepted by them is expended in evaporation and is excluded from circulation in the soil-plant-animal-soil system. Another part, flowing down the above-ground parts of plants, is enriched to a certain degree with organic and mineral matter and reaches the litter surface, and later the soil, in a modified state. Consequently atmospheric precipitation, on contact with the vegetation in phytocoenoses, acquires new properties dependent on the characteristics and structure of the phytocoenoses.

That dependence is many-faceted, being determined by the composition, density, and structure of forest phytocoenoses, and also by the structure of the plants forming the phytocoenoses (arrangement of branches, needles, and leaves; nature of trunks, especially their bark structure; etc.).

The depth, composition, and structure of forest litter are no less important, since litter is the first recipient of solutions flowing down the plants, and it not only can enrich the soil with the matter that they contain but can itself absorb part of it.

Thus, by successive interactions with the above-ground parts of plants and with litter, atmospheric precipitation undergoes continual transformation before it enters into interaction with the soil. Such changes in the composition of precipitation depend also on external environmental conditions. The latter include principally the intensity of incidence of heat and moisture on the upper surface of the biogeocoenotic layer.

Therefore we may conclude *a priori* that with increase in atmospheric precipitation and decrease in incident radiant energy the leaching of matter out of phytocoenoses will increase, and with decrease in precipitation and increase in incident heat it will diminish.

It is not possible to summarise all existing data on this kind of addition of matter and energy to forest biogeocoenoses, since they have been obtained by different methods and cannot be properly compared in order to discover their significance in the total circulation of matter and energy. We therefore quote only the most complete data (unpublished) obtained by V. N. Mina.

In spruce forests of the southern taiga subzone (Yaroslav province) precipitation passing through the crowns of various tree species has the following pH values. On an open space pH is 5·1-6·1; under spruce crowns 4·6-6·0; under birch 4·5-5·9; and under pine 3·7-5·5.

In the mixed-forest subzone (Moscow province) the pH in an open space is 5·7-6·1; under pine crowns 4·7-5·3; under linden 5·4-5·9; and under mountain ash 5·3-5·9.

In all conditions precipitation after passing through tree crowns becomes more acid than precipitation falling directly on the soil surface, because of the increased concentration of soluble organic matter. That acidification is highest with pine, followed by birch, spruce, mountain ash, and linden.

Precipitation flowing down tree trunks changes its chemical composition even more markedly (Table 137).

The high acidification of water flowing down tree trunks is also due to the presence of soluble organic compounds. The rougher and more corrugated the bark of the trunks, the greater is the acidification. The flow from spruce, pine, and linden is most acid, that from birch and mountain ash less so. According to the data of Ehwald *et al.* (1959), the flow from oak trunks is less acid than that from pine trunks.

Water flowing down trunks carries away much more calcium, magnesium, potassium, and ammonia nitrogen from them in the central part of the taiga zone than in the southern part.

This phenomenon accounts for the highly-podzolised soils around pine and spruce trunks, which receive precipitation with higher acidity and a higher content of organic compounds that attack the mineral part of the soil. The downflow also carries a certain amount of mineral matter and nitrogen leached out of the tree bark.

Data regarding different types of biogeocoenoses (Table 138) show that leaching of mineral elements and nitrogen from tree crowns depends not only on the amount of precipitation passing through but also on the composition of the stand.

The leaching of calcium and potassium is most noticeable, that of magnesium and nitrogen being less pronounced. The relationship between types of spruce biogeocoenoses and the amount of leaching is clearly evident. The more diverse the floristic composition of the forest, the

Table 137. *Chemical composition of precipitation flowing down tree trunks, from 13.VI to 12.IX (per tree in mg/litre).*

Tree species	Amount of water flowing down, litres	pH variation	Content					
			H	Ca	Mg	K	Ammoniacal N	Oxidisability, mg O_2
Southern taiga (Yaroslav province)								
Spruce	4·5-6·5	2·6-3·5	22-32	733-1085	54-65	229-299	50-84	Not determined
Pine	22	3·3-4·7	20	528	86	165	66	Not determined
Birch	14-16	3·7-5·0	5-6	42-48	6-17	18	10-14	Not determined
Mixed forests (Moscow province)								
Pine	—	2·8-3·5	—	37	6	9	15	437
Linden	—	3·3-3·6	—	58	10	36	19	406
Birch	—	4·6	—	4	1	1	2	112
Mountain ash	—	4·5-4·9	—	10	2	4	3	113

greater is the leaching. That relationship disappears in other types of biogeocoenoses.

As water seeps through litter it generally loses mineral elements on account of absorption of some of the salts and acids by the decomposing organic mass. In only a few types of biogeocoenoses is there an increase in the content of mineral matter, which results from leaching of the litter and is governed by the depth of the litter and its richness in certain elements. For instance, calcium, magnesium, and nitrogen are leached from the litter of an *Oxalis*-fern spruce stand during the vegetative period, but potassium is absorbed by it. The litter of a bilberry spruce stand absorbs all elements except nitrogen, which is leached from it.

These phenomena also appear in varying degree in other forest types.

According to the data of Ehwald *et al.* (1961), the relationship between the leaching of nitrogen and mineral elements and the amount of precipitation is clearly displayed (Table 139).

Although these data do not cover zonal variations in the amounts of nitrogen and mineral elements reaching the soil, they show that it is of

Table 138. *Leaching of mineral elements and nitrogen from tree crowns and litter in different types of forest biogeocoenoses from 13.VI to 12.IX (in kg/ha).*

Type of biogeocoenose	Tree layer					Litter				
	Precipitation passing through (mm)	Leached				Precipitation passing through (mm)	Leached			
		Ca··	Mg··	K·	N		Ca··	Mg··	K·	N
Oxalis-fern spruce	188	3·0	1·1	8·3	1·8	111	22·4	3·9	1·0	1·7
Bilberry spruce	153	6·5	2·4	9·5	0·9	95	1·3	0·5	2·5	2·9
Sphagnum spruce	196	5·3	0·8	3·0	0·1	—	—	—	—	—
Mixed-herbage birch	217	3·1	0·5	1·3	1·0	103	3·7	0·9	3·9	0·2
Bilberry birch	175	6·4	1·0	2·3	1·4	94	2·1	0·1	0·9	−0·1
Bilberry pine	177	8·4	1·6	7·0	3·3	109	2·4	1·3	1·0	−0·8
Open area	273	6·6	1·6	5·1	1·0	—	—	—	—	—
Hazel-bilberry-mixed-herbage pine	154	5·5	1·5	3·2	1·9	115	3·7	0·4	2·6	0·2

Table 139. *Relationship between leaching of nitrogen and mineral elements and amount of precipitation.*

Type of biogeocoenose	Year	Precipitation (mm)	Leaching			
			N	Ca	K	P
Open area	1957	489	6·5	8·6	1·9	0·4
	1958	613	11·2	12·8	4·7	0·6
Pine-beech forest	1957	293	85	23·2	10·5	0·7
	1958	387	12·0	24·0	19·0	0·6
Pine forest	1957	324	9·4	9·4	11·7	0·9
	1958	410	12·0	12·0	12·4	0·3

definite significance in determining the characteristics of plant-soil interactions in forest biogeocoenoses.

We must assume that with the onset of drought during the vegetative
period the amount of organo-mineral compounds carried into the soil by
precipitation will decrease. But in these conditions, just as in the vicinity
of large industrial and populated areas, it is replaced by aerial transportation of the same elements and also of others (often injurious for the
life of forest biogeocoenoses—gases, radioactive compounds, matter of
organic origin, etc.). Inadequate attention, however, has as yet been paid
to these phenomena and their effect on forest biogeocoenose development.

Thus the precipitation that enters the soil of forest biogeocoenoses is
enriched with mineral elements and nitrogen leached from the above-
ground parts of plants and partly from litter. Therefore atmospheric precipitation cannot be regarded merely as a factor that adds moisture to the
soil and helps to remove nutrients from it. To some extent it stimulates
accumulation of biogenic elements, which reduces and sometimes eli-
minates its leaching affect.

Further investigations should include determination of the amounts of
mineral elements and nitrogen that enter the soil in this way, and of their
significance in the circulation of matter and energy in forest biogeocoenoses.

ROLE OF TRACE ELEMENTS AND RADIO-ISOTOPES IN THE CIRCULATION OF MATTER AND ENERGY IN FOREST BIOGEOCOENOSES

Very little attention has as yet been given to trace elements in forest soils.
Therefore the questions of the content of trace elements in these soils,
their significance in the life of trees, and their role in the circulation of
matter and energy have been little studied. Trace elements may have been
neglected because forest economy has not come into contact with their
positive or negative effects in natural conditions. It is known, however,
that many trace elements not only stimulate the growth and raise the
productivity of plants, but also form part of many metabolic products
excreted by animals and micro-organisms during life or formed by decomposition of their dead remains. Forest soils with a litter horizon (A_0)
and fairly well humified upper layers have greater stocks of trace elements
than cultivated soils. That is because soil-forming strata are the chief
source of trace elements. Forest vegetation returns trace elements to the
soil in litter fall, as a result of which large quantities accumulate in litter
and in the upper soil layers.

Underground water plays a major part in their accumulation, as it
contains many trace elements in solution, and when stagnant helps to
concentrate them both in soils and in trees (Table 140).

All trace elements are concentrated in horizon A, which is typical also
for other soils. Still greater amounts of iodine accumulate in litter, as is
shown by data on iodine content in horizons A_0 and A_1 (Bykova 1961) (in
mg/kg of dry soil):

	Depth (cm)		Amount
Dark-grey sandy soil under deciduous stands	A_0	0-5	1046
	A_1	11-21	753
Soddy-strongly-podzolised soil under birch stand	A_0	0-6	788
	A_1	8-18	265

As may be seen, the iodine content also depends on the degree of humification of horizon A_1. It is greater in dark-grey than in soddy-podzolic soil. A similar situation is observed regarding the content of chromium, nickel, cobalt, copper, vanadium, etc. in the same soils.

Almost no data exist on the use of trace elements by tree species. M. A. Glazovskaya et al. (1961) give data on the content of trace elements in birch leaves, branches, and roots. In some cases they report higher copper content in leaves (up to 0·128% of the ash content) than in the soil (up to 0·040%); as for nickel, cobalt, lead, gold, no differences are apparent between their content in the soil and in the separate parts of the birch trees. For the present that is all the information available on trace element content in forest soils and trees.

The role of radio-isotopes in the soils of forest biogeocoenoses may be positive or negative.

When radio-isotopes enter forest biogeocoenoses in relatively large amounts they may have indirect negative effects. Without considering all the possible consequences of radio-active pollution of the biosphere, we shall dwell only on those that affect the life and development of forest biogeocoenoses.

Radio-active pollution of forest biogeocoenoses from the atmosphere may occur in two ways: (i) direct absorption of radio-isotopes by the green parts of plants; (ii) their subsequent arrival in the soil with litter fall or root exudates, or on the soil surface beneath forest canopy.

According to existing data (Aleksakhin 1963), the forest litter covering the soil of forest biogeocoenoses accumulates very great amounts of radio-isotopes falling from the atmosphere with precipitation. These may seriously affect the activity of animals and micro-organisms (to the point of poisoning them), which in turn may alter not only soil processes but also the circulation of matter between these organisms and vegetation.

The entrance of radio-isotopes into plants both directly and through the soil may, as some investigators (Aleksakhin 1963) have shown, induce severe pathological changes in plants (chlorosis, defoliation, withering of

Table 140. *Content and distribution of trace elements in humus-gleyed soils, with underground water near by, under alders (in mg/kg of calcined batches)* (*Vinnik 1961, Bykova 1961*).

Horizon	Depth (cm)	V	Cr	Ni	Co	Mo	Cu	Li	Sr	Mn	I
A_1	0-10	210	93	120	40	8	120	800	<5	3500	27,776
B_2	39-49	110	52	41	13	7	74	500		830	8475
C	87-97	110	48	17	7	5	33	100		200	751

tips, etc.), leading to their death. With concentrated fall-out of radio-isotopes such action may also produce radiation damage, the struggle against which may last for many years. In this way the detrimental effect of radio-isotopes may show itself in varied forms and to varying degrees, and alter the normal life-activity of forest biogeocoenoses in different ways. Knowledge of these processes is essential in order to forestall them and devise effective means of combating them.

INTERACTIONS OF ANIMALS AND MICRO-ORGANISMS WITH SOILS

Micro-organisms and animals constitute an indispensable part of all forest biogeocoenoses, but their activities vary greatly. Some fulfil their functions mainly in the above-ground layer of a biogeocoenose, others in the underground (soil) layer, where they are represented both by specialised species and by species common to both layers (especially bacteria and fungi).

Birds dwell mainly in the above-ground layer, and only a few of them nest in the soil (sand martins, etc.), but they all fulfil the same soil-forming functions by taking part in the exchange of matter between plants and soil. Their participation, however, is on a very small scale, and as yet has not been quantitatively assessed.

The role of other vertebrates in the process of matter and energy exchange is more substantial, but because of their mobility and their widely dispersed places of abode it also cannot be precisely evaluated.

We can merely assume that in a temperate climate animals cannot (on account of the small amount of organic matter produced by them) alter the character and type of circulation of matter as determined by vegetation. In the tropics, perhaps, their role in separate types of forest biogeocoenoses is greater, but data on that aspect of animal life have not yet been collected. We do not even possess data on the composition of organic remains of animals and its differences from the composition of plant remains.

In some cases, especially in the event of irruptions of insects or other animals, their role in the circulation of matter and energy may be considerable and sometimes even decisive. We may also assume that animal activity is manifested not only in direct changes in the composition of organic remains (resulting from their consumption of plant matter) but also through the indirect (biochemical) action of their excretions, such as urea and a number of specific substances of enzyme type, etc. In this way the soil may be enriched with various compounds (nitrogenous and other), and a number of biochemical processes of decomposition of organic matter and synthesis of humic and other compounds may be accelerated. In addition, soil that passes through the alimentary tracts of insects, earthworms, wireworms, and other animals is enriched with micro-organisms and acquires some new properties through its interaction with digestive secretions.

Equally important is the role of some groups and species of animals in distributing organic matter on the soil surface (e.g., ant-hills and local piles of organic matter connected with them). Burrowing animals cause increased non-capillary porosity by their tunnelling, and by throwing out soil from deeper layers on the surface they enrich litter and the upper soil horizons with mineral compounds (carbonates, iron compounds, soluble salts, etc.).

As a result the circulation of matter becomes more complex, and also separate structural elements or parcels (as N. V. Dylis calls them) are created, with differences in soil conditions and in exchange of matter between plants and soil.

These changes in microrelief themselves produce the prerequisites for dispersal of matter and energy within biogeocoenoses. This leads to the formation of varying meso- and micro-complexity of soil cover in forest biogeocoenoses. More detailed investigation will not only enable us to discover the role and significance of these phenomena, but possibly also to devise finer classifications of forest biogeocoenoses in order to solve a number of scientific and practical problems of forest economy.

From the above, we may at this stage regard the role of animals in soil formation mainly from the following aspects: enriching the soil with micro-organisms; accelerating the transformation of plant remains into humic matter; altering the physical and chemical properties of the soil; stimulating and complicating the circulation of matter and energy; dispersing matter in connection with changes in meso- and micro-relief; converting some nutritive elements (nitrogen, phosphorus, potassium, etc.) into more available compounds; and a certain amount of retention of organo-mineral matter in their own composition.

The activities of animals (like those of plants) keep the circulation of matter and energy in operation and enrich the surface soil layers with nutrients. Soil-dwelling micro-organisms and animals are adapted to an inter-related existence. That has been demonstrated by Sekera and Franz (1955) and other authors, who hold that fauna is a 'natural reservoir' of microflora that lives in animal intestines. Sekera has proved that artificial exclusion of soil fauna from the processes of decomposition of organic remains almost inhibits the formation of humic matter. He points out that only with simultaneous activity by fungi, bacteria, and soil-dwelling fauna is part of organic remains converted into humus and used by plants as it undergoes further decomposition.

In the opinion of other investigators, the chief agents of decomposition of organic matter in litter and of formation of humic matter are micro-organisms and mould fungi (Kononova 1951, Pushkinskaya 1954). V. Ya. Chastukhin (1962) believes that the chief agents of decomposition of complex ligno-cellulose compounds in forest litter are Basidiomycetes, which form part of the group of litter saprophytes. Chastukhin remarks on the well-known incompleteness of the bacteriological methods of investigation directed towards study of the limited group of spore-forming micro-organisms on the supposition that the composition of the latter defines the type of soil formation. A. A. Kozlovskaya (1959), while aware of the close link between the life-activity of soil-dwelling fauna and microflora, still gives preference to mesofauna. She believes that particular groups of micro-organisms cannot bring the process of litter breakdown to the point of formation of soft humus, i.e., to a definite quality of humic matter.

We must remark that at present a new trend of thought is developing, according to which micro-organisms participate indirectly in the nutrition of higher plants (Krasil'nikov 1958). Not only do their metabolic products act upon the organo-mineral part of the soil, but the excretion by micro-organisms of various vitamins, growth compounds, etc., stimulates the processes of exchange of matter.

From this brief survey it is seen that investigators do not agree on evaluation of the biogeocoenotic role of animals and micro-organisms. Perhaps there is some over-valuation of the 'work' of the latter, in spite of the inadequate study of their role in the processes of decomposition and transformation of matter and energy.

The inter-relationships and interdependence of the activities of separate groups of animals and micro-organisms especially need clarification.

Existing experimental data enable us to suggest that the role of meso-fauna and higher fungi may not consist solely in the mechanical preparation of organic matter for subsequent treatment and synthesis of humic matter from it by micro-organisms. Mesofauna and higher fungi, to judge from the investigations of Kozlovskaya, Chastukhin, and others, are able to convert the organic matter in litter fall and litter into humic substances characteristic of the so-called soft humus.

The question of the role of micro-organisms in that conversion arises. Considering that most of them are specialised for a diet of food that is more soluble and has undergone preliminary treatment by other organisms, we may suggest that their participation in all processes of transformation of the organo-mineral part of the soil are indissolubly linked with mesofauna and possibly also with macrofauna. In particular, the colonisation of litter and soil by micro-organisms takes place through the medium of fauna. Judging from the same data, mesofauna and higher fungi are to a great degree adapted to existence and activity independent of micro-organisms. That is confirmed by the data of K. A. Gavrilov (1950), M. S. Gilyarov (1953), and especially L. S. Kozlovskaya (1959), who have discovered fairly clear correlations between specific complexes of soil-dwelling mesofauna and definite types of forest biogeocoenoses.

Kozlovskaya takes a very firm stand on that subject. She believes that we must distinctly separate the qualitative and quantitative aspects of decomposition of organic remains. According to her views, with which we cannot disagree, the quality of decomposition, or the stage reached by the process of alteration of litter fall and litter, depends not only on the total composition of the fauna but also on the characteristics of the dominant groups in it. For instance, oligochaets bring organic remains to the soft-humus stage, and earthworms enrich the latter with calcium (S. M. Ponomareva 1948). Among humus-formers Kozlovskaya includes Apterygota, nematodes, and Acarina. The mechanical trituration of litter is performed mainly by Myriopoda and higher insects and their larvae.

The role of insects has been very well described by A. P. Travleev (1961). His data relate to the numbers of larvae of St. Mark's fly (*Bibio hortulanus*) in the litter of steppe plantations on dark-chestnut soils. In an oak plantation their numbers reach 12,000 per m^2, and they can consume up to two tons of litter per hectare. In black locust plantations, with the same number of larvae per m^2, they consume 5·6 ton/ha, and in a honeylocust plantation (numbering 9600 per m^2) only 1·6 ton/ha. In another experiment the same larvae consumed 93-96% of ash litter (with moisture equal to field water capacity), 63% of oak litter, 33% of honeylocust litter, and 93% of black locust litter. It was observed that the excrement of the larvae contained up to 20·5% of humic matter.

Black wireworms (Myriopoda) are even more important in steppe plantations on chernozems. According to D. F. Sokolov (1957) their numbers fluctuate from year to year from 210,000 million to 2,110,000 million per hectare, and their weight from 56 to 560 kg/ha. The amount of excreted products of their life-activity may reach 30-50 kg/ha per day.[1] Black wireworms feed on undecomposed organic matter, and the products of their life-activity contain a considerable amount of compounds of composition close to humus. The formation of these compounds, as stated by I. V. Tyurin and M. M. Kononova (1962), is due to enzymes of the phenol oxide type and others secreted by the intestinal epithelium.

The combination of two groups of mesofauna, mechanically-triturating and humus-forming, is most effective. At the same time the amount of organic matter worked upon by fauna, as Kozlovskaya well states, depends not only on the abundance of the animals but also on the weight of their biomass as an index of the intensity of their feeding. The number of individuals may be high but their biomass weight low, when the fauna is represented mostly by very small animals (e.g., Collembola).

The intensity of life-activity of animals also is important, depending on their ecological living conditions. The same number of individuals of any group will consume a large mass of organic matter in favourable, and a smaller mass in unfavourable, conditions.

Fig. 75 (Kozlovskaya 1959) illustrates the qualitative and quantitative differentiation of mesofauna in a number of types of forest biogeocoenoses, arranged according to their degrees of moisture supply. Whereas in some types (shrub-sphagnum pine, sphagnum spruce, herbage-green-moss

[1] If we take their active period as 90 days, they may consume 1800 to 4500 kg/ha, or the entire amount of the year's offal.

spruce, and bilberry spruce) the soil fauna is represented mostly by Enchytraeidae, Collembola, and Acarina, in a boggy-herb spruce stand the most hygrophilous complexes of mesofauna develop—Diptera larvae, aquatic oligochaets, and aquatic larvae of Diptera, etc.

The numbers of mesofauna increase from sphagnum pine stands through sphagnum spruce to herbage-green-moss and bilberry spruce stands, corresponding to changes in the kind of diet and in the water-air regime of the soil. The soils poorest in mesofauna are those of shrub-sphagnum pine and boggy-herb spruce stands. In the latter soil, along with general impoverishment, there is expansion of faunistic complexes containing hygrophilous groups.

In the above series of types of forest biogeocoenoses Kozlovskaya has discovered a clear relationship between the biomass of the principal humus-formers and the degree of peat decomposition.

In this way, many of the most recent investigations point to a more

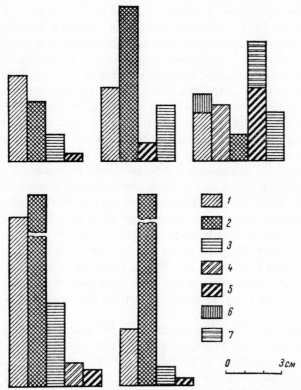

Fig. 75. Complexes of soil-dwelling fauna in different types of forest. 1—Enchytraeidae; 2—Collembola; 3—Nematoda; 4—Acarina; 5—Diptera larvae; 6—aquatic Oligochaetae; 7—aquatic Diptera larvae. Vertical scale; 1 cm (see inset 3-cm scale)—1000 specimens. Broken column on left—7635 specimens; on right—1182 specimens.

important role for animals and higher fungi than was formerly credited to them.

Especially indicative in this regard are Chastukhin's (1962) data on the role of higher fungi in breaking down fallen oak leaves. It is useful to compare them with data on the breakdown of similar plant material by micro-organisms, obtained by E. V. Runov and D. F. Sokolov (1956).

These laboratory experiments were so arranged as to exclude, or almost exclude, the influence of other organisms. Over a period of almost a year the losses in weight during the decomposition of oak leaves by cultures of litter-dwelling saprophytes reached the following figures: *Collybia dryophila*, up to 82%; *Lepiota procera*, up to 76%; *Phallus impudicus* up to 73%; *Trichotoma album*, up to 69%; *Clitocybe inversa*, up to 62%; *Cl. nebularis*, up to 53%; and *Mycena mucor*, up to 48%.

The maximum figures for loss of organic matter during bacterial decomposition in optimal (thermostatic) conditions are: for oak leaves in the district of the Derkul' forestry experimental station (Lugansk province), 43%; for oak and maple leaves in the same district, 45%; for oak and maple leaves in the Tellerman experimental forest (Voronezh province), 41%; and for the same leaves, but with the midrib removed, 48%.

As may be seen, saprophytic fungi show more decomposing activity than bacteria. Moreover, these data enable us to conclude that present-day tendencies to ascribe the greatest importance in the decomposition of organic matter to bacterial flora need revision rather than refinement. It seems to us that the process of breaking down organic matter and converting it into humic matter in the soil consists of a number of stages, determined by the activity of different groups of living organisms and by variation in the relationships between them. It is scarcely possible to give preference to any one of them, since all interact upon one another, and fluctuations in their activity are due not only to the kind of food but also to the degree of its preparation for acceptance by any group. It is difficult to believe that bacteria can feed directly on leaf, and especially on needle, litter. Requiring food that is more edible by them—or, more correctly, that is prepared to a certain extent by other organisms—they are not only closely interlinked with mesofauna but probably are dependent on the latter, since they infest the bodies of insects and other animals.

The intensity of their life-activity is no less due to saprophytic vegetation, since the latter is able to convert fresh organic matter into compounds more soluble and therefore more easily used by micro-organisms.

Micro-organisms, it appears to us, have on the one hand the faculty of synthesising specific humic substances from the products of breakdown of organic matter by mesofauna and saprophytic plants, and on the other hand the ability to break them down when the intensity of the 'work' of saprophytes in the soil is inadequate and the organic remains are mineralised by heat and physico-chemical processes. That phenomenon is most characteristic of chestnut, grey, and tropical savannah and treeless soils.

In northern soils under coniferous forests, as is well known, there is very little humification and accumulation of humus in spite of the abundance of organic matter. That is apparently due to inadequate development and activity of micro-organisms, on account of the unfavourable living conditions for them. With predominance of saprophytic flora in the litter, the acid decomposition products formed by them are leached out of the soil more rapidly than they can be accumulated by micro-organisms and converted into less soluble humic compounds.

That condition is probably also typical of tropical forests, where the processes of mineralisation take place vigorously, unlike the situation in northern forests.

Saprophytes have an important role in forest soils not only as agents of decomposition but also as regulators of the life-activity of micro-organisms. As is well known, much attention has recently been given to the study of the toxic compounds found in organic matter (Runov and Enikeeva 1955, Egorova 1962) and in micro-organism excretions.

These investigations have shown that tree litter contains soluble substances that in varying degree suppress Actinomycetes and bacterial flora. Therefore the picture of possible exhaustion of soils where tree stands (especially spruce) have long been growing becomes complicated. Such data, however, obtained in 'pure' laboratory experiments, are not confirmed by observations in natural conditions.

V. Ya. Chastukhin and M. A. Nikolaevskaya (1962) believe that Basidiomycetes (in the group of litter saprophytes) are able to reduce the toxicity of extracts from oak litter, thus having a beneficial effect on bacterial flora. At the same time mutual antagonism has been observed, especially in combined cultures, between microscopic fungi and cellulose-decomposing bacteria.

In this way the mutual aid of saprophytes and bacteria benefits not only the development of the latter but also the humification process. The latter, we believe, consists of three stages: (i) treatment by animals and saprophytic plants, with formation of soluble nitrogen and carbon compounds; (ii) development of a bacterial flora, which assimilates the soluble compounds and manufactures various metabolic products of a humic and growth-stimulating character; (iii) breakdown of the humic compounds by bacteria when soluble compounds run short. These stages may be concurrent or consecutive, depending on the composition of the organic matter and the phytoclimatic conditions of its decomposition.

In the same atmospheric conditions, but with differing phytoclimatic regimes beneath the canopy of stands of different composition, the rates of decomposition of organic matter are far from uniform. That is a result of the characteristics of the organic matter, as well as of the causes mentioned above, and is confirmed by the following data of E. V. Runov and I. E. Mishustina (1960):

Stand	Total no. of micro-organisms ('000 per g of dry matter)
Pine	35,351
Spruce	45,414
Oak	84,388
Larch	185,944
Birch	240,764

In spite of the large amount of litter in spruce and pine stands, they contain many fewer micro-organisms than do larch and birch stands. Oak litter occupies an intermediate place. The distribution of bacteria in litter is reflected in the soils (Table 141).

Table 141. *Number of micro-organisms in soils under forest stands* (*10^3 per g of dry soil*).*

Stand	Bacteria	
	in meat-peptone agar	in starch-ammonia medium
Pine	1700	3100
Spruce	1550	4300
Oak	2880	7400
Larch	2950	6400
Birch	3680	5700

* Mokhovoe, Moscow province.

Comparison of these data (as of many others) confirms that the quantity and composition of bacterial flora depends to a very great extent on the kind of litter and on the phytoclimatic conditions of its decomposition.

Numerous data indicate that the same species of bacteria occur in the soil in both taiga and forest-steppe biogeocoenoses, and their numbers fluctuate only within narrow limits.

Censuses of bacterial flora by total numbers and by proportions of separate groups, made by microbiologists, unfortunately do not permit determination of their biogeocoenotic role as a whole or by groups in the general circulation of matter and energy.

A. A. Rode (1954) remarked a long time ago that the chief shortcoming of work of this type is defective methods, even in purely floristic studies of soil microflora. Unfortunately that criticism is still valid, despite some advances by A. V. Rybalkina (1957), T. V. Aristovskaya and O. M. Parinkina (1961), and others, using different investigational methods.

Nevertheless even now we do not possess the data required for a quantitative evaluation of the activity of micro-organisms in the treatment of the organic matter that reaches the surface of forest soils.

The extensive activity of micro-organisms must not be regarded in isolation from that of animals and saprophytic plants. Only with a definite synchronism of the work of these groups of living organisms can the treatment of organic matter and its conversion into humic matter proceed most effectively. With any disruption of the relationships between them, the processes of breakdown of organic matter may be depressed or accelerated independently of external conditions.

Micro-organisms (especially bacteria) are of very great significance in the processes of transformation of humic matter. We must remember that bacteria feed mainly on humic matter, which for them is the principal source of energy (Vinogradskii 1952). While consuming humic matter bacteria at the same time synthesise new, more active compounds: various enzymes, growth substances, acids, etc., valuable in plant nutrition and still more so in the weathering and transformation of the mineral part of the soil. Bacteria break down minerals and convert the elements contained in them into more soluble forms. In that way they prepare the mineral nutrients essential for plant nutrition. Therefore bacteria should be regarded as participants in, and stimulators of, matter circulation in the soil-vegetation system.

Equally valuable is the role of bacteria in the processes of nitrification, ammonisation, denitrification, and fixation of atmospheric nitrogen. In forest biogeocoenoses these processes have their own peculiar features on account of the intense activity of ammonising, and the slight activity of nitrifying, bacteria.

Up to the present it has been ascertained that in forest biogeocoenoses the mineralisation of nitrogen into ammoniacal compounds predominates, and that into nitrates is very slight. It has thence been deduced that trees are adapted for preferential consumption of ammoniacal, not nitrate, forms of nitrogen compounds. That deduction, however, is contradicted by microbiological data on the presence of varying numbers of nitrifying bacteria in forest soils. Therefore another suggestion has gained ground: that nitrates may be rapidly absorbed by tree roots and so are not observed in the soil. This all indicates how little we know as yet of the mechanism of formation and consumption of soluble nitrogenous compounds in the soils of forest biogeocoenoses. We must assume that there cannot be a universally-applicable solution to this problem. It is possible that in conditions of accumulation of deep, and especially of peaty, litter the nitrifying processes are suppressed, since nitrogen exists there in combination with fresh organic matter that is unavailable for nitrifying bacteria.

With intensive soft-humus decomposition of the organic matter in litter and with sufficient accumulation of humus in the soil, the nitrifying process may be stimulated.

Works have appeared recently (Aref'eva 1963, Khrenova 1963, Aref'eva and Kolesnikov 1964) showing that the processes of nitrification and ammonisation in the soils of forest biogeocoenoses are very complex. In particular, it has been discovered that in the Pripyshminsk pine forests microbiological and biochemical activity in forest soils does not cease in winter, as at that time there is intensive accumulation of ammoniacal forms of nitrogen, reaching substantial figures (in mg/kg):

Horizon	March	October		December	
	1960	1959	1960	1959	1960
A_0	207	63	74	489	143
A_1	130	17	74	36	32
A_2	30	9	10	—	17

When litter is burned the accumulation of ammoniacal forms of nitrogen does not increase, but nitrate forms appear. The year after the 1959 fires the following amounts of nitrogen in nitrate form were found in the soil (in mg/kg):

Horizon	March	October	December
A_0	18	165	153
A_1	21	83	107
A_2	8	23	47

These figures support the view that nitrifying processes in forest soils are depressed when organic matter is only slightly decomposed. Only with mineralisation of organic remains does nitrification become more intensive.

It follows from these data that a substantial increase in the productivity of forest biogeocoenoses may be achieved by methods and measures aimed at stimulating mineralisation of litter in types with a considerable accumulation of it, especially when it is peaty. In the light of new investigations (Aref'eva 1963, Aref'eva and Kolesnikov 1964), the burning of litter (at a temperature not over 400°C) not only does not reduce the total amount of nitrogen, but accelerates its conversion into ammoniacal and nitrate forms, aiding the regeneration and growth of trees. In these cases microbiological activity is increased and chemical processes are accelerated, with release of ammonia from the crystalline framework of minerals.

We must also draw attention to the role of nodule bacteria in fixing atmospheric nitrogen. In many forest types (especially those rich in leguminous plants) that means of accumulating nitrogen and bringing it into circulation may be important.

So far we have dealt with the life-activity of micro-organisms, but we must not undervalue the role of their dead cells. They enrich the soil with more-available organo-mineral and nitrogenous compounds and also, probably, with metabolic products that directly and indirectly affect soil formation and stimulate and accelerate the circulation of matter and energy. Little investigation has as yet been made in this field, and the development of such investigations with extensive use of isotopes is an urgent task. Use of isotopic 'tagging' will permit deeper understanding of the biogeocoenotic 'work' of both animals and micro-organisms.

SUBDIVISION OF SOILS BY TYPES OF MATTER AND ENERGY CIRCULATION IN FOREST BIOGEOCOENOSES

The above-mentioned separate processes of biological accumulation of organic matter and of mineral elements contained in it cause variations in the circulation of matter and energy among plants, animals, and soils as components of forest biogeocoenoses.

In determining types of matter and energy circulation we take into account all accompanying phenomena. These include: forms of accession of materials involved in matter and energy circulation; intensity of circulation of atmospheric elements (C, O, H, N); intensity of circulation of terrestrial elements (Ca, Mg, K, Na, Fe, Al, Si, etc.) with relation to the species of organisms that create living matter and to the migration of elements due to the characteristics of the water regime of forest biogeocoenoses as a whole and of their soils in particular.

From these factors it is possible to classify all the soils of forest biogeocoenoses into types, series, and groups of types of circulation.

1. Groups of types—according to the accumulation of organic matter as determined by the varying composition of vegetation.
2. Series of types—according to the forms of accession and movement of material, with which is linked the migration of elements within the living organisms-soil system.
3. Types—according to circulation of atmospheric and terrestrial elements, linked with the rate of decomposition of organic remains and the types of water regime.

The intensity of accumulation of atmospheric elements in soils is subject to zonal-climatic laws, in consequence of which all soils under forests are combined in the following groups of types:

(*i*) forest-tundra, (*ii*) taiga-forest, (*iii*) southern-taiga (coniferous-broadleaved forest), (*iv*) forest-steppe, (*v*) steppe, (*vi*) semi-desert, (*vii*) desert, (*viii*) tropical and subtropical.

Within each group of types it is possible to subdivide forest soils into three series of types of matter and energy circulation (by forms of accession and movement of material): (*i*) biological, (*ii*) biogenic-volcanic, and (*iii*) biogenic-alluvial.

Soils of the biogenic-volcanic series are distinguished by periodic accession of mineral matter overlaying the accession of biological matter, which also affects the characteristics of migration of matter and energy.

We give below brief descriptions of the separate groups of soils of forest biogeocoenoses.

The group of forest-tundra types of soils is formed with very slight accumulation of atmospheric elements. That results in low increment of living organic matter (mostly in the form of mosses and lichens) and intensive accumulation on the soil surface of dead (peaty) organic remains, because of their slow decomposition. The circulation capacity is small; outflow of soluble products is rapid. Biological accumulation of humic matter and mineral elements in the soil is on a small scale.

The group of taiga-forest soils (including taiga-frozen) is distinguished by the ability to produce a large total mass of living matter (mainly C, O, H, N) with wood predominating over green parts. The formation of these soils is affected by the presence of dead organic remains from coniferous and small-leaved tree species. The intensity and capacity of matter and energy circulation are very high. There is great accumulation of dead organic matter; although decomposition is vigorous, conservation of the matter is clearly evident. There is considerable outflow of decomposition products, with partial accumulation of them in the lower soil horizons.

The group of southern-taiga soils of coniferous-broad-leaved forests accumulates a still greater total mass of organic matter, with trunk wood predominant. They develop under the influence of trees, with high participation by shrubs and herbage. The intensity and capacity of matter and energy circulation are still higher and nearer the optimum; prolonged conservation is not so evident. Decomposition products are held in the soil and only partly carried into the lower parts of their biogeocoenotic layer. There is well-marked accumulation of humic and mineral matter.

The group of forest-steppe soils under broad-leaved forests produces a very large mass of living organic matter. The intensity and capacity of matter and energy circulation are at the highest level. Rapid decomposition of dead organic matter results in very small accumulation of litter and intensive accumulation of humic and mineral matter. Iron and aluminium compounds are subject to migration, and carbonates of calcium and magnesium to leaching.

The group of steppe soils (under forest plantations) shows less accumulation of atmospheric elements and of living organic matter (especially roots) than the preceding group, but more than soils under herbage. Dead organic matter is subject to rapid decomposition and partial mineralisation, which leads to predominance of accumulation of humic and mineral elements and leaching of organogens ($CaCO_3$, $MgCO_3$) and gallogens ($CaSO_4$, NaCl, etc.).

The group of semi-desert soils (under artificial forest biogeocoenoses) produces a still smaller mass of living organic matter; dead matter is subject to almost complete mineralisation. Matter and energy circulation is impeded by excessive heat and deficiency of moisture. It is mainly mineral elements that are accumulated, with very little humic matter. Gallogens are leached to a very small depth, and organogens even less.

The group of desert soils (under forest vegetation—saxaul, etc.) is characterised by slow and small production of living organic matter and patchy distribution of dead matter. Circulation of matter and energy is very slight and distinctively gallogenic, with accumulation of carbonates of sodium and potassium on the surface. Leaching is very slight and irregular.

The group of subtropical and tropical soils produces a very large mass of living organic matter. Biological circulation is distinguished by great capacity and high rate of accession of atmospheric elements (with a relatively low rate for terrestrial elements). Accession of dead organic matter is at the highest level, but accumulation of it in the form of litter is at a minimum because of intensive decomposition and mineralisation. The products of mineralisation saturate soil waters with CO_2 and organic acids. Accumulation of humic and mineral elements is small (mostly iron and aluminium; very little calcium, magnesium, etc.).

Soils are subdivided according to types of exchange of matter mainly by the prevalence of groups of organo-mineral compounds that accumulate in the soil to varying extents. In this field the exchange of energy is least understood, because we still lack objective criteria for evaluating it. Investigations have to be made in this direction. The proposed division of forest soils by types of exchange of matter and their nomenclature must be regarded as provisional, requiring refinement and more thorough analysis.

Soils are classified according to the following types of matter and energy exchange:
1. Fulvatic, H-Fe-Al,
2. Fulvatic, H-Fe,
3. Humatic-fulvatic, H-Ca,
4. Fulvatic-humatic, Ca,
5. Humatic, Ca,
6. Fulvatic-allitic, Al-Fe,
7. Humatic-allitic, Ca-Al.

We give below a brief description of these types of exchange of matter and energy.

Fulvatic, H-Fe-Al type

Fulvatic compounds predominate—free and combined with aluminium and iron. Under their influence humic-illuvial ferruginous horizons are often formed. With a stagnant-flowing water regime, a large amount of the residual salts and mineral matter (except SiO_2) migrates through the profile and is often leached out. Two subtypes are defined: frost-free and seasonally or permanently frozen. In the latter, leaching is impeded by the presence of a frozen waterproof barrier; in the former, there is more intensive accumulation of organic matter on the surface. This type is characteristic of the soils of spruce and pine groups of biogeocoenose types formed in eluvial and transitional conditions.

Fulvatic, H-Fe type

Free and iron-combined fulvatic compounds predominate, stimulating outflow of the majority of bases with formation of illuvial-ferruginous horizons and partial outflow of iron and aluminium from the soil-forming layer. Two subtypes are proposed: frost-free, and seasonally-frozen with a temporarily-stagnant water regime and ferruginisation of the soil, often from the surface. This type is characteristic of taiga-forest biogeocoenoses of the north-east and east of the U.S.S.R.

Humatic-fulvatic, H-Ca type

Characterised by increased decomposition and lower accumulation of organic remains. Energy exchange is higher, with formation of free and calcium-combined humic and fulvic acid compounds, which impede the aggressive action of free fulvic acid. This type is formed in conditions of slightly-leaching and periodically-leaching water regimes, which cause less outflow of bases and marked accumulation of humic and mineral matter. It is most apparent in the southern-taiga subzone under coniferous broad-leaved forests.

Fulvatic-humatic, Ca type

Characterised by predominant accumulation of products of decomposition of organic matter, wide dispersal of terrestrial mineral matter over the soil layer but less dispersal of it through the soil profile, and strong retention of humic substances in the soil. Among these substances humates and fulvates of calcium and iron predominate, making the soil highly structural and improving aeration. This type is formed in conditions of periodically-leaching water regimes, under broad-leaved steppe forests.

Humatic, Ca type

Accumulation, decomposition, and mineralisation of organic remains are balanced, and litter forms periodically and is found on all soils. Accumulation of mineral and humic matter predominates over their dispersal in the soil layer, because of the high buffering property of steppe soils against the effects of forest plantations. That restricts utilisation of the soils in depth, as a result of the presence in them of residual organogenic ($CaCO_3$, $MgCO_3$) and gallogenic ($CaSO_4$, NaCl, etc.) compounds. This type is characteristic of the soils of natural forests and of artificial forest plantations on steppes and semi-deserts.

Fulvatic-allitic, Al-Fe type

Formed in conditions where the mineralisation of organic remains predominates over their decomposition, with release into the soil of large amounts of CO_2 and free fulvic acids, which cause intensive decomposition of the mineral part of the soil, allitisation of it, and accumulation of aluminium and iron in compounds with little exchange possibility. All other bases are largely washed away. Humic compounds of the iron and aluminium fulvate type predominate. Shortage of bases (especially calcium and magnesium) determines the characteristics of mineral matter circulation, and in particular more vigorous exchange of aluminium and iron. This type is characteristic of the soils of humid subtropical and tropical forests.

Humatic-allitic, Ca-Al type

Distinguished by considerable removal of ferro-organic compounds from circulation by leaching, and their conversion because of inadequacy of moisture into compounds of low solubility. Exchange of calcium and aluminium is at a higher level, causing formation of humic compounds combined with these elements. The soils are distinguished by higher humus accumulation and saturation with bases, resulting in predominance of biogenic accumulation. They are formed under savannahs and subtropical and tropical forests.

EFFECT OF FOREST BIOGEOCOENOSE INTERACTIONS ON SOILS AND UNDERGROUND WATERS

The mutual effects of biogeocoenoses on each other include all the phenomena caused by the penetration of roots of one biogeocoenose into another, and the passing (in different forms) of products of the life-activity of one biogeocoenose into another.

That is most clearly seen in connection with the locations of separate biogeocoenoses according to relief.

On that circumstance is based (as stated above) the division of soils of forest biogeocoenoses into eluvial (developing on watersheds independently or automorphically, i.e., without receiving any matter or energy from adjacent parts of the surface occupied by other biogeocoenoses); transitional (located on watershed slopes and bringing into circulation matter and energy not only from the layers occupied by them but also received from higher-situated biogeocoenoses, and themselves naturally passing matter and energy on to biogeocoenoses situated at a lower level). (*Translator's Note*: Apparently some lines have been omitted here in the original text.)

And, finally, these latter soils bring into circulation all matter and energy accumulated by the soil layer as a result of migration from higher-situated biogeocoenoses. They may be either accumulative-emergent or submerged.

Such interactions may occur in various ways: mechanical or physical and chemical transportation along soil surfaces of products of the life-activity of biogeocoenoses in solid or dissolved form; transportation of organic remains in the form of litter by wind, water, animals, or other agents; transportation of organic and mineral compounds in solution by surface-flowing waters. Still more important is the dispersal of life-activity products by intra-soil flow of excess moisture arriving in eluvial biogeocoenoses. That dispersal may occur at different depths and involve soil layers of varying thickness. Soil and subsoil waters moving from higher to lower parts of the relief provide a medium for very clear manifestation of these interactions.

Such phenomena are widely distributed in nature, and become more marked where there are greater differences in altitude.

Up to the present, however, these processes have been little studied either qualitatively or quantitatively. Nor do we know the laws governing such interactions in different zonal conditions, with different combinations of separate types of biogeocoenoses, and determining the qualitative composition and amount of transported matter and of the energy contained in it.

These wide gaps in our knowledge of the nature of forests should be filled by investigations in the immediate future, using all the newest methods, especially using isotopes.

These problems have been touched upon in the works of T. A. Rozhnova and L. S. Schastnaya (1959) and T. A. Rozhnova (1962), who have indicated some elements of interaction in the case of some types of forest biogeocoenoses.

The arrangement of forest biogeocoenoses according to topography, changes in soil structure, and location of underground water is shown in Fig. 76.

Analysis of underground waters (Table 142) has shown that the content of all elements and oxides increases regularly from cross-section 2 through cross-section 4 to cross-section 5. That increase is considerably higher in spring than in autumn, indicating differences in rates of leaching and of movement of cations and anions. The increase in content of these substances from cross-section 2 to cross-section 5 is a gross amount, which reflects differences in their content in the soil and inflow through leaching from higher-situated biogeocoenoses.

Table 142. *Variation in composition of underground waters in adjacent types of forest biogeocoenoses (mg/litre) (Rozhnova and Schastnaya 1959).*

Number of cross-section	Salt pH	Dry residue	$Ca^{..}$	$Mg^{..}$	$HCO_3{}'$	Cl'	$SO_4{}''$	$Na^{.}-K^{.}$	SiO_2
Spring									
2	5·35	56·0	5·0	2·0	7·3	11·4	5·8	3·5	8·5
4	5·20	51·5	2·1	1·8	4·9	5·2	5·4	4·6	12·0
5	6·35	462·5	4·79	30·4	222·0	117·2	12·4	53·1	34·5
Autumn									
2	5·01	44·0	4·5	0·9	4·8	4·6	2·0	Not determined	
4	5·22	95·0	3·5	0·6	4·2	5·2	—	Not determined	
5	5·76	111·0	13·2	4·1	59·4	5·2	2·0	Not determined	

In the soil horizons affected by underground water, pH, exchange acidity, and total exchange cations (Table 143) change from the bilberry-green-moss spruce stand to the herbaceous spruce stand with birch. This example of interactions between forest biogeocoenoses and their reflection in the composition of underground water and in some properties of the soils that interact with underground waters is very typical. It shows how important it is to study the manifestations of biogeocoenotic processes in different zonal conditions in order to understand these processes.

Having no direct data on the interactions of forest biogeocoenoses and soils, we can only note some possible variations in them arising out of the general laws of biogeocoenotic exchange.

Interactions in the soils of adjacent types of forest biogeocoenoses of the mixed-forest type will be intensified by increased input of calcium and magnesium into underground waters, by reduced input of SiO_2, Fe_2O_3, and Al_2O_3, and by even greater increase in the content of exchange cations and decrease in exchange acidity in lower-lying soils, as compared with biogeocoenoses located at a greater altitude.

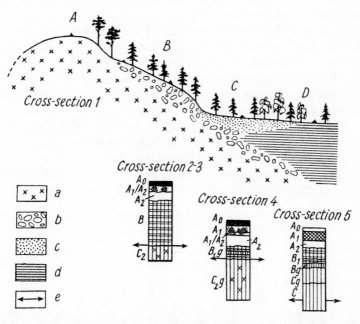

Fig. 76. Diagram of interactions of biogeocoenoses in soils along the profile from the top of a hill to a glacial-lake depression (Rozhnova 1962). Cross-section 1—primitive-accumulation soil; cross-section 2-3—peaty-strongly-podzolised ferruginous-humus soil; cross-section 4—peaty-strongly-podzolised gleyed soil; *a*—crystalline rock; *b*—coarse stony sandy loam; *c*—sand with admixture of gravel; *d*—heavy stratified loam; *e*—underground water; A—outcrop of granite with cover of fruticose lichens; B—bilberry-green-moss spruce stand; C—bilberry-sphagnum spruce stand; D—herbaceous birch stand with spruce.

In forest-steppe conditions similar interaction will be governed by the transportation by underground waters of large amounts of calcium and magnesium, and partly of Cl and SO_4, and slight transportation of SiO_2, Al_2O_3, and Fe_2O_3.

In the soils of steppe biogeocoenoses interactions will be decided by the transportation of $CaCO_3$, $MgCO_3$, $CaSO_4$, and partly of chlorites and sulphates of sodium. All other compounds are not subject to transportation because of their stable bonds with the mineral part of the soil.

In the soils of subtropical and tropical biogeocoenoses, transportation due to interactions may be extended to SiO_2, Fe_2O_3, and Al_2O_3, and partly to water-soluble humic matter. Other oxides, contained in small amounts, remain in biological circulation or may be washed out in negligible quantities.

SOIL DYNAMICS IN FOREST BIOGEOCOENOSES

Questions concerning the dynamics, or evolution, of soils in forest biogeocoenoses are still insufficiently investigated, in spite of their great scientific and practical interest. The existing scattered material and observations have up to the present been almost entirely neglected, both for correlation and for discussion.

Some general views on soil evolution have been expressed by A. A. Rode (1947), whose work has not yet been evaluated properly. He believes that soil evolution is closely linked with biogeocoenose evolution, since soil is an indispensable component of a biogeocoenose, and that the most important dynamic force of that evolution is the activity of living organisms, especially plants.

A. A. Rode distinguishes three groups of factors causing soil evolution. The first is connected with changes in factors outside of the entire biogeocoenotic envelope and the biogeocoenoses that compose it. It includes the succession of atmospheric processes in various parts of the earth and the total of changes that they cause in vegetation and animal life; changes

Table 143. *Variation in properties of lower soil horizons under the action of underground water in adjacent biogeocoenoses (Rozhnova and Schastnaya 1959).*

Bilberry-green-moss spruce stand				Bilberry-sphagnum spruce stand				Herbaceous spruce stand with birch			
Depth (cm)	Salt pH	Exchange acidity mg-equivalent	Exchange cations mg-equivalent	Depth (cm)	Salt pH	Exchange acidity mg-equivalent	Exchange cations mg-equivalent	Depth (cm)	Salt pH	Exchange acidity mg-equivalent	Exchange cations mg-equivalent
42-55	4·5	0·13	1·09	53-63	4·1	0·65	*	60-70	4·81	0·09	15·10
61-73	5·0	0·09	1·49	80-90	4·1	0·67	*	80-90	5·21	0·02	11·70
								115-200	5·24	0·03	8·30

* Not determined.

in soil erosion; and the general system of drainage, peneplain formation, etc.

The evolution of soils so produced may be considered to have a geological aspect. But even with such a series of changes, the role of living organisms as a vital force is not eliminated. To reconstruct past levels and stages in soil evolution (as in evolution of the biogeocoenotic envelope as a whole) is extremely difficult, because data are scarce and contradictory. Therefore the question of palaeobiogeocoenotic reconstruction, in spite of its importance for the understanding of contemporary soil evolution in forest biogeocoenoses, remains very obscure.

The second and third groups of causative factors are connected, according to Rode, with 'remote action of adjacent biogeocoenoses' and with 'self-development of biogeocoenoses'. That division does not seem clear to us. In the first case, as Rode says, the 'initiative' belongs to vegetation and sometimes to animals. In the second, changes in the composition and properties of soils as a result of gradual development of soil formation lead to changes in the composition of vegetation, which later in turn affect soil changes.

In both cases the leading role is evidently taken by vegetation, especially when it is considered that the soil-forming process (as A. A. Rode correctly holds) is not an independent process. Therefore it remains unclear in what sense the terms 'remote action of adjacent biogeocoenoses' and 'elementary biogeocoenose' are used. The latter probably describes separate parts, or (as N. V. Dylis says) parcels, of a biogeocoenose, in which the role of vegetation in soil evolution is the same as in 'self-development of biogeocoenoses'.

Therefore we see no basis for such a division of the evolution of soils. Understanding by 'self-development of biogeocoenoses' changes due to time (age), we cannot deny that the role of the causative factors is always displayed against the background of the dynamics of all biogeocoenotic processes. Soil evolution takes place as a result of development of all biogeocoenose components and is most intimately linked with their seasonal, annual, and secular dynamics. Therefore at this stage it is more useful to divide the causes of soil evolution into two groups: (i) those determined by constant (or uninterrupted for long periods) action of forest biogeocoenoses on soils; and (ii) those determined by interrupted or alternating action of plant formations on soils.

In the first case soil development proceeds in conditions of uniform matter and energy circulation, which change in the course of time mainly in their quantitative parameters. As a result, soil evolution does not go beyond the limits of a single type of soil formation. In the second case, diversity in matter and energy circulation (e.g., replacement of steppe by forest) leads to change in both quantitative and qualitative parameters, which causes the earlier type of soil formation to be overlaid by another type.

In other words, in the first type we are dealing with an unbroken natural series of phases of soil evolution, determined by the dynamics of the living components of forest biogeocoenoses with their characteristic processes of accumulation and migration of matter and energy within the biogeocoenose-soil system. In such cases soils evolve from primitive to more advanced stages of the same type of soil formation. In the second, one type of soil formation is replaced by another under the influence of evolution of soil properties determined by changes in vegetation or (more rarely) in animals. In this instance the soils retain properties characteristic of two or more types of soil formation.

Whereas the first type of evolution is provisionally described as constant-normal, the second is described as composite.

Another (a fourth, according to Rode) group of causes of soil evolution is connected with the development of new plant species possessing new biochemical characteristics of use and return of matter and energy, and is of value for understanding not only the geological past but also the present. It appears to us that this explains the evolution of soils in modern tropical regions. In these regions soil evolution takes place as a result both of plant phylogenesis and of features of weathering and soil formation that retain characteristics of earlier geological epochs (including the Tertiary). There forest and other plants are adapted to the resulting geochemically-different soils, and also induce types of matter and energy circulation differing qualitatively and quantitatively from those in other zones. As a result, the action of vegetation on the formation and evolution of soils differs substantially from that in other zonal conditions.

The differences, in our view, consist primarily in the prolonged retention of the mineral skeleton of the soil in the form of a thick lateritised layer, which results in complete, or almost complete, elimination of the mineral resources for soil formation—primary minerals. In this case the soils pass into a somewhat inert state, expressed in the retention and apparent invariability of soil features and properties. That state of inertness, however, is merely apparent, evoked by the short-comings of our present methods of study of soil. These methods were devised for soils in other zonal conditions, in which they are justified by making it possible to identify changes occurring as a result of mobilisation of existing reserves of primary minerals. In this case also the influence of vegetation as a factor in soil evolution is directed in the greatest degree, perhaps, towards transformation of elements contained in the soil and constant introduction of them into the matter and energy cycle, in association with fluctuations in the atmospheric component of forest biogeocoenoses. In such conditions not only the mineral composition but also the plasma[1] of the soil

---

[1] By plasma of the soil it is customary to understand the most active colloidal organo-mineral part of it, possessing physico-chemical energy and deciding the character and course of soil formation.

are altered qualitatively and quantitatively. The plasma is then in an active state, with slight evidence of elements of 'aging'.

In tropical conditions the mineral skeleton not only is deprived of the reserves that produce plasma, but is represented by secondary argillaceous minerals of the ferruginous and kaolin groups. Although the amount of plasma formed by them is considerable it is there inert, 'aging', and to some extent passing into a solid phase (ferruginisation, lateritic crust, etc.). In these conditions the activity of vegetation (especially forest vegetation) is directed to a considerable (if not a preponderant) degree to the 'revival' of plasma, to restoration of its activity and to concentration of the elements whose content in tropical soils is very small and thinly dispersed, especially calcium, magnesium, and potassium. At the same time the vegetation induces the conversion of iron (included in unavailable, almost insoluble compounds in secondary argillaceous minerals, or in various kinds of concretions) into more soluble ferro-organic compounds, very active and fulfilling the same, or almost the same, role as calcium in the soils of other zones.

In this way the role of vegetation and animals as factors in soil evolution in tropical conditions is very complex. It is largely determined by the composition and nature of the organic matter produced by them, since the products of decomposition are vectors of the energy accumulated in soil plasma. It is that energy that causes the evolution of soils and the maintenance of their fertility at a definite level. At the same time one may find a certain analogy between the formation and evolution of soils in tropical and in temperate regions. Thus, with changes in vegetation (in time and space) from humid tropical, through deciduous, to savannah forests, an evolution of soils towards 'steppe-ness' takes place (transition from yellow-earth tropical forest soils into red-earth forest soils and cinnamon-red savannah-forest soils). This, to a certain extent, repeats the pattern of transition from podzolic taiga-forest soils through podzolised-leached to leached forest-steppe soils. These questions, however, require more thorough and independent examination, which is beyond our present scope.

We have touched on this subject because humid-tropical forest biogeocoenoses, and consequently their soils, represent a clear example of their transformation into a dynamically-unstable system, or the so-called climax stage. Later that stage will be replaced by a stage of natural breakdown of such biogeocoenoses and as a result soil development will cease. In fact, when such forests are destroyed by man they do not regenerate naturally and the soils are often transformed into laterite crusts. Formation of the latter merely indicates that as yet we have not understood sufficiently the laws of development of natural formations, believing that the results of our actions (of any intensity) upon the components of natural forest biogeocoenoses (to the point of complete destruction of the living organisms composing them) can be overcome, and the biogeocoenoses restored. The example of tropical forests is visual proof of the necessity of knowing the laws of development of natural resources before we exploit them. If we lack that knowledge, formations of no value to agriculture or forest economy may develop, instead of the original biogeocoenoses.

The above statement is true not only for tropical but for all other zonal conditions. The differences consist only in the nature and direction of soil evolution. In the tropics destructive changes are more obvious and more difficult to overcome; in other zones restoration of the original conditions is less difficult, but it still requires more effort and resources than are expended thereon by nature, especially with intelligent co-operation by man. In this connection we should start from the premise that soil development proceeds uninterruptedly in forest biogeocoenoses, although at different rates in different stages. We should strive to avoid checking natural soil development—on the contrary, we should always assist it. Interference in soil development should be based on knowledge of the general laws of soil evolution in forest biogeocoenoses. Identifying ourselves with this trend in science as the most progressive, and denying the existence of a climax stage in soil formation, we (following V. N. Sukachev and A. A. Rode) hold that the evolution of forest soils is in all cases both a cause and a result of the evolution of forest biogeocoenoses.

On the basis of the above statement about soil evolution we shall try to survey in greater detail some aspects of soil evolution in forest bio-geocoenoses in the boreal zone. In this connection we deal separately with the following very important groups of phenomena: (*i*) the dynamics of soils on the first occupation of territory by forest vegetation, and (*ii*) the dynamics of forest soils resulting from the development of forest biogeocoenoses.

SOIL DYNAMICS ON FIRST OCCUPATION OF TERRITORY BY FOREST VEGETATION

As remarked above, the formation and development of soils on their first occupation by vegetation of soil-forming species takes different courses on friable and on compact strata.

On friable strata, homogeneous and relatively rich in nutrients, soil de-velopment may begin at the moment of colonisation by pioneer trees. A good example is the evolution of volcanic ash (Zonn *et al.* 1963).

The first stages of soil formation on such friable strata begin at the moment of their occupation by single larch trees (*Larix kurilensis*). In these conditions and generally on friable soils, especially with some shortage of moisture, the pioneers of soil formation are tree species with root systems that develop rapidly and deeply, and therefore are better able to satisfy their requirements of moisture and nutrients.[1]

[1] When the strata are rich in nutrients, and especially in moisture, colonisation not by single trees but by large numbers of seedlings is possible.

The next stage begins with colonisation of the ground under the tree crowns by mosses and lichens, which arrive after a shallow (not more than 0·5-1 cm) layer of larch litter is formed under the crowns. Its high moisture content makes the establishment and growth of mosses and lichens possible. Later the depth of litter is increased by the dying-off of parts of these, which improves the conditions for development of the larch population in the occupied area. An open forest canopy is created, changing the atmospheric component into a phytoatmosphere. After the mosses, another pioneer—cowberry—becomes established under the tree canopy, while the litter increases to a depth of 2-3 cm. More moisture accumulates in the litter, aiding the formation of a denser larch stand of higher quality class. As time goes on *Ledum* (wild rosemary) invades the

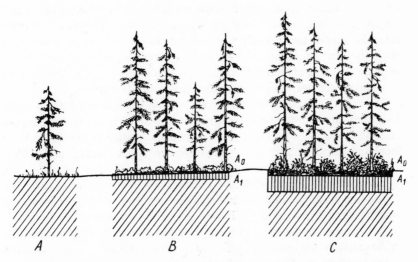

Fig. 77. Stages of formation of larch biogeocoenoses on volcanic ash (Kamchatka). A—scattered larches; B—cowberry larch stand; C—*Ledum* larch stand.

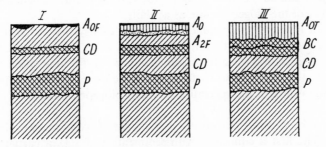

Fig. 78. Diagram of development of soil profile on volcanic ash in Kamchatka when occupied by larch. Horizon A_{0F}—fragment litter; horizon A_{2F}—fragments; horizon A_{0T}—dry peaty litter; P—ash layer; I-III—stages of soil development.

area under the canopy, producing a large mass of above-ground organic matter and helping to form still deeper (up to 4-6 cm) litter of coarse-humus or dry-peaty type. That corresponds to the complete development of a typical larch-*Ledum* forest biogeocoenose, in which all components enter the phase of full interaction.

The stages of soil development in such conditions are diagrammatically shown in Fig. 77. From it we may see, in the first place, that soil development is intimately linked with formation and accumulation of forest litter as a distinctive forest component ensuring improvement of forest-growing conditions (increasing moisture supply and the availability of mineral elements through the action of organic acids in the litter). In the second place it is clear that soil development consists of definite stages. The first stage, which involves the establishment of scattered larch trees, may be called fragmentary-biogeocoenotic, when the action of living organisms on soil formation is just beginning. The second is the formation of forest litter, leading to the formation of incipient forest soil. Then in the third stage the final formation of a forest biogeocoenose takes place, with its appropriate cycle of matter and energy, which determines the development of a soil profile corresponding to the type of soil formation.

The rate of soil development increases from the first to the third stage. Its future rate of progress is governed by the amount of moisture supply and the intensity of decomposition of organic matter in the litter. With increased moisture and heat, soil formation is accelerated, and with a decrease in moisture it is slowed down.

Development of the soil profile corresponds to these phases (Fig. 78).

The first stage corresponds to volcanic-fragmentary soil, the second to volcanic slightly-podzolised soil, and the third to volcanic coarse-humus slightly-leached forest soil. The incipient podzolisation of the soil, arising in the second stage under the action of mosses and lichens, is not observed in the third stage. That is due to two causes: (*i*) weakening of the action of mosses and lichens; and (*ii*) arrival of new layers of ash, which enrich the soil.

Further soil development may follow various trends, depending on the dynamics of the biogeocoenose as a whole.

Such a type of soil dynamics shows very clearly that its chief moving force is vegetation and the litter formed by it, as a component that not only acts as an intensive transformer of the soil-forming strata but also improves the supply of water to plants and ensures greater complexity in the composition of the vegetation in forest biogeocoenoses.

On compact strata there are often two types of evolution or evolutionary series. The first is connected with sedimentary strata, in particular with limestones. Occupation of the strata begins with herbage, with formation of fragmentary soddy-carbonate soil (Fig. 79, I). In the second stage scattered pines take root and the soil becomes shallow soddy-humus-

carbonate soil (Fig. 79, II). Later, under the influence of developing cowberry-pine stands, a transition to typical humus-carbonate soil takes place,
which (depending on the thickness of the fine-grained layer) is somewhat
leached (Fig. 79, III). The thicker the layer, the greater is the leaching.
As the thickness increases to 60-70 cm the soil may evolve into brown

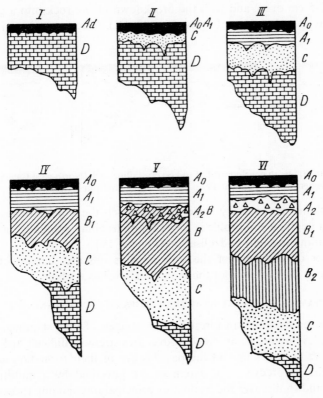

Fig. 79. Stages in soil development on compact sedimentary strata
(limestones). Explanation in text.

leached or podzolised soil, on account of invasion of the pine stands by
spruce. That stage may vary in duration, increasing from the taiga-forest
to the forest-steppe zone (Fig. 79, IV). In the following stage transition
to zonal soil, podzolic or grey forest soil, takes place (Fig. 79, V). Later,
evolution slows down and tends towards stronger leaching or podzolisation (Fig. 79, VI). But these processes are regulated by the evolution of
the biogeocoenoses, i.e., by the succession of vegetation in different layers
or throughout the biogoecoenoses. Such a series in soil development is
characteristic of all districts where soil formation takes place on carbonate
strata, especially on limestones (Silurian plateaux, etc.).

The second type is found on crystalline strata (granites, gneisses, etc.).

We know particularly of a development series beginning with a flat granite surface and ending with formation of brown forest soil under spruce, in the North-west Caucasus. It begins with formation of cover of crustose lichens (Fig. 80, stage I). That first stage passes into a second—a moss-lichen carpet with scattered birch, poplar, mountain ash, etc. (Fig. 80, stage II). With the formation in that fragmentary biogeocoenose of soil 2-5 cm deep, and with the breakdown of the rock into a gravelly mass, invasion by spruce begins (Fig. 80, III). The formation of mossy

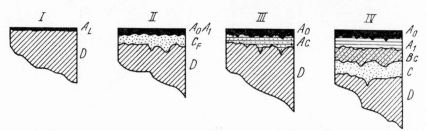

Fig. 80. Stages in development of soil on crystalline rock (North-west Caucasus). A_L—lichen; C_F —fragmentary (skeletal). Explanation in text.

spruce stands is a lengthy stage, during which the soil continues to increase in depth, a soil profile is formed, and leaching gradually increases, often leading to slight podzolisation (Fig. 80, IV).

Further development of the soils remains obscure, but doubtless is linked with the evolution of the biogeocoenoses as a whole.

SOIL DYNAMICS IN FOREST BIOGEOCOENOSES OF LONG STANDING

To trace the dynamics in forest biogeocoenoses of long standing, which are still developing, is, as stated above, an extremely difficult task, since usually we do not know all the past history of their formation and development. Moreover, soil dynamics are governed by a multitude of factors, and to discover the leading or principal ones among them is also very difficult. Therefore we limit ourselves here to a brief survey of some trends in soil dynamics in forest biogeocoenoses.

With increase in the age of biogeocoenoses and simultaneous slight change in stand composition, certain soil characteristics also change, gradually altering and complicating the profile. As a rule, however, there is no overlaying or gradual transformation of one soil type into another.

Changes are mostly restricted to formation of new subtypes, genetically interlinked, and reflecting stages of the development of biogeocoenoses as they grow older.

We may quote two examples of such changes: one from the forest (taiga) zone, the other from the forest-steppe zone.

In the forest zone, with podzolic soils washed by an invariable or long-lasting water regime, podzolisation is intensified and the thickness of the structural (illuvial) horizon B increases. For instance, the soddy-podzolic soils of composite spruce stands may pass into coarse-humus podzolic soils of green-moss spruce stands. Further increases in the degree of podzolisation and in the thickness of horizon B are due to the evolution of spruce stands of the green-moss group into green-moss-bilberry spruce stands.[1] That kind of change in soil is illustrated in Fig. 81. Later, as horizon B becomes more water-permeable, the water supply in the upper soil layers (including horizons A_0 and A_2) increases, leading to improvement in the abundance of the bilberries and change in the cover from

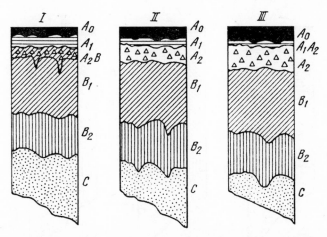

Fig. 81. Stages in soil development resulting from succession of types of forest biogeocoenoses. I—soddy-podzolic soil of composite spruce stands; II—podzolic soil of green-moss spruce stands; III—podzolic soil of green-moss-bilberry spruce stands.

green mosses to sphagnum. In such a case the podzol or strongly-podzolised soil may evolve into gleyed-podzolic; the depth of litter increases, it becomes peaty, and the thickness of horizon A_2 and partly that of horizon B decrease on account of the gleying.

In the forest-steppe zone dark-grey soils developed on carbonate loam may evolve in two directions. With normal soil drainage and a relatively constant water regime the upper boundary of the carbonate horizon is lowered, the thickness of structural horizon B increases, and the thickness of horizon A is reduced, with a simultaneous rise in its humus content. In other words, the soils evolve from less-leached to more-leached, to the point of being weakly-podzolised, as is seen from the diagram of soil-profile development (Fig. 82). In this case the depth of the carbonate horizon changes extremely slowly, because the depth of soil wetting is constant or may even decrease slightly because of deterioration in the filtration of moisture through horizon B. Further changes take place in

[1] With forest-cutting and transformation of forest into meadow biogeocoenoses, or with cultivation, evolution into soddy-podzolic soils takes place.

the upper (mostly humus) layer of the soil, resulting from the dynamics of the biogeocoenose composition.

With better drainage, and watering of the soils by the same amount of precipitation as in the preceding case, soil development may involve more intensive leaching of humic matter and podzolisation of the upper soil layer above the carbonate, as far as it is subject to drainage, without change in the depth at which the carbonate horizon lies. In this case dark-grey soil may be converted into grey podzolised forest soil, with an

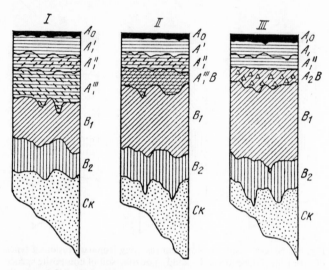

Fig. 82. Stages in development of dark-grey forest soils on account of evolution of forest biogeocoenoses. I—sedge oak stand; II—goutwort oak stand; III—birch-linden oak stand.

accompanying change in the composition of the biogeocoenose—the numbers of linden, birch, and elm rise while the proportion of oak falls.

Similar evolution is observed on narrow watersheds between valleys or ravines, and may be illustrated by diagrams of profiles (Fig. 83).

In the taiga zone evolution of soils is fairly widespread, because of changes in biogeocoenose composition under the influence of individual components. For instance, increase in water supply may induce not only changes in soil characteristics but also thinning-out of tree stands by various natural methods. It may take place with additional inflow of moisture from adjacent areas occupied by other biogeocoenoses, including non-forest biogeocoenoses.

In such cases soil evolution results in the superimposition of gleying on podzolisation, with formation of podzolic-gley soils that are gleyed and podzolised to different extents.

Soil evolution is usually accompanied by a succession of phytocoenoses and ultimately by formation of new types of forest biogeocoenoses. This trend of evolution may be illustrated (Fig. 84) by the transition from the soddy-weakly-podzolised soil (I) of a herbage-fern spruce stand into the medium- or strongly-podzolised soil (II) of a bilberry spruce stand and the strongly-podzolised-gley soil (III) of a bilberry-sphagnum spruce stand. We do not, however, possess any reliable data with which to answer the question of how rapidly the transition from one stage to another takes place.

With relative constancy of the tree layer and changes in the shrub and

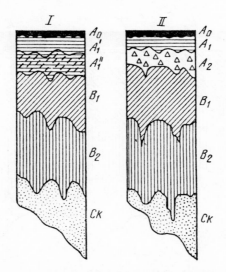

Fig. 83. Transformation of dark-grey forest soil (I) into grey podzolised forest soil (II) with increase in natural drainage.

herbage layers, soil dynamics have their own specific features, which very often are not taken into account in forest practice.

Some people think that in forest biogeocoenoses the leading and even the only role in soil dynamics is played by tree stands. Their opinion is based largely on visual appraisal of the (possibly powerful) action of tree stands on soil, and not on factual data.

That view of tree stands resulted in many objections to the typological trend towards classifying forest types not only by trees but also by their accompanying vegetation. These objections originated in lack of knowledge of the laws of development of biogeocoenoses and especially of their soils. Existing data (admittedly few) on soil development in interrelated series of biogeocoenoses indicate that very often not only the tree layer but also the low-growing and shrub layers substantially change the characteristics of soils and ultimately their development. For instance, N. P. Remezov (1952) allotted more importance to moss vegetation than to spruce litter in soil formation. Weakening of podzolisation, on the other hand, is due to invasion by shrub and herbaceous vegetation, etc. According to Zonn (1960), with the same tree canopy but with change in the composition of the low-growing and shrub layers, a regular and concomitant variation in the composition of humic matter and exchange cations takes place (Table 144). These variations are so substantial beneath high-mountain spruce and beech stands that we cannot disregard them as indicators of soil development. In all types, an increase in the herbaceous and shrub vegetation lowers acidity and raises the content of

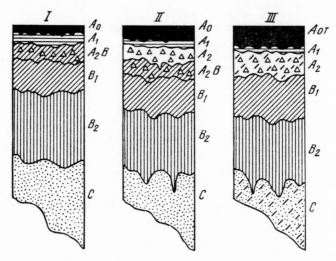

Fig. 84. Stages in transition from soddy-podzolic to strongly-podzolised gley soil with change in surface water supply and succession of spruce stands. Explanation in text.

humic acids, which are the acids most stably combined with the mineral part of the soil. Therefore in study of the dynamics of soils as components of forest biogeocoenoses it is essential to take into account changes in the vegetation of the low-growing and shrub layers.

Soil evolution in biogeocoenoses that are unforested but are affected by forest phytocoenoses was circumstantially described by T. I. Popov (1914). He was one of the first to point out that, at a definite stage of the desalinisation of solonetz on the lower slopes of watersheds, poplar colonises them and transforms solonetz into solodised-podzolised soil. Later, when poplar is replaced by birch, elm, and finally by oak, zonal grey and dark-grey forest soils are formed.

Similar evolution of solonetz soils on the old terraces of the Khoper river has been described by Zonn (1947). The following stages are seen there: solodisation, podzolisation, and formation of dark-grey residual-podzolised soils. I. V. Tyurin (1930) has interestingly and circumstantially described the transformation of grey forest soils into leached chernozem and the reverse process. Finally, artificial creation of forest strips on different soils (from leached chernozem even to chestnut soil), widespread during the last 90-100 years, is another clear example of evolution of the soils of steppe biogeocoenoses under the influence of forest vegetation into forest-biogeocoenose soils, unique in their origin and in their evolution.

In spite of views held earlier, soils typical of natural forest biogeocoenoses (in the sense of their podzolisation) have not been formed there, but there is reduction in leaching of carbonates, some decrease in the thickness of humus horizons, increased humus content, and improvement in the structural and physical properties of the soil layer occupied by roots. Soil evolution therefore is directed towards formation of unique chernozem-forest, chestnut-forest, and similar soils. It is difficult to forecast the direction of their future development, but, taking into account the characteristics of matter and energy circulation, we can scarcely suppose that it can lead to podzolisation.

The above examples show clearly how multiform and diverse the paths of soil development in forest biogeocoenoses can be, and how important their study is as a scientific foundation for measures to transform forests and increase their productivity, and equally to diminish the transforming role of the entire biogeocoenotic cover of our country.

We must add that anthropogenic factors are constantly becoming more

Table 144. *Variation in humus composition and in content of exchange cations in soils under the influence of vegetation in the low-growing and shrub layers (horizon A).*

Type of forest biogeocoenose	Total Carbon	Carbon (% of total carbon taken as 100)					$C_H:C_F$	Content of exchange cations, mg-equivalent				Degree of Unsaturation (%)
		Wax-resin	Soluble in 1·0 g Na_2SO_4	Humic Acids	Fulvic Acids	Unhydrolysable residue		C_A	M_G	H	Total	
Spruce stands on dark-coloured soils (Bulgaria)												
Grassy	8·41	0·8	1·9	22·7	28·3	45·1	0·80	—	—	—	—	—
Green-moss bilberry	5·69	6·5	1·4	20·2	29·2	41·3	0·63	2·1	0·4	14·2	16·8	85
Bilberry	8·67	8·5	1·8	17·9	28·5	41·5	0·63	2·0	0·6	20·2	22·3	89
Juniper	6·44	1·0	2·8	18·8	35·8	40·6	0·53	0·6	1·0	6·4	8·0	80
High-mountain spruce stands on brown forest soils (Eastern Tibet)												
Moss-bamboo	5·89	3·7	1·2	41·0	15·4	33·5	2·64	10·7	4·7	9·2	24·6	35
Herbaceous	12·81	4·4	1·6	22·5	25·4	44·9	0·88	11·0	3·6	3·6	18·2	19
Moss-sphagnum	4·40	8·0	1·4	10·4	38·0	37·0	0·28	2·7	1·2	13·8	17·7	77
Beech stands on brown forest soils (Bulgaria)												
Dead-cover	2·30	5·2	2·2	13·0	35·5	46·1	0·39	4·7	1·8	4·5	11·0	40
Fern	2·98	4·0	2·7	18·5	34·2	40·6	0·54	8·9	1·1	None	10·0	—
Laurelcherry	2·25	2·7	2·2	20·4	32·0	42·7	0·63	7·4	1·7	None	9·1	—

influential in soil evolution: the irreversible destruction of forest biogeo-coenoses, their artificial restoration, fires of human origin, etc. That aspect of soil evolution has been little studied, because (among other reasons) very often separate stages of brief duration leave no traces in the soil.

Questions regarding soil evolution under the influence of anthropogenic factors require independent study and discussion.

GENERAL APPRAISAL OF THE ROLE OF SOIL IN A FOREST BIOGEOCOENOSE

The soil of forest biogeocoenoses is not only a substrate—a habitat for plants and animals—but also a very active biogeocoenotic component. Its activity is primarily due to the fact that the growth and formation of the 'crop' of forest vegetation depends largely on the properties of the soil, which govern interactions established between higher plants, litter, soils, and the animals and micro-organisms living in them. Whereas the annual supply to the soil surface of an enormous mass of dead organic matter—a potential source of heat energy, mineral matter, and organic compounds of various kinds—is provided by vegetation, its conversion into positively-acting matter and other forms of energy is accomplished by animals and micro-organisms.

Therefore the role of soils in forest biogeocoenoses cannot be assessed solely on the basis of their properties and their nutrient content, as is widely done with regard to soils used in agriculture. Forest soils may potentially possess adequate favourable properties and the requisite amounts of nutrients, and yet not provide the conditions required to produce a forest 'crop'. That may be due to various causes, including weakness of the activity of animals and micro-organisms, which fail to give proper treatment to the products of plant life-activity; these products then begin to have unfavourable effects on the soil and to cause un-desirable changes in it.

All of these processes are observed, to some degree, in unforested bio-geocoenoses. They are halted annually, however, by natural means (in annual phytocoenoses) and can be comparatively rapidly and correctly altered by human interference. The soils of forest biogeocoenoses are subject to secular action by forest vegetation and therefore acquire distinctive tendencies and manifestations of interaction processes peculiar to themselves.

In this way forest soils reflect in themselves and in their characteristics the many-faceted processes that take place in forest biogeocoenoses during not one but many generations of forest and give rise to a succession of biogeocoenotic types. That succession takes place not only as a result of natural development of forest vegetation, but also as a result of ele-mentary phenomena (fires, etc.) and of human interference (cutting, burning, etc.). Residual features of these phenomena exist in soils in a 'preserved' state. Discovery of them and revelation through them of the chief stages of development and succession of various processes constitute one of the still-unturned pages of biogeocoenotic knowledge of soils. Such investigations may bring to light valuable features of the complex and multiform paths of soil evolution linked with successions of forest vegetation.

It is no less important to discover the significance of the separate components in creating fertility in forest soils. Fertility is not a direct function of the mineral composition of the soil. The latter is only one component of fertility, brought into action by the activity of living organisms in conversion of mineral into organo-mineral matter.

Thus the fertility of forest soils is a function of complex biogeocoenotic processes, and knowledge of it is indissolubly linked with knowledge of the history and development of biocoenoses. Soil, consequently, stands out not only as a factor governing the composition and structure of biocoenoses, but still more as a natural formation reflecting in its characteristics and its fertility the principal stages in changes in the life-activity of living organisms.

That role of soil is very clearly demonstrated in establishing types of circulation of matter and energy. Soil takes a leading part in them, and all the features of consumption, return, and outflow of the compounds and substances that characterise every type of matter and energy circulation are connected with the soil.

It follows that soil is in no way a stable and passive component of forest biogeocoenoses. It continually changes under the action of other biogeocoenose components, and itself actively affects them.

That property of soil gives it special significance in the life and formation of biogeocoenoses. Soil is the source of various synthetic growth substances manufactured by soil micro-organisms and representing products of a metabolic character. They are created and accumulated in the soil as a result of varying trends in matter and energy circulation. In this respect the circulation of organic matter is most important as the source of various carbon compounds that constantly act upon the mineral part of the soil, with participation by animals and especially by micro-organisms.

As we see, the biogeocoenotic representation of the processes that take place in forest soils is not limited to formal knowledge of their separate features; it calls for study of the processes whereby the soil is enriched with metabolic products. The latter have varying effects on forest vegetation and play no small role in the production of its organic mass.

We have remarked that the soils of forest biogeocoenoses, although they have features of similarity with the soils of biogeocoenoses of other plant formations (steppes, deserts, and bogs) also show great differences from them, resulting from the nature of the interactions between soils and other biogeocoenose components, especially differences in biogeocoenose structure and dimensions.

The soils of forest biogeocoenoses are formed in conditions in which the atmosphere is transformed by plants and acquires the characteristics of a phyto-atmosphere. In all other biogeocoenose formations such changes are only slightly manifested, and soil formation is connected mainly with general atmospheric conditions, so that differences occur in the water, heat, salt, and other regimes. Another distinction is that soils of forest biogeocoenoses are as a rule characterised by a combination of processes of transformation of organic matter, in which the leading role is taken by decomposition regulated not only by forest canopy but also by soil-dwelling animals and micro-organisms. In the soils of steppe and desert biogeocoenoses mineralisation of organic remains predominates, precluding their accumulation on the surface in the form of litter. In the soils of boggy biogeocoenoses, on the contrary, accumulation of organic mass is predominant because of the weakening of decomposition processes.

As a result of such differences in the nature of accession of organic matter to the soil, there is substantial variation in types of matter and energy circulation.

In the soils of steppe and semi-desert biogeocoenoses the exchange of mineral matter, and in boggy biogeocoenoses the exchange of organic compounds of an acid nature, are intensified. Destruction of forests leads, as a rule, to changes in the character of circulation of matter and energy; in the taiga zone factors of bog formation, and in the forest-steppe and more southerly zones mineralisation factors, become more prominent therein.

The above are the principal differences in the conditions governing soil formation in different types of biogeocoenoses in the biogeosphere.

They show that in forest biogeocoenoses soil processes and soil fertility are in the highest degree inter-related with the interactions of all components. In all other biogeocoenoses these interactions are to some extent weakened, some being stimulated and others depressed. In boggy biogeocoenoses the activity of micro-organisms is decreased and the effects of water are heightened; in steppe and desert biogeocoenoses, on the other hand, water supply is reduced and the effects of radiant energy are stronger, causing periodic suppression of the 'work' of animals, micro-organisms, and even plants. All these factors are reflected in one way or another in the soils by altering their properties, especially their formation and fertility.

At the same time any disruption of the biogeocoenotic interactions determined and regulated by vegetation has a substantial effect on the soils of forest biogeocoenoses, since any destruction of forest cover requires a long time for its restoration. The duration of the interactions creates such mutually-dependent, profound, complex, and specific inter-relationships that it markedly differentiates forest soils from the soils of other types of biogeocoenoses.

That is why soil study is so important not only for the development of forest biogeocoenology but also for the solution of the practical problems facing sylviculture and forest economy.

These problems may be formulated as follows:

1. Soils, which express the characteristics and types of matter and energy circulation, form an objective criterion for determining types of forest biogeocoenoses and a basis for delimiting them.

2. Classification of soils by differences in matter and energy circulation is of special value. Such an approach provides better knowledge of the processes and phenomena that decide the course of development of forest biogeocoenoses and their production of organic matter of varying quality and quantity.

3. The very complete expression in soils of the types and peculiarities of matter and energy circulation in forest biogeocoenoses permits more thorough study of such important sylvicultural problems as succession of species, forest regeneration, creation of forest plantations on soils with different types of matter and energy circulation (steppes, deserts, bogs), application of various forestry measures, selection of tree species, etc.

4. The above-mentioned possibilities of understanding the nature of forests in order to manage them call for radical revision of present ideas on the role of the various natural components in forest life. This refers, above all, to soils. Their study should be systematic and obligatory, in contrast to the present attitude towards it; only in exceptional cases does it receive attention in sylviculture, and then only in order to discover certain features of soils by using methods that often fail to correspond to our present level of knowledge. The study of soils should be based on recognition of their outstanding role among forest biogeocoenose components.

5. Soil study should be directed primarily towards the discovery and classification of the features and properties that determine variations in the circulation of matter and energy, and of the role of the soils themselves in these variations. Without knowledge of such processes and of the soil characteristics created by them, it is impossible to achieve management of biogeocoenotic processes and thereby rational utilisation of forest resources.

6. Investigations should be directed, not so much to the study of individual features of soils and their effect on tree stands in various types of biogeocoenoses, as to systematic study within forest biogeocoenoses of (i) elements and types of matter circulation; (ii) the energy aspect of matter circulation; (iii) the dynamics of soil processes; (iv) mutual influences of separate biogeocoenoses.

Study of all these questions should be carried on with use of all the latest methods, including the use of isotopes, which offers the highest possibilities for discovery of the secrets of natural processes and for revealing the intimate aspects of the interactions of soils with plants and animals.

CHAPTER VII

Dynamics of forest biogeocoenoses

CONCEPT OF SUCCESSIONS AND OF DEVELOPMENT OF THE BIOGEOCOENOTIC COVER OF THE EARTH

A forest biogeocoenose, as we have seen, is a complex unity: all of its components interact upon one another, and thereby exist in a state of interdependence. By their direct action these inter-relationships alter each component to some extent and change its characteristics in one way or another, and consequently the interactions of the biogeocoenose components are altered. The result is that no forest biogeocoenose remains, even for a brief period, perfectly stable and unchanged. Forest biogeocoenoses are always changing to some degree; sometimes these changes are so rapid that we can easily record them, sometimes so slow that we do not notice them.

We must also remember that the environment, whether forest or non-forest, surrounding each forest biogeocoenose is also constantly changing. These environmental changes in turn act upon every component of the forest biogeocoenose and therefore upon their interactions, i.e., they alter the biogeocoenose as a whole to a greater or lesser extent.

As early as 1891 S. I. Korzhinskii wrote that the contemporary state of the vegetation in any country is merely one of the stages in continual alteration of its plant cover, the outcome of past conditions, the beginning of future ones; that statement of his, in view of what we have said above, can be applied in full to any biogeocoenose as a whole.

Materialistic dialectics, however, teach us that not every change in an object or a phenomenon is a development of it. In this connection V. I. Lenin has written: 'A condition of our knowledge of all processes in the world in their "*self-motion*", in their spontaneous development, in their vital life, is knowledge of them as contradictory unities. Development is a "conflict" of contradictions. Two fundamental (or two possible? or two observed-in-history?) concepts of development are: development as a decrease and increase, as a repetition, *and* development as a unity of contradictions (a bifurcation of a single unit into mutually-exclusive contradictions and the inter-relationship between them).

'The first concept of movement leaves in obscurity its *self*-movement, its *moving* force, its source, its motive (or that source exists *outside*—god, subject, etc.). With the second concept, it is to knowledge of the *source* of "self"-movement that attention is chiefly directed.

538

'The first concept is dead, pallid, arid. The second is alive. *Only* the second provides the key to "self-movement" of all existing things; it also provides the key to "leaps", to "a break in gradual continuity", to "conversion into contradiction", to annihilation of the old and emergence of the new.' (V. I. Lenin, *Works*, vol. 38, p. 358.)

Therefore we may describe the changes in any object or phenomenon of nature only as 'development' when they are the consequence of contradictory processes taking place within the object or phenomenon itself. Although a biogeocoenose is an open system, all of its components together still form a certain integral dialectical unity, characterised by internal contradictory interactions, which never produce a state of equilibrium within that unity (system). On the contrary, these internal contradictions are a force that produces constant movement in it, which is its development. If a change takes place in a biogeocoenose as a result of action upon it by any force from outside, from its surroundings, from the environment in which it exists or from more distant objects or phenomena, that change, in accordance with the principles of materialistic dialectics, cannot be called development of it. Changes in biogeocoenoses under the influence of such forces are usually to some degree disruptions of its normal development. Like the processes of development, however, they are subject to definite laws, which we shall discuss below. Knowledge of them is quite indispensable.

The normal process of development of forest biogeocoenotic cover does not always lead to increase in its economic value, in its productivity. If we know the laws of development, we can often guide them along paths leading to a biogeocoenose more productive or, on the whole, more valuable for forest economy.

Before discussing the different forms of the dynamics of forest biogeocoenoses, we must take the following into consideration.

As soon as the bare surface of the earth appears and begins to be occupied by organisms, interactions between them and inanimate nature arise. Inanimate nature (air, water, rock strata) forms the organisms' environment, and at the same time (in certain parts of it) an essential condition of their existence. As the population density of the organisms increases in a given territory, there also arises a process of interactions of the organisms among themselves, whose intensity constantly increases, in accordance with the continuing increase in population density. All of these interactions are accompanied, as we know, by a process of matter and energy exchange both between the organisms and their environment and among the organisms. Even before the appearance of living matter a process of interaction among the elements of inanimate nature had been taking place in that territory, a process that also consisted in the exchange of matter and energy among these elements on that part of the earth's surface; but now it acquires much greater complexity and usually greater intensity, and becomes qualitatively different. This new distinctive type of matter and energy exchange, which is characteristic of a biogeocoenose, may be called the biogeocoenotic process.[1] That process continues uninterruptedly as long as organisms exist in the territory.

One of the most important features of the biogeocoenotic process is the fact that the interactions of biogeocoenose components lead to migration of matter (and with it, inseparable energy) both within a biogeocoenose and beyond its borders, producing, as stated above, changes in the components themselves, and concurrently changes in the biogeocoenotic process. While organisms exist, energy (being inexhaustible) is a source of constant movement, which ensures incessant movement also in the biogeocoenotic cover. The biogeocoenotic process includes not only interactions and exchange of matter and energy between biogeocoenose components, but also interactions and exchange of matter and energy between biogeocoenoses and their surroundings—the environment in which they exist, and other biogeocoenoses (both those immediately adjacent and those more remote). Such interactions extend to distances where there are no living organisms, where there is only inanimate nature. Since the process of interaction of a biogeocoenose with its environment is partly expressed in incessant outflow of energy into space, it has, as it were, an entropic character. But at the same time new matter and energy are constantly entering the biogeocoenose. Solar radiation is the most important item in the influx of energy. It is the ultimate cause of the whole biogeocoenotic process; a biogeocoenose receives only part of its energy directly from solar radiation, and the rest through the entrance into it of components of other adjacent or distant biogeocoenoses, which in their own time and place have received energy from the sun. Since the forms and rates of action of that multiform exchange of matter and energy depend on the characteristics of the biogeocoenose components, on the characteristics of living organisms, and also on air, water, and geological strata (mineral matter), and on the spatial combinations and distribution of all these, i.e., on the structure of the biogeocoenose and its environment, in each biogeocoenose the biogeocoenotic process takes place in its own way and has its own specific character. Biogeocoenoses similar in composition and structure (which usually means similar also in environment) have similar biogeocoenotic processes. In other words we may say that each type of biogeocoenose has its own appropriate type of biogeocoenotic process; and since there are very many types of biogeocoenoses in the world, the types of biogeocoenotic processes also are very diverse.

As stated above, the biogeocoenotic process continually induces changes in biogeocoenose components, leading to conflicting inter-

---

[1] Sometimes the term 'physico-geographical process' is used with a similar meaning. It is better, however, to confine that term only to the aggregate of processes taking place within inanimate nature on the earth's surface, without participation by organisms.

relationships between them, since the action of one component upon another breaks down an established inter-relationship and creates a new one, resulting in the continual reconstruction of biogeocoenoses, the destruction of some and the creation of others. This internal contradiction in the biogeocoenotic process is manifested also in the continual development of the biogeocoenotic cover of the earth. In order to be able to guide that development we must study all of its aspects and discover the mechanism of the changes in biogeocoenose components that originate in it.

If we consider merely the forest cover of the earth, then (as may be seen from the preceding chapters) although much has been discovered about forest-biogeocoenose components and their inter-relationships, a huge task still lies ahead. Therefore we cannot yet present a sound system of developmental types for forest cover on the earth and demonstrate all the successive stages of changes in biogeocoenoses and the moving forces behind them. The task is made still more difficult because knowledge of the components of forest biogeocoenoses and their inter-relationships requires research in diverse sciences; and because of recent advances in science, the rapid rate of scientific development, and specialisation, biogeocoenotic study is a very complex subject. We shall mention only the main landmarks as a forest cover develops and explain the chief tasks confronting us in the solution of the basic problem, which has great theoretical and practical significance.

Both the earliest stages of development of biogeocoenotic cover and the methods and rates of later successions of biogeocoenoses depend (ignoring human influence for the present) on two kinds of factors: on the characteristics of the biogeocoenoses themselves, and on the nature of the surrounding biogeocoenoses.

Independently of the varying effects of these factors we may note the following general features of the biogeocoenotic process. The first colonisation of the earth's surface by micro-organisms, higher plants, and animals is largely determined by the characteristics of the process of biocoenose (principally phytocoenose) formation, which is linked with the constantly-increasing occupation of territory by organisms, the process called *syngenesis* in phytocoenology. That term may be applied also to the formation of a biogeocoenose as a whole, and at that stage of the biogeocoenotic process we may speak of syngenetic successions of biogeocoenoses. The successions of phytocoenoses and zoocoenoses are determined mainly by the composition of the plants and animals in them, by their reproduction and distribution through the substrate and by the conditions of their co-existence, which are created by two mutually-contradictory processes: (*i*) interspecific struggle for existence between organisms and (*ii*) favourable effects of one organism upon another when living together or, to use a conventional term, their mutual aid. Although these two processes are always present in a biogeocoenose, the struggle for existence usually predominates. At the same time interactions by the components belonging to inanimate nature (i.e., air, rock strata, and water) are taking place. These interactions produce weathering of the rock strata. At this early stage of the biogeocoenotic process, alteration of its characteristics by the organisms inhabiting it is still not great enough to make substantial changes in it, or to change a biogeocoenose so much that we can speak of a succession of biogeocoenoses.

As soon as the first organisms appear on the surface of mineral strata a biogeocoenose exists, although in a primitive form. Usually the first participants are bacteria, algae, lichens, mosses, Protozoa, and a few pioneer higher plants and higher animals. Soon there appear either more-or-less-complex herbaceous communities with their corresponding fauna or, less often, tree and shrub vegetation, i.e., a forest biogeocoenose with its own fauna and micro-organisms. In the first stages of development of biogeocoenoses, the inanimate components (rocks, water, and atmosphere) undergo changes caused by animal organisms and their remains. As a result of their interactions a new natural formation begins to take shape, i.e., soil (called by V. I. Vernadskii a bio-inanimate substance), with its own distinctive features. With the start of soil formation, the new biogeocoenose component takes part in the mutual interactions of the components. The first stages of the soil-formation process have already been discussed in the chapter dealing with soil as a biogeocoenose component.

Further development of the biogeocoenotic process is accompanied by further soil-formation. These two processes proceed in parallel, and although the soil is a product of the biogeocoenotic process it also takes part in it, interacting with the other biogeocoenose components.

S. D. Muraveiskii, regarding a biogeocoenose as a natural geographical complex of varying extent, briefly but correctly describes the general composite process thus: 'The earth's surface . . . is the stage for three fundamental natural processes, which determine the existence of geographical complexes in general: the process of weathering, the process of development of the organic world, and the process of soil formation. All of these processes develop automatically (spontaneously) on the basis of laws inherent in them' (1948, p. 98). The origin of geographical complexes is connected with the successive development of these three processes, leading to the creation of a single whole, within which appear new regular relationships between phenomena, broader than those between the phenomena in the three processes. These relationships determine the further development of the whole.

The place of the soil-forming process in biogeocoenotic development is unique, and differs from that occupied by other biogeocoenose components.

Although each component of a forest biogeocoenose changes during its development and thereby alters the whole biogeocoenose, the roles of the separate components are not of equal value. The atmosphere, changing during the process of its development as a global phenomenon, also changes as a component within a given biogeocoenose. Fluctuations in solar radiation originate from general cosmic causes and affect also the proportion of solar energy utilisable by plants. Global changes in the humidity, movement, and composition of the air also change the characteristics of the air within a forest biogeocoenose. Morphological changes in the whole surface of the earth or in local areas, as a result of tectonic movements and other geological processes, produce microclimatic changes and changes in the water regime of the substrate and at the same time alter these components within a given forest biogeocoenose. Finally, all evolutionary change in the organic world, the development of taxonomic units and the corresponding changes in their ecological characteristics, their dispersal through new territories and, in general, changes in their geographical and topographical distribution, are reflected in the organic life of each biogeocoenose.

In all these cases alteration in the components of a forest biogeocoenose affects its dynamics and produces varying changes in its whole. In such cases a forest biogeocoenose is altered as a result of changes in any natural unit affecting its formation.

The role of soil in altering a forest or any other biogeocoenose as a whole is essentially different from that of other components. Any soil, as we have seen, is to some extent a product of the interaction of all other components of the biogeocoenose, mainly of their interaction with the mineral strata of the earth's surface. The evolution of the soil of a biogeocoenose does not show separate self-development according to its own laws, but is a result of the interaction of other developing components of the biogeocoenose. Soil, being formed in each separate biogeocoenose and acquiring its peculiar qualities, continues to develop according to the specific laws of the biogeocoenose. These laws are local, not global, in character; they are closely connected with the laws of development of the whole biogeocoenose, being in essence a part of them. That also decides the role of the soil in forest-biogeocoenose dynamics. Whereas changes in a biogeocoenose as a whole due to changes in other components (resulting from general fluctuations in climate, movements of the earth's surface, and the total distribution of living organisms and their evolutionary development) should be classified as exogenous successions, the succession of biogeocoenoses due to alterations in the organic world connected with variations in the soil and in its physico-chemical features produced by the entire biogeocoenotic process should be classified as endogenous (endodynamic) successions.

The development of the soil and of the entire biogeocoenose have much in common, while at the same time differing from the development of other biogeocoenose components. As a biogeocoenose is not merely the sum of its components but is qualitatively a new natural phenomenon, developing according to its own new laws, so the soil also develops according to its own laws. These laws are, however, relevant only to a certain soil, originating from a combination of other components of a particular biogeocoenose, if we except the most general laws—such as that the development of any soil is a result of the development of all other interacting components. The soil-formation process (or, more precisely, the soil-evolution process: usually soil already exists in a forest biogeocoenose as an already-formed bio-inanimate structure) is quite inseparable from the total biogeocoenotic process, actually constituting an inalienable part of it.

A similar role in the forest-biogeocoenotic process is played by changes in forest litter.[1] In every biogeocoenose the shedding of dead parts of plants (and partly of animals) takes place. In some biogeocoenoses the shed matter is small in amount and decomposes rapidly, not forming an independent layer; in others, in forest and some steppe biogeocoenoses, it accumulates in considerable volume, acquires its own physico-chemical characteristics, possesses its own world of micro-organisms, and develops according to its own laws. For that reason, as shown in the chapter on soil, forest litter is to some extent a unique component of a forest biogeocoenose, being a transitional link between the components of the organic world and the soil. We may therefore speak not only of the litter-formation process but also, after the litter is formed, of the litter-evolution process. That process (like that of the soil) is very specific but is not distinct from the forest-biogeocoenotic process, being actually a part of it. Therefore the biogeocoenotic process includes soil-evolution and litter-evolution.

Sooner or later (but in conditions favourable for tree growth, which in many cases are created by the biogeocoenoses themselves) the original primitive moss-lichen or herbaceous biogeocoenoses are replaced by more complex forest biogeocoenoses.

As our task consists in the study of succession of forest biogeocoenoses, we shall discuss these in greater detail.

There cannot be a single biogeocoenotic process any more than there can be a single soil-formation process. V. P. Williams once tried to develop the concept of a single soil-formation process, but those who approached the subject without preconceived ideas saw the complete fallacy of that concept. Scarcely any pedologist now accepts it. Sometimes the term 'forest-formation process' is used (Kolesnikov 1956 et al.). It also cannot be considered a single process. It is extremely multiform, depending on the internal and external conditions in which the life of the forest runs its course. The forest-formation, or, more precisely, forest-evolution, process is a particular aspect of the biogeocoenotic process.

---

[1] Although the term 'forest litter' is in almost universal use, it cannot be considered a happy one. It would be desirable to have a special word for this concept, as for the analogue of forest litter on the steppes, which is sometimes called by the equally unhappy names of 'steppe felt' or 'vetosh', and on the steppes of Askania-Nova in Khersonshchina has received the name 'kaldana'.

Recently P. D. Yaroshenko (1961) spoke of a single process of development of plant cover, relating it to 'the single soil-formation process'. Both of these processes, together with the development of 'bioclimate', merging into one another, produce (as Yaroshenko says) the total process of development of the earth's biosphere. In Yaroshenko's view neither the process of plant-cover development nor that of soil formation is a single process in the sense given to the term by Williams. In each case both plant cover and soil cover, like the whole biosphere, are constantly developing and to avoid ambiguity in the meaning of the term 'single', it should not be used at all in such cases.

If we disregard possible changes in the general conditions of both the immediate and the distant surroundings of a given biogeocoenose, as discussed earlier, and consider as constants both the inflow of energy in solar radiation and the characteristics of the organisms included in the biogeocoenose, then the biogeocoenotic process may be defined as consisting of four processes: (*i*) the interactions of biogeocoenose components among themselves, constantly altering them and the biogeocoenotic process to some degree; (*ii*) the introduction of plants and animals by wind, water, or other means, and in general colonisation by new organisms from the surroundings, which by changing the composition of the biogeocoenose also change the biogeocoenotic process; (*iii*) the introduction, with dust or inflowing water, of mineral and some organic matter; (*iv*) the carrying-away of mineral and some organic matter from the biogeocoenose by water and other agents.

The first process, being a purely internal one, may be given the name of *endocoaction*; the second is called *inspermation*; the third was called *inpulverisation* by G. N. Vysotskii, who had in mind only introduction of matter with dust, but the meaning of the term may be broadened to include also introduction of matter with water; the fourth process may accordingly be called *expulverisation*.

The endocoaction process is in turn composed both of interactions of biogeocoenose components among themselves and of interactions among the separate elements of each component. We have seen in the relevant chapters that the different physical and chemical properties of the atmosphere not only affect other biogeocoenose components—mineral strata, soil, plants, animals, and micro-organisms—and in turn depend on them, but also affect each other. The gaseous composition of the atmosphere, its water content, air temperature, and air movement, not to mention the solar radiation penetrating the biogeocoenose, are interdependent. The physical and chemical properties of mineral strata and soil are also interdependent. Plants, animals, and micro-organisms affect each other in the struggle for existence, and by mutual benefit through mechanical action and through chemical secretions, through aid in the distribution of diaspores, etc. The process of distribution of chemical substances in the soil by means of the life-activity of plants and, to some extent, of animals is particularly important, being connected with the so-called biological cycle of matter, which is also linked with the transformation of energy within a biogeocoenose. All of these multiform and complex endocoactional processes are continually—now rapidly, now more slowly— altering the biogeocoenose components, and at the same time causing replacement of one biogeocoenose by another. The process of internal interaction never ceases, but generally has a tendency to slacken its pace. That is due mainly to the slowing down of inspermation because of exhaustion of the source from which seeds or entire organisms of new plant, animal, and micro-organism species are derived, and because of the formation of biogeocoenoses with organisms more adapted to each other, with firmer structure, more stable and relatively less penetrable by new arrivals.

The second and third processes, assuming that the general environmental conditions of the biogeocoenose remain unchanged, may themselves remain unchanged; but when the conditions of the surrounding area change these processes do not long remain the same. The general tendency to slowing-down of the biogeocoenotic process, causing also retardation in the succession of phytocoenoses, has led some foreign scientists to take the view that ultimately plant cover becomes stable and, unless the climate changes, may exist indefinitely in the same phase of development.

This was the origin of Sernander's concept of a terminal formation, and later the theory of climax of vegetation, worked out by Clements[1] and a number of other botanists. The theory of climax of vegetation relates essentially to phytocoenoses; but since at present not only our own, but also many foreign phytocoenologists regard plant succession as being linked with soil evolution, we must briefly discuss that theory from the biogeocoenological point of view.

Much has been written about the climax of vegetation, especially in foreign literature. The original concept as devised by Clements was later considerably altered by himself and other authors. Originally Clements applied the term 'climax' to vegetation covering wide areas in which, having passed through a number of successive stages of development, it has finally acquired a composition and structure appropriate to the local climate, has become stable, and does not change further until the climate changes. In that sense the climax, often called monoclimax, corresponds (to a certain extent) to zonal vegetation, if one considers only plakor territories in the sense of many of our authors. In recent years such a climax has often been called a climatic climax. But since, in any place with varying topographical conditions and therefore with varying soils, even fully-developed vegetation is very diversified, the concept of mono-

---

[1] For the climax theory see especially the works of Clements (1916, 1936). A summary of the latest views on this subject has been presented by V. D. Aleksandrova (1962).

climax is unacceptable. In its place the theory of polyclimax has been developed. In particular, we have begun to speak of edaphic climaxes (pedoclimaxes). Until recently, however, 'climax' has often been defined as the final stage in development of the vegetation that occupies a certain territory while the climate remains unchanged. It is usually stressed that natural changes in vegetation lead it to a final stage of equilibrium, i.e., to a dynamic equilibrium among climate, geomorphology, soil, and vegetation. In giving that definition to 'climax' Braun-Blanquet (1951, p. 462) adds that the concept of a single form of vegetation is inadequate. Thus Braun-Blanquet, in fact, understands 'climax' almost from the biogeocoenotic standpoint.

Up to the present foreign authorities have suggested very many different types of climaxes. Whittaker (1953) lists 35 terms containing the word 'climax', including ten introduced by Clements. Matters have reached a stage where not only radical vegetation[1] but also that resulting from the action of man or of catastrophic agents, i.e., the so-called derivative vegetation, has begun to be called 'climax'; for instance, some speak of 'biotic climax' produced by constant pasturing, or 'fire climax' due to forest fires.

Therefore many protests have been made in the literature against the climax concept. The work of Selleck is notable in this respect. Summarising the present state of the subject and quoting the results of field geobotanical investigations, he writes 'in its development vegetation rarely reaches a climax: the latter is affected by so many obscure factors that its theoretical significance is doubtful'. Later he writes: 'It would be desirable not to use a term expressing any degree of "culmination" of successions, and in defining "climax" we must avoid terms indicating "conclusion" of successions' (Selleck, 1960, pp. 535, 543).

Nevertheless even now the term 'climax' is widely used in foreign literature; it has its advocates even in our own country. A. Ya. Gordyagin (1900), studying the dynamics of forests in Western Siberia and Tartary, spoke of some of their formations (understanding the term 'formation' in a sense close to the modern concept of plant association) as being ter-

---

[1] It is sometimes held that the expression 'radical association', occasionally used in our literature, corresponds to 'climax'. That is not so. By 'radical association' we usually mean any natural association not altered by man or by any catastrophic agent, regardless of whether it is serial or climactic in Clements' sense.

V. B. Sochava (1963, p. 8) distinguishes between strictly radical associations, which correspond to the concept of a zonal type of plant grouping, and quasi-radical stable associations, 'whose structure is due to the prevailing action of some edaphic factor (as well as to zonal characteristics) that in some form or other modifies zonal features (e.g., excessive moisture, lack of development of the soil-forming process on a stony substrate, etc.)'. Sochava notes that Guinochet (1955) names the first category 'climactic' and the second 'permanent'. It is hardly worth while following up either of these suggestions. There is no need to change the term 'radical association', long in use in our country, and to do so would merely cause confusion. It is enough to speak, as is usually done, of zonal and non-zonal (azonal) vegetation. Guinochet's 'permanent vegetation' corresponds to the term 'edaphic climax', in use abroad.

minal, following Sernander. B. N. Gorodkov (1944) very strongly supported the climax theory. In recent years some Soviet authors, although not sharing the view of 'climax' as vegetation that has ceased to develop and is able to remain unaltered indefinitely while the climate does not change, regard it as being fully developed and able to remain without substantial change for a fairly long period. Thus E. M. Lavrenko writes 'in our daily work we should use the concept, introduced by F. E. Clements, of succession series and climax, not, of course, giving the latter term an absolute meaning' (1959, p. 59).

P. D. Yaroshenko (1961) gives the name of 'nodal stage' to a stage in the succession process in any area of plant cover when the latter remains in practically the same state for a fairly long period; that stage corresponds to the climax of foreign phytocoenologists. But we see no particular advantage in the term. It tells nothing.

The question arises: in what relation does the climax theory stand to biogeocoenology in general and forest biogeocoenology in particular? As we have seen above, Braun-Blanquet remarks that climax is characterised by a state of dynamic equilibrium among climate, geomorphology, soil, and vegetation. Actually the biogeocoenotic process is reflected in alteration in a biogeocoenose as a consequence of the uninterrupted mutual action of the components upon each other, and for various reasons that process slows down when a 'fully-developed' biogeocoenose has been formed.

The theory of climax vegetation could have arisen only with insufficient appreciation of the great effects produced by animals, plants, and microorganisms on other components of a biogeocoenose. Since these effects never cease, and since the other three processes (inspermation, inpulverisation, and expulverisation) are always operative, the biogeocoenotic process, as stated above, never halts. It constantly produces replacement (succession) of one biogeocoenose by another, which excludes the possibility of existence of a really climactic biogeocoenose. One can only speak of the biogeocoenotic process slowing down with the passage of time, when the plant cover (and with it the animal population) acquires some relative stability, and that is all. Therefore if the term 'climax' is used, it should be only in the sense that vegetation or a biogeocoenose as a whole has become relatively very slow in development; we must remember that, even when all the other above-mentioned agents that are able to alter it are absent, replacements (successions) of biogeocoenoses can only be much retarded, and that a biogeocoenose may appear to us to be substantially unaltered for a very long time, but in fact the processes preparing it for replacement (succession) by another are continuing incessantly.

The process of retardation is usually uneven. It may at times be comparatively rapid, depending considerably on the nature of the resident organisms. The organisms that are able to multiply vigorously and propagate rapidly by vegetative methods, and at the same time possess strong environment-transforming power, may accelerate the biogeocoenotic process.

Two stages may be distinguished in the biogeocoenotic process. The first stage is characterised by the occupation of new territory by organisms; by their struggle for existence (with environmental conditions only); and by natural selection of the species that can survive in the climatic conditions and with the physico-chemical properties of the substrate. Later, as the plant cover becomes denser and competition among the plants for living requirements becomes stronger, species are selected that not only can live in the given physico-geographical conditions but can survive competition. That stage in plant-cover development has been called syngenesis. This term can be applied also to development of biogeocoenotic cover produced by the occupation of a territory by organisms and by their selection on the basis of the physico-geographical conditions and of interspecific competition resulting from the struggle for existence. Although the organisms produce a certain amount of change in the physico-geographical environment, that is not enough to be the driving force in replacement of the vegetation and of the entire biogeocoenose.

After some time these changes become so great and begin to act so strongly upon the whole organic world that the latter also is changed, and the driving force of the biogeocoenotic process is then the assemblage of all the changes in the biogeocoenose resulting from the interactions of all its components. We may give the name of *endogenesis* to that prolonged period in the life of biogeocoenotic cover.

Thus the biogeocoenotic process, having begun with syngenesis, later acquires the characteristics of endogenesis. Sometimes the elements of syngenesis continue to exist.

American authors, developing the theory of climax vegetation, have stated that plant cover approaching the climax acquires, as a rule, a more mesophytic character, regardless of whether its first stages of development were hygrophytic or xerophytic. In its general outlines that law applies to the development of all biogeocoenotic cover.

In this way the biogeocoenoses observed by us, including forest biogeocoenoses, are merely links in an uninterrupted biogeocoenotic process that determines the self-development in which the biogeocoenotic cover of the earth is always engaged.

Recently Western European literature has been using the term 'potential vegetation', denoting the vegetation that would most closely correspond to the entire complex of historical and physico-geographical conditions of a locality. Sochava (1963) attaches great weight to that concept, believing it to be of value for the future development of geobotanical cartography. As an example of a case where use of the concept is necessary he mentions the Siberian taiga, which has changed its appearance with the elimination of forest fires, many taiga plant associations differing from the former ones now occupying a permanent place in it. In this case we have the so-called 'demutation' process (to use G. N. Vysotskii's term), whereby vegetation on recovery after regression does not assume exactly the same form as the pre-catastrophic vegetation. Sochava also suggests broadening the concept of potential vegetation to include the consequences of changes taking place in plant cover as the result of its transformation through the action of other components of the geographical environment. Such broadening, however, would greatly change the meaning of the concept of potential vegetation as proposed by Tyuksen.

EXOGENOUS SUCCESSIONS OF FOREST BIOGEOCOENOSES

Since all objects and phenomena in nature are subject to interactions and are inter-related, every unit, especially in biogeocoenotic cover, enters into the composition of other units. These alter while developing according to their own laws and thereby affect the course of development of the units linked with them. Units whose existence runs parallel to that of others are included in larger complexes; these parallel units interact upon one another, and during their development they themselves change and cause each other to change. In this way endogenous processes, which determine the development of biogeocoenotic cover and the successions of bio-geocoenoses, are from time to time disrupted by exogenous influences arising from changes in other units, both those whose existence is parallel to that of the units concerned and those of a higher order in which the latter are included.

Units comprising several forest biogeocoenoses interact, in the course of their development and change, with these forest biogeocoenoses and alter them, i.e., they cause one forest biogeocoenose to be replaced by another. We may give a few examples of such replacements (successions).

1. A rock stratum and the soil developing upon it depend on the de-velopment of the entire surface of the earth and on geomorphological processes that may vary in nature and affect a given forest biogeocoenose in different ways. For instance, if we have a willow-aspen forest on a river flood-plain, as the valley develops with deepening of the river bed the flood-plain gradually changes from a depression to relatively high ground. That process, which is connected with the general geomorpho-logical development of the area, leads first to changes in its water regime and the process of accumulation of alluvial deposits (i.e., to alteration of the conditions of water and mineral nutrition and of living conditions for micro-organisms and many animals), to alteration of soil processes and consequently of the soil, and in some degree also to alteration of the lowest atmospheric layer. All these factors create the foundation for changes in the whole of the higher vegetation. The willow-aspen bio-geocoenose, in the forest-steppe, for instance, begins to be replaced by elm (mostly *Ulmus scabra*) or elm-oak forest. In this case the replacement

of one forest biogeocoenose by another is the result of the general geo-morphological development of the locality.

2. Another example is the replacement of a forest biogeocoenose as a result of suffosion causing karst phenomena and the sinking of a part of the earth's surface. If, for instance, in the broad-leaved forest-subzone such an area has been occupied by a sedge-goutwort oak stand, when the land sinks the soil-moisture regime deviates towards damper conditions. That also alters the water and mineral nutrition of plants and the living conditions of micro-organisms and many higher animals. The sedge-goutwort oak stand is converted into one of the better-watered types of oak stand, including more hygrophilous tree, shrub, and herbage vegeta-tion. A spiraea-birch-poplar oak stand may develop. In this case also the general geomorphological development of the earth's surface, due to geological development, has produced a succession of biogeocoenoses.

3. In the same category of biogeocoenose successions we must include those resulting from landscape changes during the Holocene period, although the rate of such changes is incomparably slower. The develop-ment of phenomena of a cosmic order, which has led to the onset of ice ages during the period of human existence, has caused tremendous changes in the biogeocoenotic cover of the earth, destroying it in the glaciated areas and greatly altering it in adjacent regions. The recession of the most recent glaciation and the general climatic change that followed, which in fact were consequences of a particular phase of global develop-ment, led to changes not only in vegetation but in the entire organic world over vast areas, and were reflected in soil-evolution processes in all types of biogeocoenoses and also, of course, in their biogeocoenotic processes. We may mention the changes in landscape and in forest-bio-geocoenose successions in the central parts of the Russian plain, which were covered with ice. Palaeogeographical, palaeobotanical, palaeozoo-logical, and partly also archaeological investigations have shown that during the postglacial period plakor areas in that territory, which formerly possessed their own types of landscape in which spruce forests pre-dominated, became covered by biogeocoenoses of various types of birch-pine forests approximately 10,000 years ago. Some time later broad-leaved tree species began to invade them—elms, oak, and linden, and also hazel. In place of the pine-birch forests there appeared mixed pine-broad-leaved forests, and in places only broad-leaved and occasionally oak stands. The general climate of that territory was then more favourable for broad-leaved forests than it is now, and in suitable soil conditions they spread farther north than they do now. Later spruce began to invade these forests, and in many places it gradually forced out the broad-leaved species, apparently as the result of the onset of a somewhat colder and much damper climate. Palaeozoological investigations show that the forest fauna also varied during the Holocene. There is no doubt that the evolutionary processes of both litter and soil were affected by changes in

climate, vegetation, animal life, and micro-organisms, and were diverted from their own endogenous course of development. The result, on the whole, was that during the Holocene the biogeocoenotic process underwent changes in its evolutionary path and produced successions of forest biogeocoenoses differing greatly from each other.

Although the replacement of forest biogeocoenoses (or, better, of biogeocoenotic cover) in the central part of the Russian plain was basically caused by climatic fluctuations, doubtless the changes that took place in other biogeocoenose components and their interactions played a definite part in the trend of the biogeocoenotic process. The whole mechanism of these successions of biogeocoenoses is not yet known, and their immediate causes are obscure, especially the replacement of landscape containing spruce (*Picea obovata*) by pine-birch biogeocoenoses in the early Holocene. It is difficult to attribute that change simply to climatic fluctuations.

In order to discover the immediate causes of the above succession of biogeocoenoses more data are required, for more thorough study of the changes in all biogeocoenose components in historical perspective, taking into account their constantly-changing interactions.

4. Among the units that affect the development of plant cover while developing in parallel with it, we must include human society. It develops according to its own unique laws, as revealed by K. Marx, and at the same time substantially affects the development of natural biogeocoenoses. Marx, however, stated that society, while acting upon the surrounding world and altering it, thereby also alters its own character. Therefore in this case also we may speak of interaction of these units. In view of the distinctive nature of human action upon forest biogeocoenotic cover we shall dwell in greater detail upon anthropogenic changes in it.

In all stages of development human society has exerted, and exerts, great influence on nature and on the whole biogeocoenotic cover of the earth. The higher human society has risen in its historic course of development, the more strongly that influence has been exerted and the more complex, diverse, and purposeful it has become. Whereas in earlier stages of human society its action consisted in direct use of the products of nature required for human existence (apart from occasional disruption of natural processes), in the more advanced stages the efforts of society have become more and more directed towards planned changes in nature, transformation of it, and increase of its resources in the interests of mankind. This process has, however, included various forms of casual and ignorant interference with nature. All of these actions affect the biogeocoenotic cover of the earth and the biogeocoenotic process and have produced definite replacements (successions) of certain biogeocoenoses by others.

All biogeocoenoses are subjected to human intervention, but its effect on the course of the biogeocoenotic process in the forest—on the succession of forest biogeocoenoses, because of their complexity and internal interdependence—is especially great and many-faceted. Although we must formally include successions due to human action among those produced by the development of adjacent units, i.e., among exodynamic successions, they are sometimes placed in a separate category because of the above-mentioned unique features.

Human influence on forest biogeocoenoses may take two forms: either man acts directly upon a forest biogeocoenose as a whole, or he acts upon it indirectly through other factors that can—independently, or affected in turn by other agents—affect the biogeocoenose as a whole. We shall examine the first of these categories in greater detail.

Human activities may have incidental effects on biogeocoenoses, or even destroy them in their natural form, by crushing vegetation or injuring other components (and sometimes the whole) of a forest biogeocoenose. The structure of the biogeocoenose is altered, and the biogeocoenotic process takes a new direction. As a rule human action upon the vegetation in a forest biogeocoenose is accompanied by alterations in other components, e.g., the soil is compacted; its air regime and other physical and chemical properties are changed; birds and other vertebrates are driven away, etc. All of these forms of human intervention are complex in character, and it is difficult to segregate the separate factors therein. It is necessary, however, to analyse these effects of human action upon a forest biogeocoenose and to distinguish the separate causes of succession of biogeocoenoses, in order to be able to avert those successions that run counter to human interests.

An example of that kind of succession is seen in forest biogeocoenoses in the environs of populated places, especially large towns. In forests near towns the lower layers of vegetation are usually much altered. When a forest is visited frequently and the herbaceous and moss-lichen cover is trampled down, the cover changes in composition and may even be partly obliterated. The change in cover is due, on the one hand, to damage to plants and general deterioration in their growing conditions, and on the other hand often to the introduction of new, so-called 'weed' species and their subsequent growth. Change in the herbaceous layer may not be immediately obvious; but in such cases changes often take place also in other layers of vegetation, in the shrub layer, and in the first and second tree storeys.

Because the problem of forest management in forest-park zones is very urgent nowadays, studies of changes in forest biogeocoenoses there are of great practical value, but as yet little work has been done on them. A good example of such work is that of R. A. Karpisonova (1962 a, b) who, studying the so-called Ostankinsk oak stand in the Central Botanical Garden in Moscow and using available published information, defined four phases of changes in that oak stand under human influence (without tree-felling). In its undamaged state the oak stand has oak in its first, upper dense storey; linden and maple in the second storey; hazel, spindle-tree, alder buckthorn, and other species in the shrub layer; and oak-forest elements with a large number of ephemeroids in the herbage. In the first phase of human influence, when human action is still on a small scale, only oak remains in the tree stand; the shrub layer is still complete and the herbage is little changed, only the number of ephemeroids being reduced. In the second phase, when human influence is increasing, the shrubs are thinned out; meadow species appear in the herbage and rhizomatous plants predominate there, being well able to endure the increased illumination (goutwort, hairy sedge, etc.). In the third phase, with human influence still stronger, the shrub layer becomes very thin (density 0·2); loose bushy meadow grasses are predominant in the herbage; and forest species survive only under bushes and near trees. Finally, in the fourth phase shrubs are entirely absent, and there are no oak-forest plants in the herbage. Dense, bushy meadow grasses (hairgrass, etc.) predominate in damp conditions, and annual grasses where moisture is moderate or inadequate. The trees begin to die at the tops at the age of only 100-120 years. It is interesting to note that while the effects of human activity are still slight there are considerable numbers of oak shoots and the oak undergrowth thrives, since elimination of the second-storey trees and thinning-out of the shrub layer favours that situation. But intensive human interference (fourth phase) practically puts an end to tree regeneration. All of these phenomena are due mainly to changes in soil and light conditions in the oak stand resulting from human action, but they are also affected by changes in other components of the oak-stand biogeocoenose. As a result Karpisonova was able to identify distinctive forest-vegetation associations for each phase: for the first phase, yellow archangel (*Lamium galeobdolon*) and lungwort oak stands; for the second, sedge oak stands; for the third and fourth, herbaceous oak stands. Thus under the influence of human activity and without tree-felling a succession of vegetation takes place, corresponding to a succession of types of forest biogeocoenoses. It is very probable that, if a detailed study of all components of these biogeocoenoses were made, the third and fourth phases, in spite of the similarity of their herbaceous cover, would have to be regarded as different types of biogeocoenoses.

In the southern belt of the forest zone of the European part of the U.S.S.R. mountain ash grows vigorously in the undergrowth of forests near cities, which are affected by human activity. In forests untouched or little touched by man it occurs in comparatively small numbers, but under human influence it thrives very well. In such conditions entirely new plants sometimes appear in the undergrowth, substantially changing the forest biogeocoenose. Red-berried elder, for instance, grows strongly, although it is not found in the native forests of the territory. In some parts of forests in the Moscow district and some other European-Russian areas the same role is played by the Canadian serviceberry, *Amelanchier canadensis* (L.) Medik. That berry does not generally occur in forests of the U.S.S.R. forest zone, and is brought accidentally from gardens and parks by people and birds.

The immediate cause of successions of forest biogeocoenoses in such cases may be very complex, and discovery of it is not a simple matter. Even with such a common event as the vigorous growth of mountain ash in undergrowth and the second tree storey, the cause is not clear. Most probably the reason is that man, by destroying the natural density of the herbage-shrub layer and breaking up the moss cover, creates an environment more favourable for germination of mountain ash seeds and growth of its shoots. In this case human activity produces compaction of the soil and alters its gas regime and its animal and microbial population, which lessens the competitive ability of the natural vegetation. Therefore the succession of forest biogeocoenoses is a consequence of change in almost all their components, but human activity is the basic cause.

Human influence is very important in successions of forest biogeo-coenoses due to forest exploitation. Different methods of tree-felling may lead to replacement of forest biogeocoenoses, but clear-cutting, of course, plays an exceptionally important role.

In other chapters of this book we have explained how open areas and forests have different atmospheric and soil regimes, and how their vegetation and their animal and micro-organism populations are also different. Therefore elimination of tree stands or heavy thinning of them induces changes in all components of forest biogeocoenoses, i.e., replacement (succession) of them. Sylviculturists and phytocoenologists have made fairly thorough studies of the way in which vegetation changes in clearings, of the rate at which it changes, of the cases in which the original vegetation is restored, and of the cases in which it is not restored at all. Data also exist regarding changes in the atmosphere and in the physical and chemical properties of the soil on cleared lands after tree-felling, and regarding the effect produced on these components, and on the appearance and development of new growth of various tree species, by different methods of cutting and hauling timber. These changes lead to changes in other forest-biogeocoenose components—animal life and micro-organisms—but the latter have not been adequately investigated. In any case we can definitely speak of changes in the entire biogeocoenotic process in the forest as a result of tree-felling. The study of tree-felling from all aspects in relation to types of forest biogeocoenoses is necessary to achieve the replacement of a clearing with a forest of desirable composition and structure in the shortest possible time. Every tree-felling system introduces some kind of change into a forest biogeocoenose. Clear-cutting changes it immediately and catastrophically. Concentrated clear-cutting has a particularly marked effect. Selective, gradual, and other cutting methods produce lesser changes, these sometimes being so slight that one cannot describe them as successions of forest biogeocoenoses.

Generally, after clear-cutting a herbaceous biogeocoenose develops on the cleared area, and if left alone sooner or later gives place to a forest biogeocoenose. Then either a biogeocoenose of the former composition is regenerated, or a biogeocoenose appears containing other tree species. In this connection the classification of types of cutting developed by I. S. Melekhov (1959) is worthy of note; it derives from the correct premise that clearings, being specific biogeocoenoses, are subject in their continued development to conversion into forest biogeocoenoses according to definite laws, which can be discovered and which it is possible to control only by thorough study of all the components of the biogeocoenoses on clearings.

Since the process of replacing a forest biogeocoenose with a clearing biogeocoenose, and later the process of restoring the forest biogeocoenose (degressive-demutational replacement, in G. N. Vysotskii's terminology) depends essentially on the destruction and regeneration of a tree stand, we may to a certain extent use the known types of vegetation successions after tree-felling and their nomenclature to study the successions of biogeocoenoses that take place on cleared areas, and to study the regeneration of forests on these areas.

Thus we may speak of radical forest biogeocoenoses, which have developed without human interference, and of derivative ones, which arise as a result of the above-mentioned human activities. The latter are sometimes called temporary, which we cannot consider to be correct, since even radical biogeocoenoses are, in a way, temporary. If tree seeds are not blown over a cleared area, or if after trees are felled conditions are created on the cleared area preventing development of trees there (e.g., swamping of the clearings, which often happens after heavy cutting in the taiga zone), then in place of the removed forest biogeocoenoses there develop meadow or bog biogeocoenoses, which, developing according to their own laws, may hold the area for a prolonged period.

Usually a forest biogeocoenose is re-established on the cleared area after some time has passed. The process of forest regeneration without human intervention, i.e., the so-called natural forest-regeneration process, may take place rapidly, within a few decades, but sometimes it stretches out over a long period, many decades. In the first case we may speak of the appearance of quickly-produced, in the second case of slowly-produced forest biogeocoenoses.

An example of the first type is the succession observed in pine or spruce forests where, after cutting, a profusion of birch or poplar shoots appears on the cleared area. Within two or three decades they form birch or poplar biogeocoenoses, which are distinguished not only by changed composition but also by changed atmospheric conditions: the light regime and the composition, temperature, humidity, and movement of the air are changed, and so is all the other vegetation; soil processes take place differently, and the animal life and micro-organisms are also more or less different. In such a birch or poplar forest the development of dense herbage, and to some extent other conditions, impede the development of seedlings of these tree species. On the other hand, pine or spruce seedlings appear beneath the canopy, and their development still further prevents the growth of birch or poplar seedlings. Therefore within a single generation of these trees the pine or spruce stand is regenerated and the former type of biogeocoenose is restored.

An example of a slowly-produced forest biogeocoenose is the appearance of a new oak biogeocoenose when pine has been cut out from a forest bio-geocoenose with a first storey of pine, second storey of oak, and under-growth of hazel. With such cutting, even when the cleared strip is not too wide and there is adequate blowing-in of pine seeds, natural regeneration of pine does not take place. In this case removal of the pine favours in-creased growth of the oak and hazel, and partly of herbs, which prevents pine seedlings from developing. The oak biogeocoenose thus created may occupy the area for a very long time.

In these two categories of forest successions the effect of human influence consists in disruption of the integrity, or partial destruction, of vegetation as a component of a forest biogeocoenose, inducing substantial changes in its other components; human action also causes the appearance of a new type of forest biogeocoenose, which with the passage of time may be replaced by the original type.

If man does not interfere in the forest-regeneration process on cleared areas, that process takes the form of occupation of a herbaceous biogeo-coenose by trees and other forest vegetation with their appropriate animals and micro-organisms, which, together with the residual forest life, gradually forms a forest biogeocoenose with distinctive, partly-new characteristics of soil and atmosphere.

In this case the demutational successions are essentially syngenesis, on which endogenesis is overlaid. Therefore these regenerating, demutational successions are to be classified as syngenetic-endogenous.

Man may also assist in the replacement of one forest biogeocoenose by another by sowing or planting trees on a cleared area. But he may do the same in a meadow, steppe, tundra, or desert biogeocoenose. In such cases man usually creates a new environment to enable the tree stands created by him to grow successfully, which usually involves altering the soil and giving new characteristics to it.

Since we give the name 'anthropogenic' to successions resulting from human activity, we must use that term only when human activity directly affects the biogeocoenotic process and governs its direction. Such cases include successions resulting from the beating down of vegetation, from various types of tree-felling, from the creation of new tree stands by sowing or planting, etc. Therefore whereas degressive successions must be called anthropogenic, demutational ones are not so. As we have stated, they are syngenetic-endogenous.

Successions caused by the activities of domestic animals or by forest fires of human origin also cannot be called anthropogenic. In the first case they must be classified as zoogenic successions, which also include successions caused by wild animals; in the second case they may be called pyrogenic, a type often produced by natural causes (e.g., lightning).

We cannot use such terms for biogeocoenotic successions produced by changes in water regime due to artificial drainage or irrigation of an area. They have to be included in the anthropogenic category, although similar biogeocoenotic successions may be caused by changes in water regime through geological or soil processes. By drainage or irrigation man directly affects that regime, altering the characteristics of the most important component of a biogeocoenose—the soil.

Exogenous influences disrupt the process of self-development of biogeocoenotic cover and cause the appearance of new succession-series of biogeocoenoses. Actually all definite successions of biogeocoenoses usually take place under the influence of both endogenous and various exogenous processes.

These processes, acting simultaneously and overlapping each other, make biogeocoenotic successions extremely diverse and complex. In order to analyse them, to understand their dynamic forces and ways of development and to direct them in the interests of mankind, we must segregate the effects of the three different processes when we are investigating biogeocoenotic successions, and we must discover their interactions and respective roles.

CYCLIC (PERIODIC) CHANGES IN FOREST BIOGEOCOENOSES

When a forest biogeocoenose develops by itself, or changes under the influence of other more inclusive or neighbouring units, its successions are irreversible. If external operative factors disappear or their action ceases, the biogeocoenose may return to practically its former state. In such cases we may speak of reversible changes. Only the first-mentioned changes in biogeoceonoses are called replacements (successions). The second are called *cyclic (periodic)* changes.

The latter may be due to various causes. A biogeocoenose does not remain unchanged *even for a single day*. During that period not only does the inflow of solar energy fluctuate cyclically, but other elements of the atmosphere also vary: the temperature, humidity, movement, and even composition of the air. These properties vary under the influence of changes in solar radiation and albedo, and also of the plants themselves. Whereas on a clear sunny summer day photosynthesis (together with respiration and exhalation of carbon dioxide) proceeds intensively, with considerable exhalation of oxygen, by night only respiration continues. These processes—like other life processes in plants and also those taking place in other biogeocoenose components, soil, animals, and microorganisms—are affected by the temperature, humidity, and movement of the air.

The chief factor governing the daily dynamics of a biogeocoenose is certainly fluctuation in the properties of the atmosphere, its climate throughout the day; but changes induced by that factor in other components in turn alter the interactions between these components and their effect on the atmosphere. In fact all components and their interactions fluctuate during the day, producing the daily-cyclic dynamics of a forest biogeocoenose. Although during the course of a day some irreversible changes always occur in a forest biogeocoenose, and it is never exactly the same as it was the day before, these changes are usually so comparatively small that we can consider the forest biogeocoenose as being practically unchanged. If the development of a forest (as of any other) biogeocoenose is likened not to a circle but to a spiral, as is always stressed by materialistic dialectics, then in its daily dynamics the successive coils of the spiral lie very close together. Although these changes are cyclic (reversible), in different types of forest biogeocoenoses they take place in different ways specific to these types. Therefore in the study of forest biogeocoenoses it is essential to take account of these different dynamics and it is necessary to study the daily regime of all components of the biogeocoenoses.

Another feature of the cyclic dynamics of biogeocoenoses is their nature *at different seasons of the year*. In temperate-climate countries marked changes in biogeocoenoses are caused by the succession of seasons (spring, summer, autumn, winter) as a result mainly of differences in the temperature regime; in more southerly countries, such changes depend mainly on variations in atmospheric humidity and the occurrence of specific dry periods. In these different periods the appearance of the biogeocoenoses may be radically altered. Such seasonal changes in biogeocoenoses may be called successions of aspects, caused mainly by phenological successions in aspects of vegetation.

The variability of a forest biogeocoenose may also be displayed in its dependence on alternation of climatic conditions *from year to year*, e.g., in drier and wetter years. In this case tree stands and undergrowth generally do not change, but herbaceous cover may be severely affected, and changes may also take place in such components as the animal and micro-organism populations, or at least in their life-activity. Soil processes may take a different course. The structure and aspect of the phytocoenose, and in consequence the aspect of the biogeocoenose as a whole, may then be changed. If in some years the climatic conditions change very much or the periods of successive dry or wet years are prolonged, a forest biogeocoenose may be altered as a whole. In such cases the alteration may be irreversible, and we may speak of successions of forest biogeocoenoses.

Among reversible biogeocoenose changes we include those due to the natural regeneration of tree stands. This process may take widely-differing forms, but in all cases it induces some changes in other biogeocoenose components. Therefore a biogeocoenose changes as a whole with changes in the tree stand and the phytocoenose structure, as has been stated in Chapter I. Up to the present, however, very few studies have been made of the changes in all components connected with this form of the dynamics of forest biogeocoenoses. Rather more has been done by forest surveyors in evaluating the changes in young trees during their growth under the maternal canopy and their gradual replacement of it. Some of the other changes in the forest that accompany the growth of young trees have been noted at the same time. In this connection the investigations of B. P. Kolesnikov (1956) and other authorities on regeneration in Far Eastern forests are of interest.

In natural conditions of forest life, when nothing catastrophic happens to it, we may differentiate two principal forms of stand regeneration. With the first form, undergrowth of tree species appears beneath the canopy of a natural (usually dense) stand, mostly in the better-lighted places, in gaps and where the trees are farther apart. With shade-tolerant species (spruce and fir) that process may also take place even in very shady places, where the undergrowth develops slowly and may remain in a severely-suppressed state for many years. When a tree standing beside a young tree dies of old age or through damage by pests (which occurs in any forest as a natural element of the biogeocoenose) the young tree begins to grow vigorously and, gradually filling up the gap, enters the first storey. This is a normal and regular process, and in the course of a longer or shorter period the composition and structure of the biogeocoenose and all the interactions of its components, although not remaining completely unchanged, change so little that the biogeocoenose practically retains the same form. In such a forest the trees are always of varying ages, which is a constant characteristic of that type of forest. All other components of the forest situated in the vicinity of the fallen tree also change to some extent on account of the growth of the young trees. The forest has, as it were, a mosaic structure, which naturally is reflected in all of its components.

Up to the present, however, no investigation has been made into the course of the biogeocoenotic process in each element of the mosaic and in each component of the forest biogeocoenose. To understand the life of the entire biogeocoenose, to discover the nature of the process of regeneration of the tree stand, it is necessary to know more precisely how the atmosphere, soil, and organic life are differentiated in the mosaic elements of such forest biogeocoenoses. Although the differences may be comparatively small they may still play a definite and sometimes a considerable role in the life of the stand and of other parts of the biogeocoenose. Therefore there is great theoretical and practical value in the organisation of studies of these mosaic elements from all aspects.

In this connection it is necessary to remember that in the above type of regeneration the mosaic structure of the forest is constantly mobile. Each element of such a mosaic is usually only a fragmentary expression of the characteristics of the biogeocoenose. At the same time the mosaic pattern is a permanent structural element of the forest biogeocoenose, and the dynamics of the latter are due to the incessant movement of the pattern within the biogeocoenose. But if we look at the whole area of the biogeocoenose we may observe that over a definite period of time it retains all of its distinctive features, and its dynamics have a reversible, and in a certain sense a cyclic, character.

It is as yet difficult to say how widespread this form of dynamics, linked with a definite form of stand regeneration, is in natural forests. We can merely state that it is observed in spruce and fir forests, and also in a number of types of pine and oak forests, but whether it exists in all types of these is not yet known. This form of biogeocoenose dynamics may be linked not only with the process of stand regeneration, but also to some (although perhaps a considerably smaller) extent with regeneration of undergrowth.

In some forest types the process of stand regeneration takes another form. An example is the regeneration of beech forests in the Crimea (Sukachev and Poplavskaya 1927). There natural regeneration (mainly by seed) proceeds successfully everywhere, and the course of replacement of old by young trees follows a definite sequence. If we observe parts of a beech forest that give the impression of pure stands, of unmixed composition and typical in their structure, we find the age of the trees usually varying from 100-200 years. In such a beech forest there are abundant shoots and seedlings up to 0·5 metres in height. There are no taller young beeches, however, because the shoots die off at the age of 1-3 years. Therefore the impression is created that in a beech forest no seedlings have any hope of growing up, and that picture may persist for 50 years. Then the old beeches, having reached the age of 250 years, begin to die off. When gaps are formed by the death of the old trees the young shoots that had been suppressed, and also new ones, begin to grow rapidly.

Now the beech forest has a mixed-age appearance, and this period lasts for another 100 years. By that time the process of dying-off of the trees composing the original stand is completed, since those that were 100 years old at the start will have reached the age of 250 years, about the limit of duration of life for beeches on the best soils of that district. But by then the oldest beeches of the new generation are 100 years old, and they are accompanied by younger beeches of various ages. An abrupt break is observed in the regeneration conditions. The older beeches do not fall, no gaps are formed, and therefore new shoots cannot succeed in growing. In this way about another 150 years pass, during which the stand grows without any new seedlings having any hope of growing.

We see that in the life of these beech forests there is an alternation of two periods. In the first (150-year) period the age of the stand changes from 100 years at the beginning to 250 years at the end, and no regeneration of the forest takes place. Although shoots and young beeches appear during that period they partly die off and partly drag along their existence, not reaching any considerable height. A spur to their growth may be provided only by the occasional fall of neighbouring trees out of the compact beech canopy as a result of exceptional happenings (violent storms, etc.). Later the second period (lasting approximately 100 years) begins, during which the old trees die off and young ones grow vigorously in their place. Thus there is an alternation of two waves lasting about 100 and 150 years respectively. During the first wave gradual regeneration takes place, to some extent following the same pattern as the first type of restoration of original forest described above; during the second wave regeneration is halted, and only growth of the trees is observed.

As yet we do not know how frequent this form of self-regeneration of a forest is in nature. It is possible that with further study of the regeneration process it will be discovered that the second type of regeneration actually exists in the forest biogeocoenoses to which we are usually inclined to attribute a gradual, continuous process of self-regeneration.

Cyclic dynamics are also characteristic of forest biogeocoenoses where stand regeneration is of the second type, but the cycles are considerably longer, covering several decades.

If regeneration of a tree stand begins after it has been completely or partially destroyed by some catastrophic factor (e.g., forest-clearing, fire, storm damage, mass pest reproduction) and if the stand is restored with the same tree species, during the ontogenetic development of the stand the composition of the other layers of vegetation and the structure of the phytocoenose also change. Consequently all other components of the biogeocoenose and the inter-relationships between them and the vegetation also change. At the same time the composition of the phytocoenose either remains practically the same, only its structure being altered, or it partially changes. In the first case we may speak only of the cyclic dynamics of the forest biogeocoenose, linked with the ontogenetic development of the stand; in the second we have short-term replacements (successions) of biogeocoenoses.

Speaking of different forms of biogeocoenose dynamics, we must mention one frequently found, due to the movement of a mosaic pattern of soil cover resulting from patchy plant growth: for instance, a type of pine forest with fern (bracken) cover. Sometimes the bracken (*Pteridium aquilinum*) forms dense cover; sometimes it grows in patches of various sizes, from a few square metres to many times as large. The older parts of the bracken patches, which usually grow along the periphery of root-stocks, gradually die off, and therefore they are constantly migrating through the area occupied by the pine stand. Such migratory patches are also often observed in pine stands with club-moss (*Lycopodium complanatum*). Usually the club-moss patches, which also grow on the periphery, die off at the centre in the course of time. They acquire the form of rings, which by constantly expanding may attain a diameter of many metres. In every such patch one finds (besides the dominant species creating it) other plants peculiar to it, i.e., differing in some respects from those outside the ring in the same type of forest. Not only do the plant communities inside and outside the ring differ in composition and structure, but the soil processes, the animal and microbial populations, the microclimatic features, and the conditions for regeneration of the tree layers are all different. Thus from the biogeocoenotic point of view these elements of the mosaic differ considerably from each other. The mosaic elements represent a very dynamic phenomenon—they migrate through the territory occupied by the forest biogeocoenose. As the faculty of forming patches or beds is possessed by many plants of forest soil cover, this type of synusial forest-biogeocoenose dynamics is not infrequent, being sometimes more, sometimes less clearly displayed, and varying in effectiveness. In some cases such dynamics are merged with the tree regeneration dynamics discussed above, but such forms need not be segregated since they are due to various causes.

Generally speaking, synusiae of biogeocoenoses are usually a spatially-mobile phenomenon (e.g., lichen synusiae on tree trunks). Especially worthy of note are the successions of the synusiae called 'parallel' by N. V. Dylis. They are included among successions of microphytocoenoses, in P. D. Yaroshenko's sense (1961). Such parallel dynamics are very interesting both theoretically and practically, and deserve further deep study from all aspects.

GENERAL CONCLUSIONS ON, AND CLASSIFICATION OF FORMS OF, THE DYNAMICS OF FOREST BIOGEOCOENOSES

Summarising what has been said about the dynamics of forest biogeocoenoses, we may draw the following conclusions.

1. Forest biogeocoenoses, like all others, are always changing, i.e., they are dynamic. That state results from the fact that their environment (both immediate and more remote) is constantly changing, but also because interactions between their components never cease. These interactions have a contradictory character. For instance, the action of atmosphere upon rock strata or upon the soil formed above them changes the properties of the latter. Therefore if a biogeocoenose is more or less adapted to certain properties of the soil it has to develop new forms of adaptation, and thereby begins to act in a new way both upon the soil and upon the atmosphere. Such relationships also exist among other components, since all of them interact upon one another.

The process is rendered still more complex by the fact that every component of a biogeocoenose (except soil and litter) is part of a general natural phenomenon, which develops according to its own laws because of the internal contradictory interactions, peculiar to itself, of its component elements. Thus the properties of the air that forms part of a biogeocoenose undergo changes connected with the development of the entire atmosphere of the earth. This applies also to the rock stratum that affects the soil and (through the soil) other biogeocoenose components, which is constantly undergoing change with the development of the earth's lithosphere. The organic world (plants, animals, micro-organisms) develops and is distributed according to its own laws, as do the phyto-, zoo-, and micro-biocoenoses formed by it. Thus every biogeocoenose is the arena of the most complex and diverse internally-contradictory processes, which are the origin, the moving force of the biogeocoenotic process, and of the processes of soil and litter evolution. These processes mostly operate slowly, but they never cease, and ultimately lead to replacement (succession) of one biogeocoenose by another.

In the development of forest-biogeocoenotic cover a major role is played by successions of phytocoenoses; the mechanisms of each phyto-coenose succession, and therefore of each succession of biogeocoenoses as a whole, is the expulsion of some species by others in the course of their interspecific struggle for existence, their competition. In the evolution of the organic world the chief moving force in the process of species formation is natural selection, a result of intraspecific competition and both intra- and inter-specific struggle for existence (for more detailed discussion of this point see Shmal'gauzen 1946, Sukachev 1956). In the replacement of one biogeocoenose by another the chief guiding role is naturally played by interspecific relationships. Intraspecific relationships, while having some effect on the latter, exert only minor influence on biogeocoenose replacements.

All of the above complex contradictory processes result in the biogeo-coenotic (including the forest) cover of the earth being always in movement, always developing. There are not, and cannot be, biogeocoenoses that are permanent, whose development has stopped. While there are no grounds for speaking of a 'climax' in the sense in which that word has previously been understood, and is still occasionally used in foreign literature with reference to vegetation, there is still less reason to use the term with regard to biogeocoenotic cover and separate forest biogeo-coenoses. But since the rates at which the dynamics of biogeocoenoses operate vary greatly, and usually the natural process of development of biogeocoenotic cover takes place rapidly at first and then decelerates, it is permissible to apply the term 'climax', in a conditional sense, to the latter stage. In this case, however, it would be better to speak of a developed biogeocoenose and of developed biogeocoenotic cover, always remembering that even developed biogeocoenoses are subject to replacements, to successions.

2. There are two directions in which plant cover develops: progressive and regressive (for more detailed discussion see Bykov 1957, p. 272, and Aleksandrova 1963).

The symptoms of progressive development are, for the most part, com-plexity of organisation and increase in mass of phytocoenose per unit of land surface area, better use of habitat, and increased amount of matter and energy brought into biological circulation by the phytocoenose (linked with more intensive alteration of its environment), etc. The process of evolution of organisms, as is well known, is characterised by increase in their relative independence of their environment; they become more emancipated from it and acquire relatively greater autonomy in their growth and development. They still, however, do not break away from their environment, since they cannot exist without it; they merely become able to transform the environment more effectively in their own interests and to subject its properties to their uses. Phytocoenoses, as well as biogeocoenoses as a whole, also develop along these lines. This is only a general law, however, and the phytocoenotic process does not always follow the path of progressive development. Cases are observed where the progressive movement of the phytocoenotic process takes a direction opposite to that described above and to the features that we consider progressive. It then acquires a regressive character, that is, it tends towards simplification of organisation, towards decrease in plant mass, towards reduction in use of environment, and so on.

We may also speak of progression and regression with regard to the development of biogeocoenotic cover. In biogeocoenology also the signs of progression must be taken to be complexity of organisation and structure, and a more intensive biogeocoenotic process of exchange of matter among the biogeocoenose components, embracing all of these components more thoroughly and in more different ways. But we cannot speak of it in the same way regarding exchange of matter and energy. We must distinguish two forms of exchange of matter and energy with the environment: one leads to accumulation of matter and energy in the biogeocoenose and the other is accompanied by losses of them. In the first case we may speak of a progressive trend in the process, but in the second the trend is regressive. In the first case, as a rule, forest biogeocoenoses develop more usefully for man, but in the second their value diminishes.

Unfortunately there have as yet been few investigations in our country that throw full light on the process of matter and energy exchange either within a forest biogeocoenose or between it and the environment. Therefore we cannot yet establish series of progressive and regressive successions of forest biogeocoenoses. As examples of progressive successions we may mention the replacement of herbaceous by forest biogeocoenoses in syngenetic biogeocoenose successions on a new land surface where occupation by vegetation is beginning, or the replacement of a willow-poplar biogeocoenose by an oak one in a river valley. We may consider endodynamic processes, leading to swamping of forests and replacement of forest by bog biogeocoenoses, as being regressive.

I. M. Zabelin (1960, 1963) speaks of progress in the development of the entire biosphere in a fairly similar sense (calling it, moreover, the 'biogenosphere', the sphere in which life arises). Having in mind this development of the biosphere, he writes of growth of the autonomy of the bio-genosphere, of its segregation from the cosmos and from other parts of the earth, of increase in the number and distinctiveness of the features of its independence as a natural formation, and of the activation and complexity of the inter-relationships and interdependences of the components of the biogenosphere, as a result of which it is constantly becoming a more integrated natural formation.

3. The above-described replacements of certain forest biogeocoenoses by others, excluding those of a catastrophic nature, take place slowly and gradually. The question arises: at what moment during the changes does it become possible to speak of a replacement of biogeocoenoses as having taken place, or (more precisely) of the appearance of a biogeocoenose of a new, different type?

As was stated in the opening chapter, a new type of biogeocoenose arises when changes in all or most of its components are so great that a biogeocoenose has become qualitatively different and substantial alterations have occurred in the character of matter and energy exchange both within the biogeocoenose (between its components) and between the components and other natural phenomena. Although it is sometimes difficult at present to give strictly objective criteria for deciding whether a new type of biogeocoenose has or has not appeared, since the process of matter and energy exchange in biogeocoenoses is still little studied, there is no reason to exaggerate the difficulty. In field investigation, as a rule, the question can be decided with sufficient assurance by using the above criteria.

4. The forms of the dynamics of forest biogeocoenoses are consequently very diverse. They may be classified in different ways, using different characteristics as the basis of classification.

One of the most obvious differentiating features of successions of plant cover is the varying duration of the time in which they are realised. This criterion has been used by a number of authors in phytocoenology. (For more detailed discussion see the survey by Aleksandrova, 1963.) For instance, E. M. Lavrenko (1959) divides all successions into: (i) secular, including successions of geological dimensions, involved in the evolution of continents and flora; (ii) prolonged, stretching out over decades and sometimes centuries (e.g., the replacement by new spruce forests of birch or pine forests that have appeared as a result of the cutting down of spruce forests); (iii) rapid, taking place within a few years or a few decades (e.g., succession of forest biogeocoenoses due to age-changes in the tree stands); (iv) catastrophic, when because of external factors mentioned above (e.g., tree-felling, torrential floods, fires, etc.) the original forest biogeocoenose undergoes a sudden violent change or even is entirely destroyed, and syngenesis begins afresh.

The above classification may also be applied to forest biogeocoenoses. Accordingly we may distinguish the following biogeocoenotic successions of forests: secular, prolonged, rapid, and catastrophic. This classification can be of practical value.

There is also a visual classification of successions based on the soil conditions (or, as sylviculturists say, on the forest-growing conditions) in which the successions take place (Clements, G. F. Morozov, *et al.*).

It is more useful, however, to base the classification of the various forms of forest-biogeocoenose and phytocoenose dynamics on the causes that produce them. The classification then takes the following form:

I *Cyclic (periodic) dynamics of forest biogeocoenoses.* (Reversible changes in forest biogeocoenoses.)
   (*i*) Daily changes in biogeocoenoses.
   (*ii*) Seasonal changes in biogeocoenoses.
   (*iii*) Annual changes in biogeocoenoses.
   (*iv*) Changes in biogeocoenoses due to the process of regeneration and growth of arboreal and other vegetation:
      (*a*) regular regeneration of arboreal vegetation;
      (*b*) irregular (wave) regeneration of tree stands;
      (*c*) synusial dynamics, especially parcel dynamics.

II *Dynamics of the forest-biogeocoenotic cover of the earth, or successions of forest biogeocoenoses.*
   (*i*) Autogenous (irreversible) successions of biogeocoenoses (development of the forest phytogeosphere, of forest biogeocoenogenesis).
      (*a*) Syngenetic succession of biogeocoenoses.
      (*b*) Endogenous (endodynamic) successions of biogeocoenoses.
      (*c*) Phylocoenogenetic successions of biogeocoenoses:
         (*i*) phytophylocoenogenetic successions of biogeocoenoses;
         (*ii*) zoophylocoenogenetic successions of biogeocoenoses.
   (*ii*) Exogenous (reversible and irreversible) successions of biogeocoenoses.
      (*a*) Hologenetic (irreversible) successions of biogeocoenoses:
         (*i*) climatogenic successions of biogeocoenoses;
         (*ii*) geomorphogenic successions of biogeocoenoses;
         (*iii*) selectocoenogenetic, or areogenic, successions of biogeocoenoses:
            (*a*) phytoareogenic successions of biogeocoenoses;
            (*b*) zooareogenic successions of biogeocoenoses.
      (*b*) Local (reversible and irreversible) catastrophic successions of biogeocoenoses:
         (*i*) anthropogenic successions of biogeocoenoses;
         (*ii*) zoogenic successions of biogeocoenoses;
         (*iii*) pyrogenic successions of biogeocoenoses;
         (*iv*) storm-damage successions of biogeocoenoses;
         (*v*) successions of biogeocoenoses produced by torrential floods, landslides, sudden inundations, and such causes.

Cyclic changes in forest biogeocoenoses are characterised by periodic returns to a state practically identical with the original, although, of course, there is no absolute identity with it. For instance, as stated above, even after a daily change in a biogeocoenose it is not exactly the same as it was the day before, although we may think it identical. Such changes cannot be called successions in the sense given above to the term. This category includes all the changes occurring in the course of a day or a season, as well as those resulting from differences in the weather of separate years or the self-restoration process of plant cover, mainly of trees.

Successional changes in forest biogeocoenoses result in the creation of new biogeocoenoses. Such changes are usually irreversible, but sometimes, after a variable time interval, a biogeocoenose of the original type is regenerated because the causes that led to its replacement have been eliminated. The dynamic process is then of a reversible character.

A large subdivision of biogeocoenotic changes consists of those induced by internal opposing interactions of components, which are always in operation within biogeocoenoses and which are a source of change and of self-development of forest-biogeocoenotic cover. In this group we include syngenesis and endogenesis of biogeocoenotic cover.

Changes in forest biogeocoenoses due to external causes, i.e., exogenous successions, form a second large subdivision. It includes two groups of successions.

One type of succession is due to large-scale natural phenomena, sometimes planet-wide in scope. This group includes successions resulting from general changes in climate (climatic); from geological processes that alter the ecotopes of biogeocoenoses through alterations in geomorphological conditions (tectonic and other earth movements, suffosion processes, development of river valleys, erosion, drainage and irrigation measures, etc.). Such successions are irreversible, often take place slowly, and may be classified as secular. In this group we include successions due to dispersal of plants and animals, when one or more new species are extending their range into a biogeocoenose and forcing out some of the original species, thereby changing the biogeocoenose as a whole. This process is linked with the history of formation or building-up of plant communities, which was formerly called selectocoenogenesis (Sukachev 1944). It may originate either in a phytocoenose or in a zoocoenose, that is, it may be either phytoselectocoenogenesis or zooselectocoenogenesis (in combination they may be called bioselectocoenogenesis).

Successions induced by these processes and embracing whole types of biogeocoenoses may be called phyto- or zoo-areogenic. They are irreversible, and, taking place sometimes slowly, sometimes more rapidly, may often approach the secular type.

There are also cases where organisms included in a certain type of biogeocoenose (in which abiogenic components remain for a long time in an unchanged state) become so altered by species-forming processes that they are transformed into new species with distinctive new characteristics, and then they change the type of biogeocoenose. Such processes can take place only where climatic factors remain stable for centuries and in territories that keep their ecotypes practically unchanged for a very long time. In nature such conditions are apparently very rare, because over a long period of time ecotopes are altered in various ways. Successions due to these factors are called phytophylogenic or zoophylogenic. There is reason to suspect that they have occurred in a number of places in Southeast Asia. These successions must be classified as secular; they also are irreversible. They are linked with the phenomenon called phylocoenogenesis (Sukachev 1944) and depend on the phylogenic development of the species composing the biogeocoenose.

Often species change during dispersal, and then biocoenogenesis, i.e., the formation of new biocoenoses with new inter-relationships among organisms (and therefore also between them and the abiogenic components of biogeocoenoses), cannot be separated from the process of selectocoenogenesis. In this case selectocoenogenetic (areogenic) biogeocoenose successions merge into the phylocoenogenetic type.

The last large group of successions results in local replacements of biogeocoenotic cover. They usually begin suddenly and run their course rapidly, being due to some cause that acts briefly but strongly. Such changes are often segregated under the name 'catastrophic'. They act only upon a single biogeocoenose, or upon a group of biogeocoenoses located comparatively close together. After sudden destruction of, or serious damage to, a forest biogeocoenose, a derivative one develops fairly quickly from what remains, or, if the tree stand or the entire plant cover has been destroyed, a new syngenetic process begins. These successions may be due to various causes. The prime cause is human activities, whose magnitude and extent we have described (tree-felling, cultivation and general elimination of primary plant cover in order to replace it with crops of various kinds, trampling-down of vegetation, etc.), i.e., anthropogenic successions. We include in this group successions of forest cover resulting from forest fires, regardless of whether they are caused by human or natural agencies. Such successions may be produced in a forest by storm damage, torrential floods, sudden landslides, etc.

More complex is the problem of classification of successions caused by intensive animal activity, especially by pest outbreaks (e.g., pine moths, bark beetles, locusts, etc.). We include successions caused by the pasturing of domestic animals in the last-mentioned group. When at certain times animals that normally (in somewhat limited numbers) form part of the zoocoenotic component of a forest biogeocoenose multiply excessively and produce both death of tree stands and succession in forest biogeocoenoses, such successions must be considered endogenous. They must not, however, be included among successions due to the self-development of forest biogeocoenoses, because in fact they distort the natural process of development of forest-biogeocoenotic cover. Such successions, mostly due to pest outbreaks, ultimately induce the appearance of short-term forest biogeocoenoses. They are reversible. Therefore it is more correct to include them in the preceding (local, catastrophic) group of successions.

In many cases observed changes in biogeocoenoses cannot easily be referred to any of the chief subdivisions of dynamics; that is, it is difficult to say whether they should be classified as successions or not. An example is a change in a biogeocoenose due to weather fluctuations, to climatic conditions in different years (annual dynamics). If climatic changes extend over several successive years the changes in a forest biogeocoenose become so great and relatively so permanent that a forest biogeocoenose of another type is formed and retains its characteristics even if years with the former climatic conditions follow. In this case there are certainly successions of forest biogeocoenoses. It is often difficult, however, to give clear, absolute criteria for deciding whether or not a new type of forest biogeocoenose has appeared. Therefore divergence in the views of investigators is always possible.

We cannot exclude the possibility that successive (even if intermittent) identical annual changes in weather may be so strongly reflected in a forest biogeocoenose and induce such changes that during a series of such periods they lead to a succession of biogeocoenoses. In this case we may say that the annual dynamics of biogeocoenoses prepare them for replacement. That phenomenon was studied by Yaroshenko (1961), with particular reference to replacement of phytocoenoses.

We may quote another example where it is not easy to distinguish cyclic dynamics from successions. After trees have been felled or destroyed in a catastrophe the former type of forest usually regenerates. Not always are all the features of the original forest biogeocoenose restored. Some of the successional stages of the developing biogeocoenose may be similar to independent types of forest biogeocoenoses. K. V. Kiseleva (1962) describes age-changes in spruce forests in the north-west of Moscow province, where, having regenerated after cutting or fires, they pass successively in the mature state through the stages of *Oxalis* spruce stands and composite spruce stands with hazel. The latter, as is well known, frequently occur as types of spruce stands already formed and of long standing.

P. D. Yaroshenko (1961) regards successions as replacements of phytocoenoses that do not cover extensive territory and are produced by purely-local causes. He also calls local replacements 'partial replacements', and classifies them as:

I Natural replacements:
  (*i*) successive:
    (*a*) endoecogenetic;
    (*b*) hologenetic;

    (*ii*) sudden:
        (*a*) climatogenic;
        (*b*) edaphogenic;
        (*c*) biogenic;
II Anthropogenic replacements:
    (*i*) successive;
    (*ii*) sudden.

By 'general replacements' he means those connected with the general history of the plant cover, caused by geologo-historical landscape development.

By 'evolution of coenoses' he means the development of new, not-previously-existing types of plant communities, i.e., the process mentioned above and called coenogenesis. His classification of replacements of phytocoenoses, differs little from that proposed by me (Sukachev 1944). I prefer, however, to follow the system of dynamics presented above for both phytocoenotic and forest-biogeocoenotic cover.

In using this classification of dynamics of forest-biogeocoenotic cover we must bear in mind the following points:

1. Although the first category of cyclic changes includes, as stated above, reversible changes, at the same time a process of autogenic replacements is always taking place in forest biogeocoenoses, never ceasing so long as the territory contains a living population, i.e., the biogeocoenotic process. Hologenetic successions may also take place simultaneously. But since both operate slowly, we do not observe them against the background of various more obvious cyclic changes.

2. It is not always possible to draw a sharp dividing line between hologenetic and local successions.

3. We must stress particularly that in every case an observed succession usually results from several simultaneous causes, and different forms of succession may be overlaid one upon another. As a rule, however, at a given moment some one cause predominates. In order to be able to direct biogeocoenotic changes in the interests of the national economy, we must analyse an observed succession from different aspects and define the causes producing it, evaluating the significance of each of them.

CHAPTER VIII

Principles of construction of a classification of forest biogeocoenoses

Although biogeocoenology is a young science, and relatively little study is generally given to the principal features of the biogeosphere, recent literature shows that much interest is being taken in the problem of typology and classification of biogeocoenoses, with special reference to forests. That problem was discussed on a broad scale in a special symposium at the IX International Botanical Congress in 1959, in reports by Hills, Rowe, Ovington, Krajina, Dansereau, and others (*Silva fennica*, 1960, No. 105). Separate aspects of it were examined in the same year at the Conference on Problems of Classification of Ural Vegetation, in reports by V. B. Sochava, N. V. Timofeev-Resovskii, and A. G. Dolukhanov (*Problems of Classification of Vegetation*, 1961).

In the opinion of most investigators, biogeocoenotic classification of forests can provide objectively a more thorough and many-sided description of a forest than, for example, phytocoenotic classification, and can serve to reconcile the views of different schools of forest typology. The latter received much attention in the reports and discussion of the International Symposium on Forest Typology.

A practical solution of the problem of classifying ecosystems (biogeocoenoses) is seen by some investigators in a combination of the characteristics of living organisms and of the environment in which they live at local and regional levels, i.e., within the framework of division into natural districts (Halliday 1937, Hills 1952, 1961, Rowe 1959, Coaldrake 1961, Sochava 1961). Others see it in differentiation of biogeocoenotic classification—beginning with the grade 'group of types of biogeocoenoses' —into independent parts that reflect the characteristics of vegetation and ecotopes separately (Dolukhanov 1959, 1961). A third group prefer systematisation of the multiformity of the biogeosphere on the basis of typology of biogeocoenotic exchange of matter and energy (Sukachev 1947, 1949, Timofeev-Resovskii 1959, 1961, and in essence also Ovington 1961).

If we consider a biogeocoenose to be a qualitatively-unique natural phenomenon, whose principal specific feature is the interlinked metabolism of its components, the approach of Sukachev and Timofeev-Resovskii to the classification of biogeocoenoses must be considered the most correct in principle.

Applied to forests, this approach should lead ultimately to a system of classification for the forests of the world, constructed with due regard to the degree of accuracy attained in classifying them by the nature of the process of matter and energy exchange among the biogeocoenose components, or, using N. V. Timofeev-Resovskii's terminology, constructed on typology of the geochemical activity of biogeocoenotic systems. A classification should be complete and the principles of classification should be maintained throughout. We do not agree that 'a more or less complete and many-sided biogeocoenotic classification should consist of at least *three independent parts* (italics ours, N.D.): classifications of biocoenoses, classifications of succession cycles of development, and classifications of physico-geographical conditions of growth' (Dolukhanov 1959). A biogeocoenose is not the sum of a biocoenose and its environment, but an integral and qualitatively-individualised phenomenon of nature, acting and developing according to its own laws, the foundation of which is the metabolism of its components, perfected under the action of solar radiation.

For that reason a classification of biogeocoenoses can be constructed only as a classification of natural units and not of mechanical sum-totals, and on the basis of features inherent in these units and not in their separate components. Therefore there cannot be a biogeocoenotic classification of vegetation (Dolukhanov 1957), soils, animal life, etc.; there can properly be only biogeocoenotic evaluation of the roles of these components in the metabolism and development of biogeocoenoses.

Experiments in the construction of biogeocoenotic classifications on the basis of typology of biogeocoenotic metabolism have not been made, although there are some systems (of methodological interest on account of the way in which they are devised) of types of biological circulation of elements, types of physico-geographical environment, types of territory, and geochemical types of landscapes, worked out by a number of investigators (Williams 1936, Grigor'ev 1938, 1939, 1942, Gozhev 1945, 1946, 1956, Perel'man 1960) on a basis similar to biogeocoenology—exchange (balance, intensity) of matter and energy between components of the physico-geographical envelope of the globe.

A number of the classifications proposed on that basis by the above investigators can even be used directly in the construction of biogeocoenotic classifications.

Biogeocoenotic classifications based on direct indices of biogeocoenotic metabolism become more feasible because of the investigations of matter exchange between separate components of the biogeosphere, which have recently been intensified both here and abroad. Data on matter circulation in the vegetation-soil system are rapidly accumulating. In forest pedology the subject has practically taken the lead among all modern studies. Compilation and comparison of the material obtained enables us to draw clear boundaries based on the nature of the circulation of a number of substances, not only between such different types of the biogeosphere as tundra, forest, steppe, desert (Bazilevich 1955, Perel'man 1960, 1961) but even between finer subdivisions, e.g., between composite and green-moss spruce stands, different types of pine stands, etc. (Smirnova 1951 a, b, 1952, Remezov et al. 1949, Remezov et al. 1959). Separate works have appeared, devoted also to the energy side of biogeocoenotic metabolism (Ovington 1961, Molchanov 1961 a, b, Ovington and Heitkamp 1960). Ovington's investigations have shown, for instance, substantial differences in energy accumulation between the ecosystems of coniferous and deciduous forests.

As yet the data are insufficient to solve the problem of classifying the earth's biogeocoenoses, and much work still lies ahead to obtain basic material on all aspects of biogeocoenotic metabolism. The possibility of solving it in the above manner, however, is quite clearly indicated by the data already obtained and we believe that if research continues as at present rate the solution is near.

Because of the shortage of direct data on the scale, intensity, and direction of matter and energy exchange in biogeocoenoses throughout the world, it is necessary in developing a biogeocoenotic classification to use a number of indirect indices of biogeocoenotic metabolism that are visible externally and are clearly and convincingly connected with the main biogeocoenotic process. A fairly large number of these items may be selected, which, in spite of the unfortunate absence of direct metabolic data in many cases, provide extensive opportunities for cross-checking of results at each stage of classification. It is particularly appropriate to use such indices as: the radial dimensions of the biogeocoenotic layer; the duration and rhythm of biogeocoenotic exchange, and especially the characteristics of growth of vegetation and life-activity of fauna; heat-exchange; moisture-exchange; stocks of living matter; current productivity, and nature of organic production; rate and specific features of decomposition of organic matter; scale and forms of partial losses of matter and energy from current circulation by their accumulation in wood, peat, litter, and humus; relationship and regime of downflow and upflow in the soil; direction, rate, and chemical characteristics of outflow through the soil of products of biogeocoenose life-activity; soil respiration; soil aeration and its regime; soil temperature; rate and scale of accession to a biogeocoenose of volcanic products, dust, salts, fluvial and marine silt; and relief.

Some of these indices relate to gross aspects of metabolism, others to finer details, which makes it possible to compile a system of classification of biogeocoenoses using the same principle throughout all grades.

With regard to the principles of classification of the biogeosphere we must point out that the biogeosphere, unlike such concepts as physico-geographical envelope (A. A. Grigor'ev), geochora (Yu. P. Byallovich), and epigenema (R. I. Abolin), is not of global extent. It occupies only that part of the earth's surface where living matter is present, even though only for a brief period, and takes part in the processes of exchange, circulation, or migration. Unlike landscapes, choras, types of territory, epigenemas, etc., biogeocoenoses cannot exist for ever without life. We must therefore exclude from the biogeosphere: permanent snow and ice, some kinds of hot sandy deserts, very salt lakes (?), fresh lava flows, fresh landslides, fresh scree and washouts, fresh alluvium, many sandy and shingly sea beaches, quarries, railway and highway roadbeds, and urban territory (except gardens, parks, and squares). The biogeosphere thus conceived has considerable vertical extent. V. N. Sukachev (1960) believes that the vertical boundaries of the biogeosphere are the upper and lower limits of the phytogeosphere in the sense of E. M. Lavrenko (1949), which embraces the lower layer of the troposphere, including the part penetrated by plants plus some 10-30 metres above the vegetation, and also the soil and subsoil. Theoretically, perhaps, it is more correct to set the limits of the vertical dimension of the biogeosphere at those layers of the troposphere and the lithosphere where the specific characteristics of horizontal anisotropy of biogeocoenotic metabolism cease, but unfortunately we do not yet have any factual data on this subject. For the purpose of practical differentiation of the biogeosphere into homogeneous natural units—biogeocoenoses—it is perhaps more useful not to take the whole depth of the biogeosphere into account, but only the part with the most sharply defined external outlines: the phytosphere, in V. B. Sochava's sense, and the soil.

Although biogeocoenogenesis has been proceeding on earth without interruption since the most remote geological epochs, there are many places in the modern biogeosphere where we do not find fully-developed and fully-active biogeocoenoses. Even in optimal living conditions of heat and moisture exchange there are places thinly populated with plants and animals because of the 'youth' of the substrate: growing alluvial deposits, sandbanks, screes with plant growth, washouts, dunes, lava flows, reefs, etc. There the biogeocoenotic complex is underdeveloped, its structure and relationships have not yet become stabilised, and the biological side of matter and energy circulation is very weak. On that account, it is appropriate to differentiate between *mature* or built-up biogeocoenoses with well-developed and metabolically-active relationships and structure, and *young* choras, underdeveloped in that respect. It is useful to segregate, by their origins, *radical* biogeocoenoses formed in a chora in the course of natural historical development and distinguished by deep-seated and stable adaptations among their components, and *derivative* biogeocoenoses that arise in the place of the former because of various disruptions of the normal course of biogeocoenogenesis, often with thoroughgoing reorganisation of all interactions and of biogeocoenotic metabolism, as is very clearly seen in biogeocoenoses created by man by sowing and planting with the use of advanced agrotechnical methods (cultural biogeocoenoses).

Among natural biogeocoenoses we have those that are *complete*, consisting of a full assemblage of biogeocoenotic components (atmosphere, lithosphere, pedosphere, vegetation, animal and microbial populations) and are therefore distinguished by very complex biogeocoenotic metabolism and a most complex morphological structure; and those that are *incomplete*, lacking some component, with impoverished structure and simplified metabolism, e.g., without soil, without atmosphere, or without vegetation. The latter are normal for the aqueous sector of the biogeosphere, where there is no soil and (with some exceptions) no atmosphere (biohydrocoenoses), but they occur also on land, e.g., biogeocoenoses of bird bazaars, where there is no vegetation (if we exclude seaweed washed up on the rocks) and where therefore there are no balanced *in situ* autotrophic-heterotrophic cycles of biological circulation. The same category includes off-shore and peat-bog biogeocoenoses, which lack such components as soil, and perhaps some others.

Among both aquatic and terrestrial biogeocoenoses there are some in which autotrophic living matter is developed in a complex and compact mass, and where consequently biological circulation of matter and energy proceeds vigorously and all features of biogeocoenotic mutual exchange are *completely* affected and transformed by phytocoenoses. In such biogeocoenoses there are no *direct connections* and therefore no *direct exchange* between, for instance, the atmosphere and the lithosphere. Their contact and interactions are entirely controlled and altered by the action of the green plant cover. Forests, meadows, and bogs are biogeocoenoses of that type and may be regarded as *biologically closed*. Many biogeocoenoses have a very limited mass of living matter of autotrophic organisms, distributed over the chora not continuously but in patches or thinly. They are distinguished by a lower rate of biological circulation, which has very little influence on the general biogeocoenotic metabolism, and by *constant direct contact* and *direct exchange* between the atmosphere and the soil or other substrate. Such biogeocoenoses are characterised, on the one hand, by terrain that is in various ways difficult for life (hot and cold deserts, stony mountain peaks, etc.) and on the other hand by the first developmental stages of biogeocoenogenesis on geologically-young substrates. They may be called *biologically open*.

The biogeosphere is divided by its structure and metabolism into two broad sectors (divisions, fields, or systems): terrestrial and aquatic. It is possible also to distinguish an intermediate (off-shore) aquatic zone of the biogeosphere.

In view of the scope of this book we shall not dwell on possible subdivisions of the aquatic sector of the biogeosphere, which indeed are almost unstudied biogeocoenotically, but shall restrict ourselves to discussion of the terrestrial sector.

Amid the great variety of biogeocoenotic cover of dry land, the most striking feature is the great differences in depth (essential dimensions) of the layer of the earth's envelope in which most biological metabolism takes place (landscape depth, according to Perel'man) and the *density* of its occupation by autotrophic organisms (i.e., the degree of biological closure). The deeper the layer is and the more densely it is populated by organisms, the greater is its biomass and the greater the role of living matter not only in the total exchange but also in its effect on interactions between other biogeocoenose components.

The depth of the layer over dry land varies from a few millimetres or centimetres (e.g., on rocks among crustose lichens) or decimetres (e.g., on moss or lichen tundras, wasteland meadows, etc.) to metres (on steppes, meadows, beds of Alpine or other shrubbery, stunted forests), or tens or even hundreds of metres (in forests). V. I. Vernadskii (1926) drew attention to the diversified significance of this feature for the earth's 'film of life'; and it was used under the name of 'landscape depth' by A. I. Perel'-man (1960) in devising principles for the geochemical classification of landscapes. Not all subdivisions of the biogeosphere can, of course, be identified by this feature, but it very clearly and visually (or, using Yu. K. Efremov's (1960) terminology, picturesquely) distinguishes the forest division[1] in which we are now interested from all other varieties in the biogeosphere.

The forest division of the biogeosphere is also characterised, in contrast to other biogeosphere divisions, by certain distinctive features, reflecting in various degrees the characteristics of biological metabolism in forests:

1. Well-balanced heat and moisture exchange. The moisture coefficient, according to N. N. Ivanov (1948), is unity or higher.

2. Maximal and *year-round* intermediate position of living matter between contacts of radiation, water, and atmospheric gases with the lithosphere, and maximal and *year-round* alteration of the internal climate of biogeocoenoses, whereas in most other divisions of the biogeosphere there is definite periodicity in these respects.

3. Maximal removal of part of the matter and energy from the annual biological cycle (as a result of their accumulation for long periods in the wood of trunks, branches, and roots, in bark, and sometimes in leaves) in the form of the highest stocks of living plant mass in the biogeosphere, amounting to tens and hundreds of tons per hectare (dry weight), as compared with one-hundredth as much in other parts of the biogeosphere.

4. Strongly-marked predominance of above-ground over underground

---

[1] In original, 'forest type'. Here, and below in similar contexts, the author uses 'type' and 'subtype' in a sense different from their previous use, to denote divisions and subdivisions of the biogeosphere. In Russian the distinction between this 'forest type' (i.e., biogeosphere division pertaining to forests) and 'type of forest' is clear, but difficult to express in English without ambiguity. Therefore 'forest type of biogeosphere' and 'subtype' are translated 'forest division' and 'subdivision'. It is of interest that the original book contains an 'Index of Terms' in which there are nine page references to 'type' in the sense of 'kind', but none to 'type' in the sense (as here) of 'division of the biogeosphere'. (*Tr.*)

organic matter (3 or 4 times as much). In this respect forest biogeocoenoses differ sharply from steppe and desert biogeocoenoses, for instance, where the proportions of above-ground and underground masses in phytocoenoses are reversed: on the steppes the weight of roots is twice as large as that of above-ground parts, and in deserts three times as large (Bazilevich 1955).

5. Increased mobility of soil solutions, and increased transportation of easily-soluble mineral nutrients out of terrestrial biogeocoenoses by water.

6. Predominance of aerobic fungi in the processes of decomposition of organic matter and liberation of energy and mineral compounds, a feature especially stressed by V. P. Williams (1936).

7. Predominance of the ammonising process in the transformation of soil-contained nitrogen.

8. Presence of distinctive litter, due to some retardation (occurring in forest conditions) of the decomposition of organic litter reaching the soil both from current plant growth and from stocks of organic matter removed from circulation (mainly by trees and shrubs) many years earlier.

It is possible that in the course of further investigations some other distinctive features of forest biogeocoenoses will be discovered, but diligent comparative study of all the divisions of the biogeosphere is required, with synthesis of the data obtained into a single terrestrial biogeocoenotic classification.

The forest division of the biogeosphere occupies a vast area and is very heterogeneous in the conditions, character, and forms of manifestation of biogeocoenotic metabolism, but with the present status of factual material we are very far from any real possibility of classifying that diversity of forest biogeocoenoses in a well-ordered system of graduated units.

We quote below some examples of possible biogeocoenotic classification of forests at different levels on the basis of the above 'representative' features of biogeocoenotic metabolism as a whole or of separate parts of it. Among these features we must give a very high place to the *rhythm of biogeocoenotic metabolism*. This feature is most clearly displayed in the annual course of development of a phytocoenose and in the annual course of variation in heat and moisture conditions, but it can also be traced in the 'behaviour' of other components (e.g., fauna and soil). It is important systematically for all grades of biogeocoenotic classification of forests, but its differentiating value is most outstanding at the highest levels. On the basis of this feature the forest division of the biogeosphere is naturally separated into two subdivisions:

1. forests with a regular biogeocoenotic metabolism that continues without interruption throughout the annual cycle, as a result of uninterrupted and uniform reception of heat and moisture by the biogeocoenose;

2. forests with an intermittent biogeocoenotic metabolism more or less clearly displayed, resulting from seasonal lowering of the inflow of heat and moisture.

The first subdivision includes forests of the constantly-humid equatorial and tropical districts of the globe, the biogeocoenotic characteristics of which are very well summarised in the works of A. A. Grigor'ev (1938, 1939) and A. D. Gozhev (1946, 1956).

The second subdivision includes all forests of temperate latitudes and a large part of the forests of the tropical and subtropical zones of the earth; in these there are long periods without precipitation or (in mountain districts) heat.

Both subdivisions of the forested biogeosphere are very diverse in form, and may be further divided into many sections homogeneous in their metabolism. From the biogeocoenotic point of view the following categories can be distinguished in the first subdivision (forests with regular and uninterrupted annual metabolism).

1. Humid-tropical and equatorial forests, with their extremely regular and intensive heat and moisture exchange, high productivity and uninterrupted metabolic phytocoenotic activity, highly-intensive and thorough weathering of the soil because of the regularity and large amount of precipitation, and rapid and complete decomposition of organic litter so that a continuous layer of litter is never formed on the soil surface. The entire biogeocoenotic metabolism in the forest is overwhelmingly influenced by the forest vegetation.

2. Mangrove forests, distinguished by the fact that part of the daily exchange between their biogeocoenose components takes place in an aerial-terrestrial and part in an aerial-aquatic environment because of daily sea-water inundations; for the same reason there is no soil, in the strict sense of the word, the soil being replaced by a substrate of marine silt, always soaked with salt water, badly aerated, and with a high content of sulphates and chlorides of sodium and of hydrogen sulphide. In the nature of their metabolic processes mangrove forests are on the borderline between the terrestrial and aquatic sectors of the biogeosphere.

3. Swampy tropical forests or, in other terminology, forested tropical swamps. As biogeocoenotic systems they are distinguished by accumulation and little movement of very acid soil moisture, by marked shortage of oxygen in the soil and subsoil, and by slow decomposition of organic matter, which accumulates on the ground as a damp layer of arboreal-herbaceous peat, sometimes of considerable depth.

There are some other categories in the subdivision of forests whose biogeocoenotic functions proceed at a uniform rate.

In forests with intermittent biogeocoenotic metabolism two groups are clearly differentiated:

1. group one is characterised by a reduction in biogeocoenotic metabolism due to periodic cessation of moisture influx (hydroperiodic forests);

2. group two is characterised by a marked biogeocoenotic pause due to replacement of a warm by a cold period (thermoperiodic forests).

The first group includes savannah and monsoon tropical forests, and partly monsoon subtropical forests. Since the warm period of the year is uninterrupted, the cessation in biogeocoenotic metabolism is not profound and universal. Even the component most sensitive to moisture (vegetation) does not entirely discontinue its biogeocoenotic metabolism during the dry period of the year. Even at the end of the dry period, when the stocks of available moisture in the soil seem to be completely exhausted, many plants do not shed their leaves and continue to function, though at a reduced rate. At that time there may be abundant blossoming of many trees and shrubs and even growth of shoots. There is no cessation in the activities of fauna or in soil processes (respiration, movement of solutions as a result of rising moisture, etc.). During the wet periods biogeocoenotic metabolism in the forests recommences on a scale and at a rate probably no lower than in forests with regular uninterrupted metabolism.

The periodicity of biogeocoenotic metabolism has a different character in the group of forests with a pause due to a cold season replacing a warm one. The onset of long-lasting cold practically puts an end to the biogeocoenotic work even of evergreen plants, sharply reduces the activity of fauna (migration, hibernation), stops the course of almost all soil processes, the work of micro-organisms, and gas-exchange between soil and atmosphere. The biogeocoenotic characteristics of forests of this group are most clearly displayed in forests of cold and moderately-cold climates of the earth, with more or less long-lasting snow cover and mild or warm vegetative periods. Almost all of the forests of the Soviet Union fall into this subdivision. As biogeocoenotic systems they are very diverse, and among them we may segregate many biogeocoenotic classes according to the duration of the heat-pause, to the daily or seasonal rhythms of biogeocoenotic metabolism caused by local photoperiodism and amounts of heat and moisture, to biological production, and to many other factors. Among such classes are the following:

1. Northern forest-tundra, northern-taiga, and high-mountain forest biogeocoenoses, distinguished by: very short biogeocoenotic metabolism period (about three months); low annual production of organic matter, especially by trees, because of shortage of heat in the atmosphere and the soil; small amounts of organic matter in wood (or in needles), in spite of the partial 24-hour synthesis by green parts; shallow weathering of the substrate and very small outflow beyond the biogeosphere of mobile products of soil-formation, especially when permafrost is near the surface; shallowness of the air-soil zone affected by biogeocoenotic metabolism (5-15 metres).

2. Taiga biogeocoenoses, distinguished by: longer and more intense biogeocoenotic metabolism (4·5-5 months); considerably higher annual

productivity of organic matter and larger amounts of wood (or needles); deep substrate weathering and high mobility of chemical elements in the soil because of abundant atmospheric precipitation and better soil warming; considerable outflow of part of the mobile products of soil-formation beyond the biogeosphere, in spite of the fairly intensive accumulation of many of them in horizons A_1 and A_0; greater depth of the biogeocoenotic layer (20-40 metres), with much biological cover being uniform horizontally.

3. Broad-leaved biogeocoenoses, distinguished by: the longest period of biogeocoenotic metabolism (5-6 months); the highest (in this group of classes) annual productivity of organic matter and amount of wood in phytocoenoses; higher biogenic accumulation of many chemical elements in the upper soil horizons than in the taiga-biogeocoenose class, and much smaller outflow of them beyond the biogeosphere; greatest depth of the biogeocoenotic layer (30-55 metres), with highest diversity both horizontally and vertically, on account of the high biogeocoenotic variability in the phytomass.

Each class of forest biogeocoenoses can be subdivided into biogeocoenotic formations or any other units selected according to differences in biogeocoenotic metabolism arising from differences in the amount and form in which atmospheric precipitation enters into circulation, and also from the major qualitative biogeocoenotic features of the chief components of the phytomass of a forest biogeocoenose. The following examples of such groups are taken from the class of taiga biogeocoenoses in Eurasia.

1. Formations of evergreen dark-conifer taiga, located in provinces where oceanic influence is strong. As biogeocoenotic systems formations of this group are distinguished by: a 'cold photosynthesis' phase, a prolonged period of biogeocoenotic exchange in the phytocoenose-atmosphere system during early spring and late autumn; abundant moisture supply by rain in summer and autumn and by water from melting snow, which causes deep soil soaking and high mobility of solutions, and removal of part of the dissolved matter out of local biogeocoenotic circulation; uniform structure of the plant mass throughout the year, because of the evergreen nature of the principal synusia of the phytocoenoses; and marked year-round transformation by it of the internal climate of the biogeocoenoses.

2. Formations of summer-green taiga, distributed through provinces with a more extreme continental climate and with less precipitation, especially in winter. This group differs from the preceding group in: absence of a 'cold photosynthesis' phase; abrupt alternations in plant-mass structure, due to leaf-shedding by the principal synusia of the phytocoenoses, and consequently with a strongly-marked periodicity in modification of the internal climate of the biogeocoenoses, with division into summer (strong influence) and winter (very weak influence) phases; heavy soaking of the soils, substantial lowering of the mobility of soil

solutions (especially in spring and autumn) and their removal from local biogeocoenotic circulation.

These features of the biogeocoenotic metabolism of each group of formations are most typically shown in biogeocoenoses occupying the so-called 'plakor' habitats—well-drained watersheds with loamy soil without carbonates. In all other cases various deviations are observed in the course of biogeocoenotic metabolism, due to local variations in accession and distribution of heat, moisture, and mineral elements of plant nutrition and in aeration of the soil, etc. Therefore a number of biogeocoenotic formations can be differentiated within the groups. For instance, in the group of evergreen-taiga formations in flat districts the following formations occur:

1. Plakor evergreen taiga, occupying well-drained watersheds with loamy soil and subsoil and distinguished by the group's most 'normal' conditions of heat, light, moisture, air movement, migration of soil solutions, soil aeration and composition, and productivity, accumulation, and decomposition of organic matter.

2. Evergreen taiga on fairly steep slopes in river valleys, with good illumination. It differs from the plakor formation in having more heat and light, a longer period of vegetative plant activity and of general biogeocoenotic metabolism, less moisture (because of more rapid outflow and greater output in evaporation and transpiration), drier air, less soil leaching, richer composition and higher productivity of organic matter, more rapid decomposition of litter, and less-well-developed litter.

3. A formation on fairly steep slopes in river valleys, with shade conditions. It differs from the plakor formation in having less heat and light, moister air, a shorter period of biogeocoenotic activity for plants, and less rapid decomposition of organic litter, which becomes fairly deep.

4. Swamp forest or peaty forest formation in low-lying or badly-drained level parts of interfluvial areas and terraces. This group is distinguished by impaired heat and air exchange in the atmosphere-soil system, very slow movement of oxygen-poor soil moisture, low tree productivity (quality classes III-IV-V), and slow and incomplete decomposition of organic litter, with accumulation of peaty layers on the soil surface and of fulvic acids, ferrous oxide, and hydrogen in the soil.

5. Taiga forest formation on carbonate strata. It is distinguished by very high productivity and diversity of composition of organic matter, large amounts of wood in trees and roots, and rapid and complete mineralisation of organic litter because of the intensive activity of the abundant soil fauna and micro-organisms (especially bacteria) and the fact that the physico-chemical properties of the soil (structure, permeability to water and air, good heat regime) favour oxidation.

6. Valley-bottom taiga, distinguished particularly by the fact that its biogeocoenotic metabolism is complicated and enriched by inflow of mineral and organic matter and water (with high water levels in spring

or summer and with underground water). Consequently tree growth is greatly accelerated and numerous eutrophic shrubs, grasses, and many hygrophytes are present. At the same time a tendency to 'continentalisation' of ecological regimes and of the entire biogeocoenotic metabolism is clearly displayed in lowering of the erosion level and of the water table, and enhanced drainage of the surface layers of soil in the course of evolutionary development of river terraces. This is often reflected in the vegetation by the formation of plant communities that become increasingly similar to those of watershed choras (e.g., *Oxalis*-spruce stands on the higher levels of the valley bottoms).

Another well-defined formation is taiga forests on quartz sands, with their oligotrophic regime, dryness, very low productivity of organic matter by plants, and impoverished flora and fauna. In some places the formation of forests on volcanic ash and other volcanic products is clearly distinguishable, possessing (as investigations of Kamchatka forests, for example, have shown) well-marked biogeocoenotic characteristics.

Within formations the differentiation of biogeocoenotic metabolism among components takes place under the influence of the composition (at the genus, sometimes the species, level) of the dominant trees in the forests, which have different ways of using (because of genetically-caused selective absorption) and of transforming (through formation of distinctive litter, cholines, and phytoncides, and through transpiration) a similar environment.

On this basis each formation is divided into groups of types of forests of spruce, fir, stone pine, etc.

The lowest taxonomic unit in forest biogeocoenology, suggested by V. N. Sukachev and accepted by the 1950 Conference on Forest Typology, must be taken as the type of forest biogeocoenose, or type of forest that combines areas homogeneous in composition of tree species, in other layers of vegetation, and in fauna; in the complex of forest-growing conditions (climatic, soil, and hydrological); in inter-relationships between plants and environment; in regenerative processes; and in the trend of successions in these forest areas (for further details see Chapter I, p. 46).

Whereas the boundaries of a biogeocoenose usually coincide fully with the boundaries of the corresponding phytocoenose, making it easily distinguishable from adjacent biogeocoenoses, the extent of a type of forest biogeocoenose identified by the homogeneity of all of its components does not always correspond to the area of a type of forest phytocoenose or forest association, which is distinguished by the similarity of separate communities in their composition of dominant and subdominant species and in their vegetational structure. Tracts of land having similar kinds of plant cover may display great dissimilarity in other biogeocoenose components. Many cases are known where there is close similarity in forest vegetation on watersheds and on certain levels of river terraces; in sandy-loam and in loamy areas; in climate of the taiga and in climate of the forest-steppe (e.g., in moss and lichen pine stands); and so on. In geobotany such cases are combined in the concept of convergent plant communities.

A very interesting example of this is provided by the bilberry spruce stands in the Rhodope mountains of Bulgaria and in the northern taiga of Europe. From the phytocoenotic point of view they are extremely similar: the dominant producer synusia is formed of the same species of spruce (*Picea excelsa* Link.), it is equally dense, its productivity is ranked in the same quality class (III-IV), the forest structure is equally three-storeyed, and the herbaceous cover is everywhere dominated by bilberries (morphologically indistinguishable) with a number of accompanying species found in both areas (*Goodyera repens, Vaccinium vitis-idaea, Deschampsia flexuosa, Fragaria vessa, Oxalis acetosella, Ramischia secunda, Moehringia pendula,* etc.); identical plants dominate in the well-developed moss cover (*Hylocomium proliperum, Rhytidiadelphus triquetrus*) with an admixture of the same species (*Pleurozium schreberi, Dicranum scoparium, Sphagnum girgensohnii,* etc.) (Dylis *et al.* 1959).

There are, of course, some differences in the composition of secondary plant species between the Bulgarian and the northern-taiga bilberry spruce stands, but they do not affect the close similarity of their main species and general community structure. On that basis they must be classified from the phytocoenotic point of view as a single association, probably on the level of geographical variants or subassociations. From the bio-geocoenotic point of view, which embraces other components besides vegetation, primarily climate and soil, the two are very different. The Bulgarian bilberry spruce stands develop in a much milder climate (the average temperature of the coldest months, January and February, does not fall below $-4.4°C$), they receive 50% more precipitation (up to 1000 mm), the soil practically never freezes, and water circulation proceeds throughout the whole year. But the outstanding feature of the Bulgarian bilberry spruce stands is in the soil. There is no podzolisation, a dark humus horizon is strongly developed, reaching a thickness of 26-36 cm, with high humus content and a splendid powdery-cloddy-granular struc-ture. These characteristics are so notable and unusual that they have induced pedologists to classify such soils as a special type of mountain-forest, dark-coloured soils peculiar to Bulgaria (Zonn 1957).

Since the soil is the mirror of biogeocoenotic metabolism and inte-grates within itself the biogeochemical work of different components of a biogeocoenose, the above-listed features of the soils under the Bulgarian bilberry spruce stands must indicate some very deep, although as yet almost undiscovered, differences from the northern-taiga forests in their metabolism, especially with regard to decomposition of organic matter and its interaction with the soil. In making a biogeocoenotic evaluation of these forests we are compelled by these features to separate them from the taiga bilberry spruce stands not only in different forest types but in different classes of biogeocoenotic formations. Evidence of the peculiarities of metabolism of the Bulgarian spruce stands is given (as well as by the soils) by the failure of birch or poplar to replace spruce after forest-cutting. That replacement is always observed in the taiga zone of Eurasia, and always introduces substantial variations into the course and character of biogeocoenotic metabolism in the forest.

Analysis of this and of a number of other cases of close convergence of the vegetation in ecotopes differing sharply in relief, climate, or soil composition leads to the conclusion that we should not speak of convergence as applying to the biogeosphere. It is not inherent in it, and in classification of the biogeosphere we must regard biogeocoenoses that are very similar in vegetation, but differ in the condition and character of other components, as items of different, sometimes quite distant, taxonomic subdivisions.

On the other hand one sometimes observes the opposite relationship between a type of biogeocoenose and a type of phytocoenose (association), namely, the type of biogeocoenose appears to be a category of wider scope than the association. For instance, from the biogeocoenotic point of view two areas of bilberry spruce stands may be placed in different associations (and not only associations!) if in one *Picea excelsa* Link., and in the other (perhaps adjacent to it) *Picea obovata* Ledb. is dominant. From the biogeocoenotic point of view there may be scarcely any justification for differentiating between such forests: they should be combined into a single type.

Although the type of forest biogeocoenose is a very fine unit of the biogeosphere and is distinguished by high homogeneity of its components and developmental tendencies, nevertheless the areas combined in a single forest type are not absolutely identical biogeocoenotically; in other words, a certain degree of variability in its characteristics and properties is inherent in a type of forest biogeocoenose, as in any natural phenomenon. Even in virgin forests one observes, in passing from one part of a forest to another typologically similar, various deviations in the composition of the constituent plants, in their relationships, and in their lives; and the same applies to the fauna, to the depth of the soil and of its separate horizons, to the pH value, etc. Sometimes also one may observe substantial variations within a single type of forest in the plant-community structure and therefore in the character of biogeocoenotic metabolism. In the virgin forests of Tibet, for instance, we have seen spruce stands with dense undergrowth of monocarpic bamboo (*Sinarundinaria*). The external appearance of such forests is very distinctive, and they are clearly differentiated by their undergrowth from other types of forest. But in one part of them the whole bamboo population will suddenly come into flower, the plants produce fruits and then wither (including their root systems and new shoots) and fall on the ground. Nothing is left of the dense and luxuriant bamboo undergrowth but decaying debris on the soil surface. The external appearance and structure of the forest are changed beyond recognition for a time, and it resembles spruce stands of another local biogeocoenotic group—moss spruce stands. At the same time the biogeocoenotic metabolism of the forest also changes: the total biomass and increment of the phytocoenose are greatly diminished; the soil becomes better illuminated, watered, and warmed; the composition of the litter changes; and the growth of a number of ground-level plants (especially mosses) is stimulated.

Very great alterations in the state and activity of other components within a single type of forest are observed as the trees become older. In a forest the amount of organic matter in the above-ground and underground parts of the phytocoenose increase with age, the radial dimension of the biogeocoenose increases, the diversity and mass of pl ants in the lower layers of the phytocoenose change (usually increasing), the community structure is greatly altered, the mass of litter and the depth of litter increase, the phytoenvironment (light, air humidity, warming, snow melting, penetration of liquid precipitation, etc.) change; consumption of moisture, gases, and mineral elements, and the composition and abundance of fauna all change, as may be seen from the fluctuations in a number of essential characteristics of forest biogeocoenoses in the diagrammatic illustration prepared for one of the principal types of forest-steppe oak forests (Fig. 85) (see also Sakharov 1948, Molchanov 1961a). For that reason the classification of biogeocoenoses should take into account homogeneity not only in the above-listed characteristics of a forest but also in their dynamic transformations, the regimes of their components, and their changes with time. Within a single forest type all these features should be homogeneous.

Natural variability in forest biogeocoenoses is considerably increased under the influence of such factors as invasion by pests or animals and the economic activities of man (improvement cutting, amelioration, haymowing, livestock-pasturing, etc.). Changes due to these external factors may affect any component or any aspect of biogeocoenotic metabolism, although their principal and original action is usually exerted upon the phytocoenose, its work, and its links with other components. In many cases the changes in the biogeocoenoses may be so extensive that it is impossible to group them together with biogeocoenoses of the same original type but not modified by these factors. It is necessary to segregate different types of forest biogeocoenoses: radical and derivative (transformed in one way or another). In forests this happens primarily in connection with a succession of the tree species that form the main producer synusia of a forest after a fire or concentrated cutting; for instance, the replacement of coniferous by deciduous forests, which takes place over vast stretches of the taiga. Such succession of species is not a mechanical replacement of one species by another, but a very complex, intricate process of changes, affecting all features of the forest biogeo-

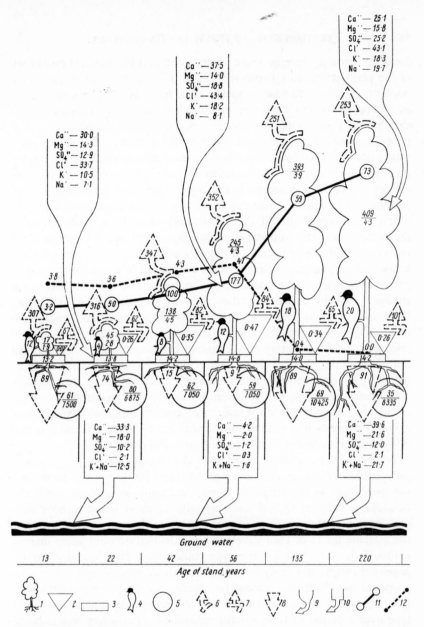

Fig. 85. Some of the most important biogeocoenotic features of a sedge-goutwort oak stand in the forest-steppe and their changes with the age of the stand (compiled by N. V. Dylis from data of A. A. Molchanov *et al.*). 1—(numerator) amount of organic matter in leaves (tons/ha, dry weight); 2—amount of organic matter in herbaceous cover (tons/ha, dry weight); 3—amount of litter (tons/ha, dry weight); 4—number of pairs of nesting birds per ha; 5—in the numerator, number of earthworms (individuals per m²), and in denominator, number of micro-organisms ('000 per g of soil); 6—transpiration of water by tree stand (mm per annum); 7—evaporation from herbaceous cover (mm per annum); 8—circulation of water in depth of soil (mm per annum); 9—accession of mineral matter with precipitation during vegetative period (kg/ha); 10—outflow of mineral matter with underground water (kg/ha); 11—lighting between crowns (% of light in an open space); 12—current increment of wood (tons/ha, dry weight).

coenoses through changes in the composition and biochemical properties
of the plant community (Morozov 1912). For instance, when birch and
poplar occupy an area where a spruce stand has been cut down or burned,
an extremely abrupt deviation in the whole course of biogeocoenotic
metabolism takes place for a long period of years; the phase of 'cold
photosynthesis' disappears, and therefore the period of multilateral biogeocoenotic metabolism is curtailed; the seasonal structure of the phytomass is greatly altered, and therefore its transformation of the internal
climate of the biogeocoenose is sharply differentiated into summer and
autumn-winter phases; there are changes in the amount, kind, and rate
of mineralisation of litter, in the floristic and synusial composition, and in
the composition and abundance of the fauna and micro-organisms; air
movement and warmth, and lighting (including that in summer) are increased; there is a large increase in the amount of litter reaching the soil;
and the chemistry, composition, and movement of soil solutions are
changed. When demutation takes place as spruce develops under the
birch canopy these changes are gradually eliminated, and finally the
original type of biogeocoenotic metabolism is re-established. Complete
restoration of the structure and regime of biogeocoenoses, however, does
not always take place in the course of degressive-demutational replacements in nature, even when this course is not checked by fresh human
interference. In particular, instances are known of irreversible swamping
of cut and burned areas in the taiga, with conversion of some varieties of
dry-valley forests into swamp forests with peaty-gley soils, and a quali-
tatively-different course of biogeocoenotic metabolism.

In view of the continuous nature of the changes taking place in biogeocoenoses with regeneration and growth, and the diversity of forms and
the magnitude of the effects of human economic activity on them, the
boundaries between radical and derivative types of forest in both time
and space are often very provisional. They are clear and obvious only at
the extreme limits of their development. On that account some investigators flatly deny that a forest type could retain morphological homogeneity throughout a fairly long period of time. Accordingly it has been
proposed to substitute a genetic (or, more precisely, ontogenetic) classification, based on consideration of all the changes taking place in a forest
type over a period of time, for the natural classification, which is based
on morphological homogeneity of the forest areas combined into one type
and does not make constructive use of dynamic characteristics (Kolesnikov
1956, 1958 a, b, 1961). The advocates of a genetic classification greatly
broaden the meaning of 'forest type' as compared with that adopted at the
1950 Conference on Forest Typology. According to B. P. Kolesnikov,
'forest type' covers all stages of its development from germination to
death, especially all growth and regeneration stages, and therefore includes
a number of types of forest biogeocoenoses.

Conversion of forest classification to a system that not only establishes similarities and differences in static features but also reflects the course and trends of development is, in principle, a very alluring project. The problem is, however, extremely difficult, as has already been noted in the literature (Dolukhanov 1961), quite apart from the question of whether we should regard a forest as a plant community or a biogeocoenose. Attempts to construct such systems are certainly rational. It appears, however, that this is impossible without a very undesirable breaking-up of the basic taxonomic unit of forest cover, the forest type. We recall, for instance, that successful solution of the problem of proper grading of radical and derivative forest types in the classification of forests was provided by the concept (introduced by S. Ya. Sokolov 1938) of a series, combining a radical association and all those derived from it. It is useful to retain that concept for the same purpose in biogeocoenotic classification of forests, narrowing it somewhat by excluding all those variants of derivative types that in the course of further development are not fully restored to biogeocoenoses of the original kind but start new diverging lines of biogeocoenogenesis. Combining types of forest biogeocoenoses into a series should ideally show series of successive demutations of derivative types of forests up to the original, radical type.

Kolesnikov's contrasting of basic and lower elementary units of forest cover also appears undesirable. They should, of course, coincide at the level of type of forest biogeocoenose, and at the same time that unit should be basic for any grouping of forests undertaken for scientific or practical purposes.

At present biogeocoenotic classification of forests is still in a very embryonic state, and in order to advance it we have to collect very extensive factual data on matter and energy exchange in forest biogeocoenoses throughout the earth. Concurrently it is necessary to devise special nomenclature for taxonomic subdivisions of the biogeosphere and special cartographic methods for them, especially for the purpose of discovering the laws of geographic variations in biogeocoenotic metabolism.

CHAPTER IX

Possibility of applying the ideas and methods of cybernetics to forest biogeocoenology

In recent years the ideas and methods of cybernetics have been applied widely and fruitfully not only in technical subjects but also in different fields of the biological and economic sciences and the humanities.

The cybernetic approach has been successfully used in biology with regard to subjects at different organisational levels, from the intra-cellular molecular level (Gamov *et al.* 1956, Krik 1956, Timofeev-Resovskii and Rompe 1959, Lyapunov and Malenkov 1962, Spirin 1962) to the super-organism level, including species-populations and biogeocoenoses (Shmal'-gauzen 1958 a, b, 1959, 1960, Takhtadzhyan 1959, Aleksandrova 1961, Starobogatov 1962, Vilenkin 1963, Margaleff 1957, 1958, 1961, Quastler 1959, Patten 1959, Odum and Pinkerton 1955, Odum, Cantlon, *et al.* 1960); work on neurocybernetics has now been extensively developed. Cybernetics has been successfully used in economics, linguistics, study of literature, etc.

The subjects to which cybernetics can be applied include systems of the most varied complexity, from simple systems like a water tap or an electric switch (Grenevskii 1959) or a formula representing a simple arithmetical operation (Lyapunov and Yablonskii 1961), or a system such as Watt's centrifugal governor (the cybernetic diagram of which is simple), to extremely complex systems or 'very large systems', to use Eshbi's term. Eshbi gives the following description of the latter: 'The words "very large system"... will denote the following: given a certain observer, with specified equipment and methods, and a system that in some respect is, in practice, large for him: for instance, he cannot observe or manage it completely, or make all the calculations required to predict its behaviour. ... Such systems, too large for our present means of observation and direction, are very common in the biological world and its social-economic parallels' (1959, p. 94). The approach used in cybernetics enables one to examine 'very large systems' using strict, scientific methods and to work out procedures for studying them, in spite of the impossibility of studying a system in all its details (Eshbi 1959, Lyapunov and Yablonskii 1961).

A forest biogeocoenose is a typical example of a 'very large system' in Eshbi's sense. The phenomena of self-regulation that are observed in the development of this complex system give us the right to look upon it as a

590

subject for cybernetics and therefore to regard a biogeocoenose as a cybernetic system.

Regulation is expressed primarily in the fact that in similar environmental conditions monotypic biogeocoenoses are observed, characterised by similarity in species-composition, in the structure and productivity of their biological components, in the morphology of the soil profile, in the chemistry and fertility of the soil, in the type of exchange of matter with the environment and of effects on the environment, etc. Only such similarity enables us to separate types of biogeocoenoses and later to classify them, and also to estimate their productivity and lay down practical norms for their use and improvement on the basis of calculations and experimental investigations carried out on test plots. Regulation is also displayed in regular successions in time, when biogeocoenoses are formed either on newly-exposed areas (sandbanks, outcrops, solidified lava flows, etc.) or after destruction of a biogeocoenose by any factor (cutting, uprooting, cultivation, fire, etc.). In these cases successions of series are observed, composed of stages regularly succeeding each other and leading to development of a 'climax' biogeocoenose, distinguished by 'mature' soil and full development of the biological component and of its species-composition and productivity. Although in the course of successions, if they are traced in different (but monotypic in their conditions) localities, certain fluctuations are observed, depending both on environmental conditions and on the most diverse incidental circumstances, the stages of succession are fairly monotypic. Regulation is also displayed by biogeocoenoses in that having reached the highest degree of development they acquire features of stability. These include the capacity to maintain equilibrium in their components for an indefinite period of time; to maintain their own typical species-composition, productivity, and structure of plant cover and animal population; to maintain the morphology, chemistry, and fertility of the soil; to maintain the characteristics of the microclimate within their borders; and to maintain the character of their exchange of matter and energy with the environment, and of their influence on the environment. These clearly-displayed phenomena show that a biogeocoenose is, undoubtedly, a natural system in which relationships producing the effect of regulation are realised, and which therefore is a suitable subject for cybernetics.

The principles of study of cybernetic systems set forth by Viner (1958) were developed in their most abstract and general aspect by U. R. Eshbi (1958, 1962) and others. They were developed in the most concrete and constructive way by the eminent Soviet cybernetic experts A. A. Lyapunov and S. V. Yablonskii in a number of works (Lyapunov 1958, Yablonskii 1959, Lyapunov and Yablonskii 1961). According to Lyapunov and Yablonskii, in studying cybernetic systems[1] the 'macro-approach' and 'micro-approach' are to be distinguished.

[1] For such complex objects as biogeocoenoses regarded as a whole, the term 'cyber-

In the 'macro-approach' (Lyapunov and Yablonskii 1961) a cybernetic system is regarded as a 'black box'.[1] In this case the investigator does not study the whole system, but restricts himself to certain problems connected with the external behaviour of the system. Here we include primarily the discovery of flows of information received by the system from the external environment. In this connection it is important to formulate clearly the purpose of the regulation attainable through treatment of the information, in order to decide what information received from outside is essential or not essential for the system. The entrances to the system are discovered, i.e., those elements in it that accept stimuli or actions from the external environment containing information. Since any system has an unlimited number of possible entrances (because of the unlimited number of possible external actions upon it), those entrances that accept essential information are segregated. In a similar way the chief, or essential, exits are discovered: those whereby the influence of the system on the environment (the result of treatment of the entering information) is realised.

Time scales are established connected with the treatment of information: 'steps' or 'measures' or 'elements of time' of the work of the system. There is established (for both incoming and outgoing information) a system of 'signals', i.e., a collection of conditions of external actions, conditions of inflow, conditions of outflow ('repertoire' of the system, according to Grenevskii 1959). The presence or absence of reverse links of the system with its environment is determined. The external reverse links are described, and their role in regulation of the system is discovered. A number of problems coming to light during the macro-approach are solved on the level of pure interest, not on a formal level (Lyapunov and Yablonskii 1961).

The macro-approach is insufficient for full study of the structure of a system, which is necessary for mathematical description of it; for this the micro-approach is required.

With the micro-approach, a cybernetic system is divided into elements. These elements, in the words of Lyapunov and Yablonskii, are 'elementary bricks' in the treatment of information. Each element has its own entrances and exits, through which the relationships among them are realised. Since a cybernetic system typically possesses a large number of elements through which regulatory information enters, there is a complex system of relationships between elements on different scales giving a hierarchy of relationships and levels. In the micro-approach the

netic system' (Poletaev 1958) is more appropriate than that of 'governing system' used by A. A. Lyapunov and S. V. Yablonskii (1961). If the latter term is accepted, a biogeocoenose, strictly speaking, must be called not a 'governing' but a 'self-governing' system, since it contains the most essential phenomena of autoregulation.

[1] The expression 'black box' is used to denote systems whose internal structure is unknown or almost unknown, or is inadequately studied, or is not taken into consideration as it is outwith the scope of the investigation.

elements of a cybernetic system are segregated by different means and on different 'scales'. Once all the elements of a cybernetic system have been found and classified, they must be studied in detail. Algorithms[1] of transformation of information are found for each class of elements of the cybernetic system and experiments may be made to determine the algorithms. The tasks of the micro-approach include study of the algorithms of synthesis of cybernetic systems, study of their evolution (i.e., of the transformation of a cybernetic system under the action of its function into a new cybernetic system), study of the reliability of cybernetic systems and of their stability on account of suitable structure (achieved in systems of homoeostat type), and search for algorithms of reliable cybernetic systems (Lyapunov and Yablonskii 1961).

The tasks of study of a biogeocoenose as a cybernetic system, in conformity with the scheme proposed by A. A. Lyapunov and S. V. Yablonskii, are presented in Table 145.

Fulfilment of these tasks has not yet been undertaken and we can only present the problem and quote examples of known attempts at cybernetic analysis of separate biogeocoenose components.

The main problem is the analysis of the structure and functions of a biogeocoenose through discovery of the streams of information reaching the system, and determination of the methods of treatment (coding) of information in the system. Another problem of great importance is the reverse links in the system of a biogeocoenose, which play a decisive role in achieving regulation.

In speaking of information with application to a biogeocoenose, it is expedient to remember the warning given by I. I. Shmal'gauzen that extension of the term 'information' to biology 'is being made without sufficient discrimination, and its use is not always justified. This leads to a situation where the term "information", of value for its precision, is fast losing its clear-cut boundaries and becoming indefinite, like the majority of biological concepts, which thereby become a source of endless and quite fruitless "discussion". To prevent this, it is absolutely necessary to use the term "information" only as accepted in the general theory of information. All the value of the concept of "information" lies in its specificity, in the possibility of measuring amounts of information and evaluating its quality. That is possible only on the basis of the statistical theory of information worked out by Shennon' (1960, pp. 121-122).

The concept of information is connected with the selection or realisation of certain possibilities or events out of a definite number. The amount of information is calculated in accordance with the probability of these events: the amount of information I_i of a single event i, or a single selection, or realisation of one of a number of possibilities is equal (according to Shennon) to the logarithm of probability of the event p_i, taken with the

[1] The rules by which transformation takes place are called algorithms. Any mathematical formula is an example of an algorithm.

Table 145. *Tasks in the study of cybernetic systems.*

In the field of calculating and controlling machines (Lyapunov and Yablonskii 1961)	In the field of biology (Lyapunov and Yablonskii 1961)			In the field of biogeocoenoses (original)
	Genetics	Theory of evolution	Physiology of nervous system and receptors	
Discovery of currents of information in the machine	Discovery of means of transmission of hereditary information (chromosome theory of inheritance)	Study of currents of information that govern evolution	Study of currents of information circulating in nervous system and receptors	Discovery of currents of information entering a biogeocoenose from the external environment and being processed in it
Devising means of coding numbers of commands, etc. in the machine	Discovery of means of coding hereditary information (segregation of types, cytology)	Study of means of coding information governing evolution	Study of means of coding information in nervous system and receptors	Discovery of the means of processing information in a biogeocoenose
Discovery of the relationship of the work of the calculating machine to the project	Study of method of production of genotype (phenogenetics)	Study of destiny of population or taxon with relation to external conditions and hereditary features	Study of reactions, reflexes, and behaviour of animals	Study of means whereby a biogeocoenose can achieve maximum use of matter and energy in a given area, and of the structural stability of the biogeocoenose

Quantitative evaluation of information in the machine. Evaluation of productivity. Evaluation of retardation. Statistics of performance of operation	1. Evaluation of amount of genetic information in individuals of a definite group of animals of a definite species. 2. Statistical analysis of genotypes of progeny, when genotypes of progenitors are known. 3. Transmission of hereditary information as a means of securing stability of living forms and their evolution 4. Selection and mutational processes (population genetics) 5. Study of genotype structure—structure of groups of linkages and interaction of genes	Evaluation of amount of information that governs evolution. Statistical problems of population dynamics. Study of characteristics of regimes of evolution process (stabilising selection, dynamic selection). Evolution as a means of ensuring permanent existence of life. Appraisal of conditions of selection according to kinetics of population	Evaluation of amount of information and transmission capacity of channels of nerve network. Problems of interaction of man and machine. Work of the organism in extreme external conditions. Study of purposeful actions of higher animals and man, physiology of active movements, complexes of reflexes. Physiology of training and accumulation of experience, hierarchy of functioning, collective behaviour. Assessment of structure of nervous system and its parts with relation to functioning	Evaluation of amount of information entering the biogeocoenose and of information processed in the biogeocoenose
Devising elements for data storage	Problem of genes. Discovery of biochemical vectors of hereditary information	Discovery of elementary acts at base of evolutionary process (factors in evolution)	Segregation of elementary components of nervous system and its parts, down to separate cells, and also of receptors and their elementary reactions	Segregation within a biogeocoenose of elements of different dimensions that ensure reception, processing, and exit of information, and elements of data storage in the biogeocoenose

Table 145—continued

In the field of calculating and controlling machines (Lyapunov and Yablonskii 1961)	In the field of biology (Lyapunov and Yablonskii 1961)			In the field of biogeocoenoses (original)
	Genetics	Theory of evolution	Physiology of nervous system and receptors	
Study of interaction of elements	Study of structure of genotypes of normal and damaged cells. Structure of groups of gene linkages and chromosome cytology. Localisation of genes in chromosomes, structure of DNA molecules	Study of interactions of factors in evolution	Study of relationships between separate organs in the nervous system	Study of interactions of biogeocoenose components at different levels of their organisation. Study of the character of their inter-relationships that ensure regulation of the stability of the biogeocoenose or of its progressive development with successions
Formalisation of description of structure and functioning of machines	Algorithmic description of processes taking place under influence of vectors of hereditary information. Study of biochemical role of DNA	Discovery of means of functioning of information governing the evolutionary process	Algorithmic description of functioning of nervous system and receptors	Formalisation of description of structure and functioning of a biogeocoenose as a cybernetic system ensuring the processing of information and regulation. Discovery and description of the complex hierarchy of structural elements in the biogeocoenose

Statistics of subplans. Statistics of work of subplans	Genetical analysis of individuals and populations	Study of population dynamics. Theory of struggle for existence	Study of algorithms of processing of information in nervous system	Discovery of algorithms of processing of information in a biogeocoenose
Devising methods of synthesis of machines (use of mock-ups and programme modelling). Problem of high speed	Methods of selective synthesis of desirable genotypes	Methods of synthesis of desirable biotopes		Use of mock-up and programme models for study of biogeocoenoses; reconstruction of biogeocoenoses, creation of new biogeocoenoses
Designing and studying self-adjusting machines	Study of evolution of genes and genotypes	Study of variations in selection regimes with changes in external conditions	Problem of the evolution of the nervous system	Problem of phylocoenogenesis and evolution of soils
Statistics of fractions. Ensuring reliability of plans. Methods of machine control (tests)	Study of means of ensuring reliability in retention and transmission of hereditary information	Study of means of ensuring permanence of populations, biotypes, taxons, etc.	Problems of study of the reliability of functioning of the nervous system. Compensation of functions of nervous system	Study of means of ensuring stability of the system of a biogeocoenose

opposite sign (probability is always expressed as a fraction, and since the logarithm of a fraction is a negative number, the change of sign makes it positive)

$$I_i = -k \log p_i \tag{1}$$

The coefficient k is a constant, reflecting the unit of measurement.

The average amount of information I, obtained by realisation of any one out of n available possibilities, according to Shennon, is equal to:

$$I = -k \sum_{i=1}^{n} p_i \log p_i \tag{2}$$

If the unit of amount of information is taken as a binary unit, or bit (contraction for 'binary digit'), equal to the amount of information that can be obtained by selection or realisation of two equally probable possibilities (e.g., in tossing coins, 'heads' or 'tails'), then the coefficient in formulae (1) and (2) is unity and the formulae take the form:

$$I = -\log_2 p_i \tag{3}$$

$$I = \sum_{i=1}^{n} p_i \log_2 p_i \tag{4}$$

Such a statistical approach to the concept of information lays down definite conditions for the discovery of information entering a biogeo- coenose and being processed in it. Thus for determination of the amount of information a discrete series of events should be extant. The flow of matter and energy pouring into the biogeocoenose and being processed in it, and also the stimuli created by variation in the environmental condi- tions of the biogeocoenose, may be regarded as carriers of information coming in at the entrance of the biogeocoenose. The majority of these phenomena are not discrete and they are constantly changing. In order to look upon them as sources of information it is necessary to transform these continuous changes into discrete phases. That may be done by defining gradients in the limits of amplitude of quantitative changes in the factor under consideration. The division into gradients is not provisional but essential, important in principle, as it should express the limits of the quantitative variations in that factor that produce transformations of substantial value in regulation. Even for discrete (not continuous) factors the question of the 'value' of an elementary member of a series for which the amount of information can be calculated is not easily solved. Thus when biological components of the environment of a bio- geocoenose in the form of macroscopic and microscopic plants and animals enter a biogeocoenose, they bring definite information into it. The biological components of the environment are discrete by their nature, since they consist of individuals or embryos of individuals. But whether we should regard the individual or the species-population as an element of an information series is a problem requiring a special solution.

The next difficulty in principle is that a separate event can always be treated as a member of one of the most diverse groups of events, depending on how we look at it. For instance, an individual or a population with hereditary features built into it, when entering a biogeocoenose and being a vector of certain information, may be regarded from different points of view: from the ecologo-physiological point of view (its relationships with moisture, heat, etc.); from the point of view of its producer properties and its competitive ability; from the point of view of its potential mass; from the point of view of the part that it can take in mineral exchange; and so on.

Depending on the particular way in which an individual or a population is regarded, it may be a member of different groups with different distribution of probabilities (determined by observation or experiment) and with different evaluation of the information that it may contain in these circumstances.

No less difficulty is connected with the necessity of determining temporal units of the 'work' of the biogeocoenose system, its 'steps' or 'measures'—the 'calendar' of the system (Grenevskii 1959). What time element constitutes a 'step' or 'measure' of a biogeocoenose system in the process of its regulation? Is its daily rhythm significant, or only the seasonal rhythm in connection with the annual cycle of circulation of matter and transformation of biomass? Or are the successions of generations in the biological component of a biogeocoenose important, as taking place in its most vital and active part? These questions have not yet been decided and call for special analysis. Only with correct definition of the time elements, or 'steps', of a biogeocoenose system—only with correct definition of the series of events carrying information—can we reveal the cybernetic structure of a biogeocoenose and construct a logical model of it, the transformations in which will adequately represent the transformations taking place in the biogeocoenose. With introduction of errors, with incorrect determination of the 'calendar' of the system and incorrect determination of the series of significant events that carry information, such requirements are not attained and over-estimation of the system is the result. Eshbi (1959) gives special attention to such essential difficulties.

Among problems soluble by the macro-approach is the discovery of reverse links; in these the changes in the system produced by certain conditions themselves exert influence on these conditions and alter them; i.e., a reverse link consists of the effect of an exit from the system on an entrance to it. In carrying out the macro-approach, positive, supporting reverse links and negative, restricting reverse links in the biogeocoenose should be discovered.

No attempts to use the macro-approach[1] with respect to a biogeo-

---

[1] Such an attempt has been made with regard to a phytocoenose (Aleksandrova 1961); for a biogeocoenose, this is one of the tasks for the micro-approach.

coenose, to describe the structural hierarchy of the cybernetic elements of a biogeocoenose, or to show the stages by which information enters a biogeocoenose and is processed in it, have yet been made. Patten (1959) attempted to estimate the amount of information contained at the molecular level in the biological component of the aquatic biogeocoenose of 'Silver Springs' lake in Florida. According to Patten's calculations, the information (negentropy)[1] contained in the biogeocoenose per cm^2 of lake surface is equal to $3 \cdot 1 \times 10^{24}$ bits per annum, of which $1 \cdot 8 \times 10^{24}$ bits are used in metabolism by plants and the remainder goes in food chains, a large part of it finally being converted by microflora into entropy (i.e., energy dissipation), and part is carried out of the biocoenose, via the lake outlet. The amount of information estimated in this way gives a measure of the changes in microconditions undergone in the biocoenose per cm^2 per annum, on the assumption that they are equally probable. Patten based his calculations on data for the energy balance for the biogeocoenose obtained by Odum (1957). According to Odum, the input of energy into the biogeocoenose was $3129 \cdot 6$ cal per cm^2 per annum. Since the lake was thermostatic with a temperature of $290°K$, Patten was able to calculate the physical entropy of the biocoenose per cm^2 per annum, which proved to be $7 \cdot 19$ kcal/deg per cm^2 per annum. The conversion from entropy to information was made by him by means of the ratio obtained by Augenstine (1953),

$$I = 0 \cdot 726 \ S,$$

according to which multiplication of the data in units of entropy by $0 \cdot 726$ and by the Avogadro number ($6 \cdot 02 \times 10^{23}$) gives the amount of information in bits.[2]

Patten's calculations give us very little assistance, primarily because they were made at the molecular level and not for biological structures. In addition they were made on the supposition that all micro-conditions are equally probable, which is not the case. Also doubtful is the direct conversion from physical entropy (and negentropy, with which Patten works) on a kcal/deg scale to information entropy as a measure of uncertainty of experience and of information, calculated in bits, since the quantitative expression for physical entropy is connected with a thermodynamic closed system, and a biocoenose is an open system with an external source of energy (Eshbi 1959).

V. D. Aleksandrova (1961) has demonstrated the presence, within a biogeocoenose, of reverse links regulating the species-composition,

---

[1] Negentropy (contraction of negative entropy) is a quantity opposite to entropy ($N = S$; Brillyuen 1960), characterising the ability of a system to perform work. It increases with the growth of an organised system and with transition of the latter into a less probable state.

[2] The bit figures show in this case, in effect, the amount of diversity (information entropy), which is calculated in the same way as information, and is reckoned as equal to it.

structure, and productivity of its vegetation component. She also pointed out the stochastic (chance-linked) character of transformations taking place in a phytocoenose at the individual level, and the determinate character of transformations taking place at the level of species-populations or of the phytocoenose as a whole. The determinate character of the system of a phytocoenose, having a statistical origin, is expressed not only in monotypy of phytocoenoses developing in similar habitat conditions but also in the fact that monotypic phytocoenoses react identically to external actions upon them. It was also shown that the determinate character displayed by a phytocoenose as well as by the species-populations that enter into its composition is not absolutely precise but is, as it were, blurred or eroded, and therefore the response of a phytocoenose to a given action cannot be foretold with absolute accuracy; it varies within certain limits.

Among the various kinds of micro-approach in the study of the cybernetic structure of a biogeocoenose, we must include the works of I. I. Shmal'gauzen (1958 a, b, 1959, 1960), which are devoted to the regulating role of a biogeocoenose in the micro-evolution of species-populations by the operation of reverse links between the latter and the biogeocoenose as a whole, and also to the amount of phenotypic information contained in a population as an elementary evolving unit. Here also are included the works of Margalef (1957, 1958, 1961), MacArthur (1955), Odum *et al.* (1960), Ya. I. Starobogatov (1962), B. Ya. Vilenkin (1963), and others, relating to the amount of information at the organism level and to its transcoding to other (population, biocoenotic) levels, and also to the use of criteria of the information balance of a biocoenose for quantitative evaluation of the dynamic state of the latter. According to MacArthur, Patten, and Starobogatov, a positive balance in the amount of information characterises the progressive stages of a succession; a zero balance, a climax state; and a negative balance, a state of regressive succession. These works are significant for showing the cybernetic structure of a biogeocoenose at several of its levels. At the present time interest in this problem is rising among biologists, ecologists, and geographers (e.g., A. D. Armand, G. F. Khelmi), and in the near future we may look for new publications relating to questions linked with the application of cybernetics to biogeocoenology.

Application of the ideas and methods of cybernetics to biogeocoenology should be one of the essential additions to the total assemblage of methods of biogeocoenose study. Analysis of the structure of a biogeocoenose, regarded as a cybernetic system, should lead to clear definition of elementary phenomena in this very complex process and provide for accuracy and clear definition of concepts. Discovery of the essential links in a biogeocoenose system on which operation of the regulatory mechanism depends should direct attention to the aspects of inter-relationships and interdependences within the biogeocoenose, and also in its links with the environment, that call for special effort in the way of ecological, physiological, and other specific investigations. It is highly important that the cybernetic approach enables us to obtain quantitative criteria for objective evaluation and comparison both of the separate aspects of the very complex system of a biogeocoenose and of the entire system as a whole. Finally, mathematical description of a biogeocoenose system makes possible wide application of models as a method of scientific investigation of various characteristics of the biogeocoenose. By the use of models, cybernetics employs experimental methods of investigation that (together with mathematical methods) play a major role in it. Therefore modelling has been called the cybernetic experiment (Lyapunov and Yablonskii 1961).

Modelling consists in substitution of an isomorphic model, which is then studied, for the object under investigation. Models may be physical, being replicas of the original, or abstract (mathematical or logical), or models made by programming on computers. The latter method is the most promising kind of cybernetic experiment.

Modelling as a method of studying a biogeocoenose has great possibilities. We must agree with N. V. Timofeev-Resovskii, who says that 'the introduction of cybernetic principles and concepts into population-genetical and biogeocoenological-biochorological levels of life . . . creates the possibility of almost unlimited use of machine models for analysis of the most complex biological phenomena' (1962, p. 44).

In conclusion we must point out another aspect of the application of cybernetics to biogeocoenology, namely, the use of modern cybernetic computers, both electronic calculators with complex programming, and analytical machines with punched cards (or punched tape). Calculating-punchcard machines, which make calculations and sort punched cards by appropriate methods, are capable of performing logical operations and thus fulfilling the functions of a cybernetic device. There are still greater possibilities with the use of electronic calculating machines, which, appropriately programmed, can not only produce an ordinary statistical analysis of material with great rapidity but also trace range maps, decipher aerial photographs, compile geobotanical maps, etc. The processing of forest-survey data by computers is already in wide use. The use of electronic and punchcard machines should not only save time and labour for biogeocoenologists but also ensure objective and uniform methods of collecting primary information relating to study of biogeocoenoses and their components and make possible increased responsibility for the quality and quantity of collected material.

The application of cybernetic methods, which are accurate methods, should be regarded as a progressive trend in biogeocoenology. This trend, combined with the wide and constantly-increasing introduction of the methods of physiology, biophysics, physics, and chemistry, should help to convert biogeocoenology into an exact science. This tendency, now clearly displayed, coincides with the general tendency in the development of sciences in our time, when (in the words of A. I. Berg, 1961) the walls standing between 'exact' and 'descriptive' sciences and methods are being broken down, and the criteria and logic of the exact sciences are ever more authoritatively demanding admittance both to the social-economic fields of knowledge and through the doors of biological laboratories.

CHAPTER X

Theoretical and practical significance of forest biogeocoenology

And so we have seen that the biogeocoenotic cover of the earth, which according to V. I. Vernadskii constitutes the most active part of the biosphere, and which is called the phytogeosphere by E. M. Lavrenko, is the arena in which organic matter is synthesised, where the most complex processes of creation of new matter take place, and where energy is constantly being accumulated and converted into matter in one form or another, and exchanged both within biogeocoenoses and between them and other phenomena of nature. That is because biogeocoenotic cover is the arena of life. Living organisms participate directly and individually in all of the above processes. Not only the terrestrial but also the cosmic role of living organisms (of 'living matter', as V. I. Vernadskii says) is performed in biogeocoenotic cover. All of man's economic activity depends directly or indirectly on processes taking place in biogeocoenoses.

Therefore these processes have to some extent been studied by man for a very long time. In the course of centuries a large number of sciences, studying nature from various aspects, have become differentiated. Although the theory that all natural phenomena are interacting and inter-related was propounded in ancient times, the idea of the universality and significance of these relationships has been most clearly expressed only in the philosophy of dialectical materialism. The idea that a special science was necessary—a science that, based on the data of the sectional sciences, would study all the inter-relationships of natural phenomena on the earth's surface, and both the units and the total composed of these inter-relationships—was originated by the ingenious Russian scientist V. V. Dokuchaev, who put it forward in the closing years of the nineteenth century. It was developed only during the present century. In a number of foreign countries these concepts led to development of the science of ecosystems, and in our country to the science of biogeocoenoses, which is constantly acquiring greater theoretical perceptive and philosophical importance and at the same time is constantly becoming a more active scientific foundation for the transformation of nature in the interests of mankind, for more rational use of its resources (principally soil, vegetation, and the animal world), for increasing their productivity, and for improving and conserving natural resources.

As we advance along the road to Communism and to a higher level of human economic activity, the role and importance of biogeocoenology will continue to grow. Life itself calls for further and more thorough working-out of the problems of biogeocoenology. That is due to the fact that when changing any part of nature in our own interests we must always know how the change affects other aspects of nature, and whether (because of the inter-relationships of all natural phenomena and processes) undesirable changes in nature, harmful to man's interests, will be initiated.

As we study the interactions of all biogeocoenose components, more profound inter-relationships and interdependences are revealed. To learn the laws governing them we must use the newest achievements and methods of physics, chemistry, biology (at all of its levels), mathematics, and cybernetics. Biogeocoenology will continue to combine the achievements of many other sciences, to process them from its viewpoint, and to create new generalisations of great theoretical and practical value.

V. I. Vernadskii most brilliantly expounded the necessity of deeper knowledge of the role of 'living matter' in the biosphere not only on a planetary but on a cosmic scale, but it has now become clear that such study can be successful only when it is closely related to the biogeocoenotic cover of the earth and takes into consideration its geographical and topographical variability.

While the biogeocoenotic cover of the earth is a laboratory in which the above-discussed biochemical reactions of living matter take place, forest cover—forest biogeocoenoses—plays an extremely important part in these processes, since nowhere in the living cover of the earth are these processes so intensive or so complex as in a forest.

It becomes clear why in recent times, when both philosophers and naturalists (including biologists) are striving simultaneously to solve the general problems of the knowledge of nature, interest in the problem of the biogeocoenose from the philosophical point of view has begun to grow. In the compendium *Outline of the dialectics of living nature* this idea has been expressed in several articles, but most definitely by N. P. Naumov. He writes: 'The relationships of a biocoenose with its environment are so intimate and profound that they constitute an indubitable unity, called a biogeocoenose by V. N. Sukachev and an ecosystem abroad (by Tansley and Dyce)' (1963, p. 123).

In noting this general theoretical importance of biogeocoenology, we must not forget that all evolution in the organic world, all processes of species-formation in plants and animals take place in populations located in definite biocoenoses, and are thereby linked with definite types of biogeocoenoses and with the whole complex of factors affecting the populations. We can state with assurance that the arena of species-formation is almost always a biogeocoenose.

I. I. Shmal'gauzen (1958, 1961) was the first to describe the mechanism of evolution of populations in biogeocoenoses in cybernetic terms, formulating it briefly as follows: 'Regulation of the evolutionary process is thus performed within a population system by means of natural selection of variants (signals) on the basis of their comparative valuation in the biogeocoenose. This denotes the presence of entrance-relationships with the environment. The result of regulation is transmitted through the medium of signals in the hereditary code of sex cells, is intensified in the reproduction process, and is transformed into signals of reverse information (phenotypes) entering the biogeocoenose by the exit channel of links for the control of fulfilment (phenotypes)' (1961, pp, 124, 125).

Giving a graphic visual sketch of this process, Shmal'gauzen remarks: 'This "cybernetic" cycle is merely a paraphrase of the Darwinian concept of evolution. Application of cybernetic principles, however, enables us to introduce greater clarity into understanding of the separate stages of this cycle, especially in determining the place occupied by the "struggle for existence" as a complex system of control, in analysing the concept, and in quantitative evaluation of the diversity of forms as material for natural selection' (1961, p. 126).

In the light of these statements we see clearly the great importance of biogeocoenology for evolutionary theory and for study of species-formation, and consequently for practical selection of plants and animals.

When considering the immediate practical importance of biogeocoenology as a theoretical science we have to remember the words (quoted at the beginning of the book) of such a great theorist and scientist, and one who at the same time did so much for the practical application of new achievements of physics in various fields of production, as Joliot Curie, namely, 'It is proper to acknowledge that almost all great technical and industrial inventions and innovations are based on knowledge and discoveries attained in theoretical laboratories.' Approximately the same idea was recently expressed by N. N. Semenov: 'What is most important in theory is always, as experience shows, later proved to be most important in practice also' (*Pravda*, 6 February 1964).

In practical sylviculture it is of primary significance that a forest, like any biogeocoenose, is regarded as a complex system in which there is great diversity in the relationships between its components. These interrelationships are expressed either in interactions mutually affecting the components and their elements or in reciprocal bonds between them. The mutual effects may be beneficial for both interacting components or elements of them (in this case both components or elements are alive), or unilaterally beneficial (when both components are alive, or only one of them is); or the effects may be beneficial for one and harmful for the other, but the benefit exceeds the harm, and therefore the system does not break down; or there may be cases where the interactions are harmful to both components, but they receive some other benefit from co-existence, and therefore the system is still preserved.

Study of all these forms of interactions and of the mechanism of their mutual effects and inter-relationships enables us to control and reconstruct them or to create new biogeocoenoses, with beneficial application of the laws discovered.

In recent years the practical importance of forest biogeocoenology has been increasingly recognised by growing numbers of sylviculturists. In the opening chapter we referred to a statement to that effect made by the Czechoslovak scientist, Krajina, when summing up the work of the forestry symposium at the International Botanical Congress held in Canada in 1962.

In 1963 a course on sylviculture was published in the German Democratic Republic (*Der Wald und die Forstwirtschaft*), compiled on the basis of forest biogeocoenology by the staff of the Higher Forestry School at Tarandt (near Dresden) under the editorship of Professors J. Blanckmeister and E. Kienitz. The first part of the book is devoted to survey of the nature of a forest as a biogeocoenose. In the second part the problems of forest economy are set forth on the same basis. The book closes with a general programme for study of a forest biogeocoenose. The authors write: 'The practical significance of forest biogeocoenology is that it discloses the regularity of the processes taking place in a biogeocoenose and makes their use possible. When the laws pertaining to the forest are discovered, it is possible to rationalise forest economy and to conduct sylviculture on a scientific basis and not merely empirically' (p. 19). Professor A. Scamoni (1960, p. 921) of the Forest Research Centre at Eberswald (near Berlin) writes: 'For the practice of agriculture (*Bodenkultur*) the biogeocoenotic approach is of great value. Although the newest agricultural techniques may create greatly-altered phytocoenoses, it is still necessary, for the highest plant production, to study a biogeocoenose in its structure and in the laws of its development in order to carry out proper agronomic measures.'

How appropriate and necessary the biogeocoenotic approach is from the point of view of practical forest economy is shown in an article by G. F. Starikov and A. A. Stepanov, in which they describe the problem of reconstruction of the forest landscape of the Amur basin. Although the editorial board of *Problems of the Geography of the Far East*, which published the article, remarks that it contains a number of disputable statements and does not cover the whole problem, the article does point out the general path of 'prognosis, regulation, and management of the process of reconstruction of nature (in many respects still in its original state) in the Far East. At the same time it shows that only complex biogeocoenotic study of the forest can provide a correct scientific foundation for economic measures for proper (from the point of view of the national economy) reconstruction of the forest landscapes of the Far East, i.e., reconstruction directed ultimately towards increasing and extending exploitation of the natural resources of the Far East, combined with measures to facilitate self-restoration of all renewable resources' (1963, pp. 72, 73, 157).

V. S. Shumakov (1963), when reviewing the effects of plantations of various tree species on the soil and its fertility, and pointing out practical ways of proper establishment of forest plantations, remarks that the biogeocoenose concept 'in its depth and dialectical essence is the most fruitful in modern sylviculture' (p. 167).

We could quote many similar statements by scientists and practical sylviculturists about the value of studying forests from the biogeocoenotic point of view. In foreign literature also high praise is often given to the science of ecosystems, which, as was shown in the opening chapter, is very close to the science of biogeocoenoses.

Much earlier, however, this very approach to the forest was stressed by G. F. Morozov. In his famous *Forest Science* he brilliantly expressed the idea that for all the principal forestry measures (clear and strip felling, forest management, aiding natural regeneration, artificial forest planting, protection against pests, fire prevention, etc.) it is necessary to know not only the biological characteristics of the tree stand and other layers of forest vegetation but also the climatic (including microclimatic), pedological, and hydrological conditions, and the fauna and microbial population. The more thoroughly we know the characteristics of these forest-forming components, the sounder will be the forestry measures and the greater their practical effect.

Human economic activity in nature consists largely in guiding biogeocoenotic processes in such a direction that maximum use may be made of biogeocoenoses. In particular, forestry measures consist essentially in altering the direction and intensity of biogeocoenotic processes so as to obtain production from the forest in greatest amount and of highest quality (wood, vegetable food products, animal products) or to increase its other useful properties (its protective, anti-erosion, water-conserving, and sanitational roles).

We must repeat that the organisation of any experiments is of value only when they are adapted to a definite type of forest, this term being understood to mean a type of forest biogeocoenose. An experiment undertaken in a forest without consideration of its type has no value, since the same experiment may give entirely different results in forests of different types. Therefore forestry measures based on a specific experiment can produce the expected result only within that forest type in which the experiment was made or in types with similar biogeocoenotic conditions. Consequently, biogeocoenotic forest typology, accompanied by compilation of maps of forest types, is a necessary foundation for the application of the results of experimental forestry work.

As was pointed out in the opening chapter, the existence of the so-called continuum of plant (including forest) cover, reflecting the fact that vegetation constantly occupies all places suitable for it and that plant

associations often merge very gradually into each other, does not prevent
the investigator from grouping associations, defining these in nature, and
charting them on a map or plan.

That applies equally to biogeocoenoses. We must not, however, exaggerate the difficulty of separating types of biogeocoenoses, since often
in nature, when environmental living conditions for plants change gradually, their communities are quite sharply defined because the interspecific
struggle for existence among plants plays a substantial role in the distribution of plant communities. In a gradually-changing environment,
therefore, a phytocoenose occupies all the territory in which its constituent
plants can win the struggle for existence against plants of other communities. A good example of such a community is a pine forest covering
sand dunes. Often the summits of the dunes are occupied by lichen pine
stands, the slopes by moss pine stands, and the hollows between the
dunes by herbaceous pine stands. The boundaries between these three
types of pine biogeocoenoses are usually sharp, although the relief (and
with it the soil and microclimatic conditions) changes very gradually.

We must not overrate the mobility, i.e., the dynamic nature, of biogeocoenoses as an impediment to definition of their types. Although the
processes of development and of changes in general never cease in biogeocoenoses, as a rule forest biogeocoenoses retain their monotypy so
long that even such long-living plant organisms as trees can pass through
many generations. That results largely from the property of self-regulation, which is to some extent inherent in all biogeocoenoses. Biogeocoenoses are, as we have seen above, self-regulating systems. That
feature enables a forest to offer considerable resistance to external destructive influences and to maintain its specific characteristics for a long time.
In a forest biogeocoenose (as in others) two mutually-contradictory
tendencies are always in operation: on the one hand, internal interactions
are always trying to change, to break down the inter-relationships built
up between components, and on the other hand there is the capacity to
resist the changing action, to correct it, and to restore what has been
destroyed. A forest biogeocoenose is particularly capable of resistance to
the penetration of new species of organisms into it. The relative stability
of developed biogeocoenoses and their self-regulating ability result from
the fact that the components of biogeocoenoses include organisms that
in the process of their evolutionary development have become adapted
to a definite environment and to living together, whereby there have been
developed (as well as certain relationships of competition for the means
of existence, or of parasitism) such mutually-beneficial influences of one
organism upon another that these influences have become a necessary
condition of existence for them.

We can therefore agree with V. K. and L. K. Arnol'di when they (understanding a biocoenose as being essentially the same as a biogeocoenose)
write: 'The most important feature of a biocoenose is monotypy, repeated

from year to year through many generations, of the process of production and general circulation of organic matter, based upon constancy of the species-composition of the producers and consumers and the capacity for self-regulation of that species-composition' (1962, p. 117).

Although complex forest-biogeocoenotic investigations are many-sided and expensive, and require the participation of highly-qualified investigators of various specialised sciences, only they can ensure all the necessary understanding for development of the forest economy. An example of the fruitfulness for the forest economy of the biogeocoenotic method of forest study, taking forest types into consideration, is the investigations organised and conducted under the leadership of A. A. Molchanov (1963) in the Tellerman experimental forest in Voronezh province. They have done very much for the theory of forest biogeocoenology and have made it possible to recommend a number of practical measures for rationalisation of the forest economy in the broad-leaved forests of the forest-steppe.[1] Although G. F. Morozov, during the first decade of this century, urged that study of forest communities should be regarded as the principal theoretical basis of forest economy, in the following decade he strongly supported the view that study of forest communities together with their whole environment is the scientific basis of forest economy, i.e., the assemblage of knowledge that to-day composes forest biogeocoenology.

We agree with N. V. Timofeev-Resovskii when he writes that 'in the general plan for solution of the problems of nature conservation, and especially in study and use of the biological resources of the earth, it is absolutely necessary to start from general biogeocoenological concepts; only when we take into account all the features of the phenomena occurring on the evolutionary and biochorological levels of life can we avoid gross errors and markedly raise human utilisation of the earth's biological productivity' (1962, p. 45).

Without touching here upon the programmes of forest biogeocoenotic investigations, we need only remark that complex investigation of natural forest biogeocoenoses should include the following main subjects.

1. Study of the composition of the plant and animal populations of biogeocoenoses. This study should not be restricted merely to determination of the species-composition of the organic world of both higher and lower plants and animals but should be in greater detail, since species units embrace a variety of vital characteristics. This study should aim at clarification of phylogenetic relationships and an experimental approach is essential.

2. Study of the ecological and biological features of plant and animal

---

[1] In A. A. Molchanov's book *Scientific foundations for management of oak stands on the forest-steppe* (1964), issued after this book was compiled, it is shown very clearly how necessary complex biogeocoenotic study of a forest is as a scientific foundation for sylvicultural measures.

organisms (their life forms). This study calls for thorough physiological and chemical study, with extensive use of the experimental method.

3. Thorough study of the properties (inherent in a given forest biogeo-coenose) of the soil, of the hydrological conditions throughout the depth of the phytogeosphere, and of the atmosphere throughout the height of the phytogeosphere.

4. The study should differentiate not only between separate biogeo-coenoses but also the synusiae composing them. Such investigations call for field work and prolonged experimental work covering a number of years, with multilateral study of the inter-relationships of all biogeo-coenose components among themselves.

5. Study of the structure of forest biogeocoenoses, of the internal interactions and inter-relationships of their components, of their inter-actions with adjacent biogeocoenoses, of their topographical and geo-graphical distribution, and of their dynamics. Here also there should be wide use of field and experimental methods, and also use of small-scale and large-scale mapping of biogeocoenoses.

On the basis of such complex study of natural forest biogeocoenoses, the processes of transformation of matter and energy in them and in the phytogeosphere generally should be clarified, and general deductions of theoretical value should be obtained as a contribution to the general theory of biogeocoenology; practical suggestions arising therefrom should be noted. As a rule, practical suggestions should be directed towards:

(*i*) more rational, multipurpose use of forest production, with regard both to wood and to other plant and animal resources; further increase in their productivity in the interests of the national economy, industry, urban development, and health preservation;

(*ii*) organisation and carrying-out of biological, chemical, and other measures for forest protection and pest control;

(*iii*) maximum use of natural and artificial forest biogeocoenoses and of forest vegetation for windbreak and water-conservation purposes; for control of soil erosion, floods, and landslides; and for control of soil, water, and air pollution, taking into consideration the capacity of plants to accumulate harmful radioactive elements;

(*iv*) devising new types of artificial, but more productive and valuable, forest biogeocoenoses; suggestion of sound practical measures for en-suring natural regeneration and generally increased restoration of forest resources; protection of forest landscapes and their regenerative power, and heightening of their aesthetic value.

Biogeocoenology is a new scientific discipline that has arisen in recent years, like a number of other disciplines, on the foundation of a combina-tion of several sciences. It is still in the earliest stages of its growth. In the future it certainly should and will use new methods of investigation created on the basis of both its own achievements and those of a number of other sciences, and will disclose new, still-unused biological resources, including those of the forest. At the same time it will point out new and more effective ways of using natural biological resources, and will aid in the general reconstruction of the biosphere in the interests of mankind, or, more precisely, in reconstruction of its most active part, the phytogeosphere, in which the most important place is now occupied, and will continue to be occupied, by forests.

Bibliography

RUSSIAN LIST

ABOLIN, R. I. 1914. Experiment in epigenological classification of bogs. *Boloto-vedenie*, 3.

ABRAMOV, K. G. 1954. *Ungulates of the Far East.* Khabarovsk.

ABRAMOVA, M. M. 1947. Seasonal variation in some chemical properties of podzolic forest soils. *Trudy pochv. Inst.*, 25.

ABRAMOVA, M. M. 1953. Water movement in the soil during evaporation. *Trudy pochv. Inst.*, 41.

ABRAMOVA, M. M., BOL'SHAKOV, A. F., ORESHKINA, N. S., and RODE, A. A. 1956. Evaporation of suspended water from soil. *Pochvovedenie*, No. 2.

AFANES'EVA, E. A., KARANDINA, S. N., KISSIS, T. YA., and OLOVYANNIKOVA, I. N. 1952. Forest-vegetational properties of southern chernozems and growth of tree plantations on them. *Trudy Kompleks. nauch. Eksped. Vopr. polezashch. Lesorazv.*, 1.

AFANES'EVA, E. A., KARANDINA, S. N., KISSIS, T. YA., and OLOVYANNIKOVA, I. N. 1955. Combined study of root systems and water regime of soils in oak-maple plantations on ordinary chernozems. *Trudy Inst. Lesa*, Moscow, 39.

AKHROMEIKO, A. I. 1936. Exudation of mineral matter by plant roots. *Izv. Akad. Nauk SSSR*, No. 1.

AKHROMEIKO, A. I., and SHESTAKOVA, V. A. 1958. Study of the role of micro-organisms in tree nutrition. *Mikrobiologiya*, 27.

AKULOVA, E. A. 1962. *Study of energy exchange in leaves of higher plants.* (Author's extract of candidate's dissert.) Petrozavodsk.

AKULOVA, E. A., KHAZANOV, V. S., TSEL'NIKER, YU. L., and SHISHOV, D. M. Light regime beneath forest canopy in relation to incident radiation and crown density. *Fiziologiya Rast.* (in press).

AL'BITSKAYA, M. A. 1960. *Principal laws of formation of herbaceous cover in artificial forests in the steppe zone of the Ukrainian S.S.R.* In: *Artificial forests of the Ukrainian steppe zone.* Khar'kov.

AL'BITSKAYA, M. A., and BEL'GARD, A. L. 1950. Interrelationships of tree-shrub and herbaceous vegetation in artificial forests in Dnepropetrovshchina. *Bot. Zh. Kyyiv.*, 35.

ALEKSAKHIN, R. M. 1963. *Radioactive pollution of soil and plants.* Publ. Acad. Sci. U.S.S.R. Moscow.

ALEKSANDROVA, I. V., and KRASOVSKII, L. I. 1957. Observations on summer diet of moose in the Pri022sko-Terrasnyi reserve. *Trudy Priok.-Terr. Zap.*, 1.

ALEKSANDROVA, V. D. 1961. The plant community in the light of some principles of cybernetics. *Byull. mosk. Obshch. Ispyt. Prir.*, 66.

ALEKSANDROVA, V. D. 1962. The problem of development in geobotany. *Byull. mosk. Obshch. Ispyt. Prir.*, 67.

ALEKSEEV, A. M. 1948. *Water regime of plants and effect of drought on it.* Tatgosizdat. Kazan.

ALEKSEEV, S. V., and MOLCHANOV, A. A. 1938. *Clear-cutting of forests in the north.* Vologda.

ALEKSEEVA, T. A. 1957a. *Investigation of photosynthesis and respiration of some tree species in Tellerman experimental forest.* (Author's extract of dissert.) Publ. Inst. Forest. Acad. Sci. U.S.S.R. Moscow.

ALEKSEEVA, T. A. 1957b. *Photosynthesis and respiration of tree species in different ecological conditions in Tellerman forest.* Publ. Inst. Forest. Acad. Sci. U.S.S.R. Moscow.

ALPAT'EV, A. M. 1954. *Moisture circulation in cultivated plants.* Gidrometizdat. Leningrad.

613

ANOKHIN, P. K. 1961. Physiology and cybernetics. In: *Philosophical problems of cybernetics*. Moscow.

ANOKHIN, P. K. 1962. Theory of a functional system as a prerequisite for construction of philsophical cybernetics. In coll. *Biological aspects of cybernetics*. Publ. Acad. Sci. U.S.S.R. Moscow.

ANTIPOV, V. G. 1957. Effects of smoke and gas emitted by industrial undertakings on seasonal development of trees and shrubs. *Bot. Zh. SSSR*, **42**.

AREF'EVA, Z. N. 1963. Effects of fire on some biochemical processes in forest soils. *Trudy Inst. Biol.*, Sverdlovsk, **36**.

AREF'EVA, Z. N., and KOLESNIKOV, B. P. 1964. Effects of fire on nitrate and ammonia regime of soils. *Pochvovedenie*, No. 3.

ARISTOVKAYA, T. V., and PARINKINA, O. M. 1961. New methods of studying micro-organism communities. *Pochvovedenie*, No. 1.

ARISTOVKAYA, T. V., and PARINKINA, O. M. 1962. Study of the microbial landscape of the soils of Leningrad province. *Mikrobiologiya*, **31**.

ARKHANGEL'SKII, M. P. 1929. Effect of the work of earthworms on crops of oats and barley in relation to the addition of certain fertilisers to the soil. *Nauchno-agron. Zh.*, No. 11.

ARMAND, D. L. 1949. Functional and correlational relations in physical geography. No. 1. *Izv. vses. geogr. Obshch.*, **81**.

ARMAND, D. L. 1950. Experiment in mathematical analysis of connections between vegetation types and climate. *Izv. vses. geogr. Obshch.*, **82**.

ARNOL'DI, K. V. 1957. Theory of range in connection with ecology and origin of species-populations. *Zool. Zh.*, **36**.

ARNOL'DI, K. V., and ARNOL'DI, L. V. 1962. Some basic ecological concepts applicable to biocoenose studies. In: *Voprosy ekologii*, IV. Izd. kiev. Univ. Kiev.

ARNOL'DI, K. V., and ARNOL'DI, L. V. 1963. The biocoenose as one of the basic concepts of ecology, its structure and scope. *Zool. Zh.*, **42**.

ARNOL'DI, L. V. 1960. Brief methodical instructions on study of consortial relation-ships of insects in biocomplex investigations. In: *Programme-method notes on bio-complex and geobotanical study of the steppes and deserts of Central Kazakhstan*. Publ. Acad. Sci. U.S.S.R. Moscow.

ARNOL'DI, L. V., and BORISOVA, I. V. 1962. Some features of consortial relationships between vegetation and insects. *Paper delivered at conference on problems of inter-actions of plants and animals in a biocoenose, Moscow branch of Geographical Society*.

ARNOL'DI, L. V., and LAVRENKO, E. M. 1960. Brief programme notes on study of consortial relationships between animals and higher plants in plant communities. In: *Programme-method notes on biocomplex and geobotanical study of the steppes and deserts of Central Kazakhstan*. Publ. Acad. Sci. U.S.S.R. Moscow.

Bacterial fertilisation. 1961. Izd. sel'sk.-khoz. Lit. Zh. Plak. Moscow and Lenin-grad.

BANNIKOV, A. G., and LEBEDEVA, L. S. 1956. Significance of deer in the forests of the Belovezh bush. *Byull. mosk. Obshch. Izpyt. Prir.*, No. 4.

BARABASH-NIKIFOROV, I. I. 1957. *Wild animals of the south-eastern part of the cher-nozem centre*. Voronezh.

BARAKOV, P. 1910. Carbon dioxide content in soils at different periods of plant growth. *Russ. J. exp. Agric.*, **11**.

BARANEI, A. V. 1940. Experiments in introduction of mycorrhiza fungi into the soil. *Les. Khoz.*, No. 10.

BARANOV, P. A. 1960. Experiment in analysis of adaptational evolution of climbing plants. *Trudy mosk. Obshch. Ispyt. Prir.*, **3**.

BARASKINA, E. M. 1959. Interactions of oak and pine root systems with soil. *Trudy voronezh. gos. Zapov.*, No. 8.

BASHENINA, N. V. 1951. Material on the dynamics of small-rodent numbers in the forest zone. *Byull. mosk. Obshch. Ispyt. Prir.*, No. 2.

BASURMANOVA, O. K. 1958. Biological forms of flat-headed pear-borers (*Agrilus viridis* L.). *Zool. Zh.*, **37**.

BAZILEVICH, N. I. 1955. Features of the circulation of mineral elements and nitrogen in some soil-vegetational zones of the U.S.S.R. *Pochvovedenie*, No. 4.

BAZILEVICH, N. I., and RODIN, L. E. 1954. Features of forest biological circulation in different soil-vegetational zones. *Dokl. Akad. Nauk SSSR*, **97**, No. 6.

BEI-BIENKO, G. YA. 1930. The problem of zonal-ecological distribution of grasshoppers in the Western Siberian and Zaisan lowlands. *Trudy Zashch. Rast.*, **1**.

BEI-BIENKO, G. YA. 1962. Succession of stages of terrestrial organisms as an ecological principle. In: *Voprosy Ekologii*, IV. Izd. kiev. Univ. Kiev.

BEILIN, I. G. 1950. Mistletoe in Western Europe and the U.S.S.R. *Trudy Inst. Lesa*, Moscow, **3**.

BEKLEMISHEV, V. N. 1931. Basic concepts of biocoenology in application to the animal component of terrestrial communities. *Trudy Zashch. Rast.*, **1**.

BEKLEMISHEV, V. N. 1951. Classification of biocoenological (symphysiological) relationships. *Byull. mosk. Obshch. Ispyt. Prir.*, **56**.

BEL'GOVSKII, M. L. 1955. Relationship between the systematic position of Russian elm (*Ulmus laevis* Pall.) and infestation of it by insects. *Soobshch. Inst. Lesa*, No. 4.

BELKIN, V. D. 1961. Cybernetics and economics. In coll. *Cybernetics in the service of Communism*. Ed. by Acad. A. I. Berg. Publ. Acad. Sci. U.S.S.R. Moscow & Leningrad.

BEREZINA, V. M. (1936) 1937. Distribution of soil entomofauna in sandy and chestnut soils of the Kamyshinsk forest-improvement area. *Itogi nauchno-issled. Rab. vses. Inst. Zashch. Rast.*, 1936, part 1.

BEREZOVA, E. F. 1950. Interrelationships of plants and soil microflora. *Agrobiologiya*, No. 5.

BERG, A. I. 1961. Cybernetics in the service of Communism. In coll. *Cybernetics in the service of Communism*. Ed. by Acad. A. I. Berg. Publ. Acad. Sci. U.S.S.R. Moscow & Leningrad.

BERG, L. S. 1913. Experiment in dividing Siberia and Turkestan into landscape and morphological zones. *Vol. Jub. D. N. Anuchin*, Moscow.

BERG, L. S. 1915. Subject and tasks of geography. *Izv. russk. geogr. Obshch.*, **1**.

BERG, L. S. 1931. *Landscape-geographical zones of the U.S.S.R.*, part 1. Sel'khozgiz. Moscow.

BERG, L. S. 1936. *Physico-geographical (landscape) zones of the U.S.S.R.*, part 1. Geografgiz. Moscow.

BERG, L. S. 1945. Facies, geographical aspects, and geographical zones. *Izv. russk. geogr. Obshch.*, **77**.

BERNAL, D. D. 1960. *World without war*. Moscow.

BESKARAVAINYI, M. M. 1955. Root coalescence in some tree species in the district of the Kamyshin river. *Agrobiologiya*, No. 3.

BIBIKOV, D. I. 1948. Ecology of the nutcracker. *Trudy pech.-ilych. Zap.*, **4**.

BKHUVANASVARI, SUBBO-RAO. 1959. Root exudates and the rhizosphere effect. *Abstr. J. Biologiya*, No. 4.

BLAUBERG, I. V. 1960. Problem of the integrity of the geographical envelope. *Philosophical questions of natural history*, part III. Geologo-geographical sciences. Moscow State Univ. Moscow.

BOBRITSKAYA, M. A. 1962. Entrance of nitrogen into the soil with atmospheric precipitation in different zones of the European part of the U.S.S.R. *Pochvovedenie*, No. 12.

BODROV, V. A. 1951. *Forest improvement*. Goslesbumizdat. Moscow.

BOL'SHAKOV, A. F. 1950. Water regime of the deep chernozems during the drought period of 1946-1947. *Trudy pochv. Inst.*, **32**.

BOL'SHAKOVA, V. S. 1964. Microflora of the rhizosphere of tree species on the soddy-podzolic soils of Moscow province. In: *Field biogeocoenotic investigations in the southern taiga subzone*. Nauka. Moscow.

BRAINES, S. N. 1961. Neurocybernetics. Coll. *Cybernetics in the service of Communism*. Ed. by Acad. A. I. Berg. Publ. Acad. Sci. U.S.S.R. Moscow & Leningrad.

BREZHNEV, I. E. 1950a. Parasitic and saprophytic microflora of tree and shrub species in windbreaks. *Trudy lesostep. Sta. Les na Vorskle*.

BREZHNEV, I. E. 1950b. Soil microflora of the 'Les na Vorskle' reserve. *Trudy lesostep. Sta. Les na Vorskle*.

BRILLYUEN, L. 1960. *Science and theory of information*. Moscow.

BUDYKO, M. I. 1956. *Heat balance of the earth's surface*. Gidrometizdat. Moscow.

BUDYKO, M. I., and DROZDOV, O. A. 1950. Moisture circulation on a limited territory of dry land. In: *Problems of the hydrometeorological effectiveness of windbreak plantations.*

BURNASHEVA, N. YA. 1955. Varying infestation of aspens by hornet clearwing moths. *Trudy Inst. Lesa,* Moscow, **25.**

BURSOVA, A. I. 1955. Study of soil microflora of spruce stands growing in Leningrad province. *Trudy vses. zaoch. lesotekh. Inst.,* **1.**

BYALLOVICH, YU. P. 1947. *Method of phyto-amelioration. Scientific report for 1945.* Ukr. nauch.-issled. Inst. Agrolesomel. les. Khoz. Kiev & Khar'kov.

BYALLOVICH, YU. P. 1954. Practical significance of differences in times of spring foliation of different tree and shrub species. *Nauch. Trudy Ukr. nauch.-issled. Inst. les. Khoz. Agrolesomel.,* No. 16.

BYALLOVICH, YU. P. 1960. Biogeocoenotic horizons. Coll. of works on geobotany, botanical geography, plant systematics, and palaeogeography. *Trudy mosk. Obshch. Ispyt. Prir.,* **55.**

BYKOV, B. A. 1957. *Geobotany.* Alma-Ata.

BYKOVA, L. N. 1961. Iodine in the forest soils of the Voronezh national reserve. *Trudy voronezh. gos. Zap.,* **13.**

CHASOVENNAYA, A. A. 1954. Effect of volatile organic exudates of plants on seed germination and growth and development of some species of herbaceous plants. *Vest. leningr. gos. Univ.*

CHASOVENNAYA, A. A. 1961. Problem of the mechanism of the chemical interaction of plants. *Vest. leningr. gos. Univ.*

CHASTUKHIN, V. YA. 1945. Ecological analysis of decomposition of plant remains in spruce forests. *Pochvovedenie,* No. 2.

CHASTUKHIN, V. YA. 1948. Ecological analysis of decomposition of plant remains in young pine plantations. *Pochvovedenie,* No. 2.

CHASTUKHIN, V. YA. 1952. Decomposition of plant remains and role of fungi in the soil-formation process. *Agrobiologiya,* No. 4.

CHASTUKHIN, V. YA. 1962. Decomposition of forest litter by pure cultures of Basidiomycetes. *Uchen. Zap. leningr. gos. Univ.,* No. 313 (*ser. biol.* No. 49).

CHASTUKHIN, V. YA., and NIKOLAEVSKAYA, M. A. 1953. Investigations of the decomposition of organic remains under the action of fungi and bacteria in oak stands, steppes, and windbreaks. *Trudy bot. Inst. Akad. Nauk SSSR,* Ser. 2.

CHASTUKHIN, V. YA., and NIKOLAEVSKAYA, M. A. 1962. Interrelationships among the micro-organisms that decompose plant remains, in composite cultures. *Uchen. Zap. leningr. gos. Univ.,* No. 313 (*ser. biol.* No. 49).

CHEKANOVSKAYA, O. V. 1960. *Earthworms and soil formation.* Publ. Acad. Sci. U.S.S.R. Moscow.

CHELYADINOVA, A. I. 1941. Amount and character of needle development in a pine plantation. *Trudy vses. nauchno-issled. Inst. les. Khoz.*

CHERNOBRIVENKO, S. I. 1956. *Biological role of plant exudates and interspecific relationships in mixed plantations.* Sovetskaya Nauka. Moscow.

CHERNOV, YU. I. 1961. Study of animal population of Arctic tundra soils in Yakutia. *Zool. Zh.,* **10.**

CHETYRKIN, V. M. 1960. *Central Asia. Experiment in complex geographical description and division into districts.* Izd. Sredn.-Aziatsk. Univ. Tashkent.

CHISTYAKOV, F. M., and BOCHAROVA, Z. Z. 1938. Effect of low temperatures on development of moulds. *Mikrobiologiya,* **7.**

CHISTYAKOV, F. M., and NOSKOVA, G. L. 1938. Effect of low temperatures on development of bacteria and yeasts. *Mikrobiologiya,* **7,** No. 5.

CHIZHOV, B. A. 1940. Problem of utilisation of nutrients from dry soil. *Izv. Akad. Nauk SSSR (ser. biol.),* No. 4.

CHUGAI, N. S. 1960. Phytoclimatic features of artificial forests in the steppe zone of the Ukraine. In: *Artificial forests in the steppe zone of the Ukraine.* Khar'kov.

DADYKIN, V. P. 1950. Study of root systems of plants growing on permanently-cold soils. *Byull. mosk. Obshch. Ispyt. Prir.,* **55.**

DANILEVSKII, A. S. 1961. *Photoperiodism and the seasonal development of insects.* Leningrad.

DANILEVSKII, A. S., and BEI-BIENKO, I. G. 1958. The green oak roller moth (*Tortrix viridana* L.) and the problem of resistance by oak to infestation. *Uchen. Zap. leningr. gos. Univ.*, No. 240.

DANILOV, D. N. 1934. *Game habitats*. Moscow.

DANILOV, D. N. 1937. The spruce seed crop and its use by squirrels, crossbills, and woodpeckers. *Byull. mosk. Obshch. Ispyt. Prir.*, **42**.

DANILOV, D. N. 1944. Food resources of spruce forests and their use by squirrels. *Trudy tsentr. lab. biol. okhot. prom.*, No. 6.

DANILOV, M. D. 1953. Laws of development of monoculture tree stands in relation to leaf-mass dynamics. *Les. Khoz.*, No. 6.

DARWIN, C. 1941. *Collected works*, vol. 8. Publ. Acad. Sci. U.S.S.R. Moscow & Leningrad.

DAVYDOVA, YU. A. 1964. Transpiration of some tree and herbaceous species in a maple-hazel linden-oak stand and an oak-hazel pine stand. In: *Field biogeocoenotic investigations in the southern taiga subzone*. Nauka. Moscow.

DAVYDOVA, YU. A., PAVLOVA, N. N., and TSEL'NIKER, YU. L. 1964. Productivity and transpiration coefficient of photosynthesis of different layers of vegetation in a maple-hazel linden-oak stand and an oak-hazel pine stand. In: *Field biogeocoenotic investigations in the southern taiga subzone*. Nauka. Moscow.

DEKATOV, N. E. 1957. Problems of hay-mowing and cattle-pasturing in forests. *Les. Khoz.*, No. 9.

DEKATOV, N. E. 1959. *Forest pasturage and hay-mowing*. Moscow & Leningrad.

DENISOV, I. A. 1962. Agricultural use of tropical soils in Africa. *Pochvovedenie*, No. 7.

DIANOVA, N. V. 1959. Damage by ungulates in leskhozy (managed forests) in the U.S.S.R. *Soobshch. Inst. Lesa.*, No. 13.

DIANOVA, N. V., and VOROSHILOVA, A. A. 1925. Absorption of bacteria by soil and its effect on microbiological activity. *Nauchno-agron. Zh.*, No. 11.

DINESMAN, L. G. 1959a. Damage by ungulates in leskhozy (managed forests) in the U.S.S.R. *Soobshch. Inst. Lesa*, No. 13.

DINESMAN, L. G. 1959b. Effect of blue hares on larch forest regeneration in Central Yakutia. *Byull. mosk. Obshch. Ispyt. Prir.*, No. 5.

DINESMAN, L. G. 1961. *Effect of wild mammals on tree-stand formation*. Moscow.

DINESMAN, L. G., and SHMAL'GAUZEN, V. I. 1961. Role of moose in circulation and transformation of matter in a forest biogeocoenose. *Soobshch. Lab. Lesov.*, No. 3.

DOBROGAEV, I. I. 1939. Microbiological life of forest soil in relation to its swamping. *Pochvovedenie*, No. 5.

DOKUCHAEV, V. V. 1898a. Place and role of contemporary pedology. *Ezheg. Geol. Miner. Ross.*, No. 10.

DOKUCHAEV, V. V. 1898b. *Study of zones in nature*. St. Petersburg; 2nd ed., 1948. Geografgiz. Moscow.

DOLUKHANOV, A. G. 1957. Problems of coenotic classification of forests in relation to appearance of convergent vegetation. *Tez. Dokl.*, No. IV, flora and vegn. sect., part 2. Publ. Acad. Sci. U.S.S.R. Moscow.

DOLUKHANOV, A. G. 1959. Problems of natural classification of forest coenoses. *Trudy tbilis. bot. Inst.*, **20**.

DOLUKHANOV, A. G. 1961. Principles of classification of plant communities. *Trudy Inst. Biol.*, Sverdlovsk, **27**.

DOPPEL'MAIR, G. G. 1915. Biological differentiation of areas in a forest and study of forest fauna. *Les. Zh.*, No. 3.

DRABKIN, B. S., and BALOVNEV, V. M. 1952. Effect of phytoncides on earthworms. In: *Phytoncides, their role in nature and significance for medicine*. Publ. Acad. Sci. U.S.S.R. Moscow.

DROBOT'KO, V. G., and AISENMAN, P. E. 1958. *Anti-microbe substances in higher plants*. Publ. Acad. Sci. Ukr. S.S.R. Kiev.

DUBININ, N. P., SIDOROV, V. N., and SOKOLOV, N. N. 1961. *Physico-chemical and structural bases of biological phenomena*. Publ. Acad. Sci. U.S.S.R. Moscow.

DUBININ, N. P., and TOROPANOVA, T. A. 1960. Some laws of distribution of birds in the forest zone. *Ornitologiya*, No. 3.

DUSHECHKIN, A. I. 1911. *Results of laboratory experiments in study of the effects of temperature and moisture on the process of nitrate accumulation in the soil.* Vseross. Obshch. Sakharovod. St. Petersburg.

DYLIS, N. V., UTKIN, A. I., and USPENSKAYA, I. M. 1964. Horizontal structure of forest biogeocoenoses. *Byull. mosk. Obshch. Ispyt. Prir.*, **69.**

DZHAFAROV, B. A. 1960. Seasonal dynamics of litter accumulation and decomposition of litter in beech forests on the southern slopes of Great Caucasus. *Izv. Akad. Nauk azerb. SSR (ser. biol.),* No. 6.

DZHEIMS (JAMES), B. 1956. *Plant respiration.* Inostr. Lit. Moscow.

EDEL'MAN, N. M. 1956. Biology of the gypsy moth in the conditions of Kubinsk district in Azerbaidzhan s.s.r. *Zool. Zh.,* **35.**

EDEL'MAN, N. M. 1963. Age-changes in the physiological state of larvae of some xylophagous insects in relation to their nutrition conditions. *Ent. Obozr.,* **42.**

EDEL'SHTEIN, V. I. 1946. Nutrition area as one of the conditions of high vegetable production. *Proc. jub. Conf. Timiryazev agric. Acad. Moscow.* Selkhozgiz. Moscow.

EFREMOV, YU. K. 1960. Two logical stages in the process of physico-geographical division into districts. *Vest. mosk. gos. Univ.,* No. 4.

EGLITE, A. K. 1958. Difficultly-soluble minerals and organic matter as a source of food for mycorrhiza fungi. *Microbiol. Inst. zinat. Rak.,* **2.**

EGLITE, A. K., and YAKOBSON, Z. A. 1961. Relationship between microbiological processes in the soil and productivity of plantations on sandy soils. *Mežsaimn. Probl. Inst. Rak.,* **22.**

EGOROV, N. N., RUBTSOVA, N. N., and SOLOZHENIKINA, T. N. 1961. The green oak roller in Voronezh province. *Zool. Zh.,* **40.**

EGOROV, O. V. 1961. *Ecology and trapping of Yakutian squirrels.* Moscow.

EGOROVA, S. V. 1962. Effects of tree root systems on soil microflora. *Soobshch. Lab. Lesov.,* **6.**

EITINGEN, G. R. 1918. Effect of tree stand density on plantation growth. *Les. Zh.,* Nos. 6-8.

EITINGEN, G. R. 1925. Individual growing energy of tree species. *Trudy les. opyt. Delu Ross.,* No. 8.

ELAGIN, N. N., and MINA, V. N. 1953. Structure of oak root system in dark-grey and solonetz soils. *Trudy Inst. Lesa,* Moscow, **12.**

ELTON, C. 1934. *Animal ecology.* Biomedgiz. Moscow.

ELTON, C. 1960. *Ecology of plant and animal invasions.* Translated from English. Inostr. Lit. Moscow.

ENIKEEVA, M. G. 1947. *Soil moisture and micro-organism activity.* (Candidate's dissert.) Moscow.

ENIKEEVA, M. G. 1952. Soil moisture and micro-organism activity. *Trudy Inst. Microbiol.,* Moscow, No. 2.

EREMEEV, G. N. 1936. Role of root system in plant resistance to excessive and inadequate soil moisture. Work on applied botany, genetics, and selection, ser. A. *Trudy pribl. Bot. Genet. Selek. Ser. A,* No. 18.

EREMEEV, G. N. 1938. Drought resistance and desiccation resistance in plants. *Dokl. Akad. Nauk SSSR,* **18.**

ERPERT, S. D. 1962. Moisture consumption in transpiration by Chinese elm in optimal moisture conditions in North-west Prikaspiya. Physiology of trees. In: *Vol. Jubil. L. A. Ivanov.* Publ. Acad. Sci. u.s.s.r. Moscow.

ESHBI, U. R. 1959. *Introduction to cybernetics.* Inostr. Lit. Moscow.

ESHBI, U. R. 1962. *Brain structure.* Inostr. Lit. Moscow.

FAGELER, P. 1935. *Fundamentals of soil study in subtropical and tropical countries.* (Translated from German.) (Soviet section of International Assocn. of Pedologists.) Moscow.

FEDORAKO, V. I. 1940. Damage by brown hares to various tree and shrub species. *Vest. Zashch. Rast.,* No. 5.

FEDOROV, M. V. 1952. *Biological fixation of atmospheric nitrogen.* Sel'khozgiz. Moscow.

FEDOROV, M. V. 1954. *Soil microbiology.* Sovetskaya Nauka. Moscow.

FEL'DMAN, YA. I. 1950. Role of oases in the deserts of Central Asia and formation of local weather with sukhoveis. *Trudy Inst. Geogr.,* Leningrad, **48.**

FETT, V. 1961. *Atmospheric dust.* (Translated from German.) Inostr. Lit. Moscow.

FOLITAREK, S. S. 1947. Material on the economic biology of the Barguzinsk sable. *Trudy vses. nauchno-issled. Inst. Okhot. Prom.*

FOMICHEVA, L. I. 1962. Effect of light on distribution of tree aphids in plantations. *Uchen. Zap. mosk. gosud. pedag. Inst.*, **186.**

FORMOZOV, A. N. 1933. Stone pine seed crop, flight of Siberian nutcrackers to Europe, and fluctuations in squirrel numbers. *Byull. nauchno-issled. Inst. Zool. mosk. gosud. Univ.*

FORMOZOV, A. N. 1950. Some features of bird biology in relation to problems of protection of forests and plantations. In: *Birds and forest pests.* Mosk. Obshch. Ispyt. Prir. Moscow.

FRIDERIKS, K. 1932. *Ecological bases of applied zoology and entomology.* Sel'khozgiz. Moscow.

GALKIN, G. I. 1960. Some problems in the creation of reserves in primary breeding foci of Siberian pine moths in forests of the Krasnoyarsk region. In: *Material on Siberian pine moth problems.* Publ. Acad. Sci. U.S.S.R., Sib. Br. Novosibirsk.

GALKIN, G. I. 1963. The Siberian pine moth in the forests of the Krasnoyarsk region. *Trudy Vost.-sib. nauchno-issled. proekt Inst. les. derev. prom.*

GAMOV, G., RICH, A., and IKAS, M. 1956. Problems of transmitting information from nucleic acids to proteins. *Khim. Nauka Prom.*

GAR, K. A. 1954. Light intensity beneath oak canopy of different ages and different forest types. *Soobshch. Inst. Lesa.*, Moscow.

GARDER, L. A. 1927. Effect of temperature on development of *Azotobacter* and consumption of molecular nitrogen by it. *Trudy Otd. sel.-khoz. Mikrobiol. gos. Inst. opyt. Agron.*, **2.**

GAUZE, G. F. 1944. Some problems of chemical biocoenology. *Usp. sovrem. Biol.*, **17.**

GAUZE, G. F., PREOBRAZHENSKAYA, T. P., *et al.* 1957. *Problems of classification of antagonistic Actinomycetes.* Medgiz. Moscow.

GAVRILOV, K. A. 1950. Effect of the composition of a tree stand on the microflora and fauna of forest soils. *Pochvovedenie.*

GAVRILOV, K. I. 1962. Role of earthworms in enriching the soil with biologically-active substances. *Vop. Ekol. Biocen.*, **7.**

GEIGER, R. 1960. *Climate of the ground layer.* Moscow.

GELLER, I. F. 1961. Methods and problems of natural-territorial (physico-geographical) classification of the lands of Brandenburg and Altmark (East Germany). *Vest. mosk. gos. Univ.*

GENKEL', P. A., and BUTYLIN, E. I. 1935. Process of nitrification by Vaksman's method in specimens of soil with unbroken structure. *Mikrobiologiya*, **4.**

GERASIMOV, I. P. 1954. *Condition and tasks of Soviet geography at the present stage of its development.* Pre-published abstracts of contributions for 2nd session of the U.S.S.R. Geog. Society. Moscow.

GILYAROV, M. S. 1939. Effect of soil conditions on soil-dwelling pest fauna. *Pochvovedenie.*

GILYAROV, M. S. 1942. Comparative invertebrate populations of dark-coloured and podzolic soils. *Pochvovedenie.*

GILYAROV, M. S. 1947. Distribution of humus, root systems, and soil-dwelling invertebrates in soils of the nut-producing forests of the Fergansk range. *Dokl. Akad. Nauk SSSR*, **55.**

GILYAROV, M. S. 1949. *Characteristics of soil as a habitat and its significance in insect evolution.* Publ. Acad. Sci. U.S.S.R. Moscow & Leningrad.

GILYAROV, M. S. 1953a. Soil fauna and soil fertility. *Proc. Conf. problems of soil microbiology.* Publ. Acad. Sci. U.S.S.R. Moscow.

GILYAROV, M. S. 1953b. Soil fauna of broad-leaved forests and its value for soil diagnosis. *Zool. Zh.*, **32.**

GILYAROV, M. S. 1954. Species, population, and biocoenose. *Zool. Zh.*, **33.**

GILYAROV, M. S. 1956a. Soil fauna of forest plantations and open steppe areas in the Derkul' river basin. *Trudy Inst. Lesa*, Moscow, **30.**

GILYAROV, M. S. 1956b. Investigation of soil entomofauna as a method of soil-type diagnosis. *Ent. Obozr.*, **35.**

GILYAROV, M. S. 1957. Black wireworms (Iuloidea) and their role in soil formation. *Pochvovedenie.*

GILYAROV, M. S. 1959a. Soil-zoological investigations and their tasks. *Vest. Akad. Nauk SSSR*

GILYAROV, M. S. 1959b. Problems of modern ecology and theory of natural selection. *Usp. sovrem. Biol.*, **48**.

GILYAROV, M. S. 1960. Soil invertebrates as indicators of characteristics of the soil and plant cover of the forest-steppe. *Trudy tsent.-chernoz. gos. Zapov.*

GIRNIK, D. V. 1955. Water regime of trees in winter, and winter drought. *Trudy Inst. Lesa*, Moscow, **27**.

GLAZOVSKAYA, M. A. 1953. Biological absorption of mineral elements and possibilities of using plants for soil improvement. *Vop. Geogr.*, **33**.

GLAZOVSKAYA, M. A., MAKUNINA, A. A., et al. 1961. *Geochemistry of landscapes and the search for useful minerals in the Southern Urals.* Izd. mosk. gos. Univ. Moscow.

GLEZER, V. D., and TSUKKERMAN, I. I. *Information and vision.* Moscow & Leningrad.

GOLOVYANKO, Z. S. 1926. Method of estimating the degree of infestation of pines by bark beetles. *Trudy les. opyt. Delu Ukr.*

GOLOVYANKO, Z. S. 1952. Secondary pine pests in the Lower Dnieper sands. *Les i Step'*.

GOL'TSBERG, A. A. 1957. *Microclimate and its significance in agriculture.* Gidrometeoizdat. Leningrad.

GORDYAGIN, A. YA. 1900. Material on study of soils and vegetation in Western Siberia. *Trudy Obshch. Estest. imp. kazan. Univ.*, **34**.

GORLENKO, M. V. 1946. Toxins in mould fungi. *Dokl. Akad. Nauk SSSR*, **54**.

GORODKOV, B. N. 1916. Observations on the life of stone pine in Western Siberia. *Trudy bot. Mus. Akad. Nauk.*

GORODKOV, B. N. 1944. The study of successions and climax in botany. *Priroda*, Moscow.

GORSHENIN, K. P. 1960. The problem of soil classification. *Pochvovedenie.*

GORSHENIN, N. M. 1962. Contradictions and their role in the development of plant formations. Higher school scientific reports. *Filosof. Nauki.*

GORTINSKII, G. B. 1963. Allelopathy and biogeocoenology (biogeocoenological approach to problems of allelopathy). *Byull. mosk. Obshch. Ispyt. Prir.*

GOZHEV, A. D. 1930. Types of sands in the western part of the Tersko-Dagestan massif and their economic use. *Izv. russk. geogr. Obshch.*, **62**.

GOZHEV, A. D. 1934. Methodology of physical geography. *Izv. russk. geogr. Obshch.*, **66**.

GOZHEV, A. D. 1945. On the nature of terrestrial surface. *Izv. vses. geogr. Obshch.*, **77**.

GOZHEV, A. D. 1946. Some laws of the development of the nature of terrestrial surface. *Uchen. Zap. leningr. gos. pedag. Inst.*, **49**.

GOZHEV, A. D. 1948a. *South America.* Geografgiz. Moscow.

GOZHEV, A. D. 1948b. Variations in types of territory. *Uchen. Zap. leningr. gos. pedag. Inst.*, **73**.

GOZHEV, A. D. 1956. Zonal-provincial nature of terrestrial surface. *Uchen. Zap. leningr. gos. pedag. Inst.*, **116**.

GRENEVSKII, G. 1959. *Elements of cybernetics, explained non-mathematically.* Warsaw (in Polish). Quoted by Povarov, 1960.

GRIGOR'EV, A. A. 1938-42. Experiment in description of the main types of physico-geographical environment. *Problems of physical geography*, part I, 1938, No. 5; part II, 1938, No. 6; part III, 1938, No. 6; part IV, 1939, No. 8; part V, 1942, No. 11.

GRIGOR'EV, A. A. 1956. Interrelationships and dependences of components of geographical environment and the role played in them by matter and energy exchange. *Izv. Akad. Nauk SSSR (ser. geogr.).*

GRIGOR'EVA, O. Z. 1950. Role of plant cover in formation of soil fauna. *Pochvovedenie.*

GRIMAL'SKII, V. I. 1961a. Reasons for resistance of pine plantations to needle-eating pests. *Zool. Zh.*, **40**.

GROMYKO, E. P. 1960. Reasons for toxicity of podzolic soils for *Azotobacter. Izv. Akad. Nauk SSSR (ser. biol.).*

GRUDZINSKAYA, I. A. 1956. Tree root systems in Derkul' forest plantations. *Trudy Inst. Lesa*, Moscow, **30**.

GRUMMER, G. B. 1957. *Mutual influence of higher plants—allelopathy.* (Translated from German by A. N. Boyarkina.) Inostr. Lit. Moscow.

GUBAREVA, V. A. 1962. Relationships of circulation of water to its chemical composition, root and leaf exudates, and growth of oaks. *Soobshch. Lab. Lesov.*

GUKASYAN, A. B. 1960. Microbiological method of control of Siberian pine moths. *Zashch. Rast. Vredit. Bolez.*

GUKASYAN, A. B. 1963a. Bacteriological method of control of Siberian pine moths. In: *Microbiological methods of control of insect pests.* Publ. Acad. Sci. U.S.S.R. Moscow.

GUKASYAN, A. B. 1963b. Bacteriological method of control of Siberian pine moth in Tuva. *Izv. Akad. Nauk SSSR (ser. biol.).*

GUKASYAN, A. B., and DOMB, N. S. 1961. Retention (on needles) of virulence by organisms causing disease in Siberian pine moths. In: *Proceedings of the Planning and Method Conference on plant protection in the Ural zone and Siberia, 1960.* Novosibirsk.

GULIDOVA, I. V. 1955. Transpiration by trees and shrubs in the southern chernozem subzone. *Trudy Inst. Lesa,* Moscow, **27.**

GULIDOVA, I. V. 1958. Transpiration by trees and herbage in the central taiga zone and its dependence on meteorological conditions. *Trudy Inst. Lesa,* Moscow, **41.**

GULIDOVA, I. V., and TSEL'NIKER, YU. L. 1962. Exchange of matter and growth processes in birch and spruce. Physiology of trees. In: *Vol. Jubil. L. A. Ivanov.* Publ. Acad. Sci. U.S.S.R. Moscow.

GULISASHVILI, V. Z. 1962. Reserve substances and their conversion in tree species. Physiology of trees. In: *Vol. Jubil. L. A. Ivanov.* Publ. Acad. Sci. U.S.S.R. Moscow.

GURSKII, A. V. 1939a. Tree root systems in steppe and desert soils. *Dokl. vses. Akad. sel.-khoz. Nauk.*

GURSKII, A. V. 1939b. Experiment in tree planting in the desert without irrigation. *Dokl. vses. Akad. sel.-khoz. Nauk.*

GUR'YANOVA, N. I. 1954. Effect of the composition of its diet on the physiological condition of the field maybeetle (*Melolontha hippocastani* Fabr.). *Trudy vses. Inst. Zashch. Rast.*

GUSEINOV, B. Z. 1952. *Physiology of drought-resistance in tree species in Apsheron.* (Doct. dissert.) Moscow.

IL'INSKAYA, S. A. 1963. Study of synusial structure of communities. In: *Siberian forest types.* Moscow.

IL'INSKII, A. I. 1928. Laws of reproduction of the lesser pine-shoot beetle and theoretical basis for methods of control of it in forests. *Trudy les. opyt. Delu Ukr.*

IL'INSKII, A. I. 1949. The pine beauty moth, the bordered white moth (pine looper), and the gypsy moth. *Les. Khoz.*

IL'INSKII, A. I. 1958. Secondary pests of pine and spruce and measures for their control. *Sb. Rab. les. Khoz.* Goslesbumizdat. Moscow.

IL'INSKII, A. I. 1959. *The gypsy moth and measures for its control.* Goslesbumizdat. Moscow.

IMSHENETSKII, A. A. 1953. *Microbiology of cellulose.* Publ. Acad. Sci. U.S.S.R. Moscow. *Interaction of earth-study sciences.* 1963. Publ. Acad. Sci. U.S.S.R. Moscow.

IOGANZEN, B. G. (JOHANSEN). 1962. Identity of biotope and biocoenose. In: *Problems of zoological investigations in Siberia.* Gorno-Altaisk.

ISACHENKO, A. G. 1953. *Basic problems of physical geography.* Leningrad.

ISACHENKO, B. L., and SIMAKOVA, T. N. 1934. Bacteriological investigations of Arctic soils. *Bull. of VIEM.*

ISAEV, E. M. 1959. State of wild ungulate population in the territories of the R.S.F.S.R. *Soobshch. Inst. Lesa.*

ISHCHEREKOV, V. 1910. Soil solutions. *Uchen. Zap. imp. kazan. Univ.,* 77.

IVANOV, L. A. 1916a. Evaluation of evaporation by tree species. *Les. Zh. S-Peterb.,* **16.**

IVANOV, L. A. 1916b. Anatomical structure of root tips of pine. *Izv. les. Inst.,* **30.**

IVANOV, L. A. 1929. Solar energy and its use by plants. *Nauch. Slovo.*

IVANOV, L. A. 1932. Laws of distribution of light in forest associations. *Bot. Zh. SSSR,* **17.**

IVANOV, L. A. 1936. *Physiology of plants.* Goslestekhizdat. Leningrad.

IVANOV, L. A. 1941a. Photosynthesis and crop production. Coll. of works on plant physiology. *Vol. Jubil. Timiryazev.* Publ. Acad. Sci. U.S.S.R. Moscow.

IVANOV, L. A. 1941b. Changes in transpirational capacity of trees in winter. *Bot. Zh. SSSR*, **26**.

IVANOV, L. A. 1946. Light and moisture in the life of our tree species. *Timiryazev lectures, V.* Publ. Acad. Sci. U.S.S.R. Moscow & Leningrad.

IVANOV, L. A. 1953. Suction apparatus of the chief tree species. *Dokl. Akad. Nauk SSSR*, **18**.

IVANOV, L. A. 1956a. Transpiration by windbreak tree species in the Derkul' steppe. *Trudy Inst. Lesa*, Moscow, **30**.

IVANOV, L. A. 1956b. Fall in transpiration by trees during a sukhovei as a result of transpiration resistance. In: *Vol. Jubil. Sukachev.* Publ. Acad. Sci. U.S.S.R. Moscow.

IVANOV, L. A., GULIDOVA, I. V., TSEL'NIKER, YU. L., and YURINA, E. V. 1963. Photosynthesis and transpiration by tree species in different climatic zones. In: *Water regime of plants in relation to exchange of matter and productivity.* Publ. Acad. Sci. U.S.S.R. Moscow.

IVANOV, L. A., and KOSSOVICH, N. L. 1930. Work of accumulative apparatus of different tree species. I. Pine. *Zh. russk. bot. Obshch.*, **15**.

IVANOV, L. A., and KOSSOVICH, N. L. 1932. Work of accumulative apparatus of different tree species. II. *Bot. Zh. SSSR*, **17**.

IVANOV, L. A., and ORLOVA, M. M. 1931. Problem of winter photosynthesis of our coniferous species. *Zh. russk. bot. Obshch.*, **16**.

IVANOV, L. A., and SILINA, A. A. 1951. Actinometric method of determining transpiration by a forest. *Bot. Zh. SSSR*, **36**.

IVANOV, L. A., SILINA, A. A., ZHMUR, D. G., and TSEL'NIKER, YU. L. 1951. Determination of transpirational output by tree stands in a forest. *Bot. Zh. SSSR*, **36**.

IVANOV, L. A., SILINA, A. A., and TSEL'NIKER, YU. L. 1953. Transpiration by windbreaks in the Derkul' steppe. *Bot. Zh. SSSR*, **38**.

IVANOV, L. A., and YURINA, E. V. 1961. Effect of spectral composition of light on intensity of transpiration by tree species. *Soobshch. Lab. Lesov.*

IVANOV, N. N. 1948. Landscape-climatic zones of the earth. *Zap. vses. geogr. Obshch.*

IVANOV, S. L. 1961. *Climatic theory of formation of organic matter.* Moscow.

IVANOV, V. V., and SHAUMYAN, S. K. 1961. Linguistic problems of cybernetics and structure of linguistics. In: *Cybernetics in the service of Communism.* Ed. by Acad. A. I. Berg. Publ. Acad. Sci. U.S.S.R. Moscow & Leningrad.

IVANOVA, E. N., and ROZOV, N. N. 1959. Experiment in systematics of steppe-zone soils, parts I, II. *Pochvovedenie.*

IVANOVA, N. E. 1950. Natural regeneration of ash and oak in Tellerman experimental forest. *Trudy Inst. Lesa*, Moscow, **3**.

IVLIEV, L. A. 1960. The Siberian pine moth in the forests of the Far East. In: *Material on the problem of the Siberian pine moth.* Publ. Acad. Sci. U.S.S.R., Sib. Br. Novosibirsk.

IVLIEV, V. S. 1962. Energy bases of problems of biological productivity. *Voprosy ekologii*, 4. Izd. kievsk. Univ. Kiev.

KALESNIK, S. V. 1940. Tasks of geography and field geographic investigations. *Uchen. Zap. leningr. gos. Univ. (ser. geogr.).*

KALESNIK, V. S. 1955. *Fundamentals of general agriculture.* Uchpedgiz. Moscow.

KALETSKAYA, M. L. 1959. Moose damage to young pines in the Darwin reserve. *Soobshch. Inst. Lesa.*

KALITIN, N. N. 1938. *Actinometry.* Gidrometeoizdat. Moscow.

KAMENSKII, F. M. 1886. Symbiotic union of fungal mycelia with roots of higher plants. *Trudy S-Peterb. Obshch. Ertest.*, **17**.

KAPLANOV, L. G. 1948. *Tiger, moose, and Manchurian red deer.* Izd. mosk. Obshch. Ispyt. Prir. Moscow.

KARANDINA, S. N. 1950. Tree root systems in the broad-leaved forests of the forest-steppe. *Uchen. Zap. leningr. gos. Univ. (ser. biol.)*, **25**.

KARANDINA, S. N. 1953. Growth of oak seedlings in relation to number of acorns sown in seed-holes. *Soobshch. Inst. Lesa*, No. 1.

KARANDINA, S. N. 1962. Ecological differentiation of oaks in dense plantations. *Soobshch. Lab. Lesov.*

KARANDINA, S. N. 1963. *Features of the growth of English oak in the Caspian lowlands.* Publ. Acad. Sci. U.S.S.R. Moscow.

KARANDINA, S. N., and ERPERT, S. D. 1961. Effect of herbaceous cover and parent rock on growth of box elder. *Soobshch. Lab. Lesov.*

KARPINSKAYA, N. S. 1925. Problem of absorption of bacteria in soil. *Nauchno-agron. Zh.*

KARPISONOVA, R. A. 1962. Changes in plant cover in the Ostankinsk oak forest. *Byull. glavn. bot. Sada,* Leningrad, **46.**

KARPOV, V. G. 1952. Ecology of tree growth in the desert-steppe zone. *Bot. Zh. SSSR,* **37.**

KARPOV, V. G. 1954. Effect of environment of steppe pine forests on the drought-resistance of pine seedlings. *Uchen. Zap. leningr. gos. Univ.,* **166.**

KARPOV, V. G. 1955a. Competition between trees and seedlings in plantations on the arid steppe. *Bot. Zh. SSSR,* **40.**

KARPOV, V. G. 1955b. Competition among tree roots in plantations on the arid steppe. *Dokl. Akad. Nauk SSSR,* **104.**

KARPOV, V. G. 1956. Factors governing interrelationships between trees and herbage in plantations on the arid steppe. In: *Vol. Jubil. Sukachev.* Publ. Acad. Sci. U.S.S.R. Moscow.

KARPOV, V. G. 1958. Competition by tree roots and structure of herbage-shrub layer in taiga forests. *Dokl. Akad. Nauk SSSR,* **119.**

KARPOV, V. G. 1959a. Competition for nutrients and regenerative processes in steppe-zone plantations. *Dokl. Akad. Nauk SSSR,* **125.**

KARPOV, V. G. 1959b. Competition by tree roots and nutrient content of leaves of plants in the herbage-shrub layer in taiga forests. *Dokl. Akad. Nauk SSSR,* **129.**

KARPOV, V. G. 1960a. Experimental solution of some problems of spruce-forest phytocoenology. *Dokl. Akad. Nauk SSSR,* **182.**

KARPOV, V. G. 1960b. Principal results of experimental investigations into inter-relationships between plants in forests in the central taiga. *Bot. Zh. SSSR,* **45.**

KARPOV, V. G. 1962. Experimental clarification of some problems of spruce-forest phytocoenology. *Soobshch. Lab. Lesov.*

KARTSEV, G. K. 1903. *The Belovezh forest.* St. Petersburg.

KASHIN, K. I., and POLOSYAN, KH. P. 1950. Water exchange in the atmosphere. *Meteorologiya Gidrol.*

KASHKAROV, D. N. 1945. *Animal ecology.* Moscow.

KAUTSIS, A. R. 1956. Significance for forestry of trophobiosis of ants and aphids. *Sb. Trud. Zashch. Rast.* Riga.

KAZNEVSKII, P. F. 1958. Problem of conservation of deer numbers in U.S.S.R. reserves. *Okhr. Prir. tsentr.-chern. Obl.*

KAZNEVSKII, P. F. 1959. Interrelationships of forests and red deer in U.S.S.R. reserves. *Soobshch. Inst. Lesa.*

KERZINA, M. N. 1956. Role of tree-felling and fires in formation of forest fauna. In: *Role of animals in the life of the forest.* Moscow.

KHALABUDA, T. V. 1948. Results of soil microflora investigations. *Mikrobiologiya,* **17.**

KHALIFMAN, I. A. 1961. Use of ants for forest protection. *Les. Khoz.*

KHARITONOVICH, F. N. 1951. *Oaks in steppe conditions and their cultivation.* Gos-lesbumizdat, Moscow.

KHIL'MI, G. F. 1957. *Theoretical forest biogeophysics.* Publ. Acad. Sci. U.S.S.R. Moscow.

KHLEBNIKOVA, N. A. 1958a. Transpiration and photosynthesis of trees and shrubs in the Caspian lowlands. *Trudy Inst. Lesa,* Moscow, **38.**

KHLEBNIKOVA, N. A. 1958b. Features of photosynthesis in different parts of a planta-tion. *Trudy Inst. Lesa,* Moscow, **10.**

KHLEBNIKOVA, N. A. 1962. Physiological characteristics of trees with different growth rates in young Scots pine stands. Physiology of trees. In: *Vol. Jubil. Ivanov.* Publ. Acad. Sci. U.S.S.R. Moscow.

KHLEBNIKOVA, N. A., and MARKOVA, M. I. 1955. Features of growth and water regime of trees in the Caspian lowlands. *Trudy Inst. Lesa,* Moscow, **25.**

KHODASHOVA, K. S. 1960. *Natural environment and animal life in the Zavolzh'e clay semi-desert.* Publ. Acad. Sci. U.S.S.R. Moscow.

KHOLODNYI, N. G. 1944a. Exhalation of volatile organic compounds by living organisms and consumption of them by soil microbes. *Dokl. Akad. Nauk SSSR*, **41**.

KHOLODNYI, N. G. 1944b. Volatile exhalations by flowers and leaves as a source of nutriment for micro-organisms. *Dokl. Akad. Nauk SSSR*, **43**.

KHOLODNYI, N. G. 1944c. The atmosphere as a possible source of vitamins. *Dokl. Akad. Nauk SSSR*, **43**.

KHOLODNYI, N. G. 1948. Physiological action of volatile organic substances on plants. *Dokl. Akad. Nauk SSSR*, **12**.

KHOLODNYI, N. G. 1951a. Volatile organic compounds exhaled by the soil. *Pochvovedenie*.

KHOLODNYI, N. G. 1951b. Aerial nutrition of roots. *Dokl. Akad. Nauk SSSR*, **76**.

KHOLODNYI, N. G. 1951c. Volatile organic substances in the soil. *Dokl. Akad. Nauk SSSR*, **80**.

KHOLODNYI, N. G. 1957a. *Selected works.* Publ. U.S.S.R. Acad. of Sci. Kiev.

KHOLODNYI, N. G. 1957b. Oak distribution in natural conditions. *Selected works.* Publ. Acad. Sci. U.S.S.R. Kiev.

KHOLODNYI, N. G., ROZHDESTVENSKII, V. S., and KIL'CHEVSKAYA, A. A. 1945. Consumption of volatile organic substances by soil bacteria. *Pochvovedenie*.

KHRENOVA, G. S. 1963. Effect of fire on microflora of forest soils in industrial pine forests in Transuralia. *Trudy Inst. Biol.*, Sverdlovsk, **36**.

KHUDYAKOV, N. N. 1926. Adsorption of bacteria by soil and its effect on microbiological processes in the soil. *Pochvovedenie*.

KHUDYAKOV, YA. P. 1951. Physiology of mycotrophism in trees. *Trudy kompleks. nauch. Eksped. Vopr. polerzashch. Lesorazv.*, **2**.

KIRSHENBLATT, YA. D. 1962. Telergons as a means of action by animals upon biotic factors in the external environment. In: *Voprosy ekologii*, **4**. Izd. kievsk. Univ. Kiev.

KISELEVA, K. V. 1962. Problem of interrelationships of spruce and oak in Moscow province. *Vest. mosk. gos. Univ.*

KISHKO, YA. G. 1961. Distribution of bacterial flora in the air above inhabited places and at a distance from them. *Mikrobiologiya*, **30**.

KITREDZH, DZH. (KITTREDGE, J.) 1951. *Effect of a forest on climate soil, and water regime.* Inostr. Lit. Moscow.

KLESHNIN, A. F. 1954. *The plant and light.* Publ. Acad. Sci. U.S.S.R. Moscow.

KLESHNIN, A. F., and SHUL'GIN, I. A. 1963. Transpiration and temperature of plant leaves in sunlight. In: *Water regime of plants in relation to exchange of matter and productivity.* Publ. Acad. Sci. U.S.S.R. Moscow.

KLYUCHNIKOV, L. YU., and PETROVA, A. N. 1960. Effect of frequent application of herbicides on soil microflora. *Mikrobiologiya*, **29**.

KOCHMAREK, V. 1962. Biocoenotic relationships among predators of soil fauna. *Voprosy ekologii*, **4**. Izd. kievsk. Univ. Kiev.

KOLCHEVA, B. 1954. *Microbiological characteristics of the chief soil types in Bulgaria.* (Cand. dissert.) Moscow.

KOKINA, S. I. 1926. Problem of effect of soil moisture on plants. *Izv. glav. bot. Sada SSSR*, **26**.

KOLDANOV, V. YA. 1958. Seed-hole sowing of trees and growth of their root systems. *Bot. Zh. SSSR*, **43**.

KOLESNICHENKO, M. V. 1960. Effect of exudates of white birch (*Betula verrucosa* Ehrh.) on photosynthesis in English oak (*Quercus robur* L.). *Dokl. Akad. Nauk SSSR*, **132**.

KOLESNICHENKO, M. V. 1962. Biochemical interactions of pine and birch. *Les. Khoz.*

KOLESNIKOV, B. P. 1956. Stone pine forests of the Far East, vol. 2. *Trudy dalnevost. Fil. Akad. Nauk SSSR*, **14**.

KOLESNIKOV, B. P. 1958a. Present state of Soviet forest typology and the problem of genetic classification of forest types. *Izv. sib. Otdel. Akad. Nauk SSSR*.

KOLESNIKOV, B. P. 1958b. Genetic classification of forest types and tasks of forest typology in the eastern districts of the U.S.S.R. *Izv. sib. Otdel. Akad. Nauk SSSR*.

KOLESNIKOV, B. P. 1961. Genetic classification of forest types and its problems in the Urals. *Trudy Inst. Biol.*, Sverdlovsk, **27**.

KOLOMIETS, N. G. 1957. The Siberian pine moth—a pest of the level taiga. *Trudy les. Khoz. Zap. Sib.*

KOLOMIETS, N. G. 1960a. The Siberian pine moth in Tuva. *Trudy les. Khoz. Zap. Sib.*

KOLOMIETS, N. G. 1960b. Study of the Siberian pine moth in Western Siberia, and prospects for use of its parasites. In: *Materials on the problem of the Siberian pine moth.* Publ. Acad. Sci. U.S.S.R., Sib. Br. Novosibirsk.

KOLOSOV, I. I. 1962. *Absorbent action of plant root systems.* Publ. Acad. Sci. U.S.S.R. Moscow.

KOL'TSOV, V. F. 1954. *Water regime of oak plantations on the chernozems of the Donsk leskhoz.* (Author's extract of Cand. dissert.) Moscow.

KOMAROV, V. L. 1961. *Origin of plants.* Publ. Acad. Sci. U.S.S.R. Moscow.

KONDAKOV, YU. P. 1963. The gypsy moth (*Ocneria dispar* L.) in the forests of the Krasnoyarsk region. In: *Protection of Siberian forests from insect pests.* Publ. Acad. Sci. U.S.S.R. Moscow.

KONDRAT'EV, P. S. 1939. Effect of planting density on growth of pine plantations. *Les. Khoz.*

KONONOVA, M. M. 1951. *The soil humus problem and modern methods of studying it.* Publ. Acad. Sci. U.S.S.R. Moscow.

KONSTANTINOV, A. R., and FEDOROV, S. F. 1960. Experiment in use of graduated poles in determining evaporation and heat exchange in a forest. *Trudy gos. gidrol. Inst.*, 81.

KORAB, I. I., and BUTKOVSKII, A. B. 1939. *Principal results of study of sugar-beet nematodes.* In coll. of works on nematodes that infest farm crops. (Ed. by E. S. Kir'yanova.) Sel'khozgiz. Moscow & Leningrad.

KORCHAGIN, A. A. 1956. Problem of the character of the interrelationships of plants in a community. *Vol. Jubil. Sukachev.* Publ. Acad. Sci. U.S.S.R. Moscow.

KORELOV, M. N. 1947. Significance of the wild boar in the life of the mossy spruce forests of Tian-Shan. *Vest. Akad. Nauk Kazakh. S.S.R.*

KOROL'KOVA, G. E. 1957. Activity of insectivorous birds in oak stands. In: *Young foresters—40th anniversary of Great October.* Moscow.

KOROL'KOVA, G. E. 1961. Attracting insectivorous birds to forest strips, broad-leaved forests, and isolated forests. *Soobshch. Lab. Lesov.*

KOROL'KOVA, G. E. 1963. *Effect of birds on numbers of forest insect pests.* Moscow.

KORZHINSKII, S. I. 1891. Northern limit of the chernozem region of the eastern part of European Russia, part 1. *Trudy Obshch. Estest. imp. kazan. Univ.*

KOSHCHEEV, A. L. 1955. Transpirational activity of regenerating spruce stands—the chief factor in the drying-up of cleared areas. *Trudy Inst. Lesa*, Moscow, 26.

KOSHEL'KOV, S. P. 1961. Formation and subdivision of litter in southern taiga coniferous forests. *Pochvovedenie.*

KOSHKINA, T. V. Comparative ecology of red-backed voles in the northern taiga. *Fauna Ekol. Gryz.*

KOSSOVICH, N. L. 1940. Effect of forest improvement cutting on assimilation, lighting, and growth of spruce in a spruce-deciduous stand. *Sb. Trud. tsent.-nauchno-issled. Inst. les. Khoz.*

KOSSOVICH, N. L. 1945. Photosynthesis by spruce and its connection with growth, with heavy thinning in forest biocoenoses. *Dokl. vses. Soveshch. Fiz. Rast.*

KOSSOVICH, N. L. 1952. Dynamics of light conditions under spruce-deciduous canopy in relation to forest improvement cutting. In: *Physiology of trees.* Publ. Acad. Sci. U.S.S.R. Moscow.

KOSTIN, I. A. 1958. Outbreak of mass reproduction of Siberian pine moths in the mountain forests of Eastern Kazakhstan. *Trudy zool. Inst.*, Alma-Ata, 8.

KOSTYCHEV, S. P. 1933. *Physiology of plants*, vol. 1. Sel'khozgiz. Moscow & Leningrad.

KOSTYCHEV, S. P., and KHOLKIN, I. 1929. Mineralisation of soil nitrogen and nitrification in the Bukharian saltpetre beds. *Trudy Otd. sel.-Khoz. Mikrobiol. gos. Inst. opyt. Agron.*, 4.

KOTELEV, V. V. 1955. Significance of soil microflora in migration of phosphorus and its absorption by plants when introduced at focal points. *Izv. moldav. Fil. Akad. Nauk SSSR.*

KOVAL'SKII, V. V. 1962. Geochemical ecology. *Voprosy ekologii*, **4**. Izd. kiev. Univ. Kiev.

KOVAL'SKII, V. V. 1963. Geochemical ecology and its evolutionary trend. *Izv. Akad. Nauk SSSR (ser. biol.)*.

KOVRIGIN, S. A. 1952. Dynamics of nitrates of ammonia and of available forms of phosphorus and potassium in soils under different tree species. *Pochvovedenie*.

KOZHANCHIKOV, I. V. 1951. Towards understanding of mass insect reproduction. *Zool. Zh.*, **32**.

KOZHANCHIKOV, I. V. 1960. Chief features of the phenology of Lepidoptera of the forest zone, and some practical problems of insect phenology. *Trudy phenol. Soveshch.* Gidrometeoizdat. Leningrad.

KOZHEVNIKOV, V. A. 1950. *Spring and autumn in the life of plants.* Izd. mosk. Obshch. Ispyt. Prir. Moscow.

KOZLOV, K. A. 1962. Problem of landscape microbiology. *Dokl. Inst. Geogr. Sib. Dal'nevost.*

KOZLOVA, E. I. 1955. Study of genus and species composition of micro-organisms in oak rhizosphere. *Mikrobiologiya*, **24**.

KOZLOVA, E. I. 1958. Microbiological characteristics of oak rhizosphere. *Vest. mosk. gos. Univ. (ser. biol.)*.

KOZLOVSKAYA, L. S. 1957a. Fauna of forest soils of Kotlassk leskhoz. *Trudy Inst. Lesa*, Moscow, **36**.

KOZLOVSKAYA, L. S. 1957b. Comparative characteristics of soil fauna of the Arctic part of the Usa basin. *Trudy Inst. Lesa*, Moscow, **36**.

KOZLOVSKAYA, L. S. 1959. Role of soil fauna in decomposition of organic remains in boggy forest soils. *Trudy Inst. Lesa*, Moscow, **49**.

KOZLOVSKAYA, L. S., and ZHDANNIKOVA, E. V. 1961. Concurrent activity of earthworms and microflora in forest soils. *Dokl. Akad. Nauk SSSR*.

KOZLOVSKII, A. A. 1960. *Moose and forest. Protection of forests from moose damage.* VNIILM, Moscow.

KRAINOVA, L. V. 1951. Diet of bison in the Caucasus reserve. *Byull. mosk. Obshch. Ispyt. Prir. (otd. biol.)*, **56**.

KRASIL'NIKOV, N. A. 1944a. Bacterial mass in plant rhizospheres. *Mikrobiologiya*, **13**.

KRASIL'NIKOV, N. A. 1944b. Effect of plant cover on composition of microbes in soil. *Mikrobiologiya*, **13**.

KRASIL'NIKOV, N. A. 1954. Micro-organisms and soil fertility. *Izv. Akad. Nauk SSSR (ser. biol.)*.

KRASIL'NIKOV, N. A. 1958. *Soil micro-organisms and higher plants.* Publ. Acad. Sci. U.S.S.R. Moscow.

KRASIL'NIKOV, N. A., and GARKINA, N. R. 1946. Microbiological factors in soil exhaustion. *Mikrobiologiya*, **15**.

KRASIL'NIKOV, N. A., KORENYAKO, A. I., and MIRCHINK, T. G. 1955. Toxicosis of podzolic soils. *Izv. Akad. Nauk SSSR (ser. biol.)*

KRASIL'NIKOV, N. A., and KOTELEV, V. V. 1956. Effect of soil bacteria on absorption of phosphorus compounds by plants. *Dokl. Akad. Nauk SSSR*, **110**.

KRASIL'NIKOV, N. A., KRISS, A. E., and LITVINOV, M. A. 1936. Microbiological characteristics of rhizospheres of cultivated plants. *Mikrobiologiya*, **5**.

KRASIL'NIKOV, N. A., RYBALKINA, A. V., GABRIELYAN, M. S., and KONDRAT'EVA, T. M. 1934. Microbiological characteristics of soils of Zavol'zhe. *Trudy Kom. Irrig.*, **3**.

KRASOVSKAYA, I. V., and SMIRNOVA, A. D. 1950. Use of mycorrhiza when sowing oak acorns in arid conditions in Saratov province. *Les i Step'*.

KRASOVSKAYA, N. V. 1931. Problem of competition between primary and secondary crops sown in the same place. *Trudy prikl. Bot. Genet. Selek.*, **25**.

KRAVCHINSKII, D. M. 1903. *Forest growth.* 2nd ed. St. Petersburg.

KRIK, F. 1956. Structure of hereditary substance. *Khim. Nauka Prom.*

KROKER, V. 1951. *Plant growth.* Inostr. Lit. Moscow.

KRUGLIKOV, G. G. 1939. Damage to forest economy by squirrels and woodpeckers. *Les. Khoz.*

KRUPENIKOV, I. A. 1951. Observations on effect of insects on soil. *Byull. mosk. Obshch. Ispit. Prir. (otd. biol.)*.

KRYSHTEL', A. F. 1955. Present state and tasks of ecological terminology. *Proc 3rd ecol. conf.*, part IV, **54.** Kiev.

KUCHERUK, V. V. 1960. Types of mammal shelters and their distribution in natural zones of non-tropical Eurasia. *Vop. Geog.*, **48.**

KUCHERYAVYKH, E. G. 1954. Tree root systems and transpiration. *Nauch. Trudy ukr. nauchno-issled. Inst. Lesov. Agrolesomel.*, **16.**

KUDRYAVTSEVA, A. A. 1925. Accumulation of nitrates in soil by means of cultivation. *Nauchno-agron. Zh.*

KULIK, M. S. 1957. Criteria of sukhoveis. In: *Sukhoveis, their origin, and combating them.* Publ. Acad. Sci. U.S.S.R. Moscow.

KUNTS, D. E., and RAIKER, A. D. 1956. Use of radioactive isotopes in study of the role of tree root growth in movement of water, nutrients, and disease-generating organisms. In: *Use of radioactive isotopes in industry, medicine, and agriculture.* Publ. Acad. Sci. U.S.S.R. Moscow.

KUPREVICH, V. F. 1947. *Physiology of diseased plants.* Publ. Acad. Sci. U.S.S.R. Moscow.

KUPREVICH, V. F. 1949. Extra-cellular enzymes of higher autotrophic plant roots. *Dokl. Akad. Nauk SSSR*, **68.**

KUPREVICH, V. F. 1952. Physiological role of mycorrhiza. *Trudy kompleks. nauch. Eksped. Vopr. polezashch. Lesorazv.*

KURCHEVA, G. F. 1960. Role of invertebrates in the breakdown of oak litter. *Pochvovedenie.*

KURSANOV, A. L., BLAGOVESHCHENSKII, V., and KAZAKOVA, M. 1933. Effect of soil moisture on physiological processes and chemical composition of sugar beets. *Byull. mosk. Obshch. Ispyt. Prir. (n. ser., otd. biol.),* **42.**

KUTUZOV, P. K., KONEV, G. I., and SAVCHENKO, A. M. 1963. Results of damage by pine loopers in the Tuva forest massif. *Trudy vost.-sib. nauchno-issled.-proekt. Inst. les. derev. Prom.,* **7.**

KUZ'MIN, S. P. 1929-30. Water balance and drought-resistance of plants in Apsheron in relation to features of their root-system structure. *Trudy prikl. Bot. Genet. Selek.,* **23.**

LABUNSKII, I. M. 1951. Some features of the structure and growth of oak root systems. *Agrobiologiya.*

LARIN, B. A. 1955. Effect of intensity of forest-cutting on productivity of game areas. *Trudy VNIO,* **14.**

LARIN, I. V. 1926. Experiment in defining soils, parent rocks, agricultural areas, and other elements of landscape in the central part of the Ural government by their plant cover. *Trudy Obshch. Izuch. Kazakhst. (Otd. Estest. Geogr.),* **7.**

LASHCHINSKII, N. N., and RAIMERS, N. F. 1959. Role of animals in the life of the Altai larch forests. *Izv. sib. Otdel. Acad. Nauk SSSR.*

LAVRENKO, E. M. 1945. Significance of the biochemical works of Acad. Vernadskii in knowledge of the plant cover of the earth. *Priroda,* Moscow, **5.**

LAVRENKO, E. M. 1949. The phytogeosphere. *Vop. Geogr.*

LAVRENKO, E. M. 1955. Study of productivity of terrestrial plant cover. *Bot. Zh. SSSR,* **40.**

LAVRENKO, E. M. 1959. *Field geobotany,* vol. 1. Publ. Acad. Sci. U.S.S.R. Moscow & Leningrad.

LAVRENKO, E. M. 1962. V. N. Sukachev's science of the biogeocoenose. *Soobshch. Lab. Lesov.*

LAVRENKO, E. M., ANDREEV, V. N., and LEONT'EV, V. L. 1955. Productivity profile of the above-ground part of the natural plant cover of the U.S.S.R. *Bot. Zh. SSSR,* **10.**

LEBEDEV, A. F. 1921. Carbon assimilation by saprophytes. *Izv. rostovsk. gos. Univ.,* **1.**

LEBEDEVA, L. S. 1956. Ecological characteristics of wild boars in the Belovezh forest. *Uchen. Zap. mosk. gorod. pedag. Inst.*

LEBLE, B. B. 1959. Changes in ungulate numbers in Archangel province due to forest-cutting. *Soobshch. Inst. Lesa,* **13.**

LEMAN, V. M. 1961. *Course in light-culture of plants.* Izd. Vysshaya Shkola. Moscow.

LEVINA, R. E. 1957. *Methods of distribution of plants and seeds.* Moscow.

LEVINA, V. I. 1960a. Annual mass of litter from above-ground parts of plant cover in two types of pine forests in Kola peninsula. *Bot. Zh. SSSR,* **45.**

LEVINA, V. I. 1960b. Characteristics of exchange of mineral elements between moss-lichen-shrub cover and soil in two types of pine forests in Kola peninsula. *Poch-vovedenie.*

LEVITSKAYA, K. I. 1961. Effect of mature trees on anatomical structure of leaves of English oak seedlings in plantations on the arid steppe. *Bot. Zh. SSSR,* **46.**

LIKHACHEV, G. N. 1959. Some features of the ecology of the badger in the broad-leaved forests in the Tula reserve. *Coll. of material on results of mammal study in reserves.* Moscow.

LIKVENTOV, A. V. 1954. Effect of windbreak-planting on distribution of leaf-eating pests. *Trudy vses. Inst. Zashch. Rast.*

LINDEMAN, G. V. 1963. Infestation of elms by trunk pests in foci of Dutch elm disease in Tellerman forest in Voronezh province. *Vop. Lesozashch.*

LINDEMAN, G. V. 1964. Infestation of deciduous species by trunk pests in forest-steppe oak stands, with relation to their weakening and dying-off (e.g. in the Tellerman forest). In: *Protection of forests from insect pests.* Nauka. Moscow.

LIPMAA, T. M. 1946. Synusiae. *Sov. Bot.*

LISIN, S. S. 1949. Growth of seedlings of Scots pine and larch with introduction of mycorrhiza into the soil. *Les i Step'.*

LITVINOV, L. S. 1939. Problem of criteria for evaluating soil moisture. *Bot. Zh. SSSR,* **17.**

LOBANOV, N. V. 1947. Method of studying tree-root growth with fluctuating soil moisture. *Dokl. Akad. Nauk SSSR,* **3.**

LOBANOV, N. V. 1952. Application of mycorrhised soil under oaks planted on the steppes. *Trudy kompleks. nauch. Eksped. Vopr. polezashch. Lesorazv.,* **2.**

LOBANOV, N. V. 1953. *Mycotrophism of trees.* Sovetskaya Nauka. Moscow.

LOBANOV, N. V. 1960. New information on use of mycotrophism of trees in forest cultivation. *Vest. sel.-khoz. Nauki.* Moscow.

LOZOVOI, D. I. 1956. Insect pests of park and forest-park plantations in Rustav'. *Vest. tbilis. bot. Sada,* **63.**

LUGOVOI, A. V. 1963. Pine bark beetles in the forest of the Kazakh hillock district. *Trudy kazaks. nauchno-issled. Inst. les. Khoz.*

LYAPUNOV, A. A. 1958. Some general problems of cybernetics. *Problemy Kibern.*

LYAPUNOV, A. A., and MALENKOV, A. G. 1962. Logical analysis of structure of hereditary information. *Problemy Kibern.*

LYAPUNOV, A. A., and YABLONSKII, S. V. 1961. *Theoretical problems of cybernetics.* (Brief summary of report to United Theoret. Conf., div. of methodol. seminars.) Publ. Acad. Sci. U.S.S.R. Moscow.

LYSENKO, T. D. 1946. Natural selection and intraspecific competition. *Agrobiologiya.*

LYU SHOU-PUE. 1958. First study of soils under a spruce forest on the northern slope of Begedoshan. *Acta pedol. sin.,* **6** (in Chinese).

LYUBIMENKO, V. N. 1909. Effect of light of different intensities on the accumulation of dry matter and chlorophyll in light-loving and shade-loving plants. *Trudy les. opyt. Delu,* **13.** St. Petersburg.

LYUBIMENKO, V. N. 1910. Chlorophyll content in a chloroplast and energy of photo-synthesis. *Trudy S-Peterb. Obshch. Estest. (Otd. bot., ser. 3),* **41.**

LYUBIMENKO, V. N. 1935. *Photosynthesis and chemosynthesis in the plant world.* Moscow.

MAKAROV, B. N. 1953. Soil respiration. *Priroda,* Moscow, **9.**

MAKSIMOV, N. A. 1952. *Selected works on drought- and cold-resistance in plants,* vol. 1. Publ. Acad. Sci. U.S.S.R. Moscow.

MALOZEMOV, YU. A. 1962. Characteristics of the entomofauna of young pine and pine-birch stands in the Il'mensk reserve. Paper delivered at 2nd Sci.-tech. Conf. of young forest-production specialists of the Urals on 1961 work. Sverdlovsk.

MALYSHEVA, M. S. 1962. The pine looper *Bupalus piniarius* L. (Lepidoptera, Geo-metridae), and its entomophagy in the Saval'sk forest in Voronezh province. *Ent. Obozr.,* **41.**

MALYSHKIN, P. E. 1951. Effect of soil micro-organisms on oak growth. *Pochvovedenie.*

MAMAEV, B. M. 1960. Stag-beetle larvae (Coleoptera, Lucanidae) as destroyers of rotting wood in oak forests in the European part of the U.S.S.R. *Zool. Zh.,* **39.**

MAMAEV, B. M. 1961. Activity of large invertebrates—one of the chief factors in natural wood decomposition. *Pedobiologia.*

MANTEIFEL', A. YA., ZHUKOVA, A. I., and DEM'YANOVA, E. K. 1950. Study of oak rhizosphere microflora. *Mikrobiologiya*, 19.

MARCHENKO, A. I., and KARLOV, E. M. 1962. Mineral exchange in spruce forests of the northern taiga and forest-tundra of Archangel province. *Pochvovedenie*.

MARKEVICH, V. P. 1957. *The concept of 'facies'*. Moscow.

MARKUS, E. 1937. State of equilibrium in landscape. *Zemlevedenie*.

MATSKEVICH, V. B. 1953. Carbon dioxide regime in soil-contained air on the Kamennaya steppe. *Problems of the grass-arable system of farming*, vol. II. Publ. Acad. Sci. U.S.S.R. Moscow.

MATVEEVA, A. A. 1954. Changes in herbaceous cover in accordance with forest type, habitat conditions, and age of tree stand. *Soobshch. Inst. Lesa*.

MATYAKIN, G. I. 1952. *Windbreaks and microclimate*. Moscow.

MAZILKIN, I. A. 1956. Microbiological characteristics of soddy-forest and humuscarbonate soils in the Olyokminsk district of the Yakut A.S.S.R. In: *Material on natural conditions and agriculture in the south-east of the Yakut A.S.S.R.*, No. 1. Publ. Acad. Sci. U.S.S.R. Moscow & Leningrad.

MAZING, V. V. 1957. Role of birds in distribution of seeds of forest and bog vegetation. *Proc. 2nd Baltic Ornithological Conf.* Moscow.

MEDVEDEV, S. I. 1936. Some ideas on post-glacial changes in climate on the Black Sea-Azov arid grassy steppe. *Vop. Ekol. i Biotsen.*

MEISEL', M. N. 1950. *Functional morphology of yeast organisms*. Publ. Acad. Sci. U.S.S.R. Moscow & Leningrad.

MEISEL', M. N., and TROFIMOVA, N. 1946. Use of volatile biocatalytic substances by micro-organisms. *Dokl. Akad. Nauk SSSR*, 53.

MEKHTIEV, S. YA. 1953. Microflora of soddy-podzolic soils and effect on it of some ploughing methods. (Cand. dissert.) Moscow.

MEKHTIEV, S. YA. 1957. Proportions of *Bac. megatherium* and *Bac. cereus* in soddy-podzolic soil in relation to its cultivation. *Mikrobiologiya*, 26.

MEKHTIEV, S. YA. 1959. Microflora of Moldavian soils. *Trudy pochv. Inst. mold. Fil. Akad. Nauk SSSR*.

MELEKHOV, I. S. 1957. Accumulation of forest litter in relation to forest type. *Trudy arkhangelsk. lesotekh. Inst.*, 17.

MELEKHOV, I. S. 1959. Fundamentals of forest-cutting typology. Coll. *Fundamentals of forest-cutting typology and its significance in forest economy*. Archangel.

MIL'KOV, F. N. 1959. Basic problems of physical geography. *Selected lectures*. Izd. voronezh. Univ. Voronezh.

MINA, V. N. 1951. Mineral exchange in oak forests on different soils. *Trudy Inst. Lesa*, Moscow, 7.

MINA, V. N. 1954a. Interaction of trees and soils in oak stands on the southern forest-steppe. *Trudy Inst. Lesa*, Moscow, 23.

MINA, V. N. 1954b. Carbon dioxide content in air in forest soils in relation to age of tree stand. *Soobshch. Inst. Lesa*.

MINA, V. N. 1955. Circulation of nitrogen and mineral elements in oak stands on the forest-steppe. *Pochvovedenie*.

MINA, V. N. 1957. Biological activity of forest soils and its dependence on physico-geographical conditions and tree-stand composition. *Pochvovedenie*.

MINA, V. N. 1960. Intensity of formation of carbon dioxide and its distribution in soil-contained air in leached chernozems, in relation to the composition of forest vegetation. *Trudy Lab. Lesov.*

MINAKOV, K. N. 1962. Accession of nitrogen and mineral elements with tree offal in the forests of Kola peninsula. *Pochvovedenie*.

MINYAEV, N. A. 1963. *Structure of plant associations*. Publ. Acad. Sci. U.S.S.R. Moscow & Leningrad.

MIROLYUBOV, I. I., and RYASHCHENKO, P. N. 1949. *Axis deer*. Vladivostok.

MISHUSTIN, E. N. 1946a. Geographical variation in soil bacteria. *Usp. sovrem. Biol.*, 22.

MISHUSTIN, E. N. 1946b. Distribution of varieties of *Bac. mycoides* in the soils of the Soviet Union. *Mikrobiologiya*, 15.

MISHUSTIN, E. N. 1947. *Ecologo-geographical variation in soil bacteria*. Publ. Acad. Sci. U.S.S.R. Moscow & Leningrad.

MISHUSTIN, E. N. 1949. Law of zonality and composition of the bacterial population of the soil. *Proc. of Jubilee session celebrating the 100th anniversary of the birth of V. V. Dokuchaev.* Publ. Acad. Sci. U.S.S.R. Moscow & Leningrad.

MISHUSTIN, E. N. 1950. *Thermophilous micro-organisms in nature and practice.* Publ. Acad. Sci. U.S.S.R. Moscow & Leningrad.

MISHUSTIN, E. N. 1954. Law of zoning and study of microbial associations in the soil. *Usp. sovrem. Biol.,* 37.

MISHUSTIN, E. N. 1955. Mycotrophy in tree species and its significance in sylviculture. *Tr. of Conf. on mycotrophy in plants.* Inst. of Microbiology of U.S.S.R. Acad. of Sc.

MISHUSTIN, E. N. 1956. *Micro-organisms and soil fertility.* Publ. Acad. Sci. U.S.S.R. Moscow.

MISHUSTIN, E. N., and MIRZOEVA, V. A. 1953. Proportions of chief groups of micro-organisms in soils of different types. *Pochvovedenie.*

MISHUSTIN, E. N., and NAUMOVA, A. N. 1955. Exudation of toxic substances by lucerne and their effects on cotton plants and soil microflora. *Izv. Akad. Nauk SSSR (ser. biol.).*

MISHUSTIN, E. N., and PERTSOVSKAYA, M. I. 1954. *Micro-organisms and self-purification of soil.* Publ. Acad. Sci. U.S.S.R. Moscow.

MISHUSTIN, E. N., and PUSHKINSKAYA, O. I. 1941. Distribution of *Bac. mycoides* (Flügge) in different types of soil. *Mikrobiologiya,* 10.

MISHUSTIN, E. N., and PUSHKINSKAYA, O. I. 1960. Ecologo-geographical laws of distribution of soil-dwelling microscopic fungi. *Izv. Akad. Nauk SSSR (ser. biol.).*

MISHUSTIN, E. N., PUSHKINSKAYA, O. I., and MIRZOEVA, V. A. 1951. Micropopulation of the soil and formation of mycorrhiza on oak. *Trudy kompleks. nauch. Eksped. Vopr. polezashch. Lesorazv.*

MISHUSTIN, E. N., PUSHKINSKAYA, O. I., and TEPLYAKOVA, Z. F. 1961. Ecologo-geographical distribution of soil-dwelling microscopic fungi. *Trudy pochv. Inst. Kazakh. Fil. Akad. Nauk SSSR,* 12.

MISHUSTINA, I. E. 1955. Oligonitrophilous soil organisms. *Trudy Inst. Microbiol.,* Moscow, 4.

MOISEENKO, F. P., and KOZHEVNIKOV, A. M. 1963. Losses of increment in pine stands infested with sawfly. *Les. Khoz.*

MOLCHANOV, A. A. 1938. Damage to spruce seed crop by birds and squirrels. *Les. Khoz.*

MOLCHANOV, A. A. 1947. Rate of decomposition of pine and spruce litter. *Dokl. Akad. Nauk SSSR,* 56.

MOLCHANOV, A. A. 1949. Amounts of needles in pine stands of different ages. *Dokl. Akad. Nauk SSSR,* 67.

MOLCHANOV, A. A. 1952. *Hydrological role of pine forests on sandy soils.* Publ. Acad. Sci. U.S.S.R. Moscow.

MOLCHANOV, A. A. 1954. Variation in biological, ecological, and hydrological factors in different types of oak forests. *Soobshch. Inst. Lesa.*

MOLCHANOV, A. A. 1960. *Hydrological role of the forest.* Moscow.

MOLCHANOV, A. A. 1961a. Experiment in biogeocoenotic description of forest types. *Soobshch. Lab. Lesov.*

MOLCHANOV, A. A. 1961b. Circulation of organic matter during process of growth of a bilberry pine stand. *Soobshch. Lab. Lesov.*

MOLCHANOV, A. A. 1961c. *Forest and climate.* Publ. Acad. Sci. U.S.S.R. Moscow.

MOLCHANOV, A. A. 1963. *Experimental complex (biogeocoenotic) study of broad-leaved forests as a scientific foundation for forestry measures. Biogeocoenotic investigations in oak stands in the forest-steppe zone.* Moscow.

MOLCHANOV, A. A., and PREOBRAZHENSKII, I. F. 1957. *Forests and forest management in Archangel province.* Moscow.

MØLLER, P. E. 1907. Effect of earthworms on germination of rhizomatous plants, mostly in beech forests. *Les. Zh.*

MORAVSKAYA, A. S. Insect damage to early- and late-leafing forms of oak and elm. *Soobshch. Inst. Lesa.*

MOROZOV, G. F. 1899. Soil moisture and natural regeneration of pines in Prussia. *Pochvovedenie.*

Morozov, G. F. 1912, 1926. *Forest science.* St. Petersburg; Moscow & Leningrad.

Morozov, G. F. 1931. *Forest science.* 6th ed. Sel'khozgiz. Moscow.

Morozov, G. F., and Okhlyabin, I. N. 1911. Experiment on effect of the root system of a pine plantation on the moisture in the soil under it. *Les. Zh.*

Moshkalev, A. G., and Spitsin, L. M. 1962. *Instructions for compiling composite forest-management tables on punchcard-computing machines.* Leningrad.

Moshkov, B. S. 1950. Significance of separate parts of the spectrum in the physiological range of illumination for growth and development of certain plants. *Dokl. Akad. Nauk SSSR*, **71**.

Muraveiskii, S. D. 1948. Role of geographical factors in the formation of geographical complexes. *Vop. Geogr.*, coll. **9**.

Muromtsev, I. A. 1940. Dynamics of development of the active part of the apple-tree root system. (Cand. dissert.) Michurinsk.

Mustafova, N. N. 1958. Effect of mineral fertilisation on the microflora of forest soils. *Vest. leningr. gos. Univ. (ser. biol.)*, **15**.

Mustafova, N. N. 1959. Microbiological observations in podzolic soils of spruce-*Oxalis* and spruce-bilberry stands. *Vest. leningr. gos. Univ. (ser. biol.)*, **15**.

Nalivkin, D. V. 1956. *Study of facies.* Publ. Acad. Sci. u.s.s.r. Moscow.

Napalkov, A. V. 1962. Processing of information by the brain. In: *Biological aspects of cybernetics.* Moscow.

Naumov, N. P. 1955. *Animal ecology.* Moscow.

Naumov, N. P. 1963. Biological systems. *Priroda*, Moscow.

Naumov, N. P. 1963. The organic world as a whole. In: *Outline of the dialectics of living nature.* Moscow.

Naumov, S. P. 1947. *Ecology of the blue hare.* Izd. mosk. Obshch. Ispyt. Prir. Moscow.

Nesterov, N. S. 1908. Effect of a forest on the strength and direction of the wind. *Trudy mosk. les. Obshch.*

Nesterov, N. S. 1931. *Outlines of forestry.* Goslestekhizdat. Moscow & Leningrad.

Nesterov, N. S. 1960. *Outlines of forestry.* Sel'khozgiz. Moscow.

Nesterov, V. G. 1954. *General forestry.* Goslesbumizdat. Moscow.

Nesterov, V. G. 1961. *Problems of modern forestry.* Sel'khozgiz. Moscow.

Nesterov, V. G. 1962. *Cybernetics and biology.* Moscow.

Nesterov, V. G., and Nikso-Nikkochio, N. V. 1951. *Dependence of the reproduction of some forest pests on fluctuations in climatic conditions and seed crops.* Moscow.

Nette, I. T. 1955. Denitrifying bacteria in oak rhizosphere. *Mikrobiologiya*, **24**.

Nichiporovich, A. A. 1955. *Light and carbon nutrition of plants (photosynthesis).* Publ. Acad. Sci. u.s.s.r. Moscow.

Nichiporovich, A. A., Strogonova, L. E., Chmora, S. N., and Vlasova, M. V. 1961. *Photosynthetic activity of cultivated plants.* Publ. Acad. Sci. u.s.s.r. Moscow.

Nikiforov, L. P., and Gibet, L. A. 1959. Effect of moose on pine regeneration in Karelia. *Soobshch. Inst. Lesa.*, **13**.

Nikitin, S. A. 1961. Forest types in the Serebryanoborsk experimental forest. *Trudy Lab. Lesov.*

Nikitin, S. A., and Grebennikova, E. F. 1961. Field investigations of the biogeocoenose of a composite pine forest. *Trudy Lab. Lesov.*

Nikolaevskaya, M. A. 1957. Study of the microflora that break down plant remains in the forest zone. Delegation to the May 1957 session of the All-Union Bot. Soc. *Abstracts, No. V. Spore-bearing plants.*

Nikolaevskaya, M. A., and Chastukhin, V. Ya. 1945. Microflora of spruce wood in different stages of decomposition. *Pochvovedenie.*

Nikol'son, A. M., and Aru, A. A. 1962. Automation of processing of survey descriptions in forest management. *Les. Khoz.*

Nikolyuk, V. F. 1956. *Soil-dwelling Protozoa and their role in cultivated soils in Uzbekistan.* Publ. Acad. Sci. Uzb. s.s.r. Tashkent.

Nikolyuk, V. F. 1962. *Soil Protista and their biological role.* Second Zool. Conf. of the Lithuanian s.s.r. (Abstracts). Vilnyus, Lith. s.s.r. Acad. of Sc., Inst. of Zool. and Parasitol.

Nomokonov, L. I. 1963. Some methodological problems of biogeocoenology. *Izv, sib. Otdel. Akad. Nauk SSSR (ser. biol. med.)*, **8**.

NORIN, B. N. 1956. Features of seed-regeneration of trees in Malyi Yamal peninsula. In: *Vegetation of the Far North of the U.S.S.R. and its utilisation*, No. 1. Publ. Acad. Sci. U.S.S.R. Moscow & Leningrad.

NOVIKOV, G. A. 1956. Spruce forests as animal habitat. In: *Role of animals in the life of the forest*. Moscow.

NOVIKOV, G. A. 1959. *Ecology of mammals and birds in forest-steppe oak stands.* Leningrad.

NOVIKOV, V. A. 1961. *Plant physiology.* Moscow.

NOVOGRUDSKII, D. M. 1936a. Investigations of ability of soils to absorb bacteria. I. Distribution of micro-organisms between solid and liquid phases of soil. *Mikrobiologiya*, 5.

NOVOGRUDSKII, D. M. 1936b. Investigations of ability of soils to absorb bacteria. II. Ability of soil to absorb different micro-organisms and its dependence on the environment. *Mikrobiologiya*, 5.

NOVOGRUDSKII, D. M. 1937. Investigations of ability of soils to absorb bacteria. III. Seasonal fluctuations in ability of soil to absorb bacteria. *Mikrobiologiya*, 6.

NOVOGRUDSKII, D. M. 1946a. Microbiological processes in semi-desert soils. I. Soil micro-organisms and hygroscopic soil moisture. *Mikrobiologiya*, 15.

NOVOGRUDSKII, D. M. 1946b. Microbiological processes in semi-desert soils. II. Lower limit of soil moisture for life-activity of bacteria. *Mikrobiologiya*, 15.

NOVOGRUDSKII, D. M. 1947. Microbiological processes in semi-desert soils. III. Categories of soil moisture and nitrification. *Pochvovedenie*.

NYUBERG, N. D. 1962. Problems of coding of information about colour in the retina. In: *Biological aspects of cybernetics*. Moscow.

OBOZOV, N. A. 1954. Forest pasturage. *Priroda*, Moscow.

OBOZOV, N. A. 1957. *Regulation of cattle-pasturing and hay-mowing in the forests of the central provinces of the R.S.F.S.R.* Bryansk.

OBRAZTSOV, B. V. 1956. Role of animals in the afforestation of the steppes. *Priroda*, Moscow.

OBRAZTSOV, B. V. 1961. Material from experiments and observations on distribution of tree and shrub seeds by wild animals in open biotopes of the forest-steppe. *Soobshch. Lab. Lesov.*

OBRAZTSOV, B. V., and SHTIL'MARK, F. R. 1957. Sylvicultural significance of mouse-like rodents in oak forests in the European part of the U.S.S.R. *Trudy Inst. Lesa*, Moscow, 35.

OGIEVSKII, V. V. 1954. Growth of root systems in plantations. *Les. Khoz.*

OLOVYANNIKOVA, I. N. 1953. *Interrelationships of trees and herbage in forest plantations on the southern chernozems of Volgograd province.* (Auth. abstr. of dissert.) Moscow.

OLOVYANNIKOVA, I. N. 1962. Plant ecology in the pine belts in the Irtysh valley. *Soobshch. Lab. Lesov.*

ORLENKO, E. G. 1955. Interrelationships of oaks in dense plantations. *Dokl. Akad. Nauk SSSR*, 102.

ORLOV, A. YA. 1955a. Role of suction roots of trees in enrichment of soil with organic matter. *Pochvovedenie*.

ORLOV, A. YA. 1955b. Method of quantitative evaluation of suction roots of trees in soil. *Byull. mosk. Obshch. Ispyt. Prir. (otd. biol.)*, 60.

ORLOV, A. YA. 1957. Observations on suction roots of spruce (*Picea excelsa* Link.) in natural conditions. *Bot. Zh. SSSR*, 13.

ORLOV, A. YA. 1959. Distribution of suction roots through waterlogged soils of spruce forests in relation to aeration conditions. *Byull. mosk. Obshch. Ispyt. Prir. (otd. biol.)*, 64.

ORLOV, A. YA. 1960a. Growth and age-changes of suction roots of spruce (*Picea excelsa* Link.). *Bot. Zh. SSSR*, 45.

ORLOV, A. YA. 1960b. Effect of soil factors on chief characteristics of some forest types in the southern taiga. *Byull. mosk. Obshch. Ispyt. Prir. (otd. biol.)*, 65.

ORLOV, A. YA. 1962. Growth and dying-off of roots of pine, birch, and spruce with periodic flooding by underground water. *Soobshch. Lab. Lesov.*

ORLOV, A. YA., and MINA, V. N. 1962. Dynamics of soil factors in some types of forest. *Trudy Inst. Lesa Drev. sib. Otd.*, 52.

ORLOVSKAYA, E. V. 1962. Use of an experimental strain of nuclear polyhedral virus to create an epizootic in gypsy moth populations. In: *Problems of ecology*, **8**. Izd. kievsk. Univ. Kiev.

OSKRETKOV, M. YA. 1953. Evaluation of needles of trees of different growth classes. *Trudy bryansk. lesokhoz. Inst.*, **6**.

OSKRETKOV, M. YA. 1954. *Variations in crowns and in wood quality in accordance with stand density and age.* (Cand. dissert. abstract.) Voronezh.

OSKRETKOV, M. YA. 1956. Variations in quantity and quality of pine needles in accordance with stand density and age. *Trudy bryansk. lesokhoz. Inst.*, **7**.

OSKRETKOV, M. YA. 1957. Effect of varying degrees of illumination on pine and spruce regeneration. *Trudy bryansk. lesokhoz. Inst.*, **8**.

OZOLS, G. E. 1961. Treatment of pine seedlings with a DDT preparation for protection against pine weevils. In: *Brief summary of scientific investigations on plant protection in the Baltic zone of the U.S.S.R.*, No. 2. Riga.

PAAVER, K. O. 1953. *Distribution and ecology of mouse-like rodents—forest pests in the Estonian S.S.R.* Scientific session on problems of biology and agriculture. Riga.

PAINTER, R. 1953. *Plant resistance to insects.* Inostr. Lit. Moscow.

PANESENKO, V. G. 1944. Ecology of soil fungi. *Mikrobiologiya*, **13**.

PANFILOV, D. V. 1961. *Insects in the tropical forests of Southern China.* Izd. mosk. gos. Univ. Moscow.

PARSHEVNIKOV, A. L. 1957. Succession of species in spruce forests of the central taiga and its effect on soils. *Soobshch. Inst. Lesa.*

PARSHEVNIKOV, A. L. 1958. Succession of species in spruce forests of the central taiga and its effect on the soil. *Soobshch. Inst. Lesa.*

PARSHEVNIKOV, A. L. 1962. Circulation of nitrogen and mineral elements in relation to succession of species in forests of the central taiga. *Trudy Inst. Lesa Drev. Sib. Otd.*, **52**.

PEREL', T. S. 1958. Dependence of numbers and species-composition of earthworms on species-composition of tree plantations. *Zool. Zh.*, **37**.

PEREL'MAN, A. I. 1960. Geochemical principles of landscape classification. *Vest. mosk. gos. Univ.*

PEREL'MAN, A. I. 1961. *Landscape geochemistry.* Geografgiz. Moscow.

PERSHAKOV, A. A. 1939. Biocoenotic method of control of forest rodents. *Trudy povolzhsk. lesotekh. Inst.*

PERSHAKOV, A. A. 1940. Mouse control in forestry. *Les. Khoz.*

PERSHAKOV, E. P. 1934. Mouse control in upland oak stands. *Izv. povolzhsk. lesotekh. Inst.*

PETERBURGSKII, A. V. 1959. *Metabolic absorption in the soil and consumption of nutrients by plants.* Moscow.

PETRENKO, E. S. 1961. Material on lepidopterous forest pests in Central Yakutia. *Uchen. Zap. krasnoyarsk. gos. pedag. Inst.*, **20**.

PETROV, A. P. 1947. Some problems of principle and method in forest phytocoenology. *Sov. Bot.*

PETROV, V. V. 1954. Extent of effect of mouse-like rodents on forest regeneration under forest canopy. *Dokl. Akad. Nauk SSSR*, **95**.

PIVOVAROVA, E. P. 1956. Distribution by biotopes, diet, and significance to forestry of mouse-like rodents in the Belovezh bush. *Uchen. Zap. mosk. gor. pedag. Inst.*

POCHON, J., and DE BARJAC, G. 1960. *Soil microbiology.* Inostr. Lit. Moscow.

PODGORNYI, V. D. 1962. Leopard moth infestation of ash trees. In: *Nature protection in the central chernozem belt*, No. 4. Voronezh.

POGORILYAK, I. M. 1962. Ecological grouping of bark beetles in relation to age of spruce forests in the Southern Carpathians. *Dokl. Soobshch. uzhgorodsk. Univ. (ser. biol.).*

POGREBNYAK, P. S. 1955. *Fundamentals of forest typology.* Publ. Acad. Sci. Ukr. S.S.R. Kiev.

POLETAEV, I. A. 1958. *Signal. Some concepts in cybernetics.* Moscow.

POLOZHENTSEV, P. A. 1952. Role of nematodes in suppression of insect pests. *Nauch. Zap. voronezhsk. les.-khoz. Inst.*, **13**.

POLYAKOVA, N. F. 1957. Relationships of leaf mass, wood increment, and transpiration. *Dokl. Akad. Nauk SSSR*, **96**.

POLYAKOVA-MINCHENKO, N. F. 1961. Leaf formation in broad-leaved plantations in the steppe zone. *Soobshch. Lab. Lesov.*

POLYNOV, B. B. 1925. Landscape and soil. *Priroda,* Moscow.

POLYNOV, B. B. 1934. *The mantle (weathering crust),* part 1. Publ. Acad. Sci. U.S.S.R. Leningrad.

POLYNOV, B. B. 1946a. Geochemical landscapes. In: *Problems of mineralogy, geochemistry, and petrography.* Publ. Acad. Sci. U.S.S.R. Moscow.

POLYNOV, B. B. 1946b. Role of pedology in landscape study. *Izv. vses. geogr. Obshch.,* 78.

POLYNOV, B. B. 1948. Problem of the role of elements of the biogeosphere in the evolution of organisms. *Pochvovedenie.*

POLYNOV, B. B. 1952. Geochemical landscapes. In: *Geographical works.* Geografgiz. Moscow.

POLYNOV, B. B. 1953. Landscape science. *Vop. Geogr.,* 33.

PONOMAREV, A. N. 1937. Elementary landscapes of the abrasion-erosion platform of the Urals in the mixed-herbage-*Stipa* steppes. *Zemlevedenie,* 39.

PONOMAREVA, A. V. 1951. Increasing the germination rate and growth energy in some tree seeds by means of steeping in suspensions of *Trichoderma* and strains of *Azotobacter. Izv. Akad. Nauk SSSR.*

PONOMAREVA, A. V. 1953. Effect of *Azotobacter* on growth of seedlings of some tree species and on accumulation of chlorophyll in their leaves. *Izv. Akad. Nauk beloruss. SSR.*

PONOMAREVA, S. I. 1948. Rate of formation of calcite by earthworms in soil. *Dokl. Akad. Nauk SSSR,* 11.

PONOMAREVA, S. I. 1949. Effect of earthworm activity on the creation of erosion-resisting soil structure. *Proc. of Jubilee session of V. V. Dokuchaev Pedol. Inst.* Moscow & Leningrad.

PONOMAREVA, S. I. 1950. Role of earthworms in creation of a stable structure in grass-arable crop rotation. *Pochvovedenie.*

PONOMAREVA, S. I. 1953. Effect of earthworm activity on creation of structure of soddy-podzolic soil. *Trudy pochv. Inst. Akad. Nauk SSSR,* 41.

PONOMAREVA, S. I. 1958. *Role of earthworms in increasing the fertility of eroded soddy-podzolic soil.* Abstracts of papers delivered at All-Union Conf. on Soil Zoology. Moscow.

PONOMAREVA, V. V. 1962. Role of humic matter in soil-formation processes (theory of formation of soddy-podzolic and grey forest-steppe soils). *Problems of pedology.* Publ. Acad. Sci. U.S.S.R. Moscow.

PONYATOVSKAYA, V. M. 1941. Present state of the problem of interrelationships of leguminous plants and grasses in herbage. *Sov. Bot.*

PONYATOVSKAYA, V. M. 1959. Two trends in phytocoenology. *Bot. Zh. SSSR,* 4.

POPLAVSKAYA, G. I. 1924. Experiment in phytocoenotic analysis of vegetation of the virgin steppe reserve of Askania-Nova. *Zh. russk. bot. Obshch.,* 9.

POPOV, T. I. 1914. Origin and development of aspen shrubs. *Trudy Uchen. pochv. Kom.*

POPOVA, N. N. 1962. Effect of small mammals on moisture distribution in soil beneath forest canopy. *Byull. mosk. Obshch. Ispyt. Prir. (otd. biol.),* 67.

POVAROV, G. N. 1960. A new look at cybernetics. *Priroda,* Moscow.

POZDNYAKOV, L. K. 1953. Light regime beneath larch canopy. *Dokl. Akad. Nauk SSSR,* 90.

POZDNYAKOV, L. K. 1956. Role of precipitation passing through forest canopy in the process of matter and energy exchange between forest and soil. *Dokl. Akad. Nauk SSSR,* 107.

PROKAEV, V. I. 1961. *The facies as the basic and smallest unit in landscape science.* Proc. of V All-Union Conf. on problems of landscape science. Moscow.

PROZOROV, S. S., KORSHUNOV, L. M., and ZEMKOVA, R. I. 1963. The vapourer moth (*Orgyia antiqua* L.)—a pest of Siberian larch. In: *Protection of Siberian forests from insect pests.* Publ. Acad. Sci. U.S.S.R. Moscow.

PUSHKINSKAYA, O. I. 1951. Material on description of soil microflora of oak stands in the Tellerman experimental forest (prelim. report). *Trudy Inst. Lesa,* Moscow, 7.

PUSHKINSKAYA, O. I. 1953. Soil microflora of the Tellerman experimental forest. *Trudy Inst. Lesa*, Moscow, **12**.

PUSHKINSKAYA, O. I. 1954a. Microflora of the forest soils of the oak stands in the Borisoglebsk forest massif. *Trudy Inst. Lesa*, Moscow, **23**.

PUSHKINSKAYA, O. I. 1954b. Microbiological characteristics of forest soils in different relief conditions. *Soobshch. Inst. Lesa.*

PUSHKINSKAYA, O. I. 1962. Effect of clear-cutting of forests on the composition of microflora in dark-grey forest soil. *Soobshch. Lab. Lesov.*

PYATNITSKII, G. K. 1929. A few words on ecological lists of forest pests. *Zashch. Rast. Vredit.*, **6**.

PYATNITSKII, S. S. 1955. Viability, longevity, and regenerability of plantations on the steppes. *Zap. Kharkov. sel.-khoz. Inst.*, **10**.

P'YAVCHENKO, N. I. 1960. Biological circulation of nitrogen and mineral elements in swampy forests. *Pochvovedenie.*

PYL'TSINA, L. M. 1956. Some data on the effect of attracted birds on the numbers of insect pests. In: *Ways and methods of attracting birds to control insect pests.* Moscow.

RABINOVICH, E. 1951. *Photosynthesis.* Inostr. Lit. Moscow.

RABOTNOV, T. A. Life cycle of perennial herbaceous plants in meadow coenoses. *Trudy bot. Inst. Akad. Nauk SSSR (ser. III, geobot.).*

RABOTNOV, T. A. 1952. Life cycle of meadow crowfoot (*Ranunculus acer* L.) and goldilocks (*R. auricomus* L.). *Byull. mosk. Obshch. Ispyt. Prir. (otd. biol.)*, **63**.

RABOTNOV, T. A. 1959. The works of G. Ellenberg on causal study of meadow vegetation. *Bot. Zh. SSSR*, **44**.

RABOTNOV, T. A. 1962. Some problems of the study of dominant producers in meadow coenoses. *Problemy Bot.*

RABOTNOV, T. A. 1963. Experiment in use of the principle of plant cover continuum in study of the vegetation of Wisconsin (U.S.A.). *Byull. mosk. Obshch. Ispyt. Prir. (otd. biol.)*, **18**.

RACHKOV, L. D. 1963. *Study of the role of plants in the migration of radio-isotopes through the components of natural biogeocoenoses and model systems.* (Dissert. abstract.) Sverdlovsk.

RAFES, P. M. 1956. Role of *Sesia apiformis* and *Poecilonota variolosa* in the death of black poplar in the Achikulaksk leskhoz (Report 2). *Soobshch. Inst. Lesa*, **48**.

RAFES, P. M. 1957. Insect pests of plantations on the Narynsk sands of the trans-Volga semi-deserts. *Zool. Zh.*, **36**.

RAFES, P. M. 1960. Insect pests of aspen, poplar, and willow growing on the Narynsk sands of the trans-Volga semi-deserts. *Trudy Inst. Lesa*, Moscow, **48**.

RAFES, P. M. 1961. Principles of investigations of mass multiplication of forest pests. *Soobshch. Lab. Lesov.*

RAFES, P. M. 1962. Length of burrows and number of offspring of bark beetles in relation to population density, in the case of Kholodkovskii's lesser bark beetle. *Soobshch. Lab. Lesov.*

RAFES, P. M. 1964. Mass multiplication of insect pests as special cases of matter and energy circulation in a forest biogeocoenose. In: *Studies in forest protection.* Nauka. Moscow.

RAIMERS, N. F. 1956. Role of mammals and birds in regeneration of stone pine forests in the Lake Baikal region. *Zool. Zh.*, **35**.

RAIMERS, N. F. 1958. Reafforestation of burned areas and forest massifs destroyed by pine moths in the stone-pine taiga of the southern Lake Baikal region, and the role of vertebrates in that process. *Byull. mosk. Obshch. Ispyt. Prir. (otd. biol.)*, **63**.

RAKHTEENKO, I. N. 1958a. Migration of mineral nutrients from one plant to another through action of their root systems. *Bot. Zh. SSSR.*

RAKHTEENKO, I. N. 1958b. Seasonal cycle of absorption and exudation of mineral nutrients by tree roots. *Fiziologiya Rast.*, **5**.

RAKHTEENKO, I. N. 1961. Growth and interaction of tree root systems in plantations. (Dissert. abstract.) Minsk.

RAKHTEENKO, I. N. 1963. *Growth and interaction of tree root systems in plantations.* Publ. Acad. Sci. Beloruss. S.S.R. Minsk.

RAMENSKII, L. G. 1910. Comparative method of ecological study of plant communities. *Proc. of XII Congress of Russian naturalists and surgeons, sect. II.* Moscow.

636 FUNDAMENTALS OF FOREST BIOGEOCOENOLOGY

RAMENSKII, L. G. 1924. Basic laws governing plant cover. *Vestn. opyt. Dela.* Voronezh.

RAMENSKII, L. G. 1938. *Introduction to complex soil-geobotanical land investigations.* Sel'khozgiz. Moscow.

RAMENSKII, L. G. 1952. Some of the fundamental principles of modern geobotany. *Bot. Zh. SSSR,* 37.

RASKATOV, P. B. 1940. Experimental estimate of transpirational losses by plantations. *Nauch. Zap. voronezhsk. lesokhoz. Inst.,* 6.

RASPOPOV, P. M. 1961. Dynamics of foci of mass reproduction of the black arches moth and other pests in the forests in the north-western part of Chelyabinsk province. *Trudy ilmensk. gos. Zapov.*

RATNER, E. I. 1950. *Mineral nutrition of plants and absorptive capacity of the soil.* Publ. Acad. Sci. U.S.S.R. Moscow & Leningrad.

RATNER, E. I., and KOLOSOV, I. I. 1954. Root nutrition of plants and new methods of investigating it. *Priroda,* Moscow.

RATS, I. I. 1938. Soil moisture and moisture consumption through absorption by tree root systems in a hornbeam plantation. *Problemy sov. Pochvov.*

RAUNER, YU. L. 1962. Components of the heat balance. In: *Heat balance of forest and field.* Izd. mosk. gos. Univ. Moscow.

RAYNER, N., and NEILSON-JONES, W. 1949. *Role of mycorrhiza in nutrition of trees.* Inostr. Lit. Moscow.

RAZUMOVSKAYA, V. G., and MUSTAFOVA, N. N. 1959. Biological activity of the soils of spruce-*Oxalis* and spruce-bilberry stands. *Vest. leningr. gos. Univ.*

REMEZOV, N. P. 1941. Ammonisation and nitrification in forest soils. *Trudy vses. nauchno-issled. Inst. les. Khoz.,* 1.

REMEZOV, N. P. 1952. *Soils, their characteristics and distribution.* Uchpedgiz. Moscow.

REMEZOV, N. P. 1953. Role of the forest in soil formation. *Pochvovedenie.*

REMEZOV, N. P. 1956. Role of biological circulation of elements in soil formation beneath forest canopy. *Pochvovedenie.*

REMEZOV, N. P. 1961. Summary of studies of interactions of oak forest and soil. *Trudy voronezh. gos. Zapov.*

REMEZOV, N. P., and BYKOVA, L. N. 1952. Consumption and circulation of nutrient elements in an oak forest. *Vest. leningr. gos. Univ. (ser. mat. estest.).*

REMEZOV, N. P., BYKOVA, L. N., and SMIRNOVA, K. M. 1959. *Consumption and circulation of nitrogen and mineral elements in forests in the European part of the U.S.S.R.* Izd. mosk. gos. Univ. Moscow.

REMEZOV, N. P., RODIN, L. E., and BAZILEVICH, N. I. 1963. Methodological instructions for study of biological circulation of mineral elements and nitrogen in terrestrial plant communities in the chief natural regions of the Temperate Zone. *Bot. Zh. SSSR,* 18.

REMEZOV, N. P., SMIRNOVA, K. M., and BYKOVA, L. N. 1949. Some conclusions from study of the role of forest vegetation in soil formation. *Vest. mosk. gos. Univ.*

RICHARDS, P. 1961. *Tropical rain forest.* (Translated from English.) Inostr. Lit. Moscow.

RICHTER, A. A. 1914. Problem of the mechanism of photosynthesis. *Izv. imp. Akad. Nauk,* 8.

RODE, A. A. 1947. *The soil-formation process and evolution of soils.* Geografgiz. Moscow.

RODE, A. A. 1950. Soil-underground-water regime in podzolic, podzolic-boggy, and boggy soils. *Trudy pochv. Inst.,* 32.

RODE, A. A. 1952. *Soil moisture.* Publ. Acad. Sci. U.S.S.R. Moscow.

RODE, A. A. 1954. Address to the Conf. on For. Soil Management (25-28 Dec. 1952). *Trudy Inst. Lesa,* Moscow, 23.

RODE, A. A. 1956. Water regime, and types of it. *Pochvovedenie.*

RODE, A. A. 1961. Water-physical constants of the soil. *Pochvovedenie.*

RODIN, L. E. 1961. *Dynamics of desert vegetation.* Publ. Acad. Sci. U.S.S.R. Moscow.

ROGACHEVA, A. I. 1947. Thermostability of spores of bacteria of different climatic zones. *Mikrobiologiya,* 16.

ROMANKOVA, A. G. 1954. Microflora of podzolic soils in different geographical districts of the U.S.S.R. *Vest. leningr. gos. Univ. (ser. biol. geogr. geol.).*

ROZANOVA, I. M. 1960. Circulation of mineral elements and variation in physico-chemical features of leached chernozems under coniferous and broad-leaved plantations. *Trudy Lab. Lesov.*

ROZHKOV, A. S., SENDAROVICH, B. P., and DANOVICH, K. I. 1962. Some interrelationships between Siberian pine moth larvae and the larch (*Larix sibirica* Ldb.) infested by them. *Trudy vost.-sib. Fil. Akad. Nauk SSSR*, 35.

ROZHNOVA, T. A. 1962. Biochemical characteristics of the landscapes of the Karelian isthmus. *Pochvovedenie,*

ROZHNOVA, T. A., and SCHASTNAYA, L. S. 1959. Study of interrelationships of vegetation and soils in Karelian isthmus conditions. *Pochvovedenie.*

RUBENCHIK, L. I., ROIZIN, M. B., and BELYANSKII, F. M. 1934. Adsorption of bacteria in salt lakes. *Mikrobiologiya*, 3.

RUBTSOV, N. I. 1950. New data on growth of root systems of some tree species. *Agrobiologiya.*

RUBTSOV, V. I. 1960. Growth of Scots pine in dense plantations. *Nauch. Zap. voronezhsk. lesokhoz. Inst.*, 18.

RUDNEV, D. F. 1958. Causes of lowering of resistance in plantations and conditions of formation of foci of pests in forests in the Ukraine. *Byull. nauchno-tekh. Inf. ukr. Inst. Zashch. Rast.*

RUDNEV, D. F. 1959. Role of herbage-covered forest-regeneration areas and some other anthropogenic factors in the reproduction of forest pests in the Ukraine. *Zool. Zh.*, 38.

RUDNEV, D. F. 1962. Effect of the physiological state of plants on mass reproduction of forest pests. *Zool. Zh.*, 41.

RUKHIN, L. B. 1962. *Fundamentals of general palaeogeography.* Leningrad.

RUNOV, E. V. 1952. Formation of mycorrhiza on oak in the arid steppe. *Soobshch. kompleks. nauch. Exped. Vopr. polezashch. Lesorazv.*

RUNOV, E. V. 1954. Microflora of chernozems under tree plantations in the Derkul' steppe. *Trudy Inst. Lesa*, Moscow, 15.

RUNOV, E. V. 1955. Experiment in mycorrhisation when sowing oak acorns on the arid steppe. *Proc. of Conf. on plant mycotrophy of U.S.S.R. Acad. of Sci.*

RUNOV, E. V. 1957. Role of microflora in the life of the forest. In: *Young foresters —40th anniversary of Great October.* Moscow.

RUNOV, E. V. 1960. Effect of forest tree species on soil microflora. *Papers given by Soviet pedologists at VII International Congress in U.S.A.* Publ. Acad. Sci. U.S.S.R. Moscow.

RUNOV, E. V., and EGOROVA, S. V. 1958. Effect of litter of coniferous and deciduous plantations and tree roots on microflora of leached chernozems. *Mikrobiol. Inst. zinat. Rak.*

RUNOV, E. V., and EGOROVA, S. V. 1962. Comparative characteristics of microflora of deciduous species in Tellerman experimental forest. *Soobshch. Lab. Lesov.*

RUNOV, E. V., and EGOROVA, S. V. 1963. Toxicosis of dark-grey soils under oak forests in the forest-steppe zone. *Pochvovedenie.*

RUNOV, E. V., and ENIKEEVA, M. G. 1955. Effect of tree-shrub vegetation litter on microflora of chernozem soils of the arid steppe. *Mikrobiologiya*, 24.

RUNOV, E. V., and ENIKEEVA, M. G. 1959. Toxicosis of chernozems under forest plantations on the arid steppe. *Soobshch. Inst. Lesa.*

RUNOV, E. V., and ENIKEEVA, M. G. 1961. Microflora of tree rhizospheres on chernozem soil on the arid steppe. *Trudy Inst. Mikrobiol.*, Moscow.

RUNOV, E. V., and ENIKEEVA, M. G. 1963. Effect of roots of trees and shrubs on growth of plant shoots and soil microflora. In: *Biogeocoenotic investigations in oak stands in the forest zone.* Publ. Acad. Sci. U.S.S.R. Moscow.

RUNOV, E. V., and KUDRINA, E. S. 1954. Effect of tree plantations on microflora of chernozems of the arid steppe. *Trudy Inst. Lesa*, Moscow, 23.

RUNOV, E. V., and MISHUSTINA, I. E. 1960. Effect of tree plantations of different composition on microbiological processes in leached chernozem. *Trudy Lab. Lesov.*

RUNOV, E. V., and SOKOLOV, D. F. 1956. Investigation of effect of litter on biochemical and microbiological processes in soils under tree plantations. *Trudy Inst. Lesa*, Moscow, 30.

RUNOV, E. V., and SOKOLOV, D. F. 1958. Changes in composition of organic matter and microflora in leached chernozem under the influence of tree plantations. *Byull. mosk. Obshch. Ispyt. Prir. (otd. biol.)*, **63**.

RUNOV, E. V., and TEREKHOV, O. S. 1960. Problem of catalase activity in some forest soils. *Pochvovedenie.*

RUNOV, E. V., and VALEVA, S. A. 1962. Microbiological characteristics of forest soils in the central taiga of Vologda province. *Trudy Inst. Lesa Drev. sib. Otd.*, **52**.

RUNOV, E. V., and ZHDANNIKOVA, E. N. 1956. Microflora of birch and spruce rhizospheres in different soils of the central taiga of Vologda province. *Soobshch. Lab. Lesov.*

RUNOV, E. V., and ZHDANNIKOVA, E. N. 1962. Effect of fires on microbiological activity in soils of the central taiga of Vologda province. *Trudy Inst. Lesa Drev. sib. Otd.*, **52**.

RYABOVA, T. I. 1938. *Effect of grazing by axis deer on plant cover of forest pastures.* Material on flora, vegetation, and soils of the Far East, No. 1.

RYASHCHENKO, Z. M., RYUMINA, T. A., and POPOVA, M. N. 1952. *Azotobacter* in trass soils of the Saratov-Kamyshin forest belt, and its effect on oak development. *Uchen. Zap. saratov. gos. Univ.*, **29**.

RYBALKINA, A. V. 1957. *Microflora of the soils of the European part of the U.S.S.R.* Publ. Acad. Sci. U.S.S.R. Moscow.

SABININ, D. A. 1925. The root system as an osmotic apparatus. *Izv. biol. nauchno-issled. Inst. biol. Sta. perm. gosud. Univ.*, **4**, Appendix 2.

SABININ, D. A. 1940. *Mineral nutrition of plants.* Publ. Acad. Sci. U.S.S.R. Moscow.

SABININ, D. A. 1949. *Significance of the root system in the life-activity of plants.* Timiryazev lecture. Publ. Acad. Sci. U.S.S.R. Moscow.

SABININ, D. A. 1955. *Physiological basis of plant nutrition.* Publ. Acad. Sci. U.S.S.R. Moscow.

SABININ, D. A., and KOLOSOV, I. I. 1935. Investigations on entrance of matter into plants. 1. Study of processes of adsorption of electrolytes by root systems. *Trudy vses. Inst. Udobr. Agrokh.*

SABLINA, T. B. Adaptive features of diet of some ungulate species, and effect of these species on plant succession. *Soobshch. Inst. Lesa.*

SAKHAROV, M. I. 1939a. Organic litter in forest phytocoenoses. *Pochvovedenie.*

SAKHAROV, M. I. 1939b. Effect of phytocoenoses on snow cover. *Les. Khoz.*

SAKHAROV, M. I. 1940a. Phytoclimates of forest phytocoenoses. *Trudy bryansk. lesokhoz. Inst.*, **4**.

SAKHAROV, M. I. 1940b. Radiation and albedo in forest phytocoenoses. *Meteorologiya Gidrol.*

SAKHAROV, M. I. 1947. Role of wind in development of forest biogeocoenoses. *Dokl. Akad. Nauk SSSR*, **8**.

SAKHAROV, M. I. 1948a. Changes in a forest biogeocoenose with age of stand. *Dokl. Akad. Nauk SSSR*, **59**.

SAKHAROV, M. I. 1948b. Effects of the separate layers of forest coenoses on radiation and illumination. *Dokl. Akad. Nauk SSSR*, **62**.

SAKHAROV, M. I. 1950. Elements of forest biogeocoenoses. *Dokl. Akad. Nauk. SSSR*, **71**.

SAKHAROV, M. I. 1951. Structure of forest coenoses in some forest types in Poles'e. In: *In the forests of Poles'e.* Publ. Acad. Sci. Beloruss. S.S.R. Minsk.

SAKHAROV, M. I., and SAKHAROVA, N. M. 1951. Effect of a forest on forest soil. *Pochvovedenie.*

SALYAEV, R. K. 1961. *Study of the physiologically-active part of root systems in the chief forest-forming species in the taiga zone.* (Dissert. abstract.) Petrozavodsk.

SAMTSEVICH, S. A. 1955a. Significance of ecto-endotrophic mycorrhiza in tree nutrition. *Proc. of Conf. on plant mycotrophy.* Publ. Acad. Sci. U.S.S.R. Moscow.

SAMTSEVICH, S. A. 1955b. Soil microflora in oak stands on the forest-steppe in the Ukrainian S.S.R. *Trudy Inst. Lesov.* Kiev.

SAMTSEVICH, S. A. 1955c. Seasonality and periodicity of micro-organism development in the soil. *Mikrobiologiya*, **24**.

SAMTSEVICH, S. A. 1956. Biodynamics of the rhizospheres of deciduous and coniferous plantations on the forest-steppe. *Mikrobiologiya*, **25**.

SAMTSEVICH, S. A. 1958. Interrelationships between soil micro-organisms and higher plants. Problems of soil microbiology. *Mikrobiol. Inst. zinat. Rak.*

SAMTSEVICH, S. A., KORETSKAYA, Z. M., and VIZIR, A. P. 1949. Effect of forest vegetation on the microflora of chernozem soils in the Black Forest. *Trudy Inst. Lesov.* Kiev.

SANADZE, G. A. 1961. *Exhalation of volatile organic matter by plants.* Publ. Acad. Sci. Georgian S.S.R. Tbilisi.

SAVINA, A. V. 1949. Ecological and physiological factors governing tree growth with improvement cutting. *Les. Khoz.*

SAVINOV, G. V., KRUSHINSKII, L. V., FLESS, D. A., and VALERSHTEIN, R. A. 1962. Experiment in use of mathematical models for study of the interrelationships of stimulation and suppression processes. In: *Biological aspects of cybernetics.* Moscow.

SAVOISKAYA, G. I. 1955. The saxaul lady-beetle. *Uchen. Zap. biol.-pochv. Fak. kirg. Univ. (Zool.).*

SELYANINOV, G. T. 1937. *World agroclimatic reference book.* Leningrad.

SEMENOV-TYAN-SHANSKII, O. I. 1960. Ecology of Tetraonidae. *Trudy lapland. gos. Zapov.*

SEMIKHATOVA, O. A. 1960. Effect of temperature on photosynthesis. *Bot. Zh. SSSR,* 45.

SEREBRYAKOV, I. G. 1954. Morphogenesis of living tree forms in forest species in the central zone of the European part of the U.S.S.R. *Byull. mosk. Obshch. Ispyt. Prir. (otd. biol.),* 59.

SEREBRYAKOVA, L. K. 1951. Transpiration by oak and Norway maple in the chestnut zone of the south-east. *Les. Khoz.*

SHALYT, M. S. 1950. The underground parts of some meadow, steppe, and desert plants and phytocoenoses. *Trudy bot. Inst. Akad. Nauk SSSR (ser. geobot.),* 6.

SHAPOSHNIKOV, F. D. 1949. Relationships between stone pine and animals of the Altai mountain taiga. *Nauchno-metod. Zap. Glav. Upr. Zapov.*

SHARAPOV, N. I. 1954. *Plant chemistry and climate.* Publ. Acad. Sci. U.S.S.R. Moscow.

SHCHERBAKOV, D. I. 1963. Features of the present state and developmental trends of earth sciences. In: *Interaction of earth-study sciences.* Publ. Acad. Sci. U.S.S.R. Moscow.

SHEMAKHANOVA, N. M. 1955a. Use of tagged atoms in study of plant mycotrophy. In: *Isotopes and microbiology.* Publ. Acad. Sci. U.S.S.R. Moscow.

SHEMAKHANOVA, N. M. 1955b. Significance of mycorrhiza for pine and oak seedlings. *Tr. of Conf. on plant mycotrophy.* Publ. Acad. Sci. U.S.S.R. Moscow.

SHEMAKHANOVA, N. M. 1962. *Mycotrophy of trees.* Publ. Acad. Sci. U.S.S.R. Moscow.

SHENNIKOV, A. P. 1939. Experimental study of interrelationships between plants. Jubilee coll. *To Acad. V. L. Komarov.* Publ. Acad. Sci. U.S.S.R. Moscow.

SHENNIKOV, A. P. 1941. *Meadow management.* Sovetskaya Nauka. Moscow.

SHENNIKOV, A. P. 1942. Natural factors of plant distribution in the light of experiments. *Zh. obshch. Biol.,* 3.

SHENNIKOV, A. P. 1948. Geographical and biological methods in geobotany. *Bot. Zh. SSSR,* 33.

SHENNIKOV, A. P. 1950. *Plant ecology.* Sovetskaya Nauka. Moscow.

SHESTAKOVA, V. A. 1962. Use of *Azotobacter* in nurseries. *Les. Khoz.*

SHILOVA, E. I. 1950. Numerical composition and mass of invertebrates in podzolic soils. *Vest. leningr. gos. Univ.*

SHILOVA, E. I. 1951. Biological characteristics of the profile of podzolic soil. *Pochvovedenie.*

SHIPEROVICH, V. YA. 1936. *Forest zoology.* Goslestekhizdat. Moscow.

SHIPEROVICH, V. YA. 1937. Soil fauna in different types of forest. *Zool. Zh.,* 16.

SHISHKOV, I. I. 1953. Spruce root systems and their significance in practical forestry. *Trudy lesotekh. Akad. S.M. Kirova,* 71.

SHIVRINA, A. N. 1961. Problem of qualitative composition of humus-like compounds formed by wood-destroying fungi. In: *Complex study of physiologically-active substances in lower plants by U.S.S.R. Academy of Sciences.* Publ. Acad. Sci. U.S.S.R. Moscow & Leningrad.

SHMAL'GAUZEN, I. I. 1946. *Factors in evolution.* Publ. Acad. Sci. U.S.S.R. Moscow.

SHMAL'GAUZEN, I. I. 1958a. Control and regulation in evolution. *Byull. mosk. Obshch. Ispyt. Prir. (otd. biol.).*

SHMAL'GAUZEN, I. I. 1958b. Regulating mechanisms in evolution. *Zool. Zh.,* **37.**

SHMAL'GAUZEN, I. I. 1959. *Amount of phenotypic information and rate of natural selection.* (Proceedings of 2nd Conf. on use of math. methods in biology.) Izd. leningr. gos. Univ., Leningrad.

SHMAL'GAUZEN, I. I. 1960. Fundamentals of evolutionary progress in the light of cybernetics. *Problemy Kibern.*

SHMAL'GAUZEN, I. I. 1961. Integration of biological systems and their self-regulation. *Byull. mosk. Obshch. Ispyt. Prir. (otd. biol.),* **16.**

SHMAL'GAUZEN, V. I. 1961. Ecology of mouse-like rodents in coniferous-broad-leaved forests. *Soobshch. Lab. Lesov.*

SHTEINKHAUZ (STEINHAUS), E. 1952. *Insect pathology.* Inostr. Lit. Moscow.

SHUBIN, V. I., and POPOV, L. V. 1959. Investigation of agrotechnics of forest culture in areas of concentrated cutting in southern Karelia. *Trudy Karel. Fil. Akad. Nauk SSSR,* **16.**

SHUMAKOV, E. A. 1954. Some results of phytopathological investigations. *Soobshch. Inst. Lesa.*

SHUMAKOV, V. S. 1941. Dynamics of decomposition of plant remains and interaction of their decomposition products with forest soil. *Trudy vses. nauchno-issled. Inst. les. Khoz.,* **24.**

SHUMAKOV, V. S. 1948. Causes of retarded nitrification in forest soils. *Pochvovedenie.*

SHUMAKOV, V. S. 1949. Form of oak root system in relation to growth conditions. *Les. Khoz.*

SHUMAKOV, V. S. 1958. Principles of classification, nomenclature, and mapping of forest litter. *Sb. Rab. les. Khoz.,* **35.** Goslesbumizdat. Moscow.

SHUMAKOV, V. S. 1963. *Types of forest culture and soil fertility.* Goslesbumizdat. Moscow.

SIBIRYAKOVA, M. D. 1949. Role of soil fauna in evolution of spruce plantations and creation of a stable spruce culture. *Les. Khoz.*

SIDEL'NIK, N. A. 1953. Stability of tree and shrub species in plantations in the Staro-Berdyansk, Altagirsk, and Rodionovsk forests. *Nauch. Zap. dnepropetrovsk. gos. Univ.,* **38.**

SIDEL'NIK, N. A. 1960. Some problems of mass forest management on the steppes, and prospects for types of plantations in the steppe zone of the Ukrainian S.S.R. In: *Artificial forests in the steppe zone of the Ukraine.* Khar'kov.

SILINA, A. A. 1955. Temperature of tree leaves in the Derkul' steppe in relation to their heat-resistance. *Trudy Inst. Lesa,* Moscow, 27.

SILINA, A. A. 1955. Transpiration of trees in Tellerman forest. *Fiziologiya Rast.,* **2.**

SIMKIN, G. N. 1961. *Features of the ecology of mouse-like rodents in different types in Perm province.* (Proc. of First All-Union Conf. on mammals, part 2.) Izd. mosk. gos. Univ. Moscow.

SISAKYAN, N. M. *Biochemical characteristics of drought-resistant plants.* Moscow.

SIZOVA, T. P., and VASIN, V. B. 1961. Microflora of oak rhizosphere. *Byull. mosk. Obshch. Ispyt. Prir. (otd. biol.),* **66.**

SIZOVA, T. P., and ITAKAEVA, E. A. 1956. Dynamics of microflora in birch rhizosphere. *Byull. mosk. Obshch. Ispyt. Prir. (n. ser., otd. biol.),* **11.**

SLEPYAN, E. I. 1962. Nomenclature and classification of galls produced by arthropods and of forms of bud teratosis in relation to their place among pathological phenomena. *Bot. Zh. SSSR,* **47.**

SLESARENKO, A. YA. 1952. Microbiological characteristics of soils under forests and in the steppes in the district where part of the Saratov-Kamyshin National Forest is cut off by the Volga. *Uchen. Zap. saratov. gos. Univ.,* **29.**

SLOVIKOVSKII, V. I. 1960. Significance for plant nutrition of mineralisation and leaching of organic litter of trees and shrubs. *Dokl. Akad. Nauk SSSR,* **132.**

SLUDSKII, A. A. 1935. Game animals and stone pine. *Okhotnik Sibiri.*

SMIRNOV, V. V. 1957. Foliation of poplar stands in the Tellerman bush. *Trudy Inst. Lesa,* Moscow, **33.**

SMIRNOV, V. V. 1961. Variation with age in needle and leaf formation in spruce and spruce-deciduous stands in the central taiga. *Trudy Lab. Lesov.*

SMIRNOV, V. N. 1953. Comparative characteristics of the biological activity of soils in the southern forest zone. *Les. Khoz.*

SMIRNOV, V. N. 1954. Problem of the biological activity of soils under forests in the southern part of the taiga zone. *Trudy Inst. Lesa*, Moscow, **23**.

SMIRNOV, V. N. 1955. Problem of the interrelationships between production of soil CO_2 and productivity of forest soils. *Pochvovedenie.*

SMIRNOV, V. N. 1958. Dynamics of nutrients and biological activity of podzolic soils in the southern belt of the forest zone. *Pochvovedenie.*

SMIRNOVA, E. A. 1928. Effect of phyto-social conditions on the course of the struggle for existence between cultivated plants and weeds. *Izv. glav. bot. Sada*, **27**.

SMIRNOVA, K. M. 1951a. Circulation of nitrogen and mineral elements in a green-moss spruce stand. *Vest. mosk. gos. Univ.*

SMIRNOVA, K. M. 1951b. Circulation of nitrogen and mineral elements in mixed spruce stands. *Vest. mosk. gos. Univ.*

SMIRNOVA, K. M. 1952. Consumption and circulation of nutrients in linden stands in the Mordovsk reserve. *Vest. mosk. gos. Univ.*

SMIRNOVA, K. M. 1956. Accumulation of nutrient elements in spruce in accordance with soil conditions. *Vest. mosk. gos. Univ.*

SNIGIREVA, A. V. 1936. Experimental study of the struggle for existence in nature. *Sov. Bot.*

SNIGIREVSKAYA, E. M. 1954. *Ecology and economic importance of mouse-like rodents in the broad-leaved forests of the Zhigulevsk uplands.* (Cand. Dissert., State Library, Moscow.)

SOCHAVA, V. B. 1926. Experiment in phytosociological analysis of the interrelationships between individuals of some meadow plant species. *Zh. russk. bot. Obshch.*, **2**.

SOCHAVA, V. B. 1947. *Landscape and phytocoenose.* (Contributions from Biol. Sci. Branch Acad. Sci. U.S.S.R. for 1945.) Publ. Acad. Sci. U.S.S.R. Moscow.

SOCHAVA, V. B. 1959. Classification of vegetation and typology of phytogeographical facies. *Material on classification of Ural vegetation.* Publ. Acad. Sci. U.S.S.R., Ural Br.

SOCHAVA, V. B. 1961. Problems of classification of vegetation and typology of physico-geographical facies and biogeocoenoses. *Trudy Inst. Biol.*, Sverdlovsk, **27**.

SOCHAVA, V. B. 1962. Geographical aspects of the scientific basis for planned exploitation of the taiga. *Dokl. Inst. geogr. Sib. Dal. Vost.*

SOCHAVA, V. B. 1963. *Geobotanical cartography.* (Collected papers.) Publ. Acad. Sci. U.S.S.R. Moscow & Leningrad.

SOKOLOV, A. A. 1956. *Significance of earthworms in soil formation.* Alma-Ata.

SOKOLOV, D. F. 1957. Significance of black wireworms and ants in the transformation of organic matter under forest plantations in arid-steppe conditions. *Byull. mosk. Obshch. Ispyt. Prir.* (*otd. biol.*), **12**.

SOKOLOV, S. YA. 1938. Advances in Soviet forest geobotany. *Sov. Bot.*

SOKOLOV, S. YA. 1956. Types of struggle for existence among plants. *Vol. Jubil. Sukachev.* Publ. Acad. Sci. U.S.S.R. Moscow.

SOLNTSEV, N. A. 1948. Natural geographical landscape and some of its general laws. *Proc. of II All-Union Geog. Conf.*, **1**.

SOLNTSEV, N. A. 1960. Interrelationships of live and 'dead' nature. *Vest. mosk. gos. Univ.* (*ser. V, geogr.*).

SOLNTSEV, N. A. 1961a. Organization of landscape expeditions and fixing of norms for field work. *Vest. mosk. gos. Univ.* (*ser. V, geogr.*).

SOLNTSEV, N. A. 1961b. Some additions and refinements to the subject of landscape morphology. *Vest. mosk. gos. Univ.* (*ser. V, geogr.*).

SOZYKIN, N. F. 1939. Hydrological importance of forest litter and physical properties of forest soils. In: *Water regime in forests*, No. 18. Goslesbumizdat. Moscow.

SPIRIN, A. S. 1962. Nucleic acid and problems of protein biosynthesis. In: *Biological aspects of cybernetics.* Publ. Acad. Sci. U.S.S.R. Moscow.

STADNICHENKO, V. G. 1960. Soils of artificial forests in the steppe zone of the Ukrainian S.S.R. In: *Artificial forests in the steppe zone of the Ukraine.* Khar'kov.

STARIKOV, G. F., and STEPANOV, A. A. 1963. Reconstruction of the geographical landscapes of the Amur basin. *Vop. Geogr. Dal. Vost.*

STARK, V. N. 1926. Need for organization of scientific research work on forest entomology. *Zashch. Rast.*, **2**.

STARK, V. N. 1931. *Forest insect pests*. Sel'khozgiz. Moscow.

STARK, V. N. 1952. Bark beetles. *Fauna of U.S.S.R. Coleoptera*, vol. 31. Publ. Acad. Sci. U.S.S.R. Moscow.

STARK, V. N. 1961. Resistance of forest plantations to pest damage. In: *Protection of forests from pests and diseases*. Moscow.

STAROBOGATOV, YA. I. 1962. *Use of the information theory in ecology*. Thesis reports. Zool. Inst. of U.S.S.R. Acad. of Sci. Leningrad.

STEBAEV, I. V. 1958a. Role of soil-dwelling invertebrates (e.g. Tipulidae and Diptera larvae) in development of microflora in sub-Arctic soils. *Dokl. Akad. Nauk. SSSR*, **122**.

STEBAEV, I. V. 1958b. Animal population of first soils on rock surfaces, and its role in soil formation. *Zool. Zh.*, **32**.

STEBAEV, I. V. 1959. Soil-dwelling invertebrates of the Salekhard tundras and changes in their groupings as a result of cultivation. *Zool. Zh.*, **33**.

STEBAEV, I. V. 1962. Zoological characteristics of tundra soils. *Zool. Zh.*, **41**.

STEBAEV, I. V., and POLIVANOVA, E. N. 1959. Features of the damage to elms by elm leaf beetles and elm springtails in the environs of Volgograd. *Uchen. Zap. geogr. Fak. mosk. gos. Univ.*, **189**.

STEPANETS, I. T. 1962. Features of moisture consumption by birch and elm plantations on the dark-chestnut soils of the Ural region. *Pochvovedenie*.

STEPANOV, N. N. 1929. Chemical properties of forest litter as a basic factor in natural regeneration. *Trudy les. opyt. Delu*.

STEPANOV, N. N. 1940. Process of mineralisation of shed leaves and needles of trees and shrubs. *Pochvovedenie*.

SUDNITSYN, I. I. 1961. *Use of the thermodynamic potential method* (*moisture pressure*) *in the study of the movement of moisture in soils*. (Dissert. abstract.) Moscow.

SUDNITSYN, I. I., and TSEL'NIKER, YU. L. 1960. Relationship between pressure of soil moisture and suction power of tree leaves. *Dokl. Akad. Nauk SSSR*, **151**.

SUKACHEV, V. N. 1922. *Plant communities* (*Introduction to phytosociology*). 1st ed. Kniga. Petrograd & Moscow.

SUKACHEV, V. N. 1926. *Plant communities* (*Introduction to phytosociology*). 2nd ed. Moscow & Leningrad.

SUKACHEV, V. N. 1928. *Plant communities* (*Introduction to phytosociology*). 3rd ed. Moscow.

SUKACHEV, V. N. 1930. *Handbook on study of forest types*. 2nd ed. Moscow & Leningrad.

SUKACHEV, V. N. 1931. Forest types of the Buzuluksk pine forest. *Trudy buzuluk. Eksped*.

SUKACHEV, V. N. 1933. Experimental study of inter-biotype struggle for existence by plants. *Trudy petergof. biol. Inst.*, **15**.

SUKACHEV, V. N. 1940. Development of vegetation as an element of geographical environment in relation to development of a plant community. In: *Geographical environment in forest production*. Lesotekh. Akad. Leningrad.

SUKACHEV, V. N. 1941. Effect of the intensity of struggle for existence between plants on their development. *Dokl. Akad. Nauk SSSR*, **30**.

SUKACHEV, V. N. 1942. The concept of development in phytocoenology. *Sov. Bot.*

SUKACHEV, V. N. 1944. Principles of genetic classification in biogeocoenology. *Zh. obshch. Biol.*, **6**.

SUKACHEV, V. N. 1945. Biogeocoenology and phytocoenology. *Dokl. Akad. Nauk SSSR*, **47**.

SUKACHEV, V. N. 1946. Problem of struggle for existence in biocoenology. *Vest. leningr. gos. Univ.*

SUKACHEV, V. N. 1947. *Fundamentals of the theory of biogeocoenology*. Jubilee coll. of U.S.S.R. Ac. of Sc., commemorating the 30th anniversary of the Great October Socialist Revolution, I. Moscow & Leningrad.

SUKACHEV, V. N. 1948a. The Soviet trend in phytocoenology. *Vest. Akad. Nauk SSSR*.

Sukachev, V. N. 1948b. Phytocoenology, biogeocoenology, and geography. *Proc. of 2nd All-Union Geog. Conf.*, **1**. Geografgiz. Moscow.

Sukachev, V. N. 1949. Relation between the concepts of geographical landscape and biogeocoenose. *Vop. Geogr., coll.*, **16**.

Sukachev, V. N. 1950. Some basic problems of phytocoenology. In: *Botanical problems*, **1**. Publ. Acad. Sci. u.s.s.r. Moscow.

Sukachev, V. N. 1951. Fundamental principles of forest typology. *Proc. of Conf. on forest typology*. Publ. Acad. Sci. u.s.s.r. Moscow.

Sukachev, V. N. 1953. Intra- and interspecific relationships among plants. *Soobshch. Inst. Lesa*.

Sukachev, V. N. 1954a. Some general theoretical questions of phytocoenology. In: *Botanical questions*, **1**. Ed. for VIII International Bot. Congress in Paris.

Sukachev, V. N. 1954b. Forest types and their significance for forest economy. In: *Problems of sylviculture and forest management*. Ed. for IV World For. Congress. Publ. Acad. Sci. u.s.s.r. Moscow.

Sukachev, V. N. 1955. Forest biogeocoenology and its principal tasks. *Bot. Zh.*, **40**.

Sukachev, V. N. 1956a. Some modern problems of plant-cover study. *Bot. Zh.*, **41**.

Sukachev, V. N. 1956b. Intraspecific relationships in the plant world. *Byull. mosk. Obshch. Ispyt. Prir.*, **61**.

Sukachev, V. N. 1957. Development of forest typology in the u.s.s.r. during 40 years. In: *Scientific achievements in forestry in the U.S.S.R. during 40 years*. Goslesbumizdat. Moscow.

Sukachev, V. N. 1958. *Forest biogeocoenology and its significance in forest economy*. Publ. Acad. Sci. u.s.s.r. Moscow.

Sukachev, V. N. 1959a. Notes on the history of the origin and development of Soviet phytocoenology. *Annaly Biol.*

Sukachev, V. N. 1959b. New data on experimental study of the interrelationships of plants. *Byull. mosk. Obshch. Ispyt. Prir. (otd. biol.)*, **64**.

Sukachev, V. N. 1960a. Forest biogeocoenology as the theoretical foundation of forest management and forest economy. In: *Problems of sylviculture and forest management*. Contribution to V World For. Congress. Publ. Acad. Sci. u.s.s.r. Moscow.

Sukachev, V. N. 1960b. Relation between the concepts of biogeocoenose, ecosystem, and facies. *Pochvovedenie*.

Sukachev, V. N. 1961. Complex forest-biogeocoenotic investigations. *Trudy Lab. Lesov.*, **2**.

Sukachev, V. N., et al. 1934. *Dendrology, with fundamentals of forest geobotany*. Goslestekhizdat. Leningrad.

Sukachev, V. N., and Zonn, S. V. 1961. *Methodological instructions for study of forest types*. Publ. Acad. Sci. u.s.s.r. Moscow.

Sukachev, V. N., and Poplavskaya, G. I. 1927. Vegetation of the Crimea National Reserve. In: *The Crimea National Reserve, its nature, history, and significance*. Glavnauka. Moscow.

Sukharev, I. P. 1956. Water-regulating role of forest strips on the Kamennaya steppe. In: *Windbreak plantations*. Sel'khozgiz. Moscow.

Sushkina, N. N. 1931. Microbiology of forest soils in relation to the effect of fire upon them. In: *Forest Management Investigations*. Sel'khozgiz. Moscow.

Sushkina, N. N. 1933. Nitrification in forest soils in relation to tree stand composition, felling, and slash-clearing by fire. *Izv. Akad. Nauk SSSR*.

Sviridenko, P. A. 1940a. Diet of mouse-like rodents and its significance in the problem of forest regeneration. *Zool. Zh.*, **19**.

Sviridenko, P. A. 1940b. Diet of mouse-like rodents in conditions of natural and artificial forest regeneration. *Les. Khoz.*

Sviridenko, P. A. 1951. Significance of rodents in the problems of forest management and protection of nurseries and windbreaks from them. *Trudy Inst. Zool.*, Leningrad, **6**.

Symposium on complex study of artistic creation. Theses and annotations. 1963. Ed. by B. S. Meilakh. Leningrad.

Syroechkovskii, E. E. 1959. Role of animals in the formation of primitive soils in the conditions of the earth's polar region (in this case, the Antarctic). *Zool. Zh.*, **38**.

SZENT-GYÖRGYI, A. 1960. *Bioenergy.* Inostr. Lit. Moscow.

TAKHTADZHYAN, A. L. 1959. Evolution in terms of cybernetics and the general theory of tolerance. *Proc. of 2nd Conf. on application of mathematical methods in biology.* Izd. leningr. gos. Univ. Leningrad.

TALALAEV, E. V. 1956. Septicaemia in Siberian pine moth larvae. *Mikrobiologiya,* **25.**

TALIEV, V. I. 1930. *General diagnosis of plant diseases.* Sel'khozgiz. Moscow.

TAL'MAN, P. N. 1959. The moving force of intra- and interspecific relationships of plants and animals, and of species formation. *Trudy leningrad. lesotekhn. Akad.,* **90.**

TARABRIN, A. D. 1957. Effect of mycorrhiza on phosphorus absorption by oak seedlings. *Dokl. Akad. Nauk SSSR,* **112.**

TARABRIN, A. D. 1961. Absorption of P^{32} phosphorus by mycorrhiza and non-mycorrhiza fungi, depending on previous phosphorus nutrition. *Les. Zh.*

TARASOVA, D. A. 1962. Characteristics of the formation of breeding foci of trunk pests of Scots pine in the south-western part of Amur province. In: *Problems of zoological investigations in Siberia.* Knigoizdat. Gorno-Altaisk.

TELENGA, N. A. 1953. Role of insectivores in mass insect reproduction. *Zool. Zh.,* **32.**

TEPLYAKOVA, Z. F. 1952. Aerobic cellulose-decomposing micro-organisms in the soil of Kazakhstan. *Trudy Inst. Pochv.,* Alma-Ata, No. 1.

TEPLYAKOVA, Z. F. 1955. Factors governing the dynamics of cellulose-decomposing micro-organisms in the soil. *Izv. kazakh. Fil. Akad. Nauk SSSR (ser. biol.).*

TEREKHOV, O. S., and ENIKEEVA, M. G. 1964. *Comparative microbiological characteristics of some forest soils of Serebryanoborsk and Podushkinsk forests in Moscow province.* Nauka. Moscow.

TIKHOMIROV, B. A. 1959. *Interrelationships of the animal world and plant cover on the tundra.* Moscow & Leningrad.

TIMIRYAZEV, K. A. 1957. *Selected works,* vol. 1. Moscow.

TIMOFEEV, V. P. 1947. *Larch in plantations.* Publ. Acad. Sci. U.S.S.R. Moscow.

TIMOFEEV-RESOVSKII, N. V. 1958. Microevolution. Elementary phenomena, material, and factors in the microevolutionary process. *Bot. Zh. SSSR,* **43.**

TIMOFEEV-RESOVSKII, N. V. 1961. Some principles of the classification of biochorological units. *Trudy Inst. Biol.,* Sverdlovsk, **27.**

TIMOFEEV-RESOVSKII, N. V. 1962. *Some problems of radiational biogeocoenology.* (Doct. dissert.) Sverdlovsk.

TIMOFEEV-RESOVSKII, N. V., and ROMPE, R. R. 1959. Statistics and the amplifier principle in biology. *Problemy Kibern.*

TITOV, I. A. 1952. *Interactions between plant communities and environmental conditions.* Sovetskaya Nauka. Moscow.

TKACHENKO, M. E. 1939. *General forest management.* Goslestekhizdat. Moscow & Leningrad.

TKACHENKO, M. E. 1955. *General forest management.* 2nd ed. Goslesbumizdat. Moscow & Leningrad.

TOKIN, B. P. 1954. *Killers of microbes—phytoncides.* Goskul'tprosvetizdat. Moscow.

TOLMACHEV, A. I. 1954. *History of the origin and development of dark-conifer taiga.* Publ. Acad. Sci. U.S.S.R. Moscow & Leningrad.

TOL'SKII, A. P. 1938. Pine plantations in the Buzuluksk pine forest. *Trudy povolzhsk. lesotekhn. Inst.*

TOPCHIEV, A. G. 1960. Some data on the distribution of soil-dwelling invertebrates in the Ratsinsk forest area in Nikolaev province. *Nauch. Zap. dnepropetrovsk. gos. Univ.,* **62.**

TRAVLEEV, A. P. 1961. Role of St. Mark's fly (*Bibio hortulanus*) larvae in decomposition of forest litter on the steppes. *Nauch. Dokl. vyssh. Shk. (biol. nauka).*

TRIBUNSKAYA, A. YA. 1950. Use of *Azotobacter* fertilisation in raising tree seedlings. *Sb. lesorazv.*

TRIBUNSKAYA, A. YA. 1955. Study of rhizosphere microflora of pine seedlings. *Mikrobiologiya,* **24.**

TRUBETSKOVA, O. M. 1927. Effect of concentration of an extraneous solution on entrance of mineral matter into plants. *Izv. biol. nauchno-issled. Inst. biol. Sta. perm. gosud. Univ.,* **5.**

TRUBETSKOVA, O. M. 1935. Investigations on entrance of water and mineral matter into plants. *Uchen. Zap. mosk. gos. Univ.*

TRUBETSKOVA, O. M. 1962. Effect of temperature on active and passive absorption of water by plant root systems. Physiology of trees. In: *Vol. Jubil. L. A. Ivanov.* Publ. Acad. Sci. U.S.S.R. Moscow.

TRUBETSKOVA, O. M., MIKHALEVSKAYA, O. B., and NOVICHKOVA, I. D. 1955. Physiological role of tree mycorrhiza. *Proc. Conf. on plant mycotrophy.* Publ. Acad. Sci. U.S.S.R. Moscow.

TSEL'NIKER, YU. L. 1955a. Rate of water loss by isolated tree leaves and their resistance to dehydration. *Trudy Inst. Lesa,* Moscow, **27**.

TSEL'NIKER, YU. L. 1955b. Water regime of oak and red-leaved ash on the Derkul' steppe and effect of irrigation on it. *Trudy Inst. Lesa,* Moscow, **27**.

TSEL'NIKER, YU. L. 1956. Drought-resistance of tree plantations in steppe conditions. *Trudy Inst. Lesa,* Moscow, **30**.

TSEL'NIKER, YU. L. 1957. Effect of the moisture in ordinary chernozem on tree transpiration. *Pochvovedenie.*

TSEL'NIKER, YU. L. 1958a. Water regime of tree plantations on the steppes during their first years of life. *Trudy Inst. Lesa,* Moscow, **41**.

TSEL'NIKER, YU. L. 1958b. Indexes of the water regime of tree leaves in the steppe zone. *Trudy Inst. Lesa,* Moscow, **41**.

TSEL'NIKER, YU. L. 1959. Effect of improvement cutting on transpirational consumption of moisture by plantations in the steppe zone. *Soobshch. Inst. Lesa,* **11**.

TSEL'NIKER, YU. L. 1960a. Methods of adapting tree species to drought-resistance in steppe conditions. *Proc. Conf. on physiology of resistance in plants.* Publ. Acad. Sci. U.S.S.R. Moscow.

TSEL'NIKER, YU. L. 1960b. Growth of English oak and red-leaved ash seedlings in relation to phosphorus nutrition and irrigation. *Trudy Inst. Lesa,* Moscow, **52**.

TSEL'NIKER, YU. L., and MARKOVA, M. I. 1955. Transpiration of trees and shrubs in the conditions of the north-western part of the Caspian depression. *Trudy Inst. Lesa,* Moscow, **25**.

TSIOPKALO, V. L. 1940. Physiological characteristics of the gypsy moth (*Ocneria dispar* L.) in different stages of development, in relation to the species of its food plant. *Proc. Ecol. Conf. on Mass reproduction of animals and the forecasting of it.* Kiev.

TUMANOV, I. I. 1959. Present state of knowledge of, and urgent problems concerning, the physiology of cold-resistance in plants. *Proc. Conf. on physiology of resistance in plants.* Publ. Acad. Sci. U.S.S.R. Moscow.

TUMANOV, I. I. 1960. *Principal achievements of science in the study of frost-resistance in plants.* 11th Timiryazev lecture. Publ. Acad. Sci. U.S.S.R. Moscow & Leningrad.

TUMANOVA, D. F., and CHOCHIA, N. S. 1961. The place and significance of field investigations of landscape dynamics. *Vest. mosk. gos. Univ. (geogr.).*

TURUNDAEVSKAYA, T. M. 1963. Infestation of pine seedlings by trunk pests in locations of root fungus infection. In: *Problems of forest protection.* Moscow.

TVOROGOVA, A. S. 1959. Microflora of the upper soil horizons of hairgrass and willow-weed-burnt clearings. In: *Basic typology of clearings and its significance in forest economy.* Archangel.

TYURIN, I. V. 1930. Problem of genesis and classification of forest-steppe and forest soils. *Uchen. Zap. kazan. gos. Univ.,* **90**.

TYURIN, I. V. 1946. Quantitative proportion of living matter in the composition of the organic part of soils. *Pochvovedenie.*

TYURIN, I. V., and KONONOVA, M. M. 1962. *Biology of humus and problems of soil fertility.* Contribution to Intl. Conf. on Pedology in New Zealand. Izd. vses. Obschch. Pochv. Moscow.

TYURLYUN, I. A. 1958. Nature of the process of gas and vapour movement in soil. *Pochvovedenie.*

USPENSKAYA, L. I. 1929. Problem of the effect of intensity of competition among plants on their development. *Zap. leningr. sel.-khoz. Inst.,* **5**.

USPENSKII, V. A. 1956. *Carbon balance in the biosphere in relation to the question of carbon distribution in the earth's crust.* Gostoptekhizdat. Moscow.

UTENKOVA, A. P. 1956. Process of mineralization of leaves of oak and accompanying species. *Trudy voronezh. gos. Univ. (vyp. pochvenno-bot.),* **36**.

VAKIN, A. T. 1954. Phytopathological state of oak stands in the Tellerman forest. *Trudy Inst. Lesa,* Moscow, **16**.

VANIN, S. I. 1948. *Forest phytopathology*. 2nd ed. 1955. Goslestekhizdat. Moscow.

VARMING, E. 1901. *Ecological geography of plants*. (Introduction to study of plant communities.) Moscow.

VASIL'EV, I. S. 1950. Water regime of podzolic soils. *Trudy pochv. Inst.*, **32**.

VASIL'EV, N. G., and KOLESNIKOV, B. P. 1962. *The black-fir-broad-leaved forests of the southern Maritime region*. Publ Acad. Sci. U.S.S.R. Moscow & Leningrad.

VASIL'EV, YA. YA. 1935. Scope of the concept 'forest type' and scheme of classification of forest types. *Sov. Bot.*

VASSOEVICH, N. B. 1963. Living matter and petroleum production. *Contribution to scientific session celebrating the 100th anniversary of the birth of Acad. V. I. Vernadskii.* Izd. geogr. Obshch. SSSR. Leningrad.

VERNADSKII, V. I. 1927. *The biosphere*. Moscow & Leningrad.

VERNADSKII, V. I. 1934a. *Biochemical outlines*. Moscow.

VERNADSKII, V. I. 1934b. *Problems of biochemistry*, part 1. Importance of biochemistry for study of the biosphere. Publ. Acad. Sci. U.S.S.R. Leningrad.

VERNADSKII, V. I. 1940. *Biochemical outlines*. Moscow & Leningrad.

VERNADSKII, V. I. 1944. Some words on the ionosphere. *Usp. sovrem. Biol.*, **18**.

VIKTOROV, G. A. 1960. The biocoenose and problems of insect numbers. *Zh. obshch. Biol.*, **21**.

VIKTOROV, G. A. 1962. Content of the concept 'biocoenose' in modern ecology. *Problems of ecology*, IV. Izd. kievsk. Univ. Kiev.

VILENKIN, B. YA. 1963. Some new aspects of the study of communities of aquatic organisms. (Thesis reports.) Sevastopol biol. Sta.

VILLI (VILLEE), K. 1964. *Biology*. (Translated from English.) Mir. Moscow.

VIL'YAMS (WILLIAMS), V. R. 1936. *Soil management*. Sel'khozgiz. Moscow.

VINBERG, G. G. 1956. *Intensity of metabolism and nutritive requirements of fish*. Izd. beloruss. Univ. Minsk.

VINBERG, G. G. 1960. *Primary products of lakes*. Minsk.

VINBERG, G. G. 1962. Energy principle of study of trophic links and productivity of ecosystems. *Zool. Zh.*, **41**.

VINER, N. 1958. *Cybernetics, or control and communication in animal and machine*. Moscow.

VINER, N., and ROSENBLYUT, A. 1961. Conduction of impulses in cardiac muscle. Mathematical formulation of conduction of impulses in a network of connected stimulable elements, in particular cardiac muscle. *Cybern. symposium*, **3**. Inostr. Lit. Moscow.

VINNIK, M. A. 1961. Biological accumulation of microelements in soils beneath forest canopy. *Trudy voronezh. gos. Zapov.*, **13**.

VINOGRADSKII, S. N. 1875. Use of free atmospheric nitrogen by microbes. *Arkh. biol. nauk*, **3**.

VINOGRADSKII, S. N. 1892. Morphology of organisms that form saltpetre in the soil. *Archs. Sci. biol. St. Peterb. (Arkh. biol. Nauk)*, **1**.

VINOGRADSKII, S. N. 1952. *Microbiology of the soil*. Publ. Acad. Sci. U.S.S.R. Moscow.

VINOKUROV, M. A., DAUTOV, R. K., and KOLOSKOVA, A. V. 1959. *Effect of windbreaks on the soil*. Tatarskoye knizh. Izd. Kazan.

VIZIR, A. P. 1956. Toxicity of products leached from forest litter as a factor suppressing the development of *Azotobacter* in forest soils. *Proc. of Conf. on For. Pedology of For. Inst. of Ukr. S.S.R. Ac. of Sc.* Kiev.

VLASOV, A. A. 1955. Significance of tree mycorrhiza and means of stimulating it. *Tr. of Conf. on plant mycotrophy*. Moscow.

VOEIKOV, A. I. 1957. Effect of vegetation on climate. *Selected works*. Gidrometeoizdat. Moscow.

VOLCHANETSKII, I. B. 1950. Formation of the bird and mammal fauna of young windbreaks in arid districts in left-bank Ukraine. *Ucheni Zap. kharkyiv. derzh. Univ.*, **44** (*Trudy nauchno-issled. Inst. Biol. khar'kov. gos. Univ.*, **16**).

VOLOBUEV, V. R. 1953. *Soils and climate*. Baku.

VOROB'EV, D. V. 1959. *Method of typological investigations*. Khar'kov.

VOROB'EVA, M., and SHCHEPETIL'NIKOVA, A. 1936. Clover sickness and partial sterilisation. *Khimiz. sots. Zeml.*

VORONIN, M. S. 1961. *Selected works*. Izd. sel.-khoz. Lit. Zh. Plak. Moscow.

VORONOV, A. G. 1959. Interrelationships of animals and plants in different geographical zones. *Uchen. Zap. mosk. gos. Univ. (biogeogr.)*, **189.**

VORONOV, A. G. 1960. Biocoenological observations in the subtropical forests of Yunnan (Chinese People's Republic). *Trudy mosk. Obshch. Ispyt. Prir. (otd. biol.).*

VORONOV, N. P. 1953. Observations on the digging activities of rodents in the forest. *Pochvovedenie.*

VORONOV, N. P. 1958. Effect of the digging activities of mammals on the life of the forest. *Izv. kazan. Fil. Akad. Nauk SSSR (ser. biol.).*

VORONTSOV, A. I. 1955. Little-known lepidopterous pests of young tree plantations and nurseries. *Nauchno-tekh. Inf.*

VORONTSOV, A. I. 1960. *Biological basis of forest protection.* Moscow.

VORONTSOV, N. N. 1956. Food supplies for the northern red-backed vole. *Byull. mosk. Obshch. Ispyt. Prir. (otd. biol.).*

VOROSHILOV, V. N. 1960. *Development rhythm in plants.* Moscow.

VOSKRESENSKAYA, N. P. 1962. Dependence of oxygen absorption by green and non-green leaves on the intensity and spectral composition of light. Physiology of trees. In: *Vol. Jubil. L. A. Ivanov.* Publ. Acad. Sci. U.S.S.R. Moscow.

VOTCHAL, E. F., and KENUKH, A. M. 1928. Transpiration coefficient of assimilation. *Proc. of All-Union Conf. of botanists.* Leningrad.

VRUBLEVSKII, K. I. 1912. Theoretical differentiation of some ungulates into twig-eaters and herbivores, and its practical significance. *Arkh. vet. Nauk*, **8.**

VYSOTSKII, G. N. 1898. The earthworm. *Zeml. Gaz.*, **42-44.**

VYSOTSKII, G. N. 1912a. Forest culture in steppe experimental forests from 1898 to 1907. *Trudy les. opyt. Delu Ross.*, **16.**

VYSOTSKII, G. N. 1912b. Reasons for desiccation of tree plantations on steppe chernozem. *Trudy les. opyt. Delu Ross.*, **16.**

VYSOTSKII, G. N. 1915. Ergenya. A cultural-phytological sketch. *Trudy Byuro prikl. Bot.*, **10-11.**

VYSOTSKII, G. N. 1925. Cover management. *Zap. belorussk. gos. Inst. sel. les. Khoz.*

VYSOTSKII, G. N. 1928. The forest (problems of its study). In: *Natural productive resources of the Ukrainian S.S.R.* Khar'kov.

VYSOTSKII, G. N. 1929. Reminder to steppe foresters about mycorrhiza. *Les. Khoz.*

VYSOTSKII, G. N. 1938. *Hydrological and meteorological effects of forests.* Moscow.

VYSOTSKII, G. N. 1950. *Study of the effect of a forest in altering the area occupied by it and the surrounding area.* (Study of forest significance.) 2nd ed., revised. Goslesbumizdat. Moscow & Leningrad.

VYSOTSKII, G. N. 1952. *Hydrological and meteorological effects of forests.* Moscow.

VZNUZDAEV, N. A., and KARPACHEVSKII, L. O. 1961. Description of the hydrophysical features and water regime of forest soils of the central part of the Kamchatka river valley. *Pochvovedenie.*

YABLONSKII, S. V. 1959. Basic concepts of cybernetics. *Problemy Kibern.*

YACHEVSKII, A. A. 1933. *Fundamentals of mycology.* Moscow.

YAKHONTOV, I. A. 1909. Development of pine seedlings beneath the canopy of old stands. *Trudy les. opyt. Delu Ross.*, **20.**

YAROSHENKO, G. D. 1929. *Pine and oak in Armenia.* Erevan.

YAROSHENKO, G. D. 1953. *Fundamentals of plant-cover study.* 2nd ed. Geografgiz. Moscow.

YAROSHENKO, G. D. 1961. *Geobotany.* Publ. Acad. Sci. U.S.S.R. Moscow & Leningrad.

YUNASH, G. G. 1940. Regeneration of oak in the Shipov forest. *Les. Khoz.*

YUNOVIDOV, A. P. 1951. Growth of root systems in the forest. *Agrobiologiya.*

YURGENSON, P. B. 1959. Ungulate population density and its normalisation. *Soobshch. Inst. Lesa*, **13.**

YURINA, E. V. 1957. Photosynthesis by trees in adequate and inadequate moisture conditions. *Fiziologiya Rast.*, **4.**

YURINA, E. V., and ZHMUR, D. G. 1962. Changes in physiological radiation in tree stands of different density and composition. In: *Physiology of trees.* Publ. Acad. Sci. U.S.S.R. Moscow.

ZABELIN, I. M. 1960. Physical geography, astrogeography, and evolutionary study of the world. *Philosophical problems of nature knowledge.* III. *Geologo-geographic studies.* Izd. mosk. gos. Univ. Moscow.

ZABELIN, I. M. 1963. *Physical geography and science of the future.* Moscow.

ZABLOTSKAYA, L. V. 1957a. Scattering of coniferous and linden seeds by shrews. *Trudy priokso-terras. Zapov.*

ZABLOTSKAYA, L. V. 1957b. Nutrition and natural food of bison. *Trudy priokso-terras. Zapov.*

ZALENSKII, O. V. 1940. Distribution and ecological features of pistachio (*Pistacea vera*) and almond (*Amygdalus communis*) in Western Kopet-Dag. *Bot. Zh. SSSR,* **25.**

ZALENSKII, O. V. 1954. Photosynthesis by plants in natural conditions. *Vop. Bot.*

ZALENSKII, O. V. 1956. Ecologo-physiological study of productivity factors of wild perennial plants. In: *Vol. Jubil. V. N. Sukachev.* Publ. Acad. Sci. U.S.S.R. Moscow.

ZALENSKII, O. V. 1957. Relationship between photosynthesis and respiration. *Bot. Zh. SSSR,* **42.**

ZGUROVSKAYA, L. N. 1958. Anatomo-physiological investigations of tree roots. *Trudy Inst. Lesa,* Moscow, **41.**

ZGUROVSKAYA, L. N. 1961. Structure and growth of pine root systems on mountain slopes. *Trudy Inst. Lesa Drev. sib. Otd.,* **50.**

ZGUROVSKAYA, L. N. 1962. Permeable cells in suctorial roots existing in a state of winter or summer quiescence. Physiology of trees. In: *Vol. Jubil. L. A. Ivanov.* Publ., Acad. Sci. U.S.S.R. Moscow.

ZHARKOV, I. V. 1938. Material on study of forest mice in the forests of the Caucasus reserve. *Trudy kavk. Zapov.*

ZHDANNIKOVA, E. N., and POPOVA, ZH. P. 1961. Effect of drainage on the microflora of forest soils. *Trudy Inst. Lesa Drev. sib. Otd.,* **50.**

ZHOKHOV, P. I., GRECHKIN, V. P., KOLOMIETS, N. G., VYSOTSKAYA, A. V., and LONSHCHAKOV, S. S. 1961. *Siberian pine moth and measures for its control.* Moscow & Leningrad.

ZHOLKEVICH, V. N. 1961. Energy balance in respiration of plant tissues in different water-supply conditions. *Fiziologiya Rast.,* **8.**

ZHUKOVA, R. A. 1959. Toxicity of Kola peninsula soils for aerobic cellulose bacteria. *Mikrobiologiya,* **28.**

ZHUKOVA, R. A. 1960. Role of the biological factor in toxicity of Kola peninsula soils for aerobic cellulose bacteria. *Mikrobiologiya,* **29.**

ZHURAVLEVA, M. V. 1953. *Absorption of mineral nutrients by tree seedlings.* (Dissert. abstract.) Publ. Acad. Sci. U.S.S.R. Moscow.

ZINOV'EV, G. A. 1958. Structure, dynamics, and typology of bark beetle breeding foci. *Zool. Zh.,* **37.**

ZINOV'EV, G. A. 1959. *Significance of the complex of insectivores in restricting reproduction of bark beetles in coniferous forests.* N. A. Kholodkovskii memorial lectures, 1956-7. Publ. Acad. Sci. U.S.S.R. Moscow & Leningrad.

ZINOV'EVA, L. A. 1955. Soil fauna in different types of forest in the Belorussian Poles'e. *Zool. Zh.,* **34.**

ZONN, S. V. 1950a. *Mountain-forest soils of the North-west Caucasus.* Publ. Acad. Sci. U.S.S.R. Moscow.

ZONN, S. V. 1950b. Geomorphological and soil conditions of forest growth in the Tellerman experimental forest. *Trudy Inst. Lesa,* Moscow, **3.**

ZONN, S. V. 1951. Water regime of oak-forest soils. *Trudy Inst. Lesa,* Moscow, **7.**

ZONN, S. V. 1954a. *Effect of forest on soil.* Publ. Acad. Sci. U.S.S.R. Moscow.

ZONN, S. V. 1954b. Material on study of the water regime of chernozems under tree plantations. *Trudy Inst. Lesa,* Moscow, **15.**

ZONN, S. V. 1959. *Soil moisture and tree plantations.* Publ. Acad. Sci. U.S.S.R. Moscow.

ZONN, S. V. 1960. Interactions of forest vegetation with soils in the light of new biogeocoenotic investigations. *Trudy mosk. Obshch. Ispyt. Prir.,* **3.**

ZONN, S. V. 1961. *Mountain-forest soils of coniferous and beech forests in Bulgaria.* Bulgarian Acad. Sci. Sofia.

ZONN, S. V. 1964. *High-mountain forest soils of Eastern Tibet.* Nauka. Moscow.

ZONN, S. V., and ALESHINA, A. K. 1953. Decomposition of oak-forest litter and interactions of its mineral elements with soils. *Trudy Inst. Lesa,* Moscow, **12.**

ZONN, S. V., GEORGIEV, A., NAUMOV, Z., and STOYANOV, ZH. 1958. Some problems of the development of mountain-forest soils under spruce forests in Western Rhodope. *Izv. Inst. Pochvov.*, book V.

ZONN, S. V., KARPACHEVSKII, L. O., and STEFIN, V. V. 1963. *Volcanic forest soils of Kamchatka.* Publ. Acad. Sci. U.S.S.R. Moscow.

ZONN, S. V., and KUZ'MINA, E. A. 1960. Effect of coniferous and deciduous species on the physical properties and water regime of leached chernozems. *Trudy Lab. Lesov.*, 1.

ZONN, S. V., and LI CHEN-KVEI. 1960. Characteristics of biological energy processes in tropical forest soils. *Pochvovedenie.*

ZONN, S. V., and LI CHEN-KVEI. 1962. Dynamics of the decomposition of litter and seasonal changes in its mineral composition in two types of tropical forest. *Soobshch. Lab. Lesov.*

ZRAZHEVSKII, A. I. 1957. *Earthworms as a factor in the fertility of forest soils.* Publ. Acad. Sci. Ukr. S.S.R. Kiev.

ZVYAGINTSEV, D. G. 1959. Activity of bacteria adsorbed by soil particles. *Mikrobiologiya*, **28**.

Non-Russian List

Aaltonen, V. T. 1926. On the space arrangement of trees and root competition. *J. For.*, **24**.

Aaltonen, V. T. 1942. Einige Vegetationsversuche mit Baumpflanzen. *Acta for. fenn.*, **50**.

Aaltonen, V. T. 1948. *Boden und Wald*. Berlin and Hamburg.

Achby, K. R. 1959. Prevention of regeneration of woodland by field mice and voles. *Q. J. For.*, No. 3.

Adlung, K. G. 1957. Zur Anlockung des Schwammspinners (*Lymantria dispar* L.) an seine Wirtspflanzen. *Z. angew. Zool.*, **44**.

Aichinger, E. 1933. *Vegetationskunde der Karawanken*. Jena.

Aichinger, E. 1951. Vegetationsentwicklungstypen als Grundlage unserer land- und forstwirtschaftlichen Arbeit. *Angew. Pfl. Soziol.*, **1**.

Aichinger, E. 1960. Können wir eine gemeinsame Plattform für die verschiedenen Schulen in der Waldtypenklassifikation finden? Forest types and forest ecosystems during the IX Internat. Bot. Congr. Montreal, 1959. *Silva fenn.*, **105**.

Al-Asawi, A. F., and Norris, D. M. 1959. Experimental prevention of bark beetle transmission of *Ceratocystis ulmi* (Buis.) Moreao with the systemic insecticide Chipman R-6199. *J. econ. Ent.*, **52**.

Allee, W., Park, O., et al. 1949. *Principles of animal ecology*. Philadelphia and London.

Anderson, J. M., and Fisher, K. G. 1956. Repellency and host specificity in the white pine weevil. *Physiol. Zool.*, **29**.

Anstett, A. 1951. Sur l'activation microbiologique des phénomènes d'humification. *C. r. hebd. Séanc. Acad. Agric. Fr.*, **37**.

Arens, K. 1934. Die kutikuläre Exkretion des Laubblätters. *J. wiss. Bot.*, **80**.

Arnborg, T. 1960. Can we find a common platform for different schools of forest type classification? *Silva fenn.*, **105**.

Arthur, A. P. 1962. Influence of host tree on abundance of *Itoplectis conquisitor* . . . *Can. Ent.*, **94**.

Atger, P., and Chastang, S. 1961. Conservations du pouvoir infectieux des polyèdres cytoplasmiques de *Thaumatopoea pityocampa*, Schiff. *Phytiat.-Phytopharm.*, **10**.

Auer, Ch. 1961. Ergebnisse zwölfjähriger quantitativer Untersuchungen des Populationsbewegung des grauen Lärchenwicklers *Zeiraphera griseana* Hübner (*diniana* Guenée) im Oberengadin (1959/1960). *Mitt. schweiz. Anst. forstl. Verswes.*, **37**.

Balch, R. E. 1958. Control of forest insects. *Rev. Ent.*, **3**.

Balogh, J. 1959. Über die Bedeutung der Collembolen und Milben in der Zoozönose der ungarischen Waldtypen. *Zentbl. Bakt. Parasitkde, Abt.* 2, **112**.

Baltensweiler, W. 1962. Die zyklischen Massenvermehrungen des grauen Lärchenwicklers (*Zeiraphera griseana* Hb.) in den Alpen. *XI Int. Congr. Ent.*, 1960, **2**.

Barley, K. P., and Jennings, A. C. 1959. Earthworms and soil fertility. III. The influence of earthworms on the availability of nitrogen. *Aust. J. agric. Res.*, **10**.

Barney, C. W. 1947. *A study of some factors affecting root growth of loblolly pine trees*. D. Duke Univ. School of Forestry.

Bartlova, D., Kozderkova, V., and Venclikova, E. 1955. Vplyv Kotlikového hospodarstvi na zmenu pudni mikroflory. *Sb. čsl. Akad. Věd., Rada C*, **28**.

Bautz, E. 1953. Einwirkung verschiedener Bodentypen und Bodenextrakte auf die Keimung von *Picea excelsa*. *Z. Bot.*, **41**.

Becker, Gyillemat. 1951. Sur les toxines racinaires des sols incultes. *C. r. hebd. Séanc. Acad. Sci.*, Paris, **232**.

Bennett, Wm. H. 1958. *The Texas leaf-cutting ant*. Forest Pest Leafl. U.S. Dept. Agric., No. 23.

Benz, G. 1962. Untersuchungen über die Pathogenität eines Granulosis-Virus des grauen Lärchenwicklers—*Zeiraphera diniana*. *Agron. Glasn.*, **12**.

Bergold, G. H. 1953. Insect viruses. In: *Advances in Virus Research*, vol. 1. N.Y.

Bernat, I., and Novotna, V. 1955. Vplyv presvetlenia porastu. *Sb. čsl. Akad. Věd., Rada C*, **28**.

Bernat, I. 1955. Vplyv smrekovédo porastu a porastu cerveneho smreka na mikrobiologické pomery v pôde. *Sb. čsl. Akad. věd. Rada C*, **28**.

BEVAN, D. 1958. Some aspects of forest entomology. *Biology Hum. Affairs*, **24**.

BIRCH, CLARK. 1953. Forest soil as an ecological community with special reference to the fauna. *Q. Rev. Biol.*, **28**.

BJEGOVIC, S. 1963. Laboratorijsko ispituvanje redukcione moci i nekih bioloskih osobina jajnih parazita gubara. *Arh. poljopr. nauke*, **16**.

BJORKMAN, E. 1942. Über die Bedingungen der Mykorrhizaabbildung die Kiefer und Fichte. *Symb. bot. upsal.*, **6**.

BJORKMAN, E. 1944. Forest planting and soil biology. *Svenska Skogsvför. Tidskr.*, **42**.

BJORKMAN, E. 1944-1945. On the influence of light on the height-growth of pine plants on pine-heaths in Norrland. *Meddn. St. Skogsförsanst.*, **34**.

BJORKMAN, E. 1956. Über die Natur der Mycorrhizabbildung unter besonderer Berücksichtigung der Waldbäume und die Anwendung in der forstlichen Praxis. *Forstwiss. Zentbl.*, **75**.

BLACKMAN, G. E., and BLACK, J. N. 1959. Physiological and ecological studies in the analysis of plant environment, II. The role of the light factor in limiting growth. *Ann. Bot.*, **23**.

BLACKMANN, F. 1905. Optima and limiting factors. *Ann. Bot.*, **19**.

BLANCKMEISTER, J., KIENITZ, E., *et al.* 1963. *Der Wald und die Forstwirtschaft*. Berlin.

BODE, H. R. 1940. Über die Blattausscheidungen des Wermuts und ihre Wirkungen auf anderer Pflanzen. *Planta*, **30**.

BODENHEIMER, F. S. 1938. *Problems of animal biology*. Oxford Univ. Press.

BOND, I. 1941. Symbiosis of leguminous plants and nodule bacteria. *Ann. Bot.*, No. 3.

BONNER, J. 1950. The role of toxic substances in the interactions of higher plants. *Bot. Rev.*, **16**.

BONNER, J., and GALSTON, A. W. 1944. Toxic substances from the culture media of guayule which may inhibit growth. *Bot. Gaz.*, **106**.

BORNEBUSCH, C. H. 1930. The fauna of forest soil. *Forst. ForsVaes. Danm.*, **11**.

BORNEBUSCH, C. 1932. Tierleben der Waldböden. *Forstwiss. Zentbl.*, **54**.

BÖRNER, H. 1955. Untersuchungen über phenolische Verbindungen aus Getreidestroh und Getreiderückstanden. *Naturwissenschaften*, **42**.

BÖRNER, H. 1958. Untersuchungen über den Abbau von Phlorizin in Boden. Ein Beitrag zur Problem der Bodenmüdigkeit bei Obstgehölzen. *Naturwissenschaften*, **45**.

BÖRNER, H. 1958. Experimentelle Untersuchungen zum Problem der gegenseitigen Beeinflussung von Kulturpflanzen und Unkräutern. *Biol. Zbl.*, **77**.

BÖRNER, H. 1959. The apple replant problem. I. The excretion of phlorizin from appleroot residues. *Contr. Boyce Thomson Inst.*, *Pl. Res.*, **20**.

BÖRNER, H. 1960a. Über die Bedeutung gegenseitiger Beeinflussung von Pflanzen in landwirtschaftlichen und forstlichen Kulturen. *Angew. Bot.*, **34**.

BÖRNER, H. 1960b. Experimentelle Untersuchungen über die Bodenmüdigkeit beim Apfel (*Pyrus malus*, L.). *Beitr. Biol. Pfl.*, **36**.

BÖRNER, H. 1962. Fragen einer gegenseitigen Beeinflussung höherer Pflanzen. *Naturw. Rdsch.*, Stuttgart, **14**.

BÖRNER, H., MARTIN, P., CLAUSS, H., and RADEMACHER, B. 1959. Experimentelle Untersuchungen zum Problem der Bodenmüdigkeit am Beispiele von Zein und Roggen. *Z. PflKrankh. PflSchutz*, **66**. H. 11/12.

BOSWELL, I. G. 1956. Decay of plant litter. *Nature*, **178**.

BOVEY, P., and MAKSYMOV, J. K. 1959. Le problème des races biologiques chez la Tordeuse grise du Mélèze *Zeiraphera griseana* (Hb.). Note prelim. *Vischr. naturf. Ges. Zürich*, **104**.

BOYSEN-JENSEN, P. 1932. *Die Stoffproduktion der Pflanzen*. Jena.

BRAUN-BLANQUET, J. 1951. *Pflanzensoziologie*. 2nd ed. Wien.

BRAUNS, A. 1953. Ist die Förderung ökologischer Untersuchungen 'indifferenter' und 'wirtschaftlich kaum beachtenswerter' Insektenarten wirtschaftlich vertretbar? *Anz. Schädlingsk.*, **26**.

BRINKMAN, M. 1956. Von der Begleitlebewelt in Reiheswäldern. *Orn. Mitt.*, Göttingen, **8**.

BRONSART, H. 1949. Der heutige Stand unseres Wissens von der Bodenmüdigkeit. *Z. PflErnähr. Düng., Bodenk.*, **45**.

BROWN, C. E. 1963. Damage to spruce by the spruce budworm. *Rep. Dep. For. Can.*, **19**.

BRUNS, H. 1955. Fortschritte im forstlichen Vogelschutz. *Anz. Schädlingsk.*, 28.

BRÜSEWITZ, G. 1959. Untersuchungen über den Einfluss des Regenwurms auf Zahl, Art und Leistungen von Mikroorganismen im Boden. *Naturwissenschaften*, 46.

BUBLITZ, W. 1952. Über organische Hemstoffe und ihre Wirkung auf Rohhumus. *Forstarchiv*, 30.

BUBLITZ, W. 1953. Über die keimhemmende Wirkung der Fichtenstreu. *Naturwissenschaften*, 40.

BUBLITZ, W. 1954. Über keimhemmende und antibacterielle Substanzen im Bodenwasser der Fichtenstreu. *Naturwissenschaften*, 42.

BUCHER, G. E. 1953. Biotic factors of control of the European Fir Budworm, *Choristoneura murinana* (Hbn.) (N. Comb.), in Europe. *Can. J. agric. Sci.*, 33.

BÜTTNER, H. 1956. Die Beeinträchtigung von Raupen einiger Forstchädlinge durch mineralische Düngung der Futterpflanzen. *Naturwissenschaften*, 43.

BÜTTNER, H. 1959. Über die Auswirkung von Düngenmassnahmen auf forstliche Schadinsekten. *Naturwissenschaften*, 46.

BÜTTNER, H. 1961. Der Einfluss von Düngemass nehmen auf forstliche Schadinsekten durch Veränderung der Nahrungsqualität ihrer Wirtspflanzen. *Meded. Landb-Hoogesch., Gent.*, 26.

CAJANDER, A. K. 1925. Der gegenseitige Kampf in der Pflanzenwelt. *Veröff. geobot. Inst.*, Zürich, 3.

CALLAHAM, R. Z. 1955. Sapwood moisture associated with galleries of *Dendroctonus valens. J. For.*, 53.

CAPEK, M. 1962a. Über den Einfluss des Kahlfrasses von *Zeiraphera diniana* Guen. auf den jährlichen Zuwachs der Fichte. *Schweiz. Z. Forstwes.*, 113.

CAPEK, M. 1962b. Synparasitische Beziehungen der Tannenlepidopteren in der Slowakei. *XI Int. Congr. Ent.*, 1960, 2.

CARTER, W. 1961. Ecological aspects of plant virus transmission. *Ann. Rev. Ent.*, 6.

CHARARAS, C. 1959a. Relations entre la pression osmotique des conifères et leur attaque par les Scolytidae. *Revue Path. vég. Ent. agric. Fr.*, 38.

CHARARAS, C. 1959b. Précisions sur l'efficacité des arbres-pièges en fonction des particularités biologique des scolitides. *Revue for. fr.*, No. 8-9.

CHARARAS, C. 1960a. L'attractivité exercée par *Fraxinus excelsior* L. à l'égard de *Leperesinus fraxini* Panz. (Coléoptères Scolytidae) et les modifications physiologiques de la plante-hôte. *C. r. hebd. Séanc. Acad. Sci.*, Paris, 250.

CHARARAS, C. 1960b. Recherchs sur la biologie de *Pityogenes chalcographus* L. *Schweiz. Z. Forstwes.*, 111.

CHARARAS, C. 1961. Recherches sur la spécifité de *Xyloterus lineatus* Ol. *C. r. hebd. Séanc. Acad. Sci.*, Paris, 252.

CHARARAS, C. 1962. Relations entre les variations de la pression osmotique des onifères et l'extension des coléoptères Scolytidae. *XI Int. Congr. Ent.*, 1960, 2.

CHARARAS, C., and BERTON, A. 1961. Nouvelle méthode d'analyse des exhalaisons terpénique de *Pinus maritima* et comportements de *Blastophagus piniperda. Revue Path. vég. Ent. agric. Fr.*, 40.

CHARARAS, C., and DECHAMPS, P. 1962. Le chimiotropisme chez les Scolytidae et le rôle des substances terpéniques. *XI Int. Congr. Ent.*, 1960, B. 2.

CHARARAS, C., DESVEAUX, R., and KOGANE-CHARLES, M. 1960. Relations entre l'installation des coléoptères Scolytidae et le teneur des conifères en glucides solubles et en acides organiques hydrosolubles libres. *C. r. hebd. Séanc. Acad. Sci.*, 250.

CHASE, F. E., and BARKER, C. A. 1954. A comparison of microbial activity in an Ontario forest soil under pine, hemlock and maple cover. *Can. J. Microbiol.*, 1.

CHILSON, E. 1955. Economical aspect of livestock-big game relationships as viewed by a livestock producer. *J. Range Manag.*, 4.

CLARK, R. C., and BROWN, N. R. 1962. Studies of predators of the balsam woolly aphid, *Adelges piceae* (Ratz.) (Homoptera-Adelgidae). XI. *Cremifania nigrocellulata* Cz. (Diptera-Chamaemyiidae), an introduced predator in Eastern Canada. *Can. Ent.*, 94.

CLARKE, G. L. 1954. *Elements of ecology.* N.Y., London.

CLEMENTS, F. E. 1916. Plant succession: an analysis of the developments of vegetation. *Carnegie Inst. Publ.*, 242.

CLEMENTS, F. E. 1936. Nature and structure of the climax. *J. Ecol.*, 24.

CLEMENTS, F. E., WEAVER, J. E., and HANSON, H. C. 1929. Plant competition. *Carnegie Inst. Publ.*, **398**.

COALDRAKE, J. E. 1961. The ecosystem of the coastal lowlands ('Wallum') of Southern Queensland. *Bull. Commonw. scient. ind. Res. Org.*, **283**.

COLLINS, S. 1961. Benefits to understory from canopy defoliation by gypsy moth larvae. *Ecology*, **42**.

COLLISON, R. S. 1935. Lysimeter investigations. V. Water movement, soil temperatures and root activity under apple-trees. *Agric. Extl. Techn. Bull.*, **237**.

COOK, F. D., and LOCHHEAD, A. G. 1959. Growth factor relationships of soil microorganisms as effected by proximity to the plant root. *Can. J. Microbiol.*, **5**.

CORBET, A. S. 1934. The bacterial numbers in the soil of the Malay peninsula. *Soil Sci.*, **38**.

COURTOIS, J. E., CHARARAS, C., and CHARITOS, N. 1960. Recherches sur les possibilités d'attaque de *Pseudotsuga douglasii* par *Ips typographus* L. *C. r. hebd. Séanc. Acad. Sci.*, **251**.

CRAIB, I. J. 1929. Some aspects of soil moisture in forests. *Bull. Yale Univ. School Forest.*, **25**.

CRIDER, P. S. 1945. Winter root growth of plants. *Science*, **68**.

CRINGAN, A. 1958. Influence of forest fires and fire protection on wildlife. *For. Chron.*, **34**.

CZAPEK, FR. 1924. *Biochemie der Pflanzen* vols. 1-3. 3rd ed. Jena.

DANSEREAU, P. 1960. A combined structural and floristic approach to the definition of forest ecosystem. *Silva fenn.*, No. 105.

DAXER, H. 1934. Über der Assimilationsökologie der Waldbodenflora. *Jb. wiss. Bot.*, **80**.

DAY, G. M. 1950. Influence of earthworms on soil micro-organisms. *Soil Sci.*, **69**.

DE CANDOLLE, A. P. 1832. *Physiologie végétale*, v. III. Paris.

DE CANDOLLE, A. P. 1913. *Théorie élémentaire de la botanique*. Paris.

DE LEON, D. 1954. Is host condition the cause of insect outbreaks? *J. For.*, **52**.

DELFS SORGEN. 1955. *Die Niederschlagszurückhaltung im Walde*. Koblenz.

DENGLER, A. 1935. *Waldbau auf ökologischer Grundlage*. 2nd ed. Berlin.

DOBBS, C. G., and HINSON, W. 1953. A widespread fungistasis in soils. *Nature*, **172**.

DRIFT, J. 1949. De bodemfauna in onze bossen. *Ned. Boschb. Tijdschr.*, **21**.

DRIFT, J. 1951. Analysis of the animal community in a beach forest floor. *Tijdschr. Ent.* **94**.

DRIFT, J. 1959. Field studies on the surface fauna of forests. *Meded. Inst. toegep. biol. Onderz. Nat.*, N 41.

DRIFT, J., and WITKAMP, M. 1960. The significance of the breakdown of oak litter by *Enoicyla pusilla*, Burm. *Archs. néerl. Zool.*, **13**.

DROOZ, A. T. 1961. *Mesoleius tenthredinis* Morl., in Pennsylvania and Michigan. *Can. Ent.*, **93**.

DUCHAUFOUR, PH., and ROUSSEAU, Z.-J. 1959. Les phénomènes d'intoxication des plantules de résineux par le manganèse dans les humus forestiers. *Rev. for. fr.*, **11**.

DUDICH, E., BALOGH, I., and LOKSA, I. 1952. Produktionsbiologische Untersuchungen über die Arthropoden der Waldboden. *Acta biol. hung.*, **3**.

DUNCAN, D. P., and HODSON, A. C. 1958. Influence of the forest tent caterpillar upon the aspen forests of Minnesota. *For. Sci.*, **4**.

DUNGER, W. 1958a. Über die Zersetzunge der Laubstreu durch die Boden-Makrofauna im Auenwald. *Zool. Jb., Abt. Syst.*, **86**.

DUNGER, W. 1958b. Über die Varänderung des Fallaubes im Darm von Bodentieren. *Z. PflErnähr., Düng., Bodenk.*, **82**.

DUNGER, W. 1960. Zu einigen Fragen der Leistung der Bodentiere bei der Umsetzung organischer Substanz. *Zentbl. Bakt. Parasitkde, Abt.* 2, **113**.

DUVIGNEAUD, P. 1962. Ecosystèmes et biosphère. L'écologie. *Sci. mod. synt.*, **11**.

EBERHARDT, F. 1954. Ausscheidung einer organischen Verbindung aus den Wurzeln des Hafers (*Avena sativa* L.). *Naturwissenschaften*, **41**.

EBERHARDT, F. 1955. Über fluoreszierende Verbindungen in der Wurzel des Hafers. Ein Beitr. zum Problem der Wurzelausscheidungen. *Z. Bot.*, **43**.

EBERHARDT, F., and MARTIN, P. 1957. Das Problem der Wurzelausscheidungen und seine Bedeutung für die gegenseitige Beeinflussung hoherer Pflanzen. *Z. PflKrankh. PflPath. PflSchutz.* **64.**

EHWALD, E. 1957. *Über den Nährstoffkreislauf des Waldes.* Leipzig.

EHWALD, E., GRUNERT, F., SCHULZ, W., and VETTERLEIN, E. 1961. Zur Ökologie von Kiefern-Buchen-Mischbeständen. *Arch. Forstw.*, **10.**

EIDMANN, F. E. 1943. Untersuchungen über die Wurzelatmung und Transpiration unserer Hauptholzarten. *Schiftenreihe H. Göring Akad. Dtsch. Forstwiss.*, **5.**

EIDMANN, F. E. 1950. Investigations on root respirations on 15 European tree species characterizing their pioneer quality in virgin soils. *Proc. III Wld For. Congr.*, 1950.

EIDMANN, H. 1949. Das Problem der Indifferenz. *Naturwissenschaften*, **36.**

ELLENBERG, H. 1950. Kausale Pflanzensoziologie auf physiologischer Grundlage. *Ber. dt. bot. Ges.*, **63.**

ELLENBERG, H. 1953. Physiologisches und ökologisches Verhalten derselben Pflanzenart. *Ber. dt. bot. Ges.*, **65.**

ELLENBERG, H. 1954a. Über einige Probleme der kausalen Vegetationskunde. *Vegetatio*, **5-6.**

ELLENBERG, H. 1954b. Über einige Fortschritte des kausalen Vegetationskunde. *Vegetatio*, **8.**

ELLENBERG, H. 1958. Über die Beziechungen zwischen Pflanzengesellschaft, Standort, Bodenprofil und Bodentyp. *Angew. PflSoziol.*, **15.**

ELLENBERG, H. 1964. *Vegetation Mitteleuropas in kausaler, dinamischer und historischer Sicht.* Stuttgart.

ELTON, C. 1927. *Animal ecology.* N.Y.

ENGELMANN, M. D. 1961. The role of soil arthropods in the energetics of an old field community. *Evol. Monogr.*, **31.**

ESCHERICH, K. 1940-1941. *Die Forstinsekten Mitteleuropas*, vol. 5. Berlin.

ESENTHER, G. R., ALLEN, T. C., CASIDA, J. E., and SHENEFELT, R. D. 1961. Termite attractant from fungus-infected wood. *Science*, **134.**

FABRICIUS, L. 1927. Der Einfluss des Wurzelwettbewerb und Lichtentrug des Schirmstandes und den Jungwuchs. *Forstwiss. Zbl.*, **62.**

FABRICIUS, L. 1929. Neue Versuche zu Feststellung des Einflusses von Wurzelwettbewerb und Lichtenzug des Schirmstandes auf den Jungwuchs. *Forstwiss. Zbl.*, **51.**

FEHER, D. 1933. *Untersuchungen über die Microbiologie des Waldbodens.* Berlin.

FEHER, D., and FRANK, M. 1937. Experimentelle Untersuchungen über den Einfluss der Temperatur und des Wassergehaltes auf die Tätigkeit der Mikroorganismen des Bodens. *Archs. Microbiol.*, **8.**

FENTON, G. R. 1947. The soil fauna: with special reference to the ecosystem of forest soil. *J. Anim. Ecol.*, **16.**

FORBES, S. A. 1887. The lake as a microcosm. *Bull. Soc. A. Peoria.*

FOSBERG, E. R. 1959. *Bodenzoologie als Grundlage der Bodenpflege.* Berlin.

FRANCKE-GROSSMANN, H. 1953. Lassen sich schädlingsresistente Wälder erziehen. *Umschau*, **53.**

FRANCKE-GROSSMANN, H. 1963. Some new aspects in forest entomology. *A. Rev. Ent.*, **8.**

FRANK, A. B. 1888. Über die physiologische Bedeutung der Mycorrhizen. *Ber. dt. bot. Ges.*, **6.**

FRANZ, H. 1950. *Bodenzoologie als Grundlage der Bodenpflege.* Berlin.

FRANZ, H. 1955. Die Bedeutung der Kleintiere für die Humusbildung. *Z.PflErnähr., Düng., Bodenkde*, **69.**

FRENZEL, B. 1957. Zur Abgabe von Aminosäuren und Amiden an das Nahrmedium durch die Würzeln von Heleantus. *Planta*, **49.**

FREREKS, W. 1954. Die Bodenatmung als Mittel zur Erfassung der Mikroorganismentätigkeit in Moor und Heides und Böden. *Z. PflErnähr., Düng., Bodenk.*, **66.**

FRICKE, K. 1904. 'Licht und Schattenholzarten', ein wissenschaftlich nicht begrundetes Dogma. *Zentbl. ges. Forstw.*, **30.**

FRIEND, W. C. 1958. Nutritional requirements of phytophagous insects. *A. Rev. Ent.*, **3.**

FÜHRER, E. 1963. Studien über den Fichtennestwickler *Epiblema* (*Epinotia*) *tedella* Cl. in Deutschland. *Anz. Schädlingsk.*, **36.**

FURR, J. R., and REEVE, J. O. 1945. Range of soil moisture percentage through which plants undergo permanent wilting in some soils from semiarid irrigated areas. *J. Agric. Res.*, **71**.

GABRIELSEN, E. 1960. Beleuchtungsstarke und Photosynthese. *Handbuch der Pflanzenphysiologie*, vol. 5, pt. 2. Berlin.

GAMS, H. 1918. Prinzipienfragen der Vegetationsforschung. *Vjschr. naturf. Ges. Zürich*, **63**.

GAMS, H. 1961. Erfassung und Darstellung mehrdimensionaler Verwandtschaftsbeziehungen von Sippen und Zebeusgemeinschaften. *Ber. geobot. Forsch. Inst. Rubel*, **32**, 1960.

GARFINKEL, D. 1962. Digital computer simulation of ecological systems. *Nature*, **194**.

GAUSS, R. 1960. Über Nahrungspflanzen-Wechsel bei Insekten. *Z. angew. Ent.*, **45**.

GEIGER, R. 1942. *Das Klima der bodennahen Luftschicht*. 2nd ed. Braunschweig.

GEORGESCU, C. C., NITU, CH., and TUTUNARU, V. 1960. Cercetari asupra circulatiei apei la molizii defoliati de *Lymantria monacha* L. In: *Probl. actuale biol. si stiinte agric*, Bucuresti. Acad. RPR.

GERE, G. 1956. The examination of the feeding biology and the humification function of Diplopoda and Isopoda. *Acta biol. hung.*, **6**.

GESSNER, F. 1960. Die Assimilationsbedingungen im tropischen Regenwald. *Handbuch der Pflanzenphysiologie*, vol. 5, pt. 2. Berlin.

GHENT, A. W. 1958. Studies of regeneration in forest stands devastated by the spruce budworm. II. Age, height, growth, and related studies of balsam fir seedlings. *For. Sci.*, **4**.

GILBERT, O., and BOCOCK, K. L. 1960. Changes in leaf litter when placed on the surface of soils with contrasting humus types. II. Changes in the nitrogen content of oak and ash leaf litter. *J. Soil Sci.*, **11**.

GOETHE, R. 1895. Einige Beobachtungen über Regenwürmer und deren Bedeutung für das Wachstum der Wurzeln. *Jahrb. nassauisch. Verein Naturkunde*, **48**. Wiesbaden.

GOETLIEB, S., and PELSZAR, M. I. Microbiological aspects of lignin degradation. *Bact. Rev.*, **15**.

GOLLEY, F. B. 1960. An index to the rate of cellulose decomposition in the soil. *Ecology*, **41**.

GRAHAM, A. K. 1949. *79th Annual Rept. Entomol. Soc. Ontario*.

GREAVES, I. E., and CARTER, E. G. 1920. Influence of moisture on the bacterial activities of the soil. *Soil Sci.*, **10**.

GREENLAND, D. J., and KOWAL, J. M. L. 1960. Nutrient content of the moist tropical forest of Ghana. *Pl. Soil*, **12**.

GRÜMMER, G. 1955. *Gegenseitige Beeinflussung höherer Pflanzen. Allelopathie*. Jena.

GRÜMMER, G. 1961. The role of toxic substances in the interrelationships between higher plants. *Sympos. Soc. Exptl Biol.*, **15**.

HADORN, CH. 1933. *Recherches sur la morphologie, les stades évolutifs et l'hivernage du bostryche liseré* (Xyloterus lineatus *Oliv.*). Berne.

HALLIDAY, W. E. D. 1937. A forest classification for Canada. *Forest Serv. Bull.*, **89**.

HARLEY, J. L. 1937. Ecological observations on the mycorrhiza of beech. *J. Ecol.*, **25**.

HARLEY, J. L. 1952. Associations between micro-organisms and higher plants (*Mycorrhiza*). *A. Rev. Microbiol.*

HARLEY, J. L., and BRIERLEY, I. K. 1955. The uptake of phosphate by excised mycorrhizal roots of the beech. VII. Active transport of P^{32} from fungus to host during uptake of phosphate from solution. *New Phytol.*, **54**.

HARLEY, J. L. 1959. *The biology of mycorrhiza*. London.

HARPER, J. L. Approaches to the study of plant competition. *Sympos. Soc. Exptl Biol.*, **15**.

HÄRTEL, O. 1959. Der Erwerb von Wasser und Mineralstoffen bei Hemiparasiten. *Handbuch der Pflanzenphysiologie*, vol. 11. Berlin.

HARTMANN, F. 1960. Vorläufige Ergebnisse einiger Waldkalkungsversuche in Osterreich. *Allg. Forstztg*, **71**.

HATCH, A. B. 1936. The role of mycorrhizae in afforestation. *J. For.*, **34**.

HAYNES, D. L., and BUTCHER, J. W. 1962. Studies on host preference and its influence on European pine shoot moth success and development. *Can. Ent.*, **94**.

HENSON, W. R. 1960. Weather and insect epidemics. *Rept. Fifth World Forestry Congr.* Seattle, U.S.A.

HENSON, W. R., and STARK, R. W. 1959. The description of insect numbers. *J. econ. Ent.* **52.**

HESSE, G., KAUTH, H., and WÄCHTER, R. 1955. Frasslockstoffe beim Fichtenrüsselkäfer *Hylobius abietis. Z. angew. Ent.*, **37.**

HESSELMAN, H. 1916-1917. On the effect of our regeneration measures on the formation of saltpetre in the ground and its importance in the regeneration of coniferous forests. *Meddn. St. SkogsförsAnst.*, **2.**

HESSELMANN, H. 1926. Studien über die Humusdecken des Nadelwaldes, ihre Eigenschaften und deren Abhängigkeit vom Waldbau. *Meddn. St. SkogsförsAnst.*

HILLS, G. A. 1959. *A ready reference to the description of the land of Ontario and its productivity.* Canada.

HILLS, G. A. 1960. Comparison of forest ecosystems (vegetation and soil) in different climatic zones. *Silva fenn.*, **105.**

HILLS, G. A. 1952. The classification and evaluation of site for forestry. *Ont. Dept. Lands Forest Res. Rept,* **24.**

HILTNER, L. 1904. Über neuere Erfahrungen und Probleme auf dem Gebiete der Bodenbakteriologie und unter besonderes Berücksichtigung der Grundungung und Brache. *Zbl. Bact.*, **2.**

HINZE, G. 1950. Der Biber. Körperbau und Lebenswiese. *Verbreitung und Geschichte.* Berlin.

HIRSCHMANN, W. 1954. Das Lückensystem der Bäume—ein wenig beachteter Lebensraum. *Mikrokosmos,* **43.**

HOFFMANN, K., and BOLLAND, G. 1959. Über Lochkartenverfahren und ihre Anwendungsmöglichkeiten in der Forstwissenschaft. *Arch. Forstw.*, **8.**

HÖFLER, K., MIGSCH, H., and ROTTENBURG, W. 1941. Über die Austrocknungsresistenz landwirtschaftlicher Kulturpflanzen. *Forschungsdienst,* **12.** H. 1.

HOSKINS, L., and DALKE, Y. 1955. Winter browse on the Pocatello big game range in Southeastern Idaho. *J. Wildl. Mgmt,* **2.**

HÜBER, B. 1952. Tree physiology. *Rev. Pl. Physiol.*, **3.**

HÜBER, B. 1960. Die CO_2-Konzentration in Pflanzengesellschaften. *Handbuch der Pflanzenphysiologie,* vol. V, pt. 2. Berlin.

HUSSON, R. 1954. Cas de polyédrose chez le géométrie de *Ennomos quercinaria* Hufn. et considérations générales sur les polyédroses. *Revue Path. vég. Ent. agric. Fr.*, **33.**

HUTCHINSON, S. A., and KAMEL, M. 1956. The effect of earthworms on the dispersal of soil fungi. *J. Soil Sci.*, **7.**

IVARSON, K. C., and KATZNELSON, H. 1960. Studies on the rhizosphere microflora of yellow birch seedlings. *Pl. Soil,* **12.**

JACOT, P. A. 1936. Soil structure and soil biology. *Ecology,* **17.**

JAHN, E. 1958. Fragen zur chemischen und biologischen Fortschädlingsbekämpfung bzw. biozönotischen Regelung. *Allg. Forsztg,* **69.**

JAHN, E. 1959. Waldränder und ihrer Auswirkung auf Boden, Bodentierleben und Wiederinbestandsbringung. *Allg. Forsztg,* **70.**

JAHN, E. 1962. Hinweise zur Erkennung des Schlagalters von Lärchen- und Kiefernstöcken an Befallsfolgen durch Insekten und dem allgemeinen Zustand in verschiedenen Altersstufen. *Allg. Forsztg,* **73.**

JAMNICKY, J. 1958. Ucast hmyzu (*Insecta*) na odumierani brezy bradavicnatej. *Biologia,* **13.**

JANKOVIC, L. 1958. Prilog poznavanju biljki hraniteljki gubara u privodi u toku poslednje gradacije (1953-1957 god.). *Zashtita bil'a.*

JULANDER, O. 1955. Deer and cattle range relations in Utah. *For. Sci.*, **2.**

KANGAS, E. 1963. Über das schädliche Auftreten der Diprion-Arten (Hym. Diprionidae) in finnischen Kiefernbeständen in diesem Jahrhundert. *Z. angew. Ent.*, **51.**

KARCZEWSKI, J. 1962. Znaczenie borowki czernicy (*Vaccinium myrtillus*) dla entomocenozy lesnej. *Folia forestalia polon.* A.

KATZNELSON, H. 1940. Survival of azotobacter in soil. *Science,* **49.**

KATZNELSON, H., and STEVENSON, L. 1956. Observation of metabolic activity of the soil microflora. *Can. J. Microbiol.*, **2.**

KEMMER, E. 1935. Über die Auswirkung der Bodenmüdigkeit bei Obstgehölzen. *Gartenwelt,* **39.**

KENDRICK, W. B. 1958. Micro-fungi in pine litter. *Nature,* **181.**

KENDRICK, W. 1959. The time factor in the decomposition of coniferous leaf litter. *Can. J. Bot.*, **37**.

KERN, H. 1959. Parasitismus und Simbiose. *Handbuch der Pflanzenphysiologie*, vol. 11. Berlin.

KEVAN, K. 1962. *Soil animals*. London.

KITTREDGE, J. 1948. Forest influences. *The effects of woody vegetation on climate, water and soil*. N.Y.

KLEPAC, D. 1959. Izracunavanje gubitka na prirastu u sastojinawa koje je napao gubar (*Lymantria dispar*). *Sum. List*, **83**.

KNAPP, R. 1954. *Experimentelle Soziologie des höherer Pflanzen*. Stuttgart.

KNIGHT, C. B. 1961. The Tomocerinae (Collembola) in old field stands of North Carolina. *Ecology*, **42**.

KNIGHT, F. 1958. The effects of woodpeckers on the populations of the Engelmann spruce beetle. *J. econ. Ent.*, **51**.

KNÖSEL, D. 1958. Über die Wirkung aus Pflanzenresten freiwerdender phenolischer Substanzen auf Mikroorganismen des Bodens. I. Der Einfluss von p-Oxybenzosäure auf die Entwicklung von Pilzen, Actinomyceten und Bakterien. *Z. PflErnähr., Düng., Bodenkde*, **80**.

KNOWLES, R., and LAISHLEY, E. 1959. Factors in forest-tree litter extracts affecting the growth of soil micro-organisms. *Nature*, **184**.

KOCH, A. 1914. Über die Einwirkung des Laub und Nadelwaldes auf den Boden und die ihn bewohnenden Pflanzen. *Zbl. Bakteriol.*, **9**.

KOCH, A., and OELSNER, A. 1916. Einfluss von Fichtenharz und Tannin auf den Stickstoffhaushalt des Bodens. *Zbl. Bakteriol.*, **11**.

KORSTIAN, C. P., and COILE, T. 1938. Plant competition in forest stands. *Bull. Duke Univ. School For.*, **3**.

KOSZUBIAK, HENRYK. 1961. Vplyw streptomycyny na mikrocenoze glebowa. *Pr. Kom. Nauk roln. leśn. Poznań.*, **7**.

KOVACEVIC, Z. 1956. Die Nahrungswahl und das Auftreten der Pflanzenschädlinge. *Anz. Schädlingsk.*, **29**.

KRAEMER, G. D. 1963. Die kritischen Grenzen der Brutbaumdisposition für Borken-käferbefall an Fichte (*Picea excelsa* L.). *Z. angew. Ent.*, **34**.

KRAJINA, V. I. 1960. Can we find a common platform for the different schools of forest type classification? *Silva fenn.*, No. 105.

KRAMER, P. J., and DECKER, J. P. 1944. Relation between light intensity and rate of photosynthesis of loblolly pine and certain hardwoods. *Pl. Physiol.*, **19**.

KRAMER, P. J., and KOZLOWSKI, T. T. 1960. *Physiology of trees*. N.Y.

KRAMER, P. J., and WILBUR, K. M. 1949. Absorption of radioactive phosphorus by mycorrhizal roots of pine. *Science*, **110**.

KRIEBEL, H. B. 1954. Bark thickness as a factor in resistance to white pine weevil injury. *J. For.*, **52**.

KRISHNAMURTI, K., and SOMON, S. 1951. Studies in the absorption of bacteria. *Proc. Indian Acad. Sci., ser. B*, **34**.

KUDLER, J. 1961. K moznostem obrany proti kozlicku osikovemu (*Saperda populnea* L.). *Sb. čsl. Akad. zemed. véd, Lesn.*, **1**.

KUDLER, J., and LUSENKO, O. 1963. Pokusy s hubenim bekyne vrbove (*Leucoma salicis* L.) patogennimi mikroorganismy. *Lesn. Case.*, **9**.

KUDLER, P. 1962. Lessives (Parabraunerde, Fahlerde) aus Geschiebemergel der Würm-Eiszeit im norddeutschen Tiefland. *Z. PflErnähr., Düng., Bodenkde.*, **95**.

KÜHNELT, W. 1944. Über Beziehungen zwischen Tier- und Pflanzengesellschaften. *Biologia gen.*, **17**.

KULMAN, H. M., HODSON, A. C., and DUNCAN, D. P. 1963. Distribution and effect of jack-pine budworm defoliation. *For. Sci.*, **9**.

KURIR, A. 1952. Vegrösserung der Zahl der Raupenstadien und Verlängerung des Raupenlebens durch die Nahrung. *Bodenkultur*, **6**.

KURIR, A. 1953. Die Frasspflanzen des Schwammspinners (*Lymantria dispar* L). *Z. angew. Ent.*, **34**.

LARCHER, W. 1961. Jahresgang des Assimilations- und Respirationsvermögen von *Olea Europaea* L. ssp. *sativa* Hoff et Link. *Quercus ilex* L., und *Quercus pubescens* Willd., aus dem Nordlichen Gardaseegebiet. *Planta*, **56**.

LAUSBERG, TH. 1953. Quantitative Untersuchungen über die kutikuläre Exkretion des Laubblätters. *Jb. wiss. Bot.*, **81**.

LEJEUNE, R. R., and HILDAHL, V. 1954. A survey of parasites of the larch sawfly (*Pristiphora erichsonii* Hartig) in Manitoba and Saskatchewan. *Can. Ent.*, **86**.

LINDQUIST, B. 1941. Untersuchungen über die Bedeutung einiger skandinavischer Regenwürmer für die Zersetzung der Laubstreu und für die Struktur des Mullbodens. *Svenska Skogsv. Tidskr.*, 179-242.

LIPMAN, I. C., and BROWN, P. E. 1908. Report of the soil chemist and bacteriologist. *29th A. Rpt. N.Y. Agric. Exptl. Sta.* (1907-08).

LOUB, WALTER, 1959. Walddüngungsmassnahmen im Lichte der Bodenmikrobiologie. *Allg. Forstz.*, **14**, N 20.

LUNDEGÅRDH, H. 1924. *Der Kreislauf der Kohlensäure in der Natur.* Jena.

LUNDEGÅRDH, H. 1951. *Die Nährstoffaufnahme der Pflanze.* Jena.

LUNDEGÅRDH, H. 1957. *Klima und Boden in ihrer Wirkung auf das Pflanzenleben.* 4th ed. Jena.

LUTZ, H. J. 1945. Vegetation on a trenched plot twenty-one years after establishment. *Ecology*, **26**.

LUTZ, H. J., and CHANDLER, R. F. 1947. *Forest soils*, 2nd ed. N.Y., London.

LYFORD, W. H. 1943. The palatability of freshly fallen forest tree leaves to millipeds. *Ecology*, **24**.

LYONS, L. A. 1962. The effect of aggregation on egg and larval survival in *Neodiprion swainei* Midd. (Hymenoptera: Diprionidae). *Can. Ent.*, **94**.

MACARTHUR, R. 1955. Fluctuation of animal population as a measure of community stability. *Ecology*, **36**.

MCCOMB, A. L. 1943. Mycorrhizae and phosphorus nutrition of pine seedlings in a prairie soil nursery. *Res. Bull. Iowa Agric. Exptl. Stat.*, **314**.

MCCOMB, A. L., and GRIFFITH, I. 1946. Growth stimulation and phosphorus absorption of mycorrhizal and non-mycorrhizal northern white pine and Douglas fir seedlings in relation to fertilizer treatment. *Pl. Physiol.*, **21**.

MCCORMICK, J. 1959. *The living forest.* N.Y.

MACFADYEN, A. 1961. Metabolism of soil invertebrates in relation to soil fertility. *Ann. appl. Biol.*, **49**.

MACFADYEN, A. 1963. *Animal ecology.* 2nd ed. London.

MARGALEFF, R. 1957. La teoria de la informacion en ecologia. *Mems. R. Acad. Cienc. Artes*, Barcelona, **32**.

MARGALEFF, R. 1958. Information theory in ecology. *Gen. Syst.*, **3**.

MARGALEFF, R. 1961. Communication of structures in planktonic populations. *Limnol. Oceanogr.*, **6**.

MARKUS, E. 1925. Naturkomplex. *Protok. Obshch. Estest.*, Yur'ev, **32**.

MARKUS, E. 1932. Chorogenese und Grenzverchiebung. *Acta Comment. Univ. tartu.*, **23**.

MARTIN, I. P. 1948. Effect of fumigation, fertilization and various other soil treatments on growth of orange seedlings in old citrus soils. *Soil Sci.*, **66**.

MARTIN, I. P., and ERVIN, I. O. 1958. Greenhouse studies on influence of other crops and of organic materials on growth of orange seedlings in old citrus soil. *Soil Sci.*, **85**.

MARTIN, P. 1956. Qualitative und quantitative Untersuchungen über die Ausscheidung organischer Verbindungen aus den Keimwurzeln des Hafers (*Avena sativa* L). *Naturwissenschaften*, **43**.

MARTIN, P. 1957. Die Abgabe von organischen Verbindungen, insbesondere von Scopoletin aus den Keimwurzeln des Hafen. *Z. Bot.*, **45**.

MARTIN, P. 1958. Einfluss der Kulturfiltrate von Mikroorganismen auf die Abgabe von Scopoletin aus den Keimwurzeln des Hafers (*Avena sativa* L). *Arch. Mikrobiol.*, **29**.

MARTIN, P., and RADEMACHER, B. 1960. Untersuchungen zur Frage der Wurzelallelopathie von Kulturpflanzen und Unkrälern. *Beitr. Biol. Pfl.*, **35**.

MEIKLEJOHN, J. 1955. The effect of bush burning on the microflora of a Kenya upland soil. *J. Soil Sci.*, **6**.

MEINECKE, TH. 1927. *Die Kohlenstoffernährung des Waldes.* Berlin.

MELIN, E. 1925. *Untersuchungen über die Bedeutung der Baummykorrhize.* Jena.

MELIN, E. 1955. Nyare undersökningar over skogsträdens mykorrhizasvamper och det fysiologiska vuxelspelet mellan dem och tradens rotter. *Acta Univ. Uppsala*, 3.

MELIN, E. 1959. Micorrhiza. *Handbuch der Pflanzenphysiologie*, vol. XI. Berlin.

MELIN, E., and NILSON, H. 1950. Transfer of radioactive phosphorus to pine seedlings by means of mycorrhizal hyphae. *Physiologia. pl.*, 3.

MELIN, E., and NILSON, H. 1953. Transfer of labelled nitrogen from glutamic acid to pine seedlings through the mycelium of *Boletus variegatus* (Sw.). Tr. *Nature*, 171.

MELIN, E., and NILSON, H. 1955. Ca^{45} used as indicator of transport of cations to pine seedlings by means of mycorrhizal mycelium. *Svensk bot. Tidskr.*, 49.

MELIN, E., and NILSON, H. 1957. Transport of C^{14} labelled photosynthate to the fungal associate of pine mycorrhiza. *Svensk bot. Tidskr.*, 51.

MERKER, E. 1956a. Die Abhängigkeit der Waldverderber von ihrer Frasspflanze. *Forschn. Fortschr.*, 30.

MERKER, E. 1956b. Der Widerstand von Fichten gegen Borkenkäferfrass. *Allg. Forst- u. Jagdz.*, 127.

MERKER, E. 1958. Forstschutz gegen Insekten durch Düngung der Baumbestände? *Allg. Forstz.*, 13.

MERKER, E. 1960. Der Einfluss des Baumzustandes auf die Übervermehrung einiger Waldschädlinge. *Z. angew. Ent.*, 46.

MERKER, E. 1961. Welche Ursachen hat die Schädigung der Insekten durch Düngung im Walde? *Allg. Forst- u. Jagdztg*, 132.

MERKER, E. 1962a. Augenblicklicher Stand der Untersuchingen über die schädigende Wirkungsweise von Düngestoffen auf Waldschädlinge. *Allg. Forst- u. Jagdztg*, 133.

MERKER, E. 1962b. Zur Biologie und Systematik der mitteleuropäischen Tannenläuse (Gattung *Dreyfusia*). *Allg. Forst- u. Jagdztg*, 133.

MERKER, E. 1962. Über die Ursachen der Vernichtung von Waldschädlingen durch Düngung. *Meded. LandbHoogesch. OpzoekStats Gent.*, 30.

METZ, H. 1955. Untersuchungen über die Rhizosphäre. *Arch. Mikrobiol.*, 23.

MEYER, F. H. 1960. Vergleich des mikrobiellen Abbaus von Fichten- und Buchenstreu auf verschiedenen Bodentypen. *Arch. Mikrobiol.*, 35.

MICCOLA, P. 1952. Effect of forest humus on parasitic fungi causing damping of disease of coniferous seedlings. *Phytopathology*, 42.

MICCOLA, P. 1960. Comparative experiment on decomposition rates of forest litter in Southern and Northern Finland. *Oikos*, 11.

MILLER, R., and RISCH, I. 1962. Zur Frage der Kohlensäureversorgung des Waldes. *Forstwiss. Zbl.*, 79, 1-2.

MILTHORPE, F. L. 1961. The nature and analysis of competition between plants of different species. *Sympos. Soc. Exptl. Evol.*, 15.

MOGREN, E. W. 1955. A study of some aspects of susceptibility of ponderosa pine to attack by Black Hills beetle. (Doct. Dissert. Univ. Michigan.) *Diss. Abstr.*, 15.

MOLISCH, H. 1937. Der Einfluss einer Pflanzen auf die andere. *Allelopathie*. Jena.

MØLLER, C. M. 1945. *Untersuchungen über Laubmenge Stoffverlust und Stoffproduction des Waldes*. Copenhagen.

MOREAU, R. 1959. Sur la microflore de quelques terres de Sapinières. *C. r. hebd. Séanc. Acad. Sci.*, 249.

MORRISON, T. M. 1961. Absorption of phosphorus from soils by mycorrhizal plants. *New Phytol.*, 61.

MORRISON, T. M. 1962a. Uptake of sulfur by mycorrhizal plants. *New Phytol.*, 61.

MORRISON, T. M. 1962b. Absorption of phosphorus from soils by mycorrhizal plants. *New Phytol.*, 61.

MOTHES, K. 1931. Zur Kenntnis des N-Stoffwechsels höherer Pflanzen. *Planta*, 12.

MULDREW, J. A. 1953. The natural immunity of the larch sawfly, *Pristiphora erichsonii* (Htg.) to the introduced parasite *Mesoleius tenthredinis* in Manitoba and Saskatchewan. *Can. J. Zool.*, 31.

MÜLLER, H. 1958. Zur Kenntnis der Schäden, die Lachniden an ihren Wirtsbäumen hervorrufen können. *Zangew. Ent.*, 42.

MUNNICHKE, K. O., and VOGEL, I. C. 1958. *Naturwissenschaften*, 45.

MUNTER, F., and ROBSON, W. P. 1913. Über den Enfluss der Boden und des Wassergehaltes auf die Stickstoffumsetzungen. *Zbl. Bakteriol.*, *Abt. II*, 39.

MURPHY, P. W. 1955. The biology of forest soils with special reference to the meso- or mediofauna. *J. Soil Sci.*, 4.

Nature Conservancy. 1962. *Report for the year ended 30.IX.1962.* London.

NEF, L. 1957. Etat actuel des connaissances sur le rôle des animaux dans la décomposition des litières de forêts. Extr. *Agricultura*, Ser. 2, V, No. 3.

NEF, L. 1959. Protection des bourgeons de pins contre les attaques du tétrax lyre. *Bull. Soc. r. for. Belg.*, 66.

NEMEC, A. 1930. Untersuchungen über die chemischen Veränderungen der organischen Substanz bei der natürlichen Zersetzung der Humusauflage in Wäldern. *Z. PflErnähr., Düng., Bodenkde*, 28.

NEMEC, A., and KVAPIL, K. 1927. Über Einfluss verschiedener Waldbestände auf den Gehalt und die Bildung von Nitraten in Waldboden. *Z. Forst- u. Jagdw.*, 59.

NEUEGEBAUER, W. O. 1951. Das Problem der Indifferenz von Forstinsekten unter besonderer Berücksichtigung der Ökologie des Kieferntriebwickelers. *Verh. d. Ges. angew. Ent.*, 2.

NEUWIRTH, G. 1959. Der CO_2-Stoffwechsel einiger Koniferen während des Knospenaustreiben. *Biol. Zbl.*, 78.

NIECHZIOL, W. 1958a. Biologisch-ökologische Studien zur Kalamität der kleinen Fichtenblattwespe (*Lygaeonematus pini* Ratz) im Mooswald bei Freiburg. *Dt. ent. Z.*, 5.

NIECHZIOL, W. 1958b. Die Bekämpfung der kleinen Fichtenblattwespe als forstlicher Dauerschädling auf ökologischem Wege. *Allg. Forstz.*, 13.

NIELSEN, C. O. 1955. Studies on the Enchytraeidae. 2. Field studies. *Natura jutl.*, 4.

NEILSON-JONES, W. 1941. Biological aspects of soil fertility. *J. Agric. Sci.*, 31.

NOSEK, J. 1954. *Vyzkum pudni zvireny jako soucast vyzkumu biocenosy lesa.* Praha.

NYKVIST, N. 1959a. Leaching and decomposition of litter. I. Experiments on litter of *Fraxinus excelsior. Oikos*, 10.

NYKVIST, N. 1959b. Leaching and decomposition of litter. II. Experiments on needle litter of *Pinus silvestris. Oikos*, 10.

ODUM, E. P. 1954. *Fundamentals of ecology.* Philadelphia.

ODUM, E. P. 1959. *Fundamentals of ecology.* 2nd ed. Philadelphia and London.

ODUM, E. P. 1962. *Fundamentals of ecology.* Philadelphia and London.

ODUM, E. P., CONNEL, C. E., and DAVENPORT, L. B. 1962. Population energy flow of three primary consumer components of old-field ecosystems. *Ecology*, 43.

ODUM, H. T. 1956. Efficiencies, size of organisms, and community structure. *Ecology*, 37.

ODUM, H. T., CANTLON J. E. and KORNICKER L. S. 1960. An organizational hierarchy postulate for the interpretation of species-individual distribution, species entropy ecosystem evolution and the meaning of a species-variety index. *Ecology*, 41.

OECHSSLER, G. 1962. Studien über die Saugeschäden mitteleuropäischer Tannenläuse im Gewebe einheimischer und ausländischer Tannen. *Z. angew. Ent.*, 50.

OHNESORGE, B. 1958. Das Austreiben der Fichten und ihr Befall durch die kleine Fichtenblattwespe *Pristiphora abietina* (Christ.). *Forstwiss. Zbl.*, 77.

OLDIGES, H. 1958. Waldbodendüngung und Schädlingsfauna des Kronenraumes. *Allg. Forstz.*, 13.

OLDIGES, H. 1959. Der Einfluss der Waldbodendüngung auf das Auftreten von Schadinsekten. *Z. angew. Ent.*, 45.

OLDIGES, H. 1960. Walbodendüngung und Kronenfauna. *Z. angew. Ent.*, 47.

OLMSTEAD, CH. E. 1941. The roles of light and root competition in an oak-maple forest. *Bull. ecol. Soc. Am.*, 22.

O'NEIL, L. C. 1963. The suppression of growth rings in jack pine in relation to defoliation by the Swaine jack-pine sawfly. *Can. J. Bot.*, 41.

OOSTING, H. J., and KRAMER, P. J. 1946. Water and light in relation to pine reproduction. *Ecology*, 27.

OVERGARD, C. 1949. Freeliving nematodes and soil microbiology. *Rept. Proc. IV Int. Congr. Microbiol.*, 1947, 483-484.

OVINGTON, J. D. 1960. The ecosystem concept as an aid to forest classification. *Silva fenn.*, 105.

OVINGTON, J. D. 1961. Some aspects of energy flow in plantations of *Pinus sylvestris* L. *Ann. Bot.*, N.S., 25.

OVINGTON, J. D., and HEITKAMP, D. 1961. The accumulation of energy in forest plantations in Britain. *J. Ecol.*, **48.**

OVINGTON, J. D., and MADGWICK, H. A. I. 1959. Distribution of organic matter and plant nutrients in a plantation of Scots pine. *For. Sci.*, **5.**

PASSARGE, S. 1912. Physiologische Morphologie. *Mitt. geogr. Ges.*, **26.**

PASSARGE, S. 1920. *Die Grundungen der Landschaftskunde.* III. Hamburg.

PASSARGE, S. 1921. Vergleichende Landschaftskunde. In: *Aufgaben und Methoden der vergleichenden Landschaftskunde.* Berlin.

PASSARGE, S. 1929. Länder, reale Landschaften, ideale Landschafts-typen. *Naturwissenschaften*, **17.**

PATRICK, I. A. 1955. The peach replant problem in Ontario. II. Toxic substances from microbial decomposition products of peach root residues. *Can. J. Bot.*, **33.**

PATTEN, B. C. 1959. An introduction to the cybernetics of the ecosystem: the trophic-dynamic aspect. *Ecology*, **40.**

PAVAN, M. 1960. Works for the protection of the Italian woods by means of ants. *Rept. Fifth Wld For. Congr.*, 1958.

PEELE, I. 1936. Adsorption of bacteria by soils. *Cornell Univ. Agric. Stat. Mem.*, **197.**

PELZ, E. 1958. Beobachtungen zur Rauchhärte der Kiefer. *Allg. Forsztg.*

PETERSEN, B. B., and SØEGAARD, B. 1958. Studies on resistance to attacks of *Chermes cooleyi* (Gill) on *Pseudotsuga taxifolia* (Poir.) Britt. *Det forstlige forsøgsvaesen i Danmark*, N 1.

PFEFFER, A. 1963. Insektenschädlinge an Tannen im Bereich der Gasexhalationen. *Z. angew. Ent.*, **51.**

PFEIFER, S., and RUPPERT, H. 1953. Versuche zur Steigerung der siedlungsdichtem Höhlen und buschbrütender Vogelarten. *Biol. Abh.*, **2.**

PISEK, A. 1960. Immergrüne Pflanzen (einschliesslich Coniferen). *Handbuch für Pflanzenphysiologie*, vol. 5, pt. 2. Berlin.

PISEK, A. 1960. Pflanzen der Arktis und des Hochgebirge. *Handbuch für Pflanzenphysiologie*, vol. 5, pt. 2. Berlin.

PISEK, A., and CARTELLIERI, E. 1939. Zur Kenntnis des Wasserhaushaltes der Pflanzen. IV. Bäume und Sträucher. *Jb. wiss. Bot.*, **88.**

PISEK, A., and KNAPP, R. 1959. Zur Kenntnis der Respirationintenstität von Blattern verschiedener Blutenpflanzen. *Ber. dt. bot. Ges.*, **72.**

PLAISANCE, G. 1959. Les formations végétales et paysages ruraux. *Lexique et guide bibliographique.* Paris.

POLLOCK, B. M. 1953. The respiration of *Acer* buds in relation to the inception and termination of the winter rest. *Physiologia pl.*, **61.**

POLSTER, H. 1950. *Die physiologischen Grundlagen der Stofferzeugung im Walde.* München.

POOLE, T. B. 1959. Studies on the food of Collembola in a Douglas fir plantation. *Proc. Zool. Soc. Lond.*, **132.**

POSTNER, M. 1961. Schadfrass des Wollkäfers *Lagria hirta* L. (Lagriidae, Coleopt.) an Jungfichten. *Anz. Schädlingsk*, **34.**

POSTNER, M. 1963. Insektenschäden an der Lärche ausserhalb ihres natürlichen Verbreitungsgebietes. *Forstwiss. Zbl.*, **82.**

POTOCKA, J., ČAPEK, M., and CHORVAT, K. 1962. Prispevok k poznanju korunovej fauny clankovcov na duboch slovenska, predovsetkym so zretel'om na rad Lepidoptera. *Biol. pr.*, **8.**

PSCHORN-WALCHER, H. 1958. Forstschutz durch naturgemässe Waldwirtschaft. *Allg. Forsztg*, **69.**

PSCHORN-WALCHER, H., and ZWÖLFER, H. 1956. Neuere Untersuchungen über die Weisstannenläuse der Gattung *Dreyfusia* C.B. und ihren Verteilerkreis. *Anz. Schädlingskde*, **29.**

QUASTLER, H. 1959. Information theory of biological integration. *Am. Nat.*, **93.**

QUISPEL, A. 1959. Lichens. *Handbuch der Pflanzenphysiologie*, vol. 11. Berlin.

RABELER, W. 1957. Die Tiergesellschaft eines Eichen-Birkenwaldes in nordwestdeutschen Altmoränengebiet. *Mitt. flor.-soc. ArbGemein.*, **6/7.**

RABELER, W. 1962. Die Tiergesellschaften von Laubwäldern (Querco—*Fagetea*) im oberen und mittleren Wesergebiet. *Mitt. flor.-soc. ArbGemein.*, **9.**

RADEMACHER, R. 1957. Die Bedeutung allelopathischer Erscheinungen in der Pflanzen-pathologie. *Z. PflKrankh.*, **64**.

RADEMACHER, R. 1959. Gegenseitige Beeinflussung höherer Pflanzen. *Handbuch der Pflanzenphysiologie*, vol. 11. Berlin.

RAMANN, E. 1911. Regenwürmer und Kleintiere im deutschen Waldboden. *Int. Mitt. Bodenkde*, 3, Berlin.

RAWLINGS, G. B. 1961. Entomological and other factors in the ecology of a *Pinus radiata* plantation. *Proc. N. Z. Ecol. Soc.*, **8**.

RICHARDS, L. A., and WEAVER, L. R. 1943. Fifteen atmosphere percentage as related to the permanent wilting percentage. *Soil Sci.*, **56**.

RICHARDS, T. O. 1958. Concealment and recovery of food by birds with some relevant observations on squirrels. *Brit. Birds*, **51**.

RIPPEL, K. 1936. Über Begriff und Wesen der Bodenmüdigkeit. *Phytopath. Z.*, **9**.

ROMELL, L. G. 1935. An example of Myriapods as mull formers. *Ecology*, **16**.

ROMELL, L. 1938. A trenching experiment in spruce forest and its bearing on problems of mycotrophy. *Svensk. bot. Tidskr.*, **32**.

ROMELL, L. G. 1939. The ecological problem of mycotrophy. *Ecology*, **20**.

ROMELL, L., and MALMSTRÖM, C. 1944-1945. The ecology of lichen-pine forest. *Meddn. St. SkogsförsAnst.*, **34**.

RONDE, G. 1951. Vorkommen, Häufigkeit und Arten von Regenwürmern in ver-schiedenen Waldböden und unter verschiedenen Bestockungen. *Forstwiss. Zbl.*, **70**.

RONDE, G. 1953. Untersuchungen in einem Forstbetrieb des Würtenbergischen Alpes-vorlandes. *Forstwiss. Zbl.*, **72**.

ROWE, J. S. 1959. *Forest regions of Canada*. Ottawa.

RUBNER, K. 1953. *Die pflanzengeographischen Grundlagen des Waldbaus*. 4th ed. Berlin.

RUSCHMANN, G. 1953. Über Antibiosen und Symbiosen von Bodenorganismen und ihre Bedeutung für die Bodenfruchtbarkeit. Regenwurm-Symbiosen und Antibio-sen. *Z. Acker- u. PflBau*, **96**.

RYAN, R. B., and RUDINSKY, J. A. 1962. Biology and habits of the Douglas-fir beetle parasite, *Coeloides brunneri* Viereck (Hymenoptera: Braconidae), in Western Oregon. *Can. Ent.*, **94**.

SALISBURY, E. J. 1929. The biological equipment of species in relation to competition. *Ecology*, **18**.

SCAMONI, A. 1960. Can we find a common platform for the different schools of forest type classification? *Silva fenn.*, **105**.

SCAMONI, A. 1960. Biogeozönose-Phytozönose. *Fortschr. 150-Jahrfeier der Humboldt-Univ.*, vol. 2. Berlin.

SCHENK, H. 1892-1893. *Beitrage zur Biologie und Anatomie der Lianen*. Jena.

SCHIMITSCHEK, E. 1952-1953. Forstentomologische Studien im Urwald. *Z. angew. Ent.*, **34, 35**.

SCHIMITSCHEK, E. 1962. Über Zusammenhänge zwischen Massenvermehrungen von *Evetria bioliana* und *Diprion sertifer* und den Boden sowie Grundwasserverhältnissen. *Anz. Schädlingskde*, **35**.

SCHIMITSCHEK, E., and WIENKE, E. 1963. Untersuchingen über die Befallsbereit-schaft von Baumarten für Sekundärschädlinge. I. Teil. *Z. angew. Ent.*, **51**.

SCHIMPER, A. F. W. 1935. *Pflanzengeographie auf physiologischer Grundlage*. 3rd ed. Jena.

SCHMID, E. 1944. Kausale Vegetationsforschung. *Ber. geobot. Forschinst. Rubel.*

SCHMUCKER, TH. 1959a. Höhere Parasiten. *Handbuch der Pflanzenphysiologie*, vol. 11. Berlin.

SCHMUCKER, TH. 1959b. Saprophytismus bei Kormophyten. *Handbuch der Pflanzen-physiologie*, vol. 11. Berlin.

SCHÖNBECK, F., and BRÜSEWITZ, G. 1957. Untersuchungen über den Abbau phyto-toxischer Substanzen durch *Lumbricus terrestris*. *Naturwissenschaften*, **44**.

SCHÖNWIESE, H. 1958. Das Rotwildvorkommen in Österreich in Beziehung zur gegeben natürlicher Ässung. *Öst. Vjschr. Forstw.*, **99**.

SCHREMMER, F. 1958. Bibiolarven als Verarbeiten von Nadelstreu. Ein Beitrag zur Ökologie Bibioniden. *Anz. Schädlingskde*, **31**.

SCHUSTER, R. 1955. Untersuchungen über die bodenbiologische Bedeutung der Oribatiden (Acari). *Naturwissenschaften*, **42**.

SCHÜTTE, F. 1957. Untersuchungen über die Populationsdynamik des Eichenwicklers (*Tortrix viridana* L.). T. II. *Z. angew. Ent.*, **40**.

SCHÜTTE, F. 1959. Untersuchungen über die Populationsdynamik des Eichenwicklers (*Tortrix viridana* L.). *Z. PflKrankh.*, **66**.

SCHWENKE, W. 1957. Über die räuberische Tätigkeit von *Formica rufa* L. und *Formica nigricans* Emery ausserhalb einer Insekten-Massenvermehrung. Hymenoptera: Formicidae. *Beitr. Ent.*, **7**.

SCHWENKE, W. 1960. Über die biozönotischen Grundlagen der Forstzoologie. *Forstwiss. Zbl.*, **79**.

SCHWENKE, W. 1961. Walddüngung und Schadinsekten. *Anz. Schädlingskde*, **34**.

SCHWENKE, W. 1962a. Neue Erkenntnisse über Entstehung und Begegnung von Massenvermehrungen an Kiefern und Fichtennadeln fressender Schadinsekten. *Z. angew. Ent.*, **50**.

SCHWENKE, W. 1962b. Walddüngung und Insektenvermehrung. *Ber. 13th Congr. int. Un. Forest. Res. Org.*, Vienna, pt. 2.

SCHWENKE, W. 1963. Über die Beziehungen zwischen den Wasserhaushalt von Bäumen und der Vermehrung blattfressender Insekten. *Z. angew. Ent.*, **51**.

SCHWERDTFEGER, F. 1954. Forstinsekten im Ur- und Nutzwald. *Allg. Forstz.*, **9**.

SCHWERDTFEGER, F. 1956. Zum Begriff der Populationsdynamik. *Beitr. Ent.*, **6**.

SCHWERDTFEGER, F. 1957. *Die Waldkrankheiten*. Hamburg. Berlin.

SCHWERDTFEGER, F. 1962. Über Dr. Schwenke's Bericht 'Walddüngung und Insektenvermehrung'. *Ber. 13th Congr. int. Un. Forest Res. Org.*, Vienna, pt. 2.

SCOTTER, G. 1960. Productivity of arboreal lichens and their possible importance to barren-ground caribou (*Rangifer arcticus*). *Arch. Soc. zool. bot. fenn.*, **16**.

SEN-SARMA, P. K. 1963. Studies on the natural resistance of timber to termite attack. III. Results of accelerated laboratory tests of 9 species of Indian woods against the Mediterranean dry wood termite (yellow-necked dry wood termite) *Kalotermes flavicollis* Fabr. *Holzforsch. Holzverwert.*, **15**.

SEYBOLD, A. 1932-1934. Über die optischen Eigenschaften der Laubblätter. *Planta*, **16, 18, 20, 21**.

SEYBOLD, A. 1934. Über den Lichtgenuss von Sonnen- und Schattenpflanzen. *Ber. dt. bot. Ges.*, **52**.

SEYBOLD, A. 1936. Über den Lichtfaktor photophysiologischer Prozesse. *Jb. wiss. Bot.*, **82**.

SHELFORD, V. E. 1951. Fluctuation of forest animal population in East Central Illinois. *Ecol. Monogr.*, **21**.

SHIRLEY, H. L. 1945. Reproduction of upland conifers in the Lake States as affected by root competition and light. *Am. Midl. Nat.*, **33**.

SHOCKLEY, D. E. 1955. Capacity of soil to hold moisture. *Agric. Engng.*, London, **32**.

SIEGEL, H. 1956. Anwendung von mechanischen und chemischen Mitteln zur Verhütung von Wildverbisschäden. *Forst- u. Jagdwes.*, **6**.

SIIVONEN, L. 1941. Über die Kausalzusammenhänge der Wanderung beim Seidenschwan. *A. zool. Soc. 'Vanamo'*, **8**.

SIMKOVER, H. G., and SHENEFELT, P. J. 1951. *J. econ. Ent.*, **44**.

SIMMONDS, F. J. 1959. Biological control—past, present and future. *J. econ. Ent.*, **52**.

SIREN, G. 1955. The development of spruce forest on raw humus sites in Northern Finland and its ecology. *Acta for. fenn.*, **62**.

SMIRNOFF, W. A., FETTES, J. J., and HALIBURTON, W. 1962. A virus disease of Swaine's jack-pine sawfly, *Neodiprion swainei* Midd. sprayed from an aircraft. *Can. Ent.*, **94**.

SMITH, G. D. 1960. *Soil classification, 7th approximation*. Soil Survey Staff, Soil Conservation Service. U.S. Dept. Agriculture.

SOUTHWOOD, T. K. E. 1961. The number of species of insect associated with various trees. *J. Anim. Ecol.*, **30**.

STÅLFELT, M. 1960. Die Abhängigkeit von zeitlichen Faktoren. *Handbuch der Pflanzenphysiologie*, vol. 5, pt. 2. Berlin.

STARK, R. W. 1961. Methods of improving biological evaluation procedures used in reaching forest insect control decisions. *J. For.*, **59**.

STEGER, O. 1962. Zum Auftreten der Kieferneule an Strobe. *Allg. Forstz.*, **17**.

STOCKER, O. 1929. Das Wasserdefizit von Gefässpflanzen in verschiedenen Klimazonen. *Planta*, **7**.

STOCKER, O. 1956. Die Abhängigkeit der Transpiration von den Umweltfaktoren. *Handbuch der Pflanzenphysiologie*, vol. 3. Berlin.

STOKLASE, I., and STOKLASE, E. 1902. Beiträge zur Lösung der Frage der chemischen Natur der Wurzel-Sekretes. *Jb. wiss. Bot.*

STOLINA, M. 1959. Les unités typologiques et leur utilisation dans l'étude de l'écologie des insectes. In: *Ontogeny Insects*, Prague.

STONE, E. L. 1949. Some effects of mycorrhiza on the phosphorus nutrition of Monterey pine seedlings. *Proc. Soil Sci. Soc. Am.*, **14**.

STRUGGER, S. 1948. Fluorescence microscope examination of bacteria in soil. *Can. J. Res.*, sec. C, **26**.

SUBRAMONEY, N. 1960. Sulphur bacterial cycle: probable mechanism of toxicity in acid soils of Kerala. *Sci. Cult.*, **25**.

SUKACHEV, V. 1928. Principles of classification of the spruce communities of European Russia. *J. Ecol.*, **16**.

SUKACHEV, V. N. 1960. The correlation between the concept 'forest ecosystem' and 'forest biogeocoenose' and their importance for the classification of forest. *Silva fenn.*, **105**.

SUKATSCHEW, W. N. 1954. Die Grundlagen der Waldtypen. *Festschrift für Erwin Aichinger zum 60. Geburtstag*, vol. 2. Wien.

TAMM, C. O. 1950. Growth and plant nutrient concentration in *Hylocomium proliferum* (L.) Lindb. in relation to tree canopy. *Oikos*, **2**.

TAMM, C. O. 1951. Removal of plant nutrients from tree crowns by rains. *Physiologia Pl.*, **4**.

TAYLOR, W., and GORSUCH, D. 1932. A test of some rodent and bird influences on western yellow pine reproduction at Fort Valley, Flagstaff, Arizona. *J. Mammal.*, **13**.

TEUCHER, G. 1955. Die Anfälligkeit des Douglasienrassen gegenüber der Douglasien-Wollaus (*Gilletella cooleyi* (Gill) C.B.). *Forst- u. Jagdw.*, **5**.

THALENHORST, W. 1951. Die Koinzidenz als gradologisches Problem. *Z. angew. Ent.*, **32**.

THALENHORST, W. 1956. Biologischer Forstschutz: Therapie und Hygiene. *Forstarchiv*, **27**.

THALENHORST, W. 1960. Ökologische Freilandarbeit in der Forstentomologie. *Z. PflKrankh.*, **67**.

THALENHORST, W. 1963. Komponenten des Schädlingsbefalls auf forstlichen Düngungs-versuchsflächen. *Forstarchiv*, **34**.

THIELE, H. U. 1956. Die Tiergesellschaften der Bodenstreu in den verschiedenen Waldtypen des Niederbergischen Landes. *Z. angew. Ent.*, **39**.

THORSTEINSON, A. J. 1960. Host selection in phytophagous insects. *A. Rev. Ent.*, **5**.

TIMONIN, M. I. 1935. The micro-organisms in profiles of certain virgin soils in Manitoba. *Can. J. Res.*, **13**.

TISCHLER, W. 1955. *Synökologie der Landtiere*. Stuttgart.

TOUMEY, J. W. 1929. The vegetation on the forest floor, light versus soil moisture. *Proc. Int. Congr. Plant. Sci.*, **1**.

TOUMEY, J. W., and KIENHOLZ, R. 1931. Trenched plots under forest canopies. *Bull. Yale Univ. School For.*, **30**.

TRANQUILLINI, W. 1960. Das Lichtklima wichtiger Pflanzengesellschaften. *Handbuch der Pflanzenphysiologie*, vol. 5, pt. 2. Berlin.

TRAPPMAN, W. 1949. *Pflanzenschutz und Vorratsschutz*, vol. 1. Stuttgart.

TRAUTMANN, W. 1957a. Die Wald- und Forstgesellschaften des Forstamtes Neuen-heerensee. *Allg. Forst- u. Jagdztg*, **128**.

TRAUTMANN, W. 1957b. Natürliche Waldgesellschaften und nacheiszeitliche Wald-geschichte des Eggegebirges. *Mitt. flor. soc. ArbGemein.*, **6/7**.

TROLL, C. 1950. Die geographische Landschaft und ihre Erforschung. *Studium gen.*, **3**.

TROLL, C. 1956. Das Wasser als pflanzengeographischer Factor. *Handbuch der Pflanzenphysiologie*, vol. 3. Berlin.

TRUHLÁŘ, I. 1958. Srovnáni mikrobiologiských poměrů v akátových a dubových pařezinách. *Lesn. pr.*, **37**.

TurČek, F. J. 1961. Über einige Wechselbeziehungen zwischen Gehölzen, Vögeln und Forstschädlingen. Z. angew. Zool., 48.

TurČek, F. J. 1962. Eine mittelbare Beeinflussung von Forstschädlingspopulationen durch die Vögel. Festschrift der Vogelschutzwarte für Hessen, Rheinland, Pfalz und Saarland.

TurČek, F. J. 1963. Výsnam a godnota podrastových drevin, sozsirovanych vtákmi, v ekologickej ochrane leca pred hmyzovými škodcami. Lesn. Casop., 9.

Türke, F. 1956. Chemische Waldschadenverhütungsmittel im Lichte neuer Forschenergebnisse. Z. Jagdwiss., 2.

Uvarov, B. 1931. Insects and climate. Trans. Ent. Soc. Lond., 79.

Vaclav, V., Georgijevic, E., and Lutesek, D. 1959. Problem zarista gubara u Bosni i Hercegovini. Anastatus disparis kao indikator gubarevih lokalitea. Zashtita bil'a, 52-53.

Valenta, V. 1960. Pusu liemenu kenkéju ekologiniu grupiu susidrymas, priklausomai nuo medziu apmirimo tipo. Liet. žemdir. moks. tyrimo Inst. Darb., 5.

Vaněk, I. 1959. Pouzitelnost pudni zoologie pro lesnickou typologii. Sb. véd. Prac. české vys. Učeni tech. Praze, 2.

Vasiljevic, Lj. 1959. Dvogodisnje kretanje populacije gubara po zavrsetku epizootije tipa poliedrije u nekim reonima NR Srbije (Saopstenje). Zashtita bil'a, 4.

Verduin, J. 1953. A table of photosynthetic rates under optimal, near natural conditions. Am. J. Bot., 40.

Vietinghoff-Riesch, A. F. 1957. Untersuchungen über Verbreitung und Schadwirkung des Lärchenblasenfusses (Taeniothrips laricivorus Krat) in den Randzonen seines Verbreitungsgebietes in Norddeutschland, der Schweiz und Frankreich. Z. angew. Ent., 41.

Virtanen, A., and Torniainen, M. 1940. A factor influencing nitrogen excretion from leguminous host nodules. Nature, 145.

Vité, J. P., and Rudinsky, J. A. 1962. Investigations on the resistance of conifers to bark beetle infestation. Verhandl. XI Int. Congr. Ent., 1960, 2.

Volk, O. H. 1937. Untersuchungen über das Verhalten des osmotischen Werte von Pflanzen aus steppenartigen Gesellschaften und lichten Wäldern des mainfrankischen Trockengebietes. Z. Bot., 32, 3.

Volz, P. 1951. Untersuchungen über die Mikrofauna des Waldbodens. Zool. Jb. Abt. Syst., 79.

Voute, A. D. 1960. Cultural control of forest insects. Rept. Fifth Wld Forestry Congr., Seattle, u.s.a.

Vukicevic and Milosevic. 1960. Dinamika vegetacije i mikrobua populacija nekih sumskih pozarista. Hrv. šum. List, 13.

Wallihan, E. F. 1940. Factors affecting the response of forest vegetation to trenching. J. For., 38.

Walter, H. 1936. Nährstoffgehalt des Bodens und natürliche Waldbestand. Silva, 24.

Walter, H. 1951. Standortslehre (analytisch-ökologische Geobotanik). Stuttgart.

Walter, H. 1962. Die Vegetation der Erde in ökologischer Betrachtung, vol. 1. Jena.

Ward, G. M., and Durkee, A. B. 1956. The peach replant problem in Ontario. III. Amygdalin content of peach tree tissues. Can. J. Bot., 34.

Wartenberg, H. 1953. Über pflanzenphysiologische Ursachen des Massenwechsels der Apfelblutlaus (Eriosoma lanigerum) auf Malus pumila. Mitt. biol., Reichs anst. Ld- u. Forstw., 75.

Wassink, E. C. 1959. Efficiency of light energy conversion in plant growth. Pl., Physiol., 34.

Watanabe, T., and Casida, J. E. 1963. Response of Reticulitermes flavipes to fractions from fungus-infected wood and synthetic chemicals. J. econ. Ent., 56.

Weaver, J. E., and Clements, F. E. 1938. Plant ecology. 2nd ed. N.Y., London.

Weck, Y. 1957. Die Wälder der Erde. Berlin, Göttingen, Heidelberg.

Weetman, G. F. 1961. The nitrogen cycle in temperate forest stands. Montreal.

Weingärtner, H. 1962. Neue Erkenntnisse über Schad- und Nutzinsekten im Walde. Lebend Erde, 4.

Weiser, J. 1956. Die Krankheiten des Tannentriewicklers, Cacoecia murinana Hb., in der Mittelslowakei (CSR). Z. PflKrankh., 63.

WEISER, J. 1957. Mikrosporidien des Schwammspinners und Goldafters. *Z. angew. Ent.*, **40**.

WEISSENBERG, H. 1954. Die Mikroorganismentätigkeit auf rohhumushaltigen Heidesand, gemessen an der CO_2-Production. *Z. PflErnah., Düng., Bodenkde*, **66**.

WELLENSTEIN, G. 1953. Ergebnisse 25-jähriger Grundlagenforschung zur forstlichen Bedeutung der Roten Waldameise (*Formica rufa* L.). *Mitt. biol. Reichs anst. Ld- u. Forstw.*, **75**.

WENT, F. W. 1940. Soziologie der Epiphyten eines tropischen Urwaldes. *Ann. Jard. bot. Buitenz*, **50**.

WESTVELD, M. 1954. A budworm vigor-resistance classification for spruce and balsam fir. *J. For.*, **52**.

WEYLAND, H. 1912. Zur Ernährungsphysiologie micotropher Pflanzen. *Jb. Wiss. bot.*, **51**.

WHITTAKER, R. H. 1953. A consideration of climax theory; the climax as a population and pattern. *Ecol. Monogr.*, **23**.

WHITTAKER, R. H. 1962. Classification of natural communities. *Bot. Rev.*, **28**.

WIEDEMANN, A., and THOMMEN. 1959. Das maschinelle Lockhartenverfahren als Rationalisierungsmittel in der Forstwirtschaft. *Mitt. schweiz. Anst. forstl. Vers. Wes.*, **4**.

WIELSTATTER, R., and STOLL, A. 1918. *Untersuchungen über die Assimilation der Kohlensäure.* Berlin.

WILDE, S. A. 1954. Mycorrhizae fungi, their distribution and effect on tree growth. *Soil Sci.*, **78**.

WILL, G. M. 1955. Removal of mineral nutrient from tree crowns by rains. *Nature*, **176**.

WILLIAMS, W. T., and LAMBERT, J. M. 1960. Multivariate method in plant ecology. *J. Ecol.*, **48**.

WILSON, L. F. 1963. Host preference for oviposition by the spruce budworm in the Lake States. *J. econ. Ent.*, **56**.

WINSTON, P. W. 1956. The acorn microsere, with special reference to arthropods. *Ecology*, **37**.

WINTER, A. G. 1952. Die Bodenmüdigkeit im Obstbau. *Zeitfragen der Baumschule*, **7**.

WINTER, A. G. 1955. Untersuchungen über Vorkommen und Bedeutung antimikrobiellen und antiphytotischen Substanzen in natürlichen Böden. *Z. PflErnähr., Düng., Bodenkde*, **69**.

WINTER, A. G. 1960. Allelopathiales Stoffwanderung und Stoffwandlung. *Ber. dt. bot. Ges.*, **73**.

WINTER, A. G. 1961. New physiological and biological aspects in the interrelationship between higher plants. *Symp. Soc. Exptl Biol.*, **15**.

WINTER, A. G., and BUBLITZ, W. 1953a. Über die Keim- und Entwicklungshemmende Wirkung der Buchenstreu. *Naturwissenschaften*, **40**.

WINTER, A. G., and BUBLITZ, W. 1953b. Untersuchungen über die antibakteriellen Wirkungen im Bodenwasser der Fichtenstreu. *Naturwissenchaften*, **40**.

WINTER, A. G., and SCHÖNBECK, F. 1953. Untersuchungen über die Beeinflussung des Keimung und Entwicklung von Getreidesamen durch Kaltwasserauszüge aus Getreidestroh. *Naturwissenschaften*, **40**.

WINTER, A. G., and WILLEKE, L. 1951. Untersuchungen der Antibiotica aus höheren Pflanzen. *Naturwissenschaften*, **38**.

WINTER, A. G., and WILLEKE, L. 1952. Untersuchungen über Antibiotica aus höheren Pflanzen. IV. Hemmstoffe im herbstlichen Laub. *Naturwissenschaften*, **39**.

WITTICH, W. 1952. Der heutige Stand unseres Wissens vom Humus und neue Wege zur Lösung des Rohhumusproblems im Walde. *SchrReihe forstl. Fak. Univ. Göttingen*, **4**.

WITTICH, W. 1953. Untersuchungen über der Verlauf der Streuzersetzung auf einem Boden mit starker Regenwurmtätigkeit. *Mitt. forstl. Fak. Univ. Göttingen*, **9**.

WITTICH, W. 1958. Bodenkundliche und pflanzenphysiologische Grundlagen der mineralischen Düngung in Walde und Möglichkeiten für die Ermittlung der Nährstoffbedarfes. *Allg. Forstz.*, **10**.

YENUG, HANS. 1961. Derivation of state factor equations of soils and ecosystems. *Soil Sci. Soc. Am. Proc.*, **25**.

ZELLER, O. 1951. Über Assimilation und Atmung der Pflanzen in Winter bei tiefen Temperaturen. *Planta*, **39**.

ZIMNY, H. 1960. Charakterystyka mikrobiologiczna niektorych gleb zespotow Iesnych. *Ekol. pol. ser. B*, **6**.

ZINECKER, E. 1957. Der grosse Fichtenborkenkäfer (*Ips typographus* L.) in seiner Abhängigkeit vom Standort. *Anz. Schädlingskde*, **30**.

ZWÖLFER, H. 1953. Biologische und chemische Schädlingsbekämpfung. *Allg. Forstz.*, **8**.

ZWÖLFER, H., and KRAUS, M. 1957. Biocoenotic studies on the parasites of two fir- and two oak-tortricids. *Entomophaga*, **2**.

ZWÖLFER, W. 1957. Ein Jahrzehnt forstentomologischer Forschung. 1946-1956. *Z. angew. Ent.*, **40**.

ZWÖLGER, W. 1963. Über Abwehreinrichtungen unserer Waldbäume gegen Insektenschäden. *Z. angew. Ent.*, **51**.

Index of Terms

(Only terms related directly to biogeocoenology are included. The pages mentioned are, in general, those where the terms are defined, classified, or discussed in some detail. Where discussion is continued over two or more pages only the first page is listed.)